Introduction To

ELECTROMAGNETIC FIELDS, RADIATION AND ANTENNA THEORY

Nirod K. Das
Tandon School of Engineering, New York University

Fundamental Concepts and Engineering Applications

US Copyright
Registration Number: TXu 2-130-837
Effective Date: December 19, 2018

First Published Edition, February 2021
Last Update, October 2022
ISBN: 978-1-7340-6300-4
UEGM Publishing

(See http://wp.nyu.edu/ndasemthant for any supplementary material)

Preface

The material in this book have been developed by the author over recent years of teaching electromagnetics and antenna courses at the Tandon School of Engineering, New York University. Material from earlier chapters of the book can be used for a one-semester course on introduction to electromagnetics, at an undergraduate level. Suitable topics from different parts of the book can be arranged, in order to accomplish different level of emphasis on the waves and antenna radiation aspect of the above course. Material from the later part of the book would be well suited for a senior-level elective course, which focuses only on antennas and radiation. This elective course would normally follow a standard one-semester course on introductory electromagnetics, which presumably does not adequately emphasize on the antenna theory and radiation concepts. Suitable parts of the material in the book, covering more advanced concepts, may also be selected for an introductory graduate level antenna or electromagnetics course.

The style and content of different parts of the book may distinguish in many ways from those covered in other current texts. They were developed in response to current academic and engineering needs as well as students interests. The material has evolved, particularly in response to perceived gaps in available texts and teaching methods. This is based on the authors experience, reflecting back when he was learning the subject as a undergraduate or graduate student, as well as on continual feedback from students the author has taught over recent years.

Some of the significant highlights of the book include:

(1) Maxwell's equations first, not statics. Minimal or no dependence on electro- and magneto-statics, before Maxwell's equations are introduced.

(2) From cicuit theory to electromagnetics. Maxwell's equations are introduced as a generalization from simple circuit theory.

(3) From relativity to electromagnetics. Simple derivation of the electomagnetic theory from relativity, taking some mystery away from Maxwell's equations

(4) Physical conceptualization without differential operators. No dependence on curl, divergence and gradient in order to develop key concepts. Integral principles are used instead, which are much more effective for basic physical conceptualization.

(5) Radiation analysis using plane-wave synthesis. Fields of elementary radiation sources are developed by superposing current sources for plane-wave radiation.

(6) Electrodynamics of radiation. Fields of a moving charge are developed, providing an alternate view point for radiation as an electrodynamic phenomenon.

(7) Antenna in a general scattering environment. Modeling of antenna performance in a scattering environment is introduced, in addition to conventional modeling in an ideal open environment.

(8) Near-field antenna modeling using Fourier analysis. Antenna input reactance, resistance, resonant condition, and mutual coupling between elements are formulated based on plane-wave synthesis of near fields, using basic Fourier analysis.

The significant new features are further expanded in the following, discussed in general sequential order with the associated chapters of the book:

- Electrostatics and magnetostatics are usually covered in physics, and it may not be necessary to cover these topics again thoroughly in an electrical engineering course. Only essential material on electro and magneto statics are covered in the chapter 1. This material may be quickly reviewed, in reference to a basic physics text, for continuity with developments of the electromagnetic theory in the following chapters. This would allow introduction of Maxwell's equations of chapter 2, essentially in the beginning of an undergraduate course. Electrostatics and magnetostatics may be shown in chapter 2 to be only simple specialization of Maxwell's equations, when there is no time variation (DC condition). This would save time for more advanced topics of wave propagation and antenna radiation, allowing adequate coverage of the advanced topics in a single undergraduate level electromagnetics course.

- Electromagnetic theory is introduced (chapter 2) as a generalization of circuit theory concepts. Kirchoff's Voltage Law (KVL) is generalized including induced emf, using Faraday's law. Kirchoff's Current Law (KCL) is generalized by including two types of currents, both conduction as well as displacement currents. The KVL (Faraday's law) is coupled to the KCL through a generalized Ampere's law, which relates the magnetic field in the KVL (Faraday's law) to the the conduction and displacement currents in the KCL. These generalized laws constitute Maxwell's equations of the electromagnetic theory.

This approach can be particularly effective for electrical engineering students, by linking an electromagnetics course directly to a circuit theory course. It would help students, particularly at the undergraduate level, to understand and value electromagnetic theory in relation to the conventional circuit concepts which they are assumed to be already familiar with.

The circuit concepts using summation over a finite number of discrete circuit elements is easy to appreciate and grasp in the beginning. The integral forms of Maxwell's equations are then developed as generalization of the circuit concepts to a distributed medium, which maybe treated as a generalized circuit by decomposing it into a large number of inter-linked circuit elements. The simple summation forms of the circuit relations then naturally transform into the integral forms of Maxwell's equations, in the limit of infinite number of elements. The required circuit concepts can be introduced with minimal efforts. Considering the simplicity and economy, the circuit approach of introducing electromagnatic theory can be quite effective, even for students from applied physics without any formal circuit-theory background in electrical engineering.

- It is of fundamental significance to introduce electromagnetics in relation to the theory of special relativity. Maxwell's equation are developed from Gauss' laws of electricity and magnetism using relativity, using a new approach presented in chapter 3. It uses fairly simple analysis, which can be effective even at the undergraduate level. Such an approach is being tried here for the first time in an introductory electromagnetics text. Connecting Maxwell's equations to relativity concepts would provide a modern view of the electromagnetic theory. It may help develop deeper insight and understanding of the electromagnetic theory, by taking some mystery away from Maxwell's equations. The topic can be introduced in the early or late stage of an undergraduate class. It may be skipped, depending on available time or scope of the class, without any loss of continuity with the following chapters. The topic may as well be used to supplement a graduate level electromagnetics course.

- All basic concepts are developed using only summation (in discrete circuits) or integral (in distributed fields) without need for differential operators like curl, divergence and gradient. The integral or summation forms are clearly more intuitive, as compared to the differential operations, and therefore can be more helpful in physical or conceptual reasoning. Heavy dependence on differential operators in an electromagnetics class is known to be a major hurdle in introducing basic electromagnetic concepts to beginners, as well as in developing deeper physical insight even for advanced students. The general integral approach followed in this text is aimed to address the above difficulties. However, selected key expressions are also presented using differential operators, only as an alternate approach, that may be preferred at an advanced stage of learning.

- Signal propagation in a transmission line is covered in chapter 4. This topic is introduced as a basic engineering application of the electromagnetic theory, sequenced after development of Maxwell's equations in chapter 2 from circuit principles. An alternate sequencing of the topics may be possible. By selecting suitable sections in chapter 4, the transmission line may be studied using only circuit modeling concepts, without full knowledge of Maxwell's equations developed in chapter 2. Through the simple transmission line analysis, some valuable principles of wave propagation in one dimension and associated mathematical analysis can be introduced, early on. This background may be valuable, preparing students to study more general circuit concepts and electromagnetic theory developed in chapter 2, as well as for modeling more advanced wave propagation in a 3-dimensional material medium covered in chapter 5.

- Signal propagation in a uniform material medium, in the form of a simple plane electromagnetic wave, is covered in chapter 5. This topic is introduced as one of the basic application of the electromagnetic theory developed in the chapter 2. Basic differential equations and relations for a plane wave are derived from the integral forms of Maxwell's equations developed in chapter 2. The analysis is relatively simple, established without use of differential operators. The plane-wave relations along a given direction are generalized for propagation in any arbitrary direction in a 3-dimensional space, using geometric and trigonometric transformation of the coordinate system. The generalization is extended to evanescent waves as well. Reflection and transmission of plane wave through a single or multiple material layers are analyzed using boundary conditions as well as using equivalent transmission line models.

The simple plane-wave analysis is used as a basis to introduce vector potential functions, and to relate them to different field components in rectangular coordinates. The potential-field relations are then expressed in cylindrical and spherical coordinates, using simple geometric and trigonometric transformation from the rectangular coordinates. Any arbitrary wave in a locally uniform medium can always be synthesized as a superposition of general plane waves. Accordingly, the above potential-field relations are recognized to be general in scope, applicable to any arbitrary fields in a uniform material medium. Development of the general potential-field relations using the above approach avoids use of any differential operators, making the potential-field relations much easier to understand and derive. These relations would be particularly useful to model antenna radiation in the following chapter 6.

- Radiation form a small dipole antenna element is derived in chapter 6 using a plane-wave decomposition approach. This is developed as an alternate approach to a conventional derivation, which is heavily dependent on differential-operators. The conventional approach starting from a vector potential, and following differential-operator analysis to derive fields, is considered to be heavily mathematical in nature, lacking in physical understanding or insight.

- In addition to the dipole element, other elementary radiation sources such as a loop element, an infinite line source, and a moving point charge, are covered in chapter 6. A point charge under a general motion is the basic source of all electromagnetic fields and radiation. Radiation may be interpreted and modeled in terms of a charge motion, which provides a valuable alternate model of radiation, in contrast to formulations in terms a time-varying current distribution. An accelerating charge experiences a self-force, which may be interpreted to be partly due to (a) an equivalent "electromagnetic mass" associated with the reactive fields, and partly to (b) a radiative force associated with energy escaping away from the charge as radiation. These electrodynamic characteristics supplement basic concepts of antenna impedance and radiation, as well as carry independent value as fundamental physical concepts.

- General antenna parameters are covered in chapter 7. The derivation of conventional antenna parameters in chapter 7 generally follow conventional antenna texts. There are some new contributions in presentation and composition of the material. A general antenna is studied under three separate conditions of operation as a transmitter, as a receiver, and in a communication link. Two complex-vector antenna parameters - transmit and receive vectors - are developed, and they are related to each other as well as to antenna directivity and effective area, which are the respective scalar parameters. The above presentation may provide a more comprehensive understanding of antenna performance and parameters. A reciprocity relation is developed in the Appendix using circuit principles, which may be used to establish reciprocity relations between transmit and receive antennas. Accordingly, derivation of the antenna reciprocity conditions in chapter 7 may be covered without dependence on a more involved derivation of the reciprocity theorem from Maxwell's equations in chapter 2.

- Antenna performance in a scattering environment is presented in chapter 8. The topic supplements material in chapter 7, which covers antenna performance only in a open free-space environment without any signal reflection or scattering. A scattering environment is clearly a more practical or realistic scenario for ground-based wireless communication, and a suitable modeling in such an environment would account for any effects from the surrounding natural or man-made structures such as building, mountains and trees. Such modeling is not commonly covered in standard antenna texts, but are addressed in ad-hoc forms in different wireless communication texts. It would be valuable to give due coverage to this material in an antenna text, with appropriate rigor and depth from an antenna theory perspective. Basic antenna parameters, describing performance specifically under practical scattering conditions, are established based on antenna theory and wave propagation concepts developed in the earlier chapters.

- Useful antenna elements and arrays are studied in chapters 9 and 10. Plane-wave decomposition of fields allows use of Fourier analysis for modeling of planar antennas. Thin wire antennas can be modeled essentially as equivalent planar strip elements. Near-field analysis of antenna performance, such as input impedance, resonance condition, and mutual coupling between elements, as well as moment method analysis of antenna current distribution, are covered through Fourier integration. Radiation from an antenna array is formulated as a superposition of radiation from its individual elements, by treating the elements operating in isolation from each other as a first-order approximation, or by including mutual coupling between the different elements for a rigorous model. Sufficiently large periodic arrays may be approximated as infinite periodic arrays, which may be directly modeled using Fourier series expression of its periodic current distribution, without having to model individual antenna elements and mutual couplings between each elements.

All the formulations can be implemented using MATLAB or any other numerical programming, which are now commonly accessible to university students. This material has valuable overlap with conventional signal analysis and communication classes. This can be attractive to electrical engineering students with overlapping interests in both signal communications and antenna theory, particularly in the wireless communication area.

Deficiencies in the different proposed approaches are sure to surface in the process of continual delivery to students. The material would go through its normal process of experimentation and future improvement, based on class-room experience and feedback from students. However, the contents in the present form are hoped to carry essential elements of value. They may serve as useful focal point or framework for teachers, students and engineers to debate about, build upon and reshape to their own liking.

Nirod K. Das
New York, November 2010

Contents

CHAPTER 1

FIRST PRINCIPLES

The theoretical development of the electromagnetic theory is based on some basic concepts of charge, electric and magnetic fields, voltage, current, power, and related material properties such as permittivity, permeability and conductivity. These basic concepts and parameters are first defined. Gauss' laws for electric and magnetic fields, which are the primary principles governing the respective fields, are then established based on the above definitions of charge and fields. The material in this chapter are usually studied in basic physics or electric circuit theory, but are covered here for rigor and continuity with further developments. The primary principles of Gauss' laws, and the definitions of the field and charge upon which the principles are established, constitute the basic foundation for development of more advanced concepts. Starting with this foundation, we will derive the electromagnetic theory as a gradual generalization of the circuit theory, and electro- and magneto-static principles, in the following chapter 2.

1. Basic Concepts and Definitions

1.1. Electric Charge: Electric charge is the elementary source of all electromagnetic phenomena. The unit of measurement of charge is Coulomb (C). The charge can be of two basic types, positive or negative, which are distinguished by the nature of the force experienced between each other. Two charges of the same type repel each other, whereas two charges of opposite types attract each other. The natural physical element of negative electric charge is the electron, whereas that of positive charge is the proton, each with a charge magnitude of $1.6 \times 10^{-19} C$. Electrons and protons are the elementary charges present in all material substance, constituting the basic atomic structure of matter.

A basic measure of charge may be defined in a simple way when the charge is stationary. The charge is measured in proportion to the force it applies on a unit stationary charge placed at a given distance, as compared to a reference force between two identical, stationary unit charges placed at the same distance. When the charge under test is in motion or is varying in time, it might produce new force effects governed by a general theory of electromagnetics, to be covered in the following chapter. The force measured by a nearby stationary, reference charge, could now be dependent upon the specific conditions of motion or time variation of the test charge. Accordingly, the simple measure of a stationary charge defined above would be invalid for general moving or time-varying charges.

However, under general motion or time variation, a charge may be conceptually defined based on the force it applies on a reference stationary charge, as measured at each and every point over an entire closed surface surrounding the test charge. This is established by Gauss' law, to be presented in the following section. Although this method requires measurement over an entire surface, which is practically much more difficult to implement than that at a single fixed point used in the above simple measurement method, it carries fundamental theoretical significance for most general conditions.

1.1.1. *Charge Conservation.* As mentioned, any given charge is fundamentally derived from the elementary charges in the atomic structure of materials. Because all the positive and negative charges associated with the atomic structure are known to be equal to each other, the total charge available in a *complete* electrical system is assumed to be zero. A non-zero charge can only occur locally due to redistribution of the positive and negative charges through suitable physical processes. A complete electrical process is physically valid or self-consistent, only when it ensures that all the charges are properly accounted for and they all add up to zero. This basic universal principle may be referred to as the law of conservation of charge. For example, consider a charge at a particular location which is changing in time. This itself is not a physically valid situation, unless some additional charge flow is identified which describes where the additional charge is coming from, if the total charge is increasing in time, or going toward, if the total charge is decreasing in time. The fields produced by the time varying charge and the additional charge flow would now be physically valid and complete, not those by the time-varying charge taken alone.

1.1.2. *Bound and Conduction Charge.* The charges available in different materials maybe classified into two broad classes, (a) conduction and (b) bound charges. The conduction charges are those that are not tightly bound to the electronic structure of the material, and are available for free flow as electric current in the presence of applied electric field in a medium, or of voltage across two circuit terminals. On the other hand, the bound charges are confined to the internal electronic structure of the material, and are not available for current flow. They however can be influenced by any electric or magnetic field in the medium. The positive and negative charges bound to the atomic structure may be locally displaced with respect to each other, properly aligned along any electric field in the medium. The orbiting or spinning of the charges, around the atom or around itself, respectively, may also be influenced and properly aligned by any magnetic field in the medium. Such charge displacement and alignment, referred to as polarization process, produces additional electric or magnetic fields in the material, which changes the total field distribution. Polarization of bound charge and charge rotation are discussed further in relations to the definition of electric and magnetic fields in material media in sections 1.3 and 1.5, respectively.

1.1.3. *Charge Distribution.* Depending on how given amount of charge is spread over a particular region in space, it may be viewed as a localized point charge, or as a volume, surface or line distribution of charge.

(a) Point Charge. A charge Q located at a definite point in space is called a point charge. A point charge is only a physical idealization, occupying a zero volume or surface area, but in practice does not strictly occur in this form. Charges occurring in distinct lumps may be approximated as discrete or point charges, placed at their physical centers (Fig.1a). In general, however, it may be useful to describe the charge in terms of its volume, surface or linear distribution, as defined in the following. A point charge is only a limiting situation for the following general distributions, when the volume, surface or length of the respective cases are sufficiently small.

(b) Volume Charge Distribution. When the charge is distributed over a definite, non-zero volume region (Fig.1b), it is useful to define a volume charge density ρ_v. Mark a small volume element $\triangle V$ in the vicinity of a given location. Let the elemental charge present in the volume element be $\triangle Q$. The ratio of $\triangle Q$ to $\triangle V$ is defined as the volume charge density ρ_v at the particular location.

$$\rho_v = \frac{\triangle Q}{\triangle V}, \ \triangle Q = \rho_v \triangle V. \tag{1.1}$$

The charge density ρ_v is expressed as a function of position over a given volume V. The total charge in the volume V can be derived in terms of the ρ_v, by decomposing the volume into volume elements $\triangle V_j$, expressing the charge in each element in terms of the charge density as $\triangle Q_j = \rho_v \triangle V_j$, and summing over all elements j (Fig.1b). In the limit when the volume elements $\triangle V_j$ approach zero, the summation may be expressed as a volume integral of the charge density over the entire volume V.

$$Q = \sum_j \triangle Q_j = \sum_j \rho_{vj} \triangle V_j, \ Q = \iiint_V \rho_v dV. \tag{1.2}$$

In accordance with its definition, the units of measurement for ρ_v is Coulomb per cubic meter (C/m^3).

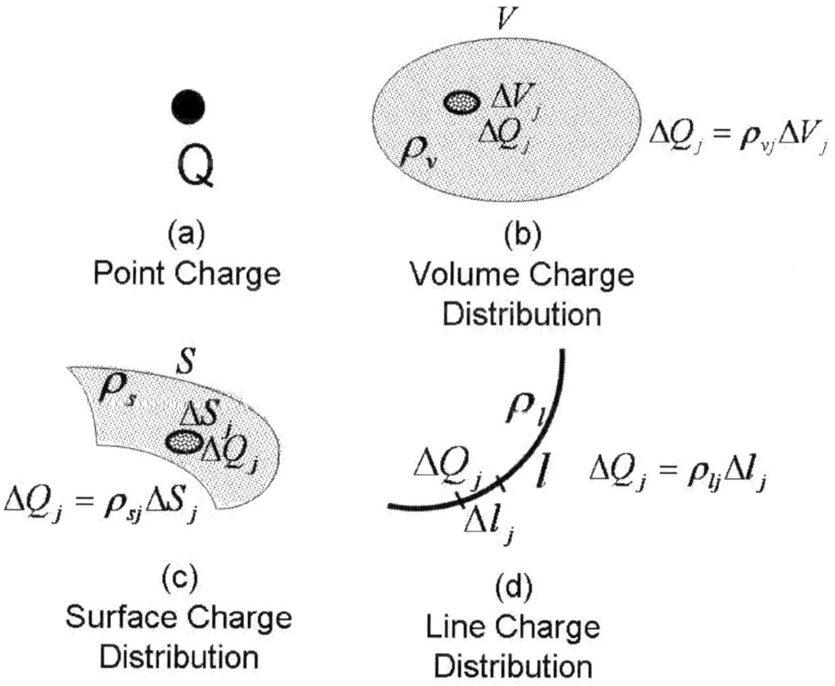

FIGURE 1

(c) Surface Charge Distribution. Sometimes, the charge may be spread over a large surface area S (Fig.1c), instead of a three-dimensional volume region. In this case, the ratio of an elemental charge $\triangle Q$ to the elemental surface area $\triangle S$ over which the charge is present may be defined as the surface charge density ρ_s. Let the distribution of ρ_s be given as a function of position over a surface S. Then, the total charge over the surface S may be derived by decomposing the surface into surface elements $\triangle S_j$, expressing the charge in each element in terms of the surface charge density as $\triangle Q_j = \rho_s \triangle S_j$, and

then summing over all elements j (Fig.1c). In the limit when the surface elements $\triangle S_j$ approach zero, the summation may be expressed as a surface integral of the charge density over the entire surface S. The concept is similar to that in equations (1.1,1.2) used for a volume distribution, except that in the present case the summation or integration is performed over a given surface region.

$$\rho_s = \frac{\triangle Q}{\triangle S}, \ \triangle Q = \rho_s \triangle S,$$
$$Q = \sum_j \triangle Q_j = \sum_j \rho_{sj} \triangle S_j, \ Q = \iint_S \rho_s dS. \tag{1.3}$$

In accordance with its definition, the units of measurement for ρ_s is Coulomb per square meter (C/m^2).

(d) Line Charge Distribution. When the cross-sectional area of a volume charge distribution, or the width of a surface charge distribution, is sufficiently small, it may be useful to described the charge in terms of a linear distribution over a given straight or curved line segment l (Fig.1d). In this case, the ratio of an elemental charge $\triangle Q$ to the elemental line segment $\triangle l$ over which the charge is present, is defined as the line charge density ρ_l. The ρ_l is in general different at different locations along the line l, and is described as a function of position along the line l. The total charge over the line l can be derived by decomposing l into small line elements $\triangle l_j$, expressing the charge in each element in terms of the line charge density as $\triangle Q_j = \rho_l \triangle l_j$, and then summing over all elements j (Fig.1d). In the limit when the line elements $\triangle l_j$ approach zero, the summation may also be expressed as a one-dimensional integral of the charge density ρ_l over the entire line l. The concept is similar to that in equations (1.1,1.2) used for a volume distribution, and in (1.3) for a surface distribution, except that in the present case the summation or integration is performed over a given line l.

$$\rho_l = \frac{\triangle Q}{\triangle l}, \ \triangle Q = \rho_l \triangle l,$$
$$Q = \sum_j \triangle Q_j = \sum_j \rho_{lj} \triangle l_j, \ Q = \int_l \rho_l dl. \tag{1.4}$$

In accordance with its definition, the units of measurement for ρ_l is Coulomb per meter (C/m).

1.2. Electric Current. Electric current is defined as the time rate of flow of free charge, and is measured in Amperes (A). A current is one Ampere (A) when there is a flow of one Coulomb of charge in every second. It is necessary to define a reference direction for the measurement of a current flow. Let $\triangle Q$ be the amount of charge flowing in a given reference direction along a conducting wire, over a time period $\triangle t$. A charge is counted positive when it is a positive charge flowing along the reference direction, or a negative charge flowing in the opposite direction, and is counted negative otherwise. The current I along the wire can then be expresses as,

$$I = \frac{\triangle Q}{\triangle t}. \tag{1.5}$$

The current along a wire may also be expressed as a vector $\overline{I} = I\hat{l}$, where $\hat{l}$ is the reference direction along the wire used for the measurement of charge flow, as discussed earlier (Fig.2a). At a given location along the wire, $\hat{l}$ is directed tangential to the conducting wire, which has two possibilities (forward or backward) that are negative of each other. One of the directions would have to be specified in order to specify proper sign of the current. Current along a conducting wire, having a small or ideally zero cross-sectional area, is sometimes referred to as a "line current," in contrast with volume or surface current to be discussed in the following.

It may be noted here, that the current is the measure of the "flow" of charge, which may not be necessarily associated with a non-zero total charge present in a given segment. A current maybe generated in a conducting wire by moving the free electrons in a particular direction in the presence of equal number of stationary positive charges. Accordingly, at any given time the total charge in a given section of wire is zero. In general, of course, one could have both a non-zero current flow as well as a non-zero net charge in a given volume. This would be true in the case of a current flow in a charged conductor, or the current flow associated in an electron beam.

1.2.1. *Volume Current Distribution.* The distribution of the current over a volume region in a conducting medium (Fig.2b) may be measured by defining a volume current density vector $\overline{J}$. At a given location, let $\overline{\triangle I}$ be an elemental current vector measured over an elemental cross-sectional area $\triangle S$. The current density vector $\overline{J}$ is defined as the ratio of $\overline{\triangle I}$ and $\triangle S$.

$$\overline{J} = \frac{\overline{\triangle I}}{\triangle S}, \ \overline{\triangle I} = \overline{J}\triangle S, \ \triangle I = J\triangle S. \tag{1.6}$$

Note that the elemental *cross-sectional* area $\triangle S$ in the above basic definition means that the surface element is on a plane which is orthogonal to direction of the current flow. For a general surface element $\triangle S$ with its normal $\hat{n}$ inclined at an angle θ with respect to the current density vector $\overline{J}$, the projection of the area orthogonal to the current flow would be $\triangle S \cos\theta$. Accordingly, the current $\triangle I$ over the general area element can be expressed by replacing $\triangle S$ in the equation (1.6) by $\triangle S \cos\theta$.

$$\overline{\triangle I} = \overline{J}\triangle S\cos\theta, \ \triangle I = J\triangle S\cos\theta = \overline{J}\cdot\overline{\triangle S}, \ \overline{\triangle S} = \hat{n}\triangle S. \tag{1.7}$$

The surface element vector $\overline{\triangle S}$ used in (1.7) is defined with a magnitude equal to $\triangle S$, with its direction along the normal $\hat{n}$ to the surface. Note that the normal $\hat{n}$ to a surface element would have two possibile directions, with respective current flows negative of each other. One of these possibilities will have to be clearly specified in order to avoid any sign ambiguity in the derivation.

Let the distribution of $\overline{J}$ be given as a function of position over a general surface S (which may be planar or non-planar). The total current I flowing through the surface S can be derived by first decomposing the surface into small surface elements $\triangle S_j$, finding the current $\triangle I_j$ over each element in terms of the current density $\overline{J}_j$ using (1.7), and then summing it over all j (Fig.2b). In the limit when the surface elements approach zero, the summation may be equivalently expressed as a two-dimensional integral over the surface S.

$$I = \sum_j \triangle I_j = \sum_j \overline{J}_j \cdot \hat{n}_j \triangle S_j = \sum_j \overline{J}_j \cdot \overline{\triangle S}_j, \ \overline{\triangle S}_j = \hat{n}_j \triangle S_j,$$
$$I = \iint_S \overline{J} \cdot \overline{dS}, \ \overline{dS} = \hat{n} dS. \tag{1.8}$$

In accordance with its definition, the units of measurement for $\overline{J}$ is Ampere per square meter (A/m^2).

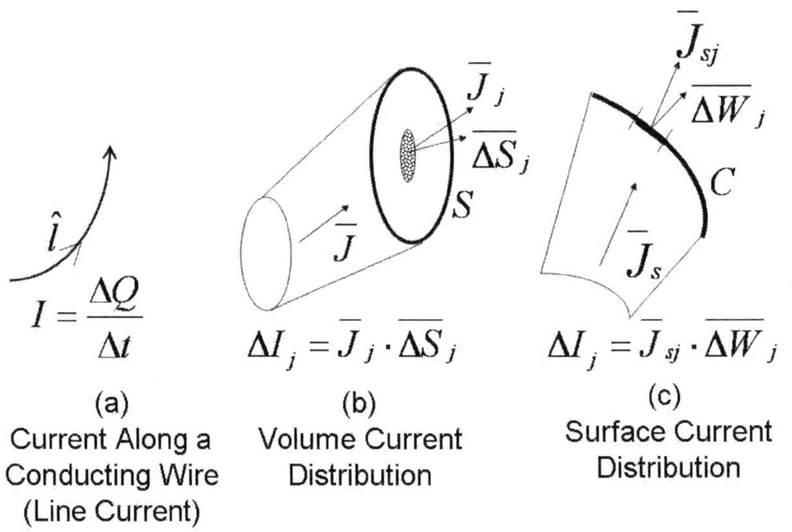

FIGURE 2

1.2.2. *Surface Current Distribution.* Sometimes the current may be flowing along a thin conducting film or strip, or may be flowing through a very thin layer on the surface of a metallic body (Fig.2c). In order to properly measure the distribution of current in such cases, it would be useful to define a surface current density $\overline{J}_s$ as the ratio of an elemental current $\overline{\triangle I}$ and an elemental transverse width $\triangle W$ over which the current flows.

$$\overline{J}_s = \frac{\overline{\triangle I}}{\triangle W}, \ \overline{\triangle I} = \overline{J}_s \triangle W, \ \triangle I = J_s \triangle W. \tag{1.9}$$

The elemental *transverse* width $\triangle W$ in the above definition means that the width element is orthogonal to the direction of the current flow. For a general width element $\triangle W$ with its normal $\hat{n}$ inclined at an angle θ with respect to the surface-current density vector $\overline{J}_s$, the projection of this width orthogonal to the current flow would be $\triangle W \cos\theta$. Accordingly, the current $\triangle I$ over the general width element can be expressed by replacing $\triangle W$ in the above equation (1.9) by $\triangle W \cos\theta$.

$$\overline{\triangle I} = \overline{J}_s \triangle W \cos\theta, \ \triangle I = J_s \triangle W \cos\theta = \overline{J}_s \cdot \overline{\triangle W}, \ \overline{\triangle W} = \hat{n} \triangle W. \tag{1.10}$$

The width element vector $\overline{\triangle W}$ in (1.10) is defined with a magnitude equal to $\triangle W$, with its direction along the normal $\hat{n}$ to the width. The normal $\hat{n}$ to a width element would have two possibile directions, with respective current flows negative of each other. One of these possibilities will have to be clearly specified in order to avoid any sign ambiguity in the derivation.

Let the distribution of $\overline{J}_s$ be given as a function of position over a general line segment C. The total current I flowing through the line C can be derived by first decomposing it into small width elements $\triangle W_j$, finding the current $\triangle I_j$ over each element in terms of the current density $\overline{J}_{sj}$ using (1.10), and then summing it over all j (Fig.2c). In the limit when the width elements approach zero, the summation may be equivalently expressed as a one-dimensional integral over the line C. The concept and procedure is similar to that followed in (1.8) for a volume current distribution, except that the surface summation or integration in (1.8) is now replaced by a similar summation or integration over the line C.

$$I = \sum_j \triangle I_j = \sum_j \overline{J}_{sj} \cdot \hat{n}_j \triangle W_j = \sum_j \overline{J}_{sj} \cdot \overline{\triangle W}_j, \ \overline{\triangle W}_j = \hat{n}_j \triangle W_j,$$

$$I = \int_C \overline{J}_s \cdot \overline{dW}, \ \overline{dW} = \hat{n} dW. \tag{1.11}$$

In accordance with its definition, the units of measurement for $\overline{J}_s$ is Ampere per meter (A/m).

1.3. Electric Field. The electric field $\overline{E}$ is described as the property of a vector field at a particular location, measured in terms of the force vector $\overline{F}$ experienced by a stationary charge Q (Fig.3). The magnitude and direction of the electric field correspond to those of the force experienced by a reference unit of positive charge placed at the particular location.

$$\overline{E} = \overline{F}/Q, \ \overline{F} = Q\overline{E}. \tag{1.12}$$

We will discuss the electric field in two different types of material media: (a) insulating or dielectric material, and (b) conducting material. An insulating material has very little or ideally no conduction charges, and the electric field changes the direction and density of alignment or polarization of its internal bound charges at the atomic level. Whereas, the electrical property in a conducting material is dominantly influenced by its conduction charges. Although some material maybe described to be more close to an ideal insulating or an ideal conducting material, all practical materials exhibit partially insulating and partially conducting behavior, having both bound and conduction charges, in different degrees.

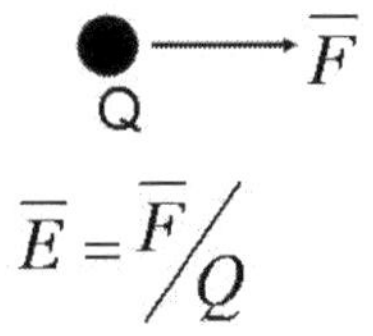

FIGURE 3

1.3.1. *Electric Field in an Insulating Material.* The electric field in an insulating material medium may be identified in two forms: the primary field $\overline{E}_0$, and the total field $\overline{E}_t$ which is the sum of the primary field $\overline{E}_0$ and an additional contribution $\Delta\overline{E}$ due to the polarization of the bound charges of the medium. An equivalent model of the material polarization is shown in Fig.4. The relative displacement of the bound charge-pairs and the relative directions of different electric field components, as shown, are determined by the basic nature/definition of charge and electric field, presented before. The primary electric field $\overline{E}_0$ displaces the positive and negative charges of the material toward and opposite to the direction of the electric field, respectively. Now, the additional field $\Delta\overline{E}$ produced by these polarized charge-pairs is pointed from the positive toward the negative ends of the bound charge-pairs, as shown, which would be opposite to the direction of the primary field $\overline{E}_0$. This results in having the total field in the material to be less than the primary field.

$$\overline{E}_t = \overline{E}_0 + \Delta\overline{E}; |\overline{E}_t| \leq |\overline{E}_0|. \tag{1.13}$$

The polarization elements are assumed to be much smaller and spaced sufficiently close to each other, compared to the dimensional scale of practical interest. Accordingly, at a macroscopic scale the field is defined to be the local average in the following sense. The total field $\overline{E}_t$ actually refers to the average as one samples along the polarization direction (x in Fig.4). Whereas, the average field is $\overline{E}_0$ as one samples on an orthogonal plane (yz in Fig.4) between two neighboring polarized elements, but is $\overline{E}_t$ if sampled on an orthogonal plane that falls right inside the polarized elements.

$\overline{E}_0$ and $\overline{E}_t$ are proportional to each other for a linear medium. Further, consider an "isotropic" medium, which means that the internal material structure of the medium looks similar as seen from any arbitrary orientation. For such a simple medium the ratio of the proportionality between $\overline{E}_0$ and $\overline{E}_t$ would be independent of the specific vector direction of the field. This ratio is defined as the relative permittivity ϵ_r of the medium. For material with a general "anisotropic" structure, the proportionality between $\overline{E}_0$ and $\overline{E}_t$ would be mathematically a bit more involved, not considered here.

With more generality, the $\overline{E}_0$ may be linearly dependent on the time-derivatives of the $\overline{E}_t$, in addition to being proportionately dependent on the $\overline{E}_t$ as assumed above. Such a medium is referred to as a dispersive dielectric medium, not considered here. With further generality, the $\overline{E}_t$ may be a non-linear function of the $\overline{E}_0$ and its time-derivatives, when the fields are sufficiently strong. Such non-linear behavior is also not considered.

$$\epsilon_r = \frac{\overline{E}_0}{\overline{E}_t} \geq 1; \; |\overline{E}_0| \geq |\overline{E}_t|. \tag{1.14}$$

ϵ_r represents the relative density or strength of polarization of the bound charge of the medium, and is greater than or equal to unity because $|\overline{E}_0|$ is greater than or equal to $|\overline{E}_t|$ as discussed earlier. For a free space medium, $\overline{E}_t$ and $\overline{E}_0$ are clearly equal to each other, leading to having $\epsilon_r = 1$ for the free space.

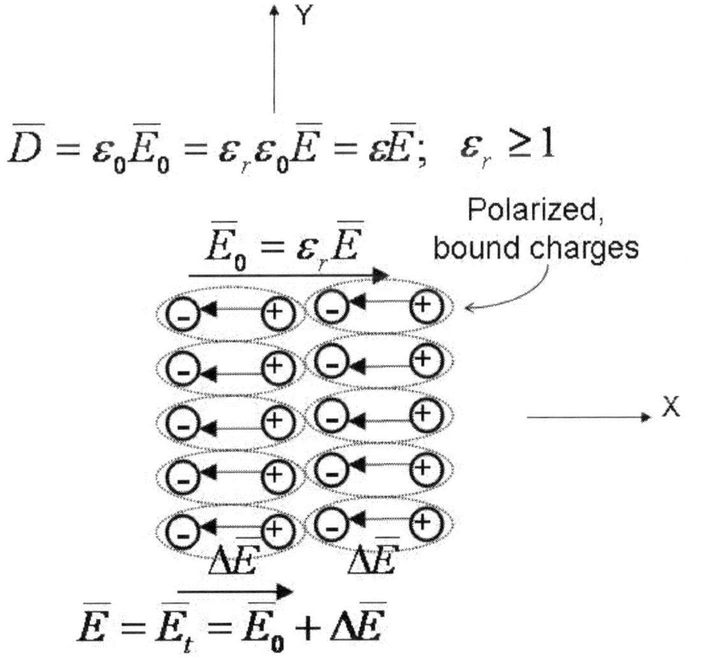

Electric field in a material medium, showing its equivalent microscopic model using polarized bound charges

FIGURE 4

Electric Field Intensity and Flux Density:

It should now be clear that the electric field inside a material medium does not have a unique meaning. The two fields, $\overline{E}_t$ and $\overline{E}_0$, have distinct physical significance. However, for analytical and notational convenience, as maybe clear in the following sections, when one refers to the electric field $\overline{E}$ in a material medium, it normally refers to $\overline{E}_t$, which may be called the electric field intensity. Whereas, $\overline{E}_0$ is defined in terms of an electric flux density $\overline{D} = \epsilon_0 \overline{E}_0$, named in reference to the concept of flux to be used in Gauss' law of section 2.1. The relationship between $\overline{E}_t$ and $\overline{E}_0$ leads to relating the flux density $\overline{D}$ and the field intensity $\overline{E}$.

$$\begin{aligned} \overline{E} = \overline{E}_t, \ \overline{E}_0 = \epsilon_r \overline{E}_t = \epsilon_r \overline{E}, \\ \overline{D} = \epsilon_0 \overline{E}_0 = \epsilon_0 \epsilon_r \overline{E}, \\ \overline{D} = \epsilon \overline{E}; \epsilon = \epsilon_0 \epsilon_r. \end{aligned} \tag{1.15}$$

The ratio between the electric flux density $\overline{D}$ and the intrinsic electric field $\overline{E}_0$ is a physical constant ϵ_0. The ratio between the flux density $\overline{D}$ and the field intensity $\overline{E}$ is $\epsilon = \epsilon_0 \epsilon_r$. The parameter ϵ for a given medium would determine how electric field interacts or is "permitted" in the medium, and is referred to as the electric permittivity or simply permittivity of the medium. When the medium is free space, then ϵ_r is unity and $\epsilon = \epsilon_0$. Accordingly, ϵ_0 is the permittivity of the free space.

The unit of measurement of the electric field intensity $\overline{E}$ is Newton per Coulomb (N/C), which is also equal to Volt per meter (V/m) based on the unit of voltage defined in a following section 1.4. The unit of measurement of the flux density $\overline{D}$ is Coulomb per square meter (C/m^2), which is based on Gauss' law relating the total flux to source charge. Based on the units of $\overline{D}$ and $\overline{E}$, the unit of measurement of permittivity ϵ is in $C/(Vm)$ which is equal to Farad per meter (F/m) using the relationship $F = C/V$. The relative permittivity ϵ_r is a dimensionless parameter.

1.3.2. *Electric Field in a Conducting Material.* In a conducting medium, the negatively-charged conduction electrons flow in a direction opposite to the direction of the electric field. This is equivalent to having positive charges flowing along the electric field, which means having an electric current along the electric field. Assuming that the electric field is sufficiently small, the conduction current density at a particular location would be proportional to the local electric field. The constant of proportionality between the current density $\overline{J}$ and the electric field $\overline{E}$ vectors is defined as the conductivity of the particular medium. One over the conductivity is defined as the resistivity (ρ_R).

$$\overline{J} = \sigma \overline{E} = \frac{\overline{E}}{\rho_R}. \tag{1.16}$$

Consider a section of the material of length Δl along the current flow, with a cross-section area ΔA (Fig.5). The voltage across its length ΔV, as defined in the following subsection 1.4, would also be proportional to the current flow ΔI over its cross section. The ratio of proportionality between voltage $\triangle V$ and current $\triangle I$ is defined as the resistance R of the material segment.

$$R = \frac{\Delta V}{\Delta I}. \tag{1.17}$$

Now, the voltage $\triangle V$ is related to the electric field magnitude E, that is $\triangle V = E\triangle l$, as established in equations (1.20,1.23). The current $\triangle I$ may be related to the current density $\overline{J}$, that is $\triangle I = J\triangle A$, as per the definition discussed earlier in section 1.2. Using the above relations, the resistance R of a material segment in (1.17) may be related to the material conductivity σ in (1.16).

$$R = \frac{\Delta V}{\triangle I} = \frac{E\triangle l}{J\triangle A} = \frac{\triangle l}{\sigma\triangle A} = \frac{\rho_R\triangle l}{\triangle A}. \tag{1.18}$$

The linearity relationship between the voltage and current of a material segment, or equivalently between the electric field and the current density for a bulk material, as established in above equations (1.16-1.18), is referred to as Ohm's law. The unit of measurement of the resistance R is called Ohm (Ω), which is equal to Volt per Ampere (V/A), by relating the unit of the resistance R to the units of voltage and current as per (1.17). The inverse of the resistance R is called the conductance $G = 1/R$, measured in mhO ($\mho = 1/\Omega = \Omega^{-1}$), which is simply the inverse of Ohm (Ω). Accordingly, the units of the resistivity ρ_R and conductivity σ are Ohm meter (Ωm) and mhO per meter ($\mho/m$), respectively, as per the relation (1.18).

1.4. Potential or Voltage: The potential difference from one given location to another is defined as the work or energy necessary to move one unit of charge from the first to the second point. If a common point can be identified for reference, and is referred to as the point of zero potential, then the potential difference from this common reference to a given point of interest is referred to as the potential or voltage of the given point. The unit of measurement for potential or potential difference is Volt (V). The potential difference between two points is one Volt, when one Joule (J) of work is necessary to move one Coulomb (C) of charge across the points. That is, $V = J/C$.

Consider now two points that are closely spaced, separated by a distance Δl. Let the potential difference from the first to the second point be ΔV_{12}. Assume also that the electric field at the mid point is $\overline{E}$. As per the definition of the electric field discussed earlier, the force $\overline{F}$ applied on a charge Q, placed between the two points, would be equal to $\overline{E}Q$. Therefore, it would require an external force $\overline{F}_e = -\overline{F} = -\overline{E}Q$ directed opposite to $\overline{F}$. Accordingly, the work ΔW_{12} necessary to move the charge from the first to the second point would be $\Delta W_{12} = \overline{F}_e \cdot \overline{\Delta l}_{12} = -Q\overline{E} \cdot \overline{\Delta l}_{12}$, as per the mechanics and definition of work or energy. Here the vector $\overline{\Delta l}_{12}$ is defined as a line vector directed from the point 1 toward 2, with a magnitude equal to the distance $\triangle l$ between the two points. This would lead to a useful relationship between the potential difference ΔV_{12}, as defined earlier, to the local electric field.

$$\triangle V_{12} = \frac{\Delta W_{12}}{Q} = \frac{\overline{F}_e \cdot \overline{\Delta l}_{12}}{Q} = -\frac{\overline{F} \cdot \overline{\Delta l}_{12}}{Q} = -\frac{Q\overline{E} \cdot \overline{\Delta l}_{12}}{Q} = -\overline{E} \cdot \overline{\Delta l}_{12}, \tag{1.19}$$

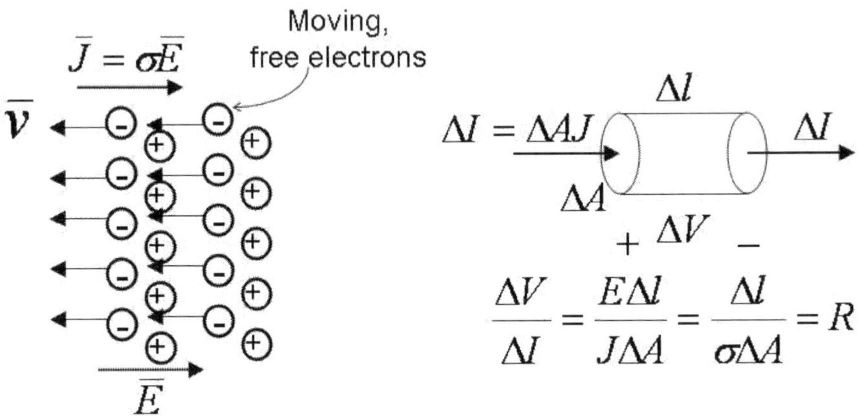

Electric field in the presence of conduction charges in a material medium

FIGURE 5

$$\triangle V_{12} = -\overline{E} \cdot \overline{\Delta l}_{12} = -E_{12}\Delta l = E_{21}\Delta l. \tag{1.20}$$

where E_{21} is the component of the local electric field directed from the point 2 towards point 1, and E_{12} is that in the reverse direction.

Potential Calculation in Material Media:

The above relationship normally works for an empty space. For a material medium, where the electric field is not uniquely defined as discussed before, one needs to find an equivalent expression for the elemental voltage difference. In reference to the polarization figure 4, an elemental path vector $\overline{\Delta l}_{12}$ maybe decomposed into two parts.

$$\overline{\Delta l}_{12} = \overline{\Delta l}_n + \overline{\Delta l}_p; \ \Delta l_n = \Delta l_{12}\sin\theta, \ \Delta l_p = \Delta l_{12}\cos\theta. \tag{1.21}$$

where θ is the angle between the path vector $\overline{\Delta l}_{12}$ and the local electric field, and subscripts p and n refer to components parallel and normal (perpendicular) to the field, respectively. Accordingly, the potential difference $\triangle V_{12}$ maybe expressed in two corresponding parts, $\triangle V_p$ and $\triangle V_n$. Note that, as per the derivation of potential difference in (1.20), $\triangle V_n$ is zero because the corresponding field is normal to the path. Further, we know that the electric field in the medium, when measured along a sample path, is $\overline{E}_t$ (see (1.13). Assuming that the total elemental potential to be equal to the sum of its two parts, $\triangle V_{12}$ may be expressed as follows.

$$\triangle V_{12} = \triangle V_p + \triangle V_n = -E_t \Delta l_p = -E_t \Delta l_{12} \cos\theta = -\overline{E}_t \cdot \overline{\Delta l}_{12}; \ \triangle V_n = 0. \tag{1.22}$$

Compare the expression of elemental potential difference (1.20) in a free space with that (1.22) in a material medium. Clearly, it would be notationally convenient to define the electric field $\overline{E}$ in a material medium to be equivalent to $\overline{E}_t$. We have proposed this definition in the last section 1.3.1, which now stands justified. With this notation, the elemental potential is expressed using the same generic relationship, valid for any general medium.

$$\triangle V_{12} = -\overline{E} \cdot \overline{\Delta l}_{12}; \ \overline{E} = \overline{E}_t. \tag{1.23}$$

The above relationships between the potential and electric field also leads to relating the units of the electric field (N/C) and voltage (V). That is, $N/C = J/(mC) = V/m$. Note that, unlike the electric field, which is a vector parameter, the voltage is a scalar quantity.

When the two points are placed at an arbitrary distance, then (1.23) needs to be integrated in order to find the potential difference. In principle, a particular path C_{12} between two points will have to be specified for the potential calculation. The path is decomposed into length elements $\overline{\triangle l}_j$, over each of which potential elements $\triangle V_j$ are calculated using (1.23). All the $\triangle V_j$'s are then added to find the total potential difference V_{12}. The sum can be expressed as an integration in the limit when the length elements approach zero.

$$V_{12} = \sum_{j,\ C_{12}} \triangle V_j = \sum_j -\overline{E}_j \cdot \overline{\Delta l}_j,$$
$$V_{12} = \int_{C_{12}} dV = \int_{C_{12}} -\overline{E} \cdot \overline{dl}. \tag{1.24}$$

If one can assume that energy is conserved in moving a charge between the two points, then calculation of the potential difference would be independent of the specific path. Equivalently speaking, under this assumption, the potential difference between the two points would be strictly a function of the two end points. Accordingly, any convenient path between the two points might be selected for the above integration. This would be analytically powerful. However, the above assumption may be strictly valid only for fields that are not time varying, and may be used only in an approximate sense for conditions of slow time variation or for relatively small path lengths. For general time-varying electromagnetic fields, where additional energy can be indirectly induced through the magnetic field, the assumption will have to be carefully evaluated. In any event, calculation of the potential or energy difference, as a charge travels between two points, is still a rigorous concept for general time-varying conditions, so long as a particular path between the points is specified. This is highlighted further in the following, in the context of an electric circuit, which underscores general limitation of the potential concept.

1.4.1. *Potential Concept in Circuits for General Time-Varying Conditions.* Energy is gained by a positive charge when it moves from a lower to a higher potential, whereas energy is released from the charge when it moves from a higher to a lower potential. Consider an electric circuit, where energy is supplied by a battery source, allowing charge to flow from a lower to a higher potential, and the energy is then released as the charge flows through other devices from a higher potential to a lower potential. This process continually maintains the current flow in the circuit. Assuming that energy is conserved, the potential gained across a battery source should be equal to the potential drop across different circuit elements as the current completes a circuit to return to the starting point at the source.

The above observation is true assuming that no additional mechanism of energy transfer (or induced energy) is available in the process. Energy induction is possible through Faraday's law if the circuit is placed inside a time-varying magnetic field, which might have been produced by time-varying currents in the circuit itself, or by other external sources. In such situations with induced energy, the potential change as a charge returns back to the same point after completing a circuit path would not be zero. In fact, it would in general depend upon the specific shape and size of the circuital path through which the charge completes the loop. Therefore, the potential energy at the given point would no longer be meaningful. Accordingly, the scalar potential or voltage concept commonly used in circuit analysis, which assigns specific potentials to nodal points, is strictly defined only for circuits with a constant current flow (direct current or DC circuits), and when any surrounding field that might exist is also time-invariant in nature. In contrast, the vector electric field is defined for general situations with a constant or time-variant processes. For slowly time-varying circuits, or sufficiently small circuit loops, any induced potential would be small compared to the primary circuit potentials. In such situations, the nodal potential concept may still be useful for understanding and approximate analysis. The issue of potential change in a circuit loop would be specifically covered in the context of Kirchoff's Voltage Law (KVL) in the following chapter 2, section 1.

1.4.2. *Power and Power Density.* Consider a current flow I in the medium, along the vector $\overline{\triangle l}_{12}$, in the presence of an electric field. Over a time period $\triangle t$ the amount of charge transferred across the length $\overline{\triangle l}_{12}$ would be $\triangle Q = I\triangle t$. As per the definition of the potential, the energy $\triangle W$ needed for the flow can be calculated by multiplying $\triangle V_{12}$ and $\triangle Q$.

$$\triangle W = \triangle Q \triangle V_{12} = -I\triangle t(\overline{E}\cdot\overline{\triangle l}_{12}) = -\triangle t\triangle l_{12}(\overline{E}\cdot\overline{I}), \quad \triangle l_{12}\overline{I} = I\overline{\triangle l}_{12}. \tag{1.25}$$

Accordingly, the power $\triangle P$ used by the current, defined to be the work needed for the charge flow per unit time, can be expressed as,

$$\triangle P = \frac{\triangle W}{\triangle t} = I\triangle V_{12} = -\triangle l_{12}(\overline{E}\cdot\overline{I}). \tag{1.26}$$

The above power would be positive when the charges gain energy, and negative when the charges dissipate energy. Therefore, if one is interested to determine the power loss in the process, the sign of the above expression would have to changed. Further, if the current is distributed over a cross section area $\triangle A$, the the current I may be expressed in terms of the current density vector $\overline{J}$.

$$\triangle P = J\triangle A\triangle V_{12} = -\triangle l_{12}\triangle A(\overline{E}\cdot\overline{J}). \tag{1.27}$$

where $\triangle A\triangle l_{12}$ is the volume element over which power is utilized. The distribution of electrical power in the medium can then be quantified by defining a power density P_d as the ratio of the power to the volume element. The total power used over an arbitrary volume can then be expressed as the volume integral of the power density.

$$P_d = \frac{\triangle P}{\triangle l_{12}\triangle A} = -\overline{E}\cdot\overline{J}. \tag{1.28}$$

The power density is conveniently expressed as the negative dot product of the electric field and the current density vectors.

1.5. Magnetic Field. The magnetic field $\overline{H}$ is described as the property of a vector field which determines the force experienced by a moving electric charge at a given location. The direction of the force vector $\overline{F}$ is anticipated to be orthogonal to the direction of the velocity $\overline{v}$ of the moving charge. This force is in addition to any non-zero force on the charge due to an electric field that may exist at the location. A static (time-independent) magnetic field is produced due a constant flow of charges or a constant current. However, for general time-varying conditions, the magnetic field can be produced due to currents as well as time-varying electric fields. The relationship between the magnetic field and the source current or local electric field is governed the by the general Ampere's law, or by the Biot-Savart's law specifically for the case of a time-invariant current element. These laws would be established in the following chapter 2 as a part of the general electromagnetic theory.

In its basic form, the magnetic field is defined in proportion the force vector $\overline{F}$ experienced by a given charge Q moving with a given velocity vector $\overline{v}$ in the free space.

$$\overline{F} = Q\mu_0\overline{v}\times\overline{H}, \tag{1.29}$$

where μ_0 is simply an appropriate constant of the proportionality. However, an equivalent parameter $\mu = \mu_0\mu_r$, properly modified for a general medium with the $\mu_r = 1$ for the specific case of a free-space medium, is also used in section 1.5.3 for modeling magnetic field in the general medium, as well as in section 2.2 for applying Gauss' law for magnetic field. The $\mu = \mu_0\mu_r$, which determines how the magnetic field interacts or "permeates" in the given medium, is referred to as the magnetic permeability, or simply the permeability, of the medium. Accordingly, the constant $\mu_0 = \mu$, when the μ_r is equal to 1 for the free space, is referred to as the permeability of the free space.

The unit of the magnetic field $\overline{H}$ is Ampere per meter (A/m), in consistency with Ampere's and Biot-Savart's laws in chapter 2, sections 3 and 3.6, which relate the magnetic field to an electric current source. Then, using this unit of the magnetic field in (1.29), the unit of the permeability μ_0 maybe be shown to be $(V/A)/(m/s) = \Omega s/m$ (Ohm second per meter), which is equivalent to Henry per meter (H/m), in reference to the unit of Henry $(H = \Omega s)$ commonly used to measure inductance in electric circuits.

Value of μ_0, the Permeability of the Free Space:

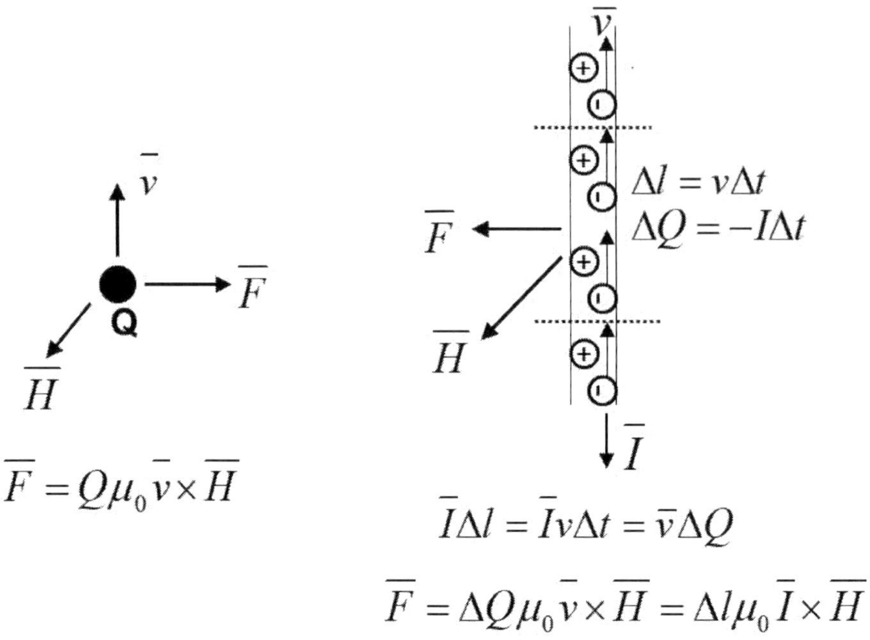

FIGURE 6

For measured values of the established mechanical parameters, force $\overline{F}$ and the velocity $\overline{v}$, the value of the constant μ_0 in an appropriate unit could, in principle, be determined as per the definition (1.29), for any given values of the electro-magnetic parameters charge Q and magnetic field $\overline{H}$, in certain units of choice. Conversely, the units of the given charge and the magnetic field have been adjusted, in consistency with units of all other physical quantities that may be related to them, as per a standard system of measurement, so that the value of the μ_0 is fixed to be exactly equal to $4\pi \times 10^{-7}$ H/m. This is for the sake of simplicity of all resulting theoretical formulations, to be used in a complete development of the electromagnetic theory.

As mentioned, the new magnetic force (1.29) is in addition to that produced by any electric field (1.12), with the new force component experienced only when the charge is in motion. Accordingly, the total force experienced by a moving charge in a free space is the sum of two parts - one due to the electric and the other due to the magnetic fields.

$$\overline{F} = Q(\mu_0 \overline{v} \times \overline{H} + \overline{E}). \tag{1.30}$$

1.5.1. *Force on a Current Element.* The above basic definition of the magnetic field may be extended in terms of force on a current element $\overline{I}$, of length Δl. As discussed before, the total charge in a current element is zero. In every segment of a current there are equal number of positive and negative charges. Therefore, the force experienced by a current segment due to the local electric field is zero. However, there is a non-zero charge "flow," which would result in a force produced due to the magnetic field. At a given time interval Δt, the amount of positive charge that flows along the current direction, through a particular cross-section of the current, is $\Delta Q = I\Delta t$. This is also equivalent to $\Delta Q = -I\Delta t$ amount of charge flowing opposite to the current direction through the same cross section. In general, the vector $\triangle Q\overline{v}$ would always be directed along the current vector $\overline{I}$. The above charge $\triangle Q$ is spread over a length $\Delta l = v\Delta t$, where v is the velocity of charge flow. Accordingly, the following vector relation can be established.

$$\overline{I}\triangle l = \overline{I}\triangle tv = \triangle Q\overline{v}. \tag{1.31}$$

One can find the force on the current element due to the magnetic field by substituting the $\Delta Q\overline{v} = \overline{I}\triangle l$ in the basic magnetic force expression (1.30).

$$\overline{F} = \triangle Q\mu_0\overline{v} \times \overline{H} = \triangle l\mu_0\overline{I} \times \overline{H}. \tag{1.32}$$

Now, because there is no additional force on the current element due to local electric field, the above force represents the total force on the current element. This relationship between the force on a current element and the magnetic field may be used as an alternate measure or definition for the magnetic field. Unlike the force on a moving charge, which includes parts due to electric and magnetic field, in this case the force on a current element is produced exclusively by the magnetic field.

1.5.2. *Torque on a Current Loop.* A current loop would experience a torque in the presence of a magnetic field. The torque may also be used to define the magnetic field in a basic way.

Consider a rectangular elemental current loop $\Delta x \times \triangle y$, placed on the $x-y$ plane, carrying a current $\overline{I}$ (Fig.7). A surface vector $\overline{\Delta S}$ may be defined with magnitude equal to the areal $\triangle S = \triangle x\triangle y$ of the current loop, and direction along normal to the surface. The normal direction to the loop is specified using a standard right-hand rule. Using this rule, the pointing direction of the thumb of the right hand is the normal direction, when the other fingers of the right hand are curled along the current direction. Applying the right-hand rule to the current loop, the normal can be seen to be along the $\hat{z}$ direction.

Unlike a linear current element $\overline{I}\Delta l$, a complete current loop would not experience any total force in a magnetic field. This is because the current elements in the opposing sides of the loop would experience forces which are equal and opposite to each other, These forces would cancel, resulting in a zero total force. However, these forces could produce a non-zero torque $\overline{\tau}$ about the center of the loop, which would lead to rotational movement of the total loop around an axial direction specified by the vector $\overline{\tau}$.

It may be useful to determine the $\overline{\tau}$ in terms of the current I and the surface vector $\overline{\Delta S}$. Toward this end, consider a general magnetic field $\overline{H} = \hat{z}H_z + \hat{x}H_x$, decomposed into components along and orthogonal to the surface vector $\overline{\Delta S}$. Here the direction of the magnetic field component normal to the surface vector (along the plane of loop) is assumed to be the reference direction $\hat{x}$ on the $x-y$ plane. The force $\overline{\Delta F}$ on the current elements $\pm I\triangle y\hat{y}$ on sides 1 and 3 of the loop may be expressed as follows. The upper and lower sign of a dual-sign notation such as $\pm$ would refer to sides 1 and 3, respectively.

$$\overline{\triangle F} = \pm\triangle yI\mu_0\hat{y} \times \overline{H} = \pm\triangle yI\mu_0\hat{y} \times (H_x\hat{x} + H_z\hat{z}) = \pm\triangle yI\mu_0(-H_x\hat{z} + H_z\hat{x}). \tag{1.33}$$

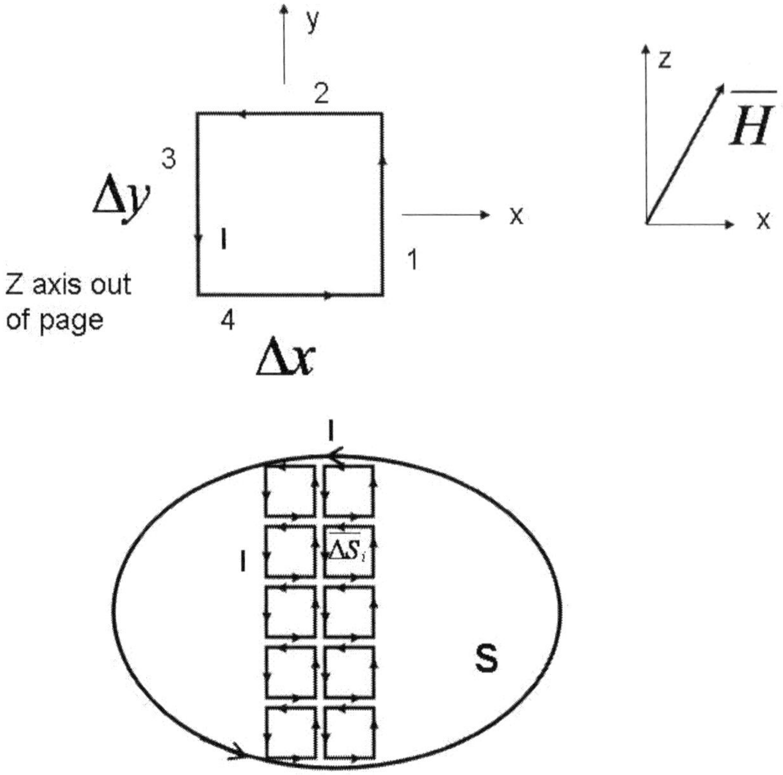

FIGURE 7

The corresponding position vectors $\overline{\triangle r}$ for the center locations of the sides 1 and 3 are $\pm\hat{x}\triangle x/2$. The resulting torque $\overline{\triangle\tau}$ may be expressed as a cross-product of the position vector and the force vector $\overline{\triangle F}$.

$$\overline{\triangle\tau} = \overline{\triangle r} \times \overline{\triangle F} = (\pm\hat{x}\triangle x/2) \times [\pm\triangle y I\mu_0(-H_x\hat{z} + H_z\hat{x})] = (\triangle x\triangle y/2)I\mu_0 H_x\hat{y}. \quad (1.34)$$

Note that the torque for both sides 1 and 3 are the same. The total torque for the two sides will therefore be twice of that for each side.

$$\overline{\triangle\tau} = \triangle x\triangle y I\mu_0 H_x\hat{y}. \quad (1.35)$$

Following similar steps, one may show that the total torque produced due to sides 2 and 4 would be zero. The total torque is therefore equal to the above contribution from the sides 1 and 3. One may express the total torque in terms of a vector cross-product of the surface vector of the loop and the magnetic field vector.

$$\overline{\triangle\tau} = \triangle x\triangle y I\mu_0 H_x\hat{y} = \triangle S I\mu_0\hat{z} \times \overline{H} - I\mu_0\overline{\triangle S} \times \overline{H}. \quad (1.36)$$

According to mechanics of torque, this would mean that the current loop, when placed in an external magnetic field, would experience a "torque action" which would turn the normal axis of the loop towards the magnetic field axis. In other words, the current loop would have a natural tendency to turn and align its normal axis with that of the magnetic field. When the normal axis remains fully aligned with the magnetic field, the torque would be zero as per the above relationship, leaving the loop in a state of equilibrium.

Torque on a General Planar Loop:

A planar current loop of general shape and size may be modeled as a superposition of small rectangular loops filling the total surface of the general loop in a "tile" form (Fig.7). The common sides shared by two neighboring tile elements would have opposing currents that would cancel by superposition. Accordingly, the superposition of the tile form can be described by the current elements only on the outer boundary, thus making the tiled superposition equivalent to the total current loop, The total torque is expressed as the sum of individual torques from the rectangular loop elements. Assuming that the magnetic field is constant over all loop elements, the total torque may be expressed by replacing $\overline{\triangle S}$ in the above expression by the total surface vector $\overline{S}$. For a general planar loop, $\overline{S}$ would have the same direction ($\hat{z}$) as the surface element $\overline{\triangle S}$, but with magnitude equal to its total surface area.

$$\overline{\tau} = \sum_i \overline{\triangle\tau_i} = \sum_i [I\mu_0\overline{\triangle S}_i \times \overline{H}\ ] = I\mu_0[\sum_i \overline{\triangle S}_i\ ] \times \overline{H} = I\mu_0\overline{S} \times \overline{H}. \tag{1.37}$$

1.5.3. *Magnetic Field in a Material Medium.* The magnetic field in a material medium is influenced by alignment of internal orbiting/spinning charges. This process is similar to the alignment of the bound charges under electric field. However, there is no magnetic process in a natural material that is equivalent to flow of conduction charges in an electric field. This is because there are are no independent "magnetic charges" observed in nature.

The magnetic field in the presence of the orbiting/spinning charges inside a material may be identified in two forms: the primary field $\overline{H}_0$, and the total field $\overline{H}_t$ which is the sum of the primary field $\overline{H}_0$ and an additional contribution $\Delta\overline{H}$ due to the alignment of the orbiting/spinning charges of the medium. An equivalent model of the material polarization, using current loops to represent orbiting/spinning charges of the medium, is shown in Fig.8. The relative orientation of the equivalent current loops and the relative directions of different magnetic field components, as shown, are determined by the basic nature/definition of the current loops and magnetic field.

The primary field $\overline{H}_0$ turns the equivalent current loops such that the normal to the loop is aligned along the direction of the magnetic field, as discussed before in section 1.5.2 (see equation (1.36)). Now, the current loops so aligned would produce an additional magnetic field $\Delta\overline{H}$. This is based on the Biot-Savart's law (which relates a current source element to the magnetic field it produces,) to be presented in the following chapter 2, section 3.6. The additional field $\Delta\overline{H}$ would be along the loop's normal axis, as shown in the Fig.8, aligned along the primary field $\overline{H}_0$. Accordingly, this results in having the total field in the material to be greater than the primary field. It may be mentioned, this situation is in contrast with the equivalent situation for the electric field in a material media in section 1.3.1, where the total electric field is less than the primary electric field.

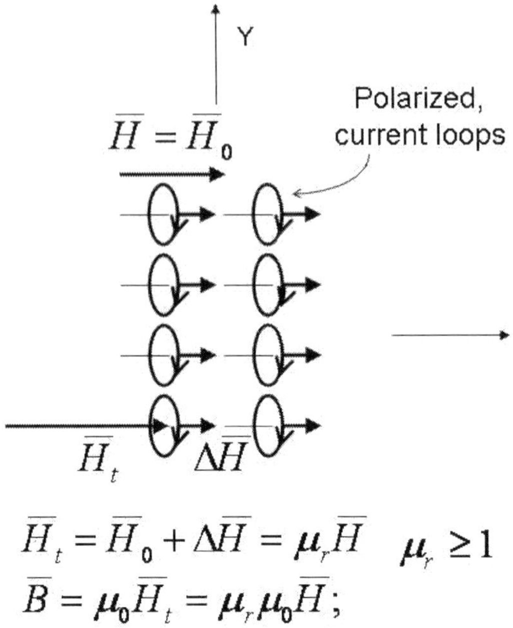

Magnetic field in a material medium, showing its equivalent microscopic model using polarized rotating charges or current loops

FIGURE 8

$$\overline{H}_t = \overline{H}_0 + \Delta\overline{H}; |\overline{H}_t| \geq |\overline{H}_0|. \tag{1.38}$$

The polarized current-loop elements are assumed to be much smaller and spaced sufficiently close to each other, compared to the dimensional scale of practical interest. Accordingly, at a macroscopic scale the field is actually measured as its local average in the following sense. If the sample path is on a plane parallel to the field (xz or xy planes in Fig.8) the average field is $\overline{H}_0$, except when the path falls just inside the polarized current loops in which case the field is $\overline{H}_t$. Whereas, the average field measured is $\overline{H}_t$ as one samples on a plane orthogonal to the field direction (yz plane in Fig.8.)

$\overline{H}_t$ and $\overline{H}_0$ are proportional to each other for a linear medium. Further, consider an "isotropic" medium for which the internal material structure of the medium looks similar as seen from any arbitrary orientation. For such a simple medium the ratio of the proportionality between $\overline{H}_t$ and $\overline{H}_0$ would be independent of the specific vector direction of the field. This ratio is defined as the relative magnetic permeability μ_r of the medium. For material with a general "anisotropic" structure, the proportionality between $\overline{H}_t$ and $\overline{H}_0$ would be mathematically a bit more involved, not considered here.

With more generality, the $\overline{H}_t$ may be linearly dependent on the time-derivatives of the $\overline{H}_0$, in addition to being proportionately dependent on the $\overline{H}_0$ as assumed above. Such a medium is referred to as a dispersive magnetic medium, not considered here. With further generality, the $\overline{H}_t$ may be a non-linear function of the $\overline{H}_0$ and its time-derivatives, when the fields are sufficiently strong. Such non-linear behavior is also not considered.

$$\mu_r = \frac{\overline{H}_t}{\overline{H}_i} \geq 1; \ |\overline{H}_t| \geq |\overline{H}_0|. \tag{1.39}$$

μ_r represents the relative strength of the alignment of the charge rotation in the medium, and is greater than or equal to unity because $|\overline{H}_t|$ is greater than or equal to $|\overline{H}_0|$, as discussed. For a free space medium, $|\overline{H}_t|$ and $|\overline{H}_0|$ are clearly equal to each other, leading to having $\mu_r = 1$ for the free space.

Magnetic Field Intensity and Flux Density:

It is now clear that the magnetic field inside a material medium does not have a unique meaning. The two fields, $\overline{H}_t$ and $\overline{H}_0$, have distinct physical significance. However, for analytical and notational convenience, when one refers to the magnetic field $\overline{H}$ in a material medium, it normally refer to $\overline{H}_0$ (see equation 2.39), which may be called the magnetic field intensity. Whereas, $\overline{H}_t$ is defined in terms of an magnetic flux density $\overline{B} = \mu_0 \overline{H}_t$, named in reference to the concept of a flux to be used in Gauss' law of magnetic field in section 2.2. It may be noted that the equivalent situation in the electric field is a bit different, in which case the primary electric field is used to define the electric flux density and the total field is used to define the field intensity. The relationship between $|\overline{H}_t|$ and $|\overline{H}_0|$ leads to relating the flux density and the field intensity.

$$\begin{aligned} \overline{H} = \overline{H}_0 &= \frac{\overline{H}_t}{\mu_r}, \\ \overline{B} = \mu_0 \overline{H}_t &= \mu_0 \mu_r \overline{H}, \\ \overline{B} = \mu \overline{H}; \mu &= \mu_0 \mu_r. \end{aligned} \tag{1.40}$$

The ratio between the magnetic flux density $\overline{B}$ and the total magnetic field $\overline{H}_t$ is the physical constant μ_0. The ratio between the flux density $\overline{B}$ and the field intensity $\overline{H}$ is $\mu = \mu_0 \mu_r$, which is referred to as the magnetic permeability of the medium. When the medium is free space, then μ_r is unity and $\mu = \mu_0$. Accordingly, μ_0 is referred as the permeability of the free space.

The unit of measurement of the magnetic field intensity $\overline{H}$ is Ampere per meter (A/m), in accordance with Ampere's law, to be presented in the following chapter 2, section 3. The unit of measurement of the magnetic flux density $\overline{B}$ is Weber per square meter (W_b/m^2), where Weber is the unit of magnetic flux. As per Gauss' law for magnetic field, presented in the following section, the total magnetic flux over a closed surface is zero, because no magnetic charge exists in nature. This is in contrast to the total electric flux over a closed surface, which may not be zero, if there is any non-zero electric charge enclosed inside the surface. Because magnetic charge does not exist, the magnetic field does not "originate" at any place. and the total magnetic flux should "flow" continuously across all material media and interfaces, and "loop back" completing a full path. Based on the units of $\overline{B}$ and $\overline{H}$, the unit of measurement of permeability μ is $W_b/(Am)$, which is equal to Henry per meter (H/m) using the relation $H = W_b/A$. The relative permeability μ_r is a dimensionless parameter.

2. Gauss' Laws - The Basic Field Laws

Two independent Gauss' laws, one for the electric field and the other for the magnetic field, constitute the most basic field principles of electromagnetic theory, upon which all electromagnetic phenomena are founded. They are presented in the following separately for the electric and magnetic fields. The two are expressed in similar forms, with some fundamental physical difference. That is, the basic source of electric field is electric charge, whereas no basic or primary source of magnetic field, so called "magnetic charge," exists in nature. Magnetic field is produced from any flow of electric charge, or any time variation in the electric field, intercoupled through a general electromagnetic theory to be developed in the following chapter 2. Accordingly, both electric and magnetic fields are one way or other produced from the electric charge, thus recognizing the electric charge to be the elemental physical source of all electromagnetic phenomena.

2.1. Gauss' Law for Electric Field: The relationship between the electric charge and the surrounding electric field is described by Gauss' law of electric field. In its basic form, Gauss' law is applicable for a free-space medium, which can then be extended to a material medium by suitable adjustments.

2.1.1. *Gauss' Law for Electric Field in a Free-Space Medium:* The basic form of Gauss' law relates the electric field to the electric charge in a free-space medium. The law is expressed in the form of an integral or summation over a closed surface S, and is based on the concept of an electric "flux." Let the closed surface S be decomposed into a number of small surface elements ΔS_i. Consider a small surface element ΔS_i, where the local electric field is $\overline{E}_i$, and the flux density vector is defined to be $\overline{D}_i = \epsilon_0 \overline{E}_i$. Now define an elemental flux $\Delta\Phi_i$ over the surface element ΔS_i as the product of the surface area ΔS_i and the component $D_{ni} = \epsilon_0 E_{ni}$ of the flux density vector along the outward normal direction to the surface element. One may also conveniently define the flux element $\Delta\Phi_i$ in the following vector form.

$$\Delta\Phi_i = D_{ni}\Delta S_i = \overline{D}_i \cdot \overline{\triangle S}_i = \epsilon_0 \overline{E}_i \cdot \overline{\triangle S}_i; \; \overline{\triangle S}_i = \triangle S_i \hat{n}_i. \tag{1.41}$$

The total flux Φ coming out of the closed surface S is calculated by adding incremental flux elements $\Delta\Phi_i$ over all surface elements ΔS_i. Gauss' law is then stated in the following simple form. The total electric flux Φ, as defined and computed above over a closed surface S, would be equal to the total electric charge Q enclosed inside S. The total charge Q may be in the form one or more number of discrete charges, or may occur in a distributed from inside the closed surface.

$$\Phi = \sum_i \Delta\Phi_i = \sum_i \overline{D}_i \cdot \overline{\triangle S}_i = \sum_i \epsilon_0 \overline{E}_i \cdot \overline{\triangle S}_i = Q. \tag{1.42}$$

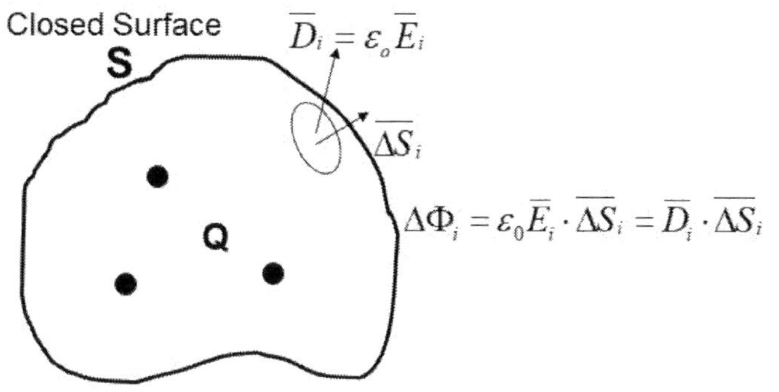

Gauss' law for electric field, applied in a free-space medium

FIGURE 9

The summation over the surface may also be expressed as a closed surface integral, in the limit when the surface area elements approach zero, or equivalently the number of surface elements on the closed surface approaches infinity.

$$\Phi = \oiint_S d\Phi = \oiint_S \overline{D} \cdot \overline{dS} = \oiint_S \epsilon_0 \overline{E} \cdot \overline{dS} = Q. \tag{1.43}$$

The fundamental concept is that, the surface-integral of the electric field $\overline{E}$ over a closed surface is a proportionate measure of the enclosed electric charge Q. The appropriate constant of the proportionality is the parameter ϵ_0, applicable for Gauss' law in a free-space medium, and is referred to as the permittivity of the free space (see section 1.3.1). Accordingly, the flux density vector $\overline{D}$ in 1.43 is defined as ϵ_0 times the electric field vector $\underline{E}$.

Value of ϵ_0, the Permittivity of the Free Space:

Consider a given spherical charge Q placed in a free-space medium. The electric field $\overline{E}$ the charge produces at a given radial distance r, in a spherical symmetry around the charge, maybe measured using the basic definition (1.12) of the electric field, from which the total surface integral of the electric field on a sphere of the radius r centered around the charge can be computed. Accordingly, the value of the constant ϵ_0 maybe deduced as per Gauss' law relation (1.43), using the above measured electric-field surface integral for the given charge Q. An accurate measured value of the ϵ_0 is $8.8541878176 \times 10^{-12}$ F/m (Farad per meter), which for analytical convenience may be approximated as $1/(36\pi) \times 10^{-9}$

F/m, within 0.2% of difference from the measured value. It may be noted, in contrast to such a measured value of the free-space permittivity ϵ_0, with inaccuracy inherent in the measurement, the value of the free-space permeability μ_0 is exactly equal to $4\pi \times 10^{-7}$ H/m, fixed as per the modern standard system of measurement (see section 1.5).

2.1.2. *Modification for a Material Medium:* Gauss' law, applicable in its basic for a free-space medium, may now be modified for a material medium with suitable adjustments. As discussed in the last section, the microscopic behavior of a material medium in the presence of an electric field may be modeled using polarized bound charges placed in a free-space medium. Because these equivalent bound charges are placed in a free-space medium, we can use the basic form of Gauss' law established for a free-space medium, as long as the bound charges are properly accounted for inside a suitable closed surface integral. Because the law applies to any arbitrary closed surface, we may deform a small surface element $\overline{\Delta S}_i$ around a bound charge-pair, such that one part $\Delta S_i \cos\theta_i$ is perpendicular to the local electric field, and the other parts are parallel to the field directions. The deformation avoids stepping inside a bound charge pair, as shown in figure 10. The figure shows the deformation as a two-dimensional illustration for convenience. The actual three-dimensional surface deformation would include four sides (one front and one back element parallel to field, in addition to the top and right-side elements shown in Fig.10), out of which only one (right side) element ($=\Delta S_i \cos\theta_i$) is perpendicular, with other three parallel to the electric field. As per the definition of the electric flux, discussed before, the flux through the three surface elements that are parallel to the field would be zero. Accordingly, the total flux in the deformed surface element is equal to that through the perpendicular element $\Delta S_i \cos\theta_i$.

$$\Delta\Phi_i = \Delta\Phi_{i1} + \Delta\Phi_{i2} + \Delta\Phi_{i3} + \Delta\Phi_{i4} = \Delta\Phi_{i1};\ \Delta\Phi_{i2,3,4} = 0, \tag{1.44}$$

$$\Delta\Phi_i = \Delta\Phi_{i1} = \epsilon_0 E_{0i} \triangle S_i \cos\theta_i = \epsilon_0\epsilon_{ri} E_i \triangle S_i \cos\theta_i = \epsilon_i E_i \triangle S_i \cos\theta_i = \epsilon_i \overline{E}_i \cdot \overline{\triangle S}_i, \tag{1.45}$$

$$\Delta\Phi_i = \overline{D}_i \cdot \overline{\triangle S}_i;\ \overline{D}_i = \epsilon_i \overline{E}_i,\ \epsilon_i = \epsilon_0\epsilon_{ri}. \tag{1.46}$$

We have now expressed the flux contribution in the deformed surface element as a flux contribution defined in the original surface element. The deformation is such that a complete, closed deformed surface would include both the positive and negative charges of a bound charge pair, all adding up to zero charge contributed by the bound charges of the material. Accordingly, the total charge inside the deformed surface would be equal to any "free" charge that may be present inside the surface. Now, let us apply the basic form of Gauss' law to the deformed surface S'.

$$\Phi = \sum_i \Delta\Phi_i = \sum_i \overline{D}_i \cdot \overline{\triangle S}_i = \sum_i \epsilon_i \overline{E}_i \cdot \overline{\triangle S}_i = Q,\ \epsilon_i = \epsilon_0\epsilon_{ri}, \tag{1.47}$$

$$\Phi = \oint\!\!\!\oint_S d\Phi = \oint\!\!\!\oint_S \overline{D} \cdot \overline{dS} = \oint\!\!\!\oint_S \epsilon \overline{E} \cdot \overline{dS} = Q,\ \epsilon = \epsilon_0\epsilon_r. \tag{1.48}$$

where, the total charge Q refers to any additional "free" charge one may have inside the surface. This excludes the internal bound charges of the medium, which add up to zero, as explained. As can be seen, the new Gauss' law in a material medium is expressed in a very

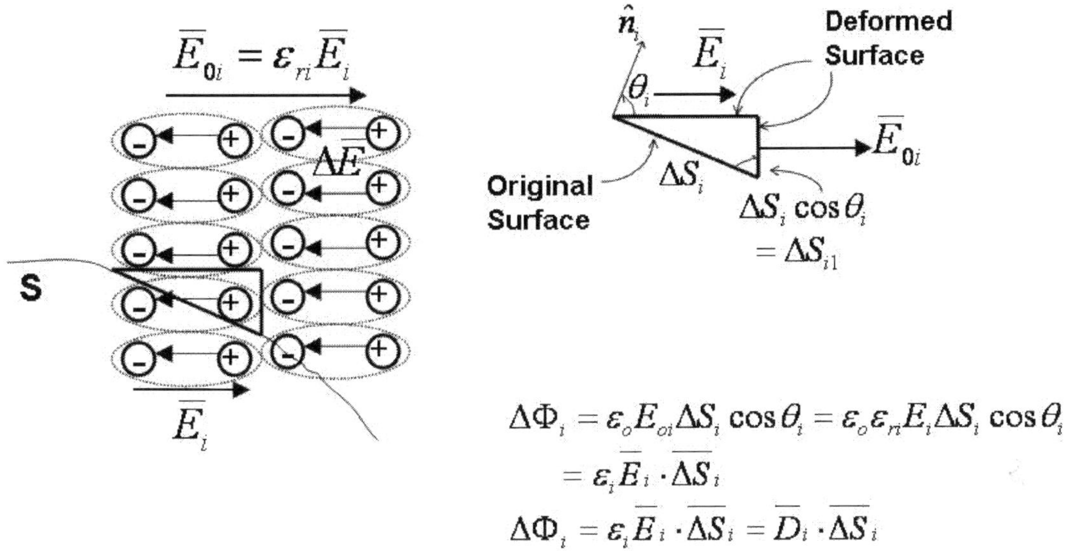

Deformation of integration surface around polarized dipoles in a material medium. The deformation is such that the total bound charge inside a closed integration surface S would be zero. Accordingly, the total charge in a closed surface S would be equal to any "free" charge that may be present.

FIGURE 10

similar form as that (1.43) in a free-space medium. The key differences is that permittivity ϵ_0 of the free space, used in Gauss' law (1.43) for an empty space, is conveniently replaced in Gauss' law (1.48) for a material medium by an equivalent permittivity $\epsilon = \epsilon_0\epsilon_r$ of the material medium.

Differential Form Using Divergence Operator:

Apply (1.48) to a surface $\triangle S$, enclosing a volume element $\triangle V$. Then apply divergence-operator concept (appendix A, section 2.3) to the left side of (1.48) and express the enclosed charge $\triangle Q$ in the right side in terms of the volume charge density ρ_v ($\triangle Q = \rho_v \triangle V$). Accordingly, (1.48) maybe shown to be equivalently expressed in a differential form using the divergence operator.

$$\oiint_{\triangle S} \overline{D} \cdot \overline{dS} = (\overline{\nabla} \cdot \overline{D})\triangle V = \triangle Q = \rho_v \triangle V, \ \overline{\nabla} \cdot \overline{D} = \rho_v. \tag{1.49}$$

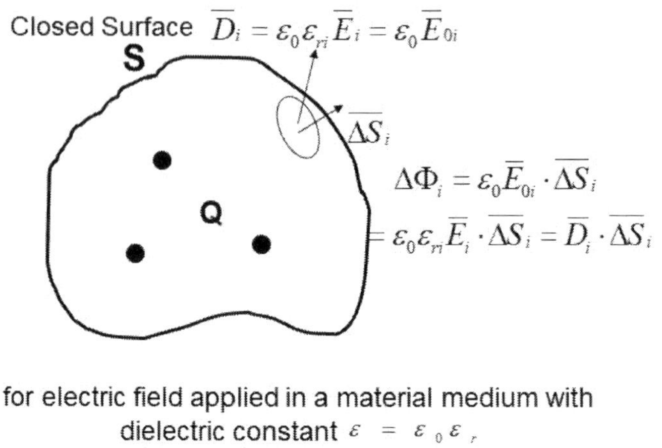

FIGURE 11

2.1.3. *Coulomb's Law.* Note that Gauss' law does not explicitly relate the electric field at every location of the surface S. It only relates the total integral of the electric field or flux over a closed surface to the total charge enclosed by the surface. However, for specific simple situations, Gauss' law may be simplified directly relating the local electric field to a specific source charge or charge distribution. The simplest case is an isolated, stationary point charge Q, placed in a uniform dielectric medium with its dielectric constant ϵ. The problem has a clear spherical symmetry, using which one may assume that the electric field is directed in the radial direction, with its magnitude dependent only on the radial distance of observation from the source charge. Now, apply Gauss' law to a spherical surface S of radius r, centered around the charge. In this case, the normal vector on each surface element is directed along the radial direction $\hat{r}$. Also, the electric field is constant over the entire surface S. This leads to the following simplification.

$$\Phi = \sum_i \overline{D}_i \cdot \overline{\triangle S}_i = \sum_i D_{ri}\triangle S_i = D_r \sum_i \triangle S_i = D_r S = \epsilon 4\pi r^2 E_r = Q, \tag{1.50}$$

$$\overline{E} = \hat{r}E_r = \frac{Q\hat{r}}{4\pi\epsilon r^2}. \tag{1.51}$$

The above simple relationship between a stationary point charge and its electric field in a uniform medium is referred to as Coulomb's law. If there are multiple number of electric charges, Coulomb's law may be applied for each charge, and the total field is then computed through superposition. For a volume charge distribution, Coulomb's law may be similarly applied to charge in a volume element, and the total field is then computed through integration. However, for certain charge distributions with suitable symmetry, as for a point charge in the above derivation, Gauss' law may be applied with a suitable choice of the closed surface S to derive the electric field directly using a relatively simple procedure (see problems).

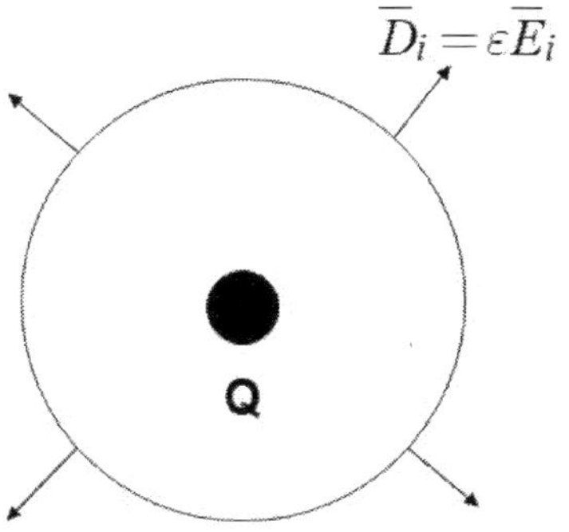

$$\Delta\Phi_i = \varepsilon \overline{E}_i \cdot \overline{\Delta S}_i = \varepsilon E_r \Delta S_i$$
$$\Delta\Phi = \sum \Delta\Phi_i = \varepsilon E_r \sum \Delta S_i = 4\pi\varepsilon r^2 E_r = Q$$
$$E_r = \frac{Q}{4\pi\varepsilon r^2}$$

Gauss' law applied to derive the electric field due to a stationary, time Invariant, point charge Q, placed in a uniform medium with a dielectric constant ε.

FIGURE 12

It may be emphasized that the above simplification in Gauss' law, in order to derive Coulomb's law, would not be possible if the charge was not stationary. For a charge moving in a given direction the spherical symmetry of the problem is intrinsically broken, biased by the specific direction of motion, thus making the above simplification and the resulting Coulomb's law invalid in this case. Similarly, consider another situation where the magnitude of the given charge at the fixed location were changing with time. This would mean that the incremental charge has to be arriving at (or exiting from) the location in a transitional state of motion, through some current flow that continually supplies the charge increment. The transitional motion of the charge in any preferential direction would break the original spherical symmetry assumed in (1.50,1.51), thus invalidating the derivation of Coulomb's law for the associated time-varying charge. In fact, a time-varying charge in isolation, without specifying the required current flow in a complete system, is an unphysical problem that violates charge conservation, making the electric field of the time-varying charge fundamentally meaningless. Therefore, seeking a derivation of the electric field for the time-varying charge in isolation, similar to the derivation in (1.50,1.51), maybe a meaningless exercise to begin with. Instead, any general time-varying field that may involve time-varying charges at different locations, should be analyzed as a complete system for physical consistency, which can be modeled as a superposition of different stationary and/or moving charges. Because the spherical asymmetry associated in an individual moving charge can not be modeled by the spherically symmetric Coulomb's field, as discussed above, any superposition of the Coulomb's field can not model a general time-varying field.

In summary, Coulomb's law explicitly models the electric field of a point charge, strictly for an "electrostatic" (stationary and time-invariant charge) condition. Whereas, Gauss' law from which we may derive Coulomb's law upon simplification, would be valid for any static as well as general time-varying fields. Due to its mathematical and conceptual simplicity, Coulomb's law can be still useful to derive exact electrostatic fields for a general charge distribution, directly through integration of the Coulomb fields of individual charge elements in the distribution. These electrostatic fields are useful as good first-order approximations for the rigorous fields, in many practical problems with sufficiently slow time variation.

2.1.4. *Coulomb's Law or Gauss' Law, Which is the First Principle of Electricity?* It may be useful to debate the relative significance of Gauss' law as compared to Coulomb's law, from fundamental and philosophical considerations. For analytical and physical understanding, the explicit expression of Coulomb's law, exhibiting its inverse-square relationship of the electric field with distance, may appear simple and intuitive. The classic inverse-square relationship naturally appears also in our day-to-day physical world of three spatial dimensions, through material and energy conservation principles. For example, the density of water captured from a spherically-symmetric lawn sprinkler, or the intensity of light seen from a household light-bulb, are known to exhibit the inverse variation with square of the distance from the source. A fixed flow of water or light energy originating respectively from the sprinkler or the light-bulb, is conserved as they pass through any spherical surface around the respective central source, independent of the radius of the sphere. In analogy, the electric flux originating from a source charge is the quantity that is conserved in Coulomb's law. The conserved electric flux in Coulomb's law, which is the product of the radial flux density $D_r = \epsilon \times E_r$ and the area of the sphere $S = 4\pi r^2$, is defined as the source charge Q. Accordingly, the inverse-square relationship is essentially established on a geometric principle, that the area of a two-dimensional spherical surface, placed in a three-dimensional space, is proportional to the square of the radius. Philosophically, if we were perhaps living instead in a two or four dimensional world, for example, Coulomb's law would have exhibited an inverse relationship with the radius or the radius-cubed, respectively.

In any case, the inverse-square relationship of Coulomb's law, which is explained as a simple conservation principle in our three-dimensional world, is also validated by experimental measurements. Like Coulomb's law for electric field, Newton's law for gravitational field also exhibits the classic inverse-square relationship for force with distance, where the flux of the gravitational field is assumed to be conserved. Accordingly, it is naturally appealing to establish the inverse-square dependence of Coulomb's law as the first principle of the electric field, produced by an elementary point charge.

Now, the explicit relationship of Coulomb's law, between a specific source charge and its electric field measured at a specific distance, maybe easier to introduce and conceptualize first, than the integral form of Gauss' law which only implicitly establishes the source-field relation, through integration of the observed field over an entire enclosed surface. The integral relationship of Gauss' law could, however, be subsequently introduced as a deduction from Coulomb's law, that is generalized for an arbitrary charge distribution and for any closed surface with any shape/size for flux calculation (see problems). The deduction would be possible by first showing that the Coulomb's field for a point charge satisfies Gauss' law for a small closed surface (or volume) element (preferably in spherical coordinates), placed around or outside the source charge. This would lead to proving validity of Gauss' law for the basic Coulomb's field, for an arbitrary shaped closed surface (or volume) by superposing the volume elements and the associated flux contributions. Validity of Gauss' law for an arbitrary charge distribution can then be established by simply superposing individual fields, and the resulting Gauss' law expressions, produced by different charge elements placed at different locations. We do not focus on the details here.

However, one would subsequently recognize that the basic Coulomb's law for a point charge, which relates the electric field at a particular location and time instant, with the distance r of the source charge from the given location measured at the same instant, could not be applicable for the charge with a general motion. Otherwise, Coulomb's law (1.51) with a time-varying distance r for a moving charge, would lead to an instantaneous change in the electric field that can be measured at the given location, which is produced due to any instantaneous change in position of the charge occurring at any distant location. Such an instantaneous cause-effect relationship between the source and field at two distant locations would be an invalid, non-physical condition. As discussed earlier, one would also recognize that the basic Coulomb's law (1.51) for a point charge, which is a spherically symmetric expression, could not be valid for a charge under motion, for which the intrinsic spherical symmetry is broken. Therefore, in order to properly model the time-varying field of a charge with a general motion, and accordingly to model any general time-varying field which maybe described as a superposition of different moving charges in general motion, the field solution for a single source charge must allow general spherical asymmetry of the field around the charge.

With the above objective, in contrast to the symmetric Coulomb's field which is expressed with explicit dependence with charge and distance, possible solutions to Gauss' law, due to its implicit integral form, would include a more general set of field distributions. For example, in the derivation of (1.50,1.51), one could establish field configurations that are not spherically symmetric, in order to accommodate any possible asymmetry in the problem, but produce the same required total flux around a given charge as the symmetric Coulomb's field for a static condition. Accordingly, the possible general solutions of Gauss' law would cover the symmetric static as well as any asymmetric dynamic conditions, superseding the simple solution of Coulomb's law, and without suffering from the inevitable contradictions of Coulomb's law discussed above. Therefore, the integral expression of Gauss' law maybe accepted as a more universal relation, valid for any general condition of asymmetry, motion or time variation.

With this advanced viewpoint, for analytical and conceptual generality, it may be inevitable to accept Gauss' law as the most fundamental principle for charge-field relationship. Coulomb's law, although it might appear conceptually and analytically simple to introduce at first, due to its limited scope and validity can only be considered as a specific solution to Gauss' law under a simple, electrostatic condition. Further, because Gauss' law is now the only general charge-field relationship available, the basic measure of charge under arbitrary, unknown conditions of motion and time variation, can be defined only through the flux integration of Gauss' law. Under the general conditions, there would be no other definitive way to measure the charge. This is a significant philosophical understanding.

2.2. Gauss' Law for Magnetic field. As mentioned earlier, Gauss' law for magnetic field takes a similar form as that for the electric field, except that the total magnetic charge is zero. Like for the electric field, in its basic form, Gauss' law for magnetic field is expressed for a free-space medium, which is then extended for a material medium by suitable adjustment using an equivalent magnetic permeability of the medium. Like in Gauss' law for electric field, Gauss' law for magnetic field maybe simplified to find the magnetic field of a hypothetical static magnetic charge (equivalent to Coulomb's law for a magnetic charge). Such a field may, however, be of little practical significance because no physical magnetic charge is known to exist.

2.2.1. *Gauss' Law for Magnetic Field in a Free-Space Medium:* Gauss' law for magnetic field in a free-space medium is expressed using the concept of magnetic flux Φ_m. The Φ_m is expressed as a surface integral over a closed surface S of the magnetic flux density vector $\overline{B}$, defined as $\overline{B} = \mu_0 \overline{H}$. μ_0 is the permeability of the empty or free space. The process is similar to defining the electric flux Φ in (1.41,1.42).

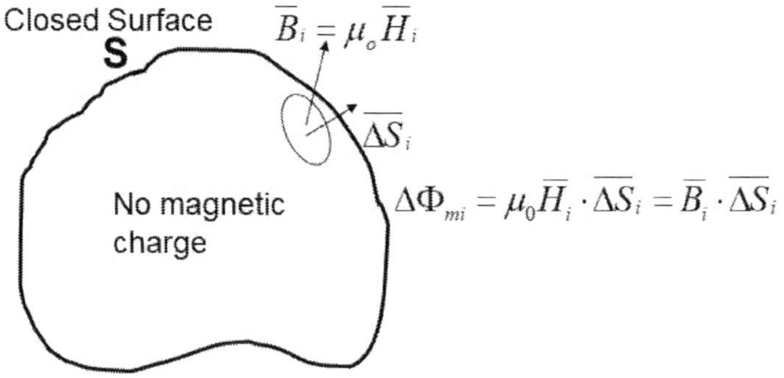

Gauss' law for magnetic field, applied in a free-space medium

FIGURE 13

$$\Delta\Phi_{mi} = B_{ni}\Delta S_i = \overline{B}_i \cdot \overline{\Delta S}_i = \mu_0 \overline{H}_i \cdot \overline{\Delta S}_i, \tag{1.52}$$

$$\Phi_m = \sum_i \Delta\Phi_{mi} = \sum_i \overline{B}_i \cdot \overline{\Delta S}_i = \sum_i \mu_0 \overline{H}_i \cdot \overline{\Delta S}_i = 0, \tag{1.53}$$

$$\Phi_m = \oint\!\!\oint_S d\Phi_m = \oint\!\!\oint_S \overline{B} \cdot \overline{dS} = \oint\!\!\oint_S \mu_0 \overline{H} \cdot \overline{dS} = 0. \tag{1.54}$$

2.2.2. *Modification for a Material Medium:* The microscopic behavior of a material medium in the presence of a magnetic field is modeled using polarized orbiting/spinning charges, or equivalent current loops, placed in a free-space medium. This is discussed in the last section. Because the equivalent polarized currents are present in a free-space medium, we can apply the basic Gauss' law for the free-space medium. Also, because the law applies to any arbitrary closed surface, we may deform a small surface element $\overline{\Delta S_i}$ around a polarized current element, such that one part $\Delta S_i \cos\theta_i$ is perpendicular to the local magnetic field, and the other parts are parallel to the field directions. This is shown in Fig.14 as a two-dimensional illustration for convenience. The actual three-dimensional surface deformation would include four sides (one front and one back element parallel to field, in addition to the top and right-side elements shown in Fig.14), out of which only one (right side) element ($=\Delta S_i \cos\theta_i$) is perpendicular, with other three parallel to the electric field. As per the definition of the magnetic flux, discussed before, the flux through the three surface elements that are parallel to the field would be zero. Accordingly, the total flux in the deformed surface element is equal to that through the perpendicular element $\Delta S_i \cos\theta_i$. Accordingly, the magnctic flux density is defined using the total magnetic field as $\overline{B} = \mu_0 \overline{H}_t$. Further, $\overline{H}_t$ may be related to the magnetic field intensity $\overline{H}$ through relative permeability of the medium.

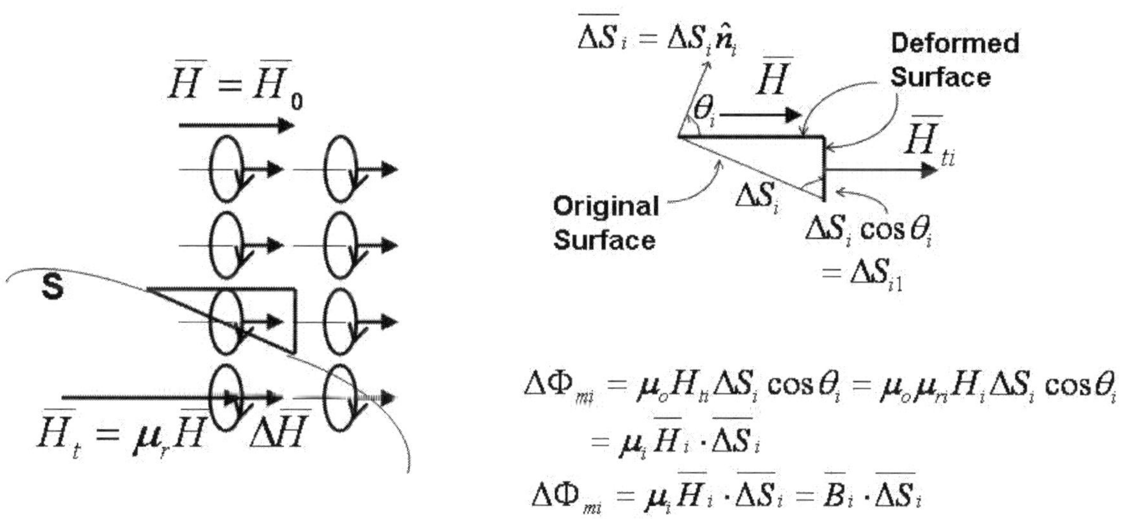

Deformation of integration surface around polarized current loops in a material medium, for application of the Gauss' Law for magnetic field.

FIGURE 14

$$\Delta\Phi_{mi} = \Delta\Phi_{mi1} + \Delta\Phi_{mi2} + \Delta\Phi_{mi3} + \Delta\Phi_{mi4} = \Delta\Phi_{mi1};\ \Delta\Phi_{mi2,3,4} = 0, \qquad (1.55)$$

$$\Delta\Phi_{mi} = \Delta\Phi_{mi1} = \mu_0 H_{ti} \triangle S_i \cos\theta_i = \mu_0 \mu_{ri} H_i \triangle S_i \cos\theta_i = \mu_0 \mu_{ri} \overline{H}_i \cdot \overline{\triangle S}_i, \tag{1.56}$$

$$\Delta\Phi_{mi} = \overline{B}_i \cdot \overline{\triangle S}_i; \ \overline{B}_i = \mu_i \overline{H}_i, \ \mu_i = \mu_0 \mu_{ri}. \tag{1.57}$$

We have now expressed the flux contribution in the deformed surface element as a flux contribution defined in the original surface element $\overline{\triangle S}_i$. Now, the original form of Gauss' law for a free-space medium may be applied.

$$\Phi_m = \sum_i \Delta\Phi_{mi} = \sum_i \overline{B}_i \cdot \overline{\triangle S}_i = \sum_i \mu_i \overline{H}_i \cdot \overline{\triangle S}_i = 0, \ \mu_i = \mu_0 \mu_{ri}, \tag{1.58}$$

$$\Phi_m = \oiint_S d\Phi_m = \oiint_S \overline{B} \cdot \overline{dS} = \oiint_S \mu \overline{H} \cdot \overline{dS} = 0, \ \mu = \mu_0 \mu_r. \tag{1.59}$$

As can be seen, the new Gauss' law (1.59) in a material medium is expressed in a very similar form as the basic Gauss' law (1.54) in a free-space medium. The simple differences is that permeability μ_0 of the free space in (1.54) is replaced in (1.59) by an equivalent permeability $\mu = \mu_0 \mu_r$ of the material medium.

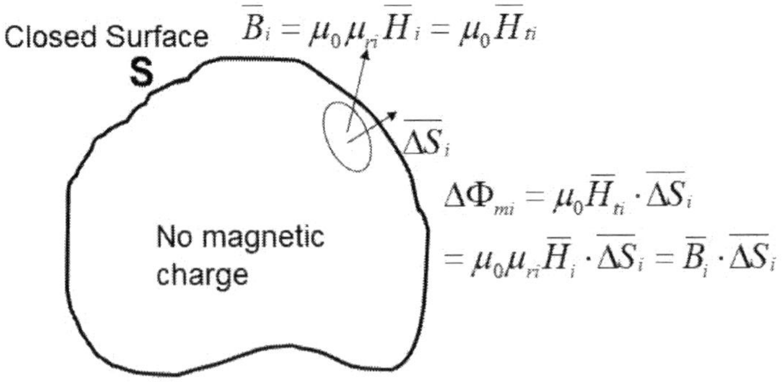

Gauss' law for magnetic field applied in a material medium with magnetic constant μ

FIGURE 15

Differential Form Using Divergence Operator:

Apply (1.59) to a surface $\triangle S$, enclosing a volume element $\triangle V$. Then applying divergence-operator concept (appendix A, section 2.3), (1.59) maybe equivalently expressed in a differential form using the divergence operator.

$$\oiint_{\triangle S} \overline{B} \cdot \overline{dS} = (\overline{\nabla} \cdot \overline{B}) \triangle V = 0, \ \overline{\nabla} \cdot \overline{B} = 0. \tag{1.60}$$

CHAPTER 2

FROM CIRCUIT TO ELECTROMAGNETIC THEORY

The theory of electromagnetic fields and waves is the most general theory available for modeling all phenomena involving electric charge, flow of charge, and their related effects. These include a wide range of concepts and applications, starting form the operation of electrical circuits, to functioning of magnetic machines and devices, propagation of radio waves and wireless communication through open space or material media, as well as optical and thermal phenomena and related applications. For a student of electrical engineering, who may have already been familiar with the basic concepts of electricity and electrical circuits, it may be natural and easier to start with the basic electrical and circuit concepts. The electromagnetic theory may then be viewed as a gradual generalization of the basic circuit principles covered in introductory classes. The circuit principles can be formulated using field equations, and may be approached from a field viewpoint as well with similar simplicity and rigor.

(a) The basic Kirchoff's Voltage Law (KVL), which governs conservation of voltage or electric energy around circuit loops, is to be generalized by taking into account Faraday's induced potential produced when the circuit is operating in the presence of a time-varying magnetic field. The magnetic field might be created by a source external to the circuit, or by the currents in the circuit itself.

(b) The basic Kirchoff's Current Law (KCL), which governs conservation of charge flow at different circuit nodes, is to be modified by including additional currents that may be needed to supply for temporary charge storage on different terminals. The currents that supply for charge storage at some points are matched by equal currents needed to "displace" equal and opposite amount of charge from other terminals. These additional currents are referred to as "displacement currents," and may be viewed to flow, or rather "jump," across terminals with a material medium in between. The displacement current is distinct from the conventional flow of "conduction current" through a resistive medium produced due to physical flow of conduction electrons. The displacement current is like a current flowing across the plates of a capacitor, and can flow or jump across any two areas of charge storage (wires or nodes).

(c) Now, the currents in KCL would produce magnetic fields as per Ampere's law, which would then produce the Faraday's induced potentials in the above KVL equations. Accordingly, it would result in inter-coupling between KCL and KVL equations. Ampere's law, which is responsible in producing the inter-coupling magnetic fields, is also to be modified in consistency with the modifications in KCL. Both currents - conduction and displacement currents - in KCL are to be used in Ampere's law, not just the conduction currents used in a basic or simple form of Ampere's law.

A rigorous circuit model, accomplished through the above generalizations, would account for distributed coupling through electromagnetic fields produced by the circuit itself. The coupled KCL and KVL equations, modified using Faraday's law, displacement current concept and Ampere's law, provide the basis for a general electromagnetic theory.

The modified version of the circuit laws are general, and can model the effects of a free-space or other material media in which a circuit is embedded. The surrounding material media contribute to the mechanism of interaction in the circuit. Like the basic circuit laws, the general circuit laws apply along conventional circuit paths through metal wires and discrete circuit elements. The general laws would apply as well along arbitrary paths through the surrounding media, by decomposing the media into distributed circuit elements. The circuit models can be formulated in the form of field equations for the distributed environment. The field equations may be used for understanding directly from a field viewpoint as well.

Now, all the free-space and materials around a circuit become integral parts of the circuit itself. In a general situation, the fields produced by a circuit in the space surrounding itself can extend far away from the circuit. In regular circuits such a situation is ignored and may not be desired. In contrast, a situation may arise where the effect of the surrounding can play a dominant role under suitable conditions, resulting in driving significant power away from the circuit in the form of electromagnetic waves. This can be used to our advantage for communicating with another circuit at a distant location. Such a special circuit, operating under the rigorous versions of circuit laws, can be seen as a unique practical device called an antenna.

The general mechanism of radiation and interaction, based on the rigorous circuit laws as discussed above, is the foundation of all electromagnetic phenomena. The rigorous versions of the circuit laws carry profound theoretical, practical as well as philosophical significance, and are described using Maxwell's equations, named after the scientist James Clerk Maxwell (1831-1879), who recognized and established the fundamental generalization.

1. Generalized Kirchoff's Voltage Law (KVL)

1.1. KVL in the Standard Form. Kirchoff's Voltage Law (KVL) in its standard form may be expressed in the following simple form.

$$\sum_{Loop\ C} V_i = 0. \tag{2.1}$$

In applying KVL, one needs to select a particular loop path C in a circuit, that originates at any point in the circuit and finally ends up at the same point after visiting several circuit elements or signal sources along the path. V_i is the voltage across the ith element visited in the loop. Consistency in the relative signs of the voltages V_i must be observed. The sign of a voltage element V_i maybe considered positive if the voltage is assumed to increase as one marches forward along the loop, and negative if otherwise.

We illustrate the application of the standard form of KVL to an example circuit, shown in Fig.1. The circuit in the Fig.1 consists of resistances, capacitances and inductances, connected to a voltage source, as shown. Application of KVL to the loop 1, for example, with the reference signs of the voltage elements indicated in the Fig.1, will result in the following equation.

$$V_1 + V_2 + V_3 + V_4 = 0. \tag{2.2}$$

Note that the signs of the different voltages indicated in the Fig.1 can be selected arbitrarily. But once selected, they should be used as reference consistently all through

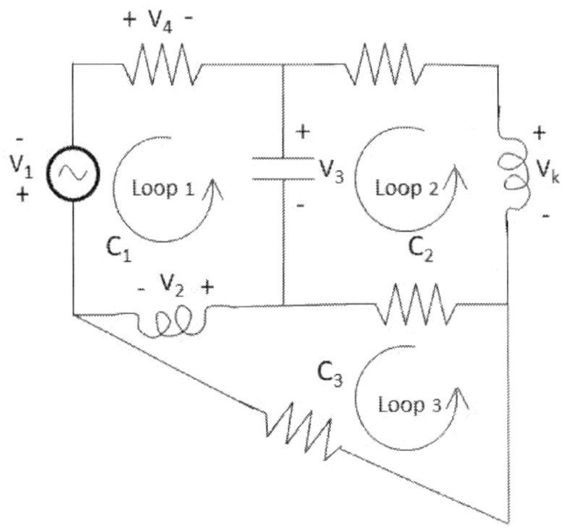

FIGURE 1

the analysis. Similar equations may be obtained by applying KVL to other possible closed loops in the Fig.1, which is left to the reader.

1.1.1. *Standard form of KVL and Energy Conservation.* We have defined voltage V_i across two points as the energy W_i gained by a unit charge when moved across the two points. If a charge of amount Q Coulombs is moved across the two points, the energy W_i it acquires is proportional to Q.

$$W_i = QV_i. \tag{2.3}$$

If a charge of Q Coulombs is moved around a particular loop path C, then the energy gained at each element added together is equal to the total energy gained.

$$\sum_{Loop\ C} W_i = \sum_{Loop\ C} QV_i = Q \sum_{Loop\ C} V_i = 0. \tag{2.4}$$

The sum of all voltages around the loop is zero, as dictated by the standard form of KVL in (2.3). As a result, the total energy gained through moving around the loop is also equal to zero, as expressed in (2.4). This means that when the charge returns back to its original position its energy does not change, or the charge energy is conserved in the process. In other words, the standard form of KVL, is an expression of energy conservation around a circuit loop.

In the above discussion, the equivalence between energy conservation and the standard form of KVL assumes that the potentials across circuit elements, that are produced by the battery or signal sources in the circuit, are the only form of energy exchange that are possible when a charge moves around a loop. If there is an additional mechanism through which energy is transferred to/from the charge, then the standard form of KVL will, of course, not account for the total energy change, and will therefore be incorrect. Such situation may arise due to interaction of the circuit on itself or from an external source. Consider a radio receiver circuit, in which we know energy is received or induced from a distant transmitter. In this case, one can not apply the standard form of KVL to the receiver circuit. Similar situation will also arise for a transmitter circuit, where energy is driven away from the circuit through interaction of the circuit on itself. In the following section, we will formulate such induced energy or potential via Faraday's law.

1.2. Faraday's Induced EMF. Let us apply the standard form KVL to a simple circuit, consisting of a resistance R with its terminals connected by a wire loop (Fig.2). However, consider the loop to be placed in a time-varying magnetic field directed normal to the plane of the loop. For example, imagine someone holding a permanent magnet above or below the loop, with one of the magnetic poles facing at the loop, and the person is moving the magnet up and down. This is a simple example where energy is induced into a circuit, not by an explicit battery or signal source, but remotely through a magnetic field from an external device. If we apply the standard form of KVL, following the prescribed procedure discussed above, we would get

$$V_1 = -V = -IR = 0; \; I = 0. \tag{2.5}$$

We see that the standard form of KVL results in no current excited in the loop. However, as per Faraday's law, there would be induced voltage $V_{induced}$ excited by the magnetic field, thus producing a non-zero current in the loop.

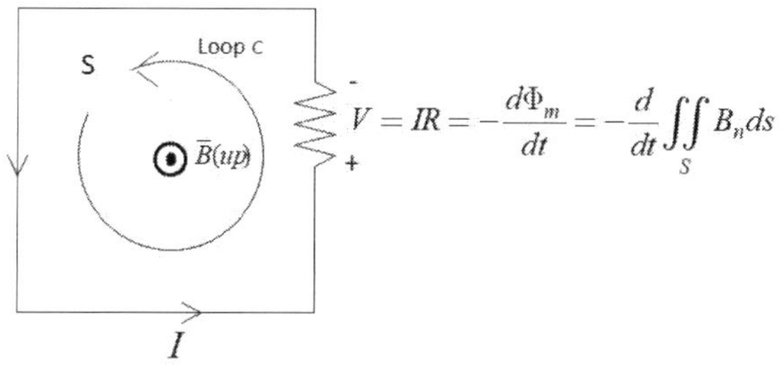

FIGURE 2

According to Faraday's law, the above loop voltage would actually be equal to the rate of change of the normal magnetic flux Φ_m passing through the cross-section area of the loop.

$$-V = -IR = -V_{induced} = \frac{d\Phi_m}{dt}. \tag{2.6}$$

where the magnetic flux Φ_m is expressed as a surface integral of the magnetic field intensity $\overline{H}$ over the surface area S enclosed by the loop path C. Computation of the magnetic flux Φ_m would follow a similar procedure in the last chapter 1, section 2.2, defined in relation to Gauss' law for magnetic field.

1.3. Adding Induced Voltage in KVL. The above problem as an example suggests that the standard form of KVL will not be applicable when the circuit is embedded in a time-varying magnetic field. The time-varying field would excite an induced voltage that may be considered to be an additional "invisible" or implicit voltage source. This induced voltage must be added to the standard form of KVL in order to be rigorous in general situations with a surrounding magnetic field.

The standard version of KVL in (2.3) may now be generalized in the following form, in order to account for any time-varying magnetic field in/around the circuit.

$$\sum_{Loop} V_i = -V_{induced} = \frac{d\Phi_m}{dt}, \tag{2.7}$$

$$\sum_{Loop} V_i + V_{induced} = \sum_{Loop} V_i - \frac{d\Phi_m}{dt} = 0. \tag{2.8}$$

As mentioned, this magnetic flux may be injected due to magnetic field produced by a source external to the given circuit. It may also be produced by the currents flowing in the different wires of the circuit itself. In other words, the induced voltage, in general, may be externally induced or self-induced by the circuit.

1.3.1. *Magnetic Flux and Right-Hand Rule.* We have defined the magnetic flux Φ_m before in (1.53) over a closed surface S, in context of Gauss' law for magnetic field. In contrast, for the present situation, we have an open surface S, enclosed by a loop path C. For a closed surface, the direction of the surface normal vector is defined to point "outward," which is clearly understood with respect to a physically closed surface. However, it would be ambiguous for an open surface S that we have in the present situation. The normal for an open surface would have two physical possibilities. In order for the expressions for the flux Φ_m, and consequently the induced voltage expressed before, to be correct, one must properly choose the correct normal direction. As a common rule, the normal direction $\hat{n}$ to an open loop surface S is selected with reference to the orientation of the loop path C using a "right hand rule." This uses the fingers of the right hand of a person as the guiding reference. As per this rule, the direction of the thumb points to the correct surface normal $\hat{n}$, when the orientation of all other fingers holding in a "fist form" points to the orientation of the loop path C (Fig.3). Once the normal direction is properly defined with respect to the loop, then accordingly the normal component of the magnetic field $\overline{H}$, $\overline{B}$, and normal flux Φ_m are definitively defined. This would follow the same procedure used in relation to the flux calculation (1.53) in Gauss' law.

$$\Phi_m = \sum_{Surface\ S} B_{ni}\Delta S_i = \sum_S \overline{B}_i \cdot \overline{\Delta S}_i = \iint_S B_n dS = \iint_S \overline{B} \cdot \overline{dS}, \tag{2.9}$$

where the elemental surface vector $\overline{\Delta S}$ or $\overline{dS}$ is defined with magnitude equal to its elemental area, with its direction pointed along the surface normal $\hat{n}$ as defined above by the right-hand rule.

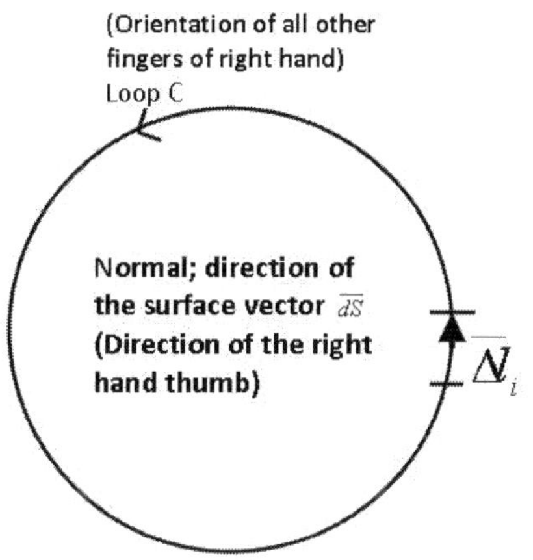

FIGURE 3

1.4. Generalized KVL Using Electric Field.

1.4.1. *A Material Media as Discretized Elements.* We now conclude that the generalized form of KVL (2.8,2.9) would work for any arbitrary circuit with explicit resistance, capacitance and inductance defined, in the presence of a magnetic field. Such a general circuit law, due to its rigor, would also work as well for any open space or any surrounding material media, by including them as a part of the circuit. The material medium can be treated as circuit components in a general sense, by breaking into smaller pieces, each with certain conductivity and dielectric permittivity. The conductivity will result in an equivalent resistance for each segment, and the dielectric permittivity will result in an equivalent capacitance. "Displacement" current can flow across the material medium equivalent to that between a capacitor, and conduction current can flow through the material medium like that in a resistor. Of course, it would result in a much complex circuit with interconnected or intercoupled segments in a three-dimensional space. We shall attempt here to extend the general KVL (2.8,2.9), by treating the media as a discretized circuit configuration. This may be an useful way to gain insight and understand the mechanism of interaction through empty space or material media.

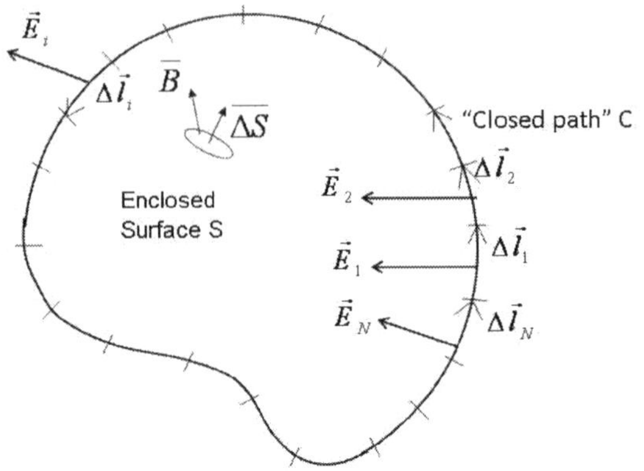

FIGURE 4

In order to treat any material medium as equivalent to interconnected circuit segments, we need to express the voltage ΔV_i across a material segment in terms of the local electric field intensity $\overline{E}$. We will use the relationship already established in (1.19).

$$\Delta V_i = \frac{\Delta W_i}{Q} = -\overline{E}_i \cdot \overline{\Delta l_i}. \tag{2.10}$$

Once we define the voltage across segments, then the general KVL (2.8,2.9) can be applied to an arbitrary loop path through the material medium (Fig.4).

$$\sum_{Loop\ C} \Delta V_i = -\sum_{Loop\ C} \overline{E}_i \cdot \overline{\Delta l_i} = -V_{induced} = \frac{d\Phi_m}{dt} = \frac{d}{dt} \sum_{Surface\ S} \overline{B}_i \cdot \overline{\Delta S}_i. \tag{2.11}$$

1.4.2. *Loop- and Surface-Integral Representation of the Generalized KVL.* In the limit as the path element $\overline{\Delta l_i}$ and the surface element $\overline{\Delta S_i}$ approach zero, the above expression for the generalized KVL may be expressed in the form of integral operations.

$$\sum_{Loop\ C} \overline{E}_i \cdot \overline{\Delta l_i} = -\frac{d}{dt} \sum_{Surface\ S} \overline{B}_i \cdot \overline{\Delta S}_i, \tag{2.12}$$

$$\oint_C \overline{E} \cdot \overline{dl} = -\frac{d}{dt} \iint_S \overline{B} \cdot \overline{dS}. \tag{2.13}$$

1.5. Extension to 3-Dimensional Circuits and Non-Planar Loops. In the above analysis it is assumed for simplicity that the entire circuit is placed on one common plane. Therefore, it was assumed that any circuit loop also exists on one plane. However, that is not always the case. A circuit can physically be structured in three dimensions, where a closed circuit path may not infact exist in one plane. In this case, however, the theory that we have developed can be applicable with some change in interpretation and extension of the mathematical formulation.

Imagine an arbitrary continuous 3-dimensional surface S, of which the non-planar loop path C defines the boundary (Fig.5). In the simple case we have already covered, when the loop C is a planar loop, one may select S to be the planar surface enclosed by the planar loop. However, in the case of a non planar loop, the choice of S will have to be non-planar.

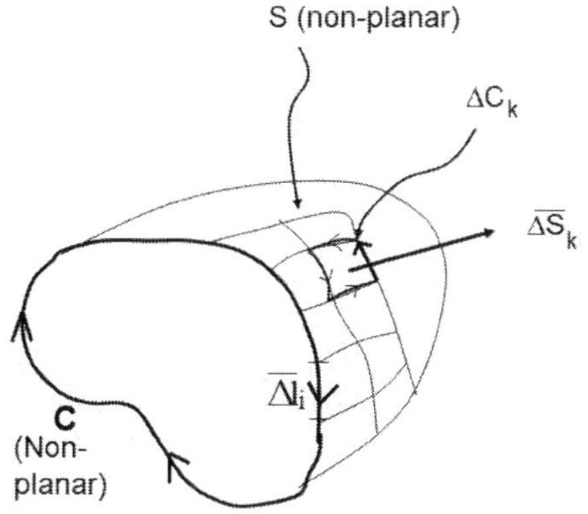

FIGURE 5

Let us segment the continuous non-planar surface S into small units ΔS_k that are approximately planar (Fig.5). Apply the standard KVL to each planar segment area ΔS_k bounded by a small closed loop ΔC_k. Remember that while applying KVL to each small loop ΔC_k, we need to decompose it into small linear segments Δl_{ik}. The planar area ΔS_k is already small enough to safely assume that the magnetic field is about uniform over it.

$$\sum_{Loop\ \Delta C_k} \Delta V_{ik} = -\sum_{Loop\ \Delta C_k} \overline{E}_{ik} \cdot \overline{\Delta l}_{ik} = \frac{d\Delta\Phi_{mk}}{dt} = \frac{d}{dt}(\overline{B}_k \cdot \overline{\Delta S}_k), \tag{2.14}$$

$$\sum_{k}\sum_{Loop\ \Delta C_k} \Delta V_{ik} = -\sum_{k}\sum_{Loop\ \Delta C_k} \overline{E}_{ik} \cdot \overline{\Delta l}_{ik} = \sum_{k} \frac{d}{dt}(\overline{B}_k \cdot \overline{\Delta S}_k). \tag{2.15}$$

The right-hand rule (Fig.3) is used for each surface element ΔS_k to determine its normal direction in reference to the path direction for the loop ΔC_k. Now, observe that neighboring ΔC_k share common path segments that have opposing path directions. The voltage increments in these path segments will be negative of each other. Therefore, when we add the voltages around all loop paths ΔC_k, the contributions from all common segments will cancel out, leaving only the contributions from the line segments that constitute the non-planar loop path C.

$$\sum_{Loop\ C} \Delta V_i = -\sum_k \sum_{Loop\ \Delta C_k} \overline{E}_{ik} \cdot \overline{\Delta l}_{ik} = \sum_k \frac{d}{dt}(\overline{B}_k \cdot \overline{\Delta S}_k) = \sum_{Surface\ S} \frac{d}{dt}(\overline{B}_k \cdot \overline{\Delta S}_k), \quad (2.16)$$

$$\sum_{Loop\ C} \Delta V_i = -\sum_{Loop\ C} \overline{E}_i \cdot \overline{\Delta l}_i = \frac{d}{dt} \sum_{Surface\ S} (\overline{B}_k \cdot \overline{\Delta S}_k), \quad (2.17)$$

$$\oint_C \overline{E} \cdot \overline{dl} = -\frac{d}{dt} \iint_S \overline{B} \cdot \overline{dS}. \quad (2.18)$$

The direction of the path integral for the non-planar boundary loop C is determined by the directions of the path integrals of all the loop elements (as per the right-hand rule) with which it shares common line segments. Note that the same expressions (2.12,2.13) of the generalized KVL for a planar loop apply as well in (2.17,2.18) for a non-planar loop. Note also that for a given loop C there exist many possible non-planar surfaces S that are bounded by the C. This means, the surface integral on the right hand side of KVL (2.17,2.18) for all such possible S will have to be the same. The generalized KVL (2.17,2.18) may also be referred to as the general integral form of Faraday's law, recognizing that Faraday's law is the fundamental basis in the development of the equation.

1.6. Gauss' Law for Magnetic Field and the General KVL. An interesting deduction is possible from the general form of KVL as expressed above for a non-planar surface. Consider a general non-planar surface S, which is bounded by a general non-planar loop C. Think of a limiting situation when the boundary loop C shrinks to an infinitesimally small loop around a point. This may happen by having the surface S enclosed by the C approach its limit in two possible ways. One trivial possibility is that the enclosed surface S gradually shrinks and approaches the point with zero area. The other possibility is to have the surface enclosed by the loop C to approach a closed surface configuration, as the loop C approach the point. Consider the second possibility. Because the loop path approaches a point, any loop integral would approach zero. In this limit, the general form of KVL (2.17,2.18) may be expressed in the following form.

$$\sum_{Loop\ C \to 0} \overline{E}_i \cdot \overline{\Delta l}_i = 0 = -\frac{d}{dt} \sum_{Closed\ S} (\overline{B}_k \cdot \overline{\Delta S}_k), \quad (2.19)$$

$$\oint_{Loop\ C \to 0} \overline{E} \cdot \overline{dl} = 0 = -\frac{d}{dt} \oiint_S \overline{B} \cdot \overline{dS}, \quad (2.20)$$

$$\sum_{Closed\ S} (\overline{B}_k \cdot \overline{\Delta S}_k) \approx \oiint_S \overline{B} \cdot \overline{dS} = 0; \frac{d}{dt} \neq 0. \quad (2.21)$$

Observe that the above integral expression (2.21) over a closed surface is Gauss' law of magnetic field we have presented before, in section 2.2 of chapter 1. The associated equivalent magnetic charge Q_m is proved to be zero, assuming general time-varying signals with non-zero time derivative. Accordingly, we have established that Gauss' law for magnetic field (1.53,1.54) can be deduced from the general form of KVL (2.17,2.18), or equivalently from Faraday's law, assuming that all natural phenomena are time variant. Faraday's law generalizes the basic form of KVL, and therefore includes in it the concept of energy conservation (2.1,2.4) of the basic KVL, together with the additional concept of induced voltage or energy from the surrounding magnetic field. Further, it also implicitly includes the concept of Gauss' law of magnetic field, as we have now discovered. This is an important development. This means that, although we established in the beginning Gauss' law of magnetic field as a fundamental basis for general electromagnetic fields (chapter 1, section 2.2), it is now superseded by Faraday's law (2.17,2.18). In other words, Faraday's law makes Gauss' law of magnetic field (1.53,1.54) analytically redundant.

2. Generalized Kirchoff's Current Law (KCL)

2.1. KCL in the Standard Nodal Form. For the application of KCL, one first selects an arbitrary node in a circuit, with which N number of current wires join. The standard form of KCL dictates that the sum of all the N branch currents is equal to zero. Kirchoff's Current Law (KCL) in the standard form may thus be expressed in the following simple form.

$$\sum_{Node} I_i = 0, \tag{2.22}$$

where I_i is the current in a particular wire connecting to the selected node (Fig.6). Correct use of KCL requires that one uses proper signs for the branch currents. As a standard practice, we assume that the currents flowing out of the node is positive, and negative otherwise. Note that the directions of currents used for analysis serve only as reference directions. If the value of a particular current is calculated to be a positive number, it means the actual current flows along the reference direction originally selected. On the other hand, the actual current flows opposite to the reference direction originally selected, if the final calculated value turns out to be negative. Once selected, however, the reference directions must be consistently used all through the analysis.

2.1.1. *Standard form of KCL and Charge Conservation.* Let us re-examine the standard form of KCL by counting the flow of charges in/out of the node. As discussed, a positive current in a particular branch refers to an outgoing current. Following this standard convention, a particular branch current I_i is equal to the number of charges leaving the node per unit time. The sum of all branch currents is, therefore, equal to the rate of total number of charges leaving the node over a unit time.

$$\sum_{Node} I_i = \sum_{Node} \frac{\Delta Q_i}{\Delta t} = \frac{\sum_{Node} \Delta Q_i}{\Delta t} = \frac{\Delta Q}{\Delta t} = 0; \ \sum_{Node} \Delta Q_i = \Delta Q = 0. \tag{2.23}$$

We see that the incremental change in charge ΔQ at the node point over a given increment of time Δt is zero. Equivalently, the amount of charge entering into a node balances with that leaving the node. In other words, the standard form of KCL is an equivalent expression of charge conservation.

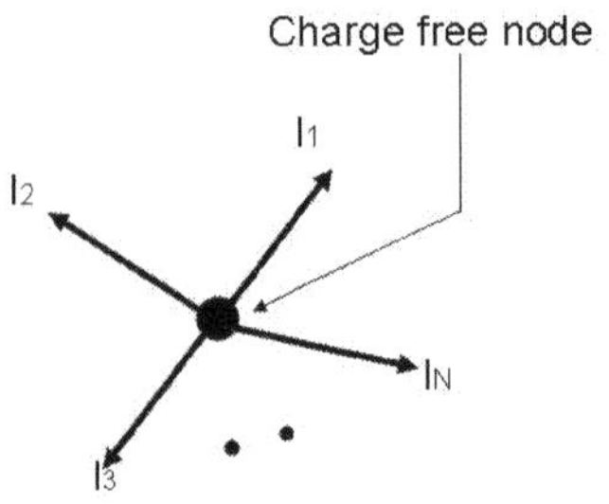

FIGURE 6

2.1.2. *Extension to a "Closed Block" of Circuit.* The above form of KCL established for a circuit node may be extended to a complete or "closed" circuit block (Fig.7). A "closed" block refers to a part of the circuit contained inside a physical volume (or a "closed" surface). The charge conservation principle which we established for a particular node should also apply to a closed physical structure. Accordingly, the sum of currents on all wires connected to the closed circuit should add up to zero.

$$\sum_{Closed\ S} I_i = 0. \tag{2.24}$$

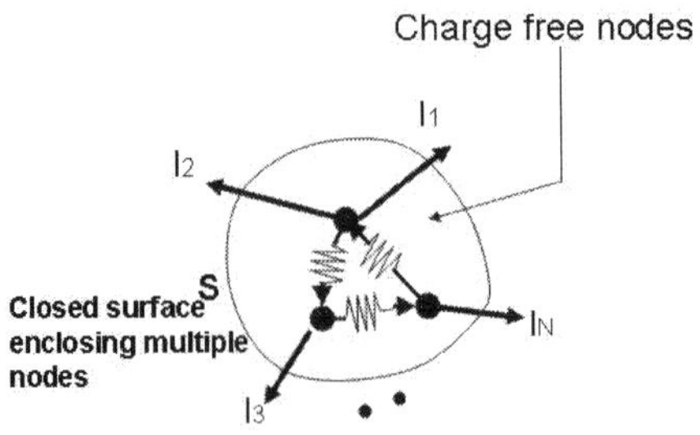

FIGURE 7

2.1.3. *Extension to a Distributed Circuit.* A material medium maybe considered to be a distributed circuit. Therefore, the above form of KCL applied to a closed circuit block can be as well extended to a material medium bounded inside a closed surface S (Fig.8). Decompose the entire surface S into surface elements $\overline{\triangle S_i}$. The current $\triangle I_i$ flowing through the surface element $\overline{\triangle S_i}$ can expressed as the product of the normal current density J_{ni} at the location and the surface area $\triangle S_i$. Now, as in the case of the closed circuit block, the sum of all the current elements $\triangle I_i$ over the boundary surface S of a distributed circuit block would be zero.

$$\sum_{Closed\ S} J_{ni} \triangle S_i = \sum_{Closed\ S} \overline{J}_i \cdot \overline{\triangle S_i} = 0. \tag{2.25}$$

The above expression can be expressed as a surface integral over a closed surface S, in the limit when the surface elements $\overline{\triangle S_i}$ approach zero.

$$\lim_{\triangle S_i \to 0} \sum_{Closed\ S} \overline{J}_i \cdot \overline{\triangle S_i} = \oiint_S \overline{J} \cdot \overline{dS} = 0. \tag{2.26}$$

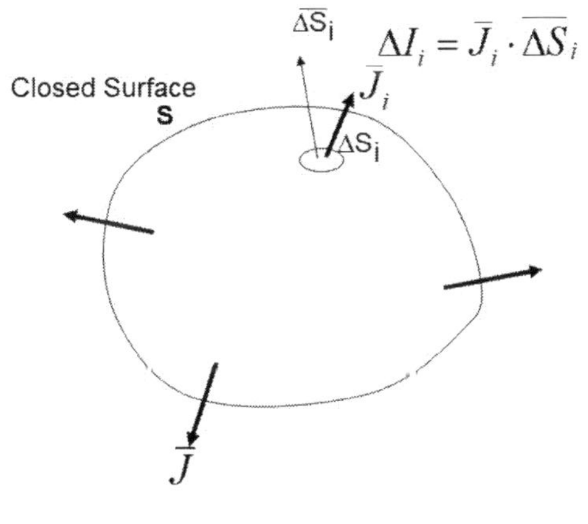

FIGURE 8

2.2. Modified KCL including Charge Storage. In the above discussion there is an implicit assumption that no charge can be stored inside the closed surface under consideration. This includes a metal node, metal structure of the connecting wires, or any material element inside the closed surface. If a metal node or any metal element has the ability to store charge temporarily, then the charge conservation relations, and consequently the standard form of KCL, need to be modified. A circuit node or a metal wire is ideally assumed to have no dimensional area or volume where any charge can be retained. Under this ideal assumption, therefore, the standard form of KCL would be valid. However, in a practical circuit every metal element has a non-zero size, and would behave like a plate of a capacitor, with any other metal structure in the circuit serving as a second plate of the capacitor. Accordingly, one may understand that the metal element can accumulate charge temporarily, like the plates of a capacitor. As discussed above, under such situations of charge storage, the standard form of KCL will have to be modified in order to account for any charge storage.

When a circuit node, such as in Fig.6, contains stored charge $Q_s(t)$, one would require the following simple modification in KCL. The rate of total outflow of charge through all the currents must equal to the rate of reduction in the total stored charge at the node.

$$\sum I_i = \frac{-dQ_s}{dt}. \tag{2.27}$$

Similar modifications can be established for a closed block of circuit (Fig.7) or material medium (Fig.8), when there is stored charge $Q_s(t)$ inside the closed surface S.

$$\sum_{Closed\ S} I_i = \frac{-dQ_s}{dt}, \tag{2.28}$$

$$\sum_{Closed\ S} J_{ni} \triangle S_i = \sum_{Closed\ S} \overline{J}_i \cdot \overline{\triangle S}_i = \frac{-dQ_s}{dt}, \tag{2.29}$$

$$\lim_{\triangle S_i \to 0} \sum_{Closed\ S} \overline{J}_i \cdot \overline{\triangle S}_i = \oiint_S \overline{J} \cdot \overline{dS} = \frac{-dQ_s}{dt}. \tag{2.30}$$

$Q_s(t)$ includes charges on all nodes, metal wires or material media present inside the surface S.

2.3. Modified KCL using Displacement Current and Electric Field. We know that the total charge will have to be conserved in an entire system. Therefore, when there is certain charge Q_s at a certain node or circuit block (Figs.9,10), at some other locations outside the circuit block there would be equal amount of charge displaced, (or opposite amount of charge $-Q_s$ stored.) Accordingly, Q_s is called the stored or displacement (as opposed to flowing or conduction) charge, and $\frac{dQ_s}{dt}$ is referred to as the total capacitive or displacement current I_d.

It would be further useful, if we can interpret the displacement current not just as the rate of change of stored or displaced charges at distant locations, but some form of continuous "flow" between the distant locations. We know electric field lines continuously connect between opposite charges in the form of "electric flux" lines, as shown in the Fig.9. Also, as per Gauss' law of electric field (chapter 1, section 2.1), the total electric flux diverging from any stored charge Q_s is equal to the stored charge (Fig.10).

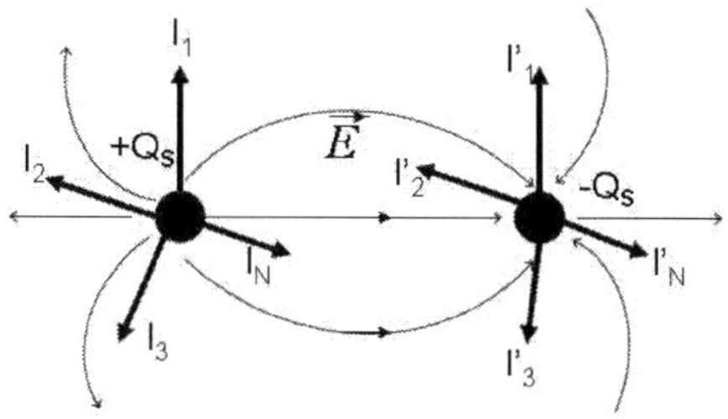

FIGURE 9

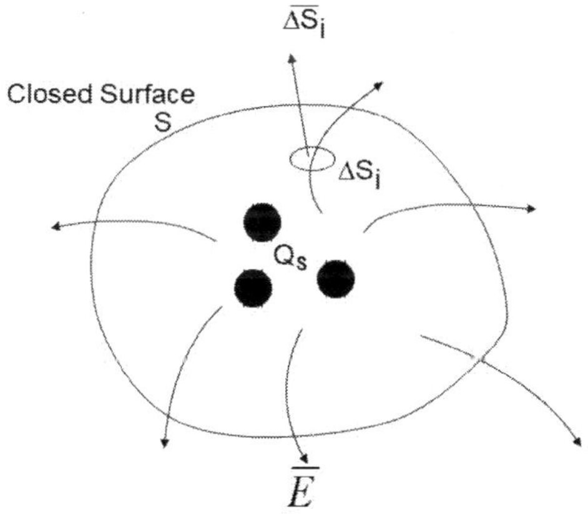

FIGURE 10

$$\sum_S \overline{D}_i \cdot \overline{\triangle S}_i = Q_s; \overline{D} = \epsilon \overline{E},$$

$$\sum_S \frac{d\overline{D}_i}{dt} \cdot \overline{\triangle S}_i = \frac{dQ_s}{dt} = I_d. \qquad (2.31)$$

Accordingly, it may be meaningful to interpret the electric field lines in Figs.9,10 to represent "flow" of displacement current, and $\frac{d\overline{D}_i}{dt}$ as the associated displacement current density $\overline{J}_{di}$.

$$\sum_S \overline{J}_{di} \cdot \overline{\triangle S}_i = I_d = \frac{dQ_s}{dt}, \quad \overline{J}_{di} = \frac{d\overline{D}_i}{dt}. \tag{2.32}$$

Now, the displacement current I_d, as expressed above in terms of electric flux density $\overline{D}$, can be used together with the modified or rigorous expression of KCL in (2.27-2.30). This would be valid for a single circuit node, a closed-block of circuit, or a material medium inside a closed surface, as illustrated in the Figs.11,12.

$$\sum I_i = -I_d = -\frac{dQ_s}{dt}, \quad I_t = \sum I_i + I_d = 0, \tag{2.33}$$

$$\sum_{Closed\ S} \overline{J}_i \cdot \overline{\triangle S}_i + \sum_{Closed\ S} \overline{J}_{di} \cdot \overline{\triangle S}_i = 0, \tag{2.34}$$

$$\oiint_S \overline{J} \cdot \overline{dS} + \oiint_S \overline{J}_d \cdot \overline{dS}$$
$$= \oiint_S \overline{J} \cdot \overline{dS} + \oiint_S \frac{d\overline{D}}{dt} \cdot \overline{dS} = 0. \tag{2.35}$$

Equations (2.32-2.35) constitute a rigorous version of KCL. They include both the conduction and displacement currents, with the displacement currents now expressed in the form of electric fields or flux density. The total current $I_t = I + I_d$, defined as the sum of the regular conduction current I and the displacement current I_d is zero. In other words, the incoming total current (conduction and displacement) to a particular three-dimensional closed region is equal to the outgoing total current from the region. In the limit, when a three-dimensional region enclosed by a surface S is compressed or flattened into a region of zero volume, with the enclosing surface S now consisting of two sides of a two-dimensional surface, the above condition would mean that the total current entering in one side of the two-dimensional surface would be equal that leaving the other side of the same surface. Accordingly, this condition maybe referred to as the continuity of the total current, literally meaning that the total current "flows" continuously through a cross sectional surface. Note that, as per the standard form of KCL (2.25,2.26), this would have meant that the conduction current flows continuously, as in a conducting wire with no charge storage. As we have learned now, that is no longer true in general conditions when the conducting wire supports charge storage. Part of the current may flow as conduction current through physical movement of charge along the conductor, and a part may "flow" as displacement current in the form of time-varying electric field lines emerging away from, or across two cross-sections of, the charged wire. The sum of the two parts, however, would always maintain continuity in general conditions.

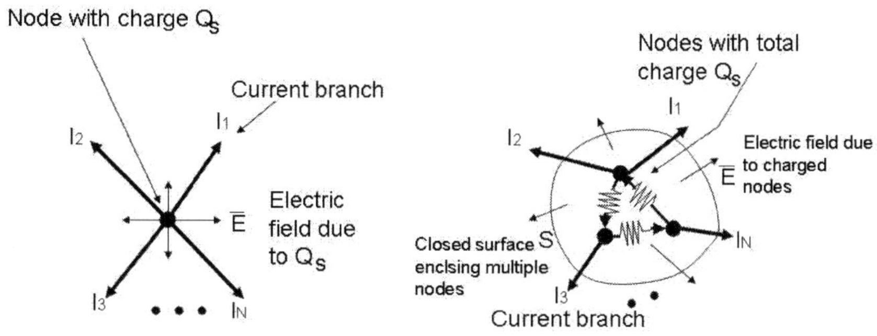

FIGURE 11

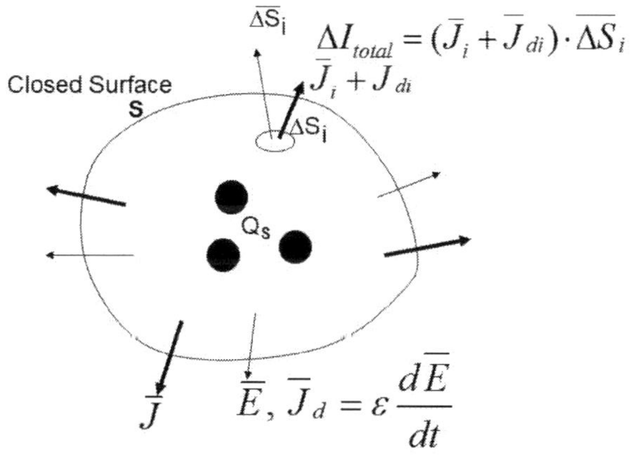

FIGURE 12

2.4. Modified KCL, Charge Continuity, and Gauss' Law of Electric Field. The continuity of the total current (conduction and displacement) in the general form of KCL (2.34,2.35) ensures that charge conservation is maintained at every location along a current flow. The concept of the local charge conservation, maintained continuously along a current flow, may also be be referred to as charge continuity. The general form of KCL (2.34,2.35) ensures the continuity of the total current, and consequently the charge continuity, based on Gauss' law of electric field. Conversely, the continuity of the total current in the general KCL (2.34,2.35) would automatically satisfy Gauss' law of electric field, if charge continuity is independently maintained.

Gauss' law of electric field simply relates the surrounding electric field or flux to the total internal charge. However, when the charge inside a closed surface is changing in time, Gauss' law does not specify where the incremental charge would be coming from (or going to). In the absence of that specification, the problem could be physically incomplete or potentially invalid. Whereas, the general form of KCL together with Gauss' law, would ensure the required charge flow through continuity of the charge. Accordingly, the general KCL (2.34,2.35), together with Gauss' law, not Gauss' law by itself, would model a physically realistic and complete problem. Alternately, the general KCL (2.34,2.35), together with maintaining charge continuity, would also model a physically realistic and complete problem. The combined two conditions in this case would make Gauss' law redundant, because they will automatically satisfy Gauss' law as discussed above. This is a significant understanding.

3. Ampere's Law

The conduction currents in different circuit branches or material media were originally thought to be the basic source of the magnetic field. The standard form of Ampere's law was first established to relate the magnetic field to the conduction currents. The general form of KCL developed in the above section has extended the concept of currents to include both the conduction as well as the displacement currents. One needs to add the displacement currents to the conduction currents in order to satisfy charge conservation and continuity of the total (conduction plus displacement) current flow. To be consistent with the general definition of currents, as defined by the general form of KCL, the standard form of Ampere's law may also have to be revised to include both the conduction as well as the displacement currents as the source of the magnetic field. Magnetic field modeled using a generalized Ampere's law can be used to derive induced potentials, which has to be added in the general form of KVL through Faraday's law. Accordingly, a general form of Ampere's law would serve as an essential link to connect or couple the general versions of KCL and KVL. This would constitute a complete set of relations that can be used to model general electromagnetic fields.

3.1. Basic Form of Ampere's Law. As mentioned, the standard form of Ampere's law relates the magnetic field to the conduction currents. It may be expressed in the following integral form. Consider a closed line path C. The line integral of the magnetic field $\overline{H}$ over the close path C is equal to the total conduction current through the surface S enclosed by the boundary path C.

$$\sum_C \overline{H}_i \cdot \overline{\triangle l}_i = I = \sum_S \overline{J}_i \cdot \overline{\triangle S}_i, \tag{2.36}$$

where I or $\overline{J}_i$ refer to conduction current and conduction current density, respectively. The sign of the current I is determined in reference to the direction of the integration path C using a right-hand rule. Accordingly, a positive direction for the current I is defined by the direction of the thumb of the right hand, while the other fingers of the right hand are curled along the direction of the loop path C (Fig.3). For now we may assume that the loop C and the enclosed surface S are on a common plane, but theory can then be extended to general loops and surfaces.

<u>*Generalization for Material Medium:*</u>

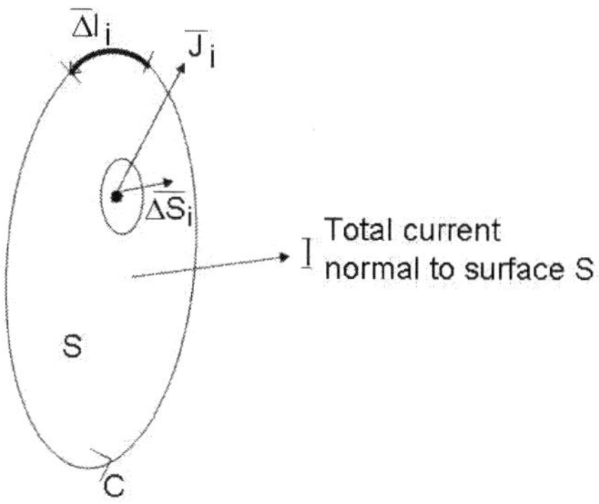

FIGURE 13

The above expression of Ampere's law is at first established for a free-space medium. The magnetic field in a material medium would polarize the orbiting/spinning charges present in the material. These polarized orbiting/spinning electrons would behave as equivalent conduction current loops which would produce additional magnetic fields, thus changing the total field in the medium. The magnetic field in a material medium may be modeled using equivalent current loops, distributed over the material, as discussed in the chapter 1, section 1.5.3. Ampere's law established for a free-space medium can be extended to general media through suitable choice of the magnetic field $\overline{H}$ and modification of the integration path C. A path element $\overline{\triangle l_i}$ on the loop C may be deformed such that it does not pass right through a polarized current loop. The deformed path C would pass between current loops, and would not enclose inside the path the equivalent conduction currents due to polarized orbital/spin charges of the material. Accordingly, the conduction current enclosed by the deformed loop path C is contributed only by the conduction currents in the medium due to flow of the conduction charges, and excludes the equivalent orbital/spin currents due to bound charges.

In reference to the magnetic polarization figure 14, chapter 1, let us decompose a line segment $\overline{\triangle l_i}$ into two parts.

$$\overline{\Delta l_i} = \overline{\Delta l}_{ni} + \overline{\Delta l}_{pi};\ \Delta l_{ni} = \Delta l_i \sin\theta_i,\ \Delta l_{pi} = \Delta l_i \cos\theta_i \tag{2.37}$$

where θ_i is the angle between the path vector $\overline{\Delta l_i}$ and the local magnetic field, and subscripts p refers to a path component parallel to the field (not passing through a loop) and n refer to a path component perpendicular to the field. Accordingly, the path integration in Ampere's law $\overline{H}_i \cdot \overline{\triangle l_i}$ maybe expressed in two corresponding parts. The contribution along $\overline{\triangle l}_{ni}$ is zero because the corresponding field is normal to the path. Further, we know that the magnetic field in the medium, when measured along a sample path parallel to the field (not passing through a current loop), is $\overline{H}_0$. Assuming such deformation in the microscopic level would not change the total conduction current I enclosed inside the loop, Ampere's law for the deformed loop may be expressed as follows.

$$\overline{H}_i \cdot \overline{\triangle l}_i = \overline{H}_i \cdot \overline{\triangle l}_{ni} + \overline{H}_i \cdot \overline{\triangle l}_{pi} = H_{oi} \Delta l_i \cos(\theta_i) = \overline{H}_{oi} \cdot \overline{\triangle l}_i, \tag{2.38}$$

$$\sum_C \overline{H}_{oi} \cdot \overline{\triangle l}_i = I = \sum_S \overline{J}_i \cdot \overline{\triangle S}_i. \tag{2.39}$$

Now compare Ampere's law (2.39) in a material medium to that (2.36) in a free-space medium. The magnetic field intensity $\overline{H}_i$ used for a free-space medium in (2.36) is replaced by the primary magnetic field $\overline{H}_{oi}$ in (2.39) for a material medium. Clearly, it would be notationally convenient to define the magnetic field $\overline{H}$ in a material medium to be equivalent to the $\overline{H}_o$. This notational definition we have already adopted in chapter 1, section 1.5.3, the consistency of which now stands justified. With this notation, Ampere's law is expressed using the same generic relationship, valid for a free-space as well as any general medium.

3.2. Modified Ampere's Law using Displacement Current. Consider a current flow across a capacitor as shown in Fig.14. If the standard form of Ampere's law is applied to a circular loop C of radius ρ around the conduction current I, we have

$$\sum_C \overline{H}_i . \overline{\triangle l}_i = H_\phi \times 2\pi\rho = I, \quad H_\phi = \frac{I}{2\pi\rho}. \tag{2.40}$$

Note that across the capacitor there is no conduction current (I=0). Therefore, if the basic form of Ampere's law would be applied to a loop C of radius ρ around the center of the capacitor. We have

$$\sum_C \overline{H}_i . \overline{\triangle l}_i = H_\phi \times 2\pi\rho = 0, \quad H_\phi = 0. \tag{2.41}$$

This would mean that the magnetic filed would have a sudden discontinuity from $H_\phi = \frac{I}{2\pi\rho}$ to $H_\phi = 0$ as the z coordinate of the observation point slides into the capacitor gap region of length d. This would be contrary to a reasonable expectation. One would expect continuity of the magnetic field as the z-coordinate of the point of observation slides down along the wire through the capacitor gap.

The physical significance of the above reasoning may be appreciated further if one imagines a current across a single capacitor or series of capacitors, each with a separation gap d, that are connected in series with conducting wire segments. The physical structure is essentially a discontinuous conducting wire, broken by the capacitive gaps. Now consider that each gap distance is reduced, approaching an infinitesimally small but non-zero value, $d \to 0$, in which case the broken structure would essentially appear to be a continuous wire. Accordingly, in this limiting case, one may be compelled to believe that the magnetic field produced by the current in the circuit would also be continuous along all z, as if the same current was flowing along a continuous wire. However that would not be the case if one applies the basic form of Ampere's law, which relates the magnetic field to only

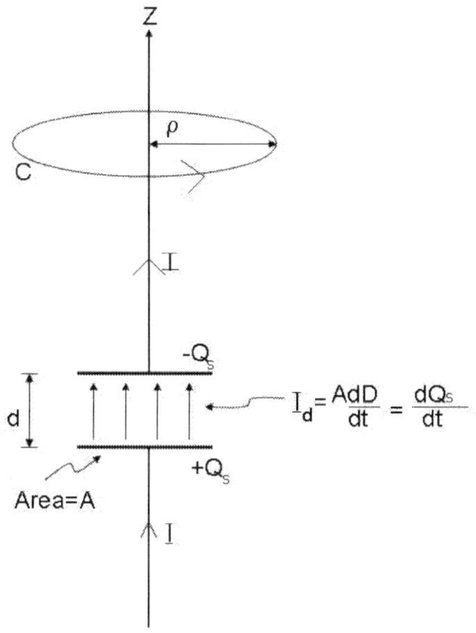

FIGURE 14

conduction currents. It would instead produce magnetic fields abruptly changing to zero when the place of observation z slides through the infinitesimally small gaps, because there is strictly no conduction current across the small, non-zero gap. This would intuitively appear to be an unacceptable physical condition.

This unreasonable situation can be overcome if the basic form of Ampere's law is modified by including both the conduction and the displacement current.

$$\sum_C \overline{H}_i \cdot \overline{\Delta l}_i = I + I_d = I_t,$$

$$= \sum_S (\overline{J}_i + \overline{J}_{di}) \cdot \overline{\Delta S}_i = \sum_S (\overline{J}_i + \frac{d\overline{D}_i}{dt}) \cdot \overline{\Delta S}_i. \quad (2.42)$$

This modification is suggested in consistency with the modification of Kirchoff's Current Law, where the conduction current I needs to be replaced by the total current $I_t = I + I_d$, including both the conduction and the displacement currents, in order to maintain current continuity (see section 2.3). Continuity of the magnetic field would be ensured by associating the field with the continuous total current, not with the conduction current alone whose continuity is not ensured.

For direct current (DC) circuits where there is no time variation, the displacement current $\overline{J}_d = \frac{d\overline{D}}{dt} = 0$. In this case only conduction current would exist, and the basic form of Ampere's law (2.36) would be valid. In the other words, the basic form of Ampere's law (2.36) would be applicable strictly for DC circuits only. The modified version of Ampere's law (2.42) would be needed for all time-varying phenomena.

3.3. Integral form of the Generalized Ampere's Law. In the limit as the path elements $\overline{\Delta l_i}$ and the surface elements $\overline{\Delta S_i}$ approach zero, the above expression (2.42) for the generalized Ampere's law may be expressed in the form of integral operations.

$$\sum_C \overline{H}_i \cdot \overline{\Delta l}_i = \sum_S (\overline{J}_i + \frac{d\overline{D}_i}{dt}) \cdot \overline{\Delta S}_i, \tag{2.43}$$

$$\oint_C \overline{H} \cdot \overline{dl} = \iint_S (\overline{J} + \frac{d\overline{D}}{dt}) \cdot \overline{dS}. \tag{2.44}$$

3.4. Line Current Across a Capacitive Gap with Spherical Terminals: We will apply the general form of Ampere's law to a line current with a spherical capacitive gap, as shown in Fig.15. Here we have conduction current along the wire, and displacement current or electric field across the capacitive gap. The conduction current is strictly confined in the wire, whereas the displacement current is spreading over the entire surrounding space, but would be mostly confined in the vicinity of the capacitive gap for a small gap separation δ.

We selected the spherical capacitive terminals because it would be analytically simple to evaluate the electric field or displacement current. It would be useful to evaluate and interpret the magnetic field produced in the presence of both conduction and displacement currents.

One may use Coulomb's law of electrostatics to express the electric fields produced due to the charges $\pm Q(t)$. The charge pair $\pm Q(t)$ is varying in time. Therefore, it may not necessarily be obvious that the resulting electric field $\overline{E}(\overline{r}, t; Q(t))$ can be expressed in terms of the $Q(t)$ using Coulomb's law, which is strictly valid only for time-invariant problems. However, the $Q(t)$ here is linearly time varying with $\frac{dQ(t)}{dt}$ equal to a constant current I. Under this special condition of a linear time variation of the charge, one may show through simple reasoning that Coulomb's law may still be used to model the time-variant electric field in a "derivative sense."

Because the surrounding medium is assumed to be time-invariant, the observed field produced due to a time-shifted source charge function $Q(t-t_0)$, would be the same as the time-shifted version of the original observed field produced due to the unshifted source charge function $Q(t)$.

$$\overline{E}(\overline{r}, t; Q(t-t_0)) = \overline{E}(\overline{r}, t-t_0; Q(t)). \tag{2.45}$$

The time-shifted function of the charge pair $\pm Q(t-t_0)$ can be expressed as a superposition of the unshifted function $\pm Q(t)$ and a static charge pair $\pm Q(t_0) = \pm It_0$. Therefore, the observed field due to the time-shifted charge pair $\pm Q(t-t_0)$ can be expressed as a superposition of the original field due to the $\pm Q(t)$ and a static Coulomb's field $\overline{E}_c(\overline{r}, Q(t_0))$ produced due to a fixed charge pair $\pm Q(t_0) = \pm It_0$.

$$Q(t) = It; \;\; Q(t-t_0) = It - It_0 = Q(t) - It_0, \tag{2.46}$$

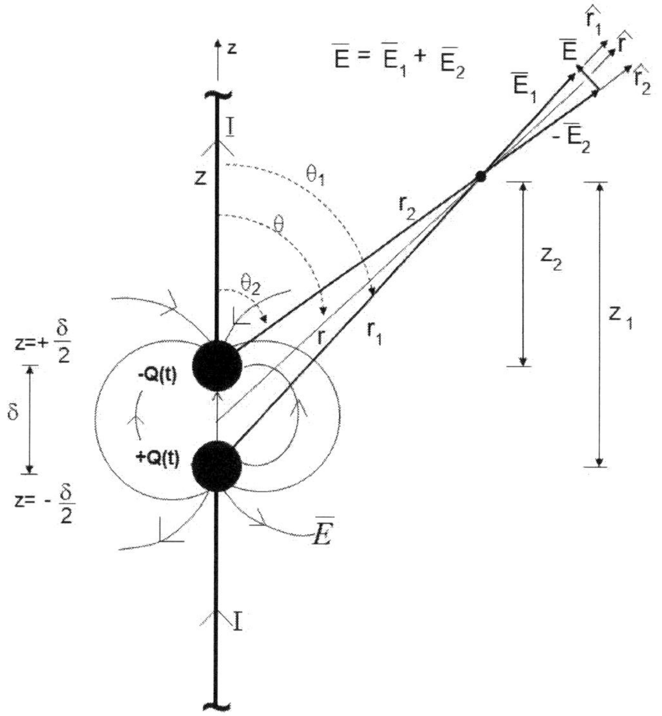

FIGURE 15

$$\overline{E}(\overline{r}, t; Q(t-t_0)) = \overline{E}(\overline{r}, t; Q(t)) - \overline{E}_c(\overline{r}, It_0). \tag{2.47}$$

Combining (2.45) and (2.47) we have,

$$\overline{E}(\overline{r}, t; Q(t)) - \overline{E}(\overline{r}, t-t_0; Q(t)) = \overline{E}_c(\overline{r}, It_0) = t_0\overline{E}_c(\overline{r}, I) = t_0\overline{E}_c(\overline{r}, \frac{dQ(t)}{dt}). \tag{2.48}$$

The above derivation also assumes that the Coulomb field $\overline{E}_c$ is a linear function of the charge amplitude. One may now relate the time derivative of the original field to the time derivative of the charge $\frac{dQ(t)}{dt} = I$ using the Coulomb function $\overline{E}_c$.

$$\lim_{t_0 \to 0}[\frac{\overline{E}(\overline{r}, t; Q(t)) - \overline{E}(\overline{r}, t-t_0; Q(t))}{t_0}] = \frac{d\overline{E}(\overline{r}, t; Q(t))}{dt} = \overline{E}_c(\overline{r}, \frac{dQ(t)}{dt}). \tag{2.49}$$

The derivative of the electric field (or equivalently the displacement current density) is now related in (2.49) to the derivative of the charge $\frac{dQ}{dt} = I$ through Coloumb's law. In the following, we derive the field due to the charge pair $\pm Q(t)$ using Coulomb's relations, which would then be used to express the displacement current in the "derivative sense," as established in (2.49).

$$\overline{E}_1 = \frac{Q}{4\pi\epsilon r_1^2}\hat{r}_1 = E_1\hat{r}_1,\ \overline{E}_2 = -\frac{Q}{4\pi\epsilon r_2^2}\hat{r}_2 = -E_2\hat{r}_2,$$
$$E_{z_1} = E_1\cos\theta_1,\ E_{z_2} = -E_2\cos\theta_2$$
$$E_{r_1} = E_1\cos(\theta-\theta_1),\ E_{r_2} = -E_2\cos(\theta_2-\theta),$$
$$E_{\theta_1} = -E_1\sin(\theta-\theta_1),\ E_{\theta_2} = -E_2\sin(\theta_2-\theta). \tag{2.50}$$

By superposition,

$$E_z = E_{z_1} + E_{z_2} = \frac{Q}{4\pi\epsilon r_1^2}\cos\theta_1 - \frac{Q}{4\pi\epsilon r_2^2}\cos\theta_2,$$
$$E_r = E_{r_1} + E_{r_2} = \frac{Q}{4\pi\epsilon r_1^2}\cos(\theta-\theta_1) - \frac{Q}{4\pi\epsilon r_2^2}\cos(\theta_2-\theta),$$
$$E_\theta = E_{\theta_1} + E_{\theta_2} = -\frac{Q}{4\pi\epsilon r_1^2}\sin(\theta-\theta_1) - \frac{Q}{4\pi\epsilon r_2^2}\sin(\theta_2-\theta). \tag{2.51}$$

Now apply Ampere's law to a loop of radius ρ_0 centered around the z-axis at $z = z_0$. The radial (r) and angle (θ) parameters (see Fig.15), associated with (ρ_0, z_0) are $(r_{10}, r_{20}, \theta_{10}, \theta_{20})$.

Case-I: $z > \frac{\delta}{2}, 0 < (\theta_1, \theta_2, \theta) < \frac{\pi}{2}$

$$H_\phi \times 2\pi\rho_0 = I + \frac{d}{dt}\int_{\rho=0}^{\rho_0}\int_{\phi=0}^{2\pi} \epsilon E_z\rho d\rho\, d\phi = I + 2\pi\epsilon\frac{d}{dt}\int_{\rho=0}^{\rho_0} E_z\rho d\rho$$
$$= I + 2\pi\epsilon\frac{d}{dt}\left[\int_{\rho=0}^{\rho_0} E_{z_1}\rho d\rho + \int_{\rho=0}^{\rho_0} E_{z_2}\rho d\rho\right]$$
$$= I + 2\pi\epsilon\frac{d}{dt}\left[\int_{\theta_1=0}^{\theta_{10}} \frac{Q}{4\pi\epsilon r_1^2}\cos\theta_1\cdot z_1\tan\theta_1\cdot z_1\sec^2\theta_2 d\theta_1\right.$$
$$\left. - \int_{\theta_2=0}^{\theta_{20}} \frac{Q}{4\pi\epsilon r_2^2}\cos\theta_2\cdot z_2\tan\theta_2\cdot z_2\sec^2\theta_2 d\theta_2\right]$$
$$= I + \frac{1}{2}\frac{d}{dt}\left[\int_0^{\theta_{10}} Q\ \sin\theta_1\ d\theta_1 - \int_0^{\theta_{20}} Q\ \sin\theta_2\ d\theta_2\right] = I + \frac{1}{2}\frac{dQ}{dt}[\cos\theta_{20} - \cos\theta_{10}]$$
$$H_\phi = \frac{I}{2\pi\rho_0} + \frac{dQ}{dt}\frac{1}{4\pi\rho_0}[\cos\theta_{20} - \cos\theta_{10}]. \tag{2.52}$$

Case-II: $z < -\frac{\delta}{2},\ \frac{\pi}{2} < (\theta_1, \theta_2, \theta) < \pi$

$$H_\phi \times 2\pi\rho_0 = I + \frac{1}{2}\frac{d}{dt}\Big[\int_\pi^{\theta_{10}} Q\ \sin\theta_1 d\theta_1 - \int_\pi^{\theta_{20}} Q\ \sin\theta_2 d\theta_2\Big]$$

$$= I + \frac{1}{2}\frac{dQ}{dt}\Big[\cos\theta_{20} - \cos\theta_{10}\Big]$$

$$H_\phi = \frac{I}{2\pi\rho_0} + \frac{dQ}{dt}\frac{1}{4\pi\rho_0}\Big[\cos\theta_{20} - \cos\theta_{10}\Big]. \tag{2.53}$$

Case-III: $-\frac{\delta}{2} < z < \frac{\delta}{2},\ \frac{\pi}{2} < \theta_2 < \pi,\ 0 < \theta_1 < \frac{\pi}{2}$

$$H_\phi\ \times 2\pi\rho_0 = \frac{1}{2}\frac{d}{dt}\Big[\int_0^{\theta_{10}} Q\ \sin\theta_1 d\theta_1 - \int_\pi^{\theta_{20}} Q\ \sin\theta_2 d\theta_2\Big]$$

$$= \frac{1}{2}\frac{dQ}{dt}\Big[-\cos\theta_{10} + \cos 0 + \cos\theta_{20} - \cos\pi\Big] = \frac{1}{2}\frac{dQ}{dt}\Big[2 + (\cos\theta_{20} - \cos\theta_{10})\Big]$$

$$= \frac{dQ}{dt} + \frac{1}{2}\frac{dQ}{dt}(\cos\theta_{20} - \cos\theta_{10})$$

$$H_\phi = \frac{dQ}{dt}\frac{1}{2\pi\rho_0} + \frac{dQ}{dt}\frac{1}{4\pi\rho_0}(\cos\theta_{20} - \cos\theta_{10}). \tag{2.54}$$

All the above cases result in the same functional expression for the H_ϕ, because $\frac{dQ}{dt} = I$.

$$H_\phi = \frac{I}{2\pi\rho_0} + \frac{I}{4\pi\rho_0}(\cos\theta_{20} - \cos\theta_{10}). \tag{2.55}$$

One may choose to apply Ampere's law to a cylindrical surface of height $\triangle z$ ($z_0 < z < z_0 + \triangle z$), radius ρ_0 and angular spread $\triangle\phi$ ($\phi_0 < \phi < \phi_0 + \triangle\phi$). It can be shown (problem) that the same H_ϕ and $\overline{E}$ as derived above would consistently satisfy Ampere's law on this surface as well.

One may note that the H_ϕ in the above three cases are continuous as one transitions between the region $|z_0| < \frac{\delta}{2}$ and the region $|z_0| > \frac{\delta}{2}$, as we have expected. This is because the common functional expression (2.55) for H_ϕ, valid in all regions, is a continuous function with respect to all variables including z_0.

3.5. Magnetostatic Field Due to a Finite-Length Line Current. The structure we are analyzing now may be considered as a continuous conduction current $I\hat{z}(-\infty < z < \infty)$ from which a finite segment of length $\delta(-\frac{\delta}{2} < z < \frac{\delta}{2})$ is "missing." This missing or complementary line current segment is associated with charge accumulation $\pm Q$ at its ends $z = \pm\frac{\delta}{2}$ which are negative of those of the original structure at the respective ends. We know that for a continuous line of constant conduction current (referred to with a subscript l) the magnetic field $\overline{H}_l = \hat{\phi}H_{l\phi} = \hat{\phi}\ \frac{I}{2\pi\rho_0}$ (it maybe shown (problem) that $H_{lz} = H_{l\rho} = 0$.) The $H_{l\phi}$ is equal to the first part of the magnetic field H_ϕ we have derived above for the total structure. Therefore, the second part of the magnetic field H_ϕ we have derived above should be negative of the magnetic field component $\triangle H_\phi$ produced by the line segment $I\hat{z}(-\frac{\delta}{2} < z < \frac{\delta}{2})$ of length δ.

$$\overline{H} = \overline{H}_l - \triangle\overline{H},$$
$$\triangle H_\phi = \frac{I}{4\pi\rho_0}\big(\cos\theta_{10} - \cos\theta_{20}\big). \tag{2.56}$$

In addition, because an infinite constant line current does not produce any electric field ($\overline{E}_l = 0$), the electric field (2.51) produced by the total structure must be negative of that ($= \triangle\overline{E}$) due to the line segment of length δ. We will express the $\triangle\overline{E}$ at the location coordinates $(\rho_0, z_0) \equiv (r_0, \theta_0)$.

$$\overline{E} = \overline{E}_l - \triangle\overline{E} = -\triangle\overline{E}; \overline{E}_l = 0,$$
$$\triangle\overline{E} = -E_r\hat{r} - E_\theta\hat{\theta}$$
$$= \frac{Q}{4\pi\epsilon}\Big[-\frac{\cos(\theta_0 - \theta_{10})}{r_{10}^2} + \frac{\cos(\theta_{20} - \theta_0)}{r_{20}^2}\Big]\hat{r}$$
$$+\frac{Q}{4\pi\epsilon}\Big[\frac{\sin(\theta_0 - \theta_{10})}{r_{10}^2} + \frac{\sin(\theta_{20} - \theta_0)}{r_{20}^2}\Big]\hat{\theta}. \tag{2.57}$$

Let us examine any $\triangle H_z$ or $\triangle H_\rho$ components that might be produced by the line segment of current I. As per equation (2.56), these would be negative of the respective fields produced by the total structure, because we know H_{lz} and $H_{l\rho}$ are zero.

$$H_\rho = -\triangle H_\rho, H_z = -\triangle H_z. \tag{2.58}$$

Note that the electric field $\overline{E} = -\triangle\overline{E}$, or equivalently displacement current density $\overline{J}_d = \epsilon\frac{d\overline{E}}{dt}$, does not have any $\hat{\phi}$ component. That is, $E_\phi = 0 = J_{d\phi}$. The E_ϕ or $J_{d\phi}$, which couples to H_ρ and H_z through Ampere's law, is therefore expected to produce $H_\rho = H_z = 0$. However, one may like to establish the proof rigorously, as follows.

3.5.1. *Special Proof, $H_\rho = H_z = 0$:* Consider an elemental rectangular closed loop $\triangle C(\triangle\rho \times \triangle z)$ (Fig.16), with normal current $\triangle I_{d\phi} = J_{d\phi}\triangle\rho\triangle z = 0$. Applying Ampere's law to the $\triangle C$, one can show,

$$\sum_{i,\triangle C}\overline{H}_i \cdot \overline{\triangle l}_i = \sum_{i,\triangle C} H_{ti}\triangle l_i = 0 = \Big(\frac{\partial H_z}{\partial z} - \frac{\partial H_\rho}{\partial \rho}\Big)\triangle\rho\triangle z,$$
$$\frac{\partial H_\rho}{\partial z} = \frac{\partial H_z}{\partial \rho}. \tag{2.59}$$

The above relation can be satisfied by defining a suitable magnetic potential ψ such that

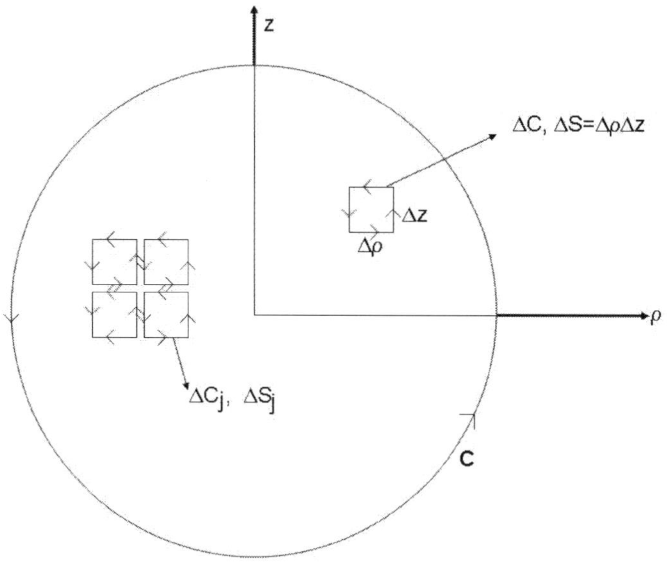

FIGURE 16

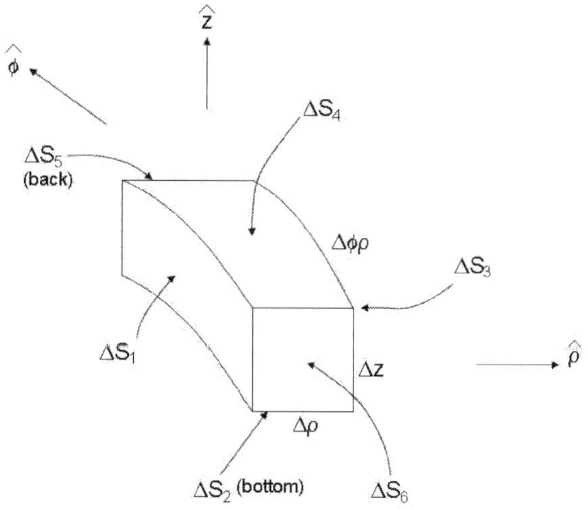

FIGURE 17

$$H_\rho = \frac{\partial \psi}{\partial \rho}, H_z = \frac{\partial \psi}{\partial z}. \tag{2.60}$$

Gauss' law for the magnetic field would also have to be satisfied. Apply Gauss' law to an elementary volume $\triangle V \equiv (\triangle \rho \times \triangle z \times \rho \triangle \phi) \equiv \triangle C \times \rho \triangle \phi$ enclosed by six side surfaces $\triangle S_i$ (Fig.17). Unlike electric charges, there are no known magnetic charges. We also know that existing non-zero magnetic field H_ϕ is independent of ϕ.

$$H_{n5} = -H_{n6} = H_\phi,$$

$$\sum_{i,\ \triangle V} \overline{H}_i \cdot \overline{\triangle S}_i = \sum_{i,\ \triangle V} H_{ni} \triangle S_i$$

$$= \sum_{i=1}^{4} H_{ni}\ \triangle S_i = 0;\ H_{n5}\triangle S_5 + H_{n6}\triangle S_6 = 0,$$

$$\rho\triangle\phi \sum_{i=1}^{4} H_{ni}\triangle l_i = \rho\triangle\phi \sum_{i,\triangle C} H_{ni}\triangle l_i,$$

$$\sum_{i,\triangle C} H_{ni}\triangle l_i = 0 = (\frac{\partial H_\rho}{\partial \rho} + \frac{\partial H_z}{\partial z})\triangle\rho\triangle z, \tag{2.61}$$

where the subscript n represents field component outwardly normal to a particular line segment $\triangle l_i$ of $\triangle C$.

Now consider the following line integral, and simplify it using the above results from Ampere's and Gauss' laws.

$$\sum_{i,\triangle C} \psi_i H_{ni}\triangle l_i = [\frac{\partial}{\partial \rho}(\psi H_\rho) + \frac{\partial}{\partial z}(\psi H_z)]\triangle S$$
$$= \left[\psi(\frac{\partial H_\rho}{\partial \rho} + \frac{\partial H_z}{\partial z}) + H_\rho\frac{\partial \psi}{\partial \rho} + H_z\frac{\partial \psi}{\partial z}\right]\triangle S$$
$$= \left[0 + (H_\rho^2 + H_z^2)\right]\triangle S \geq 0. \tag{2.62}$$

Consider a large surface S enclosed by a circular closed loop $C\big(r \to \infty\big)$ (Fig.16), decomposed into elemental loops $\triangle C_j, j = 1, \infty$, enclosing corresponding surface elements $\triangle S_j$.

$$\sum_j \sum_{\triangle C_j} \psi_{ij} H_{nij}\triangle l_{ij} = \sum_{k,C} \psi_k H_{nk}\triangle l_k; k \equiv ij\ on\ C$$
$$= \sum_j \big(H_{\rho j}^2 + H_{zj}^2\big)\triangle S_j. \tag{2.63}$$

Recall that $H_\rho = -\triangle H_\rho$, and $H_z = -\triangle H_z$ are negative of the fields of a time-independent (static) current element $(\overline{I}\delta)$ of finite length δ. For any finite-sized, static current source, with a finite source amplitude, the total stored energy in the magnetic field would be finite. Further the volume density of the magnetic field energy is proportional to the square of the total magnetic field magnitude $\left|\overline{H}\right|$. Accordingly the magnetic stored energy $\triangle W_m$ in a volume element with angular spread $\triangle\theta$ and $\triangle\phi$, and $0 < r < \infty$, can be expressed with the following constraint.

$$\triangle W_m = \int_0^\infty \alpha \left|\overline{H}\right|^2 r^2 dr \triangle\theta\triangle\phi \; < \infty, \tag{2.64}$$

where α is a proportionality constant. $\triangle W_m$ would be finite only if $\left|\overline{H}\right|$ would reduce with $\frac{1}{r^a}$ dependence for large $r \to \infty$, where $a \geq 1.5$.

$$\begin{aligned}
\int_0^\infty \left|\overline{H}\right|^2 r^2 dr &< \infty, \\
\left|\overline{H}\right| &\sim \frac{1}{r^a}, \; r \to \infty; \; a \geq 1.5, \\
|\psi| &\sim \frac{1}{r^{a-1}}, \; r \to \infty, \\
|\psi_k H_{nk}| &< \frac{b}{r^2}, \; r \to \infty, \\
\lim_{r\to\infty} \sum_{k,C} \psi_k H_{nk} \triangle l_k \; &\leq \; \lim_{r\to\infty} \frac{b}{r^2} \sum_{k,C} \triangle l_k \\
&= \lim_{r\to\infty} (\frac{b}{r^2} \times 2\pi r) = 0,
\end{aligned} \tag{2.65}$$

where b is a finite positive real number. Using the above condition of the line integral on C in (2.63), we would have,

$$\begin{aligned}
\lim_{r\to\infty} \sum_{k,C} \psi_k H_{nk} \triangle l_k &= 0 \\
= \sum_{j=1}^{\infty} \left(H_{\rho j}^2 + H_{zj}^2\right) \triangle S_j.
\end{aligned} \tag{2.66}$$

Because each of H_ρ^2 and H_z^2 is non-negative at every location, the only way the above condition can be satisfied is by having $H_\rho = H_z = 0$ on the entire surface S of Fig.16, on the $\rho - z$ plane. Considering the symmetry of the physical problem, the fields would be independent of the ϕ variable, and therefore the $H_\rho = H_z = 0$ in the entire three-dimensional space.

$$\begin{aligned}
H_\rho = H_z = \triangle H_\rho &= \triangle H_z = 0, \\
\overline{H} = H_\phi \hat{\phi} &= \left(\frac{I}{2\pi\rho_0} - \triangle H_\phi\right) \hat{\phi}, \\
\triangle\overline{H} = \triangle H_\phi \hat{\phi} &= \frac{I}{2\pi\rho_0} \left(\cos\theta_{10} - \cos\theta_{20}\right) \hat{\phi}.
\end{aligned} \tag{2.67}$$

In the above derivation, the $z-$ and $\rho-$components of the total magnetic field $\overline{H}$ are proved to be zero, by first assuming the $z-$ and $\rho-$components of the field $\overline{H}_l$ due to the infinite-line current to be zero. Otherwise, we have essentially proved that the $z-$ and $\rho-$components of the field $\triangle\overline{H}$, produced only due to the finite current segment of length δ, are zero. Now, the field $\triangle\overline{H}$ of the $z-$directed current segment, with a small segment length $\delta \to 0$, can be used to derive the field $\overline{H}_l$ due to the $z-$directed infinite-line current, by breaking the infinite line into small segments, and then integrating the $\triangle\overline{H}$ due to each segment over the entire infinite length of the line. Accordingly, in turn, we have effectively proved that $z-$ and $\rho-$components of the magnetic fields of the infinite-line current would be zero, because each of its elemental segments is now proved not to contain the $z-$ and $\rho-$directed components. This validates our initial assumption in the derivation, and therefore the final results in (2.67).

3.6. Magnetostatic Field Due to an Elemental Line Current: Biot-Savart's Law. For the special case, when $\delta << (r, r_1, r_2)$, different location parameters may be approximated.

$$\begin{aligned}
\theta_{10} \simeq \theta_0 - \frac{\triangle\theta_0}{2},\ \theta_{20} \simeq \theta_0 + \frac{\triangle\theta_0}{2},\ \triangle\theta_0 &\simeq \frac{\delta\sin\theta_0}{r}, \\
r_{10} \simeq r_0 + \frac{\delta}{2}\cos\theta_0,\ r_{20} \simeq r_0 - &\frac{\delta}{2}\cos\theta_0, \\
\cos\theta_{10} - \cos\theta_{20} \simeq \triangle\theta_0\sin\theta_0 \simeq &\frac{\delta\sin^2\theta_0}{r_0}.
\end{aligned} \tag{2.68}$$

Using these approximations in the field expressions, we get

$$\begin{aligned}
H_\phi \simeq \frac{I}{2\pi\rho_0} - \frac{I}{4\pi\rho_0}\frac{\delta\sin^2\theta_0}{r_0} &= \frac{I}{2\pi\rho_0} - \frac{I\delta\sin^2\theta_0}{4\pi r_0^2}, \\
\triangle\overline{H} = \hat{\phi}\triangle H_\phi &= \frac{I\delta\sin\theta_0}{4\pi r_0^2}\hat{\phi}, \\
E_r \simeq \frac{Q}{4\pi\epsilon r_{10}^2}\cos\frac{\triangle\theta_0}{2} &- \frac{Q}{4\pi\epsilon r_{20}^2}\cos\frac{\triangle\theta_0}{2} \\
\simeq \frac{Q}{4\pi\epsilon}\left(\frac{1}{r_{10}^2} - \frac{1}{r_{20}^2}\right) &= \frac{Q}{4\pi\epsilon}\frac{r_{20}^2 - r_{10}^2}{r_{10}^2 r_{20}^2} \\
\simeq -\frac{Q}{4\pi\epsilon}\frac{4r_0\frac{\delta}{2}\cos\theta_0}{r_0^4} &= -\frac{Q}{4\pi\epsilon}\frac{2\delta\cos\theta_0}{r_0^3}, \\
E_\theta \simeq -\frac{Q}{4\pi\epsilon r_0^2} \times \frac{\delta\sin\theta_0}{r_0} &= -\frac{Q}{4\pi\epsilon}\frac{\delta\sin\theta_0}{r_0^3}, \\
\triangle\overline{E} &= -E_\theta\hat{\theta} - E_r\hat{r}.
\end{aligned} \tag{2.69}$$

Notice that when $\delta \to 0$, the observation location $z = z_0$ with $|z_0| < \frac{\delta}{2}$ (or, equivalently $z_0 \to 0$,) and the other two cases with $z_0 > \frac{\delta}{2}$ or $z_0 < -\frac{\delta}{2}$ (equivalently $|z_0| > 0$,) result in the same expression for $H_\phi = \frac{I}{2\pi\rho_0}$, which is the magnetic field H_ϕ for a continuous line current at the radial distance ρ_0. This happens, although for $|z_0| > 0$ we have strictly conduction current, whereas for $z_0 = 0$ we have strictly displacement current, used in Ampere's law analysis of the field. The displacement current across the small gap ($\delta \to 0$) behaves as if it is an ideal continuation of the conduction line current.

It may be observed that the magnetic field of a line current $\overline{I} = I\hat{z}$ of small length δ may also be expressed in the following general form.

$$\triangle\overline{H}(r_0, \theta_0, \phi_0) = \hat{\phi}\triangle H_\phi = \hat{\phi}\frac{I\delta\sin\theta_0}{4\pi r_0^2} = \frac{(\overline{I}\delta)\times\overline{r}_0}{4\pi r_0^3}, \tag{2.70}$$

$$\triangle\overline{H}(r, \theta, \phi) = \frac{(\overline{I}\delta)\times\overline{r}}{4\pi r^3} = \frac{(\overline{I}\delta)\times\hat{r}}{4\pi r^2}. \tag{2.71}$$

The above general relation (2.71) is the Biot-Savart's law of magneto-statics, which would be applicable for any general current vector $\overline{I}$, located at the reference origin of the coordinate system, and for any general location (r, θ, ϕ) of observation represented by the general position vector $\overline{r} = r\hat{r}$. This is obtained here as a direct consequence of generalization of Ampere's law by including a time-invariant displacement current, and using Coulomb's law of electrostatics and Gauss' law of magnetic field.

The fields in the above derivations are valid only when I is a constant independent of time, or equivalently when $Q(t)$ is linearly changing with time. Accordingly, the resulting magnetic field H_ϕ in this case is independent of time. For general time variations, the H_ϕ we have obtained here from Ampere's law would feed back into Faraday's law producing additional E_θ and E_r. These additional E_θ and E_r will again feed back into Ampere's law in terms of displacement current producing additional magnetic fields, and so on. In this case, a complete coupled formulation of Faraday's and Ampere's laws will have to be considered to obtain a rigorous solution of the electromagnetic fields, as studied in the following chapters. In practical cases when the current I and the related magnetic fields are slowly changing in time, the present results may be used as good first-order approximations.

As mentioned above, Biot-Savart's law (2.71) follows as a consequence of other known electromagnetic principles, under the condition of a time-invariant (constant or DC) current. In other words, Biot-Savart's law is not an independent law, and would not be needed once other known laws are enforced. A general expression of the magnetic field for a time-invariant current is also derived in (6.32) using a different approach based on vector potential function. The same magnetic field expression (2.71) of Biot-Savart's law is shown in (6.34) to be deduced from the general expression of (6.32), as a specific limiting condition for a time-invariant current. It may be convenient to use the simple form of Biot-Savart's law in order to determine the magnetic field produced by a given DC-current distribution. This is possible by segmenting the current distribution into small elements $(\overline{I}\delta)$, finding the magnetic field $\triangle\overline{H}$ produced due to each segment using Biot-Savart law, and then superposing or integrating the $\triangle\overline{H}$ over the entire distribution.

Using the above process of superposition for a practical DC-current distribution, which maintains a closed path in order to ensure a continuous charge flow, each of the positive or negative charge accumulation that exist at the head- or tail-end of individual elemental currents $(\overline{I}\delta)$ would sequentially cancel with an opposite charge, respectively at the tail- or head-end of a neighboring segment. Accordingly, the superposition of the electric field from each DC-current element, which results from the individual charges at the two ends of the element, would also cancel to zero, resulting in the total electric field of any closed-circuit DC-current distribution to be zero. Therefore, the electric field $\triangle\overline{E}$ we have derived for a current element $(\overline{I}\delta)$ would not be needed to model closed-circuit DC currents. Therefore, the $\triangle\overline{E}$ of the complete fields (2.69,2.71) for the current element $(\overline{I}\delta)$ maybe ignored, retaining only the magnetic field $\triangle\overline{H}$ (2.71) of Biot-Savart's law, which would be sufficient for modeling any distribution of closed-circuit DC currents.

3.7. Ampere's Law for a Non Planar Surface. For simplicity, we have expressed Ampere's law (2.43,2.44) as applied to a closed loop C placed on a given plane, enclosing a planar surface S. It can be further generalized to a planar or non-planar closed loop C, enclosing a planar or non-planar surface S (Fig.18). The treatment would be similar to application of Faraday's law for a non-planar surface, presented in section 1.5. The non-planar surface S may be decomposed into small segments of planar surfaces $\triangle S_i$, each enclosed by a planar loop $\triangle C_i$, to which the planar version of Ampere's law may be used. Now summing the contribution for all closed loop segments $\triangle C_i$ would result in the line integral over the non-planar closed boundary C of the total non-planar surface. This is because the contribution from the overlapping parts of two neighboring loop segments $\triangle C_i$ would cancel due to their opposing path directions. This would therefore keep non-zero contribution from only the parts that overlap with the bounding loop C in the sum total, leading to a line integral over the loop C. On the other hand, the summation of the surface integrals over the segments $\triangle S_i$ would clearly be the surface integral over the total non-planer surface S.

$$\begin{aligned}\sum_i \sum_{\triangle C_i} \overline{H}_{ij} \cdot \overline{\triangle l}_{ij} &= \sum_{i,S} \left(\overline{J}_i + \frac{d\overline{D}_i}{dt}\right) \cdot \overline{\triangle S}_i \\ &= \sum_{ij\ on\ C} \overline{H}_{ij} \cdot \overline{\triangle l}_{ij} = \sum_{k=(ij)\ on\ C} \overline{H}_k \cdot \overline{\triangle l}_k,\end{aligned} \tag{2.72}$$

$$\sum_C \overline{H}_k \cdot \overline{\triangle l}_k = \sum_S (\overline{J}_i + \frac{d\overline{D}_i}{dt}) \cdot \overline{\triangle S}_i, \tag{2.73}$$

$$\oint_C \overline{H} \cdot \overline{dl} = \iint_S (\overline{J} + \frac{d\overline{D}}{dt}) \cdot \overline{dS}. \tag{2.74}$$

In the above indexing, j represents a segment of any close loop $\triangle C_i$, and k represents a line segment with index (ij) that is on the boundary loop C. The same form of Ampere's law (2.43,2.44) for a planar loop is now applicable in (2.73,2.74) as well for a general non-planar loop.

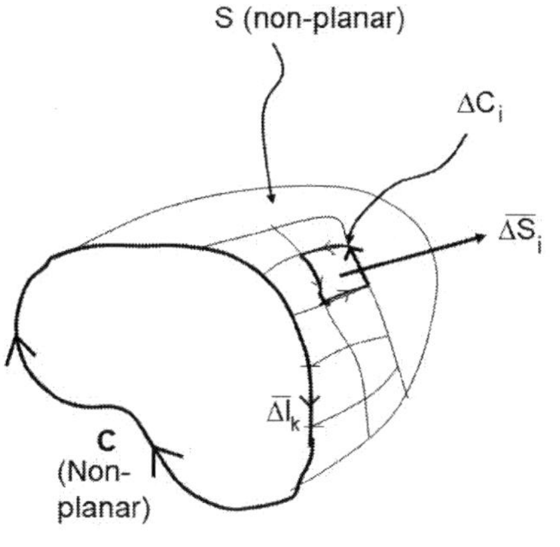

FIGURE 18

3.8. Ampere's Law, KCL, and Gauss' Law of Electric field. Interestingly, if the boundary loop C to a non-planar surface S is reduced to an arbitrarily small size, then the surface S approaches to be a completely closed surface. Because the loop C is now infinitesimally small in size, the line integral of $\overline{H}$ along the C would approach zero.

$$\sum_{C\to 0} \overline{H}_k \cdot \overline{\triangle l}_k = 0. \tag{2.75}$$

In this limit, Ampere's law (2.73,2.74) would take the following form.

$$\sum_{i,closed\ S} \left(\overline{J}_i + \frac{d\overline{D}_i}{dt}\right) \cdot \overline{\triangle S}_i = \sum_{C\to 0} \overline{H}_k \cdot \overline{\triangle l}_k = 0, \tag{2.76}$$

$$\oiint_S (\overline{J} + \frac{d\overline{D}}{dt}) \cdot \overline{dS} = 0. \tag{2.77}$$

This is essentially the general KCL of (2.34,2.35). In other words, once general form of Ampere's law (2.73,2.74) is satisfied, the general KCL of (2.34,2.35) is automatically satisfied, making the general KCL redundant. Further as discussed in section 2.4, the general KCL of (2.34,2.35) is as well an equivalent expression of Gauss' law of electric field, if charge continuity is assumed. Consequently, once the general Ampere's law (2.73,2.74) is satisfied, which already makes the general KCL of (2.34,2.35) redundant as per (2.76,2.77), would also make Gauss' law of electric field redundant, if charge continuity (local charge conservation) is independently maintained. This is a significant observation on the theoretical generality of Ampere's law.

4. Maxwell's Equations

Starting from the first principles of electric and magnetic fields, and circuit concepts such as Kirchoff's Voltage Law (KVL) and Kirchoff's Current Law (KCL), we have established two significant relations in general forms: Faraday's law and Ampere's law. Faraday's law generalized the concepts of KVL, induced potential and energy conservation, and also implicitly includes Gauss' law of magnetic field. On the other hand, Ampere's law in its general form puts together the concepts of KCL, displacement current and Gauss' law of electric field, conditional upon charge conservation, and in a static limit also includes Coulomb's law of electrostatics and Biot-Savart's law of magnetostatics. These two laws establish the inter-play between the electric and magnetic fields and material media, forming two complementary relations that supersede all other existing concepts and laws of electricity and magnetism. The generalization of the two laws was established by James Clerk Maxwell (1831-1879), who recognized that the fundamental scope of these general laws covered all existing theories and concepts of electricity and magnetism. He also recognized that these two laws apply in a complementary way to a diverse set of physical phenomena that were originally thought to be unrelated - starting from electrical circuits, magnetic machines, radio waves, to light as well as heat radiation. In recognition to his contribution, Ampere's and Faraday's laws are together referred to as Maxwell's equations of the electromagnetic theory. As mentioned, the two laws are fundamentally complimentary to each other, and the resulting theory is referred to as the electromagnetic theory in reference to the close complementarity of the electric and magnetic fields,

The two equations may be expressed in discretized summation or integral forms, as follows.

$$\sum_C \overline{E}_i \cdot \overline{\triangle l}_i = -\sum_S \frac{d\overline{B}_i}{dt} \cdot \overline{\triangle S}_i = -\sum_S \mu_i \frac{d\overline{H}_i}{dt} \cdot \overline{\triangle S}_i,$$
$$\oint_C \overline{E} \cdot \overline{dl} = -\iint_S \frac{d\overline{B}}{dt} \cdot \overline{dS} = -\iint_S \mu \frac{d\overline{H}}{dt} \cdot \overline{dS}, \tag{2.78}$$

$$\sum_C \overline{H}_i \cdot \overline{\triangle l}_i = I_d + I = I_t = \sum_S (\overline{J}_{di} + \overline{J}_i) \cdot \overline{\triangle S}_i = \sum_S \overline{J}_{ti} \cdot \overline{\triangle S}_i$$
$$= \sum_S (\frac{d\overline{D}_i}{dt} + \overline{J}_i) \cdot \overline{\triangle S}_i = \sum_S (\epsilon_i \frac{d\overline{E}_i}{dt} + \overline{J}_i) \cdot \overline{\triangle S}_i,$$
$$\oint_C \overline{H} \cdot \overline{dl} = \iint_S (\overline{J}_d + \overline{J}) \cdot \overline{dS} = \iint_S \overline{J}_t \cdot \overline{dS}$$
$$= \iint_S (\frac{d\overline{D}}{dt} + \overline{J}) \cdot \overline{dS} = \iint_S (\epsilon \frac{d\overline{E}}{dt} + \overline{J}) \cdot \overline{dS}. \tag{2.79}$$

The conduction current density $\overline{J}$ in Ampere's law may consists of two parts. Part of it may be induced in the medium by the electric field governed by the conductivity of the medium. This part may be referred to as the $\overline{J}_c$, which is related to the electric field using the conductivity of the medium. That is, $\overline{J} = \sigma\overline{E}$. The second kind of the conduction current may have been impressed in the medium in the form of an independent source. The current density of this independent source may be referred to as the source current density $\overline{J}_s$. All the currents in Ampere's law, which includes the source current, conduction current and the displacement current, are together referred to as the total current I_t. The corresponding current density is the total current density $\overline{J}_t$.

$$\overline{J}_t = \overline{J}_d + \overline{J} = \overline{J}_d + \overline{J}_c + \overline{J}_s = \frac{d\overline{D}}{dt} + \sigma\overline{E} + \overline{J}_s. \tag{2.80}$$

4.1. Symmetry in the Maxwell's Equations. The two equations appear remarkably similar to each other in the following sense. The left side of Faraday's law is a closed-loop integral of the electric field, while that in Ampere's law is a closed-loop integral of the magnetic field. The right side of Faraday's law contains the negative derivative of the magnetic flux density, while that in Ampere's law also contains the derivative of the electric flux density but with a positive sign. However, the right side of Ampere's law has the conduction current density, while there is no equivalent term in Faraday's law. This is a distinct difference. This difference arises because there is no known magnetic charge, and consequently no magnetic current, that exist in nature. This fundamental issue was also discussed in relations to Gauss' laws in chapter 1, sections 2.1,2.2. For the same above reason, Gauss' law for the electric field equates the total electric flux to the electric charge, while that for the magnetic field equates the magnetic flux to zero.

The above asymmetry between the two equations arising due to the absence of the magnetic charge may be overcome by introducing a hypothetical magnetic current I_m, and a corresponding magnetic current density represented as $\overline{M}$. Accordingly, a new term $-\overline{M}$ may be added to the right side of Faraday's law, in symmetry with the electric conduction current density term $\overline{J}$ in Ampere's law. The $-$ sign before the magnetic current density $\overline{M}$ to be introduced as a second term in the right side of Faraday's law is to maintain symmetry with the $-$ sign already associated with the first term. The first term in the right side of Faraday's law may be referred to as magnetic displacement current density $\overline{M}_d$, in symmetry with corresponding term in Ampere's law which is referred to as the electric displacement current density $\overline{J}_d$. Further, the two terms together in the right side of Faraday's law may also be referred to as the total magnetic current density $\overline{M}_t$, in symmetry with that in Ampere's law which is referred to as the total current density $\overline{J}_t$. In addition, the magnetic current density $\overline{M}$ in Faraday's law may also be decomposed into magnetic conduction and source current densities, $\overline{M}_c$ and $\overline{M}_s$, in symmetry with the electric current density $\overline{J}$ in Ampere's law decomposed into $\overline{J}_c$ and $\overline{J}_s$, respectively. Further, in symmetry with the conservation and continuity of electric charge Q, associated with the total electric current density $\overline{J}_t$ in Ampere's law, any hypothetical magnetic charge Q_m also needs to be conserved maintaining continuity, established in relation with the total magnetic current density $\overline{M}_t$ of the generalized Faraday's law.

$$\oint_C \overline{E}\cdot\overline{dl} = \iint_S \left(-\frac{d\overline{B}}{dt} - \overline{M}\right)\cdot\overline{dS} = \iint_S \left(-\mu\frac{d\overline{H}}{dt} - \overline{M}\right)\cdot\overline{dS} = \iint_S -\overline{M}_t\cdot\overline{dS}, \tag{2.81}$$

$$\oint_C \overline{H} \cdot \overline{dl} = \iint_S (\frac{d\overline{D}}{dt} + \overline{J}) \cdot \overline{dS} = \iint_S (\epsilon\frac{d\overline{E}}{dt} + \overline{J}) \cdot \overline{dS} = \iint_S \overline{J}_t \cdot \overline{dS}. \quad (2.82)$$

$$\overline{M}_t = \overline{M}_d + \overline{M} = \overline{M}_d + \overline{M}_c + \overline{M}_s = \tfrac{d\overline{B}}{dt} + \overline{M}_c + \overline{M}_s,$$
$$\overline{J}_t = \overline{J}_d + \overline{J} = \overline{J}_d + \overline{J}_c + \overline{J}_s = \tfrac{d\overline{D}}{dt} + \overline{J}_c + \overline{J}_s. \quad (2.83)$$

The magnetic-current density term introduced in Maxwell's equations, although is hypothetical and does not exist in reality, carries useful analytically significance. Some practical problems may be analytically modeled with significant simplicity when modeled as an equivalent problem using magnetic current sources. This is the principal practical motivation behind the introduction of the magnetic currents. We will include the magnetic current in future analytical treatments, as needed, particularly for derivations that carry general scope and significance.

Differential Form of the Maxwell's Equations Using Curl Operator:

Each of the above two Maxwell's equations (2.81,2.82) maybe applied to any small, flat contour $\triangle C$, enclosed by a flat surface element $\overline{\triangle S}$. Then, Stokes theorem maybe applied to the left sides of (2.81,2.82) using the curl operator (see appendix A). This would be valid for any orientation of the normal axis to the surface element $\overline{\triangle S}$. Accordingly, the above two Maxwell's equations maybe shown to be equivalently expressed in the differential form using the curl operator.

$$(\overline{\nabla} \times \overline{E}) \cdot \overline{\triangle S} = -\overline{M}_t \cdot \overline{\triangle S},$$
$$(\overline{\nabla} \times \overline{H}) \cdot \overline{\triangle S} = -\overline{J}_t \cdot \overline{\triangle S}, \quad (2.84)$$

$$\overline{\nabla} \times \overline{E} = -\overline{M}_t = -\tfrac{d\overline{B}}{dt} - \overline{M}_c - \overline{M}_s, \quad (2.85)$$
$$\overline{\nabla} \times \overline{H} = \overline{J}_t = \tfrac{d\overline{D}}{dt} + \overline{J}_c + \overline{J}_s. \quad (2.86)$$

The two Maxwell's equations, expressed in the integral forms (2.81,2.82) or the differential forms (2.85,2.86), when solved for a given material medium and for a particular source structure, would provide rigorous solutions of the electric and magnetic fields in the structure, which can then be used to derive any other practical parameters such as input impedance, power, efficiency. In the following, we will derive some special relations from Maxwell's equations that are often useful for physical interpretation and analytical simplification of general problems. In the following chapters we will apply Maxwell's equations and the special relations to specific problems of electromagnetic waves and antenna radiation, of particular interests in electrical-engineering applications.

4.2. The Poynting Theorem and Radiation Power. Consider a volume element $\Delta V = \triangle x \triangle y \triangle z$, with point O at its center, as shown in Fig.19. Consider the x component J_{xto} of the total current density at O, which includes the conduction current density J_{xco}, displacement current density J_{xdo} and any source current density J_{xso} that might be independently impressed. Apply Ampere's law to a closed loop $\triangle C$ around J_{xto}. The closed loop $\triangle C$ is on the yz plane, passing through O, as shown in Fig.20.

$$J_{xt} = J_{xc} + J_{xd} + J_{xs}, \quad J_{xto} = J_{xco} + J_{xdo} + J_{xso},$$
$$J_{xto}\triangle y\triangle z = \sum_{\triangle C} H_{ti}\triangle l_i \;= H_{y1}\triangle y + H_{z2}\triangle z - H_{y3}\triangle y - H_{z4}\triangle z$$
$$= \sum_{\triangle C} \hat{l}_i \cdot \overline{H}_i\triangle l_i = \sum_{\triangle C}(\hat{x}\times\hat{n}_i)\cdot\overline{H}_i\triangle l_i$$
$$= \sum_{\triangle C} -\hat{n}_i\cdot(\hat{x}\times\overline{H}_i)\triangle l_i; \quad \hat{l}_i = \hat{x}\times\hat{n}_i. \tag{2.87}$$

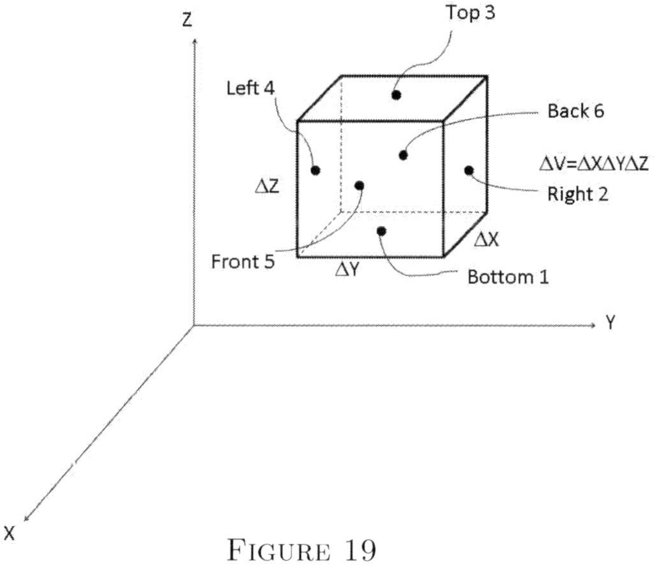

FIGURE 19

The vector relationship, $\overline{A}\times\overline{B}\cdot\overline{C} = -\overline{B}\cdot\overline{A}\times\overline{C}$, is used in the above simplification. Now multiply the above equation with the x component of the electric field, E_{xo}, at the center point O, and simplify as follows.

$$E_{x0}J_{xt0}\triangle y\triangle z = E_{x0}\sum_{\triangle C} -\hat{n}_i\cdot(\hat{x}\times\overline{H}_i)\triangle l_i,$$
$$(\hat{x}E_{x0})\cdot\overline{J}_{t0}\triangle y\triangle z = \sum_{\triangle C} -\hat{n}_i\cdot\left[(\hat{x}E_{x0})\times\overline{H}_i\right]\triangle l_i,$$
$$(\hat{x}E_{x0})\cdot\overline{J}_{t0}\triangle y\triangle x\triangle z = \sum_{\triangle C} -\hat{n}_i\cdot\left[(\hat{x}E_{x0})\times\overline{H}_i\right]\triangle l_i\triangle x,$$

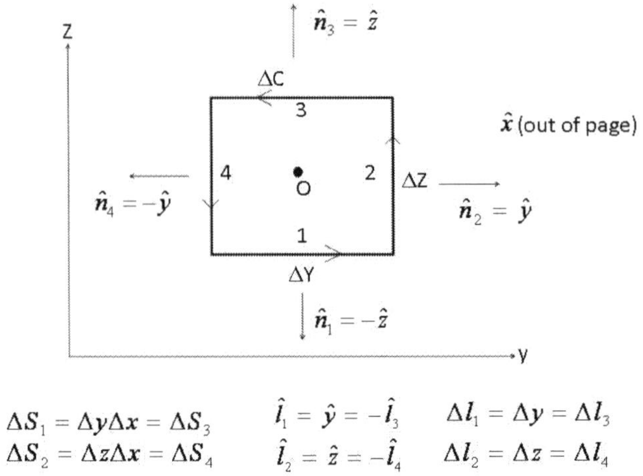

FIGURE 20

$$(\hat{x}E_{x0}) \cdot \overline{J}_{t0}\triangle V = \sum_{\triangle C} -\hat{n}_i \cdot \big[(\hat{x}E_{x0}) \times \overline{H}_i\big]\triangle S_i$$
$$= \sum_{i=1}^{4} -\overline{\triangle S}_i \cdot \big[(E_{x0}\hat{x}) \times \overline{H}_i\big],\ \overline{\triangle S}_i = \triangle S_i \hat{n}_i. \tag{2.88}$$

The above surface integration is performed over the sides $i = 1, .., 4$. Similar surface integration over the sides $i = 5, 6$ can be shown to be zero.

$$\overline{\triangle S}_{5,6} \cdot \big[(E_{x0}\hat{x} \times \overline{H}_i\big] = \pm\triangle S_{5,6}\hat{x} \cdot \big[(E_{x0}\hat{x} \times \overline{H}_i\big] = 0. \tag{2.89}$$

The vector relationship, $\overline{P} \cdot \overline{P} \times \overline{R} = 0$, is used in the above simplification. Adding the equations (2.88, 2.89), we can deduce the surface integration over all six faces of the closed surface $\triangle S$.

$$(\hat{x}E_{x0}) \cdot \overline{J}_{t0}\triangle V = \sum_{i=1}^{6} -\overline{\triangle S}_i \cdot \big[E_{x0}\hat{x} \times \overline{H}_i\big] = \sum_{\triangle S} -\overline{\triangle S}_i \cdot \big[E_{x0}\hat{x} \times \overline{H}_i\big]. \tag{2.90}$$

The above derivation may be repeated by starting with the y or z component of the total electric current, and then multiplying with the corresponding components of the electric field at the center point O.

$$
\begin{aligned}
(\hat{y}E_{y0}) \cdot \overline{J}_{t0}\triangle V &= \sum_{\triangle S} -\overline{\triangle S}_i \cdot \left[E_{y0}\hat{y} \times \overline{H}_i\right], \\
(\hat{z}E_{z0}) \cdot \overline{J}_{t0}\triangle V &= \sum_{\triangle S} -\overline{\triangle S}_i \cdot \left[E_{z0}\hat{z} \times \overline{H}_i\right].
\end{aligned} \tag{2.91}
$$

Adding the equations (2.90, 2.91),

$$
\overline{E}_0 \cdot \overline{J}_{t0}\triangle V = \sum_{\triangle S} -\overline{\triangle S}_i \cdot \left[\overline{E}_0 \times \overline{H}_i\right], \quad \overline{E}_0 = E_{x0}\hat{x} + E_{y0}\hat{y} + E_{z0}\hat{z}. \tag{2.92}
$$

Similar relations can be derived by starting with the negative of the magnetic current instead of the electric current, multiplying with the respective magnetic fields, and using Faraday's law instead of Ampere's law. Due to symmetry between Ampere's law and Faraday's law as discussed before, the new relationship can be expressed by substituting the electric current in the above relation (2.92) by negative of the magnetic current, and then exchanging the electric and magnetic fields. The total magnetic current in general includes the magnetic displacement current, and any equivalent magnetic source or conduction current that may be present.

$$
\begin{aligned}
-\overline{H}_0 \cdot \overline{M}_{t0}\triangle V &= \sum_{\triangle S} -\overline{\triangle S}_i \cdot \left[\overline{H}_0 \times \overline{E}_i\right], \\
\overline{H}_0 \cdot \overline{M}_{t0}\triangle V &= \sum_{\triangle S} -\overline{\triangle S}_i \cdot \left[\overline{E}_i \times \overline{H}_0\right], \\
\overline{M}_t = \overline{M}_d + \overline{M}_s + \overline{M}_c &= \frac{d\overline{B}}{dt} + \overline{M}_s + M_c.
\end{aligned} \tag{2.93}
$$

Adding the two equations (2.92, 2.93) we derived for the electric and magnetic currents,

$$
\begin{aligned}
\left(\overline{E}_0 \cdot \overline{J}_{t0} + \overline{H}_0 \cdot \overline{M}_{t0}\right)\triangle V &= \sum_{\triangle S} -\overline{\triangle S}_i \cdot \left[\overline{E}_i \times \overline{H}_0 + \overline{E}_0 \times \overline{H}_i\right] \\
&= \sum_{\triangle S} -\overline{\triangle S}_i \cdot \left[(\overline{E}_0 + \overline{\triangle E}_i) \times \overline{H}_0 + \overline{E}_0 \times (\overline{H}_0 + \overline{\triangle H}_i)\right].
\end{aligned} \tag{2.94}
$$

Note that the subscript i and O refer to parameters computed at the surface S_i and center location O, respectively. For an elemental surface, with $\Delta x, \Delta y, \Delta z$ approaching zero, the fields at the center and on the surface are incrementally different. In this limit, the product of incremental fields would approach zero. Accordingly, the following relationship may be established.

$$\overline{E}_i = \overline{E}_0 + \overline{\triangle E}_i\ , \overline{H}_i = \overline{H}_0 + \overline{\triangle H}_i,$$
$$\overline{E}_i \times \overline{H}_i = \overline{E}_0 \times \overline{H}_0 + \overline{\triangle E}_i \times \overline{H}_0 + \overline{E}_0 \times \overline{\triangle H}_i + \overline{\triangle E}_i \times \overline{\triangle H}_i,$$
$$(\overline{E}_i \times \overline{H}_0) + (\overline{E}_0 \times \overline{H}_i) = 2\overline{E}_0 \times \overline{H}_0 + \overline{\triangle E}_i \times \overline{H}_0 + \overline{E}_0 \times \overline{\triangle H}_i,$$
$$(\overline{E}_i \times \overline{H}_0) + (\overline{E}_0 \times \overline{H}_i)|_{\triangle V \to 0} = \overline{E}_i \times \overline{H}_i + \overline{E}_0 \times \overline{H}_0. \tag{2.95}$$

Also note that the surface integral of any constant vector over the six surfaces of a rectangular volume is zero. The normal component of the constant vector would be equal and opposite on any two opposing sides of the volume. Accordingly, the contribution of the surface integral over any two opposing faces would cancel each other, adding to zero over the total surface.

$$\sum_{\triangle S} -\overline{\triangle S}_i \cdot \left(\overline{E}_0 \times \overline{H}_0\right) = 0 = \left(\overline{E}_0 \times \overline{H}_0\right) \cdot \sum_{\triangle S} \overline{\triangle S}_i. \tag{2.96}$$

Using the above relations (2.95, 2.96) in (2.94) the following simplification is possible.

$$\begin{aligned}\left(\overline{E}_0 \cdot \overline{J}_{t0} + \overline{H}_0 \cdot \overline{M}_{t0}\right)\triangle V &= \sum_{\triangle S} -\overline{\triangle S}_i \cdot \left[\overline{\triangle E}_i \times \overline{H}_0 + \overline{E}_0 \times \overline{\triangle H}_i\right] \\ &= \sum_{\triangle S} -\overline{\triangle S}_i \cdot \left[\left(\overline{E}_i \times \overline{H}_i\right) + \left(\overline{E}_0 \times \overline{H}_0\right)\right] \\ &= \sum_{\triangle S} -\overline{\triangle S}_i \cdot \left(\overline{E}_i \times \overline{H}_i\right).\end{aligned} \tag{2.97}$$

One may extend the above result to any arbitrary volume V, enclosed by a surface S, by segmentation and superposition (see Fig.21).

$$\begin{aligned}\triangle S_j \equiv \sum_{i=1}^{6} \triangle S_{ij}, \\ V \equiv \sum_j \triangle V_j;\ S \equiv \sum_k \triangle S_k \equiv \sum_j \sum_i \triangle S_{ij},\quad k = (ij)\ on\ S,\end{aligned} \tag{2.98}$$

$$\begin{aligned}\sum_V \left(\overline{E}_j \cdot \overline{J}_{tj} + \overline{H}_j \cdot \overline{M}_{tj}\right)\triangle V_j &= -\sum_j \sum_{\triangle S_j} \overline{\triangle S}_{ij} \cdot \left(\overline{E}_{ij} \times \overline{H}_{ij}\right) \\ &= -\sum_S \overline{\triangle S}_k \cdot \left(\overline{E}_k \times \overline{H}_k\right).\end{aligned} \tag{2.99}$$

The total currents may be explicitly decomposed into the source, conduction and displacement terms. The contribution from the source currents may be separated from the other contributions. The equations in the form of summation over discretized volume or surface elements may be equivalently expressed in integral forms, and rearranged as follows.

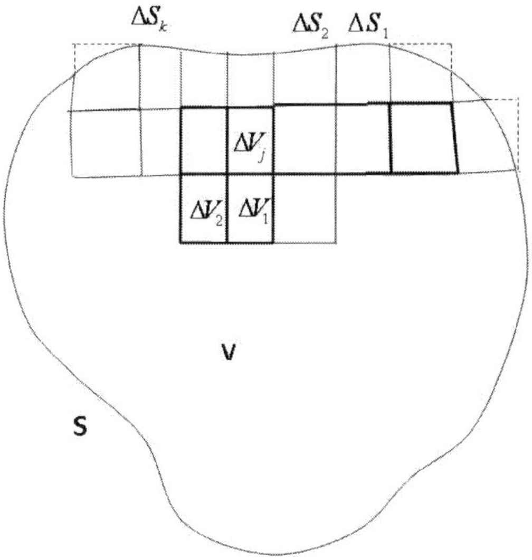

FIGURE 21

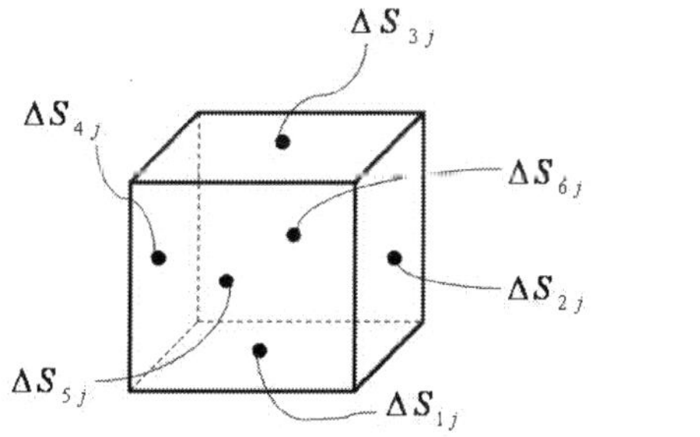

FIGURE 22

$$\sum_V \left(- \overline{E}_j \cdot \overline{J}_{js} - \overline{H}_j \cdot \overline{M}_{sj} \right) \triangle V_j$$

$$= \sum_V [\overline{E}_j \cdot (\overline{J}_{cj} + \overline{J}_{dj}) + \overline{H}_j \cdot (\overline{M}_{cj} + \overline{M}_{dj})] \triangle V_j + \sum_S \overline{\triangle S}_k \cdot (\overline{E}_k \times \overline{H}_k), \qquad (2.100)$$

$$\iiint\limits_V \left(- \overline{E} \cdot \overline{J}_s - \overline{H} \cdot \overline{M}_s \right) dV$$
$$= \iiint\limits_V \left[\overline{E} \cdot \left(\overline{J}_c + \overline{J}_d \right) + \overline{H} \cdot \left(\overline{M}_c + \overline{M}_d \right) \right] dV + \oiint\limits_S (\overline{E} \times \overline{H}) \cdot \overline{dS}. \quad (2.101)$$

The above relationship between the different field integrals is referred to as the Poynting theorem. The theorem is expressed in a basic analytical form, which may need further analysis and physical understanding.

Poynting Theorem Derived Using Differential Forms of Maxwell's Equations:

The above derivation of the Poynting theorem, based on integral or summation forms of Maxwell's equations, may provide useful physical as well as analytical insight through the individual steps. However, it may be mathematically convenient to derive the same final results from the differential forms (2.85,2.86) of Maxwell's equations, using a useful relation for the divergence operator of the cross product of two vectors (appendix A, section 4.3).

$$\begin{aligned} \overline{\nabla} \cdot (\overline{E} \times \overline{H}) &= \overline{H} \cdot \overline{\nabla} \times \overline{E} - \overline{E} \cdot \overline{\nabla} \times \overline{H} = -\overline{H} \cdot \overline{M}_t - \overline{E} \cdot \overline{J}_t \\ &= -\overline{H} \cdot \overline{M}_d - \overline{H} \cdot \overline{M}_c - \overline{H} \cdot \overline{M}_s - \overline{E} \cdot \overline{J}_d - \overline{E} \cdot \overline{J}_c - \overline{E} \cdot \overline{J}_s, \\ -\overline{H} \cdot \overline{M}_s - \overline{E} \cdot \overline{J}_s &= \overline{E} \cdot (\overline{J}_c + \overline{J}_d) + \overline{H} \cdot (\overline{M}_c + \overline{M}_d) + \overline{\nabla} \cdot (\overline{E} \times \overline{H}). \end{aligned} \quad (2.102)$$

Performing a volume integral of the above expression, and applying divergence theorem (appendix A, section 4.5) to the last term, would result in the same final expression (2.100,2.101) of the Poynting theorem.

4.2.1. *Different Kinds of Power in an Electromagnetic System.* The individual parts in the basic expression of the Poynting theorem in (2.101) represent different kinds of powers. (2.101) establishes a conservation principle among the different power contributions. We may group similar kinds of powers together for closer examination and interpretation.

$$\begin{aligned} P_s &= P_l + P_f + P_p, \\ P_s &= \iiint\limits_V \left(- \overline{E} \cdot \overline{J}_s - \overline{H} \cdot \overline{M}_s \right) dV, \\ P_l &= \iiint\limits_V \left(\overline{E} \cdot \overline{J}_c + \overline{H} \cdot \overline{M}_c \right) dV, \\ P_f = \iiint\limits_V \left(\overline{E} \cdot \overline{J}_d + \overline{H} \cdot \overline{M}_d \right) dV &= \iiint\limits_V \left(\overline{E} \cdot \epsilon \frac{d\overline{E}}{dt} + \overline{H} \cdot \mu \frac{d\overline{H}}{dt} \right) dV, \\ P_p &= \oiint\limits_S (\overline{E} \times \overline{H}) \cdot \overline{dS}. \end{aligned} \quad (2.103)$$

Source Power:

The P_s in (2.103) is contributed by the source current densities, which include the conventional electrical source current density $\overline{J}_s$ and any equivalent magnetic source current density $\overline{M}_s$ one may have. Accordingly, P_s may be recognized as the source power. We may decompose the total source power P_s into two parts, P_{se} and P_{sh}, contributed from the electric and the equivalent magnetic sources, respectively.

$$P_s = P_{se} + P_{sh},$$
$$P_{se} = -\iiint_V (\overline{E} \cdot \overline{J}_s) dV, \quad P_{sh} = -\iiint_V (\overline{H} \cdot \overline{M}_s) dV. \tag{2.104}$$

Power Loss in the Material Medium:

The P_l in (2.103) is contributed due to the conduction current densities in the medium which, like the source currents, also consist of two different types: the conventional electrical conduction current density $\overline{J}_c$, and any equivalent magnetic conduction current density $\overline{M}_c$ one may have. P_l is recognized as the power lost or dissipated in the material medium due to the conduction currents. P_l may be decomposed into two parts, P_{le} and P_{sh}, contributed from the electric and the equivalent magnetic conduction currents, respectively.

$$P_l = P_{le} + P_{lh},$$
$$P_{le} = \iiint_V (\overline{E} \cdot \overline{J}_c) dV, \quad P_{lh} = \iiint_V (\overline{H} \cdot \overline{M}_c) dV. \tag{2.105}$$

The relative sign in the expressions for the power loss terms in (2.105), as compared with those of the source power in (2.104), may be noticed. The sign difference is due to a basic distinction between the two power contributions, as discussed in chapter 1, section 1.4.2. The source power P_s refers to the power *generated* by the sources, whreas P_l refers to the power *dissipated* in the conduction currents.

Power Delivered to the Stored Field Energy:

The P_f in (2.103) is expressed in a form similar to that of P_l. The conduction currents in P_l are substituted in P_f by the corresponding displacement currents. In analogy with the power P_l, which is the power delivered to the conduction currents, the P_f may be thought of, in mathematical terms, as the power delivered to the displacement currents. However, the displacement currents are physically different from the conduction currents, as discussed in section 2.3. Therefore, the physical meaning of P_f may have to be viewed in a fundamentally different manner. P_f may be interpreted as the power utilized to build a form of energy which is temporarily stored in the electromagnetic fields.

Let us decompose P_f into two parts, P_{fe} and P_{fh}, which are contributed by the electric and magnetic displacement currents, respectively. P_{fe} and P_{fh} may be expressed as time derivatives of respective parameters, W_e and W_h. Conversely, the parameters W_e and W_h are the time integrals of the respective powers P_{fe} and P_{fh}, and therefore be seen as the field energies, *stored* in the associated fields. The powers P_{fe} and P_{fh}, are needed in order to incrementally change the field energies W_e and W_h. The powers are positive as the respective energies increase in time, and negative as the energies reduce. The field energies appear only temporarily, and are therefore referred to as stored energies They build up from their zero initial values, which is the case when there are no fields, to positive values as the fields increase with time. The stored energies reduce back to zero, when the system returns back to its initial state, with no fields.

$$P_f = P_{fe} + P_{fh},$$

$$P_{fe} = \iiint\limits_V (\overline{E} \cdot \overline{J}_d) dV = \iiint\limits_V (\overline{E} \cdot \epsilon \frac{d\overline{E}}{dt}) dV = \iiint\limits_V \frac{\epsilon}{2} \frac{d}{dt} (\overline{E} \cdot \overline{E}) dV$$

$$= \iiint\limits_V \frac{\epsilon}{2} \frac{d}{dt} |\overline{E}|^2 dV = \frac{d}{dt} \iiint\limits_V \frac{\epsilon}{2} |\overline{E}|^2 dV = \frac{d}{dt} W_e, \ W_e = \iiint\limits_V \frac{\epsilon}{2} |\overline{E}|^2 dV,$$

$$P_{fh} = \iiint\limits_V (\overline{H} \cdot \overline{M}_d) dV = \iiint\limits_V (\overline{H} \cdot \mu \frac{d\overline{H}}{dt}) dV = \iiint\limits_V \frac{\mu}{2} \frac{d}{dt} (\overline{H} \cdot \overline{H}) dV$$

$$= \iiint\limits_V \frac{\mu}{2} \frac{d}{dt} |\overline{H}|^2 dV = \frac{d}{dt} \iiint\limits_V \frac{\mu}{2} |\overline{H}|^2 dV = \frac{d}{dt} W_h, \ W_h = \iiint\limits_V \frac{\mu}{2} |\overline{H}|^2 dV. \quad (2.106)$$

An electromagnetic system may be seen as analogous to an electric circuit. The stored field energies W_e and W_h in an electromagnetic system, and the respective powers P_{fe} and P_{fh} associated with their time derivatives, are analogous to the energies, and the associated powers, stored in the capacitor and inductor elements in an electric circuit, respectively. In contrast, P_s is analogous to the power delivered by all signal sources in the circuit, and P_l to the power lost or dissipated in the resistive elements in the form of heat.

Radiation Power and Poynting Vector:

Now, let us interpret the power conservation relation (2.103), in terms of the different power contributions. The power P_s of (2.103), is generated from the source currents. It is then utilized in three different forms, constituting the three power terms in the right side of the conservation relation in (2.103). The first term, P_l, is the part which is dissipated in the form of power loss in the conducting medium. The second term, P_f, is the part which is used to build the stored energy in the electromagnetic fields. Any remaining power appears as the third term, P_p, in the right side of the conservation relation (2.103). The contribution due to P_p is conceptually different from the other power contributions, P_s, P_l and P_f, which we have discussed. In contrast to P_s P_l and P_f, which are expressed as integrals over a given volume V, P_p is expressed as an integration over its boundary surface S. Accordingly, P_p may be interpreted as the power that escapes from the region inside the closed surface S, flowing through the closed surface into the outside region. Consider an open problem extending to infinite distances. In this case, any non-zero value of P_p, computed over a closed surface which is sufficiently large in expanse, may be recognized as the "radiation" power. This power escapes from a local source, and propagates to far-away distances. It appears in the form of the electromagnetic energy, spreading in the open space through infinite distances.

There is no direct analogy to the radiation power P_p in practical electric circuits. This is because the practical electric circuits are finite in size. Whereas, an open space, in which the radiation power is considered, is a physically infinite structure. One may, however, consider an open space to be analogous to a hypothetical, infinite-sized circuit. Any power "radiated" into the electromagnetic fields at far open space may be considered analogous to any power in an infinite-sized circuit, delivered to circuit elements placed far away from the source.

As mentioned, P_p in (2.103) is expressed as a surface integral of the vector $(\overline{E} \times \overline{H})$. Accordingly, the vector $(\overline{E} \times \overline{H}) = \overline{S}_p$ may be identified as a special vector, with its direction and magnitude interpreted in relation to the power flow concept discussed above. The magnitude, P_{pd}, of the special vector $\overline{S}_p$ is interpreted as the density of the power flow over a unit surface area, placed normal to the direction of the flow. The vector direction, $\hat{p}$, of $\overline{S}_p$ is interpreted as the direction of power flow or power radiation. For a given magnitude of $\overline{S}_p$, maximum power would flow or propagate through a surface element $\overline{dS}$, placed at a given location, when the normal direction to the surface element is aligned with $\hat{p}$. The flow would be zero if the surface normal is perpendicular to $\hat{p}$, or equivalently the surface is parallel to the $\hat{p}$. This special vector, $(\overline{E} \times \overline{H}) = \overline{S}_p$, is referred to as the power-flow vector or the Poynting vector.

$$P_p = \oint\!\!\!\oint_S (\overline{E} \times \overline{H}) \cdot \overline{dS} = \oint\!\!\!\oint_S \overline{S}_p \cdot \overline{dS},\ \overline{S}_p = (\overline{E} \times \overline{H}) = P_{pd}\hat{p},$$

$$\triangle P_p = \overline{S}_p \cdot \overline{\triangle S} = P_{pd}\triangle S,\ P_{pd} = \frac{\triangle P_p}{\triangle S},\ \overline{\triangle S} = \triangle S\hat{p} \tag{2.107}$$

4.3. The Reciprocity Theorem. Consider two sets of fields, $(\overline{E}_1, \overline{H}_1)$ and $(\overline{E}_2, \overline{H}_2)$, existing in a common medium, produced by their respective sources $(\overline{J}_{s1}, \overline{M}_{s1})$ and $(\overline{J}_{s2}, \overline{M}_{s2})$. Here we will not consider any magnetic conduction current in the medium, for practical reasons. Accordingly, the total currents in the two cases, $(\overline{J}_{t1}, \overline{M}_{t1})$ and $(\overline{J}_{t2}, \overline{M}_{t2})$, consist of the following parts.

$$\overline{J}_{t1} = \frac{d\overline{D}_1}{dt} + \overline{J}_{c1} + \overline{J}_{s1} = \epsilon\frac{d\overline{E}_1}{dt} + \sigma\overline{E}_1 + \overline{J}_{s1}, \tag{2.108}$$

$$\overline{M}_{t1} = \frac{d\overline{B}_1}{dt} + \overline{M}_{s1} = \mu\frac{d\overline{H}_1}{dt} + \overline{M}_{s1}, \tag{2.109}$$

$$\overline{J}_{t2} = \frac{d\overline{D}_2}{dt} + \overline{J}_{c2} + \overline{J}_{s2} = \epsilon\frac{d\overline{E}_2}{dt} + \sigma\overline{E}_2 + \overline{J}_{s2}, \tag{2.110}$$

$$\overline{M}_{t2} = \frac{d\overline{B}_2}{dt} + \overline{M}_{s2} = \mu\frac{d\overline{H}_2}{dt} + \overline{M}_{s2}. \tag{2.111}$$

All currents other than the source component are expressed in terms of the respective field intensities and the material parameters that are common for both fields.

Now, we will adopt here essentially the same derivation steps followed in the last section, with specific differences. We will refer to necessary steps in the last section noting the essential changes. The main difference is that the derivation in the last section was dealing with only one set of fields and its associated currents. Instead, we have here two sets of fields and respective currents, from which we need to select appropriate combinations in the derivation of the last section. In the steps (2.87-2.92), substitute the electric current by $\overline{J}_{t2}$, apply Ampere's law using the corresponding magnetic field $\overline{H}_2$, and then multiply by the electric field $\overline{E}_1$. After simplification, the resulting equation which would be equivalent to (2.92), may be expressed as follows.

$$(\overline{E}_{10} \cdot \overline{J}_{t20})\triangle V = \sum_{\triangle S} \overline{\triangle S}_i \cdot (-\overline{E}_{10} \times \overline{H}_{2i}). \tag{2.112}$$

Now, repeat the above steps by starting with magnetic current $-\overline{M}_{t2}$ in place of electric current $\overline{J}_{t2}$, apply Faraday's law using the corresponding electric field $\overline{E}_2$, and then multiply by the magnetic field $\overline{H}_1$. After simplification, the resulting equation equivalent to (2.93) would be,

$$(\overline{H}_{10} \cdot \overline{M}_{t20})\triangle V = \sum_{\triangle S} \overline{\triangle S}_i \cdot (-\overline{E}_{2i} \times \overline{H}_{10}). \tag{2.113}$$

In the last section we added equations (2.92) and (2.93) to obtain (2.94). Instead, here we will subtract the corresponding equations (2.112,2.113) to obtain the following result.

$$(\overline{E}_{10} \cdot \overline{J}_{t20} - \overline{H}_{10} \cdot \overline{M}_{t20})\triangle V = \sum_{\triangle S} \overline{\triangle S}_i \cdot \big[- \overline{E}_{10} \times \overline{H}_{2i} + \overline{E}_{2i} \times \overline{H}_{10}\big]. \tag{2.114}$$

A similar equation can be obtained by exchanging the first set of fields and associated currents by the second set. This simply means to exchange the indexes 1 and 2 in (2.114).

$$(\overline{E}_{20} \cdot \overline{J}_{t10} - \overline{H}_{20} \cdot \overline{M}_{t10})\triangle V = \sum_{\triangle S} \overline{\triangle S}_i \cdot \big[- \overline{E}_{20} \times \overline{H}_{1i} + \overline{E}_{1i} \times \overline{H}_{20}\big]. \tag{2.115}$$

The above two equations (2.114,2.115) may be subtracted.

$$\begin{aligned} &(\overline{E}_{10} \cdot \overline{J}_{t20} - \overline{E}_{20} \cdot \overline{J}_{t10} + \overline{H}_{20} \cdot \overline{M}_{t10} - \overline{H}_{10} \cdot \overline{M}_{t20})\triangle V \\ &= \sum_{\triangle S} \overline{\triangle S}_i \cdot \big[- \overline{E}_{10} \times \overline{H}_{2i} - \overline{E}_{1i} \times \overline{H}_{20} + \overline{E}_{20} \times \overline{H}_{1i} + \overline{E}_{2i} \times \overline{H}_{10}\big]. \end{aligned} \tag{2.116}$$

It involves fields at the center 0 as well as those on the surface S_i. However, as simplified in the last section using equations (2.95-2.97), the fields at the center and on the surface may be related with incremental differences. This would lead to simplification such that, in the limit as $\triangle S$ and $\triangle V$ approach zero, the surface-summation in the above equation is expressed using the fields only on the surface.

$$\sum_{\triangle S} \overline{\triangle S}_i \cdot \left[\overline{E}_{10} \times \overline{H}_{2i} + \overline{E}_{1i} \times \overline{H}_{20}\right] = \sum_{\triangle S} \overline{\triangle S}_i \cdot \left[\overline{E}_{1i} \times \overline{H}_{2i}\right],$$
$$\sum_{\triangle S} \overline{\triangle S}_i \cdot \left[\overline{E}_{20} \times \overline{H}_{1i} + \overline{E}_{2i} \times \overline{H}_{10}\right] = \sum_{\triangle S} \overline{\triangle S}_i \cdot \left[\overline{E}_{2i} \times \overline{H}_{1i}\right],$$
$$(\overline{E}_{10} \cdot \overline{J}_{t20} - \overline{E}_{20} \cdot \overline{J}_{t10} + \overline{H}_{20} \cdot \overline{M}_{t10} - \overline{H}_{10} \cdot \overline{M}_{t20})\triangle V = \sum_{\triangle S} \overline{\triangle S}_i \cdot \left[-\overline{E}_{1i} \times \overline{H}_{2i} + \overline{E}_{2i} \times \overline{H}_{1i}\right]. \tag{2.117}$$

Now, decompose the total currents into specific parts using equations (2.108-2.111). Except the source currents, other parts are related to the respective fields and their time-derivatives through material parameters ϵ, μ and σ. It maybe shown to result in cancellation of contributions in (2.117) from the displacement and conduction currents in the medium, for complex harmonic signals represented with a $e^{j\omega t}$ time variation (or for DC signals with $\omega = 0$), with a given frequency ω. In this case, each time-derivative is equivalent to an algebraic multiplication by a factor $(j\omega)$. This would lead to the following result involving only the source currents.

$$(\overline{E}_{10} \cdot \overline{J}_{t20} - \overline{E}_{20} \cdot \overline{J}_{t10} + \overline{H}_{20} \cdot \overline{M}_{t10} - \overline{H}_{10} \cdot \overline{M}_{t20})\triangle V$$
$$= (\overline{E}_{10} \cdot \overline{J}_{s20} - \overline{E}_{20} \cdot \overline{J}_{s10} + \overline{H}_{20} \cdot \overline{M}_{s10} - \overline{H}_{10} \cdot \overline{M}_{s20})\triangle V = \sum_{\triangle S} \overline{\triangle S}_i \cdot \left[-\overline{E}_{1i} \times \overline{H}_{2i} + \overline{E}_{2i} \times \overline{H}_{1i}\right], \tag{2.118}$$

$$\overline{H}_{20} \cdot \overline{M}_{d10} = \mu \overline{H}_{20} \cdot \tfrac{d\overline{H}_{10}}{dt} = j\omega\mu \overline{H}_{20} \cdot \overline{H}_{10}$$
$$= \overline{H}_{10} \cdot \overline{M}_{d20} = \mu \overline{H}_{10} \cdot \tfrac{d\overline{H}_{20}}{dt} = j\omega\mu \overline{H}_{10} \cdot \overline{H}_{20},$$
$$\overline{E}_{20} \cdot \overline{J}_{d10} = \epsilon \overline{E}_{20} \cdot \tfrac{d\overline{E}_{10}}{dt} = j\omega\epsilon \overline{E}_{20} \cdot \overline{E}_{10}$$
$$= \overline{E}_{10} \cdot \overline{J}_{d20} = \epsilon \overline{E}_{10} \cdot \tfrac{d\overline{E}_{20}}{dt} = j\omega\epsilon \overline{E}_{10} \cdot \overline{E}_{20},$$
$$\overline{E}_{20} \cdot \overline{J}_{c10} = \sigma \overline{E}_{20} \cdot \overline{E}_{10} = \overline{E}_{10} \cdot \overline{J}_{d20} = \sigma \overline{E}_{10} \cdot \overline{E}_{20}. \tag{2.119}$$

Derivation Using Differential Forms of Maxwell's Equations:

The above derivation based on the integral or summation forms of Maxwell's equations, may provide useful physical as well as analytical insight through the individual steps. However, it may be mathematically convenient to the derive the above final results from the differential forms (2.85,2.86) of Maxwell's equations, using a useful relation for the divergence operator of the cross product of two vectors (see appendix A, section 4.3).

$$\overline{\nabla} \cdot (\overline{E}_1 \times \overline{H}_2) = \overline{H}_2 \cdot \overline{\nabla} \times \overline{E}_1 - \overline{E}_1 \cdot \overline{\nabla} \times \overline{H}_2 = -\overline{H}_2 \cdot \overline{M}_{t1} - \overline{E}_1 \cdot \overline{J}_{t2}$$
$$= -\overline{H}_2 \cdot \overline{M}_{d1} - \overline{H}_2 \cdot \overline{M}_{c1} - \overline{H}_2 \cdot \overline{M}_{s1} - \overline{E}_1 \cdot \overline{J}_{d2} - \overline{E}_1 \cdot \overline{J}_{c2} - \overline{E}_1 \cdot \overline{J}_{s2}, \tag{2.120}$$

$$\overline{\nabla} \cdot (\overline{E}_2 \times \overline{H}_1) = \overline{H}_1 \cdot \overline{\nabla} \times \overline{E}_2 - \overline{E}_2 \cdot \overline{\nabla} \times \overline{H}_1 = -\overline{H}_1 \cdot \overline{M}_{t2} - \overline{E}_2 \cdot \overline{J}_{t1}$$
$$= -\overline{H}_1 \cdot \overline{M}_{d2} - \overline{H}_1 \cdot \overline{M}_{c2} - \overline{H}_1 \cdot \overline{M}_{s2} - \overline{E}_2 \cdot \overline{J}_{d1} - \overline{E}_2 \cdot \overline{J}_{c1} - \overline{E}_2 \cdot \overline{J}_{s1}, \quad (2.121)$$

$$\overline{\nabla} \cdot (\overline{E}_2 \times \overline{H}_1) - \overline{\nabla} \cdot (\overline{E}_1 \times \overline{H}_2)$$
$$= \overline{E}_1 \cdot \overline{J}_{s2} - \overline{E}_2 \cdot \overline{J}_{s1} + \overline{H}_2 \cdot \overline{M}_{s1} - \overline{H}_1 \cdot \overline{M}_{s2}, \quad (2.122)$$

$$\overline{H}_2 \cdot \overline{M}_{d1} = \mu \overline{H}_2 \cdot \tfrac{d\overline{H}_1}{dt} = j\omega\mu \overline{H}_2 \cdot \overline{H}_1 = \overline{H}_1 \cdot \overline{M}_{d2} = \mu \overline{H}_1 \cdot \tfrac{d\overline{H}_2}{dt} = j\omega\mu \overline{H}_1 \cdot \overline{H}_2,$$
$$\overline{E}_2 \cdot \overline{J}_{d1} = \epsilon \overline{E}_2 \cdot \tfrac{d\overline{E}_1}{dt} = j\omega\epsilon \overline{E}_2 \cdot \overline{E}_1 = \overline{E}_1 \cdot \overline{J}_{d2} = \epsilon \overline{E}_1 \cdot \tfrac{d\overline{E}_2}{dt} = j\omega\epsilon \overline{E}_1 \cdot \overline{E}_2,$$
$$\overline{E}_2 \cdot \overline{J}_{c1} = \sigma \overline{E}_2 \cdot \overline{E}_1 = \overline{E}_1 \cdot \overline{J}_{c2} = \sigma \overline{E}_1 \cdot \overline{E}_2,$$
$$\overline{H}_2 \cdot \overline{M}_{c1} = \sigma_m \overline{H}_2 \cdot \overline{H}_1 = \overline{H}_1 \cdot \overline{M}_{c2} = \sigma_m \overline{H}_1 \cdot \overline{H}_2. \quad (2.123)$$

Consider a small volume element $\triangle V$, enclosed by the surface elements $\overline{\triangle S_i}$. Multiplying both sides of (2.122) by the volume element $\triangle V$, and then applying divergence theorem (appendix A, section 4.5) to the left side, would result in the same final expression (2.118). The subscript 'o' in (2.118) refers to evaluation at the center of the volume element $\triangle V$.

Now, the above result (2.118) can be extended to an arbitrary volume V, enclosed by a surface S, by segmentation and superposition. This is similar to the derivation of (2.98,2.99) used in the last section. Because the choice of the volume V and the corresponding surface S are arbitrary, let us use here a sphere of large radius $r \to \infty$. Also, we would use integral forms of expressions as the surface and volume elements approach zero.

$$\iiint\limits_{V\ (r\to\infty)} (\overline{E}_1 \cdot \overline{J}_{s2} - \overline{E}_2 \cdot \overline{J}_{s1} + \overline{H}_2 \cdot \overline{M}_{s1} - \overline{H}_1 \cdot \overline{M}_{s2}) \triangle V$$
$$= \oiint\limits_{S\ (r\to\infty)} \overline{dS} \cdot (-\overline{E}_1 \times \overline{H}_2 + \overline{E}_2 \times \overline{H}_1). \quad (2.124)$$

The field components on S may be decomposed into orthogonal spherical components along $\hat{\theta}$, $\hat{\phi}$ and $\hat{r}$ directions, The above surface integral on S would now involve fields at far distances. In the far space, the components of the fields produced by individual sources are expected to satisfy relations, which are independent of any specific source.

$$\frac{E_{1\theta}}{H_{1\phi}} = \frac{E_{2\theta}}{H_{2\phi}} = -\frac{E_{1\phi}}{H_{1\theta}} = -\frac{E_{2\phi}}{H_{2\theta}}. \quad (2.125)$$

Using these relations in the above surface integral (2.124), we can arrive at the following result.

$$\iiint\limits_{V\ (r\to\infty)} (\overline{E}_1 \cdot \overline{J}_{s2} - \overline{E}_2 \cdot \overline{J}_{s1} + \overline{H}_2 \cdot \overline{M}_{s1} - \overline{H}_1 \cdot \overline{M}_{s2}) dV$$

$$= \oiint\limits_{S\ (r\to\infty)} \overline{dS} \cdot \hat{r}(-E_{1\theta}H_{2\phi} + E_{1\phi}H_{2\theta} + E_{2\theta}H_{1\phi} - E_{2\phi}H_{1\theta}) = 0. \qquad (2.126)$$

This equation relates the interaction between a particular set of source currents and the field produced by a second set of source currents to that when the source and field are interchanged. In practical terms, this means that when there are two signal sources working in the presence of each other in a common environment, signal produced by the first source as sensed at the second source would be directly related to that produced by the second and sensed by the first. This is referred to as the reciprocity condition, and the above equation (2.126) is referred to as the general reciprocity equation.

As a simple case, let us consider two electric current sources $\overline{I}_{s1}$ and $\overline{I}_{s2}$ that are localized at specific points. The volume integrals of the respective current densities would be equal to $\overline{I}_{s1}\triangle l_1$ and $I_{s2}\triangle l_2$, where $\triangle l_1$ and $\triangle l_2$ are the lengths of the respective currents. The reciprocity integral for these localized "point" sources can be simplified as follows.

$$\iiint\limits_{V\ (r\to\infty)} (\overline{E}_1 \cdot \overline{J}_{s2} - \overline{E}_2 \cdot \overline{J}_{s1} + \overline{H}_2 \cdot \overline{M}_{s1} - \overline{H}_1 \cdot \overline{M}_{s2}) dV$$

$$= -V_{21}I_{s2} + V_{12}I_{s1} = 0, \qquad (2.127)$$

$$\frac{V_{21}}{I_{s1}} = \frac{V_{12}}{I_{s2}}. \qquad (2.128)$$

$-V_{21}$ is obtained by multiplying the component of the electric field from the second source, measured along the current direction of the first source, by the length of the first source. $-V_{12}$ is the equivalent term when the source and field are interchanged. V_{21} and V_{12} are the equivalent voltages seen by the respective "sensing" sources, as per the definition of terminal voltage. The above relation may be seen as a reciprocity condition for a general two-port circuit, relating terminal voltages and currents, for complex harmonic signals represented by a $e^{j\omega t}$ time dependence (or for DC signals with $\omega = 0$), with a given frequency ω. The individual ratios in (2.128) may be referred to as a mutual impedance from port 1 to port 2, and that from port 2 to port 1, that are proved to be equal to each other. An alternate proof of the reciprocity condition for circuits with discrete components is shown in Appendix G.

4.4. Boundary Conditions at Material Interfaces. Maxwell's equations (2.81,2.82) in the integral forms, or (2.85,2.86) in the differential forms, can be applicable to any spatial distribution of a material medium and current distributions. However, it may be difficult to apply Maxwell's equations (2.85,2.86) in the differential forms across an abrupt interface between two distinct media, with or without any surface current or charge distribution on the interface. Any such discontinuity in the material properties or presence

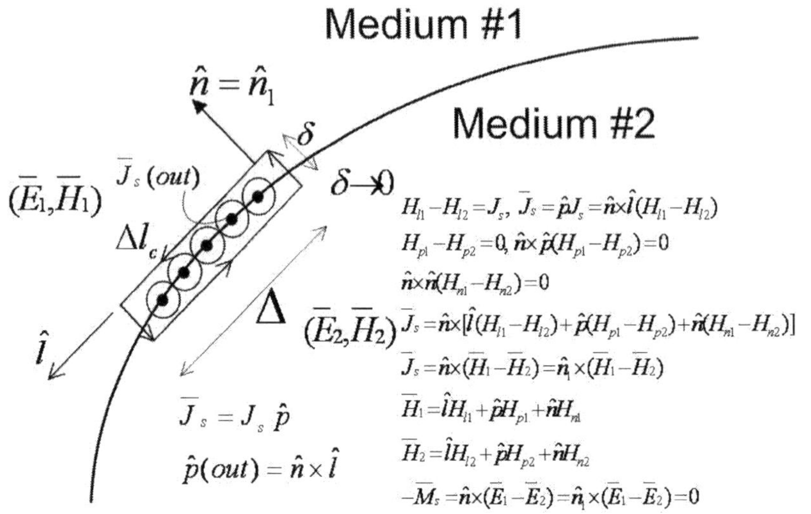

FIGURE 23

of surface current or charge distribution across the boundary, would result in possible discontinuity in different field components. This may result in difficulty to properly define field derivatives used in the differential forms (2.85,2.86) of Maxwell's equations across the boundary. For such problems, the integral forms (2.81,2.82) of Maxwell's equations, or Gauss' laws (1.48,1.59) which are implicit in Maxwell's equations, may be applied across the boundary to deduce useful relationships between fields on the two sides of the boundary, which need to be enforced. Such conditions are referred to as boundary conditions. We will separately derive in the following the boundary conditions for tangential and normal field components.

Boundary Conditions for Fields Tangential to Boundary

Apply Ampere's law (2.82) to a closed loop Δl_c across the boundary (see Fig.23). The loop is on a plane ($\hat{n} - \hat{l}$ plane) perpendicular to the direction ($\hat{p}$) of the conduction surface current density $\overline{J}_{cs} = \overline{J}_s$. The notation for the subscript s here refers to "surface" conduction current density (in A/m units) on the boundary interface, not "source" current density referenced elsewhere.

$$\oint_{\Delta l_c} \overline{H} \cdot \overline{dl} = (H_{l1}\Delta - H_{l2}\Delta) = (J_s \Delta), \tag{2.129}$$

$$H_{l1} - H_{l2} = J_s,\ \ \overline{J}_s = \hat{p}J_s = \hat{n} \times \hat{l}(H_{l1} - H_{l2}),$$
$$\overline{H}_1 = \hat{l}H_{l1} + \hat{p}H_{p1} + \hat{n}H_{n1},\ \ \overline{H}_2 = \hat{l}H_{l2} + \hat{p}H_{p2} + \hat{n}H_{n2}. \tag{2.130}$$

Similarly, applying Ampere's law to a loop on the $\hat{n} - \hat{p}$ plane, which is parallel to the current direction, a condition for the $\hat{p}$ components of the magnetic field on the two sides of the boundary maybe established.

$$H_{p1} - H_{p2} = 0, \; \hat{n} \times \hat{p}(H_{p1} - H_{p2}) = 0. \tag{2.131}$$

Combining (2.130,2.131), a general vector relationship between the surface current density $\overline{J}_s$ on the boundary and the magnetic fields $\overline{H}_1$ and $\overline{H}_2$ on the two sides of the boundary maybe established.

$$\begin{aligned} \hat{n} \times \hat{n}(H_{n1} - H_{n2}) &= 0, \\ \overline{J}_s = \hat{n} \times [\hat{l}(H_{l1} - H_{l2}) &+ \hat{p}(H_{p1} - H_{p2}) + \hat{n}(H_{n1} - H_{n2})], \\ \overline{J}_s = \hat{n} \times (\overline{H}_1 - \overline{H}_2) &= \hat{n}_1 \times (\overline{H}_1 - \overline{H}_2). \end{aligned} \tag{2.132}$$

Instead of Ampere's law (2.82), if Faraday's law (2.81) were applied across the boundary, a general vector relationship between the surface magnetic current density $\overline{M}_s$ on the boundary and the electric fields $\overline{E}_1$ and $\overline{E}_2$ on the two sides of the boundary could be established. In actual physical problems the magnetic surface current density $\overline{M}_s$ is zero.

$$-\overline{M}_s = \hat{n} \times (\overline{E}_1 - \overline{E}_2) = \hat{n}_1 \times (\overline{E}_1 - \overline{E}_2) = 0. \tag{2.133}$$

Boundary Conditions for Fields Normal to Boundary

Apply Gauss' law (1.48) for the electric field to a closed cylindrical surface (ΔS) positioned across the boundary (see Fig.24). The interface may contain a surface charge distribution ρ_s. The notation for the subscript s here refers to "surface" charge density (in C/m^2 units) on the boundary interface, not "source" charge density as may be referenced elsewhere.

$$\oiint_{\Delta S} \overline{D} \cdot \overline{dS} = \Delta A D_{n1} - \Delta A D_{n2} = \Delta A \epsilon_1 E_{n1} - \Delta A \epsilon_2 E_{n2} = \Delta q = \rho_s \Delta A, \tag{2.134}$$

$$\epsilon_1 E_{n1} - \epsilon_2 E_{n2} = D_{n1} - D_{n2} = \rho_s. \tag{2.135}$$

This establishes a relationship between the normal components D_{n1} and D_{n2} of the electric flux density, or equivalently between the normal components E_{n1} and E_{n2} of the electric fields, on the two sides of the boundary.

Instead of Gauss' law (1.48) for electric field used in the derivation (2.134), if Gauss' law (1.59) for magnetic field were applied across the boundary, a relationship similar to (2.135) but between the normal components H_{n1} and H_{n2} of the magnetic fields on the two sides of the boundary could be established. For theoretical generality, a surface magnetic charge density ρ_{sm} may also be assumed to exist on the boundary interface, which is equivalent to the electric surface charge density ρ_s in the above derivation (2.134). However, in actual physical problems ρ_{sm} is zero, because no non-zero magnetic charge exists in nature.

$$\mu_1 H_{n1} - \mu_2 H_{n2} = B_{n1} - B_{n2} = \rho_{sm} = 0. \tag{2.136}$$

Note that Gauss' laws (1.48,1.59) for the electric and magnetic fields are implicitly contained in Maxwell's equations (2.81,2.82), because Maxwell's equations are fundamentally complete (see sections 1.6, 3.8), assuming charge conservation is ensured by continuity. Accordingly, the boundary conditions (2.135,2.136) for the normal fields, that are derived from Gauss' laws, should also be implicitly satisfied once the boundary conditions (2.132,2.133) for the tangential fields are established, that are derived from Maxwell's equations. In other words, the tangential boundary conditions (2.132,2.133) constitute a complete set of boundary conditions, making the normal boundary conditions (2.135,2.136) theoretically redundant (assuming charge continuity). The normal boundary conditions are still valid, but unlike the tangential boundary conditions are incomplete. However, the normal boundary conditions maybe valuable to check for correctness of any final solutions that are established using only the tangential boundary conditions.

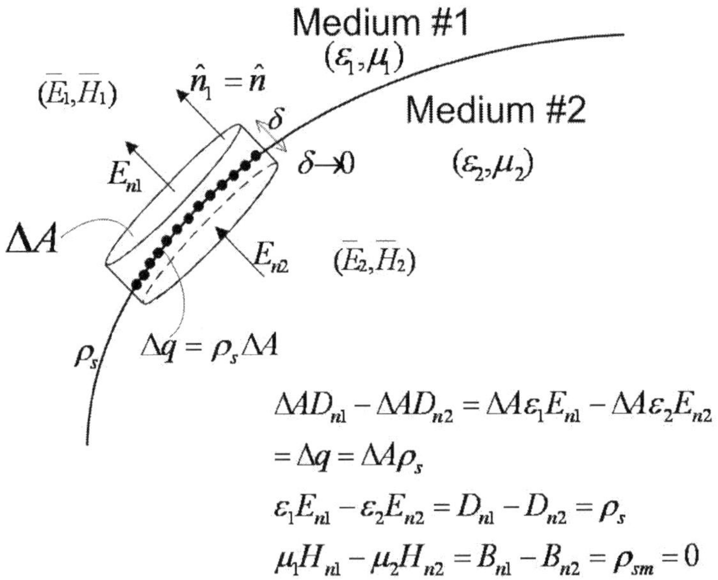

FIGURE 24

CHAPTER 3

RELATIVITY AND ELECTROMAGNETIC THEORY

The concept of Ampere's law relating the magnetic field to the conduction and displacement currents, and the concept of Faraday's law relating the electric field to the magnetic field, were historically established through practical experimentation and gradual generalization. The general findings were formalized in suitable mathematical forms for analytical use, with only limited physical and philosophical rationalization. Ampere's and Faraday's laws, in their most generalized forms, constitute Maxwell's equations, which are shown in the last chapter to supersede all other basic circuit and field laws, such as Kirchoff's Current Law (KCL), Kirchoff's Voltage Law (KVL), and Gauss' laws of electric and magnetic fields. Such consistency with existing laws, as well as their success in understanding and predicting practical experiments, develop remarkable confidence in Maxwell's equations. They have so far stood the test of time, and are now accepted to be universally valid. In spite of all the remarkable generality and success of the theory, the underlying concepts of Maxwell's equations (James Clerk Maxwell, 1831-1879), or equivalently Ampere's and Faraday's laws, may appear quite "mysterious" to basic human intuition. Without some form of logical explanation of the principles, one has to accept them solely on the basis of confidence in the experimental observations, and their consistent validity to date. This may not be fully satisfactory and insightful.

Later on, Albert Einstein's (1879-1955) theory of special relativity has been established as an independent physical principle. This is founded upon the basic assumption that the speed of light or any electromagnetic wave, propagating in a uniform free-space medium, is a fixed constant when measured from any reference frame. Further, if the reference frames are "unbiased" (referred to as inertial reference frames,) then the light would propagate with the fixed invariant speed, along any given straight-line path. The theory also assumes that any basic physical measurement, designed to verify a particular basic physical law, would lead to the same conclusion when observed or measured from any of the inertial frames of reference. This led to new transformation equations, relating the space and time coordinates of two inertial reference frames, that are consistent with the basic assumptions in the principles of relativity. In this chapter, we will use the new transformation relations to deduce Maxwell's equations from Gauss' laws of electric and magnetic fields. The deduction is intended to serve as an alternate explanation or validation for Ampere's and Faraday's laws. In basic intuitive terms, the relativistic transformation principles may not be that easy to accept or understand, as compared with the standard non-relativistic principles one gathers through common, day-to-day experiences. However, with proper interpretation and understanding, they are arguably less "mysterious" to reason than the principles of Ampere's and Faraday's laws. Similarly, Gauss' laws for electric and magnetic fields, which are fundamentally established in integral forms, at first may not seem to be readily acceptable in intuitive terms. However, they can be explained as logical and philosophical generalization from the basic concept of Coulomb's law, which is usually more easily accepted and understood (Chapter 1). Accordingly, the deduction of Maxwell's equations from Gauss' laws through relativistic transformation is expected to be less "mysterious," and therefore easier to logically accept. In this sense, the deduction would be considered philosophically or logically more fundamental.

Traditionally, the basic texts on engineering electromagnetics have used Maxwell's equations as the general starting point, from which specific electromagnetic problems are derived. Connecting Maxwell's equations to the relativity theory is not usually explored. However, introduction of the relativity concepts to engineering electromagnetics may be long overdue. It would be important for the students of engineering electromagnetics to be introduced to the theory of relativity at some useful level, and to understand its relevance to Maxwell's equations. We attempt to accomplish this in this chapter with simplicity and economy. We first introduce the space-time transformation equations of the relativity theory. They are developed reasonably quickly, using two independent approaches. One is based more on mathematical deduction, whereas the other is based more on physical analysis. Elaborate interpretation of the transformation equations, and any further application to relativistic mechanics is avoided. The space-time transformation equations are then used to derive basic transformation relations which enforce the continuity of the electric charge. This is followed by derivation of the transformation relations for Gauss' laws for electric and magnetic fields, which lead to Ampere's and Faraday's laws, respectively. The concepts maybe enhanced through suitable examples and useful problems.

1. Space-Time Transformation

Consider two frames of reference, as shown in Fig.1. The time and space coordinates of a particular event, as measured in one reference frame, are represented by four variables (x, y, z, t). The respective coordinates of the event, as measured in the second reference frame, are represented by primed variables (x', y', z', t'). Accordingly, the first and the second reference frames may be informally referred to as the "un-primed" and the "primed" reference frames or reference coordinates, respectively. The four variables may be referred to as the space-time coordinates of the event, and the event may be referred to as a particular space-time point in a coordinate system. Now, assume that the primed reference frame is moving with a uniform velocity V in the x direction, with respect the unprimed coordinates. Also assume that at time $t = 0$, the origins of the two coordinate systems are aligned. That is, at $t = 0$, the space coordinates $x' = y' = z' = 0$ in the primed system coincide with the space coordinates $x = y = z = 0$ in the unprimed system. Further, at this instant the clock in the primed coordinate system is initialized with that of the unprimed coordinate system with $t' = 0$.

1.1. Non-Relativistic Transformation. If one uses a conventional or "common-sense" coordinate transformation, referred to as non-relativistic transformation, the space-time coordinates of the event in the two reference frames can be be related as follows.

$$y = y', \; z = z', \; x = x' + Vt, \; t = t'. \tag{3.1}$$

This transformation would also lead to relating any incremental changes in the coordinates, represented by a pre-fix Δ:

$$\Delta y = \Delta y', \; \Delta z = \Delta z', \; \Delta x = \Delta x' + V\Delta t, \; \Delta t = \Delta t'. \tag{3.2}$$

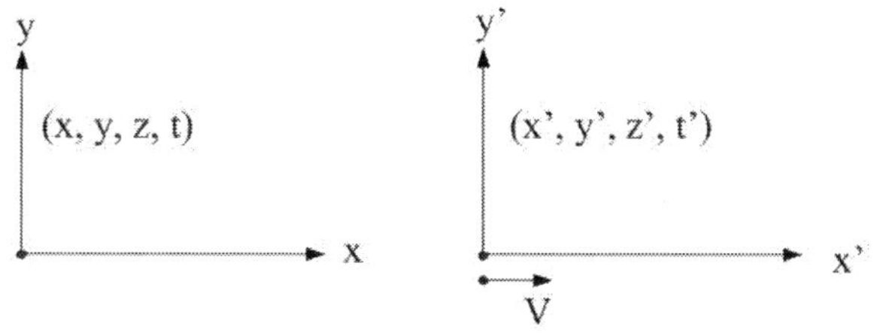

FIGURE 1

Accordingly, the components of the velocity of the event in different directions may also be related by dividing respective incremental distance by incremental time:

$$\begin{aligned} v_y &= \frac{\Delta y}{\Delta t} = \frac{\Delta y'}{\Delta t'} = v'_y, \\ v_z &= \frac{\Delta z}{\Delta t} = \frac{\Delta z}{\Delta t'} = v'_z, \\ v_x = \frac{\Delta x}{\Delta t} = \frac{\Delta x' + V\Delta t'}{\Delta t'} &= \frac{\Delta x'}{\Delta t'} + V = v'_x + V. \end{aligned} \tag{3.3}$$

Notice that the two time variables t and t' are assumed to be equal. This means, such standard transformation implicitly assume that the flow of time is "absolute," meaning it is independent of the relative motion between the frames. In addition, t is equal to t', independent of the space coordinates of the event being observed. However, the above two assumptions do not have any concrete physical basis, and should be open to review and adjustment based on any advanced understanding or experimental observation.

1.2. Special Nature of Light, and Incompatibility with Non-Relativisitc Transformation. In the non-relativistic transformation, we see that the x-directed velocity in the unprimed reference frame can be obtained from that in the primed reference frame by simply adding the relative velocity V between the frames. Accordingly, any x-directed velocity measured in one coordinate frame would not be equal to that measured in the other, for a non-zero value of the relative velocity V. Further, the magnitude of the total velocity, which is the speed, defined as the square-root of the sum of squares of the individual velocity components in the three directions x, y and z, would also not be equal when measured in the two reference frames.

Now, let us apply the standard common-sense, non-relativistic transformation to a light signal propagating in free-space. As per the transformation rules discussed above, the speed of light in free-space would be measured with different results in the two reference frames. With some insight on the special nature of light, this conclusion may be brought to question, which would lead to invalidating the non-relativistic transformation for general applications.

The speed of light is different from that of any common material substance such as a rock or a bullet. Light is not "thrown" off a light bulb like a rock thrown from someone's hand or a bullet from a gun. Therefore, unlike a rock or a bullet, the speed of light can not be fixed as a given value with respect to its source, and the given speed could not be changed as one desires by employing more strength or energy of its source. Instead, light is generated as a wave from the bulb, more like a sound wave generated from a flute, or more like the water ripples generated when a rock is dropped in a lake. Once the wave is generated by its source, it is physically detached from its source and its propagation speed is supposed to be determined by the medium, independent of the source. However, unlike a mechanical wave, such as the sound or water wave, the light wave does not require a material medium for propagation. Light can propagate in free-space. It would bring challenges in the understanding, when light propagation in an ideal free-space is considered.

The water wave propagates as a disturbance of the water particles, and the disturbance propagates with respect to the body of the water. Therefore, the body of the water, or equivalently the body of the solid ground around the lake with respect to which the water is assumed to be stationary, is the natural frame of reference. The speed of the water wave is derived from the material properties of the water, and is clearly defined with respect this natural frame of reference. Similarly, for the sound wave, the solid earth with respect to which the surrounding air is assumed to be stationary, is the natural reference frame. The speed of the sound wave is a known constant (as determined from the physical parameters of the air), clearly defined with respect to the earth or the surrounding stationary air. Once the speed of the water wave or the sound wave is definitively known, with respect to its natural, or "preferred," reference frame as discussed, pointed in any specific direction, the wave velocity with respect to any other reference frame may be determined by using velocity transformation rule. In distinct contrast, an ideal free-space in which light propagates, as per the very philosophical nature of the free-space, does not have a natural or "preferred" reference frame. Unlike the water in a lake or the air surrounding earth, the free-space is not "attached" to any definite body of reference. Therefore, to begin with, there is no preferred reference frame with respect to which we can define and fix the speed of light in free-space. Accordingly, the standard procedure of starting with a known value of the speed, pointed in any specific direction, with respect to a naturally preferred reference frame, and then calculating the velocity with respect to other references using velocity transformation, may not be as evident for light propagation in free-space. In this process, if one would start by fixing a certain value of the speed to any particular reference frame, that may appear arbitrary without proper logical basis.

1.3. Theory of Relativity and the Speed of Light in Free-Space. We may, however, look at the above situation from a new perspective. From the earlier discussion, we understand that there is no naturally preferred reference frame for propagation of light in free-space. To put it a bit differently, all reference frames should be equally preferred for treatment of the propagation of light in free-space. Accordingly, if one may start with a fixed value for the speed of light in free-space in one reference frame, arbitrarily chosen, then there would be no reason why the same value could not have been fixed in any other reference frame. This is an important philosophical argument, leading to the conclusion that, whatever is the speed of light in free-space, it may have to be the same universal constant for all reference frames of measurement. Further, if the reference frame is also "unbiased," then the light would propagate with the same speed along any given direction, with no biased preference to deflect the propagation direction to one side or other. This was postulated in Einstein's theory of relativity as a universal principle. This conclusion is also established through practical experiments.

The new theory of relativity would clearly contradict the principles of the non-relativistic transformation. The basic rule of non-relativistic transformation discussed earlier - that the speed of a given physical event would necessarily have unequal values, when measured in two reference frames moving with a non-zero relative velocity V - is clearly violated for light propagation in free-space. The transformation relations would then have to be revised in order to be consistent with the new theory of relativity. We will show in the following how this can be achieved, using two approaches: (1) mathematical analysis, and (2) physical experimentation. The first approach mainly uses mathematical arguments for the derivation of the new transformation relations. Whereas, the second approach derives the transformation relations based on a more intuitive understanding of a physical measurement process using light. The two approaches may be used equally effectively, or may be used in complementary ways, in order to develop an understanding of the new transformation relations. The new transformation relations, as derived in the following, are referred to as the relativistic transformation relations.

In order that the new relativistic transformation relations are to be universal in scope, in scientific consistency, they should be applicable not just for light in free-space, but for all physical phenomena. They would have significant implications when applied to the mechanics of material bodies. However, when the relative velocity V is much smaller than the speed of light in free-space, which is usually the case in our "common-sense" world, the new relativistic relations are seen to approach the standard non-relativistic relations. Accordingly, the non-relativistic relations are accurate enough for common day-to-day observations, as normally expected. Physical mechanics, under general conditions of relative motion, is to be revised based on the new theory of relativity and transformation relations. The reader may be referred to textbooks on modern physics for further study on this topic. Development of mechanics under the revised relativistic conditions may have indirect implications to the electromagnetic theory, but it is not the goal of the present study. Here, we would focus on the applications of the relativistic transformation specifically to the development and understanding of Maxwell's equations.

1.4. Relativistic Transformation: Mathematical Analysis. The two reference frames in Fig.1 are assumed to be moving with a relative velocity V in the x direction. The relative motion in the x direction is not expected to alter the y and z coordinates of an event, when seen from the two frames, assuming that the two frames and the free-space medium are equally unbiased in the y and z directions. That is, we have $y = y'$ and $z = z'$. This leaves the transformation of the only space coordinate along the x axis and the time coordinate. Accordingly, in the following we may represent an event with only two variables, (x, t) or (x', t').

Consider that the primed space-time coordinates (x', t') are linearly related to the respective unprimed coordinates (x, t).

$$\begin{aligned} x' &= ax + bt, \\ t' &= px + qt, \end{aligned} \tag{3.4}$$

where a, b, p and q are now unknown constants, to be determined. Let us mathematically see if transformation relations in the above form can be established, which would be consistent with the requirements of the relativity theory. The above relationships, to begin with, already satisfied the initial conditions we have assumed. From the above equations, we have $x' = 0$ when $t = 0$ and $x = 0$. That means, at time $t = 0$, the origin of the unprimed coordinate system $(x = 0)$ coincides with that of the primed coordinate system ($x' = 0$). At the same instant, that is when $x = x' = t = 0$, we would also have $t' = 0$ from the above equations. In other words, the clock in the primed coordinate system located at $x' = 0$ is initialized with $t' = 0$, when the origins of the two reference frames coincide, which is the same instant when the clock in the unprimed coordinate system is also initialized to zero $t = 0$. In addition to the initial conditions, we need other basic conditions to be enforced in order for the transformation relations to be logically valid, and meet the physical constraints of the relativity theory. These additional conditions would determine the values of the constants in the transformation equations (3.4).

At time $t = 0$ the origins of the two frames coincide. We also know, the origin of the primed coordinate system moves with a velocity V as seen from the origin of the unprimed coordinate system. Accordingly, for $x' = 0$, we have to have $x = Vt$. Using this constraint in (3.4) we get,

$$\begin{aligned} x' = 0 = ax + bt, \;\; x = -\tfrac{b}{a}t = Vt, \\ \tfrac{b}{a} = -V. \end{aligned} \tag{3.5}$$

One basic expectation in the relativity transformation is that, if one observer sees a second observer to move with a velocity V, then the second observer should see the first moving with a velocity $-V$. We just explained that the origin of the primed coordinate system moves with a velocity V as seen by an observer at the origin of the unprimed coordinate system. Therefore, the origin of the un-primed coordinate system, that is $x = 0$, should be seen by an observer at the origin of the primed coordinates to be moving with a velocity $-V$. Accordingly, for $x = 0$, we have to have $x' = -Vt'$. Using this constraint in (3.4) we get,

$$\begin{aligned} t' = qt, \;\; x' = bt; \;\; x = 0, \\ x' = bt = \tfrac{b}{q}t' = -Vt', \\ \tfrac{b}{q} = -V. \end{aligned} \tag{3.6}$$

The above constraints we have enforced are quite elementary, and are not particularly unique to the relativity theory. What would be new or unique to the relativity theory is the following. Consider a point (x, t) in the unprimed coordinate system, which is seen as a point (x', t') in the primed coordinate system. Consider that the point is moving in the x direction with a speed equal to the speed of light c in free-space, as seen in the unprimed coordinates. That is, $\frac{\Delta x}{\Delta t} = c$. Then, the speed of the same point, as measured by the primed coordinates, would also be equal to the speed of light c. That is, $\Delta x'$ and $\Delta t'$, corresponding to the incremental distance and time Δx and Δt, respectively, would be such that $\frac{\Delta x'}{\Delta t'}$ is also equal to the speed of light c. Using this condition in (3.4),

$$\Delta x' = a\Delta x + b\Delta t,$$
$$\Delta t' = p\Delta x + q\Delta t,$$
$$\frac{\Delta x'}{\Delta t'} = \frac{a\Delta x+b\Delta t}{p\Delta x+q\Delta t} = \frac{a\frac{\Delta x}{\Delta t}+b}{p\frac{\Delta x}{\Delta t}+q},$$
$$c = \frac{ac+b}{pc+q},\ pc^2 + qc = ac + b,$$
$$pc^2 + (q-a)c - b = 0. \tag{3.7}$$

Using (3.5,3.6) in (3.7),

$$pc^2 = b. \tag{3.8}$$

For $x = 0$, we have $t' = qt$. Accordingly, for $x' = 0$ we should also expect $t = qt'$. This is a basic symmetry condition between times in the two coordinates. Using this symmetry condition in (3.4),

$$x' = 0 = ax + bt,\ x = -\frac{b}{a}t,$$
$$t' = px + qt = -\frac{pb}{a}t + qt = t(q - p\frac{b}{a}) = \frac{t}{q},$$
$$q^2 - pq\frac{b}{a} = 1. \tag{3.9}$$

Now, the constants a, b, p and q can be solved from the equations (3.5-3.9).

$$a^2 - pb = 1 = a^2 - \frac{b^2}{c^2} = a^2 - a^2\frac{V^2}{c^2} = 1,$$
$$a = \frac{1}{\sqrt{1-\frac{V^2}{c^2}}} = q,$$
$$b = -Va = \frac{-V}{\sqrt{1-\frac{V^2}{c^2}}},$$
$$p = \frac{b}{c^2} = \frac{\frac{-V}{c^2}}{\sqrt{1-\frac{V^2}{c^2}}},$$
$$x' = ax + bt = \frac{x - Vt}{\alpha},$$
$$t' = px + qt = \frac{\frac{-Vx}{c^2}+t}{\alpha},$$
$$\alpha = \sqrt{1-\frac{V^2}{c^2}}. \tag{3.10}$$

1.5. Relativistic Transformation: Physical Experimentation. We will use simple physical processes employing light signals in free-space to measure time and distance. A particular process using a light signal is measured from two different reference frames. The results of measurement in the two frames can then be compared to establish transformation relations between the space and time coordinates of the two frames.

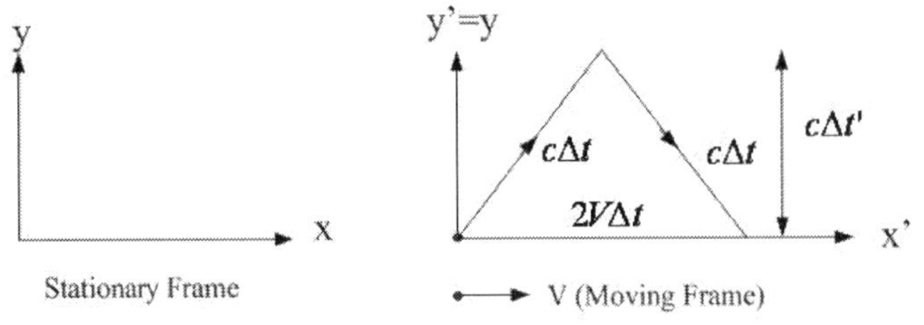

FIGURE 2

1.5.1. *Time Transformation.* First, let us measure time by bouncing a light through a fixed distance Δy in the y-direction. The round-trip delay experienced by the light to cover a fixed distance may be used as a proportionate measure of the time. Consider such a time-measurement process conducted in the primed reference frame (at the origin $x' = 0$). This same process is observed in the unprimed reference frame, and compared with another identical time-measurement process conducted at the origin of the unprimed reference frame. The relative velocity V in the x direction is assumed not to change the y dimension in the two reference frames. Accordingly, $\Delta y = \Delta y'$. Let $2\Delta t'$ be the round-trip time taken by the light, as measured in the primed reference frame, to cover a reference distance $\Delta y = \Delta y'$.

$$\Delta y = \Delta y' = c\Delta t'. \tag{3.11}$$

This light-bouncing experiment conducted in the primed frame, when observed from the un-primed frame, would appear to take a longer route (see Fig.2). The round-trip time would be measured $2\Delta t$, as compared with its local timing process conducted in the unprimed frame. Accordingly, Δt is referred to as "equivalent" to $\Delta t'$, as observed from the unprimed reference frame. As a definition, an incremental interval between a pair of time measurements, with given starting and end points, is referred to be "equivalent" or "simultaneous" to that between another pair, as observed from a particular reference frame, under the following condition. If the reference starting points of the two pairs coincide with each other (if they happen to occur simultaneously, or by introducing a suitable delay between their observations,) then the corresponding end points would also be observed in coincidence, or "simultaneously," from the given frame. In reference to the Fig.2, one can now geometrically relate Δt with $\Delta t'$ in terms of the the speed of light c and the relative velocity V.

$$c^2\Delta t^2 = c^2\Delta t'^2 + V^2\Delta t^2, \tag{3.12}$$

$$\Delta t' = \Delta t(1 - \frac{V^2}{c^2})^{\frac{1}{2}} = \Delta t\alpha. \tag{3.13}$$

The above geometric analysis assumes that the two reference frames are ideally unbiased, from which the light in free space is observed to propagate along straight lines. Also, as per the theory of relativity, we have assumed the same value c for the speed of light as observed in either of the two reference frames. Equation (3.13) is a simple relationship between an incremental time $\Delta t'$ measured at the origin of the primed coordinate system, and the corresponding incremental time Δt as measured and observed to be equivalent in the unprimed coordinate system. Now, as we have assumed before, the times t' and t are initialized to zero when the origin $x' = 0$ coincides with the origin $x = 0$. With this initialization, the above proportionality relationship between the equivalent incremental times would apply to the corresponding reference times t' and t. The time t' measured at the origin $x' = 0$ of the primed coordinates is determined to be equivalent or "simultaneous" to the time t as measured and observed in the unprimed coordinates. These equivalent times t' and t are related as follows.

$$t' = t(1 - \frac{V^2}{c^2})^{\frac{1}{2}}, \; x' = 0. \tag{3.14}$$

Note that the time t' is specified in association with a particular location of measurement $x' = 0$. This is important. If the particular location x' at which the original experiment is conducted is changed, the relationship between the equivalent times t' and t would be different. If the point x' is fixed in the primed coordinate frame (independent of t'), but not at the origin $x' = 0$, the relationship between the equivalent incremental times in (3.13) would still be valid, whereas the relationship (3.14) between the actual reference times t' and t would not (requiring some time offset). Also, as we have stated, the above relationship assumes that the t' and t are determined to be equivalent or "simultaneous" as observed from the unprimed reference frame. Instead, if the determination of equivalence or simultaneity is made as observed from the primed reference frame, the relationship would be different. For example, if the experiment is conducted at the origin of the unprimed coordinates, and the observation of simultaneity is made from the primed reference frame, it can be shown (by principle of symmetry between the frames) that the relationship between the equivalent times t and t' can be expressed using (3.14), but with the variables t' and t switched. These concepts of equivalence or simultaneity should be clear from a complete space-time transformation relationship between (x, t) and (x', t').

1.5.2. *Space-Time Transformation.* The above experiment established relationship between equivalent times t and t' in the two frames of measurement. Now we will conduct another light experiment, using which we can measure the distance and time of a particular "event" as observed in the two reference frames. Let a light signal be initiated from the origin of the unprimed coordinate system at time $t = t_1$. Let the light signal be aimed towards a particular event-point in space, as shown in Fig.3, whose space-time coordinates (x, t) and (x', t') we are interested to determine as measured from the two reference frames. The signal passes through the origin of the primed reference frame at time t_2, reaches the particular point of interest at time t_3, reflects and returns towards the starting point, passes again through the origin of the primed reference frame on its way at time t_4, and reaching finally at the origin of the unprimed reference frame at time t_5. All the above times are as measured with respect to the unprimed coordinate system. The corresponding equivalent times measured at the origin of the primed reference frame can be obtained using the above relationship (3.14). The space and time coordinates (x, t) and (x', t'), as measured with respect to the two reference frames, can be determined and related from the above timing information.

The distance x of the event is determined from the time delay $t_5 - t_1$. The time $t = t_3$ of the event is determined as the average of t_5 and t_1.

$$x = (t_5 - t_1)\frac{c}{2}, \tag{3.15}$$

$$t = t_5 - \frac{x}{c} = t_5 - \frac{t_5 - t_1}{2} = \frac{t_1 + t_5}{2} = t_3. \tag{3.16}$$

From the above two equations, the measured x and t can be related to the initial time t_1.

$$-x + ct = ct_1. \tag{3.17}$$

The distance x' of the event is determined from the time delay $t'_4 - t'_2$. The time $t' = t'_3 = \alpha t_3$ of the event is determined as the average of t'_4 and t'_2.

$$x' = (t'_4 - t'_2)\frac{c}{2}, \tag{3.18}$$

$$t' = t'_4 - \frac{x'}{c} = t'_4 - \frac{t'_4 - t'_2}{2} = \frac{t'_2 + t'_4}{2}. \tag{3.19}$$

The above (x', t') are expressed in terms of $t'_4 = t_4\alpha$ and $t'_2 = t_2\alpha$. Equation (3.17) relates (x, t) to t_1. Accordingly, in order to relate (x', t') with (x, t), we may need to relate t_2 and t_4 with t_1. This can be accomplished as follows. The distance covered by the light signal during time $t_2 - t_1$ is equal to distance Vt_2 between the reference frames at time t_2. Solving from this information, t_2 may be related with t_1.

$$c(t_2 - t_1) = Vt_2, \; t_2 = \frac{t_1 c}{c - V}. \tag{3.20}$$

Similarly, the distance covered by the light signal during time interval $t_4 - t_1$ is equal to round-trip distance $2x$ of the event point minus the distance Vt_4 between the reference frames at time t_4. Solving from this information, t_4 may be related with t_1 and x.

$$c(t_4 - t_1) = 2x - Vt_4, \; t_4 = \frac{2x + ct_1}{c + V}. \tag{3.21}$$

Using (3.17) in (3.20,3.21), we can relate t_2 and t_4 with x and t .

$$t_2 = \frac{ct - x}{c - V}, \; t_4 = \frac{2x + (ct - x)}{c + V} = \frac{ct + x}{c + V}, \tag{3.22}$$

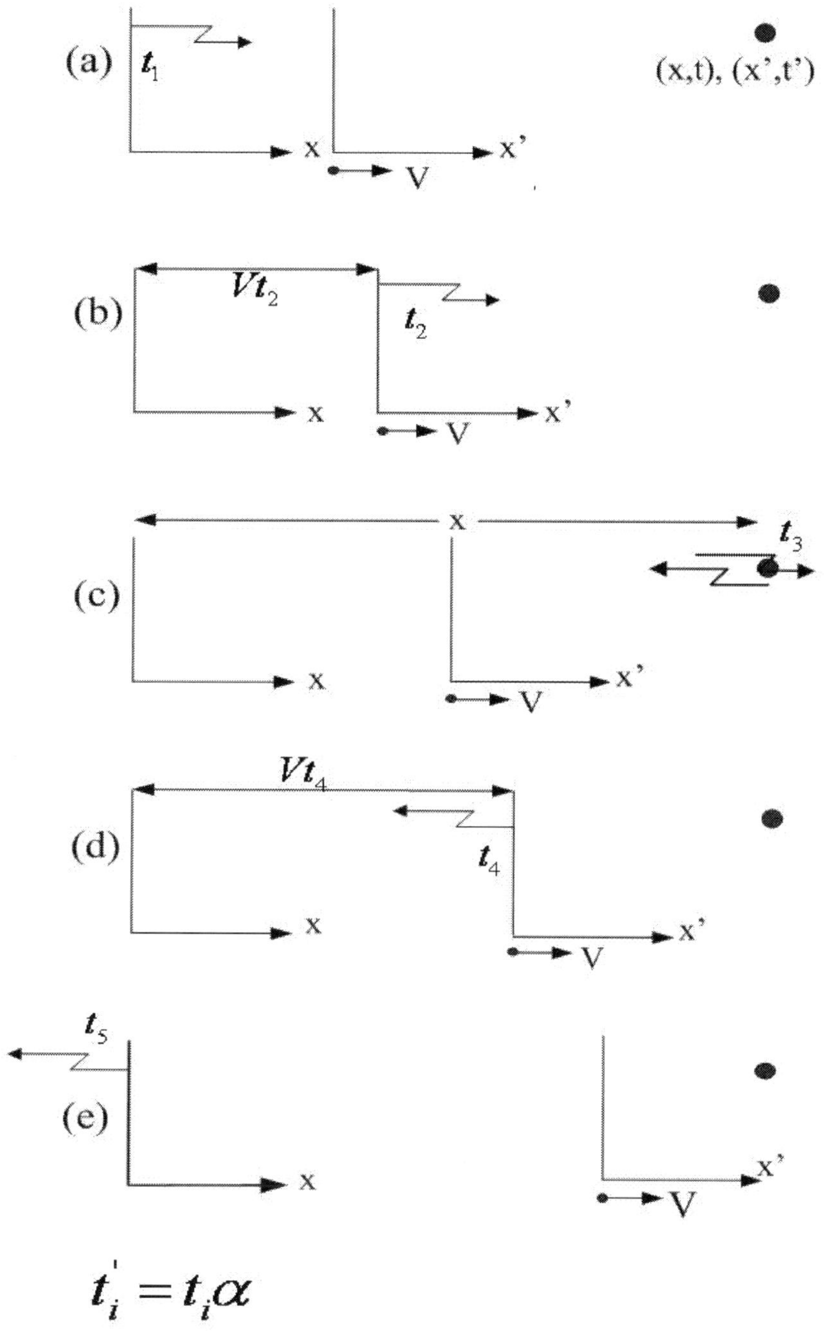

FIGURE 3

$$t_4 - l_2 = \frac{-2cVt + 2cx}{c^2 - V^2}, \; t_2 + t_4 = \frac{2c^2 t - 2Vx}{c^2 - V^2}. \tag{3.23}$$

Use (3.23) in (3.18,3.19) to relate (x', t') to (x, t).

$$x' = (t_4 - t_2)c\tfrac{\alpha}{2} = \tfrac{(c^2x - c^2Vt)\alpha}{c^2 - V^2} = \tfrac{x - Vt}{\alpha},$$
$$t' = \tfrac{t_2 + t_4}{2}\alpha = \tfrac{\alpha(c^2t - Vx)}{c^2 - V^2} = \tfrac{t - \frac{Vx}{c^2}}{\alpha}. \qquad (3.24)$$

The above transformation relations (3.24), derived here using a complete physical analysis, are the same relations in (3.10) which were derived on a mathematical basis. Both are valid and complementary approaches to establish the relativistic transformation.

Using the above transformation relations, it can be shown that a straight-line path of light propagation along any general orientation in space, as originally observed from one of the unbiased frames, would be seen with a deviated orientation in the other frame. The amount of the deviation would be dependent on the magnitude V of the relative velocity between the two frames. The relative orientation of the light path seen in the second frame, would obviously depend on the direction of the relative velocity $\overline{V}$ between the frames. Therefore, if the velocity $\overline{V}$ (magnitude and/or direction) varies in time, the resulting light path with its variable orientation in time would appear as a curved path, as seen in the second frame. This would contradict the unbiased nature of the two reference frames, that we assumed in the above analyses, which expects any observed light path to be strictly a straight line. In other words, the unbiased nature of the two frames, would require the relative velocity $\overline{V}$ between the frames to be constant in time, as a necessary fundamental condition.

2. Relativistic Transformation of Current and Charge Density

2.1. Current Along the Relative Velocity. Let us first consider only an x-directed current. An experiment is conducted in the primed reference frame to verify the continuity of current and charge conservation (Fig.4). The experiment may be mathematically expressed as follows.

$$\Delta\tau'[J'_x(x' + \Delta x'/2, t') - J'_x(x' - \Delta x'/2, t')]\Delta A = -\tfrac{\partial \Delta Q'}{\partial t'}\Delta\tau' = -\tfrac{\partial \rho'_v}{\partial t'}(x', t')\Delta\tau'\Delta A\Delta x',$$
$$\tfrac{J'_x(x' + \Delta x'/2, t') - J'_x(x' - \Delta x'/2, t')}{\Delta x'} = -\tfrac{\partial \rho'_v}{\partial t'}(x', t'),$$
$$\frac{\partial J'_x}{\partial x'} = -\frac{\partial \rho'_v}{\partial t'}. \qquad (3.25)$$

The same experiment is observed from the unprimed coordinate frame, as shown in the figure. Let us find relationships between the space $(\Delta x, \Delta x')$ and time $(\Delta t, \Delta t')$, $(\Delta\tau, \Delta\tau')$ increments observed in the two reference frames. The experiment was performed in the (x', t') coordinates with measurements done at the two ends of the volume $x' - \Delta x'/2$ and $x' + \Delta x'/2$ at the same time t'. This means $\Delta t' = 0$ at the two ends. In the (x, t) reference frame, the corresponding increments Δx and Δt) may be related with each other and with $\Delta x'$ using relativistic transform relations (3.10,3.24).

$$
\begin{aligned}
t' &= \tfrac{t - xV/c^2}{\alpha}, \; x' = \tfrac{x - Vt}{\alpha}, \\
\Delta t' &= \tfrac{\Delta t - \Delta x V/c^2}{\alpha} = 0, \; \Delta t = \Delta x V/c^2, \\
\Delta x' &= \tfrac{\Delta x - \Delta t V}{\alpha} = \tfrac{\Delta x - \Delta x V^2/c^2}{\alpha} = \alpha \Delta x, \; \Delta x = \tfrac{\Delta x'}{\alpha}, \\
\Delta x &= \tfrac{\Delta x'}{\alpha}, \; \Delta t = \Delta x V/c^2 = \tfrac{\Delta x'}{\alpha} V/c^2.
\end{aligned} \tag{3.26}
$$

The charge ΔQ and $\Delta Q'$ in the two frames would be equal. We also assume that the transverse cross-section area ΔA is the same in both frames. Because the lengths $\Delta x'$ and Δx are different when measured in the two frame, the same charge enclosed between the lengths would appear as different charge densities. Let ρ_{v0} be the charge density of the volume element as observed in the unprimed frame.

$$
\begin{aligned}
\Delta Q = \Delta Q' &= \rho_v' \Delta A \Delta x' = \rho_{v0} \Delta A \Delta x, \\
\rho_{v0} &= \rho_v' \times \tfrac{\Delta x'}{\Delta x} = \rho_v' \alpha.
\end{aligned} \tag{3.27}
$$

The time increment $\Delta\tau'$ is an arbitrarily small interval between two instants of measurements performed at a given x' $(\Delta x' = 0)$. The corresponding time interval in the (x, t) reference frame is $\Delta\tau = \Delta\tau'/\alpha$.

As per a basic principle of the theory of relativity, the outcome of the basic, physical experiment conducted in the primed reference frame at a given $t', \Delta t' = 0$, and that observed from the other reference frame should be the same. The current continuity experiment, as observed in the (x, t) frame can be expressed as follows.

$$
\begin{aligned}
&\Delta\tau[J_x(x + \Delta x/2, t + \Delta t/2) - J_x(x - \Delta x/2, t - \Delta t/2)]\Delta A \\
&\quad = -\tfrac{\partial \Delta Q}{\partial t}\Delta\tau = [-\tfrac{\partial \rho_{v0}}{\partial t}(x, t)\Delta x \Delta A]\Delta\tau, \\
&(J_x + \tfrac{\partial J_x}{\partial x}\Delta x/2 + \tfrac{\partial J_x}{\partial t}\Delta t/2) - (J_x - \tfrac{\partial J_x}{\partial x}\Delta x/2 - \tfrac{\partial J_x}{\partial t}\Delta t/2) = -\tfrac{\partial \rho_{v0}}{\partial t}(x, t)\Delta x, \\
&\tfrac{\partial J_x}{\partial x} + \tfrac{\partial J_x}{\partial t}\tfrac{V}{c^2} = -\tfrac{\partial \rho_{v0}}{\partial t}, \\
&\int \tfrac{\partial J_x}{\partial x}\partial t + \tfrac{V}{c^2}J_x = -\rho_{v0}.
\end{aligned} \tag{3.28}
$$

When the velocity $V = 0$, we know $\rho_{v0} = \rho_v$. Using this condition in the above equation, the integral term can be recognized as $-\rho_v$.

$$
\begin{aligned}
\rho_{v0} &= \rho_v' \alpha = \rho_v - \tfrac{V}{c^2}J_x = \rho_v + \Delta\rho_v, \\
\Delta\rho_v &= -\tfrac{V}{c^2}J_x, \; \tfrac{\partial \rho_v}{\partial t} = -\tfrac{\partial J_x}{\partial x}.
\end{aligned} \tag{3.29}
$$

ρ_v is the charge density the observer in the (x, t) frame normally estimates (at a given $t, \Delta t = 0$) and ρ_{v0} is the charge density the same observer would estimate in the transformed current continuity experiment (at a given $t', \Delta t' = 0$). These two parameters differ, with the difference $\Delta\rho_v$ dependent on the current density J_x. If the current is charge free as seen in the (x', t') frame, then the transformed experiment would yield

$\rho_{v0} = 0 = \alpha\rho'_v = \rho_v - J_x\frac{V}{c^2}$. This means $\rho_v = J_x\frac{V}{c^2}$. In other words, a charge free current in the (x', t') frame would look charged in the (x, t) frame. This is an important observation.

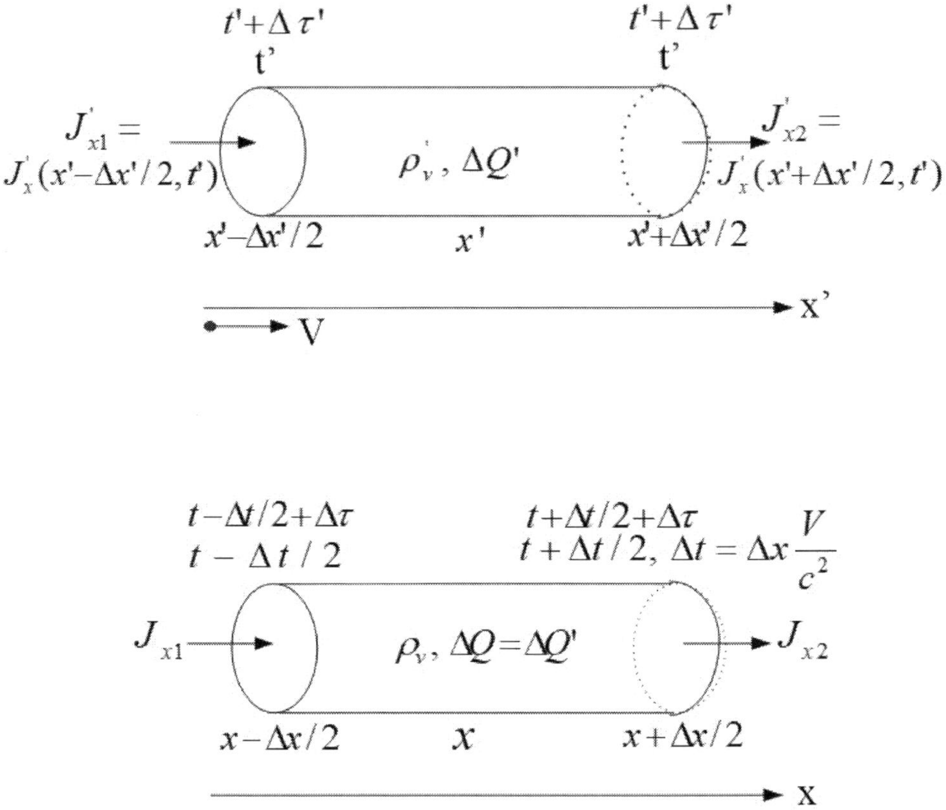

$$J_{x1}, J_{x2} = J_x(x \mp \Delta x/2, t \mp \Delta t/2)$$
$$= J_x(x,t) \mp \frac{\partial J_x}{\partial x}\Delta x/2 \mp \frac{\partial J_x}{\partial t}\Delta t/2$$

FIGURE 4

2.2. Current in Arbitrary Direction. In the above derivation, for simplicity we have considered only x-directed current (along the relative velocity). If the current is along an arbitrary direction, with x, y and z components, the above derivation may be extended. It can be shown that the same relationship between the charge density and current would be valid for the general case as well. It may not be necessary to present all derivation steps for the general case with all current directions. We will provide a conceptual picture, and discuss necessary changes in the above derivation.

For the general treatment, the volume element in the above experiment (Fig.4) may now be considered to be a rectangular box with dimensions Δx, Δy and Δz in the three rectangular coordinates. The space-time coordinates for the center of the box is (x, y, z, t). The volume element would consist of six rectangular faces, defined by the coordinates $(x \pm \Delta x/2, y, z, t \pm \Delta t/2)$, $(x, y \pm \Delta y/2, z, t)$, and $(x, y, z \pm \Delta z/2, t)$. The additional currents J_y and J_z would flow through the xz and xy faces, respectively. Let us consider changes to the equations (3.28) and (3.29), when the current continuity experiment is observed in the unprimed coordinates. The experiment is conducted in the primed reference frame ($\Delta t' = 0$). Incremental lengths $\Delta y'$ and $\Delta z'$ in the y and z directions would be observed in the unprimed frame with equal lengths ($\Delta y' = \Delta y$ and $\Delta z' = \Delta z$), at the same instant ($\Delta t = 0$). This is in distinct contrast with the dimensions in the x direction, which are observed in the unprimed frame with incremental time differences. Accordingly, the time increment Δt would be zero for treatment of currents J_y and J_z in the xz and xy surfaces. With this in mind, the modified version of the equation (3.28) can be expressed as follows.

$$\begin{aligned}
\Delta\tau\Big[\big(J_x(x+\Delta x/2, t+\Delta t/2) - J_x(x-\Delta x/2, t-\Delta t/2)\big)\Delta y\Delta z \\
+\big(J_y(y+\Delta y/2, t) - J_y(y-\Delta y/2, t)\big)\Delta x\Delta z \\
+\big(J_z(z+\Delta z/2, t) - J_z(z-\Delta z/2, t)\big)\Delta x\Delta y\Big] = -\tfrac{\partial\Delta Q}{\partial t}\Delta\tau = [-\tfrac{\partial\rho_{v0}}{\partial t}\Delta x\Delta y\Delta z]\Delta\tau, \\
\tfrac{\partial J_x}{\partial x} + \tfrac{\partial J_x}{\partial t}\tfrac{V}{c^2} + \tfrac{\partial J_y}{\partial y} + \tfrac{\partial J_z}{\partial z} = -\tfrac{\partial\rho_{v0}}{\partial t}, \\
\int(\tfrac{\partial J_x}{\partial x} + \tfrac{\partial J_y}{\partial y} + \tfrac{\partial J_z}{\partial z})\partial t + \tfrac{V}{c^2}J_x = -\rho_{v0}.
\end{aligned} \tag{3.30}$$

When the velocity $V = 0$, we know $\rho_{v0} = \rho_v$. Using this condition in the above equation, the integral term can be recognized as $-\rho_v$.

$$\begin{aligned}
\rho_{v0} = \rho'_v\alpha = \rho_v - \tfrac{V}{c^2}J_x = \rho_v + \Delta\rho_v, \\
\Delta\rho_v = -\tfrac{V}{c^2}J_x,\ \tfrac{\partial\rho_v}{\partial t} = -\tfrac{\partial J_x}{\partial x} - \tfrac{\partial J_y}{\partial y} - \tfrac{\partial J_z}{\partial z}.
\end{aligned} \tag{3.31}$$

Equation (3.31), derived here for the general case with current in arbitrary direction, establishes the same relationship (3.29) we had derived for the simple case with only J_x current. $\Delta\rho_v$, which is the difference between $\rho_{v0} = \alpha\rho'_v$ and ρ_v, depends only on the J_x component of the current. The above general equation is a useful transformation relation between current and charge densities in two reference frames. Note that ρ'_v as observed in (x', t') coordinates ($\Delta t' = 0$) depends both on ρ_v and J observed in the (x, t) reference frame. A similar expression can be written relating ρ_v to ρ'_v and J', by switching ρ_v and ρ'_v, and replacing V by $-V$. This is equivalent to switching the frame in which the current continuity experiment is conducted and the frame in which it is observed.

$$\rho_v\alpha = \rho'_v + \frac{V}{c^2}J'_x. \tag{3.32}$$

Substituting ρ'_v in the above equation using (3.31), a relationship between J'_x, ρ_v and J_x can be obtained.

$$J'_x = \frac{c^2}{V}(\rho_v\alpha - \rho'_v) = \frac{c^2}{V}(\rho_v\alpha - \frac{\rho_v}{\alpha} + J_x\frac{V}{c^2\alpha}),$$
$$J'_x\alpha = J_x - \rho_v V. \tag{3.33}$$

A similar expression can also be written relating J_x to J'_x and ρ'_v. This is obtained by replacing the primed with corresponding unprimed variables, and replacing V by $-V$. This is equivalent to switching the reference frames of measurement and observation.

$$J_x\alpha = J'_x + \rho'_v V. \tag{3.34}$$

The equations (3.31-3.34) constitute the relativistic transform relations between current and charge densities in different reference frames. Out of the above four relations, any two can be used independently, from which the other two can be derived. The relations appear very much similar to the transform relations between space-time coordinates. The current density represents timing information, and the charge density represents dimensional information (length) for charge distribution.

3. Relativistic Transformation of Gauss' Law for the Electric Field: Ampere's Law

3.1. Gauss' Law for the Electric Field, in the Primed Coordinates. Consider Gauss' law for the electric field, applied in the primed coordinate system (x', y', z', t') to a closed surface S'.

$$\sum_{S'} \overline{D}'_i \cdot \overline{\triangle S}'_i = \Delta Q', \tag{3.35}$$

where $\Delta Q'$ is the total charge inside the surface S'. In reference to Fig.(5a), the above form of Gauss' law may be applied to a closed surface S' consisting of six faces of an elemental rectangular box of dimensions $\Delta x' \times \Delta y' \times \Delta z'$. The equation (3.35) may be expressed using normal components of the flux density $\overline{D}_i$ at the center of the six faces of the box, and the internal charge density ρ'_v.

$$(D'_{x3} - D'_{x1})\Delta y'\Delta z' + (D'_{y4} - D'_{y2})\Delta x'\Delta z' + (D'_{z6} - D'_{z5})\Delta y'\Delta x'$$
$$= \Delta Q' = \rho'_v \Delta x'\Delta y'\Delta z'. \tag{3.36}$$

The above summation form of Gauss' law may also be expressed equivalently using derivatives of the flux density $\overline{D}'$ at the center of the box.

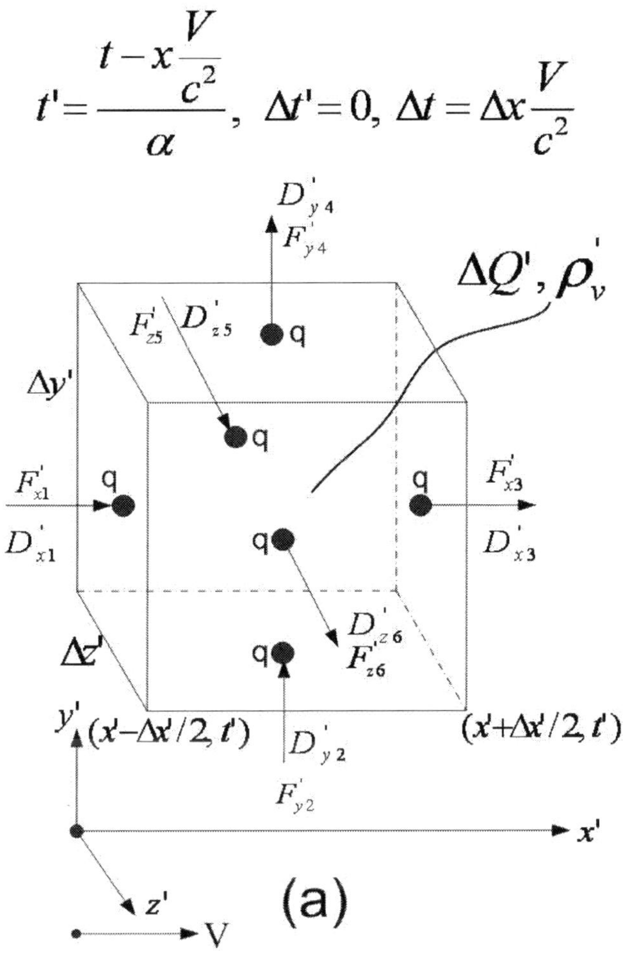

FIGURE 5

$$
\begin{aligned}
(D'_{x3} - D'_{x1})\Delta y'\Delta z' + (D'_{y4} - D'_{y2})\Delta x'\Delta z' + (D'_{z6} - D'_{z5})\Delta y'\Delta x' &= (\rho'_v)\Delta x'\Delta y'\Delta z' \\
= \tfrac{\partial D'_x}{\partial x'}\Delta x'\Delta y'\Delta z' + \tfrac{\partial D'_y}{\partial y'}\Delta y'\Delta x'\Delta z' + \tfrac{\partial D'_z}{\partial z'}\Delta z'\Delta y'\Delta x' &= \rho'_v\Delta x'\Delta y'\Delta z', \\
\tfrac{\partial D'_x}{\partial x'} + \tfrac{\partial D'_y}{\partial y'} + \tfrac{\partial D'_z}{\partial z'} &= \rho'_v. \qquad (3.37)
\end{aligned}
$$

3.2. A Basic Gauss' Law Experiment as Measured from Two Reference Frames. Let an observer in the primed coordinate system conduct an experiment to verify or validate Gauss' law for the rectangular box, placed in a free-space medium. The observer would need to measure the normal component of the electric field $\overline{E}_i$ for each of the six faces, $i = 1, .., 6$, of the rectangular box, from which the corresponding normal component of the flux density $\overline{D}_i = \epsilon_0 \overline{E}_i$ can be calculated. The electric field and the flux density are fundamentally defined in terms of the force experienced by a given stationary charge q.

$$D_{x1}, D_{x3} = D_x(x \mp \Delta x/2, t \mp \Delta t/2) =$$
$$D_x(x,t) \mp \frac{\partial D_x}{\partial x}\Delta x/2 \mp \frac{\partial D_x}{\partial t}\Delta t/2$$

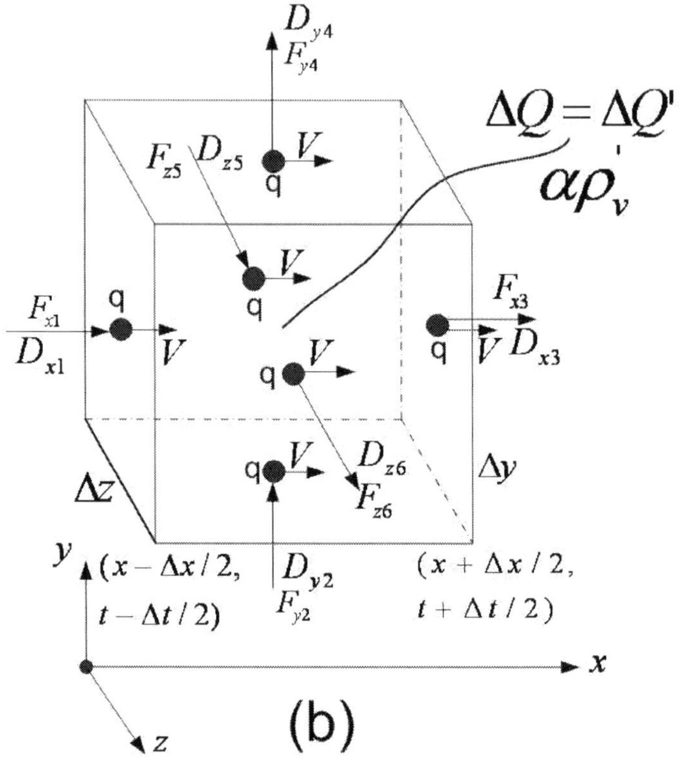

FIGURE 5. (a) Electric field divergence (Gauss' law) experiment in the (x',y',z',t') reference frame, and (b) the same experiment as seen by an observer from the (x,y,z,t) reference frame.

$$\overline{F}' = q\overline{E}' = \frac{q}{\epsilon_0}\overline{D}', \quad \overline{D}' = \frac{\epsilon_0}{q}\overline{F}'. \tag{3.38}$$

Accordingly, in order to validate Gauss' law, the observer may place a reference stationary charge q at the center of each face of the rectangular box, and then measure the force $\overline{F}_i'$ experienced by the charge using an appropriate measurement technique. Now, Gauss' law in the summation form of equation (3.36) may be expressed in terms of the measured normal components of the force and the charge density ρ_v' as follows.

$$\tfrac{\epsilon_0}{q}(F_{x3}' - F_{x1}')\Delta y'\Delta z' + \tfrac{\epsilon_0}{q}(F_{y4}' - F_{y2}')\Delta x'\Delta z' + \tfrac{\epsilon_0}{q}(F_{z6}' - F_{z5}')\Delta y'\Delta x'$$
$$= \rho_v'\Delta x'\Delta y'\Delta z'. \tag{3.39}$$

The above experiment in the primed coordinates is conducted at a given time t'. In other words, all the above force measurements are conducted simultaneously in the primed coordinate system, with time difference $\Delta t'$ between experiments at different faces equal to zero. Now, let the experiment be observed from the unprimed coordinate system. The force measurements on the six faces, performed simultaneously in the primed coordinates ($\Delta t' = 0$), would not be observed simultaneously in the unprimed coordinates ($\Delta t \neq 0$). Using the relativistic transformation relations (3.10,3.24) one may relate the time (Δt) and location (Δx) parameters of observation in the unprimed reference frame. The timing would be independent of the y and z coordinates. In other words, Δt would only depend on Δx, not on Δy or Δz.

$$
\begin{aligned}
t' &= \tfrac{t - xV/c^2}{\alpha}, \; x' = \tfrac{x - Vt}{\alpha}, \\
\Delta t' &= \tfrac{\Delta t - \Delta x V/c^2}{\alpha} = 0, \; \Delta t = \Delta x V/c^2, \\
\Delta x' &= \tfrac{\Delta x - \Delta t V}{\alpha} = \tfrac{\Delta x - \Delta x V^2/c^2}{\alpha} = \alpha \Delta x, \; \Delta x = \tfrac{\Delta x'}{\alpha}, \\
y &= y', \; z = z', \; \Delta y = \Delta y', \; \Delta z = \Delta z'.
\end{aligned} \tag{3.40}
$$

Accordingly, the forces $\overline{F}_i'$ measured simultaneously in the primed coordinate system would be measured as $\overline{F}_i$ in the unprimed coordinates at different times. The corresponding charge density, as observed from the unprimed frame, would be $\rho_{v0} = \rho_v' \Delta x'/\Delta x = \rho_v' \alpha$. This is shown in the last section, and is different from the charge density ρ_v', normally measured in the primed reference frame at a given time $t', \Delta t' = 0$. As derived in the last section, in equations (3.29,3.31), ρ_{v0} would also be different from the charge density ρ_v one would normally measure in the unprimed coordinates at a given time $t, \Delta t = 0$.

$$
\rho_{v0} = \rho_v + \Delta \rho_v = \rho_v - J_x V/c^2. \tag{3.41}
$$

As per a basic postulate of relativity, any "basic" experiment conducted in one coordinate system should lead to the same conclusion when observed from another coordinate system, moving with an uniform velocity with respect to each other. The Gauss' law experiment, which establishes the basic defining relationship for the charge, is considered one such basic experiment. This assumes that a given charge, as fundamentally defined by Gauss' law, is relativistically invariant, which means the value of the charge is independent of the relative velocity of the charge and the observer. Accordingly, the results of measurement in (3.39) should lead to the following relationship between the forces $\overline{F}_i$ and the charge density ρ_v.

$$
\begin{aligned}
&\tfrac{\epsilon_0}{q}(F_{x3} - F_{x1})\Delta y \Delta z + \tfrac{\epsilon_0}{q}(F_{y4} - F_{y2})\Delta x \Delta z + \tfrac{\epsilon_0}{q}(F_{z6} - F_{z5})\Delta y \Delta x \\
&\quad = (\rho_v + \Delta \rho_v)\Delta x \Delta y \Delta z = (\rho_v - J_x V/c^2)\Delta x \Delta y \Delta z.
\end{aligned} \tag{3.42}
$$

3.3. Recognizing the Need for a New Force Field. Let us first assume that the force on a reference charge in the unprimed frame is produced due to the electric force, in a conventional sense. With this assumption, the forces $\overline{F}_i$ and the equivalent flux densities at different faces of the test box (see Fig.5) may be related to the flux density $\overline{D}$ at the center of the box (x, y, z, t) using derivatives. Remember that the flux densities in the above experiment at different locations are measured in the unprimed coordinates at different times, as per the space-time relations (3.40).

$$\begin{aligned} F_{x1} &= \tfrac{q}{\epsilon_0}D_{x1} = \tfrac{q}{\epsilon_0}D_x(x-\triangle x/2, t-\triangle t/2) = \tfrac{q}{\epsilon_0}(D_x - \tfrac{\partial D_x}{\partial x}\Delta x/2 \\ &\quad -\tfrac{\partial D_x}{\partial t}\Delta t/2) = \tfrac{q}{\epsilon_0}(D_x - \tfrac{\partial D_x}{\partial x}\Delta x/2 - \tfrac{\partial D_x}{\partial t}\tfrac{V}{c^2}\Delta x/2), \\ F_{x3} &= \tfrac{q}{\epsilon_0}D_{x3} = \tfrac{q}{\epsilon_0}D_x(x+\triangle x/2, t+\triangle t/2) = \tfrac{q}{\epsilon_0}(D_x + \tfrac{\partial D_x}{\partial x}\Delta x/2 \\ &\quad +\tfrac{\partial D_x}{\partial t}\Delta t/2) = \tfrac{q}{\epsilon_0}(D_x + \tfrac{\partial D_x}{\partial x}\Delta x/2 + \tfrac{\partial D_x}{\partial t}\tfrac{V}{c^2}\Delta x/2), \\ F_{y2} &= \tfrac{q}{\epsilon_0}D_{y2} = \tfrac{q}{\epsilon_0}(D_y - \tfrac{\partial D_y}{\partial y}\Delta y/2), \\ F_{y4} &= \tfrac{q}{\epsilon_0}D_{y4} = \tfrac{q}{\epsilon_0}(D_y + \tfrac{\partial D_y}{\partial y}\Delta y/2), \\ F_{z5} &= \tfrac{q}{\epsilon_0}D_{z5} = \tfrac{q}{\epsilon_0}(D_z - \tfrac{\partial D_z}{\partial z}\Delta z/2), \\ F_{z6} &= \tfrac{q}{\epsilon_0}D_{z6} = \tfrac{q}{\epsilon_0}(D_z + \tfrac{\partial D_z}{\partial z}\Delta z/2). \end{aligned} \tag{3.43}$$

Note that there is no time difference between the measurements at the center of the faces 4 and 2, and similarly those of faces 6 and 5. This is because they differ only in their y and z coordinates with respect to the center of the box. The time difference Δt of observation at two locations is dependent only on the x coordinates of the locations. Now, using the force expressions of (3.43) in the equation (3.42), we get,

$$\tfrac{\partial D_x}{\partial x} + \tfrac{\partial D_x}{\partial t}\tfrac{V}{c^2} + \tfrac{\partial D_y}{\partial y} + \tfrac{\partial D_z}{\partial z} = \rho_{v0} + \Delta\rho_v = \rho_v - J_x\tfrac{V}{c^2}. \tag{3.44}$$

The above measurement for Gauss' law for the electric field was originally performed in the primed coordinates, which was then observed from the unprimed coordinates. The experiment may also be duplicated in the unprimed reference frame itself. In this case, the measurements in the different faces of the box is to be performed simultaneously in the unprimed coordinates $\Delta t = 0$. The results of this experiment may be obtained from the equivalent expression (3.37) in the primed coordinates, by simply substituting the primed variables by the corresponding unprimed variables. This would be the differential form of Gauss' law in the unprimed rectangular coordinates.

$$\frac{\partial D_x}{\partial x} + \frac{\partial D_y}{\partial y} + \frac{\partial D_z}{\partial z} = \rho_v. \tag{3.45}$$

Equation (3.44) would be clearly inconsistent with the equation (3.45) for general conditions with a non-zero current density $\overline{J}$ and/or a time-varying electric flux density $\overline{D}$. Let us examine the possible source of this inconsistency. We assume that Gauss' law for electric field applies universally in all reference frames. This means the equivalent expressions of Gauss' law in (3.35-3.39) for the primed coordinates, and in (3.45) for the unprimed coordinates, must be correct. We also assume that the basic postulate of relativity, which requires that a validation experiment for a basic theory should lead to equivalent conclusions when conducted in one reference frame, and when the experiment is observed from another frame moving with a constant relative velocity. Accordingly, the deduction of the equations (3.42) through relativistic transformation of Gauss' law in (3.39) must also be correct. This leaves equations (3.43,3.44) to be examined for any deficiency. The total forces on the reference charges, which are expected to satisfy the relationship (3.42), may not have been properly characterized in the equations (3.43,3.44).

Note that the reference charges q in the experiment are assumed to be stationary in the primed coordinates, in which the experiment was originally conducted. This experiment in the primed coordinates is described by the force and field equations (3.35-3.39). These equations are based on Gauss' law and on the definition of the electric field as the total force on a unit stationary reference charge. However, when the experiment is observed from the unprimed coordinates, the reference charges are no longer stationary, and are moving with a velocity V along the x direction. The total forces experienced by these moving reference charges in the unprimed coordinates have been established also using the electric fields through the equations (3.43,3.44). Now, the forces in (3.35-3.39), where the charges are stationary, are assumed to be valid, leading also to the validity of the force equation in (3.42). Whereas, those in (3.43,3.44), where the charges are moving, we suspect could be wrong. Therefore, one would conclude that the forces in (3.43,3.44) will have to be corrected with additional parts that must dependent on the velocity of the reference charge q.

The essential conclusion is that, the force experienced by a moving charge may have to be characterized differently from that by a stationary charge. The total force on a stationary charge is determined as the product of the charge and the electric field, as per the very definition of the electric field. Having defined the force on a stationary charge this way, the total force on a moving charge may not also be characterized in the same way - as the product of the charge and the electric field - by simply ignoring possible contributions due to its velocity. Additional force effects that depend on the velocity of the charge may have to be introduced. Otherwise, the forces on an electric charge would be relativistically inconsistent and incomplete, as we have now encountered. This is an important basic finding.

3.4. Force Correction, the Magnetic Field, and Ampere's law. As per the above discussion, we seek to adjust the forces in the equations (3.43,3.44), such that the total forces would satisfy (3.42), as well as be consistent with (3.45). Note that the differences between the equations (3.44) and (3.45) are linear with velocity V. This suggests, in order that the revised versions of (3.43,3.44) be consistent with (3.45), any new motion-dependent force that needs to be added to (3.43,3.44) should be linearly dependent on the charge velocity. Further, the new force might possibly be directed along or perpendicular to the charge velocity. However, if one would proceed by adding a new force directed along the charge velocity V, it can be shown not to resolve the above inconsistency. The new force $\overline{\Delta F}$ will have to be directed perpendicular to the charge velocity. With the charge velocity along the $\hat{x}$ direction, this means $\overline{\Delta F}$ may only have a non-zero $\hat{y}$ or $\hat{z}$ component ($\triangle F_z \neq 0$, $\triangle F_y \neq 0$), with its $\hat{x}$ component equal to zero ($\triangle F_x = 0$). Accordingly, in reference to Fig.5b, we need to revise the normal-force equations (3.43,3.44) by adding new normal forces ΔF_{y2}, ΔF_{y4}, ΔF_{z5} and ΔF_{z6}, respectively for the faces 2, 4, 5 and 6. These would be in addition to the conventional normal forces due to the electric field. Whereas, no new normal force would be added for the faces 1 and 3. This is because, for these two surfaces the normal directions are $\pm\hat{x}$, along which the new force would be zero ($\Delta F_x = 0$), as we explained.

$$
\begin{aligned}
F_{x3} &= \tfrac{q}{\epsilon_0} D_{x_3}, \\
F_{x1} &= \tfrac{q}{\epsilon_0} D_{x_1}, \\
F_{y4} &= \tfrac{q}{\epsilon_0} D_{y_4} + \Delta F_{y4}, \\
F_{y2} &= \tfrac{q}{\epsilon_0} D_{y_2} + \Delta F_{y2}, \\
F_{z6} &= \tfrac{q}{\epsilon_0} D_{z_6} + \Delta F_{z6}, \\
F_{z5} &= \tfrac{q}{\epsilon_0} D_{z_5} + \Delta F_{z5},
\end{aligned}
$$

$$
\Delta x \Delta y \Delta z (\tfrac{\partial D_x}{\partial x} + \tfrac{\partial D_x}{\partial t}\tfrac{V}{c^2}) + \Delta x \Delta z [\tfrac{\partial D_y}{\partial y} \Delta y + (\Delta F_{y4} - \Delta F_{y2}) \tfrac{\epsilon_0}{q}] \\
+ \Delta x \Delta y [\tfrac{\partial D_z}{\partial z} \Delta z + (\Delta F_{z6} - \Delta F_{z5}) \tfrac{\epsilon_0}{q}] = (\rho_v - J_x \tfrac{V}{c^2}) \Delta x \Delta y \Delta z. \tag{3.46}
$$

Combining the equations (3.45,3.46), we get,

$$
\begin{aligned}
\Delta x \Delta y \Delta z \tfrac{\partial D_x}{\partial t}\tfrac{V}{c^2} + \Delta x \Delta z (\Delta F_{y4} - \Delta F_{y2}) \tfrac{\epsilon_0}{q} + \Delta x \Delta y (\Delta F_{z6} - \Delta F_{z5}) &= -J_x \tfrac{V}{c^2} \Delta x \Delta y \Delta z, \\
\Delta y \Delta z (\tfrac{\partial D_x}{\partial t} + J_x) \tfrac{V}{c^2} &= (\Delta F_{y2} - \Delta F_{y4}) \tfrac{\epsilon_0}{q} \Delta z + (\Delta F_{z5} - \Delta F_{z6}) \tfrac{\epsilon_0}{q} \Delta y, \\
\Delta y \Delta z (\tfrac{\partial D_x}{\partial t} + J_x) &= [(\Delta F_{y2} - \Delta F_{y4}) \Delta z + (\Delta F_{z5} - \Delta F_{z6}) \Delta y] \tfrac{\epsilon_0 c^2}{qV}.
\end{aligned} \tag{3.47}
$$

For analytical convenience, one may define a new vector $\overline{H}$, using which any additional force $\overline{\Delta F}$ is indirectly expressed in terms of the charge velocity $\overline{v}$ as follows.

$$
\frac{q}{\epsilon_0 c^2} \overline{v} \times \overline{H} = \overline{\Delta F} = q \mu_0 \overline{v} \times \overline{H}, \quad \mu_0 = \frac{1}{\epsilon_0 c^2}. \tag{3.48}
$$

The force is defined proportional to the cross product of the velocity vector with $\overline{H}$. This would ensure that the magnitude of the force is proportional to the velocity, and the force is directed normal to the direction of the velocity, as per our expectation discussed earlier. The components of the new force that are normal to different faces of the test box may be related to appropriate components of the new vector $\overline{H}$.

$$
\begin{aligned}
\overline{\Delta F_2} = q\mu_0 V \hat{x} \times \overline{H}_2, \ \Delta F_{y2} &= \hat{y} \cdot \overline{\Delta F_2} = -q\mu_0 V H_{z2}, \\
\overline{\Delta F_4} = q\mu_0 V \hat{x} \times \overline{H}_4, \ \Delta F_{y4} &= \hat{y} \cdot \overline{\Delta F_4} = -q\mu_0 V H_{z4}, \\
\overline{\Delta F_5} = q\mu_0 V \hat{x} \times \overline{H}_5, \ \Delta F_{z5} &= \hat{z} \cdot \overline{\Delta F_5} = q\mu_0 V H_{y5}, \\
\overline{\Delta F_6} = q\mu_0 V \hat{x} \times \overline{H}_6, \ \Delta F_{z6} &= \hat{z} \cdot \overline{\Delta F_6} = q\mu_0 V H_{y6}.
\end{aligned} \tag{3.49}
$$

Now combine equations (3.47) and (3.49).

$$
\begin{aligned}
\Delta y \Delta z (\tfrac{\partial D_x}{\partial t} + J_x) &= -H_{z2} \Delta z + H_{z4} \Delta z + H_{y5} \Delta y - H_{y6} \Delta y \\
&= \sum_{\triangle C} \overline{H}_i \cdot \overline{\Delta l_i}\ ,
\end{aligned} \tag{3.50}
$$

$$(J_x + \frac{\partial D_x}{\partial t})\Delta S = \sum_{\Delta C} \overline{H}_i \cdot \overline{\Delta l_i} = (\overline{J} + \frac{\partial \overline{D}}{\partial t}) \cdot \overline{\Delta S}. \quad (3.51)$$

The loop ΔC refers to a closed rectangular loop parallel to the face 3 of the rectangular box in Fig.5b, passing through the center of the box. This is shown separately in Fig.6. Further, the orientation of the path of the loop ΔC is such that the normal direction to the enclosed surface vector $\overline{\Delta S}$, as per the right-hand rule, is parallel to the normal to the face 3 of the box (that is, along $+\hat{x}$).

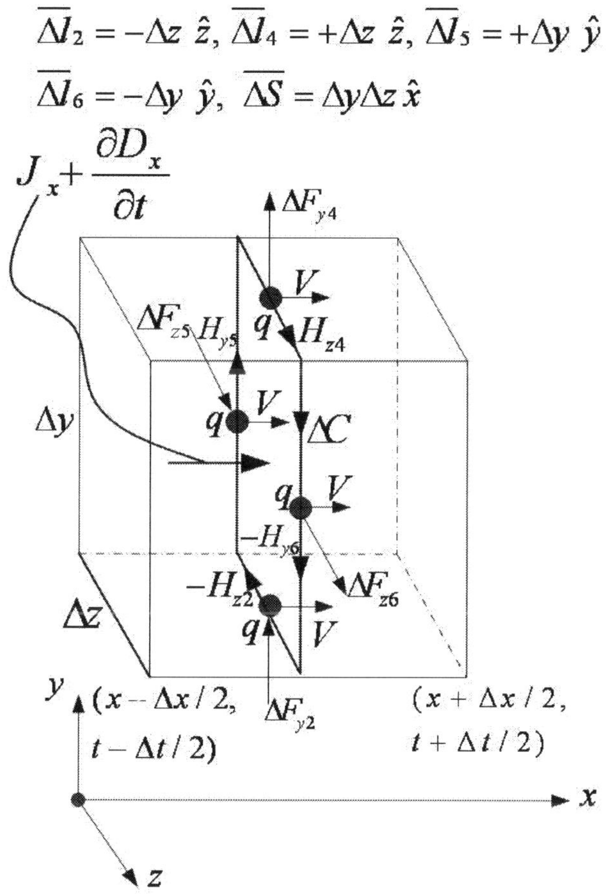

FIGURE 6. Deduction of Ampere's law from the Electric field divergence (Gauss' law) experiment of Fig.5, as seen by an observer from the (x,y,z,t) reference frame.

We have found a relationship (3.51) for the new force field $\overline{H}$, such that the equation (3.42) is satisfied for the total forces, as observed in the unprimed reference frame. This means that the Gauss' law experiment in the primed coordinates would now be consistent or equivalent when observed from the unprimed coordinates, as fundamentally required. The orientation of ΔC, and the normal to the enclosed surface $\overline{\Delta S}$, used in the above derivation of (3.51) are established by the direction of the relative motion between the primed and unprimed frames of reference. We happened to have selected the direction of the relative velocity V to be along $\hat{x}$. In general, by doing the analysis with velocity $\overline{V}$ in an arbitrary direction, it may be recognized that the equation (3.51) would apply for an elemental closed loop ΔC, and its enclosed surface $\overline{\Delta S}$, having any arbitrary orientation.

$$(\bar{J} + \frac{\partial \bar{D}}{\partial t}) \cdot \overline{\Delta S} = \sum_{\Delta C} \bar{H}_i \cdot \Delta l_i. \tag{3.52}$$

The equation (3.52) is Ampere's law expressed for an elemental loop. The new force field described by the vector $\overline{H}$, which we have defined in (3.48), may be recognized as the magnetic field. The definition in (3.48) is consistent with the standard definitions of the magnetic field used in chapter 1. The obvious implication of the derivation is that Ampere's law no longer has to be established as an independent law. It can be derived as a direct consequence of Gauss' law for the electric field, through relativistic transformation. Interestingly, the magnetic field, which relates to the electric current and flux density through Ampere's law, may be interpreted as a new "synthesized" field. This synthesized field will have to be added in order that Gauss' law for the electric field is generalized across reference frames through the principle of relativistic equivalence. This unique interpretation of the magnetic field, from the view point of the relativistic transformation, is a significant development.

3.5. Further Generalization of Ampere's Law. The equation (3.52) establishes Ampere's law for an elemental current loop. Once established for an elemental loop, it can be extended to any general loop C, enclosing a general surface S, which may be planar or non-planar. This is accomplished by decomposition of the surface S into small planar surface elements ΔS_i, with individual boundary loops ΔC_i, arranged in a "tile" form, and then superposing the results from (3.52) over all i. Such technique is commonly used in similar treatments in the last chapter.

This magnetic field in its basic form defines the force on a moving charge. It may also be equivalently defined in terms of the force on a current element $\overline{I}$ of length Δl, as covered in the chapter 1, section 1.5.

$$\overline{\Delta F} = \Delta l \mu_0 \overline{I} \times \overline{H}. \tag{3.53}$$

It may also be mentioned, that the above derivation of Ampere's law assumed a free-space medium. The derivation, in principle, is theoretically complete. However, the results may be extended to a material medium by substituting the material medium by equivalent polarization current loops and charge dipoles, and selecting suitable definitions of the electric and magnetic fields and flux densities inside the medium. Such material models were covered in the chapter 1, section 1.5, and would not be addressed here.

4. Relativistic Transformation of Gauss' Law for the Magnetic Field: Faraday's Law

The new magnetic field $\overline{H}$ was introduced in order that Gauss' law for the electric field is relativistically consistent. Now, analogous to the electric flux density and flux, we define a magnetic flux density to be $\overline{B} = \mu_0 \overline{H}$, using which magnetic flux over a given surface can be determined. Further, like Gauss' law for the electric field, we may establish a Gauss' law for the magnetic field in terms of the magnetic flux. This defines the total flux over a closed surface to the total charge enclosed by the surface. The charge in this case is referred to as the magnetic charge Q_m from which the magnetic field is supposed to originate or diverge from. However, no magnetic charge has been found to exist naturally, in which case the magnetic charge used in Gauss' law for the magnetic field would be zero. In any case, like Gauss' Law for the electric field, Gauss' Law for the magnetic field may also be considered to be the basic governing relationship for the magnetic field, which is relativistically invariant. Accordingly, let Gauss' law for the magnetic field be "observed" from different relativistic frames, in order to see what additional relationship the magnetic field will have to satisfy so that it is relativistically consistent. We will closely follow the process employed in the last section for Gauss' law for the electric field, but suitably extended to Gauss' law for the magnetic field. This is presented in the following, leading to Faraday's law as an additional relationship that must be established between the magnetic and electric fields.

4.1. Gauss' Law for the Magnetic Field, in the Primed Coordinates. Consider Gauss' law for the magnetic field, applied in the primed coordinate system (x', y', z', t') to a closed surface S'.

$$\sum_{S'} \overline{B}'_i \cdot \overline{\triangle S}'_i = 0. \tag{3.54}$$

As mentioned above, this is similar to Gauss' law for the electric field, except that the magnetic charge is assumed here to be zero for real-word phenomena. One may sometimes find hypothetical or equivalent magnetic charges included in electromagnetic modeling. Such magnetic charges are used only for useful analytical manipulations, and should not be confused to be physically realistic. Such hypothetical magnetic charges can be included in Gauss' law for analytical generality. In this case, the right side of the Gauss' law equation (3.54) is simply replaced by the total equivalent magnetic charge $\Delta Q'_m$ inside the closed surface S'. For all analytical considerations, the magnetic charge would be treated analogous to the electric charge. The magnetic charge would be assumed to be relativistically invariant. All derivations and relationships established for electric charge and current in the last sections, can also be equivalently extended for magnetic charge and magnetic current.

$$\sum_{S'} \overline{B}'_i \cdot \overline{\triangle S}'_i = \Delta Q'_m. \tag{3.55}$$

In reference to Fig.7a, the above form (3.55) of Gauss' law may be applied to a specific closed surface S', consisting of six faces of an elemental rectangular box of dimensions $\Delta x' \times \Delta y' \times \Delta z'$. For this case, the equation (3.55) may be expressed using normal

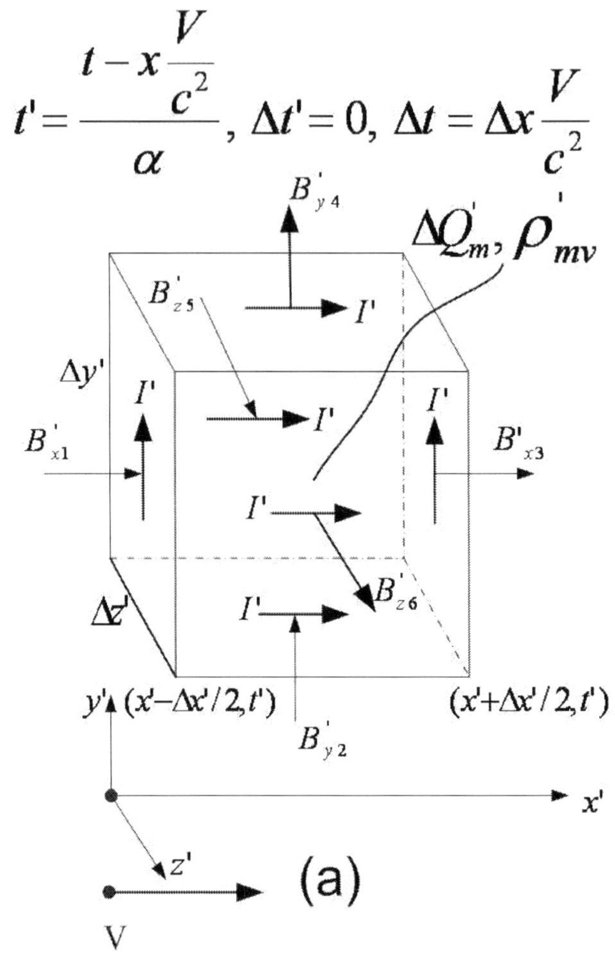

FIGURE 7

components of the flux density $\overline{B}_i$ at the center of each of the six faces of the box, and the equivalent magnetic charge density ρ'_{mv} at the center of the box. The magnetic charge $\Delta Q'_m$ is expresses as $\rho'_{mv}\Delta x'\Delta y'\Delta z'$, which is the product of the equivalent magnetic charge density and the the elemental volume of the box.

$$(B'_{x3} - B'_{x1})\Delta y'\Delta z' + (B'_{y4} - B'_{y2})\Delta x'\Delta z' + (B'_{z6} - B'_{z5})\Delta y'\Delta x' = \Delta Q'_m$$
$$= \rho'_{mv}\Delta x'\Delta y'\Delta z'. \tag{3.56}$$

The above summation form of Gauss' law may also be expressed using derivatives of the flux density $\overline{B}'$ at the center of the box.

$$B_{x1}, B_{x3} = B_x(x \mp \Delta x/2, t \mp \Delta t/2)$$
$$= B_x(x,t) \mp \frac{\partial B_x}{\partial x}\Delta x/2 \mp \frac{\partial B_x}{\partial t}\Delta t/2$$
$$\Delta l_1 = \Delta l_3,\ \Delta l_2 = \Delta l_4 = \Delta l_5 = \Delta l_6$$
$$I_1 = I_3,\ I_2 = I_4 = I_5 = I_6$$

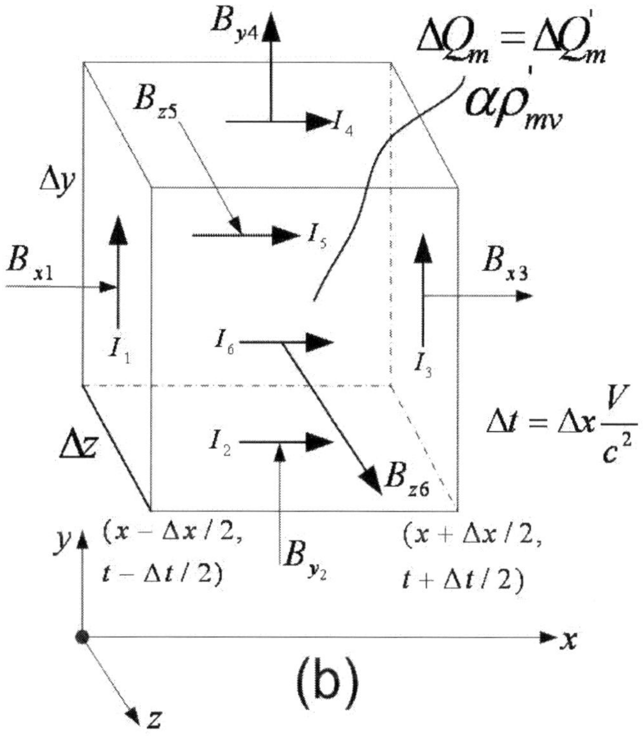

FIGURE 7. (a) Magnetic field divergence (Gauss' law) experiment in the (x',y',z',t') reference frame, and (b) the same experiment as seen by an observer from the (x,y,z,t) reference frame.

$$\begin{gathered}
(B'_{x3} - B'_{x1})\Delta y'\Delta z' + (B'_{y4} - B'_{y2})\Delta x'\Delta z' + (B'_{z6} - B'_{z5})\Delta y'\Delta x' = \\
\tfrac{\partial B'_x}{\partial x'}\Delta x'\Delta y'\Delta z' + \tfrac{\partial B'_y}{\partial y'}\Delta y'\Delta x'\Delta z' + \tfrac{\partial B'_z}{\partial z'}\Delta z'\Delta y'\Delta x' = \\
\rho'_{mv}\Delta x'\Delta y'\Delta z', \\
\tfrac{\partial B'_x}{\partial x'} + \tfrac{\partial B'_y}{\partial y'} + \tfrac{\partial B'_z}{\partial z'} = \rho'_{mv}.
\end{gathered} \tag{3.57}$$

4.2. A Basic Gauss' Law Experiment in the Primed Reference Frame. Let an observer in the primed coordinate system conduct an experiment to verify or validate Gauss' law for the rectangular box, placed in a free-space medium. With reference to the Gauss' law expression of (3.56), the observer would need to measure the normal components of the flux density $\overline{B}_i = \mu_0 \overline{H}_i$ for each of the six faces, $i = 1,..,6$, of the rectangular box S'. The magnetic field $\overline{H}_i$ or the flux density $\overline{B}_i$ are fundamentally defined in terms of the force experienced by a given current element $\overline{I}'$ of a given length $\Delta l'$, as per (3.53) applied in the primed frame.

$$\overline{F}' = \Delta l' \overline{I}' \times \overline{B}' = \mu_0 \Delta l' \overline{I}' \times \overline{H}'. \tag{3.58}$$

Accordingly, one may measure the force on a reference current element, from which a particular component of the magnetic field or flux density can be deduced. As per the equation (3.58), the component of the magnetic field which is aligned along the current element would not contribute to the total force. Therefore, one can not deduce the the component of the magnetic field along the current element from the force measurement. This implies, in order to measure the normal component of the magnetic field on a particular face of the elemental box, as needed in the validation experiment, one would like to place the reference current element parallel to the face. With this in mind, let us choose the reference currents on faces 1 and 3 to be along the y direction, but along the x direction for all other faces. For simplicity, let us select the magnitude of each reference current to be the same.

Accordingly, a conducting wire is placed centered on each face of the rectangular box, directed along the x direction for faces 2, 4, 5 and 6, but along the y direction for faces 1 and 3. Each wire is excited with a current of magnitude I'. The forces $\overline{F}'_i$ experienced by a given section $\Delta l'$ of the current wire is then measured using an appropriate technique. The actual length of the reference current, over which the force measurement is conducted, is not really critical, because we are interested in the magnetic field or flux density, which is defined in terms of the force per unit length. However, for simplicity, we have specified the reference sections of the currents over which the fore is measured on each face to be of equal length $\Delta l'$. Now, the normal magnetic field components on each face are related to appropriate force components, as per (3.58).

$$\begin{aligned}
\overline{F'}_1 = \Delta l' I' \hat{y} \times \overline{B'}_1, \; -F'_{z1} = \Delta l' I' B'_{x1}, \; B'_{x1} = -\tfrac{F'_{z1}}{\Delta l' I'}, \\
\overline{F'}_3 = \Delta l' I' \hat{y} \times \overline{B'}_3, \; -F'_{z3} = \Delta l' I' B'_{x3}, \; B'_{x3} = -\tfrac{F'_{z3}}{\Delta l' I'}, \\
\overline{F'}_2 = \Delta l' I' \hat{x} \times \overline{B'}_2, \; F'_{z2} = \Delta l' I' B'_{y2}, \; B'_{y2} = \tfrac{F'_{z2}}{\Delta l' I'}, \\
\overline{F'}_4 = \Delta l' I' \hat{x} \times \overline{B'}_4, \; F'_{z4} = \Delta l' I' B'_{y4}, \; B'_{y4} = \tfrac{F'_{z4}}{\Delta l' I'}, \\
\overline{F'}_5 = \Delta l' I' \hat{x} \times \overline{B'}_5, \; -F'_{y5} = \Delta l' I' B'_{z5}, \; B'_{z5} = -\tfrac{F'_{y5}}{\Delta l' I'}, \\
\overline{F'}_6 = \Delta l' I' \hat{x} \times \overline{B'}_6, \; -F'_{y6} = \Delta l' I' B'_{z6}, \; B'_{z6} = -\tfrac{F'_{y6}}{\Delta l' I'}.
\end{aligned} \tag{3.59}$$

The Gauss' law equation (3.56) can be expressed in terms of the measured force components in (3.59).

$$\begin{aligned}
(B'_{x3} - B'_{x1})\Delta y'\Delta z' + (B'_{y4} - B'_{y2})\Delta x'\Delta z' + (B'_{z6} - B'_{z5})\Delta y'\Delta x' = \\
\tfrac{1}{I'\Delta l'}[(-F'_{z3} + F'_{z1})\Delta y'\Delta z' + (F'_{z4} - F'_{z2})\Delta x'\Delta z'(-F'_{y6} + F'_{y5})\Delta x'\Delta y'] = \\
\Delta Q'_m = \rho'_{mv}\Delta x'\Delta y'\Delta z'.
\end{aligned} \tag{3.60}$$

The above equation (3.60) is a form of Gauss' law for the magnetic field, expressed in terms of the force components on the reference current elements. For a known charge density ρ'_{mv}, this relationship can be experimentally verified by measuring the force components, for given values of the reference currents and the length elements, It may be noted that the above experiment is conducted in the primed coordinates, at a fixed time t'. In other words, all the above force measurements are conducted simultaneously in the primed coordinate system, with time difference $\Delta t'$ between measurements at different faces equal to zero.

4.3. The Experiment as Observed from the Unprimed Frame. Now, let the above physical experiment, originally conducted in the primed frame, be "observed" from the unprimed coordinate system, as shown in Fig.7b. As per a basic postulate of relativity, any "basic" physical experiment conducted in one reference frame should lead to the same conclusion when the experiment is observed from another coordinate frame, moving with an uniform velocity with respect to each other. The Gauss' law experiment is considered one such basic experiment. It establishes a fundamental relationship between the measured forces experienced by reference current elements and enclosed equivalent magnetic charges. Accordingly, the governing relationship (3.60) for the experiment performed in the primed frame should remain invariant when the frame of observation is switched. In other words, the governing equation (3.60) should remain valid, when the different parameters in (3.60) are properly substituted by their new values of measurement, when the same physical experiment is observed from the unprimed frame.

$$\Big[(-\tfrac{F_{z3}}{I_3\Delta l_3}+\tfrac{F_{z1}}{I_1\Delta l_1})\Delta y\Delta z+(\tfrac{F_{z4}}{I_4\Delta l_4}-\tfrac{F_{z2}}{I_2\Delta l_2})\Delta x\Delta z$$
$$+(-\tfrac{F_{y6}}{I_6\Delta l_6}+\tfrac{F_{y5}}{I_5\Delta l_5})\Delta x\Delta y\Big]=\Delta Q_m. \qquad (3.61)$$

We will examine all the unprimed variables in (3.61), in contrast with the corresponding primed variables in (3.60). Fundamental and analytical implications of observing the physical experiment from the unprimed frame will have to be properly understood.

The length element $\triangle x$ would be different from $\triangle x'$, as per the transform relation (3.40). Length measurements only along the direction of relative velocity change through relativistic transformation. Whereas, those along directions orthogonal to the relative velocity remain the same. Accordingly, $\triangle y=\triangle y'$ and $\triangle z=\triangle z'$. The timing of measurement on locations with different x coordinates would be different ($\Delta t\neq 0$), whereas those with different y or z, but the same x coordinates would be simultaneous ($\Delta t=0$). Using the relativistic transformation relations (3.10,3.24), one may relate the differential time (Δt) and location (Δx) parameters of observation in the unprimed reference frame. This would lead to the relationship (3.40) deduced in the last section.

$$\triangle t=\triangle xV/c^2,\ \triangle x'=\alpha\triangle x,$$
$$\triangle y'=\triangle y,\ \triangle z'=\triangle z. \qquad (3.62)$$

Accordingly, the forces on the faces 1 and 3, which differ in their x coordinates, would be measured in the unprimed frame at different times. Whereas, those on all other faces 2, 4, 5, and 6 would be measured simultaneously, because the x coordinates at the center of all these faces have the same x coordinates. The forces $\overline{F}_i$ on a given face maybe in general assumed to be different in magnitude from $\overline{F}'_i$ on the corresponding face.

As we have indicated before, for simplicity of the original measurement in the primed frame, the reference currents I' and their length elements $\Delta l'$ were selected with equal magnitude on all faces. However, they are substituted in (3.61) with distinct variables I_i and Δl_i, $i = 1, ..., 6$, for the six faces. This is because the reference currents having equal magnitude in the unprimed frame may no longer be equal when measured in the unprimed frame. If desired, these reference currents and length elements can be related to I' and $\Delta l'$, and consequently with each other, using space-time transform relations. The final outcome of our present derivation happens to be independent of their actual magnitudes. Therefore, we choose to keep them as distinct variables just to be technically correct, without any further analysis or simplification.

The equivalent magnetic charge and current parameters, as seen in the two frames, can be treated analogous to electric charge and current parameters. The total magnetic charge $\Delta Q'_m = \rho'_{mv}\triangle x'\triangle y'\triangle z'$ is assumed to be invariant to the frame of observation, and, therefore, should be equal to the charge ΔQ_m as observed from the unprimed frame. One may like to define ΔQ_m in terms of a new equivalent charge density ρ_{mv0}, as observed from the unprimed frame, such that $\Delta Q_m = \rho_{mv0}\triangle x\triangle y\triangle z$. Accordingly, we can relate ρ_{mv0} and ρ'_{mv} by equating ΔQ_m and $\Delta Q'_m$: $\rho_{mv0} = \rho'_{mv}\triangle x'/\triangle x = \rho'_{mv}\alpha$. This derivation uses the transformation equation (3.62) to relate the dimensional variables. Now, ρ'_{mv} or ρ_{mv0} may also be related to the charge density ρ_{mv} one would normally measure in the unprimed coordinates at a given time $t, \Delta t = 0$. We may extend the derivations in the last sections (2,3), equations (3.29,3.31,3.41), to magnetic charges.

$$\rho_{mv0} = \rho'_{mv}\alpha = \rho_{mv} + \Delta\rho_{mv} = \rho_{mv} - M_x V/c^2, \tag{3.63}$$

where the parameter M represents magnetic current density, which is analogous to J used for the electric current density. By combining (3.63) with (3.61), we get,

$$\left[(-\frac{F_{z3}}{I_3\Delta l_3} + \frac{F_{z1}}{I_1\Delta l_1})\Delta y\Delta z + (\frac{F_{z4}}{I_4\Delta l_4} - \frac{F_{z2}}{I_2\Delta l_2})\Delta x\Delta z + (-\frac{F_{y6}}{I_6\Delta l_6} + \frac{F_{y5}}{I_5\Delta l_5})\Delta x\Delta y\right] =$$
$$\Delta Q_m = \rho_{mv0}\triangle x\triangle y\triangle z = (\rho_{mv} - M_x V/c^2)\triangle x\triangle y\triangle z. \tag{3.64}$$

4.3.1. *Charged Reference Currents.* Now, let us examine the forces experienced by the electric current elements $I_i\Delta l_i$, and relate them to suitable field components measured in the unprimed frame. Before we can characterize the forces on the new current elements, we need to examine the nature of these current elements $I_i\Delta l_i$ as seen in the unprimed coordinates. The reference current elements $I'\Delta l'$ used in the original experiment are assumed to be free of charge, so that the forces experienced by the current elements in the primed reference frame are contributed only due to the magnetic fields. That allowed us to directly relate the magnetic fields in Gauss' law to the total forces on the current element, which can be measured. Otherwise, the basic governing equation (3.60) for the experiment would have been invalid. However, when these charge-free currents are observed from the unprimed frame, they would look charged if the current is directed along x, but remain charge free if directed along y or z. This is governed by the derivations in section 2. An x-directed current I_x and its observed line-charge density ρ_l in the unprimed frame may be related using (3.29,3.31,3.41).

$$\begin{aligned}\rho'_v\alpha &= \rho_v - J_xV/c^2,\\ \rho'_l\alpha &= \rho_l - I_xV/c^2 = 0,\ \ \rho_l = I_xV/c^2,\\ \rho'_l = \triangle A\rho'_v &= 0,\ \ \rho_l = \triangle A\rho_v,\ \ I_x = J_x\triangle A.\end{aligned} \tag{3.65}$$

The volume distributions ρ'_v, ρ_v and J_x, in (3.29,3.31,3.41) are substituted in terms of the corresponding line distributions ρ'_l, ρ_l, and I_x, respectively, leading to the above relationship (3.65). ΔA is the small cross-sectional area of the current element, which approaches zero for an ideal line current. ρ'_v is assumed zero, because the current element is chosen to be charge free in the primed frame. This is equivalent to having ρ'_l also to be zero.

4.3.2. *Force Transformation Using Electric and Magnetic Fields.* With the understanding of the nature of the reference current elements and other relativistic considerations, discussed above, let us characterize the forces components of Gauss' law equation (3.64) using different field components measured in the unprimed coordinates. Let us classify the forces in two separate groups, based on the timing of the force measurement and the nature of their reference currents used. The two groups are, (a) forces for the faces 1 and 3, and (b) the forces for the faces 2, 4, 5 and 6. We will discuss the force modeling for these two groups separately.

For the two faces in the first group, the test current elements are directed along y, and therefore do not appear charged, as we have discussed earlier. Consequently, the forces experienced by these current elements can be expressed using magnetic field alone. This group of faces differ in their x coordinates, and therefore the forces measured on these faces in the primed coordinates at a given time (t', $\Delta t' = 0$) are observed in the unprimed coordinates at different times t, $\Delta t \neq 0$. This is in accordance with the differential space-time relationship (3.62), as discussed before. Now, the different components of the force and magnetic flux density, as measured in the unprimed frame, can be related using an extension of (3.59) as follows. The primed variables in (3.59) are replaced by the corresponding unprimed variables, and the space and time coordinates for different parameters in (3.59) are specified at the center of corresponding face (see Fig.7b). Further, the fields at the center of the faces are related to those at the center of the box using partial derivatives, and equation (3.62) is used to relate differential space ($\triangle x$) and time ($\triangle t$) variables.

$$\begin{aligned}\overline{F}_1 = \Delta l_1 I_1\hat{y}\times\overline{B}_1,\ -F_{z1} &= \Delta l_1 I_1 B_{x1} = \Delta l_1 I_1 B_x(x-\triangle x/2, t-\triangle t/2)\\ &= \Delta l_1 I_1\left(B_x(x,t) - \tfrac{\triangle x}{2}\tfrac{\partial B_x(x,t)}{\partial x} - \tfrac{\triangle t}{2}\tfrac{\partial B_x(x,t)}{\partial t}\right)\\ &= \Delta l_1 I_1\left(B_x(x,t) - \tfrac{\triangle x}{2}\tfrac{\partial B_x(x,t)}{\partial x} - \tfrac{\triangle x V}{2c^2}\tfrac{\partial B_x(x,t)}{\partial t}\right),\\ \overline{F}_3 = \Delta l_3 I_3\hat{y}\times\overline{B}_3,\ -F_{z3} &= \Delta l_3 I_3 B_{x3} = \Delta l_3 I_3 B_x(x+\triangle x/2, t+\triangle t/2)\\ &= \Delta l_3 I_3\left(B_x(x,t) + \tfrac{\triangle x}{2}\tfrac{\partial B_x(x,t)}{\partial x} + \tfrac{\triangle t}{2}\tfrac{\partial B_x(x,t)}{\partial t}\right)\\ &= \Delta l_3 I_3\left(B_x(x,t) + \tfrac{\triangle x}{2}\tfrac{\partial B_x(x,t)}{\partial x} + \tfrac{\triangle x V}{2c^2}\tfrac{\partial B_x(x,t)}{\partial t}\right).\end{aligned} \tag{3.66}$$

On the other hand, for the second group of faces 2, 4, 5 and 6, the reference currents are directed along x, and therefore they would appear charged, as discussed earlier. Equation (3.65) can be used to relate the observed line charge density ρ_l to the corresponding line current density I_x. Consequently, the force experienced by these currents are to be expressed using both the magnetic and electric fields. The centers of measurement for the four faces have the same x coordinates. Therefore, as discussed before, the experiment in the primed coordinates at a fixed time (t', $\Delta t' = 0$) are observed in the unprimed coordinates also at a fixed time t, $\Delta t = 0$. This is in accordance with equation (3.62). With these above issues in mind, the different components of the force, as measured in the unprimed frame, can now be modeled by adding two contributions: the force due to the magnetic field using an extension of (3.59), added with the force on the charges on the current elements using appropriate electric field components.

$$\begin{aligned}
\overline{F}_2 &= \Delta l_2 I_2 \hat{x} \times \overline{B}_2 + \Delta l_2 \rho_{l2} \overline{E}_2,\ F_{z2} = \Delta l_2 I_2 B_{y2} + \Delta l_2 \rho_{l2} E_{z2} \\
&= \Delta l_2 I_2 (B_y - \tfrac{\triangle y}{2} \tfrac{\partial B_y}{\partial y}) + \Delta l_2 I_2 \tfrac{V}{c^2} E_{z2}, \\
\overline{F}_4 &= \Delta l_4 I_4 \hat{x} \times \overline{B}_4 + \Delta l_4 \rho_{l4} \overline{E}_4,\ F_{z4} = \Delta l_4 I_4 B_{y4} + \Delta l_4 \rho_{l4} E_{z4} \\
&= \Delta l_4 I_4 (B_y + \tfrac{\triangle y}{2} \tfrac{\partial B_y}{\partial y}) + \Delta l_4 I_4 \tfrac{V}{c^2} E_{z4}, \\
\overline{F}_5 &= \Delta l_5 I_5 \hat{x} \times \overline{B}_5 + \Delta l_5 \rho_{l5} \overline{E}_5,\ -F_{y5} = \Delta l_5 I_5 B_{z5} - \Delta l_5 \rho_{l5} E_{y5} \\
&= \Delta l_5 I_5 (B_z - \tfrac{\triangle z}{2} \tfrac{\partial B_z}{\partial z}) - \Delta l_5 I_5 \tfrac{V}{c^2} E_{y5}, \\
\overline{F}_6 &= \Delta l_6 I_6 \hat{x} \times \overline{B}_6 + \Delta l_6 \rho_{l6} \overline{E}_6,\ -F_{y6} = \Delta l_6 I_6 B_{z6} - \Delta l_6 \rho_{l6} E_{y6} \\
&= \Delta l_6 I_6 (B_z + \tfrac{\triangle z}{2} \tfrac{\partial B_z}{\partial z}) - \Delta l_6 I_6 \tfrac{V}{c^2} E_{y6}.
\end{aligned} \tag{3.67}$$

The contribution in (3.67) from the magnetic field have been obtained by replacing the primed variables in (3.59) by the corresponding unprimed variables. The magnetic field components are measured at the center of particular faces (see Fig.7), which are then related to those at the center of the box using partial derivatives. The force contribution in (3.67) due to the charges on the current elements is expressed as product of the electric field and the charge. The equation (3.65) is used for simplification relating line charge density to the line current.

4.4. Faraday's Law, Deduced from the Force Transformation. Now, using the equations (3.66,3.67) in the force equation (3.64)), we get a relationship between the electric and magnetic fields in the unprimed reference frame.

$$\begin{aligned}
&\Big[(\tfrac{\partial B_x}{\partial x} + \tfrac{V}{c^2}\tfrac{\partial B_x}{\partial t})\triangle x \triangle y \triangle z + (\tfrac{\partial B_y}{\partial y}\triangle y + \tfrac{V}{c^2}E_{z4} - \tfrac{V}{c^2}E_{z2})\triangle x \triangle z + \\
&(\tfrac{\partial B_z}{\partial z}\triangle z - \tfrac{V}{c^2}E_{y6} + \tfrac{V}{c^2}E_{y5})\triangle x \triangle y\Big] = (\rho_{mv} - M_x V/c^2)\triangle x \triangle y \triangle z, \\
&(\tfrac{\partial B_x}{\partial x} + \tfrac{\partial B_y}{\partial y} + \tfrac{\partial B_z}{\partial z} + \tfrac{V}{c^2}\tfrac{\partial B_x}{\partial t})\triangle x \triangle y \triangle z + \tfrac{V}{c^2}\Big[(-E_{y6} + E_{y5})\triangle x \triangle y \\
&\quad + (E_{z4} - E_{z2})\triangle x \triangle z\Big] = (\rho_{mv} - M_x V/c^2)\triangle x \triangle y \triangle z.
\end{aligned} \tag{3.68}$$

We have established Gauss' law for the magnetic field in the unprimed coordinates, which is expressed in the summation form (3.56) or a differential form (3.57). The fields used in these expressions are measured simultaneously in the primed frame at a given t', $\Delta t' = 0$. Gauss' law for the magnetic field may be independently established in the unprimed frame as well, using fields that are measured simultaneously in the unprimed frame at a given t, $\Delta t = 0$. Let us rewrite the differential form of Gauss' law in (3.57) for the unprimed frame. This is accomplished by simply replacing the unprimed variables in (3.57) by the corresponding primed variables.

$$\frac{\partial B_x}{\partial x} + \frac{\partial B_y}{\partial y} + \frac{\partial B_z}{\partial z} = \rho_{mv}. \tag{3.69}$$

Using (3.69) in (3.68) we would get,

$$\begin{aligned} \frac{V}{c^2}\frac{\partial B_x}{\partial t}\Delta x \Delta y \triangle z + \frac{V}{c^2}\Big[(-E_{y6} + E_{y5})\Delta x \Delta y + (E_{z4} - E_{z2})\Delta x \Delta z\Big] \\ = -M_x \frac{V}{c^2}\triangle x \triangle y \triangle z, \\ \Big[- E_{y6}\Delta y + E_{y5}\Delta y + E_{z4}\Delta z - E_{z2}\Delta z\Big] = -(\frac{\partial B_x}{\partial t} + M_x)\triangle y \triangle z. \end{aligned} \tag{3.70}$$

The above relation (3.70) may be recognized in terms of line and surface integrals over an elemental rectangular loop ΔC and its enclosed surface $\overline{\Delta S}$, respectively.

$$\sum_{\triangle C} \overline{E}_i \cdot \overline{\triangle l}_i = -(\frac{\partial B_x}{\partial t} + M_x)\triangle y \triangle z = -(\frac{\partial \overline{B}}{\partial t} + \overline{M}) \cdot \overline{\triangle S}. \tag{3.71}$$

The loop ΔC refers to a closed rectangular loop parallel to the face 3 of the rectangular box in Fig.7b, passing through the center of the box. This is separately shown in Fig. 8. Further, the orientation of the path of the loop ΔC is such that the normal direction to the enclosed surface vector $\overline{\Delta S}$, as per the right-hand rule, is parallel to the normal to the face 3 of the box (that is, along $+\hat{x}$).

Equation (3.71) is Faraday's law as applied to an elemental loop ΔC. It is derived from Gauss' law for the magnetic field in a free-space medium, through relativistic transformation. Accordingly, Faraday's law (3.71) may be interpreted as a necessary condition, so that Gauss' law for the magnetic field is consistent through relativistic transformation across reference frames. This is a significant development. The elemental loop ΔC in the above derivation is oriented with its normal along the x direction, which is parallel to the relative velocity between the frames. The choice of the direction of the relative velocity is arbitrary. If the analysis was performed with relative velocity in a general direction, the same expression (3.71) for Faraday's law can be shown to apply to an elemental loop with any general orientation.

The above form of Faraday's law, now shown to be valid for an elemental loop in a general orientation, can also be extended for more generalized situations. This would follow the concepts in the last section, discussed for similar generalization of Ampere's law. The derivation can be extended to any loop C, small or large, planar or non-planar, enclosing a planar or non-planar surface S. This is possible through decomposition of the loop C into a number of elemental loops ΔC in a tile form, and superposition of results from (3.71) for all the elemental loops.

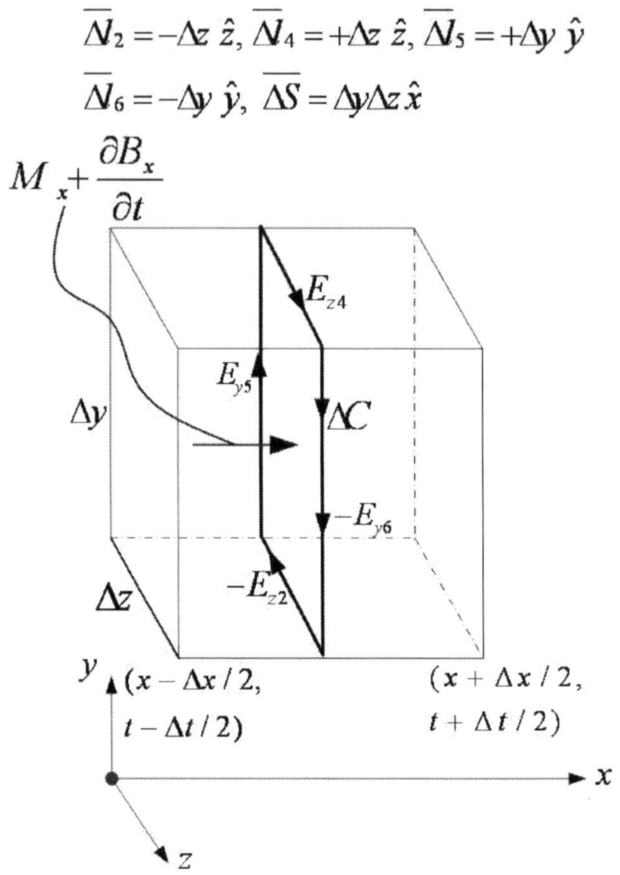

FIGURE 8. Deduction of Faraday's law from the magnetic field divergence (Gauss' law) experiment of Fig.7, as seen by an observer from the (x,y,z,t) reference frame.

$$\sum_C \overline{E}_i \cdot \overline{\triangle l}_i = -\sum_S \left(\frac{\partial \overline{B}_j}{\partial t} + \overline{M}_j\right) \cdot \overline{\triangle S}_j. \tag{3.72}$$

We have explicitly assumed a free-space medium for the derivation of Faraday's law. However, the results can be extended to a general linear medium by suitable definition of the magnetic flux density and a magnetic permeability of the material. This treatment we have presented in the first chapter. Accordingly, the derivation of Faraday's law in the elemental form of (3.71) is in principle complete.

CHAPTER 4

INTRODUCTION TO ELECTROMAGNETIC WAVES I: Signal Propagation in a Transmission Line

In this chapter, the electromagnetic theory will be applied to model propagation of electrical signals in a transmission line. A transmission line consists of two conductors running along a desired direction of signal propagation, with a suitable material medium filling the space between the boundary surfaces of the two conductors. The same relative positioning between the two conductors and the same properties of the filling material is maintained all along the propagation direction. A transmission line may take many different specific forms, mainly depending on the shape and relative configuration of the two conductors. In its basic form, it may consist of two cylindrical metal wires of the same or different diameters, running parallel to each other. This is called a two-wire transmission line, which is commonly used for house-hold electrical wiring. Other forms of transmission lines in common use include (a) a coaxial transmission line, which consists of two concentric cylindrical conductors, and (b) a printed-circuit transmission line, which consists of a planar metal strip printed on one surface of a dielectric substrate with its other surface completely coated with a conducting ground plane. We will study the signal propagation in such transmission lines applying the electromagnetic theory developed in the last chapters. An equivalent circuit description, using inductance and capacitance elements that are distributed along the propagation direction of the transmission line, can provide a simpler representation for the electromagnetic theory. The analysis establishes wave nature of the signal propagation, and the related principles would serve as an introduction to the basic concepts of an electromagnetic wave.

The electrical signal in a transmission line varies in time and distance in a manner analogous to a wave of water ripples in a lake, or like a wave of vibration along a guitar string. The speed of the signal propagation can be derived, and is shown to be finite. The signal speed is dependent ideally on the material medium filling between the two conductors of the transmission lines, but not on the shape or size of the conductors. It is also shown that the signal speed for a transmission line with an arbitrary material medium is always less than or equal to that ($\simeq 3 \times 10^8 \ m/s$) with an empty space. This upper limit for the signal speed, achieved when the transmission line is filled with an empty space, may also be recognized to be equal to the speed of light in the empty space. This significant recognition should serve as a clear indication of the similarities between the nature of electrical signal propagation in a transmission line and that of the light signal. Fundamentally, they both are known to be different forms of electromagnetic waves.

The wave nature of the signal and simple wave concepts, such as wave generation, reflection, and propagation delay, are introduced through analysis of simple transmission line circuits. An electrical signal source (a voltage or a current source) properly connected to a transmission line serves as a basic source of the electromagnetic wave in the transmission line. The amplitude of the generated wave is shown to depend on the internal impedance of the signal source and the parameters of the transmission line. Consider a simple circuit with a single section of a transmission line, connected to a single signal source (a voltage or a current source) at one end of the line. The wave generated from the source propagates through the transmission-line, eventually reaching the termination at the other end. The wave is then reflected with an amplitude smaller compared to the incident signal. The relative amplitude of the reflection is dependent on the equivalent circuit impedance seen at the termination and the transmission line parameters. The reflected signal would travel in the reverse direction, eventually reaching the termination at the source end, and would again reflect depending on the termination impedance resulting from the internal impedance of the source. The reflected signal from the source would now propagate and reflect again from the termination at the other end, and the process continues with the signal propagating and bouncing repeatedly between the opposite terminations. This forms a cascade of signal ripples, with progressively smaller amplitudes, propagating in alternating directions. The finite velocity of wave propagation results in a non-zero time delay for any particular signal ripple when observed at two different locations. Accordingly, the above train of signals would appear at different points of the transmission line with certain timing pattern, leading to a specific space-time signal pattern in the transmission line medium. In principle, the above basic concept and analysis for a simple circuit may be extended to general situations, leading to more complicated signal patterns in general circuits consisting of multiple signal sources, multiple transmission line segments and arbitrary impedances.

For simple source functions, such as a step or a pulse functions, the above space-time patterns can be kept track of in a intuitive manner. The procedure may sometimes be assisted by a "bouncing diagram" technique. For more general waveforms, although the same intuitive concept still applies, the analytical tracking procedure can be complicated, even for relatively simple circuits. Waves with sinusoidal or cosinusoidal time dependence (time-harmonic waves) carry partiular significance. In this case, the mathematical analysis of the wave propagation can be greatly simplified through uses of complex phasors. The time-harmonic waves are also associated with interesting and useful special concepts, such as standing waves and impedance transformation. Further, it is known through Fourier analysis that combinations of the harmonic signals can be used to synthesize arbitrary signal waveforms. Accordingly, the analysis of the time-harmonic waves can be used for modeling arbitrary waveforms, using Fourier decomposition and superposition. This assumes that the material properties are linear with respect to the signal amplitude, which is often the case for practical situations with a moderate signal level.

The electromagnetic effects and the wave nature of signal propagation must be included in a rigorous modeling of electrical circuits. This is in contrast to a simple circuit analysis, where signal currents and voltages are assumed to appear at two ends of a conducting wire simultaneously, with no time delay. For circuits connected by wires that are sufficiently small in lengths, such that the propagation delays are practically much smaller than the time scales of a particular application, such propagation delays and the general electromagnetic-wave effects discussed in this chapter, may be ignored. On the other hand, for general applications involving rapidly time-varying signals, where the time resolutions are comparable or larger than the time delay of signal propagation, the electromagnetic-wave effects would be critical for design and understanding.

There are further practical relevance of the transmission line theory. Based on the theory, one can design a small transmission-line segment for use as a replacement for a conventional capacitor or inductor element. This design is often practical and economic for high frequency applications, when the design lengths are relatively small for implementation. The theory may be utilized further to design other circuit functions, such as an impedance transformer or an impedance-matching circuit using simple configurations of transmission line segments. More advanced circuit functions, utilizing the wave theory of the transmission lines, may be similarly envisioned, but are beyond the scope of this chapter.

1. Basic Electromagnetic Fields for a Two-Conductor Transmission Line

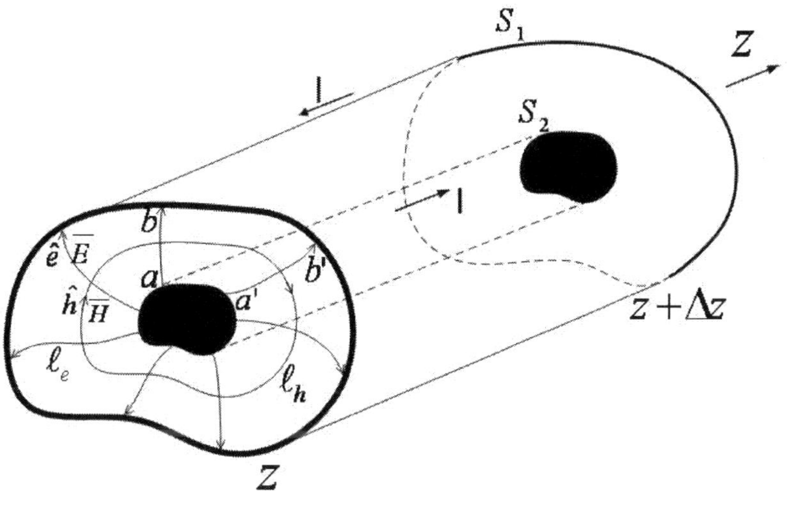

FIGURE 1

Figure 1 shows a general geometry of a transmission line. It consists of two conducting surfaces, configured in a particular form in its transverse cross section, with the transverse configuration assumed to be independent of the longitudinal or propagation direction z. The conducting surface S_1 is shown here to surround the conducting surface S_2. However, in a more general situation, one conducting surface may not enclose the other, as in the case of a two-wire transmission line.

For all rigorous analyses, we will assume the two conductors to be perfectly conducting, with an infinite conductivity. Any finite conductivity of practical conductors would be modeled only approximately. Any special treatment for such models will be discussed in relevant contexts. The material medium between the two conductors of the transmission line is ideally desired to be perfectly insulating, with its conductivity equal to zero. However, a general material medium with a non-zero conductivity can be analytically modeled in rigorous terms. Therefore, in our rigorous analytical treatments, we may allow the material medium to be in general conducting. Accordingly, the total current in the material medium may, in general, consist of displacement as well as conduction currents. The results may be simplified for the ideal case of a perfect insulating material by simply equating its conductivity to zero.

In this section we will establish certain basic properties of the electromagnetic fields of a transmission line, and derive key relationships between its electric and magnetic fields. The two conductors of the transmission line would be assumed to be perfectly conducting bodies. The analysis would be based on the electromagnetic theory developed in the last chapters, and certain assumptions for the operation mode of a transmission line. It will be followed in the next section by deriving relationships between the voltage and current signals, which are then used to develop an equivalent circuit model for simpler understanding and further analysis.

1.1. Transverse Electromagnetic (TEM) Fields. The electromagnetic effects in a general physical problem can be completely described in terms of the electric and magnetic fields, defined over all spatial coordinates of the problem. Electrical circuits constitute a special class of applications, which satisfy only a simplified version of the electromagnetic theory. For the specific problems of the electrical circuits, instead of using the electric and magnetic fields, it may be more convenient to use voltage and current parameters, defined at different circuit nodes and through different branch wirings. It would be convenient as well to model a transmission line using a voltage, properly defined across its two conductors, in relation to a current, properly defined to flow through the conductors. This could allow modeling of the transmission line using simple equivalent circuits, established based on the voltage and current concepts. However, the approach may not be always possible for a general configuration of a transmission line, where the voltage and current may not be uniquely defined. Such a situation would question the basic meaning of the voltage and current, and therefore would invalidate any subsequent use of the voltage and current in general analysis and circuit models.

In the following we will first examine the basic conditions under which the voltage and current of a transmission line can be uniquely defined. Under the required conditions, we will then establish the theory of a transmission line as a special form of the general electromagnetic theory.

1.1.1. *Voltage Definition, and Magnetic Field Vector on the Transverse Plane.* The voltage $V(z)$ on the conductor S_2 with respect to the conductor S_1, measured at a particular location z, is defined as the line integration of the electric field $\bar{E}$ across the two conductors from S_2 to S_1.

$$\int_{S_1}^{S_2} \bar{E} \cdot \bar{dl} = V \tag{4.1}$$

For a time-invariant or static signal, this voltage would be the same when computed along any arbitrary path in three dimensions, which connects an arbitrary point on the conductor S_2 to another arbitrary point on the conductor S_1. This assumes that the surfaces S_1 and S_2 are perfectly conducting, as we have indicated before. However, for a time-varying signal, we assume that the voltage is still uniquely defined when the path of the above integration is restricted to the cross-sectional plane of the the transmission line with the constant z. However, the voltage computed on two different cross-sectional planes having different z coordinates, or computed along a path which is not restricted to a constant z, may be different from each other for a general time-varying signal. The variation, when the integration path deviates from the cross-sectional plane, is attributed to additional induced voltages due to any time derivative of the magnetic field in the transverse direction. Such induced voltages would be examined in section 2.1.

Now, as per our assumption, the voltage $V(z)$ in (4.1), computed at a particular location z, is unique in the restricted sense that it is independent of of any specific path of integration on the cross-sectional plane at the given z. Accordingly, in reference to Fig.1, the voltages computed along any path from a to b should be same as that from a' to b', when the connecting paths are traced on the same cross-sectional plane.

$$\int_b^a \bar{E} \cdot \bar{dl} = \int_{b'}^{a'} \bar{E} \cdot \bar{dl} = V(z) \tag{4.2}$$

The voltage computed along a given path is negative of that when the path is reversed. Accordingly, the above condition may be equivalently expressed as follows.

$$\begin{aligned} \int_b^a \bar{E} \cdot \bar{dl} &= - \int_{b'}^{a'} \bar{E} \cdot \bar{dl} \\ \int_b^a \bar{E} \cdot \bar{dl} + \int_{b'}^{a'} \bar{E} \cdot \bar{dl} &= 0 \end{aligned} \tag{4.3}$$

It can be shown, that the above condition (4.2,4.3) would be valid if, and only if, the z-component of the magnetic field is zero ($H_z = B_z = 0$). This is accomplished by applying Faraday's law to a closed loop $l_C : abb'a'a$ (see Fig.1), under the above condition (4.2,4.3).

$$\begin{aligned} \int_{l_c} \bar{E} \cdot \bar{dl} = \int_b^a \bar{E} \cdot \bar{dl} + \int_{b'}^{b} \bar{E} \cdot \bar{dl} + \int_{a'}^{b'} \bar{E} \cdot \bar{dl} \\ + \int_a^{a'} \bar{E} \cdot \bar{dl} = \frac{d\Delta\phi_z}{dt} = \frac{d}{dt} \int_{\Delta S} B_z dS \\ = \int_{b'}^{b} \bar{E} \cdot \bar{dl} + \int_a^{a'} \bar{E} \cdot \bar{dl} \\ \Delta\phi_z = \int_{\Delta S} B_z dS = \int_{\Delta S} \mu H_z dS \end{aligned} \tag{4.4}$$

where ΔS is the surface enclosed by the closed loop l_c. Now, the tangential component of the electric field on the perfectly conducting surfaces S_1 and S_2 is zero. Accordingly, the voltage computed between two points a and a' on the surface S_2, or similarly between b and b' on S_1, along an integration path which is tangential to the respective conducting body, would be zero.

$$\begin{aligned} \textstyle\int_b^{b'} \bar{E} \cdot \bar{dl} = \int_b^{b'} E_{tan} dl = 0 \\ \textstyle\int_{a'}^{a} \bar{E} \cdot \bar{dl} = \int_{a'}^{a} E_{tan} dl = 0 \end{aligned} \tag{4.5}$$

Combining the equations (4.4,4.5) the required condition for a general, time-varying field is to have the magnetic flux $\Delta\phi_z$ enclosed inside the loop l_c to be zero. As per our earlier discussion, for a fixed set of four points a, b, b', a' in Fig.1, the loop l_c is allowed to take many arbitrary shapes on a given cross-sectional plane, depending on any specific trace of the connecting paths between a and b and between a' and b', and accordingly the area ΔS can have many possible values. The magnetic flux $\Delta\phi_z$ in (4.4) for all these arbitrary possibilities would be zero, if, and only if, the magnetic field H_z or flux density B_z is zero everywhere.

$$\Delta\phi_z = \int_{\Delta S} B_z dS = \int_{\Delta S} \mu H_z dS = 0; \; B_z = H_z = 0 \tag{4.6}$$

The above condition of having $H_z = B_z = 0$ means that the magnetic field vector for the propagation mode in a transmission line is directed only along the cross-sectional or the transverse plane.

1.1.2. *Current Definition, and Electric Field Vector on the Transverse Plane.* At a given location z, the total current $I(z)$ of a transmission line flows in the center conductor S_2 in the z direction. A current of equal magnitude $I(z)$ flows in the other conductor S_1, but in the opposite direction $-z$. This ensures continuity between the forward and reverse currents at a given plane z, which enforces charge conservation. It is assumed that the z directed currents are confined only to the body of the two conductors. This means, there is no z directed current flow anywhere in the filling material medium. Considering a general definition of current, this includes both conduction and displacement currents. Accordingly, any z directed current density J_{cz}, and z directed displacement current density J_{dz}, is zero everywhere in the material medium.

The above assumption could mean that the total current $I(z)$ of the transmission line, at a given location z, can be defined uniquely or precisely. This is in the sense that there is no ambiguity as to if any part of the z directed current, conducting or displacement type, in the filling material medium may have to be considered as a part of the current $I(z)$. When one specifies the z directed current $I(z)$, it can only refer to the z directed conduction current in the conductor S_2 or the negative of that in the conductor S_1.

We know that the conduction current density J_{cz} is equal to product of material conductivity σ and the z directed electric field E_z, and the displacement current density J_{dz} is the time derivative of the product of the dielectric permittivity ϵ and the z directed electric field E_z. As per the above assumption, both the currents can be zero for a general time-varying signal, and accordingly the current $I(z)$ would be unique, if, and only if, the z directed electric field E_z is zero everywhere in the material medium.

$$\begin{aligned} J_{cz} = \sigma E_z = 0, \; J_{dz} = \frac{\partial D_z}{\partial t} = \frac{\partial(\epsilon E_z)}{\partial t} = 0 \\ D_z = E_z = 0, \; \frac{\partial}{\partial t} \neq 0 \end{aligned} \tag{4.7}$$

The above condition of having $D_z = E_z = 0$ means that the electric field vector for the signal in a transmission line is directed only along the cross-sectional or the transverse plane.

The related condition of having no z directed current in the material medium may also be represented in terms of an equivalent condition for the magnetic field. This is possible through application of Ampere's law.

$$\oint_{l_c} \bar{H} \cdot \bar{dl} = I(z) \tag{4.8}$$

where l_c is any closed path on a particular cross-sectional plane with a given z, defined around the current $I(z)$ on the center conductor S_2. Because there is no z directed current, either conduction or displacement, in the material medium, the total current enclosed by the closed loop l_c is entirely contributed due to the current on the conductor S_2. Therefore, the total current in the loop l_c would be equal to a fixed value of $I(z)$, and would be independent of any specific form of the loop l_c.

Accordingly, the above magnetic field integral (4.8) is unique in the sense that it is independent of any specific shape or size of the closed path l_c, as long as it is confined on a particular cross-sectional plane with the given z. However, the loop integral (4.8) computed at two different cross-sectional planes with different values of z, or computed along a path l_c which is not confined strictly on a given cross-sectional plane, could produce different values. That is because, there is current flow in the material medium in the transverse direction, and the current $I(z)$ varies for a different value of z. Either of these possibilities would change the current value in (4.8) for a loop configuration which is deviated from a cross-sectional plane, thus changing the loop integral of the magnetic field. Such variation would be addressed in section 2.2.

To summarize, (1) we assume that there is no z directed current flow in the material medium. (2) The assumption leads to a unambiguous specification of the current $I(z)$ of the transmission line. (3) The above assumption is valid for a general time-varying signal, if, and only if, the z component of the electric field is zero everywhere in the material medium. This means, the electric field vector is directed along a transverse plane. (4) The condition is also equivalent to having a unique value for a loop integral of the magnetic field around the conductor S_2. This is in a limited sense that the loop integral is independent of any specific loop path on a particular cross-sectional plane, but could be different if the loop path deviates from the specific cross-sectional plane.

We have now established in the above sections (1.1.1,1.1.2) that both the electric and the magnetic field vectors for an ideal transmission line are directed only along a cross-sectional or the transverse plane, having no component along the propagation direction. Accordingly, the field in such a transmission line is a restricted form of an electromagnetic field, often referred to as the transverse electromagnetic or the TEM field. The TEM condition is shown to be valid if, and only if, the voltage $V(z)$ and current $I(z)$ are uniquely defined at a particular cross-sectional plane z.

Under this basic condition of a TEM field, we will establish additional space-time relationships between the electric and the magnetic fields in the following sections (1.2,1.3), and alternatively between voltage and current functions in section 2 using circuit concepts.

1.2. Orthogonal Directions for a TEM Field. Let the electric and magnetic field at a given location be defined by unit vectors $\hat{e}$ and $\hat{h}$, respectively. Now, apply Ampere's law to a closed loop Δl_c on the $e - z$ plane (Fig.2). Δl_c in this case is parallel to the $\hat{e}$ and $\hat{z}$ vectors.

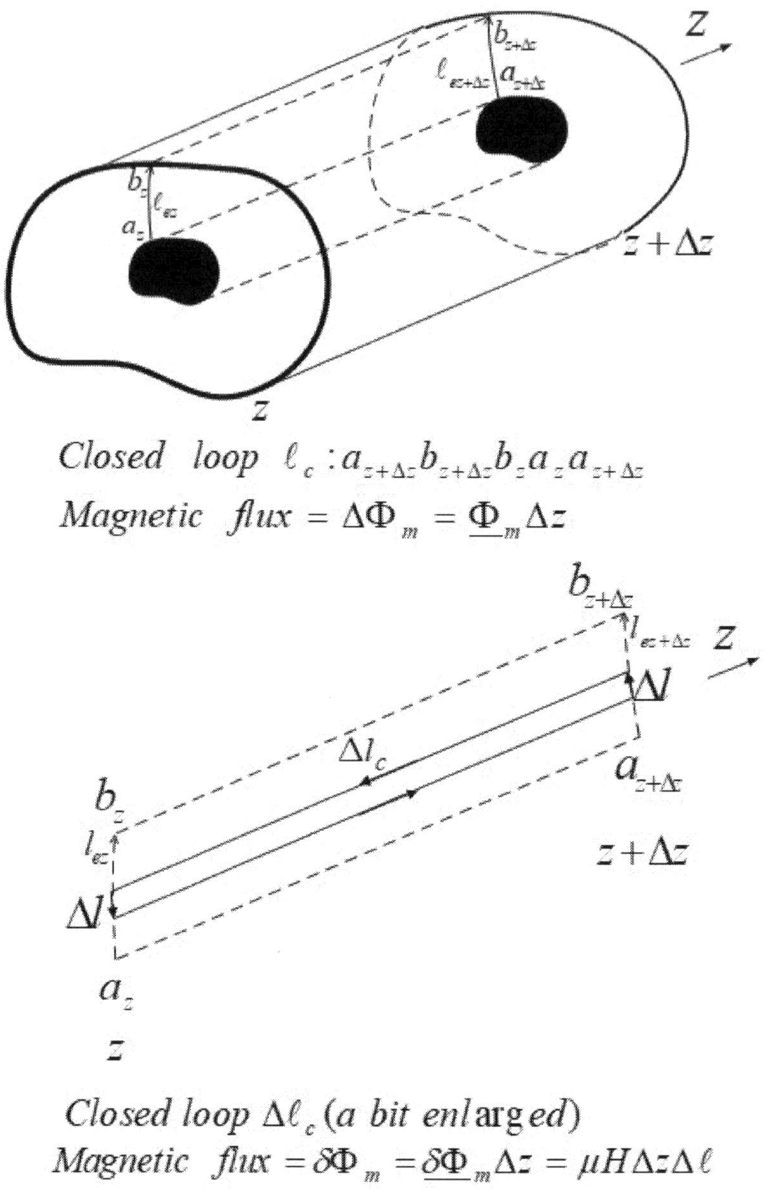

FIGURE 2

$$\oint_{\Delta l_c} \bar{H} \cdot d\bar{l} = \Delta I_c + \Delta I_d = \int_{\Delta S} \sigma E_n dS + \int_{\Delta S} \frac{d}{dt}(D_n dS) = 0 \tag{4.9}$$

Because Δl_c is parallel to the electric field $\bar{E}$, any electric field E_n normal to Δl_c is zero. Consequently, any normal component of the conduction current density $J_c = \sigma E_n$ or electric flux density $D_n = \epsilon E_n$ is zero. Accordingly, we have,

$$\oint_{\Delta l_c} \bar{H} \cdot \bar{dl} = 0 = \sum_{\Delta l_c} H_{ti} \Delta l_i \tag{4.10}$$

The tangential magnetic field H_z for the top and bottom sides of Δl_c in Fig.2 are zero, as per the TEM condition. Therefore only the left and right sides of Δl_c in Fig.2 might contribute to the above loop integration.

$$[-H_e(z) + H_e(z + \Delta z)]\Delta l = 0 = \frac{\partial H_e}{\partial z} \Delta z \Delta l \tag{4.11}$$

The subscript e refers to component along the electric field unit vector $\hat{e}$. We expect that any existing fields will have to have a non-zero variation or derivative in the z direction. This is consistent with having a propagation along the z direction. Applying this condition, that is $\frac{\partial}{\partial z} \neq 0$, in the above equation leads to having $H_e = 0$. This means, there would be no component of magnetic field along direction of the electric field. In other words, magnetic fields would be orthogonal to the electric field every where. Accordingly, the respective unit vectors $\hat{h}$ and $\hat{e}$ would be orthogonal to each other. Due to the TEM nature of the field we also have both $\hat{e}$ and $\hat{h}$ to be directed on the transverse plane. In other words, both $\hat{e}$ and $\hat{z}$ are orthogonal to $\hat{z}$. Accordingly, $\hat{e}$, $\hat{z}$ and $\hat{z}$ constitute three mutually orthogonal directions.

We need to establish a definite relationship between the reference directions of the three orthogonal vectors. In Fig.1, assuming the voltage of the conductor S_1 to be positive with respect to the conductor S_2, the electric field unit vector $\hat{e}$ is properly shown to be directed from S_1 to S_2. Further, assuming the current direction on conductor S_1 to be along the propagation direction $+z$, we have properly shown the magnetic field unit vector $\hat{h}$ to be directed clockwise, as per the right-hand rule for Ampere's law. Accordingly, one might verify from Fig.1 the following orthogonality relationship.

$$\hat{e} \times \hat{h} = \hat{z} \tag{4.12}$$

1.3. Relationships between the Transverse Fields. We have established the TEM nature of the fields of a transmission line. We have also established the orthogonality between the electric and magnetic field directions. Based on these basic conditions, we can now derive relationships between the electric and magnetic fields. Let us apply Faraday's law to the loop Δl_c on the $e - z$ plane (Fig.2).

$$\begin{aligned} \oint_{\Delta l_c} \bar{E} \cdot \bar{dl} &= (-E(z) + E(z + \Delta z))\Delta l \\ &= \frac{\partial E}{\partial z} \Delta z \Delta l = -\frac{\partial}{\partial t}(\mu H \Delta z \Delta l) \\ \frac{\partial E}{\partial z} &= -\mu \frac{\partial H}{\partial t} \end{aligned} \tag{4.13}$$

The z component of the electric field we know is zero, as per the TEM nature of the transmission line. Therefore, the contribution of the top and bottom segments of loop Δl_c (see Fig.2) to the total line integral in (4.13) are zero.

The loop Δl_c in Fig.2 is on the $e-z$ plane, which is parallel to the electric field direction. The magnetic field is normal to the electric field, as we have established earlier, and is therefore normal to the surface of the loop Δl_c.

Further, as seen from Fig.2, the magnetic field is directed positively with respect to the orientation of the loop Δl_c as per the right-hand rule. As per the above reasons, the magnetic flux through the loop Δl_c is expressed in (4.13) by multiplying the magnitude of the magnetic field H to the area element and the magnetic permeability.

The above equation (4.13) relates the time derivative of the magnetic field to the space derivative of the electric field. A similar equation, but relating the time derivative of the electric field to the space derivative of the magnetic field, may also be derived by applying Ampere's law to a loop Δl_c on the $h-z$ plane, shown in Fig.3.

$$\begin{aligned}\oint_{\Delta l_c} \bar{H}\cdot d\bar{l} &= (H(z)-H(z+\Delta z))\Delta l = -\frac{\partial H}{\partial z}\Delta z\Delta l \\ &= \frac{\partial(\epsilon E\Delta z\Delta l)}{\partial t} + J_c\Delta z\Delta l = \epsilon\frac{\partial E}{\partial t}\Delta z\Delta l + \sigma E\Delta z\Delta l \\ \frac{\partial H}{\partial z} &= -\epsilon\frac{\partial E}{\partial t} - \sigma E \end{aligned} \tag{4.14}$$

Similar arguments used in the derivation of (4.13, applying Faraday's law, is also used here in the derivation of (4.14), applying Ampere's law. The z component of the magnetic field we know is zero, as per the TEM nature of the transmission line. Therefore, the contribution of the top and bottom segments of loop Δl_c (Fig.3) to the total line integral in (4.14) are zero.

The loop Δl_c in Fig.3 is on the $h-z$ plane, which is parallel to the magnetic field direction. The electric field is normal to the magnetic field, as we have established earlier, and is therefore normal to the surface of the loop Δl_p. Any conduction current is directed along the electric field, and, therefore, is also normal to the surface of the loop Δl_p. The conduction current density $\overline{J}_c$ is proportional to the electric field $\overline{E}$, with the proportionality constant equal to the conductivity, σ, of the medium ((1.16, chapter 1).

Further, as seen from Fig.3, the electric field, and accordingly the conduction current, is directed positively with respect to the orientation of the loop Δl_p, as per the right-hand rule. As per the above reasons, the electric flux through the closed loop Δl_c is expressed in the above equation by multiplying the magnitude of the electric field E to the area element and the dielectric permittivity. And, similarly, the conduction current ΔI_c through the loop Δl_c is expressed in (4.14) by multiplying the magnitude of the current density $J_c = \sigma E$ to the area element.

When the material medium is a perfect insulator, the conductivity σ is zero. In this case there is no conduction current flowing through the material medium. Therefore, there is no power dissipation in the medium. Accordingly, the the perfectly insulating material is

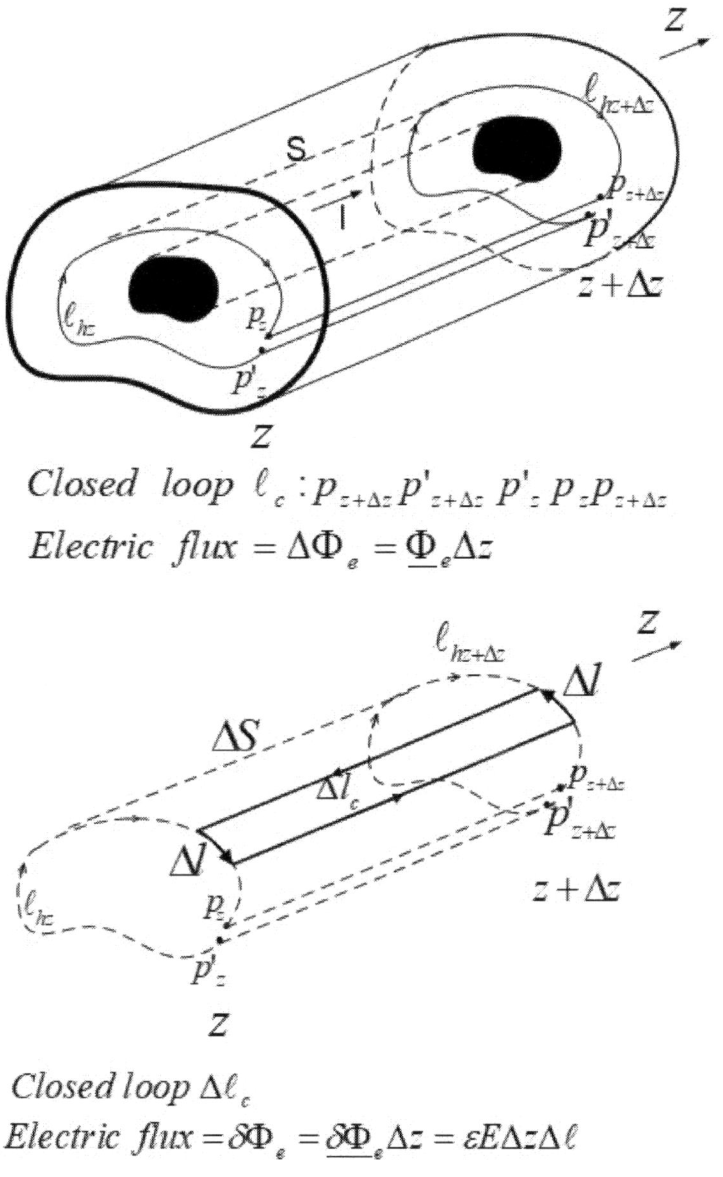

FIGURE 3

sometimes referred to as a loss-less material. Under the condition of a loss-less material, (4.14) may be expressed in a simpler form.

$$\frac{\partial H}{\partial z} = -\epsilon \frac{\partial E}{\partial t}, \ \sigma = 0 \tag{4.15}$$

2. Voltage-Current Relations

In the last section 1, we have understood the TEM nature of the fields of a transmission line, and developed relations (4.13,4.14) between the electric and magnetic fields under the TEM condition. We will now establish relationships between signal voltage and current, which are equivalent to (4.13,4.14) between the electric and magnetic fields. Similar principles used in the derivation of the field relationships (4.13,4.14), would also be used here for the derivation of the voltage-current relationships. We are now seeking integral versions of the field equations (4.13,4.14) along electric and magnetic field lines, which would lead to voltage and current expressions. We will apply Faraday's and Ampere's laws to the larger loops l_c in the Fig.2 and Fig.3, respectively. In the course of the derivation, we are particularly interested to see different parameters and relationships in terms of equivalent circuit components, such as capacitance and inductance.

2.1. Faraday's Law and Inductance per Unit Length. First, apply Faraday's law to the loop l_c, defined by interconnecting the points $a_{z+\Delta z}b_{z+\Delta z}b_z a_z a_{z+\Delta z}$ in (Fig.2). The path segments $b_{z+\Delta z}b_z$ and $a_z a_{z+\Delta z}$ are tangential to the perfectly conducting surfaces S_1 and S_2, respectively. Therefore, the tangential component of the electric field along these paths are zero, Accordingly, the contribution of the line integral of the electric field along these two segments would be zero.

$$
\begin{aligned}
\oint_{l_c} \overline{E}\cdot\overline{dl} &= \int_{a_{z+\Delta z}}^{b_{z+\Delta z}} + \int_{b_{z+\Delta z}}^{b_z} + \int_{b_z}^{a_z} + \int_{a_z}^{a_{z+\Delta z}} \overline{E}\cdot\overline{dl} \\
= \int_{a_{z+\Delta z}}^{b_{z+\Delta z}} \overline{E}\cdot\overline{dl} + \int_{b_z}^{a_z} \overline{E}\cdot\overline{dl} &= \int_{a_{z+\Delta z}}^{b_{z+\Delta z}} \overline{E}\cdot\overline{dl} - \int_{a_z}^{b_z} \overline{E}\cdot\overline{dl} \\
&= \int_{l_{ez+\Delta z}} \overline{E}\cdot\overline{dl} - \int_{l_{ez}} \overline{E}\cdot\overline{dl} \\
= V(z+\Delta z) - V(z) &= -\frac{\partial}{\partial t}\int_{\Delta S} \mu H \Delta z \Delta l = -\frac{\partial}{\partial t}\Delta\Phi_m \\
\Delta\Phi_m &= \int_{\Delta S} \mu H dl dz = \Delta z \int_{l_e} \mu H dl = \Delta z \underline{\Phi}_m
\end{aligned}
\tag{4.16}
$$

where, ΔS is the surface enclosed by the loop l_c. The loop l_c is on the $e-z$ plane, which means it is parallel to the electric field vector. Accordingly, the magnetic field is normal to the surface of the loop l_c, as per the orthogonality of the electric and magnetic fields for a TEM transmission line mode. Also, as may be verified from Fig.2, the reference direction of the magnetic field is oriented positively with respect to the loop l_c as per the right-hand rule. Accordingly, the magnetic flux in the loop is obtained by integrating the product of magnetic field magnitude H and the permeability μ over the surface ΔS.

$\Delta\Phi_m$ is the magnetic flux over the area ΔS, and $\underline{\Phi}_m$ is the magnetic flux over unit length $\Delta_z = 1$. The time derivative of this flux is responsible for the induced voltage element $\Delta V = V(z+\Delta z) - V(z)$ over a length Δz of the transmission line. Also, as seen from the above equation, $\Delta\Phi_m$ and $\underline{\Phi}_m$ are proportional to the magnetic field H, which is proportional to the current I. Following the basic circuit principle of an inductor, the transmission line segment of length Δz may be viewed as an inductor element. The inductance ΔL can be appropriately defined as the ratio of $\Delta\Phi_m$ to the current I.

$$\begin{aligned}\Delta L = \frac{\Delta\Phi_m}{I} = \frac{\int_{\Delta S}\mu H ds}{I} = \Delta z\frac{\Phi_m}{I} = \Delta z\frac{\int_{l_e}\mu H dl}{I}\\ \underline{L} = \frac{\Phi_m}{I} = \frac{\int_{l_e}\mu H dl}{I} = \frac{\int_{l_e}\mu H dl}{\int_{l_h} H dl}; \ \Delta L = \underline{L}\Delta z\end{aligned} \tag{4.17}$$

ΔL is the inductance of the transmission line of length Δz, and $\underline{L}$ is the inductance per unit length ($\Delta z = 1$). Now, combining the equations (4.16,4.17), we get the following relationship between the voltage and current signals.

$$\begin{aligned}V(z+\Delta z) - V(z) = -\frac{\partial}{\partial t}\Delta\Phi_m = -\Delta L\frac{\partial I}{\partial t}\\ \frac{\partial V}{\partial z}\Delta z = -\Delta L\frac{\partial I}{\partial t} = -\Delta z\underline{L}\frac{\partial I}{\partial t}; \ \frac{\partial V}{\partial z} = -\underline{L}\frac{\partial I}{\partial t}\end{aligned} \tag{4.18}$$

The above relationship (4.18) between the current and voltage may be recognized to be in similar form as the relationship between the magnetic and electric field in (4.13). The electric and magnetic fields in (4.13) are substituted in (4.18) by voltage and current, respectively, while the magnetic permeability μ in (4.13) is substituted by inductance $\underline{L}$ in (4.18).

2.2. Ampere's Law, and Capacitance and Conductance per Unit Length. Now, apply Ampere's law to the large closed loop l_c : $p_{z+\Delta z}p'_{z+\Delta z}p'_z p_z p_{z+\Delta z}$ in (Fig.3). The surface ΔS enclosing the loop l_c is on the $h - z$ plane. We will consider a limiting condition when the primed points approach the unprimed points. Under this condition, the integrations along the path segments $p'_{z+\Delta z}p'_z$ and $p_z p_{z+\Delta z}$ are negative of each other and would cancel to zero. The path segments $p_{z+\Delta z}p'_{z+\Delta z}$ and $p'_z p_z$ approach to be closed loops, aligned with the magnetic field lines at $z+\Delta z$ and z coordinates, respectively. These closed loops are represented as $l_{hz+\Delta z}$ and l_{hz} respectively. The surface ΔS approach to be a continuous cylindrical surface of length Δz, with its bases represented by the two closed magnetic field lines $l_{hz+\Delta z}$ and l_{hz}.

$$\oint_{l_c} \overline{H} \cdot \overline{dl} = \int_{p_{z+\Delta z}}^{p'_{z+\Delta z}} + \int_{p'_{z+\Delta z}}^{p'_z} + \int_{p'_z}^{p_z} + \int_{p_z}^{p_{z+\Delta z}} \overline{H} \cdot \overline{dl}$$

$$= \int_{p_{z+\Delta z}}^{p'_{z+\Delta z}} \overline{H} \cdot \overline{dl} + \int_{p'_z}^{p_z} \overline{H} \cdot \overline{dl} = -\int_{p'_{z+\Delta z}}^{p_{z+\Delta z}} \overline{H} \cdot \overline{dl} + \int_{p'_z}^{p_z} \overline{H} \cdot \overline{dl}$$

$$= -\oint_{l_{hz+\Delta z}} \overline{H} \cdot \overline{dl} + \oint_{l_{hz}} \overline{H} \cdot \overline{dl} = -I(z + \triangle z) + I(z)$$

$$= \frac{\partial}{\partial t} \int_{\Delta S} \epsilon E ds + \int_{\Delta S} \sigma E ds = +\frac{\partial}{\partial t} \Delta \Phi_e + \Delta I_c$$

$$\Delta \Phi_e = \int_{\Delta S} \epsilon E dl dz = \Delta z \int_{l_h} \epsilon E dl = \Delta z \underline{\Phi}_e$$

$$\Delta I_c = \int_{\Delta S} J_c dl dz = \int_{\Delta S} \sigma E dl dz = \Delta z \int_{l_h} \sigma E dl = \Delta z \underline{I}_c \tag{4.19}$$

The total current through the surface ΔS, in general, consists of displacement and conduction current. However, the conduction current would be zero when the material is loss-less (perfectly insulating). The displacement current is the time derivative of the electric flux. $\Delta \Phi_e$ is the electric flux over the area ΔS of length Δz, and $\underline{\Phi}_e$ is the electric flux over unit length $\Delta_z = 1$. ΔI_c is the conduction current over the area ΔS of length Δz, and $\underline{I}_c$ is the conduction current over unit length $\Delta_z = 1$.

The surface ΔS is on the $h - z$ plane, which is parallel to the magnetic field. The electric field is normal to the magnetic field, as we have established earlier, and is therefore normal to the surface ΔS. The conduction current density $\overline{J}_c = \sigma \overline{E}$ (1.16) is proportional to and directed along the electric field. Accordingly, like the electric field, the conduction current density is also normal to the surface ΔS of the loop Δl_p.

Further, as may be seen from the Fig.3, the electric field is directed outwardly from the surface ΔS, and is oriented positively with respect to the reference direction of the loop l_c as per the right-hand rule of Ampere's law. This would apply as well to the conduction current, which is aligned along the electric field.

As per the above reasons, the electric flux through the loop l_c is expressed as the surface integration of the electric field magnitude E multiplied by the dielectric permittivity ϵ. Similarly, the conduction current through the loop l_c is expressed as the surface integration of the magnitude of the conduction current density J_c, which is equal to the magnitude of the electric field multiplied by the conductivity σ.

Note that the path l_c could be arbitrarily selected with its enclosing surface ΔS tangential to any magnetic field line l_h around the conductor S_2. The result should be independent of such choices. Accordingly, the computation of the electric flux $\Delta \Phi_e$ or $\underline{\Phi}_e$, and electric current ΔI_c or $\underline{I}_c$ should be independent of such choices for ΔS.

2.2.1. *Charge Storage and Current Continuity.* Now, a curved cylindrical surface ΔS, together with its two side base areas that are normal to the z direction, may be referred to as a closed surface S (see Fig.3). Apply Gauss' law to this closed surface S. The normal electric flux over the two side bases would be zero, because the normal (z) component of the electric field on these surfaces we know is zero. This is due to the TEM condition we have established before. Therefore, the total electric flux in S would be equal to the electric flux $\Delta\Phi_e$ only over the curved surface ΔS. This flux is independent of the choices for the ΔS around the conductor S_2, as discussed earlier. Applying Gauss' law, this flux $\Delta\Phi_e$ would be equal to the total charge ΔQ inside S, which would be the charge located on the conductor S_2 of length Δz. By charge conservation, negative of this charge would be located on the conductor S_1 over a same length Δz. $\underline{\Phi_e}$ is the flux per unit length, and would be equal to the line charge density ρ_l, which is the charge per unit length of the line ($\Delta z = 1$).

The accumulation of the above charge ΔQ on the conductor may be interpreted in terms of current or charge continuity. As seen from the above equation (4.19), the current $I(z + \Delta z)$ at location $(z + \Delta z)$ is expected to be different from current $I(z)$ at location z. Applying current continuity, the difference is partially accounted for by the conduction current ΔI_c flowing across the material medium, and partially appears as a time-derivative of charge ΔQ accumulated on the conductor, over the length Δz.

$$-I(z + \triangle z) + I(z) = \frac{\partial}{\partial t}\Delta\Phi_e + \Delta I_c = \frac{\partial}{\partial t}\Delta Q + \Delta I_c \tag{4.20}$$

2.2.2. *Capacitance Per Unit Length.* As seen from (4.19), the flux or charge amount $\Delta\Phi_e = \Delta Q$, or that per unit length $\underline{\Phi_e} = \rho_l$, are proportional to the electric field, which would be proportional to the voltage V across the conductors. Following the basic circuit principle of a capacitor, the two conductors of length Δz, containing the positive and negative charges $\pm\Delta Q$, may be viewed as a capacitor element for temporary charge storage. The capacitance ΔC is defined as the ratio of ΔQ to the voltage V.

$$\Delta C = \frac{\Delta\Phi_e}{V} = \frac{\Delta Q}{V} = \frac{\int_{\Delta S} \epsilon E dl dz}{V} = \Delta z \frac{\int_{l_h} \epsilon E dl}{V} = \Delta z \frac{\underline{\Phi_e}}{V} = \Delta z \frac{\rho_l}{V}$$

$$\underline{C} = \frac{\underline{\Phi_e}}{V} = \frac{\rho_l}{V} = \frac{\int_{l_h} \epsilon E dl}{V} = \frac{\rho_l}{\int_{l_e} E dl} = \frac{\int_{l_h} \epsilon E dl}{\int_{l_e} E dl}; \; \Delta C = C\Delta z \tag{4.21}$$

ΔC is the capacitance of a transmission line segment of length Δz. $\underline{C}$ is the capacitance per unit length ($\Delta z = 1$).

2.2.3. *Conductance Per Unit Length.* As seen from (4.19), the conduction current amount ΔI_c, or that per unit length $\underline{I_c} = \rho_l$, are proportional to the electric field, which would be proportional to the voltage V across the conductors. The conduction current ΔI_c the across transmission line, over length Δz, may be modeled by an equivalent conductance ΔG, defined as the ratio of ΔI_c to the voltage V. ΔG may be viewed as the conductance (inverse of resistance) of a section of material of length Δz, which is measured across the two conductors of the transmission line.

$$\Delta G = \frac{\Delta I_c}{V} = \frac{\int_{\Delta S} \sigma E dl dz}{V} = \Delta z \frac{\int_{l_h} \sigma E dl}{V} = \Delta z \frac{\underline{I_c}}{V}$$

$$\underline{G} = \frac{\underline{I_c}}{V} = \frac{\int_{l_h} \epsilon E dl}{V} = \frac{\int_{l_h} \sigma E dl}{\int_{l_e} E dl}; \; \Delta G = \underline{G}\Delta z \tag{4.22}$$

ΔG is the conductance of a transmission line segment of length Δz. $\underline{G}$ is the conductance per unit length ($\Delta z = 1$).

2.2.4. *Relating Capacitance and Conductance Per Unit Length.* It may be observed, the expression for the capacitance per unit length $\underline{C}$ in (4.21) and that for the conductance per unit length $\underline{G}$ in (4.22) look very similar. The expression for $\underline{G}$ may be obtained by replacing ϵ in the expression for $\underline{C}$ with σ. This means the derivation of $\underline{G}$ and $\underline{C}$ may share the same analytical procedure. When the filling material medium is uniform, the material parameters σ and ϵ are constants that may be taken out of the integrations in the expressions for $\underline{G}$ and $\underline{C}$. Accordingly, in this case the ratio of $\underline{G}$ and $\underline{C}$ can be shown to be equal to the ratio of σ and ϵ.

$$\underline{G} = \frac{\int_{l_h} \sigma E dl}{\int_{l_e} E dl}, \quad \underline{C} = \frac{\int_{l_h} \epsilon E dl}{\int_{l_e} E dl}, \quad \underline{G} = \frac{\sigma}{\epsilon}\underline{C} \tag{4.23}$$

Now, combining the equations (4.19, 4.21,4.22), we get the following relationship between the voltage and current signals.

$$\begin{aligned} I(z+\Delta z) - I(z) &= -\frac{\partial}{\partial t}\Delta Q - \triangle I_c = -\triangle C\frac{\partial V}{\partial t} - \Delta G V \\ \frac{\partial I}{\partial z}\Delta z = -\triangle C\frac{\partial V}{\partial t} - \Delta G V &= -\triangle z\underline{C}\frac{\partial V}{\partial t} - \Delta z\underline{G}V \\ \frac{\partial I}{\partial z} &= -\underline{C}\frac{\partial V}{\partial t} - \underline{G}V \end{aligned} \tag{4.24}$$

The above relationship (4.24) between the current and voltage may be recognized to be in similar form as the relationship between the magnetic and electric field in (4.14). The electric and magnetic fields in (4.14) are substituted in (4.24) by voltage and current, respectively, while the dielectric permittivity ϵ and conductivity σ in (4.14) are respectively substituted by capacitance $\underline{C}$ and conductance $\underline{G}$ in (4.24).

When the material medium is loss-less (perfect insulator), the conductivity σ is zero. Accordingly, the conductance per unit length $\underline{G}$ in (4.24) would also be zero. Under this condition, (4.24) may be expressed in a simpler form.

$$\frac{\partial I}{\partial z} = -\underline{C}\frac{\partial V}{\partial t}, \quad \underline{G} = 0 \tag{4.25}$$

3. Distributed Circuit Model

Figure 4 shows a generic representation of a two-conductor transmission line. The two parallel wires in the figure, running along the z direction, represent the two conducting bodies of the transmission line. As a standard practice, each wire of the transmission line is shown as a "double line." This is to distinguish from a single-line connection we will use elsewhere as a standard representation for a direct electrical contact between two points. Accordingly, a double-line connection between two points would mean that there is a non-zero physical distance between the two connecting points. Whereas, a single-line connection simply refers to a direct electrical contact between two points, with zero physical distance between the points.

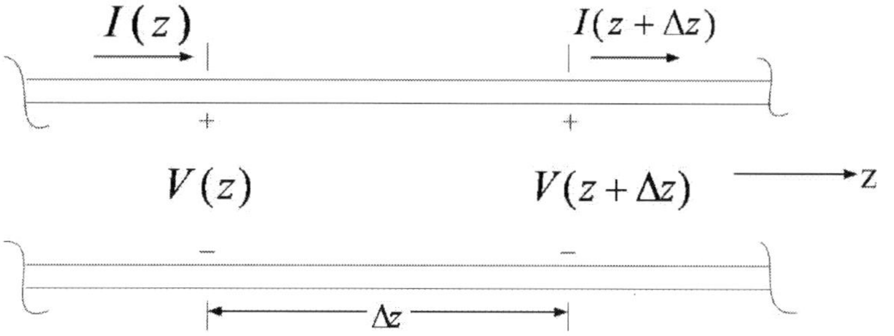

FIGURE 4

The voltage V across the two conductors is shown with a reference positive voltage for the upper wire as measured with respect to the lower wire. The current I along the transmission line refers to that in the upper wire, and the current in the lower wire is understood to be negative of that in the upper wire.

For a time-invariant signal, the voltage V across the two conductors of a transmission line would be the same for all locations z, except when the conductors have non-zero resistance (or infinite conductivity). Any variation in the voltage with z is attributed to voltage drop due to a non-zero resistance of the wire. Similarly, for a time-invariant signal, the current I along each conductor of a transmission line would be the same for all locations z, except when the filling material medium has a finite resistance (or zero conductivity). Any variation in the current with z is attributed to current leaking across the filling material due to its finite resistance (zero conductivity).

However, for a general time-varying signal, the situation is different. In this case, the voltage and current can vary with location z, even when the two conductors have zero resistance, or when the filling material medium has infinite resistance. Variation in the voltage and current at different locations can occur respectively due to induced voltage as per Faraday's law and due to capacitive or displacement current across the two conductors as per Ampere's law. These effects are analyzed in the sections (2.1,2.2). The variations can be modeled in terms of the inductance $\underline{L}$ in (4.17) and capacitance in (4.21), which are distributed along the transmission line over each unit segment of the line. We would like to model different possible causes for the variation of voltage and current along the transmission line through an equivalent circuit.

3.1. Circuit Model for a Loss-Less Transmission Line. A loss-less transmission line refers to the case where the conductors have zero resistance and the filling material medium has infinite resistance. The name loss-less means that there is no loss or dissipation of power in the zero resistance of the conductors or in the infinite resistance of the filling material medium.

The voltage-current relations derived in (4.18,4.24) are based on the ideal assumption that the two conductors of the transmission line are perfectly conducting, or loss-less. The TEM condition established in section 1.1, upon which (4.18,4.24), as well as all field relations we have derived so far, are based, is conditional upon this assumption. Any loss in the filling material medium between the conductors does not affect the voltage-current relation in (4.18) However, the filling material medium in the voltage-current relations of (4.24) is considered to be in general lossy, having a non-zero conductivity or finite resistance. The corresponding relation for the specific case when the material medium has zero conductivity, or equivalently zero conductance $\underline{G}$, is given in (4.25). Accordingly, we would refer to the voltage-current relations of (4.18,4.25) in order to establish a circuit model for a loss-less line.

A circuit model for a loss-less transmission line is shown in Fig.5, using the inductance $\underline{L}$ from (4.17) and capacitance $\underline{C}$ from (4.21). Δz is an elemental segment of the transmission line, which approaches zero. The circuit model would have to be consistent with the two voltage-current relations (4.18,4.25) for the loss-less case, as shown in the following.

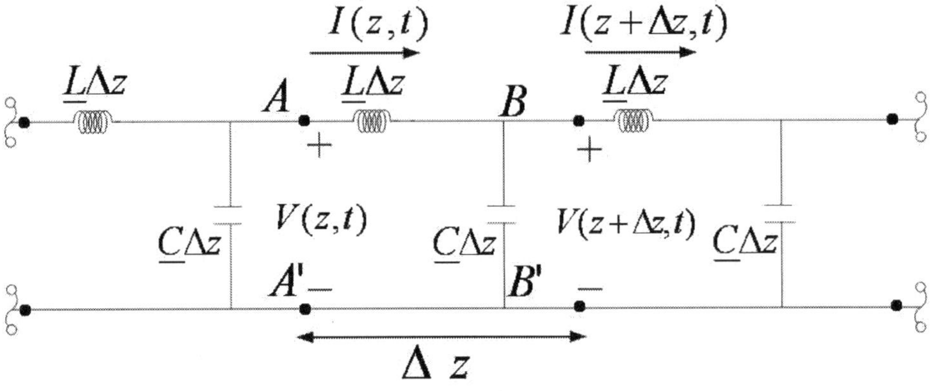

FIGURE 5

The voltage-current relation of (4.18) may be verified by applying Kirchoff's Voltage Law (KVL) to the circuit loop ($BB'A'A$) in Fig.5. Sum of all voltages around the loop should be zero.

$$-V(z+\triangle z,t)+V(z,t)-\underline{L}\triangle z\frac{\partial}{\partial t}I(z,t)=0$$

$$\lim_{\triangle z\to 0}\frac{V(z+\triangle z,t)-V(z,t)}{\triangle z}=\frac{\partial}{\partial z}V(z,t)=-\underline{L}\frac{\partial}{\partial t}I(z,t) \tag{4.26}$$

The equivalent inductance $\underline{L}$ is the self inductance of a segment of the transmission line of unit length. Voltage is induced due to the time-varying magnetic field produced by the current along the transmission line, as per Faraday's law. The induced voltage, and equivalently $\underline{L}$ is larger when the magnetic flux between the two conductors over an unit length $\Delta z = 1$ is larger for a given current I.

The other voltage-current relation of (4.25) may be verified by applying Kirchoff's Current Law (KCL) to the circuit node B in Fig.5. Sum of all currents joining the node B should be zero.

$$\begin{aligned}
-I(z+\triangle z,t) + I(z,t) - \underline{C}\triangle z\frac{\partial}{\partial t}V(z+\triangle z,t) &= 0 \\
\lim_{\triangle z\to 0}\frac{I(z+\triangle z,t) - I(z,t)}{\triangle z} &= -\underline{C}\lim_{\triangle z\to 0}\frac{\partial}{\partial t}V(z+\triangle z,t) \\
\frac{\partial}{\partial z}I(z,t) &= -\underline{C}\frac{\partial}{\partial t}V(z,t) \qquad (4.27)
\end{aligned}$$

A segment of a transmission line of length Δz, consisting of two parallel conducting surfaces S_1 and S_2, each of length Δz and certain transversal shape and dimension (see Fig.1), behaves as a capacitor element. The capacitive current across the conductors, over the length Δz, may be seen in terms of charge stored in the plates of the elemental capacitor. The voltage $V(z,t)$ across the conductors produces stored charge of equal and opposite sign in the two conductors, which are proportional to the voltage. Accordingly, the two conductors of the transmission line may be seen as charged plates. The charge over each segment Δz would change in time for a time-varying voltage $V(z,t)$. The time derivative of the charge on the conductors over length Δz would account for the variation of the current from location z to $z + \Delta z$, as per the requirement of charge conservation.

The values of the equivalent $\underline{L}$ and $\underline{C}$, for any particular transmission line can be derived using their general formulas of (4.17,4.21). This requires the field analysis of the transmission line. The values of $\underline{L}$ and $\underline{C}$ would depend on the physical dimensions and material parameters of the transmission line. Formulas of $\underline{L}$ and $\underline{C}$ for specific transmission line geometries are derived in the section 10.

3.2. Circuit Model for a Lossy Transmission Line. In contrast to a loss-less transmission line modeled in the above section, a lossy transmission line refers to the case where the conductors have non-zero resistance or finite conductivity, and the filling material medium has finite resistance or non-zero conductivity. There is loss or dissipation of signal power in the non-zero resistance of the conductors as well as in the finite resistance of the filling material medium. The voltage-current relations in (4.18,4.25) consider the material medium to be in general lossy with non-zero conductivity, but assume that the two conductors of the transmission line to be perfectly conducting, having zero resistance. We will address how to model the loss in the two conductors, and establish a circuit model which would accomodate the loss in the conductors as well as in the filling material medium.

A circuit model for a lossy transmission line is shown in Fig.6. This is essentially a modification of the model of Fig.6, by including series resistance $\underline{R}$ and parallel conductance $\underline{G}$ distributed over unit length. The series resistance $\underline{R}$ in the model is expected to account for a finite conductivity of the conductors, and the parallel conductance $\underline{G}$ is expected to account for the non-zero conductivity of the filling material medium across the conductors. Clearly, the resistance $\underline{R}$ is zero when the conductors are perfect conductors, and the conductance $\underline{G}$ is zero, or the equivalent resistance is infinity, when the filling material is a perfect insulator. Note that the parallel resistance elements are represented in terms of their equivalent conductances. This notation distinguishes the parallel resistances from the series resistances. Also, as per basic circuit analysis, it is usually convenient to model the parallel resistive elements using their equivalent conductances. This is in contrast to the series resistive elements, which are conveniently modeled using their equivalent resistances.

Two independent voltage-current relationships for a lossy transmission line can be obtained by applying Kirchoff's Voltage Law (KVL) and Kirchoff's Current Law (KCL) to the equivalent circuit of Fig.6. These derivations would essentially be modified versions of the derivations in (4.26,4.27) for a loss-less transmission line, by including the resistance $\underline{R}$ and conductance $\underline{G}$

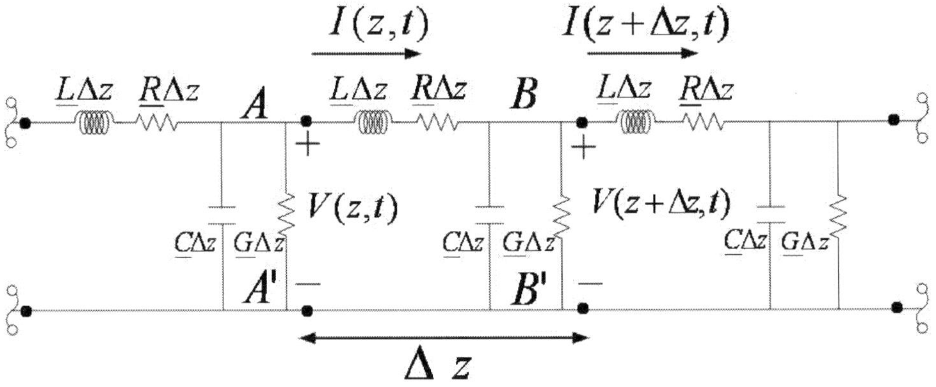

FIGURE 6

Apply the KVL to the circuit loop $(BB'A'A)$ in Fig.6. Sum of all voltages around the loop should be zero.

$$-V(z+\triangle z,t)+V(z,t)-\underline{L}\triangle z\frac{\partial}{\partial t}I(z,t)-\underline{R}\triangle z I(z,t)=0$$

$$\lim_{\triangle z\to 0}\frac{V(z+\triangle z,t)-V(z,t)}{\triangle z}=\frac{\partial}{\partial z}V(z,t)=-\underline{L}\frac{\partial}{\partial t}I(z,t)-\underline{R}I(z,t) \qquad (4.28)$$

This equation may be verified to be consistent with the voltage-current relation of (4.18), when $\underline{R}=0$. The series resistance $\underline{R}$ does not appear in (4.18), because the derivation of

(4.18) assumes perfect conductors, having zero resistance. We have discussed this issue before.

Now, apply the KCL to the circuit node B in Fig.6. Sum of all currents joining the node B should be zero.

$$-I(z+\triangle z,t)+I(z,t)-\underline{C}\triangle z\frac{\partial}{\partial t}V(z+\triangle z,t)-\underline{G}V(z+\triangle z,t)=0$$

$$\lim_{\triangle z\to 0}\frac{I(z+\triangle z,t)-I(z,t)}{\triangle z}=-\underline{C}\lim_{\triangle z\to 0}\frac{\partial}{\partial t}V(z+\triangle z,t)-\underline{G}\lim_{\triangle z\to 0}V(z+\triangle z,t)$$

$$\frac{\partial}{\partial z}I(z,t)=-\underline{C}\frac{\partial}{\partial t}V(z,t)-\underline{G}V(z,t) \quad (4.29)$$

This equation may be verified with the voltage-current relation of (4.24).

In the following we will discuss how to model the the equivalent series resistance $\underline{R}$. The inductance $\underline{L}$, capacitance $\underline{C}$, and conductance $\underline{G}$ have been modeled in (4.17,4.21,4.22). It may be noted that $\underline{L}$ and $\underline{C}$ for a lossy transmission line are derived independent of the loss parameters of the material medium and the conductors. Accordingly, $\underline{L}$ and $\underline{C}$ for a lossy transmission line would remain the same when the losses are removed by making the conductivity $(=\sigma)$ of the filling material zero and that $(=\sigma_c)$ of the conductors infinity. In other words, a lossy transmission line and its loss-less version with $\sigma=0$ and $\sigma_c=\infty$, or equivalently with $\underline{G}=\underline{R}=0$, would have the same values of $\underline{L}$ and $\underline{C}$.

3.2.1. *Series Resistance per Unit Length.* We have established the theory of an ideal transmission line in sections 1-2, assuming that the two conductors are perfectly conducting, which means that the conducting material has zero resistance. The field analysis in section 1, that led to the TEM property of the ideal transmission line, upon which the following derivations are based, would no longer be valid if that is not the case. Any non-zero resistance in a conductor would result in a non-zero voltage over any segment Δz of the conductor. Accordingly, there will be a z directed electric field on the surface of the conductor, which is equal to the voltage over a unit length. This z directed electric field appears at the interface between the conductor and the filling material medium, and would also extend into rest of the material medium by continuity. It would violate the basic TEM nature of a transmission line, which requires that the electric field along the direction of propagation (which is the z direction) is zero. However, this deviation from the TEM condition is usually small for very good conductors, having very small resistance or large conductivity. The small effect may be accounted for using an approximate analysis. This is possible by calculating the power loss in the conductors, under the assumption of large conductivity, and then selecting the equivalent series resistance in the circuit model which would result in the same power loss. Such modeling may be possible only for a time-harmonic signal. The equivalent resistance would depend on the frequency of the signal. We postpone this modeling to section 9 on the harmonic analysis of transmission line.

As discussed, the series resistance of the conductors of a transmission line invalidates the ideal TEM nature of a transmission line. In contrast, any parallel conductance due to the filling medium would not necessarily violate the basic TEM nature of a transmission line. Any conduction current in the filling medium, produced due to the voltage across the two conductors of the transmission line, is expected to flow across the two conductors, not along the propagation direction z. This does not contradict the basic condition for a TEM transmission line, which requires that there is no current or field in the filling medium along the propagation (z) direction. For a specific case when the filling material medium is uniform, having a finite resistance or non-zero conductivity of the medium would maintain the TEM condition of the transmission line, in a rigorous sense.

In any event, the voltage-current relations (4.18,4.24), which are same as (4.28,4.29), for a lossy transmission line, with a non-zero value for either the $\underline{R}$ or the $\underline{G}$, would be difficult to solve in a general form for an arbitrary time-varying signal. However, simple solutions can be possible for time harmonic signals. As discussed above, simple models for the $\underline{R}$ also happens to be possible for time time harmonic signals. Therefore, we will cover the modeling of the series resistance $\underline{R}$, and of the signal propagation in a general lossy transmission line, only for time harmonic signals in section 9. Modeling for arbitrary signals can then be indirectly performed by Fourier decomposition and superposition.

In the following sections, we will first analyse signal propagation in loss-less transmission lines and study related applications. These results apply for signals with general time variations. We will then study signal propagation specifically for harmonic signals in section 9, covering loss-less as well as lossy transmission lines.

4. Wave Propagation in a Loss-Less Transmission Line

Equations (4.26,4.27) are two independent voltage-current relations for a lossless transmission line, which may be restated in the following. The same equations would also be obtained by equating $\underline{G}$ and $\underline{R}$ to zero in the relations (4.28, 4.29) for a lossy transmission line.

$$\frac{\partial}{\partial z}V(z,t) = -\underline{L}\frac{\partial}{\partial t}I(z,t) \tag{4.30}$$
$$\frac{\partial}{\partial z}I(z,t) = -\underline{C}\frac{\partial}{\partial t}V(z,t) \tag{4.31}$$

Two independent equations for the electric and magnetic fields of a loss-less transmission line are given by (4.13,4.14), and may be restated for continuity.

$$\frac{\partial}{\partial z}E(z,t) = -\mu\frac{\partial}{\partial t}H(z,t) \tag{4.32}$$
$$\frac{\partial}{\partial z}H(z,t) = -\epsilon\frac{\partial}{\partial t}E(z,t) \tag{4.33}$$

where E and H are amplitudes of the electric and magnetic fields with respect to suitable reference directions. Their vector directions are defined by the electric and magnetic field lines, which are mutually orthogonal to each other and to the direction of propagation for a TEM transmission line. This is shown in section 1.

4.1. Equivalency Between the Voltage-Current and the Field Relations. The above equations (4.32,4.33) for the fields are recognized to be functionally equivalent to the voltage-current equations (4.30,4.31). One could obtain (4.32,4.33) from (4.30,4.31) by substituting the voltage V and current I by the electric field E and magnetic field H, respectively, and also substituting the transmission line parameters $\underline{L}$ and $\underline{C}$ by the material parameters μ and ϵ, respectively. Accordingly, any solution for the voltage-current equations (4.30,4.31) would apply for the fields as well, through the above substitution. In the following we will find solutions for the voltage and current functions from (4.30,4.31), and extend key results to the field functions using the above equivalence.

4.2. One Dimensional Wave Equation. Differentiate (4.30) with respect to z, and (4.31) with respect to t, and then simplify the first result using the second.

$$\begin{aligned}
\frac{\partial}{\partial z}[\frac{\partial}{\partial z}V(z,t)] = \frac{\partial^2}{\partial z^2}V(z,t) = \frac{\partial}{\partial z}[-\underline{L}\frac{\partial}{\partial t}I(z,t)] = -\underline{L}\frac{\partial^2}{\partial z\partial t}I(z,t) \\
\frac{\partial}{\partial t}[\frac{\partial}{\partial z}I(z,t)] = \frac{\partial^2}{\partial t\partial z}I(z,t) = \frac{\partial}{\partial t}[-\underline{C}\frac{\partial}{\partial t}V(z,t)] = -\underline{C}\frac{\partial^2}{\partial t^2}V(z,t) \\
\frac{\partial^2}{\partial t\partial z}I(z,t) = \frac{\partial}{\partial t}[\frac{\partial}{\partial z}I(z,t)] = \frac{\partial}{\partial z}[\frac{\partial}{\partial t}I(z,t)] = \frac{\partial^2}{\partial z\partial t}I(z,t) \\
\frac{\partial^2}{\partial z^2}V(z,t) =-\underline{L}\frac{\partial^2}{\partial z\partial t}I(z,t) = -\underline{L}\frac{\partial^2}{\partial t\partial z}I(z,t) \\
= -\underline{L}[-\underline{C}\frac{\partial^2}{\partial t^2}V(z,t)] = \underline{LC}\frac{\partial^2}{\partial t^2}V(z,t) \\
\frac{\partial^2}{\partial z^2}V(z,t)= \underline{LC}\frac{\partial^2}{\partial t^2}V(z,t)
\end{aligned} \tag{4.34}$$

It is assumed that the parameters $\underline{L}$ and $\underline{C}$ are independent of of the distance z and time t, and so can be taken out of any differential operation. It is also assumed that the sequence of differentiation with respect to t and z can be interchanged without changing the result.

Similarly, now differentiate (4.31) with respect to z, and (4.30) with respect to t, and then simplify the first result using the second. Follow similar simplification as earlier. We will get an equation involving only the current variable $I(z,t)$.

$$\begin{aligned}
\frac{\partial}{\partial z}[\frac{\partial}{\partial z}I(z,t)] = \frac{\partial^2}{\partial z^2}I(z,t) = \frac{\partial}{\partial z}[-\underline{C}\frac{\partial}{\partial t}V(z,t)] = -\underline{C}\frac{\partial^2}{\partial z\partial t}V(z,t) \\
\frac{\partial}{\partial t}[\frac{\partial}{\partial z}V(z,t)] = \frac{\partial^2}{\partial t\partial z}V(z,t) = \frac{\partial}{\partial t}[-\underline{L}\frac{\partial}{\partial t}I(z,t)] = -\underline{L}\frac{\partial^2}{\partial t^2}I(z,t) \\
\frac{\partial^2}{\partial t\partial z}V(z,t) = \frac{\partial}{\partial t}[\frac{\partial}{\partial z}V(z,t)] = \frac{\partial}{\partial z}[\frac{\partial}{\partial t}V(z,t)] = \frac{\partial^2}{\partial z\partial t}V(z,t) \\
\frac{\partial^2}{\partial z^2}I(z,t) =-\underline{C}\frac{\partial^2}{\partial z\partial t}V(z,t) = -\underline{C}\frac{\partial^2}{\partial t\partial z}V(z,t) \\
= -\underline{C}[-\underline{L}\frac{\partial^2}{\partial t^2}I(z,t)] = \underline{LC}\frac{\partial^2}{\partial t^2}I(z,t) \\
\frac{\partial^2}{\partial z^2}I(z,t)= \underline{LC}\frac{\partial^2}{\partial t^2}I(z,t)
\end{aligned} \tag{4.35}$$

As seen from (4.34,4.35), interestingly both $V(z,t)$ and $I(z,t)$ satisfy the same general equation of the following form.

$$\frac{\partial^2}{\partial z^2} f(z,t) = a \frac{\partial^2}{\partial t^2} f(z,t); \; a = \underline{LC}, \; f \equiv V, I \tag{4.36}$$

The above general equation (4.36) is referred to as a one-dimensional wave equation. The name is in anticipation that the solution would describe a general wave propagating in one dimension z.

4.3. General Wave Function in One Dimension. It can be shown that the general solution $f(z,t)$ for (4.36) can be written in the form $f_(z \mp vt)$, where v is a suitable positive constant. The functional form $f(z \mp vt)$ means that the variables z and t in a valid solution $f(z,t)$ can be arranged such that they occur together as $(z \mp vt)$. The symbol $\mp$ means either a $-$ or a $+$ sign would be allowed between z and vt for the function to be a valid solution. However, a particular sign, once selected, should be used in all parts of the function. It does not mean that either the $-$ or $+$ sign can be used arbitrarily in different parts of the function. Writing the $-$ sign above the $+$ sign, as we have deliberately shown here to represent either possibility, does not have particular mathematical significance. It has useful notational significance in relation to the nature and choice of the solution, as should be clear in the discussion of section 4.4. However, having this notational choice we have made, proper consistency should be followed in choice of a double sign, such as $\mp$ or $\pm$, which may appear elsewhere in the derivation process. When a double sign appears elsewhere, its upper sign should correspond to the choice of the upper sign used here, which is $-$, and its lower sign should correspond to the lower sign here, which is $+$. That would be notationally important

Consider $(z \mp vt)$ together as a dummy variable s. As per the description above, a valid solution $f(z,t)$ may be expressed as any arbitrary function $f(s)$ of a dummy variable s, where s is substituted by $(z \mp vt)$

$$f(z,t) = f(s)\Big|_{s=(z \mp vt)}; \; v > 0 \tag{4.37}$$

Further, any multiplication of the function by a constant A would also be a valid solution. Such a general form of the solution is referred to as a one-dimensional wave function. The name is in anticipation of the wave nature of the solution, which would be shown in the following. The validity of such functions as solutions for the general equation of (4.36) can be proved by substituting $f(z,t)$ in (4.36) by $Af(s) = Af(z \mp vt)$ and showing that the differential equation would be satisfied for a suitable choice for v. Towards this goal, let us find the expressions for the first and second derivatives of a function $Af(s) = Af(z \mp vt)$ with respect to z and t. These derivatives would be needed when we replace $f(z,t)$ in the equation (4.36) by $Af(s) = Af(z \mp vt)$.

$$\frac{\partial}{\partial z}[Af(z \pm vt)] = \frac{\partial}{\partial z}[Af(s)]\Big|_{s=z\mp vt} = A\frac{\partial f(s)}{\partial s}\frac{\partial s}{\partial z}|_{s=z\pm vt} = A\frac{\partial f(s)}{\partial s} = A\dot{f}$$

$$\frac{\partial^2}{\partial z^2}[Af(z \pm vt)] = \frac{\partial}{\partial z}[A\dot{f}] = A\frac{\partial \dot{f}}{\partial s}\frac{\partial s}{\partial z}|_{s=z\pm vt} = A\frac{\partial^2 f}{\partial s^2} = A\ddot{f} \quad (4.38)$$

$$\frac{\partial}{\partial t}[Af(z \mp vt)] = \frac{\partial}{\partial t}[Af(s)]\Big|_{s=z\mp vt} = A\frac{\partial f(s)}{\partial s}\frac{\partial s}{\partial t}|_{s=z\mp vt} = \mp vA\frac{\partial f(s)}{\partial s} = \mp vA\dot{f}$$

$$\frac{\partial^2}{\partial t^2}[Af(z \pm vt)] = \frac{\partial}{\partial t}[\mp vA\dot{f}] = \mp vA\frac{\partial \dot{f}}{\partial s}\frac{\partial s}{\partial t}|_{s=z\pm vt} = (\mp v)(\mp v)A\frac{\partial^2 f}{\partial s^2} = v^2 A\ddot{f} \quad (4.39)$$

where, $\dot{f}$ and $\ddot{f}$ refer to the first and second derivative of the function $f(s)$ with respect to the dummy variable s. Notice that the first derivative with respect to t is multiplied by $\mp v$, which means that the result would have a different sign when a different sign is selected in the choice of the function $Af(z \mp vt)$. However, the second derivative with respect to t is multiplied by v^2, which is a positive number. This means that the sign of the result would be independent of any particular sign selected in the choice of the function $Af(z \mp vt)$.

Now, using the derivatives (4.38) and (4.39) in the equation (4.36) we see that the general solution $Af(s) = Af(z \mp vt)$ would be valid only for a particular choice of v.

$$\frac{\partial^2}{\partial z^2}Af(z \mp vt) = A\ddot{f} = a\frac{\partial^2}{\partial t^2}Af(z \mp vt) = aAv^2\ddot{f}$$

$$av^2 = 1, \; v = \frac{1}{\sqrt{a}} = \frac{1}{\sqrt{LC}} \quad (4.40)$$

4.4. Propagation Velocity. Let us interpret the physical significance of a function of the form $Af(z \pm vt)$. For a particular function $f(s)$, at $t = 0$ we have $Af(z \mp vt)\big|_{t=0} = Af(z)$. This is shown as a function of z in the Fig.7. Now see what happens to the function $Af(z \mp vt)$ after a small time interval $t = \Delta t$. We would have $Af(z \mp vt)\big|_{t=\Delta t} = Af(z \mp v\Delta t)$. This is shown as a function of z, together with the above function for $t = 0$, plotted for both cases when the $-$ or the $+$ sign is selected. When the $-$ sign is selected, the function $Af(z - v\Delta t)$ is essentially the function $Af(z)$ for $t = 0$, but shifted to the forward or positive side of the z axis by distance $\Delta z = v\Delta t$. Whereas, when the $+$ sign is selected, the function $Af(z + v\Delta t)$ would be the function $Af(z)$ for $t = 0$, but shifted to the backward or negative side of the z axis by distance $\Delta z = v\Delta t$.

$$Af(z \mp vt)\big|_{t=0} = Af(z)$$

$$Af(z \mp vt)\big|_{t=\Delta t} = Af(z \mp v\Delta t) = Af(z \mp \Delta z)\big|_{\Delta z = v\Delta t} \quad (4.41)$$

This behavior of the function to move or shift in the distance z as time passes, while keeping the shape of the function unchanged, is a basic property of a wave. Accordingly, such general functions of the form $Af(z \mp vt)$ are recognized as wave functions in the

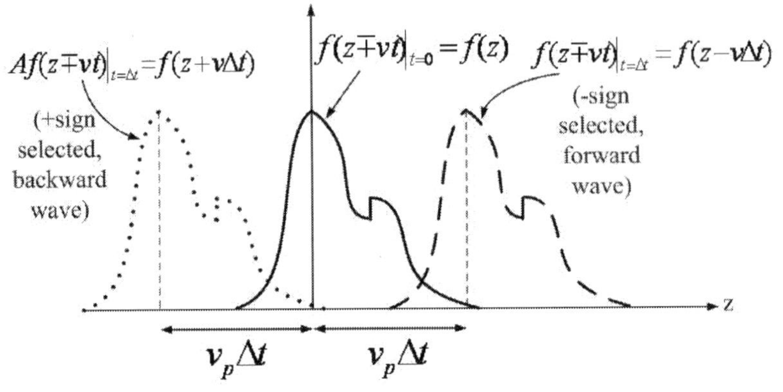

FIGURE 7

single dimension z. The movement in distance $\Delta z = v\Delta t$ is proportional to the time Δt. The time rate of the movement is the velocity of the wave, and is equal to v. This velocity at which the waveform or shape of the function appears to move would be now on referred to as the propagation velocity, v_p.

$$v_p = \frac{\Delta z}{\Delta t} = \frac{v\Delta t}{\Delta t} = v = \frac{1}{\sqrt{a}} = \frac{1}{\sqrt{LC}} \tag{4.42}$$

As shown in Fig.7, the function moves in the forward or positive direction of z, when the $-$ sign is selected in the function $Af(z \mp vt)$. Accordingly, the choice of the $-$ sign, which is the upper sign in the double sign notation $\mp$, refers to a forward wave. On the other hand, when the $+$ sign is selected in the function $Af(z \mp vt)$, the function moves in the backward or negative direction of z. Accordingly, the choice of the $+$ sign, which is the lower sign in the double sign notation $\mp$, refers to a backward wave. Any double-sign notation to be used elsewhere will have to be consistent with this. This means, the upper sign of any double-sign notation used elsewhere would as well refer to a forward or $+z$ wave, and the lower sign to a backward or $-z$ wave.

4.5. Shape Function and Amplitude. The wave function in (4.37) provides only the functional form for the solution of the general wave equation (4.36). It may represent a voltage or a current signal. The functional form of (4.37) specifies only the propagation velocity v and the direction of propagation. The function $f(.)$ represents the shape of the wave, and could take any possible form. The strength of the wave is represented by the relative amplitude of the function $f(.)$, or by a multiplying coefficient A, which also need not be specified. The shape function $f(.)$ and the amplitude factor A can be specified only when the wave is excited by connecting a particular signal source to the transmission line. The shape function $f(.)$ would then be determined by the shape or time variation of the particular signal source, and the amplitude coefficient A would be determined by the strength of a signal source as well as the specific configuration of the source connection.

Relations between a source function and the voltage or current waveforms, will be derived in a following section 5, for specific source configurations.

4.5.1. *Propagation Velocity and Material Medium.* Now, as discussed before, the relations between the electric and magnetic fields in (4.32,4.33) are equivalent to those in (4.30,4.31) for the voltage and current, respectively. Except, the parameters $\underline{L}$ and $\underline{C}$ in the voltage and current relations are to be substituted by μ and ϵ, respectively. Accordingly, the general wave equation for the electric and magnetic fields would have the same form as that in (4.36) for the voltage and current signals, with $a = \underline{LC}$ replaced by $a = \mu\epsilon$. This would lead to the same general form for the wave function (4.37) for the electric and magnetic fields, but with velocity expressed using μ and ϵ

$$v_p = \frac{1}{\sqrt{\mu\epsilon}} = \frac{1}{\sqrt{\underline{LC}}}; \quad \underline{LC} = \mu\epsilon \tag{4.43}$$

The electric and magnetic fields that exist in the filling material medium are directly associated with the voltage and current which appear at the conductor terminals of the transmission lines. Therefore, their propagations must be directly connected. The velocity of propagation of the fields will have to be the same as that for the voltage and current waves.

As derived in section 2, we expect the individual parameters $\underline{L}$ and $\underline{C}$ would depend on the parameters of the filling material medium as well as the dimensions of the particular transmission line. However, as per the above equation (4.43), the product of $\underline{L}$ and $\underline{C}$ will have to depend only on the parameters of the filling medium, and independent of the transverse dimensions of the particular transmission line. The dimensional dependence in $\underline{L}$ must therefore be inverse of that in $\underline{C}$.

Material Medium and Validity of the Circuit Model:

The equality of the velocity of the voltage-current wave to that for the electric-magnetic field waves, would fundamentally require the product of $\underline{L}$ and $\underline{C}$ to be equal to that of μ and ϵ. This is an important result, having other fundamental implications. The circuit model of Fig.5 of a loss-less transmission line, and the related derivation of the circuit parameters, are valid and meaningful only when $\underline{L}$ and $\underline{C}$, and accordingly the product $\underline{LC}$, are unique parameters for a given transmission line. Now, using the relation $\underline{LC} = \mu\epsilon$ from (4.43), the above condition would require that the product $\mu\epsilon$ has a unique value over the entire filling medium of the transmission line. In the simplest case, a unique value for the $\mu\epsilon$ could be achieved by having a single material medium, uniformly distributed over the entire filling space. Alternately, it may be achieved by having different materials in different parts of the filling space between the two conductors, but the product $\mu\epsilon$ of each medium is a constant. For example, half of the space may be filled by a material with $\epsilon = 4\epsilon_0$ and $\mu = \mu_0$, while the other half is filled by a second material with $\epsilon = 2\epsilon_0$ and $\mu = 2\mu_0$. We are referring here to the material distribution over the transverse cross-section of the transmission line. The material distribution along the propagation direction z is assumed to be uniform. We are also limiting our present discussion to loss-less transmission lines. Similar considerations for a lossy line will be presented in section 9.7.8 in the context of harmonic signals.

The "unique" value for $\mu\epsilon$, as required above, may sometimes be loosely referred to as being "constant" over the cross section of the transmission line. However, it may need to be clarified. $\mu\epsilon$ is to be "unique" or "constant" does not necessarily mean that it is required to be a constant value, defined independent of the signal waveforms. The value of $\mu\epsilon$ may have different values for two harmonic signals of different frequencies, or for two signals having different frequency bandwidths. However, whatever is the value of $\mu\epsilon$, defined for a particular signal frequency or signal band, it will have to be the same value for the same signal frequency or signal band, uniformly for all material over the cross section of the transmission line. Accordingly, the corresponding values of the parameters $\underline{L}$ and $\underline{C}$ in the equivalent circuit model of Fig.5 may have different (but unique) values for the different signals or signal bands. A parameter being unique simply means that it may not have more than one possible value, for the same device and signal conditions. That would be physically invalid.

Material Medium and the TEM Condition:

Consider a loss-less transmission line filled with an arbitrary material distribution, for which $\mu\epsilon$ is different at different locations over its cross section. As we understand from the above discussion, this transmission line can not be properly described by a circuit model of Fig.5. Let us now view this transmission line in terms of its fields. The fields at different points on the cross section would propagate with a different velocity, and appear at a distance z with different time delays for the different cross-sectional points. Accordingly, the cross-sectional field distribution would constantly change as the wave propagates over increasing distance z. Assuming that the cross-sectional fields at a given z satisfy Maxwell's equations, the changing fields may no longer satisfy Maxwell's equations at all distance $-\infty < z < +\infty$. That would be an invalid situation. Therefore, one may conclude that the basic TEM condition we have used to begin with our field analysis, leading to the derivation of the velocity expression (4.42,4.43), would have to be invalid when $\mu\epsilon$ is not unique or constant over the cross section of a transmission line.

Conversely, the basic TEM condition would be valid for a loss-less transmission line, only when $\mu\epsilon$ is a constant over the entire cross section. In this case, the field at each point on the cross-section of the transmission line would propagate with the same velocity, thus duplicating the field distribution over the entire cross section at a distance z after a time delay z/v_p. Validity of Maxwell's equations at one location z would, therefore, guarantee its validity at any other location. Having a constant value for $\mu\epsilon$, as required for a TEM transmission line discussed above, is also the condition for the validity of the circuit model of Fig.5 discussed earlier. Accordingly, the validity of the circuit model of Fig.5 of a loss-less transmission line, and the validity of the TEM condition of the transmission line, are inter-dependent and equivalent principles.

The above interrelation between the validity of the circuit model and the TEM condition may also be viewed from a different angle. In section 1.1, the TEM condition is shown to be equivalent to having unique values of the voltage $V(z)$ and current $I(z)$ for the transmission line, defined at a particular z. In other words, only under the TEM condition one may properly define the voltage and current parameters. A circuit model of a transmission line can be valid or meaningful only when the voltage and current are properly defined. Accordingly, a circuit model can be valid only when the transmission line satisfies the TEM condition. For situations where there may be small deviations from the ideal TEM condition, a circuit model may only be useful and valid in an approximate sense. Such cases may be casually referred to as quasi-TEM conditions.

As discussed before, the simplest case of satisfying an ideal TEM condition is to have a transmission line with infinite conductivity for its two conductors, and a single material uniformly filling the entire space between the two conductors. This is the case in many simple, practical designs. A common situation for a loss-less transmission line, where there is only a small deviation from the ideal TEM condition, is when one material covers regions including most of the fields of the transmission line, and a second material covers regions with only minimal fields. A specific example for this case is a microstrip line of Fig.34. Accordingly, the microstrip line may be treated as a TEM line for most practical purposes, and is modeled as an equivalent parallel-plate line with a uniform material medium. This is discussed in section 10.

4.6. Voltage and Current Wave, and the Characteristic Impedance. We know that the differential equation for voltage in (4.34), and that for current in (4.35), both satisfy the same general wave equation (4.36), with the same coefficient $a = \underline{L}\underline{C}$. Therefore, the solution for both the voltage and current would have the same basic functional form, with the same velocity v_p.

$$V(z,t) = Af(z \pm v_p t), \quad I(z,t) = Bg(z \pm v_p t), \quad v_p = \frac{1}{\sqrt{a}} = \frac{1}{\sqrt{\underline{L}\underline{C}}} \tag{4.44}$$

Clearly, the voltage and current are different quantities, and accordingly are represented by different amplitudes. We may not presume that the voltage and current waves have the same shape function. Therefore, we may choose to represent the voltage and current waves by different shape functions, $f(.)$ and $g(.)$, respectively.

The current, which is a measure of the flow of charge, is associated with a physical velocity of the charge. It may be clarified, that the propagation velocity of the current "wave" is a different concept from the charge velocity associated with the current "flow." Consider the wave function $Bg(z \mp v_p t)$ associated with the current signal. The physical velocity of a charge is governed by the amplitude of the current flow, which would be the amplitude of this function $Bg(z \mp v_p t)$ at a given point and time. Whereas, the velocity of propagation of the current wave is the velocity at which the entire function $Ag(z \mp vt)$ appears to move or shift along z as time progresses. Accordingly, one may have a particular velocity of propagation of the current wave, while the function, or equivalently the velocity of the physical charge, itself could have any arbitrary amplitude, small or large. They are distinctly unrelated quantities.

Now, using the expressions for the voltage and current waves in (4.44) in the voltage-current relationship of (4.30), we get

$$\begin{aligned} A\frac{\partial f(z,t)}{\partial z} &= A\frac{\partial f(s)}{\partial s}\frac{\partial s}{\partial z} = A\frac{\partial f(s)}{\partial s} = B\frac{\partial g(z,t)}{\partial t} = B\frac{\partial g(s)}{\partial s}\frac{\partial s}{\partial t} \\ &= -\underline{L}B\frac{\partial g(s)}{\partial s}(\mp v_p) = \pm \underline{L}Bv_p\frac{\partial g(s)}{\partial s}; \; s = z \mp v_p t \end{aligned} \tag{4.45}$$

For the above relation to be valid, the derivative functions of f and g will have to be equal for all z and t. Therefore, f and g can differ only by a constant, independent of z and t. Because we are particularly interested in time-varying functions, the above constant is assumed to be zero. Accordingly, the waveforms f and g for the voltage and current signals, respectively, would be identical.

$$\frac{\partial f(s)}{\partial s} = \frac{\partial g(s)}{\partial s}, \; g = f \tag{4.46}$$

An additional condition is required for the above relation (4.45) to be valid. This relates the amplitudes A and B, and accordingly, the ratio between the voltage and current waveforms, respectively.

$$A = \pm \underline{L} B v_p, \; \frac{A}{B} = \pm \underline{L} v_p = \pm \underline{L} \frac{1}{\sqrt{\underline{LC}}} = \pm \sqrt{\frac{\underline{L}}{\underline{C}}}$$

$$\frac{V(z,t)}{I(z,t)} = \frac{Af(z \mp v_p t)}{Bg(z \mp v_p t)} = \frac{A}{B} = \pm \sqrt{\frac{\underline{L}}{\underline{C}}} = \pm Z_c, \; Z_c = \sqrt{\frac{\underline{L}}{\underline{C}}}, \; Z_c > 0 \tag{4.47}$$

The parameter Z_c is equal to the square-root of the ratio of $\underline{L}$ and $\underline{C}$, and is a positive number. The dual sign $\pm$ before Z_c used in the expression for the ratio between the voltage and current functions should be noted. As per the notation we have been following, the upper $+$ sign refers to forward or $+z$ propagation, and the lower $-$ sign refers to backward or $-z$ propagation. Accordingly, the ratio between the voltage and current waveforms for a forward wave is a positive number equal to $+Z_c$, and that for a backward wave is a negative number equal to $-Z_c$. This simply means that, for a given voltage amplitude, the direction of the current flow would switch from $+z$ to $-z$ as the propagation direction switches from $+z$ to $-z$, as should be normally expected. The absolute value of the ratio between voltage and current functions is equal to Z_c, independent of location z, time t, and particular shape f of the waveforms. This parameter Z_c is referred to as the characteristic impedance of the transmission line. Unlike the velocity v_p, the characteristic impedance Z_c depends on the physical dimensions and the parameters of the filling material medium of a transmission line.

4.7. Field Solutions and Wave Impedance. We have recognized the equivalency between the voltage-current relations of (4.30,4.31) and the field relations of (4.32,4.33). Accordingly, any relationship for the voltage and current signals may be extended to the electric and magnetic fields, using appropriate substitutions prescribed in section 4.1. Let us apply the substitution principle to the above voltage-current relationship (4.47). The voltage and current wave functions in (4.47) are substituted by the wave functions for the electric and magnetic fields, and the inductance and capacitance parameters $\underline{L}$ and $\underline{C}$ are substituted by the material parameters μ and ϵ, respectively.

$$\frac{E(z,t)}{H(z,t)} = \pm \sqrt{\frac{\mu}{\epsilon}} = \pm \eta, \; \eta = \sqrt{\frac{\mu}{\epsilon}}, \; \eta > 0 \tag{4.48}$$

η is equal to the square-root of the ratio of μ and ϵ, and is a positive number. As per the dual sign notation we have been using, the ratio between the electric and magnetic field waveforms for a forward wave is a positive number equal to $+\eta$, and that for a backward wave is a negative number equal to $-\eta$. Similar to the situation for the voltage-current waveforms, this means that, for a given electric field amplitude, directed along its reference unit vector $\hat{e}$, the direction of the magnetic field would switch from its reference direction $\hat{h}$ to $-\hat{h}$, as the propagation direction switches from $+z$ to $-z$. Following the cross product relation (4.12) for the reference unit vectors, this would mean that the cross product for the electric and magnetic field vectors would switch from $+z$ to the $-z$ direction, as the propagation direction switches from $+z$ to the $-z$ direction. In other words, the cross product of the electric and magnetic field vectors would be aligned to the propagation direction in both cases. The absolute value of the ratio between the electric and magnetic fields is equal to η, independent of location z, time t, and particular shape f of the waveforms. This parameter η is referred to as the wave impedance. Unlike the characteristic impedance Z_c, but like the velocity v_p, the wave impedance η depends only on the parameters of the filling material medium of a transmission line, not on the physical dimensions of its conductors.

5. Excitation from a Signal Source

We will first study the simplest case of excitation of a single, infinite-long transmission line. We will extend the study to a more general situation with two, infinite-long transmission lines, connected in series or parallel with a source. Excitation in arbitrary situations can be similarly analyzed.

5.1. Source Connected to a Single Transmission Line. Figure 8 shows a transmission line connected to a voltage source, with a general time variation $V_s(t)$, and it equivalent source resistance R_s. Two terminals of the voltage source are connected to the two conductors at one end, represented as $z = 0$, of the transmission line. The transmission line is assumed here to be infinitely long, with the other end of the transmission line running to $z = +\infty$. Having an infinite-long transmission line is a physically hypothetical situation, but the assumption may be useful in the present study for the following reason. In this case, any signal wave excited by the source into the transmission line will propagate along the $+z$ direction for ever. The signal would not experience any termination or obstacle at the other end of the transmission line. Accordingly, no reflected signal will appear in the transmission line. Therefore, the signal observed in the transmission line of Fig.8 would represent only the signal excitation from the source. That is the objective of the present study.

For a practical transmission line with a finite length, which may be terminated at the other end by an arbitrary load, there would be a non-zero reflected signal in addition to the original source excitation we will study here. Modeling of the reflection would be presented in a following section 6.

The general solution for the wave equation (4.36) is expressed in a functional form of (4.37). The form of the solution in (4.37) may always be rearranged by re-writing the variable $(z \mp v_p t) = \mp v_p(t \mp z/v_p)$. Accordingly, the solution can be written in a modified functional form $f(t \mp z/v_p)$. For the present study, it may be analytically preferable to express the solution in the alternate functional form $f_(t \mp z/v_p)$.

$$f(z,t) = f(t \mp z/v_p) = f(s)\Big|_{s=t\mp z/v_p} \tag{4.49}$$

The velocity v_p here would be the same velocity we derived in (4.42). The same notation for the double sign $\mp$ used in (4.37) applies here, as well.

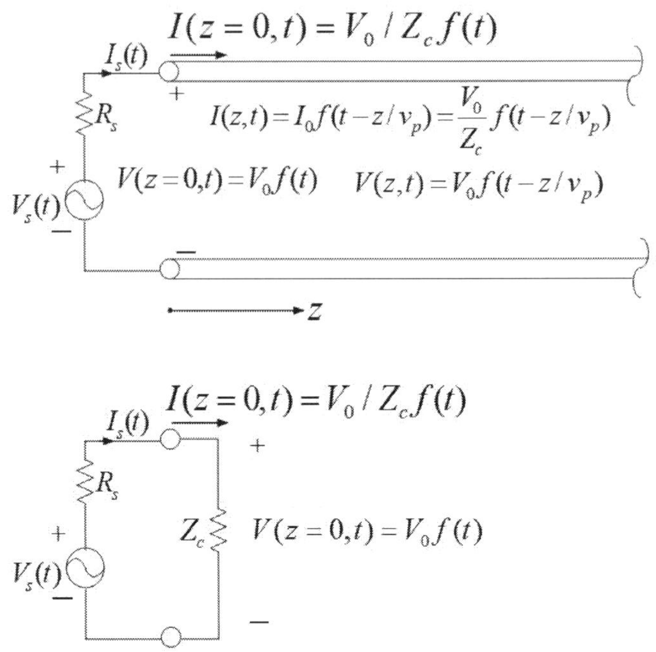

FIGURE 8

Note that the function $f(.)$ represents any generic function. It can be any specific function in order to satisfy the wave equation (4.36), as long as z and t appear everywhere together as $(z \mp v_p t)$ or as $(t \mp z/v_p)$. The actual functional shape of $f(.)$ would be dependent on a specific source of excitation, to be derived in the following . To clarify a bit further, the function $f(s)$ of the dummy variable $s = z \mp v_p t$ in (4.37) and the function $f(s)$ of the dummy variable $s = t \mp z/v_p$ in (4.49) are not meant to refer to identical functions of s, although the two functions would look closely similar. In other words, the function in (4.49) is not obtained from the function (4.37) by simply *substituting* $s = z \mp v_p t$ by $s = t \mp z/v_p$, but by appropriately *rearranging* the variables. For example, consider a function of $s = z - v_p t$ in the form $f(s) = s^2 = (z - v_p t)^2$. The function may be rearranged as $(z - v_p t)^2 = v_p^2 (t - z/v_p)^2$, which may be expressed as a new function of $s = t - z/v_p$ in the following form $f(s) = v_p^2 s^2 = v_p^2 (t - z/v_p)^2$. Notice that the first function $f(s) = s^2$ is different from the second function $f(s) = v_p^2 s^2$.

Using the functional form of the solution in (4.49), let us express the voltage waveform $V(z,t)$ excited in the transmission line of Fig.8 as

$$V(z,t) = V_0 f(t - z/v_p), \;\; V(z = 0, t) = Af(t) \tag{4.50}$$

$f(.)$ is an unknown shape function, and V_0 is an unknown amplitude of the wave, which are to be determined. We select the $-$ sign between t and z/v_p because the wave would propagate in the $+z$ direction. Because there is no reflection in the infinite long transmission line, in this case there would be no wave in the transmission line in the functional form $f(t + z/v_p)$. Using the definition of the characteristic impedance in (4.47), we may express the current waveform $I(z,t)$ in relation to the voltage waveform $V(z,t)$.

$$I(z,t) = \frac{V(z,t)}{Z_c} = \frac{V_0}{Z_c} f(t - z/v_p), \; I(z = 0, t) = \frac{V_0}{Z_c} f(t) \tag{4.51}$$

Establish continuity of current at the junction point between the source and the transmission line in Fig.8, by applying Kirchoff's Current Law (KCL) at the junction. This would result in having the current in the transmission line at $z = 0$ to be equal to the current $I_s(t)$ drawn from the source.

$$I(z = 0, t) = I_s(t) = \frac{V_0}{Z_c} f(t) \tag{4.52}$$

Now, apply Kirchoff's Voltage Law (KVL) to the complete loop in Fig.8 starting from the source, through the source resistance, and then through the input terminals of the transmission line with voltage $V(z = 0, t)$, finally returning back to the source.

$$V_s(t) - I_s(t)R_s - V(z = 0, t) = 0 = V_s(t) - \frac{V_0 R_s}{Z_c} f(t) - V_0 f(t)$$
$$V_0 f(t)(1 + \frac{R_s}{Z_c}) = V_s(t), \; V_0 f(t) = V_s(t)\frac{Z_c}{R_s + Z_c}, \; V_0 = \frac{Z_c}{R_s + Z_c}, \; f(t) = V_s(t) \tag{4.53}$$

The above result could be directly obtained from an equivalent circuit, shown with the Fig.8, where the transmission line is replaced by a resistance equal to the characteristic impedance. The ratio of the voltage and current waves, propagating in the forward $+z$ direction in the infinity transmission line, we know is equal to the characteristic impedance Z_c. Accordingly, the infinite transmission line may be replaced by an equivalent resistance equal to Z_c.

Once, the V_0 and $f(t)$ are obtained, the voltage and current waveforms in the transmission line are completely solved in terms of the source voltage $V_s(t)$, R_s and Z_c. Using (4.53) in (4.50,4.51), we get

$$V(z,t) = \frac{Z_c}{R_s + Z_c} V_s(t - z/v_p), \; I(z,t) = \frac{1}{R_s + Z_c} V_s(t - z/v_p) \tag{4.54}$$

5.1.1. *Space-Time Inter-Relations.* It may useful to illustrate the space-time relationship in a signal wave excited by an input source, through a specific example. Let us use the values $Z_c = R_s = 50\Omega$, and $v_p = 3 \times 10^8 m/s$ for the circuit of Fig.8. A specific source function V_s for the circuit is shown in Fig,9. Using the results of the above analysis, let us first sketch the voltage wave at a given time $t = 3\mu s$, as a function of the distance z. Using the above parameters in (4.54), we get,

$$V(z,t) = \frac{Z_c}{R_s + Z_c} V_s(t - z/v_p) = \frac{1}{2} V_s(3 - z/v_p) \tag{4.55}$$

where the time t is assumed to be in microseconds. The above function may be sketched, as shown in Fig.8(a) using the following simple functional steps. (i) Sketch $1/2V_s(-t)$ as a function of t by flipping $V_s(t)$ to the negative t axis and scaling the voltage axis by the factor of 1/2. (ii) Then sketch $1/2V_s(3-t) == 1/2V_s[-(t-3)]$ by shifting the function in (i) by $3\mu s$ towards positive t. (iii) The above sketch in (ii) would also be the sketch of the function $1/2V_s(3 - z/v_p)$, if the time axis is relabeled as z/v_p. We have now the desired function of (4.55), except it is sketched as a function of z/v_p, but we need to sketch as a function of z. That can be accomplished by relabeling the z/v_p axis by multiplying the scale with v_p, to obtain the corresponding z values. (iv) The sketch at this point would include values of z that are positive and negative. The negative values of z do not exist, and are non-physical. Accordingly, the sketch in the negative side of the z axis should be ignored or erased, to obtain the final sketch desired.

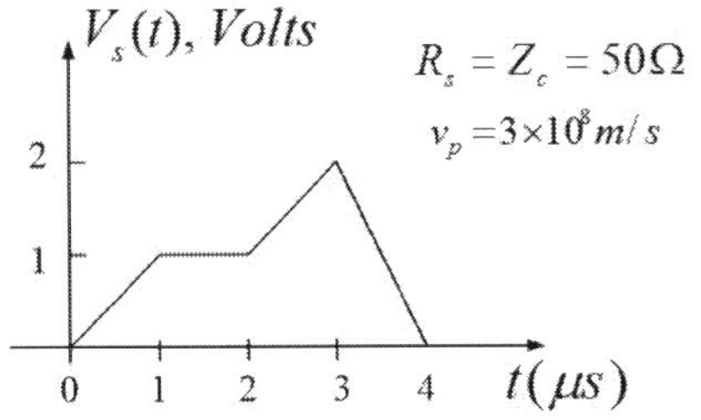

FIGURE 9

The final sketch in Fig.10(a), which we have obtained in mostly mathematical terms, may be examined for physical understanding. The scaling factor 1/2 is the relative amplitude of excitation, which depends on Z_c and R_s. Notice that the source voltage waveform $V_s(t)$, which appear for earlier times, are seen to have moved ahead as a function of distance z. This should make sense, because the source signal at an earlier time would have more time interval to propagate farther distance. Notice also that the source waveform only upto $t = 3\mu s$ affect the final sketch. This is because we are sketching the waveform in z at $t = 3\mu s$. Any source waveform for $t > 3\mu s$ refer to future events, and therefore does not yet appear as waveform in distance. This future waveform are seen in the negative z axis, which we ignored and erased in our final sketch. In this sense, the waveform in the negative z axis we erased my be interpreted as future waveform in z This part would be pushed to the positive z axis, as the time progresses, gradually evolving the waveform in z.

The above example illustrated the waveform as a function of z, observed at a given t. Now, for the same parameters and source function, let us sketch the waveform as a

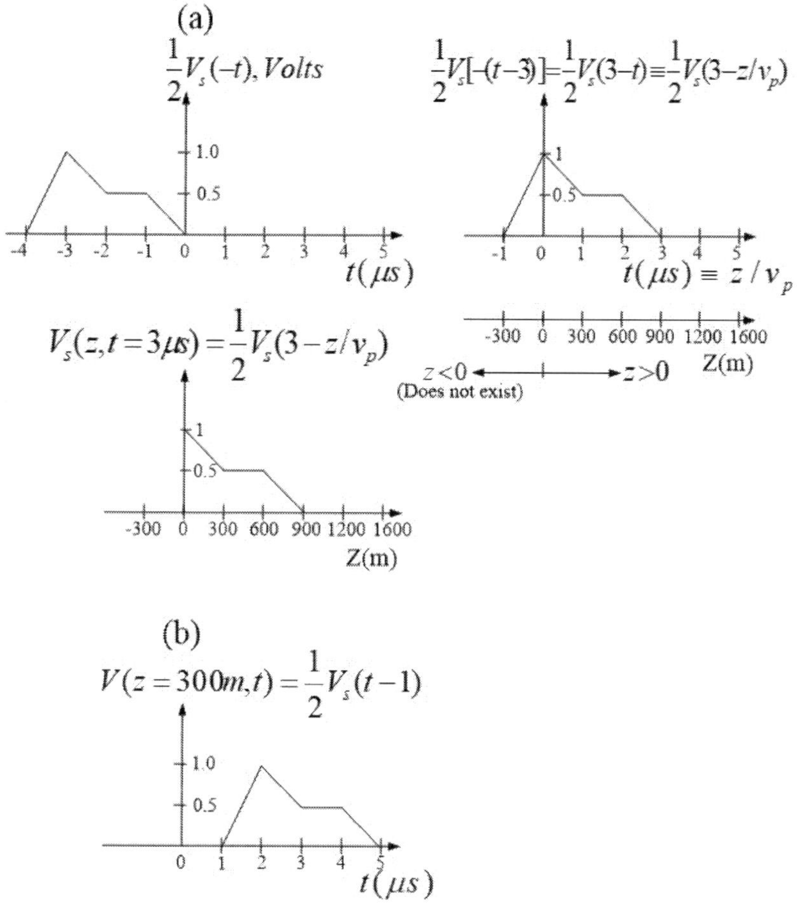

FIGURE 10

function of time t, observed at a given location $z = 300m$. Substituting $z = 300m$ in (4.55), but keeping t as a variable, we get

$$V(z,t) = \frac{Z_c}{R_s + Z_c} V_s(t - \frac{z}{v_p}) = \frac{1}{2} V_s(t - 1\mu s), \; \frac{z}{v_p} = \frac{300m}{3 \times 10^8 m/s} = 1\mu s \tag{4.56}$$

The above function may be sketched by delaying the source function V_s by $1\mu s$, and then scaling the voltage by a factor of 1/2. The sketch is shown in Fig.10b. As compared to the earlier procedure to sketch the waveform as a function of z, the procedure to sketch the waveform as a function of t at a given location z is relatively straight forward. The scaling factor 1/2 is the relative amplitude of excitation, which depends on Z_c and R_s. The signal takes time $t = z/v_p$, which is equal to $1\mu s$ for the given parameters, to arrive at the observation point. Accordingly, the entire signal waveform is delayed by $1\mu s$ when observed at the given z

5.2. Source Connected to Two Transmission Lines. We will consider two possible configurations. In the first case, two transmission lines are connected in parallel with a voltage source V_s as shown in Fig.11. Whereas in the second case, they are connected in series with the voltage source V_s, as shown in Fig.12. The transmission lines are assumed to be of infinite length, as in the case for the excitation of a single transmission line derived earlier. This means there is no reflected signal in the transmission line, propagating in the backward direction towards the source.

The basic principle in the excitation of a single transmission line will apply in present situations with two lines. As in the case with a single transmission line, one may derive the different signal waveforms from the first principles using KCL and KVL equation. Alternately, as also shown in the case with a single transmission line, one may use an equivalent circuit model, from which different waveforms may be directly obtained using basic circuit and wave-function concepts. In the present derivation, we will use the approach using circuit models, shown with the Figs.11,12. Each transmission line of infinite length is replaced at its input terminals by a resistance equal to its characteristic impedance. We will only provide the final expressions for the different voltage and current waveforms.

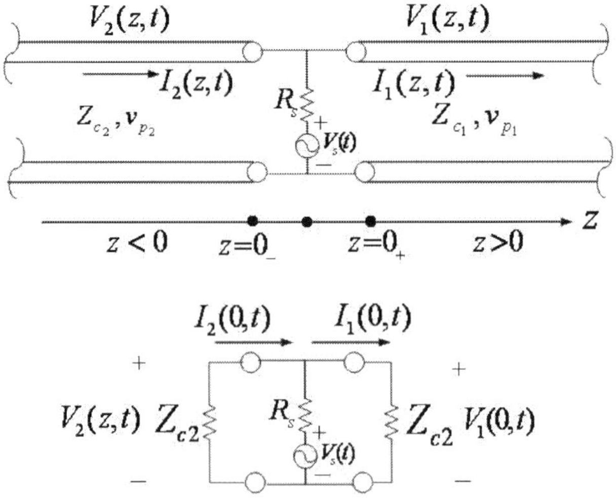

FIGURE 11

5.2.1. *Parallel Connection.*

$$\begin{aligned}
V_1(z,t) &= \frac{Z_{c1}||Z_{c2}}{Z_{c1}||Z_{c2}+R_s}V_s(t-z/v_{\rho 1}) \\
&= \frac{Z_{c_1}Z_{c_2}}{R_s(Z_{c_1}+Z_{c_2})+Z_{c_1}Z_{c_2}}V_s(t-z/v_{\rho 1});\ z>0 \\
I_1(z,t) = \frac{1}{Z_{c1}}V_1(t-z/v_{\rho_1}) &= \frac{Z_{c_2}}{R_s(Z_{c_1}+Z_{c_2})+Z_{c_1}Z_{c_2}}V_s(t-z/v_{\rho 1});\ z>0 \\
V_2(z,t) &= \frac{Z_{c1}||Z_{c2}}{Z_{c1}||Z_{c2}+R_s}V_s(t+z/v_{\rho 2}) \\
&= \frac{Z_{c_1}Z_{c_2}}{R_s(Z_{c_1}+Z_{c_2})+Z_{c_1}Z_{c_2}}V_s(t+z/v_{\rho 2}),\ z<0 \\
I_2(z,t) = -\frac{1}{Z_{c2}}V_2(t+z/v_{\rho 2}) &= -\frac{Z_{c_1}}{R_s(Z_{c_1}+Z_{c_2})+Z_{c_1}Z_{c_2}}V_s(t+z/v_{\rho 2}),\ z<0 \quad (4.57)
\end{aligned}$$

Note that the two transmission lines are aligned along the same z axis, with the source at the origin $z=0$. One transmission line is along the positive z axis, extending to $z=+\infty$, whereas the second line is along the negative z axis, extending to $z=-\infty$. For uniformity, the currents in each transmission line is considered positive when the flow is in the $+z$ direction. Accordingly, the reference current in the first transmission line is along the direction of propagation, whereas for the second transmission line it is opposite to the direction of propagation. This situation results in the $-$ sign in the expression for the current I_2 in (4.57), whereas the sign for the current I_1 is positive.

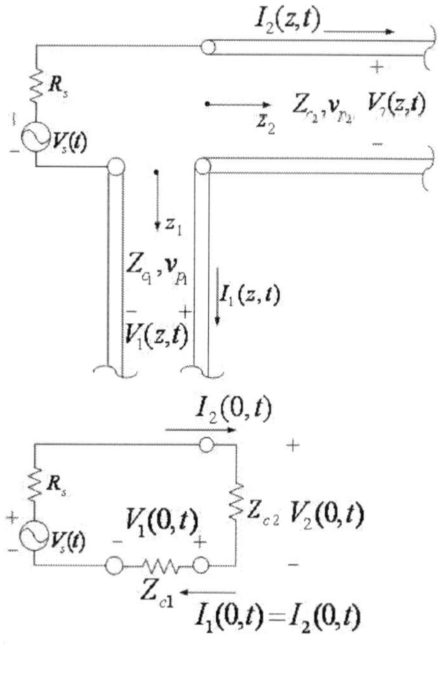

FIGURE 12

5.2.2. *Series Connection.*

$$V_1(z_1,t) = \frac{Z_{c_1}}{Z_{c_1}+Z_{c_2}+R_s}V_s(t-z_1/v_{\rho_1}),\ z_1>0$$
$$I_1(z_1,t) = \frac{1}{Z_{c1}}V_1(z_1,t) = \frac{1}{Z_{c_1}+Z_{c_2}+R_s}V_s(t-z_1/v_{\rho_1}),\ z_1>0$$
$$V_2(z_2,t) = \frac{Z_{c_2}}{Z_{c_1}+Z_{c_2}+R_s}V_s(t-z_2/v_{\rho_2}),\ z_2>0$$
$$I_2(z_2,t) = \frac{1}{Z_{c2}}V_2(z_2,t) = \frac{1}{Z_{c_1}+Z_{c_2}+R_s}V_s(t-z_2/v_{\rho_2}),\ z_2>0 \tag{4.58}$$

In this case, the two infinite-length transmission lines are aligned along two different directions, referenced with different coordinates $z_1 > 0$ and $z_2 > 0$. The source end of the transmission lines are at $z_1 = z_2 = 0$.

6. Reflection at Transmission Line Termination

Figure (13) shows a transmission line terminated at its right end ($z = 0$) with a resistance R_L. For this study, we are not particularly concerned about the left end of the transmission line, and may be considered to be at $z = -\infty$. We simply assume that, whatever source arrangement the transmission line is connected to at the left end, is producing a voltage signal $V^+(z,t)$ propagating in the forward ($+z$) direction. We use a wave function of the form (4.49), with amplitude A to express this forward voltage wave. The corresponding current signal is $I^+(z,t)$, which can be related to the voltage signal using (4.47).

$$V^+(z,t) = Af(t-\frac{z}{v_p}),\quad I^+(z,t) = \frac{A}{Z_c}f(t-\frac{z}{v_p}) \tag{4.59}$$

We have used the form $f(t - z/v_p)$, with a $-$ sign between t and z/v_p, because it is propagating in the forward ($+z$) direction.

If the transmission line was running infinitely in the $+z$ direction, then this forward wave will keep propagating in the $+z$ direction forever. However, the transmission line is now terminated by a load R_L, connected at $z = 0$. The wave will be reflected by the load, producing a voltage signal $V^-(z,t)$ propagating in the backward ($-z$) direction. The corresponding current signal is $I^-(z,t)$, which can be related to the voltage signal $V^-(z,t)$ using (4.47).

$$V^-(z,t) = Bg(t+\frac{z}{v_p}),\quad I^-(z,t) = -\frac{B}{Z_c}g(t+\frac{z}{v_p}) \tag{4.60}$$

For generality, we now allow the reflected signal to have a shape function g, which is different from the shape function f for the incident signal. We have used the form $g(t + z/v_p)$, with a $+$ sign between t and z/v_p, because the reflected signal is propagating in the

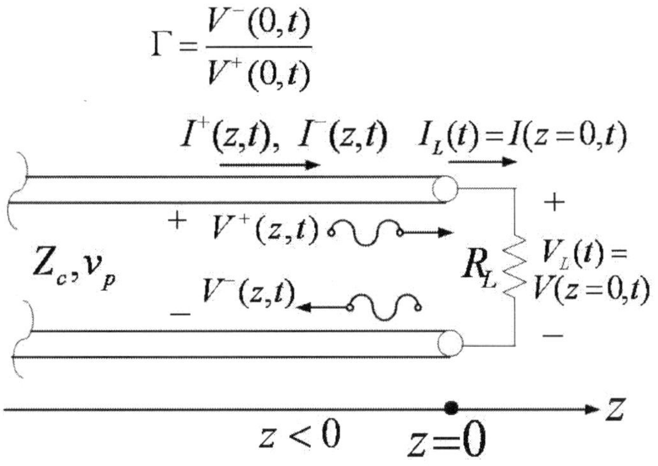

FIGURE 13

backward ($-z$) direction. Also, note that reflected current $I^-(z,t)$ in (4.60) is expressed with a $-$ sign, in contrast with a $+$ sign used in (4.59) for the forward wave. This is because, as described by (4.47), the ratio between the voltage and current signals is $-Z_c$ for a backward wave, but is $+Z_c$ for a forward wave.

The total voltage and current signals in the transmission line, represented respectively as $V(z,t)$ and $I(z,t)$, would be sum of the forward and reflected parts.

$$V(z,t) = V^+(z,t) + V^-(z,t) = Af(t - \frac{z}{v_p}) + Bg(t + \frac{z}{v_p})$$
$$I(z,t) = I^+(z,t) + I^-(z,t) = \frac{A}{Z_c}f(t - \frac{z}{v_p}) - \frac{B}{Z_c}g(t + \frac{z}{v_p}) \quad (4.61)$$

The the total voltage and current at $z = 0$ would be equal to the load voltage and current, respectively. As per Ohm's law, the load voltage is equal to the load current multiplied be the load resistance R_L. This relation would lead to relating the forward and reflected signal amplitudes and shape functions.

$$V_L(t) = V(z=0,t) = Af(t) + Bg(t), \quad I_L(t) = I(z=0,t) = \frac{A}{Z_c}f(t) - \frac{B}{Z_c}Bg(t)$$
$$V_L(t) = Af(t) + Bg(t) = R_L I_L(t) = \frac{R_L}{Z_c}(Af(t) - Bg(t))$$
$$Bg(t)(\frac{R_L}{Z_c} + 1) = Af(t)(\frac{R_L}{Z_c} - 1), \quad Bg(t) = \frac{R_L - Z_c}{R_L + Z_c}Af(t)$$
$$g(t) = f(t), \; g(s) = f(s), \; \frac{B}{A} = \frac{R_L - Z_c}{R_L + Z_c} = \frac{Bg(t)}{Af(t)} = \frac{V^+(z=0,t)}{V^-(z=0,t)} = \Gamma \quad (4.62)$$

The shape functions $f(s)$ and $g(s)$ of the forward and reflected signals are the same. The ratio of the amplitudes of the forward and reflected signals, which is B/A, is defined as the reflection coefficient Γ. This is also the ratio of the forward and reflected voltages observed at $z = 0$.

The reflection coefficient Γ is a function of the load impedance R_L and the characteristic impedance Z_c, as expressed in the above equation (4.62). We assume that the characteristic impedance Z_c, as defined in (4.47). is a positive number. The resistance R_L is also a non-negative number for a passive load. With these assumptions, let us examine the value of Γ for different conditions of R_L. Γ can be either positive or negative, depending on if R_L is larger or smaller than Z_c, respectively. Γ is zero when R_L is equal to Z_c. Accordingly, $R_L = Z_c$ may be considered as a perfect absorber of the forward incident wave. This case is equivalent to having an infinite long transmission line running to $z = +\infty$. The magnitude of $(R_L - Z_c)$ is less than or equal to that of $(R_L + Z_c)$, resulting in the magnitude of Γ to be be less than or equal to unity. The magnitude of Γ would be unity when R_L is either zero or infinity. Γ is -1 when $R_L = 0$ (short circuit). This means that the incident voltage will switch its sign, with its magnitude remaining the same, when it reflects from a short-circuited termination. Γ is $+1$ when $R_L = +\infty$ (open circuit). This means that the incident voltage will reflect, with its magnitude and sign remaining the same, when it reflects from a open-circuited termination.

$$\begin{aligned} \Gamma = \frac{R_L - Z_c}{R_L + Z_c} &> 0, \; R_L > Z_c \\ &< 0, \; R_L > Z_c \\ &= 0, \; R_L = Z_c \\ &= -1, \; R_L = 0 \\ &= +1, \; R_L = +\infty \\ |\Gamma| &\leq 1 \end{aligned} \tag{4.63}$$

7. Reflection and Transmission at a Junction Between Two Transmission Lines

Fig.14 shows a junction between two different transmission lines. Signal is assumed to be incident towards the junction from the first transmission line (referred to as the source transmission line), while the second transmission line (referred to as the load transmission line) is seen as a load. One end of the source transmission line is connected to the junction, while its other end is connected to some kind of source circuit which produces the incident signal. The present study does not concern with the specific arrangement of the source circuit. We simply need to know the incident signal waveform $V^+(z,t) = Af(t - z/v_{p1}$, with a given amplitude A and wave function $f(s)|_{s=t-z/v_{p1}}$. Similarly, one end of the load transmission line is connected to the junction, with the other end presumably connected to some kind of load circuit, which need not be specified for the present study. Soon after the given incident signal $V^+(z,t)$ arrives at the junction point $z = 0$, a reflected signal $V^-(z,t)$ and a transmitted signal $V^t(z,t)$ are generated, propagating in the source and load transmission lines in the $-z$ and $+z$ directions, respectively. We want to find the amplitudes and waveforms of these reflected and transmitted signals.

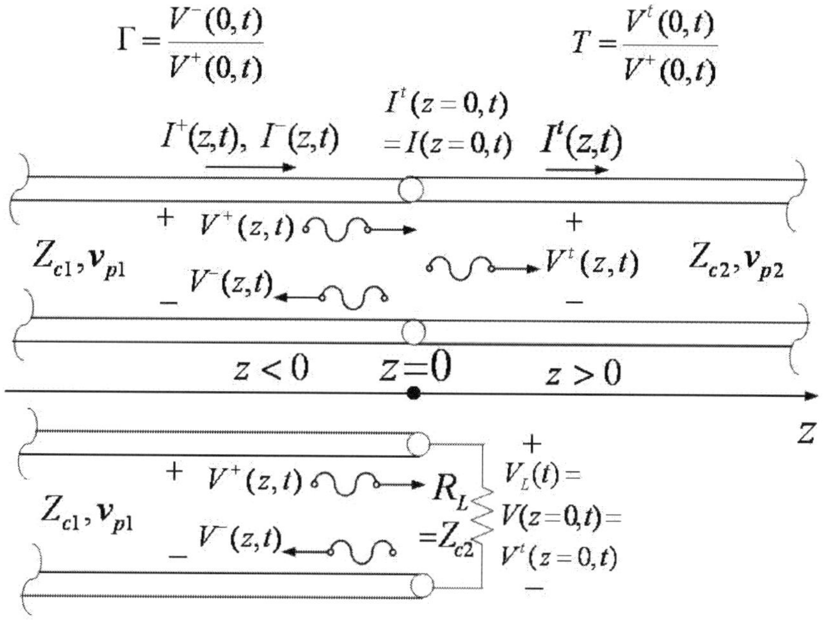

FIGURE 14

Note that the total signal in the source transmission line consists of the incident and the reflected waves, which are propagating in the forward $(+z)$ and backward $(-z)$ directions, respectively. Whereas, the transmitted signal $V^t(z,t)$ is the only wave in the load transmission line, propagating along the forward $(+z)$ direction. This assumes that there is no signal in the load transmission line, which is incident toward the junction point at $z = 0$, propagating in the backward $(-z)$ direction. In a general practical situation, there might be a non-zero incident signal in the load transmission line. This could be produced due to reflection of the transmitted signal by a specific load circuit, and/or produced by an additional signal source, connected at the other end of the load transmission line. Scattering of any such non-zero incident signal in the load transmission lines can, however, be modeled using a separate formulation, duplicating the present analysis with the parameters of the source and load sides interchanged. We are not directly concerned here with such a situation. Having no signal in the load transmission line, propagating in the $-z$ direction may be equivalently accomplished by simply assuming that the load transmission line is extended to $z = +\infty$, as shown in Fig.14.

As discussed, the total signal in the load transmission line consists of only the forward-propagating, transmitted signal $V^t(z,t)$, with no back-ward propagating wave. Therefore, as per the definition of the characteristic impedance in (4.47), the voltage-to-current ratio at any point in the load transmission line would be equal to its characteristic impedance Z_{c2}. Accordingly, the load transmission line may be treated as an equivalent resistance $R_L = Z_{c2}$, connected as a load at the end of the source transmission line. The equivalent circuit is also shown with Fig.14. Reflection in such a circuit was modeled in the last section (4.62), the results from which can as well be used here. Accordingly, the reflected signal $V^-(z,t)$ may be expressed in terms of a reflection coefficient Γ, with the Γ expressed using (4.62).

$$
\begin{aligned}
V^+(z,t) = Af(t - \frac{z}{v_{p1}}),\ I^+(z,t) &= \frac{A}{Z_{c1}} f(t - \frac{z}{v_{p1}}),\ z > 0 \\
V^-(z,t) = Bg(t + \frac{z}{v_{p1}}),\ I^-(z,t) &= -\frac{B}{Z_{c1}} g(t + \frac{z}{v_{p1}}),\ z < 0 \\
R_L &= \frac{V^t(z=0,t)}{I^t(z=0,t)} = Z_{c2} \\
\Gamma = \frac{B}{A} = \frac{V^-(z=0,t)}{V^+(z=0,t)} &= \frac{Z_{c2} - Z_{c1}}{Z_{c2} + Z_{c1}},\ B = A\Gamma,\ f(t) = g(t) \\
V^-(z,t) = A\Gamma f(t + \frac{z}{v_{p1}}),\ I^-(z,t) &= -\frac{A\Gamma}{Z_{c1}} f(t + \frac{z}{v_{p1}}),\ z < 0 \\
V(z,t) = V^+(z,t) + V^-(z,t),\ I(z,t) &= I^+(z,t) + I^-(z,t),\ z < 0
\end{aligned} \tag{4.64}
$$

The total voltage $V(z,t)$ in the source transmission line is the sum of the incident and reflected signals. The total voltage at the junction point can be obtained by replacing $z = 0$ in the expression of the total signal $V(z,t)$. The total junction voltage must be equal to transmitted signal in the load transmission line, evaluated at the junction point $z = 0$.

$$
\begin{aligned}
V^t(z,t) = Uh(t - \frac{z}{v_{p2}}),\ I^t(z,t) &= \frac{U}{Z_{c2}} h(t - \frac{z}{v_{p2}}),\ z > 0 \\
V^t(z=0,t) = Uh(t) = V(z=0,t) &= Af(t) + A\Gamma f(t) = A(1+\Gamma)f(t) \\
U &= A(1+\Gamma),\ h(t) = f(t)
\end{aligned} \tag{4.65}
$$

The shape function $h(.)$ for the transmitted signal is equal to $f(.)$, which is same as that of the incident and reflected signals. The coefficient U of the transmitted signal is related to the coefficient A of the incident signal by a factor $(1 + \Gamma)$. We define the ratio of the coefficients of the transmitted and the incident signals as the a transmission coefficient T.

$$
\begin{aligned}
T = \frac{U}{A} = 1 + \Gamma = 1 + \frac{Z_{c2} - Z_{c1}}{Z_{c2} + Z_{c1}} &= \frac{2Z_{c2}}{Z_{c2} + Z_{c1}} \\
\frac{V^t(z=0,t)}{V^+(z=0,t)} = \frac{Uh(t)}{Af(t)} &= \frac{U}{A} = T \\
V^t(z,t) = ATf(t - \frac{z}{v_{p2}}),\ I^t(z,t) &= \frac{AT}{Z_{c2}} f(t - \frac{z}{v_{p2}}),\ z > 0
\end{aligned} \tag{4.66}
$$

The transmission coefficient T is related to the reflection coefficient as $T = 1 + \Gamma$. The transmission coefficient T is seen to be the ratio of the transmitted and the incident voltages, evaluated at the junction point $z = 0$.

8. Transient Response of a Simple Transmission Line Circuit

We have developed the theories for wave propagation in a transmission line, wave generation from a source, and wave reflection from a termination load. Using these theories we can analyze a complete circuit. In the following we will study response of a simple transmission line circuit for step and pulse signals. Step and pulse functions are basic functions for transient study, which are relatively simple to model. We will illustrate the principles for a specific set of parameters.

8.1. Step Function Response. Figure 15 shows a simple transmission line circuit, showing specific values for different parameters for this study. The constant voltage source V_0, with the switch arrangement, is essentially equivalent to having a source $V_s = V_0 u(t)$, where $u(t)$ is a step function.

$$V_s(t) = V_0 u(t) \begin{aligned} &= V_0, t > 0 \\ &= 0, t < 0 \end{aligned} \tag{4.67}$$

For this excitation, let us examine the voltage waveform observed at the center of the transmission with $z = l/2$.

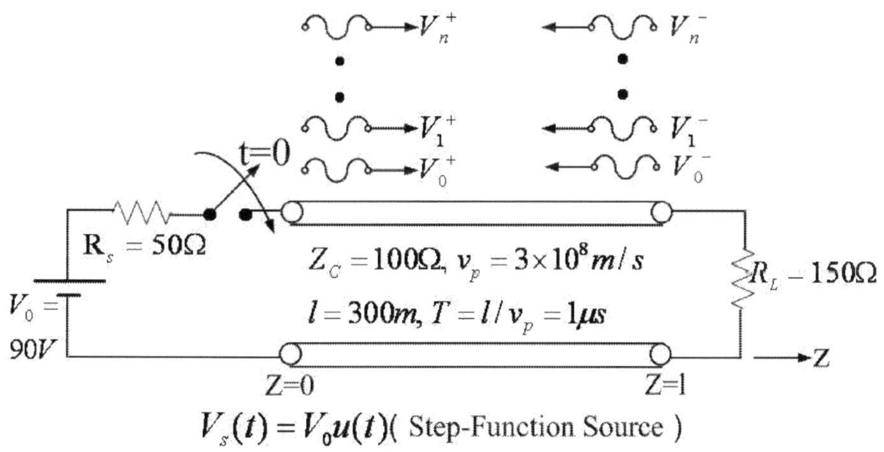

FIGURE 15

Using the theories of wave propagation, source excitation and reflection developed in past chapters, the response to the simple step-function source may sketched by simple calculations. Just after the switch is closed at $t = 0$, a step-function wave would be generated in the transmission line with amplitude $V_0^+ = 60V$, given by (4.54). This wave would propagate in the transmission line in the forward ($+z$) direction, with the given velocity $v_p = 3 \times 10^8 m/s$. The wave will reach at any particular point of observation z at time $t = z/v_p$. Before this wave arrives, the signal voltage is zero everywhere along the transmission line. Because we have here a step-function wave, the voltage will transition from the initial value of $0V$ to the amplitude V_0^+ at the arrival time t, and will continue to be V_0^+ after that. For the specific observation point $z = l/2 = 150m$, the arrival time is equal to $z/v_p = 0.5\mu s$.

$$V_0^+ = V_0 \frac{Z_c}{R_s + Z_c} \tag{4.68}$$

This wave will pass through the observation point at $z = l/2$, and reach the termination at $z = l$ at time $T = l/v_p = 1\mu s$. It will be reflected with a reflection coefficient $\Gamma_L = 1/5$, given by (4.62). This amounts to having the reflected voltage amplitude $V_0^- = V_0^+ \Gamma_L = 12V$. The reflected signal would propagate in the reverse direction $(-z)$ towards the source, and it will be added to the forward wave already existing on the transmission line. This return wave will arrive at $z = l/2$ after a delay of $0.5\mu s$, that is at time $t = 1.5\mu s$. Accordingly, the observed signal will transition from $V_0^+ = 60V$ to $V_0^+ + V_0^- = 72V$ at $t = 1.5\mu s$.

$$V_0^- = V_0^+ \Gamma_L, \; \Gamma_L = \frac{R_L - Z_c}{R_L + Z_c} \tag{4.69}$$

This return wave propagating in the $-z$ direction wave will pass through the observation point at $z = l/2$, and reach the source end at $z = 0$ at time $2T = 2l/v_p = 2\mu s$. The source impedance R_s will be seen as a load termination for this return wave. It will be reflected with a reflection coefficient $\Gamma_s = -1/3$, given by (4.62). It amounts to having a voltage wave of amplitude $V_1^+ = V_0^- \Gamma_s = -4V$, propagating in the forward direction $+z$. It will be added on top of the signals already existing, which includes the original forward wave of amplitude V_0^+, and the reflected wave from the load of amplitude V_0^-. This new wave of amplitude $V_1^+ = -4V$, which is reflected from the source end, will arrive at $z = l/2$ after a delay of $0.5\mu s$, that is at time $t = 2.5\mu s$. Accordingly, the observed signal will transition from $V_0^+ + V_0^- = 72V$ to $V_0^+ + V_0^- + V_1^+ = 68V$ at $t = 2.5\mu s$.

$$V_1^+ = V_0^- \Gamma_S = V_0^+ \Gamma_L \Gamma_S, \; \Gamma_S = \frac{R_s - Z_c}{R_s + Z_c} \tag{4.70}$$

This wave of amplitude V_1^+ propagating in the $+z$ direction will pass through the observation point at $z = l/2$, and reach the load end at $z = l$ at time $3T = 3l/v_p = 3\mu s$. It will be again reflected by the load with amplitude $V_1^-) = V_1^+ \Gamma_L$, and the reflected signal will propagate back and the procedure may be repeated until the amplitude of the reflected signal is approximately zero, which is when the total observed signal reaches its steady state value.

$$\begin{aligned} V_1^- = V_1^+ \Gamma_L = V_-^+ \Gamma_L^2 \Gamma_S, \; V_2^+ = V_1^- \Gamma_S = V_0^+ \Gamma_L^2 \Gamma_S^2 \\ V_n^+ = V_0^+ \Gamma_L^n \Gamma_S^n, \;\; V_n^- = V_0^+ \Gamma_L^{n+1} \Gamma_S^n \end{aligned} \tag{4.71}$$

The complete waveform observed at $z = l/2$, as per the above analysis, is shown in Fig.16.

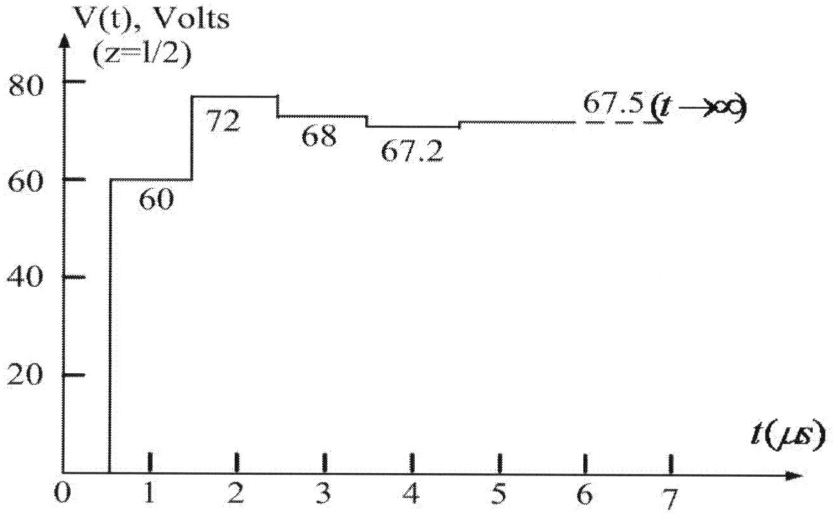

FIGURE 16

8.1.1. *Steady-State Calculation.* An expression for the steady-state signal in Fig.15, applicable for any general values of the circuit parameters, can be derived by adding all the wave amplitudes. The steady-state value would be the same at all locations on the transmission line.

$$\begin{aligned} V(z, t \to \infty) &= V_0^+ + V_0^+ \Gamma_L + V_0^+ \Gamma_L \Gamma_S + V_0^+ \Gamma_L^2 \Gamma_S + V_0^+ \Gamma_L^2 \Gamma_S^2 + \cdots \\ &= V_0^+ (1 + \Gamma_L)(1 + \Gamma_L \Gamma_S + V^+ \Gamma_L^2 \Gamma_S^2 + \cdots) = V_0^+ \frac{(1 + \Gamma_L)}{(1 - \Gamma_L \Gamma_S)} \end{aligned} \quad (4.72)$$

As discussed in last section 8, the magnitude of Γ_L and Γ_S is less than unity for practical impedances with, non-zero, finite values of R_S and R_L. Under these conditions, the magnitude of the factor $\Gamma_L \Gamma_S$ of the geometric series in the above equation is less than unity. Accordingly, the geometric series would have progressively reducing terms. In this case the sum of the geometric series converges to a finite value, which is expressed in terms of the factor $\Gamma_L \Gamma_S$.

The the above expression of the steady-state value may be simplified in terms of the source voltage and different impedances.

$$V_0^+ = V_0 \frac{Z_c}{R_S + Z_c}$$
$$1 + \Gamma_L = 1 + \frac{(R_L - Z_c)}{(R_L + Z_c)} = \frac{2R_L}{(R_L + Z_c)}$$
$$1 - \Gamma_L \Gamma_S = 1 - \frac{(R_L - Z_c)(R_S - Z_c)}{(R_L + Z_c)(R_S + Z_c)} = \frac{2Z_c(R_L + R_S)}{(R_L + Z_c)(R_S + Z_c)}$$
$$V(z, t \to \infty) = V_0^+ \frac{(1+\Gamma_L)}{(1-\Gamma_L \Gamma_S)} = V_0 \frac{Z_c}{Z_c + R_S} \times \frac{2R_L}{R_L + Z_c} \times \frac{(R_L + Z_c)(R_S + Z_c)}{2Z_c(R_S + R_L)} = V_0 \frac{R_L}{R_L + R_S} \quad (4.73)$$

As may be verified, the above steady-state value is equal to the load voltage one would obtain from a DC analysis of the circuit by connecting the load directly across the source terminals. It is independent of the parameters of the transmission line. The parameters of a particular transmission line determine the wave parameters, such as Γ_L, Γ_S and time delay, which only affect the transient response of the circuit.

The above result is theoretically significant. We have derived the steady-state value using the wave theory we have developed in this chapter. We see that all the individual wave signals add up to a steady-state value. And, this steady-state value, calculated from a wave theory, is consistent with a simple DC analysis we already know, which does not require any wave concepts. This agreement of results should develop some confidence in validity and application of the wave theory.

As should clear from the above signal bouncing analysis, that when Γ_L and Γ_s are larger in magnitude, the signal would take larger number of bounces, and therefore a longer time interval, before the stead-state value is achieved. That would be the case when R_L and R_S are very small or very large compared to Z_c. If this time interval is comparable or larger than the time scale involved in a particular application, the transient analysis, and accordingly the wave theory, would be critical for design and understanding. In a special ideal situation, when the magnitudes of Γ_L and Γ_s are unity, each bounce would be of equal magnitude. Therefore, a steady-state value would never be achieved, and the signal would keep bouncing between the source and load terminals for ever. This would be the case only under ideal situations when R_S and R_L are either zero or infinity. However, such ideal conditions are not achieved for practical, passive resistances. Further, having R_S equal to infinity may not also be physically meaningful, because in this case the source is essentially disconnected from the circuit. Accordingly, a steady state would be reached for all practical, passive resistances, although it may take considerably longer time for situations that approach the above ideal conditions.

8.1.2. *Bounce Diagram.* Using the basic principles of wave theory, discussed above, one may sketch waveforms for specific space-time conditions on a case-by-case basis. A special diagram to keep track of the complete space-time information for all signal bouncing, referred to as a bounce diagram, may assist in the process. This is intended to provide a systematic approach to sketch waveform for any general space-time conditions. Two separate bounce diagrams, one for the voltage and the second for the current signals, may be required to keep track of either the voltage or the current waveforms, as needed. The voltage and current bounce diagrams share the same basic principles, but are specifically drawn for either the voltage or the current wave, using the wave parameters for the particular signal.

Voltage Bounce Diagram:

A bounce diagram for the voltage wave, with the specific parameters of the circuit in Fig.15, is shown in Fig.17. A vertical time axis is drawn with a suitable choice of time scale, with time increasing as one traces down the axis, and its origin representing initial time $t = 0$ at the top of the scale. The horizontal axis is drawn with with a suitable scale, starting with the source end represented at the origin at $z = 0$, and the load end shown on the right at $z = l$.

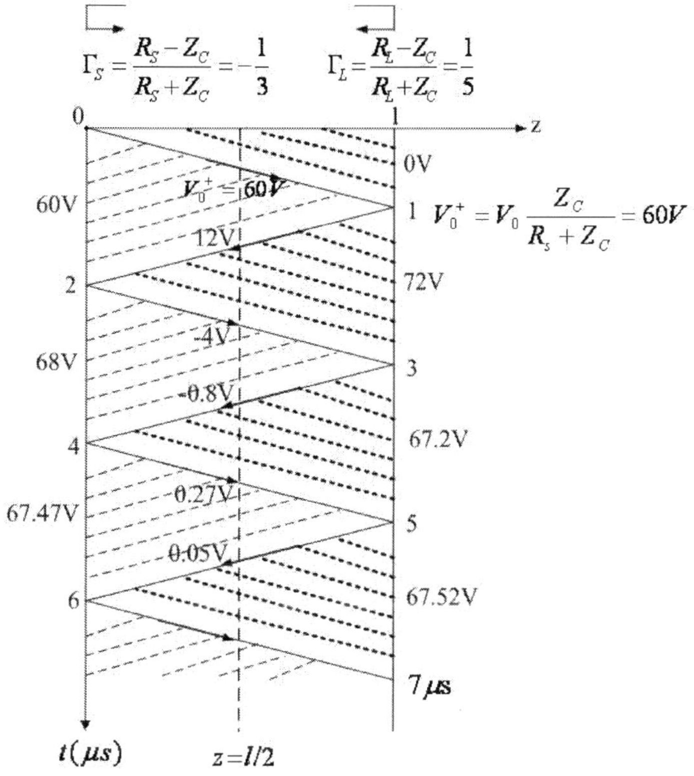

FIGURE 17

With the proper choice of the space-time axes, a set of straight lines - referred to as wavelines, are drawn, starting from the source end $z = 0$ and time $t = 0$. The wave lines join the source ($z = 0$) and load ends ($z = l$), alternately at time $t = 0,\ T,\ 2T,\ \cdots$. These lines represent the propagation of the different bouncing waves, and arrow signs on the lines show the direction of propagation of the particular wave. The first wave line, originating from $z = 0$ and $t = 0$, is labeled with $V_0^+ = 60V$, representing the first forward voltage. This wave arrives at the load end at $t = T$, and is reflected with a reflection coefficient $\Gamma_L = 1/5$. The amplitude of the reflected wave $V_0^- = V_0^+ \Gamma_L = 12V$ is labeled on the second wave line, starting from the load end at $z = l$ and $t = T$ and directed towards the source side. Similarly, all the other wave lines are labeled with the respective wave amplitude. For ready reference, the value of the reflection coefficient Γ_L is labeled at the top of the diagram in the load end, and that of Γ_S in the source end. A specific space-time point (z, t) on a wave line refer to the time t at which the particular wave arrives at the location z. Accordingly, when a space-time point (z, t) lies above or below a wave line, it means that the particular wave has not arrived or has passed the particular location, respectively. A straight line geometrically describes the linear progression of distance with time, as a wave propagates with a velocity v_p.

As can be seen in Fig.17, triangular regions are formed in the space-time plane, with the upper and lower boundaries for each region defined by two wave lines. They are shown in Fig.17 as shaded regions. Consider any space-time point (z, t) in a particular shaded region. The waves corresponding to the wave line at the upper boundary of the shaded region and all other wave lines above that, have already passed through the point. Whereas, the waves corresponding to the wave line at the lower boundary of the shaded region and all other wave lines below that have not yet arrived at the point. Accordingly, the signal voltage anywhere in a particular shaded region would differ from that in the shaded region just above it, by the voltage amplitude of the wave line separating the two regions.

As per the above principle, we can now iteratively calculate and label the voltage amplitude for all shaded regions, one by one as we go down the diagram, starting with an initial condition of $0V$ for the shaded region above the first wave line. Simply add the voltage amplitude of a shaded region, which has been already labeled, to that of the wave line passing at its lower boundary, and label the sum as the voltage amplitude for the following region just below it. These voltages are labeled in Fig.17 on the left or right sides of the particular region. For example, the voltage value for the shaded region labeled as $68V$ is obtained by adding the $72V$ labeled for the region just above it and $-4V$ labeled on the wave line passing between the regions.

Current Bounce Diagram:

The bounce diagram in Fig.17 is valid for voltage signal only. A separate bounce diagram for the current signal can be produced using a similar procedure. In this case, we will have to label the amplitudes for different current waves. Notice that the amplitude of the first wave line is labeled $0.6A$, which is the amplitude of the first, forward propagating current wave excited by the source. Also, it may be remembered that the reference direction for current switches sign as the wave reflects at a termination. This is accommodated by having the reflection coefficient for current as the negative of that for voltage. The bounce diagram for current, for the same parameters as in Fig.15, is shown in Fig.18. Notice that the reflection coefficient for current is labeled as $-\Gamma_L = -1/5$ and $-\Gamma_S = +1/3$ to account for the switch in sign discussed above. Using the first wave amplitude as I_0^+, and reflections labeled properly with an additional negative sign, other signal amplitudes for wave lines and shaded regions are labeled accordingly.

Using the Bounce Diagrams:

The bounce diagrams for a particular circuit contain complete space-time information for the signal waves in the circuit. A specific waveform for a given space-time condition can be extracted from these diagrams. Let us see how we may derive from these diagrams a waveform as a function of time t observed at a given location z, or as a function of location z observed at a given time t.

Specifically, let us sketch $V(t)$ observed at $z = l/2$ using the voltage bounce diagram of the Fig.17. We need to draw a vertical line at $z = l/2$, referred to as the observation line, representing the particular location of observation. This line will go through different shaded regions. Note the voltage value in each shaded region, and time instants at which the observation line intersects with the wave lines at the upper and lower boundaries of the particular region. The particular voltage associated with a region would be observed over the time interval between the particular time instants at the boundaries of the region. Accordingly, it may be seen from Fig17, the observation line at $z = l/2$ starts at $t = 0$ in

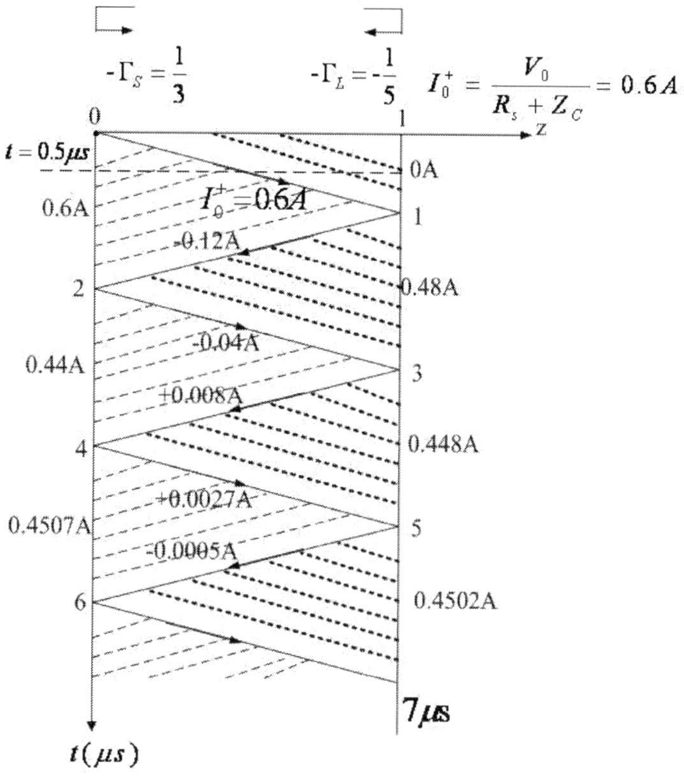

FIGURE 18

the initial region with $0V$. It meets the first wave line at $0.5\mu s$. at which point the voltage transitions from $0V$ to $60V$, which is the voltage in the next region. The observation line remains in the $60V$ region until it meets the next signal line at $t = 1.5\mu s$, at which point the signal transitions to the voltage $72V$ in the next shaded region. The process may be followed for a sufficiently long time interval, until the voltage reaches about a steady state value.

Using the above information, the required waveform can be sketched. The result may be verified with Fig.16, which we have already sketched using direct calculations for the particular case, without using the bounce diagram. Using the same procedure, we may now like to sketch a new waveform from the current bounce diagram, For example, the current waveform observed at $z = 0$, as a function of t, may be obtained from the current bounce diagram of Fig.18. The waveform is sketched in Fig.19,

The above two waveforms were sketched from the bounce diagrams, as a function of t, for a given location z. Similar procedure can be followed to sketch a waveform as a function of z, observed at a given t.

For example, we will sketch $I(z)$ observed at time $t = 0.5\mu s$. This is done by drawing a horizontal line in the current bounce diagram of Fig.18, passing through the time scale at the given time $t = 0.5\mu s$. Starting at $z = 0$, the observation line goes through the shaded region with current $0.6A$. It encounters a wave line at $z = l/2$. at which point it transitions into the shaded region with current $0A$. The observation line remains in this

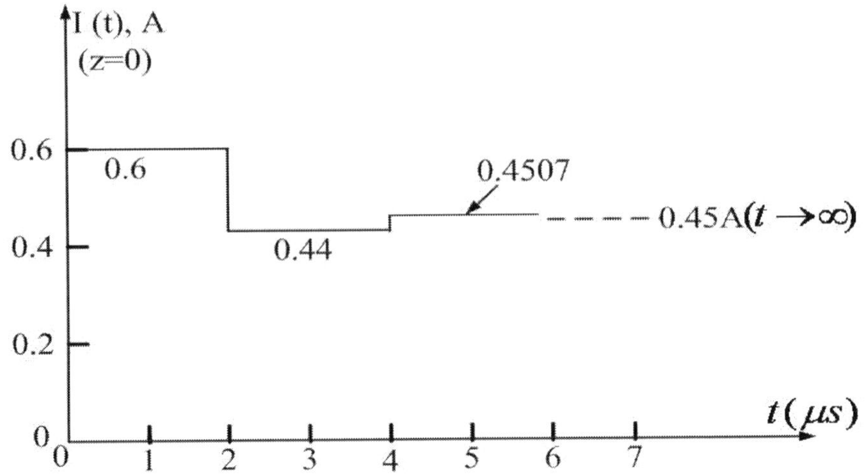

FIGURE 19

region up till the load end at $z = l$. Accordingly, the required waveform is sketched in Fig.20.

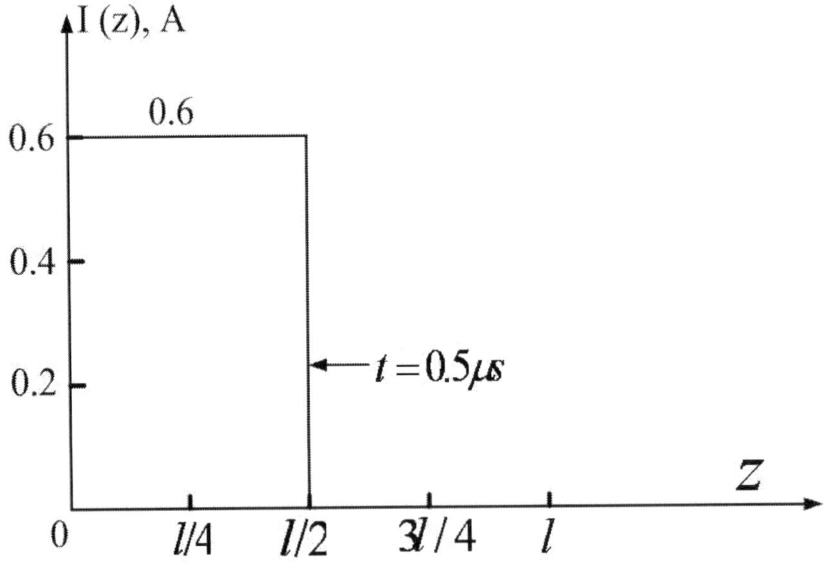

FIGURE 20

We have illustrated how to sketch waveforms using bounce diagram for selected cases. Waveforms for other specific cases can be similarly sketched (see problems). We have restricted our observations to a given location z or a given time t In principle, one may have an arbitrary curved observation line, representing the space-time coordinates of a moving observer. We may as well extend the procedure we have discussed here to sketch the waveform measured by such a moving observer.

8.2. Pulse Function Response. Fig.21 shows the same circuit of Fig.15, but with pulse source function, $V_s(t) = V_0 p(t;\tau)$, of amplitude V_0 and width τ. A pulse source of width τ may be expressed as a difference between a step function and and another step function delayed by time τ.

$$\begin{aligned} V_0 p(t;\tau) &= V_0 u(t) - V_0 u(t-\tau) \\ &= V_0, \ 0 < t < \tau \\ &= 0, \ t < 0, \ t > \tau \end{aligned} \tag{4.74}$$

In the last section, we have derived the response to a step-function source $V_0 u(t)$. The response for a delayed step-function source $V_0 u(t-\tau)$ can be obtained from that for $V_0 u(t)$ by simply delaying the time scale of the response by τ. The same bounce diagrams of Figs.17,18 would apply for the $V_0 u(t-\tau)$ source as well, except the bounce diagrams will have to start at time $t = \tau$. For this use, all labels in the time scales of Figs.17,18 will have to be increased by the time τ. Now, using the principle of superposition, the response to $V_0 p(t)$ can be obtained by simply subtracting the response to $V_0 u(t-\tau)$ from that to $V_0 u(t)$.

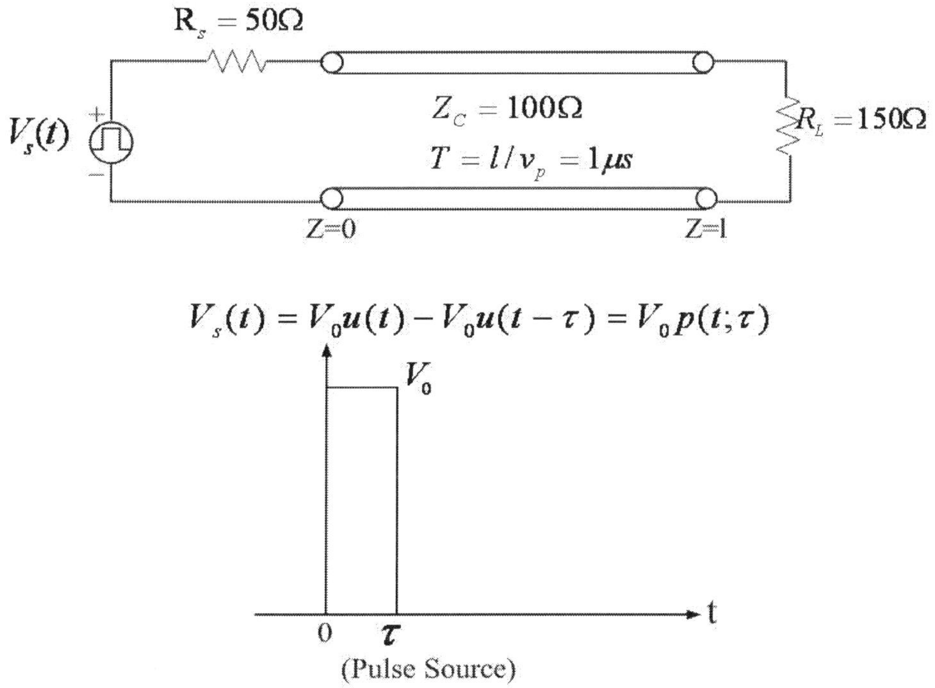

FIGURE 21

As an example, a voltage waveform for the circuit of Fig.21, excited with the pulsed source $V_0 p(t,\tau)$ of width $\tau = 0.5\mu s$ and amplitude $V_0 = 90V$, and observed at $z = l/2$, is shown as a function of t in Fig.22. This is obtained from the voltage waveform at $z = l/2$ for a step-function source $V_0 u(t) = 90u(t)$, which is already available in Fig.16. As per the procedure discussed above, the waveform for the pulsed source in Fig.22 is obtained by taking the waveform in Fig.16 and subtracting from it a delayed version of itself with a time shift of $\tau = 0.5\mu s$. It may be accomplished by observation.

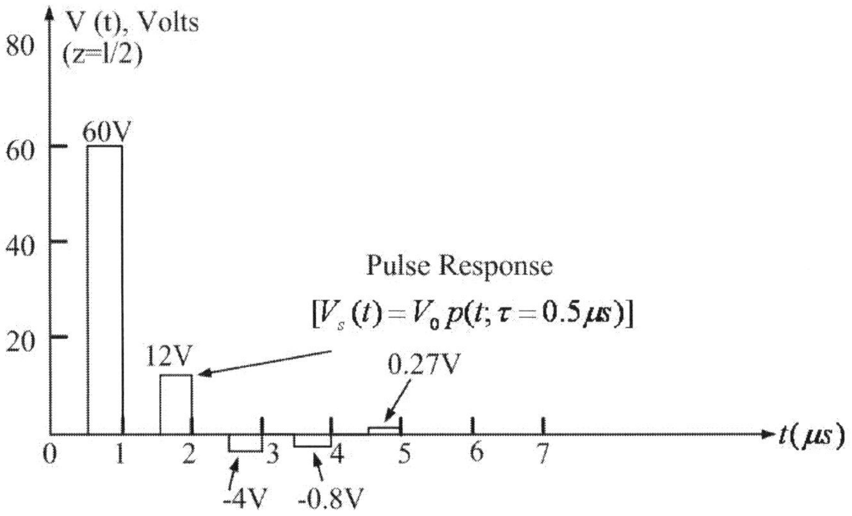

FIGURE 22

As another illustration of the procedure, for the same pulse source in Fig.21, we will now sketch the current waveform as a function of z, observed at a given time $t = 2.75\mu s$. First, the waveform $I(z)$ observed at time $t = 2.75\mu s$ is sketched for excitation $V_0 u(t)$. This is obtained from the current bounce diagram of Fig.18, using the procedure described in the last section 8.1.2. Then repeat the procedure to sketch $I(z)$ at $t = 2.75\mu s$ for excitation $V_0 u(t - \tau)$; $\tau = 0.5\mu s$). The same procedure and the bounce diagram of Fig.18 is used, except in this case the time labels are to be increased by $\tau = 0.5\mu s$. Accordingly, the observation time $t = 2.75\mu s$ of our interest would correspond to $t = 2.75 - 0.5 = 2.25\mu s$ in the original bounce diagram of Fig.18. The two waveforms for $I(z)$, observed at $t = 2.75\mu s$, with different source functions $V_0 u(t)$ and $V_0 u(t - 0.5\mu s)$, are shown together in Fig.23. The required waveform for the pulsed source $V_0 p(t; \tau = 0.5\mu s)$ is obtained by subtracting the waveforms for the $V_0 u(t - 0.5\mu s$ from that for the $V_0 u(t)$ source, and the result is shown together with the two parts in Fig.23. .

9. Time-Harmonic Waves in a Transmission Line

In this section we will study the wave propagation in a transmission line with a harmonic source function. The field and voltage-current relations we have developed in sections (1-3) are valid for any general time variation. These results were then used for the wave analysis of loss-less transmission lines in sections (4-7), which were also valid for

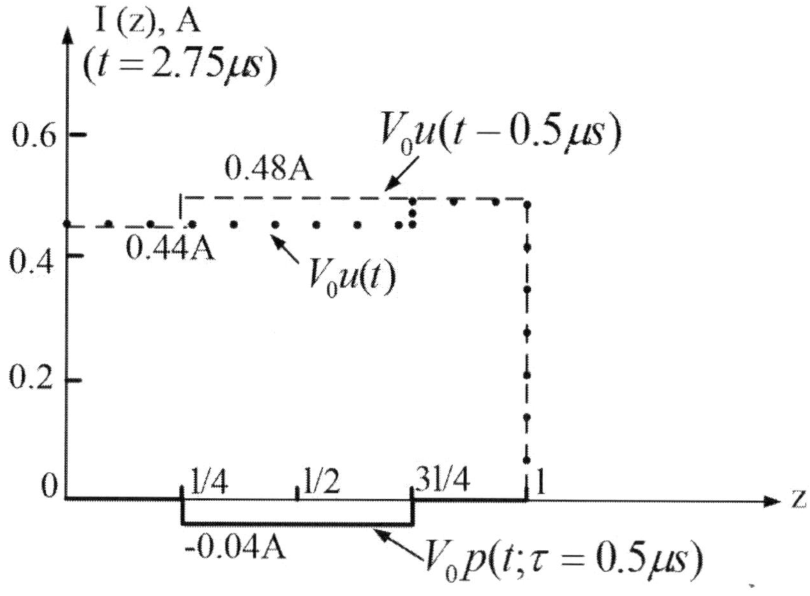

FIGURE 23

any general time variation. Accordingly, these results should also apply for the specific case of a harmonic time variation. However, for the specific situation of a harmonic time variation, significant simplification can be possible in all the general analyses by using complex phasor representation of the harmonic signals. The wave analyses in sections (4-7) were developed for general time variations, but are valid only for loss-less transmission lines and resistive source and load impedances. The analyses in sections (4-7) can not be directly applied to lossy transmission lines and complex source and load impedances that are inductive or capacitive. However, under a time-harmonic condition, the phasor analysis would allow us to extend the results in sections (4-7) to general cases with complex impedances as well as lossy transmission lines.

In the last section 8 we have studied simple transmission line problems, excited with a step- or a pulse-function source. The essential concepts of signal excitation and bouncing we have learned from the simple source excitations can be, in principle, extended to any general waveform and other circuit configurations. However, the procedure can be prohibitively difficult for understanding and analysis for general waveforms and more complex circuit arrangements, with multiple transmission line sections, multiple sources, and complex source and loads. The procedure may not also apply to lossy transmission lines. Under such circumstances, it may be useful to express any general waveform in terms of harmonic functions using Fourier decomposition. Then, the theory of the harmonic waves to be developed in this section can be applied to model each Fourier function, and the results can then be added together to obtain the total response. Accordingly, the study of the harmonic waves may be used as a essential foundation for understanding and analysis of general waves.

9.1. Time-Harmonic Wave Function for a Loss-Less Transmission Line. Using the basic phasor concept in Appendix B, the voltage-current relations (4.30,4.31) and the wave equation (4.36) may be expressed in phasor forms. We will assume that the complex phasors of different functions refer to peak phasors. For notational economy, we will use the same variable to represent the phasor and its time-dependent function. The distinction should be self evident from the context of the discussion.

$$\frac{\partial V(z)}{\partial z} = -j\omega \underline{L} I(z), \quad \frac{\partial I(z)}{\partial z} = -j\omega \underline{C} V(z)$$
$$\frac{\partial^2 V(z)}{\partial z^2} = -j\omega \underline{L}\frac{\partial I(z)}{\partial z} = -\omega^2 \underline{LC} V(z), \quad \frac{\partial^2 I(z)}{\partial z^2} = -j\omega \underline{C}\frac{\partial V(z)}{\partial z} = -\omega^2 \underline{LC} I(z)$$
$$\frac{\partial^2 f(z)}{\partial z^2} = -\omega^2 \underline{LC} f(z) = -\frac{\omega^2}{v_p^2} f(z), \ f \equiv V, I, \quad v_p = \frac{1}{\sqrt{\underline{LC}}} \tag{4.75}$$

$V(z)$, $I(z)$ and $f(z)$ are the complex phasors of the respective time-dependent functions. As per the phasor concepts in Appendix B, a time-derivative operator in the equations (4.30,4.31) and (4.36) are transformed into an algebraic multiplication by $j\omega$ in the above phasor expressions. Tbe above wave equation (4.75) is a second-order differential equation in z, which can be solved to obtain the expression for the phasor wave function $f(z)$,

$$f(z) = Be^{\mp j\beta z} = Ae^{j\phi}e^{\mp j\beta z}, \ \beta = \omega\sqrt{\underline{LC}} = \frac{\omega}{v_p}, \ A = |B|, \ \arg(B) = \phi \tag{4.76}$$

where B is a complex constant, independent of z, with magnitude A and phase ϕ. Note that the sign in the exponential factor could be either negative $(-)$ or positive $(+)$, which would correspond to wave propagation in the $+z$ and $-z$ directions, respectively. The complex coefficient B, and the corresponding magnitude A and phase ϕ, may be in general different for the cases, and represented in accordance with the direction of propagation with superscript $(+)$ or $(-)$, respectively. The above choices are represented in (4.77) using a double-sign notation, where the upper sign of all double-sign notations, and similarly all lower signs, would correspond with each other. This is consistent with the notation in (4.37), and its interpretation in (4.41,4.42), used for waves with general time dependence.

$$f(z) = B^{\pm}e^{\mp j\beta z} = A^{\pm}e^{j\phi^{\pm}}e^{\mp j\beta z}, \ \beta = \sqrt{\underline{LC}} = \frac{\omega}{v_p}$$
$$A^{\pm} = |B^{\pm}|, \ \arg(B^{\pm}) = \phi^{\pm} \tag{4.77}$$

Now, the time-dependent form of the wave function $f(z,t)$ may be expressed using the complex phasor expression $f(z)$.

$$f(z,t) = Re(A^{\pm}e^{j\phi^{\pm}}e^{\mp j\beta z}e^{j\omega t}) = A^{\pm}\cos(\omega t \mp \beta z + \phi^{\pm}), \ \beta = \frac{\omega}{v_p} \tag{4.78}$$

The above time-dependent expression for a harmonic wave may also be directly derived from the general wave function (4.37) we have established in section 4.3. The general form for a wave function in (4.37) is applicable for any arbitrary time variation. The wave function for the specific case of a time-harmonic wave may be obtained from (4.37) by selecting $f(s) = A^{\pm}\cos(\beta s \mp \phi^{\pm})$.

$$f(z,t) = f(s)\big|_{s=z\mp v_p t} = A^{\pm}\cos(\beta s \mp \phi^{\pm})\big|_{s=z\mp v_p t} = A^{\pm}\cos(\beta z \mp \beta v_p t \mp \phi^{\pm})$$
$$= A^{\pm}\cos(\omega t \mp \beta z + \phi^{\pm}),\ \beta = \frac{\omega}{v_p} \quad (4.79)$$

This is the same expression we obtained above in (4.78) from the phasor analysis.

9.2. Propagation Velocity and Wavelength. As discussed earlier, the choice of the upper or lower sign of the double-sign notations refer to propagation in the $+z$ and $-z$ direction, respectively. And, the parameter v_p is the ratio of distance Δz the wave function shifts its relative position along the propagation direction, over a given time Δt.

$$f(z, t=0) = A^{\pm}\cos(\mp\beta z + \phi^{\pm}) = A^{\pm}\cos(\beta z \mp \phi^{\pm})$$
$$f(z, t=\Delta t) = A^{\pm}\cos(\omega\Delta t \mp \beta z + \phi^{\pm}) = A^{\pm}\cos(\beta z \mp \omega\Delta t \mp \phi^{\pm})$$
$$= A^{\pm}\cos(\beta(z \mp \frac{\omega}{\beta}\Delta t) \mp \phi^{\pm}) = f(z \mp \Delta z, t=0)$$
$$\Delta z = \frac{\omega}{\beta}\Delta t,\ v_p = \frac{\Delta z}{\Delta t} = \frac{\omega}{\beta} \quad (4.80)$$

This concept of wave propagation and velocity is illustrated for a harmonic wave in Fig.24, for a simple case with $A^{\mp} = 1$, and $\phi^{\mp} = 0$. This is similar to the illustration in Fig.7 for a general waveform. Fig.24 shows how the wave shifts a distance $\Delta z = v_p\Delta t$ over time Δt, along a propagation direction, depending on the choice of the sign in the wave function.

The periodic time variation of a harmonic wave function results in also a periodic space variation, as shown in Fig.24. The periodic distance in space is called the wave length λ.

$$\beta\lambda = 2\pi,\ \lambda = \frac{2\pi}{\beta} \quad (4.81)$$

β is called the propagation constant. The periodicity λ in space is analogous to the periodicity T in time.

$$\omega T = 2\pi,\ T = \frac{2\pi}{\omega} \quad (4.82)$$

Accordingly, the propagation constant β in (4.81) for space variation may be seen as analogous to the angular frequency ω in (4.82) for time variation. In that sense, the propagation constant may be considered as the "angular space frequency."

Now, using (4.81) in (4.80), we get a relationship between v_p, ω and λ

$$v_p = \frac{\omega}{\beta} = f\lambda,\ \omega = 2\pi f,\ f = \frac{1}{T} \quad (4.83)$$

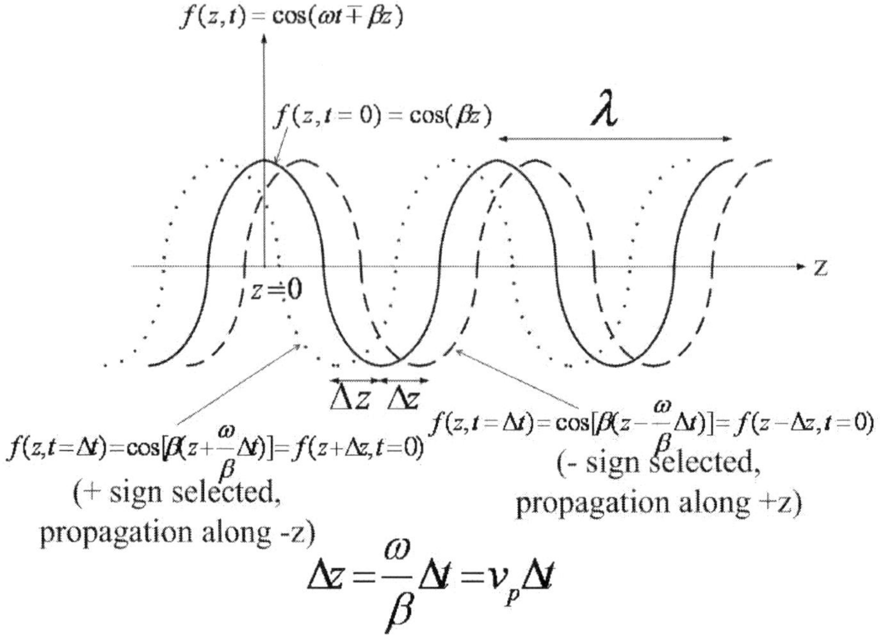

FIGURE 24

where $f = 1/T$ is the frequency of the wave.

9.3. Standing Wave in a Short Circuited Transmission Line. We will now use the phasor wave functions derived in the last section to study harmonic wave in a transmission line with a short-circuited load (see Fig.25). We have used the reference position $z = 0$ at the load end of the transmission line, The forward voltage wave, propagating in the $+z$ direction, is represented with a complex coefficient V^+ and a propagation factor $e^{-j\beta z}$. And, the backward voltage wave, propagating in the $-z$ direction, is represented with a complex coefficient V^- and a propagation factor $e^{+j\beta z}$. Accordingly, the phasor for the forward voltage wave at the load ($z = 0$) is V^+, and that for the backward voltage wave is V^-. The phasor for the total voltage wave in the transmission line is the superposition of the forward and backward components.

$$V(z) = V^+ e^{-j\beta z} + V^- e^{+j\beta z}, \; V(z = 0) = V^+ + V^- \tag{4.84}$$

The expression for the phasors for the current waves propagating in the forward and backward directions, and the phasor of the total current wave, may be expressed using the characteristic impedance Z_c, as defined in (4.47.

$$\begin{aligned} I(z) &= I^+ e^{-j\beta z} + I^- e^{+j\beta z} = \frac{V^+}{Z_c} e^{-j\beta z} - \frac{V^-}{Z_c} e^{+j\beta z} \\ I(z = 0) &= I^+ + I^-, \; \frac{V^+}{I^+} = Z_c, \; \frac{V^-}{I^-} = -Z_c \end{aligned} \tag{4.85}$$

The ratio of the forward and the backward waves at the load position, $z = 0$, is the reflection coefficient Γ_L. In reference to section 6, the reflection coefficient for a short circuit is -1. Using this relation in the above expressions for the voltage and current phasors,

$$V(z) = V^+(e^{-j\beta z} - e^{+j\beta z}) = -2jV^+\sin(\beta z), \quad |V(z)| = 2|V^+||\sin(\beta z)|$$
$$I(z) = \frac{V^+}{Z_c}(e^{-j\beta z} + e^{+j\beta z}) = 2\frac{V^+}{Z_c}\cos(\beta z), \quad |I(z)| = 2\frac{|V^+|}{Z_c}|\cos(\beta z)|$$
$$\frac{V^-}{V^+} = \Gamma_L = -1 \tag{4.86}$$

The magnitudes of the complex phasors for the voltage and current waves are plotted in Fig.25. The phasor magnitudes represent the peak amplitudes. Accordingly, in reference to Fig.25, the voltage and current waves would have zero peak amplitudes at periodic locations, with periodicity $\Delta = \lambda/2$. These locations are referred to as nodes or nulls. Because the peak amplitudes are zero, the signals at these locations would be zero at all times. This is in contrast with a propagating wave, in which any null location would shift in time, as the entire wave shifts along the propagation direction. The wave in Fig.25 may be visualized with its nodes fixed or standing at periodic locations, as time progresses. Accordingly, the wave in Fig.25 is referred to as a standing wave.

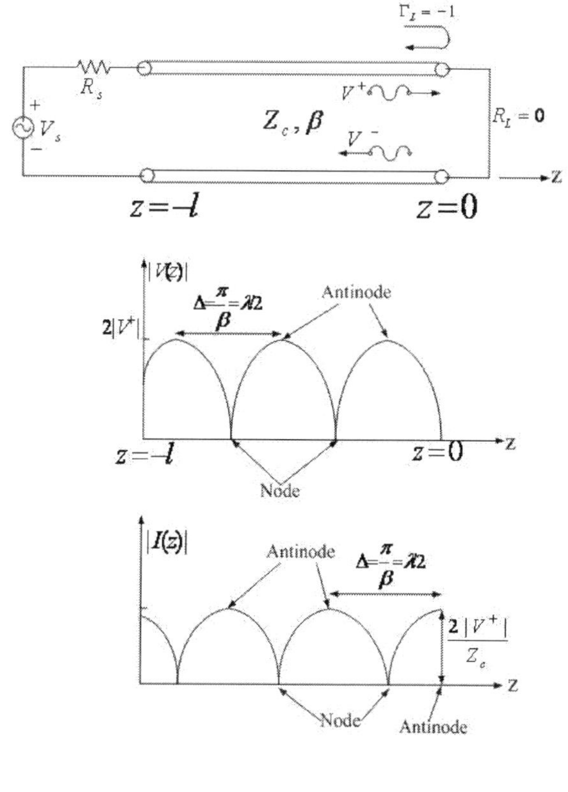

FIGURE 25

The phasor magnitude of a signal is maximum, or equivalently the peak amplitude of the signal is maximum, at locations which are equidistant from consecutive null locations for the particular signal, Accordingly, like the null locations, the locations with maximum phasor magnitude are also periodically spaced, with periodicity $\Delta = \lambda/2$. Locations with maximum phasor magnitude are referred to as anti-nodes or maximas. Notice that, the maximas for the current signal coincide with nulls for the voltage signal, and nulls for a current signal coincide with maximas for a voltage signal. As expected, there is a voltage null at the short-circuit location at $z = 0$, which is also a current maxima.

9.3.1. *Input Impedance.* The voltage and current in Fig.25 change with location z, as expressed in (4.86). The ratio of the total voltage and current phasors at a location z is the equivalent impedance $Z(z)$ seen across the two conductors of the transmission line at the location. The input impedance Z_{in} of a short-circuited transmission line of length l is the impedance $Z(z = -l)$ computed at $z = -l$

$$Z(z) = \frac{V(z)}{I(z)} = -jZ_c \tan(\beta z), \; Z_{in} = Z(z = -l) = jZ_c \tan(\beta l)$$
$$Z_{in} = jX_{in}, \; X_{in} = Z_c \tan(\beta l) \quad (4.87)$$

$Z_{in} = jX_{in}$ in this case is seen to be purely reactive in nature. The reactance X_{in} is plotted in Fig.26 as a function of length l. X_{in} is a periodic function of l, repeating at increments of $l = \lambda/2$. X_{in} is zero for $l = 0$, which should be expected because in this case we simply have the short-circuit load, with no transmission line segment connected to it. For $l = \lambda/4$ and periodic increments there on, the input reactance is infinity. This means that a short-circuited transmission line with these lengths would behave like an open circuit. Note that a short-circuited segment of a transmission line is normally expected to behave as short circuit, independent of the length of the transmission line segment, for a DC or low-frequency operation. It is interesting to know that is not the case in general situations. Contrastingly, it may even behave as an open circuit for a suitable length of the transmission line.

When the frequency of application is sufficiently high, the wave length λ can be sufficiently small. In this case, a transmission line segment of a reasonable practical length l can be designed with $0 < l < \lambda/2$. Under the situation, in reference to Fig.26, the input reactance X_{in} seen across the transmission line segment, with a short-circuit load, can be designed with any possible value $-\infty < X_{in} < +\infty$. The reactance can be either capacitive impedance in the range $\lambda/4 < l < \lambda/2$, or inductive in the range $0 < l < \lambda/4$. A transmission line segment, so designed, may be used as a reactive element in practical circuits, such as a signal filter or a impedance-matching circuit.

The voltage and current phasors in (4.86) are expressed in terms of one unknown variable V^+, which would depend on the source excitation. The transmission line segment in Fig.25 may be replaced by its equivalent input reactance X_{in}, as given by (4.87). The equivalent circuit is shown with Fig.26. Then, a circuit analysis can be used to relate the input voltage at $z = -l$ to the source voltage V_s. Accordingly, the unknown parameter V^+ can be solved, and voltage and current signals at all locations can be completely determined in terms of the source voltage.

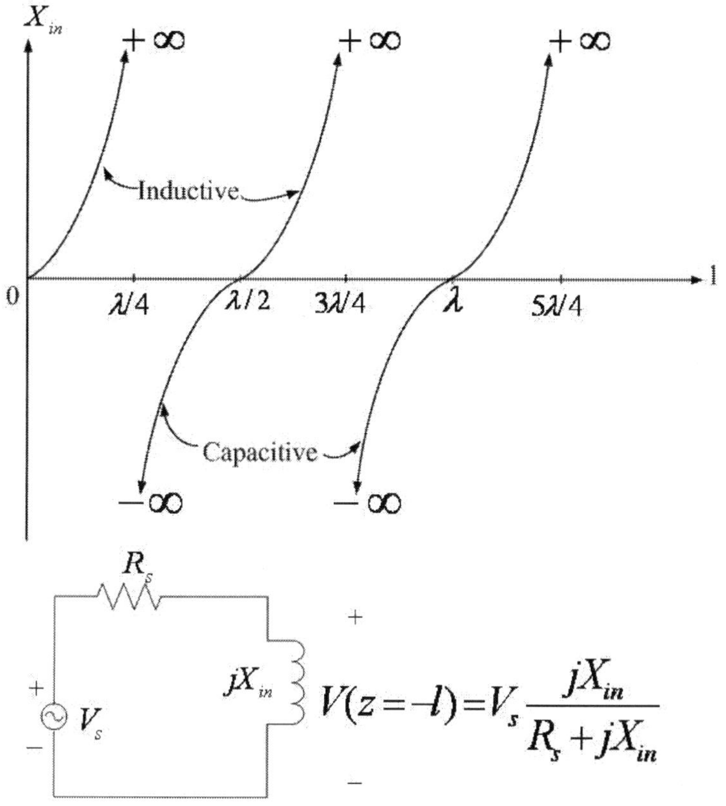

FIGURE 26

$$V(z=-l) = V_s\frac{jX_{in}}{R_s+jX_{in}} = V_s\frac{jZ_c\tan(\beta l)}{R_s+jZ_c\tan(\beta l)} = V^+2j\sin(\beta l)$$

$$V^+ = V_s\frac{Z_c}{2(R_s\cos(\beta l)+jZ_c\sin(\beta l))} \tag{4.88}$$

9.4. Standing Wave in an Open-Circuited Transmission Line. Fig.27 shows an open-circuited transmission line segment, connected to an excitation source. The analysis of the standing wave in a short-circuited transmission line, which we presented in the above section 9.3, may be repeated for the circuit of Fig.27, but with R_L equal to ∞ instead of zero. For an open-circuit load, the reflection coefficient $\Gamma_L = +1$, as compared to $\Gamma_L = -1$ for a short-circuit load. Incorporating this change in the analysis of the last section 9.3, we will write final expressions of different parameters, and discuss important observations.

$$V(z) = V^+(e^{-j\beta z}+e^{+j\beta z}) = 2V^+\cos(\beta z),\ \ |V(z)| = 2|V^+||\cos(\beta z)|$$

$$I(z) = \frac{V^+}{Z_c}(e^{-j\beta z}-e^{+j\beta z}) = -2j\frac{V^+}{Z_c}\sin(\beta z),\ |I(z)| = 2\frac{|V^+|}{Z_c}|\sin(\beta z)|$$

$$\frac{V^-}{V^+} = \Gamma_L = +1 \tag{4.89}$$

The magnitudes of the voltage and current phasors for the transmission line circuit of Fig.27, with the open-circuited load, are plotted in Fig.28. The standing wave pattern in Fig.27 may be compared with similar pattern in Fig.25 for the case with a short-circuit load. The null and maximas for voltage and current signals in Fig.25 are interchanged with corresponding maxima and null locations in Fig.27, respectively. The periodicity of the standing-wave patterns in both cases of Fig.25 and Fig.27 is $\Delta = \lambda/2$

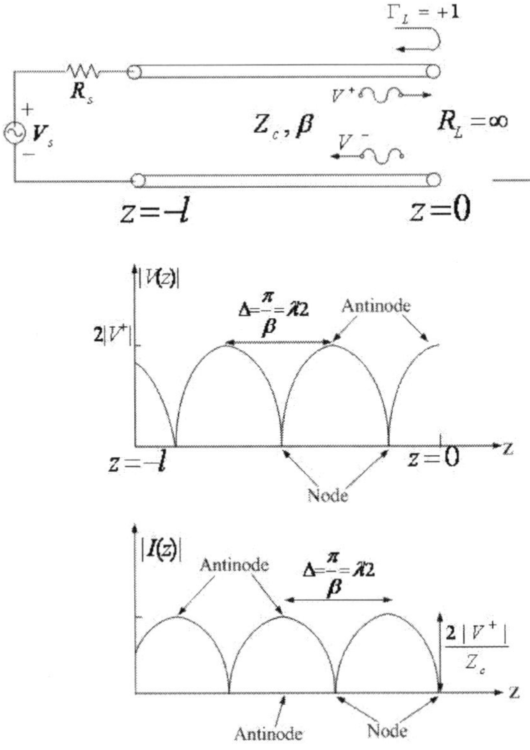

FIGURE 27

9.4.1. *Input Impedance.* The input impedance Z_{in} of an open-circuited transmission line of length l may be expressed as follows.

$$Z(z) = \frac{V(z)}{I(z)} = jZ_c \cot(\beta z),\ Z_{in} = Z(z = -l) = -jZ_c \cot(\beta l)$$

$$Z_{in} = jX_{in},\ X_{in} = -Z_c \cot(\beta l) \tag{4.90}$$

The reactance X_{in} is plotted in Fig.28 as a function of length l. X_{in} is a periodic function of l, repeating at increments of $l = \lambda/2$ X_{in} is infinity for $l = 0$, which should be expected because in this case we simply have the open-circuit, with no transmission line segment connected to it. For $l = \lambda/4$ and periodic increments there on, the input reactance is zero. This means that an open-circuited transmission line with these lengths would behave like a short circuit. X_in in Fig.28 is capacitive for a small length $l < \lambda/4$. This is in contrast with Fig.26, in which case the reactance is inductive for the same range of length l. Accordingly, a small transmission line segment with an open-circuit termination is normally used to design an equivalent capacitance, whereas that with a short-circuit termination may be used to design an equivalent inductance.

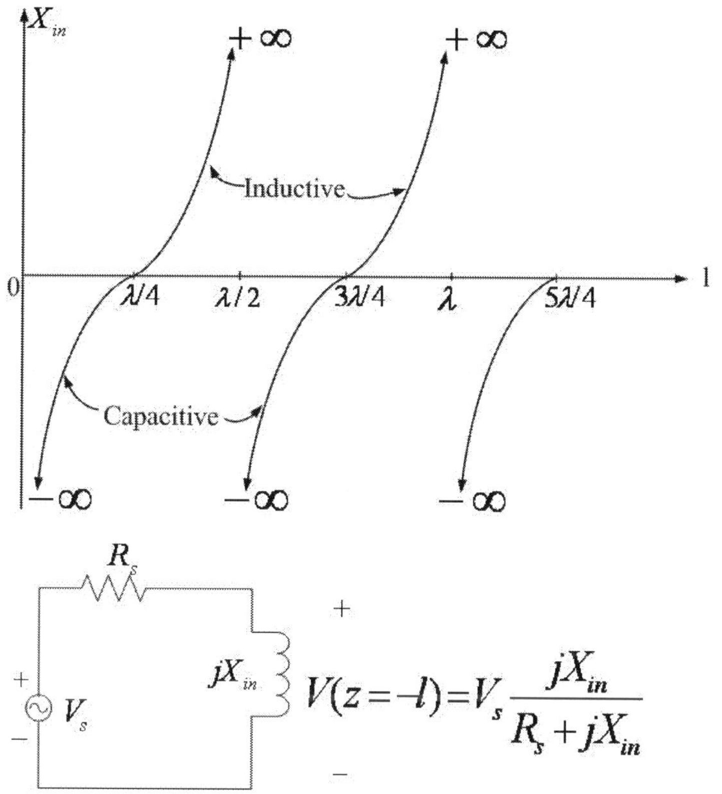

FIGURE 28

The only unknown variable V^+ in the expressions of (4.91) may be related to the source voltage V_s as follows, The procedure is similar to that used for the short-circuited transmission line in last section.

$$V(z=-l) = V_s\frac{jX_{in}}{R_s+jX_{in}} = V_s\frac{-jZ_c\cot(\beta l)}{R_s - jZ_c\cot(\beta l)} = V^+ 2\cos(\beta l)$$

$$V^+ = V_s\frac{-jZ_c}{2(R_s\sin(\beta l) - jZ_c\cos(\beta l))} = V_s\frac{Z_c}{2(jR_s\sin(\beta l) + Z_c\cos(\beta l))} \tag{4.91}$$

9.5. Standing Wave in a Transmission Line with a General Load Z_L. We will now extend the standing wave analysis, which we developed in the last sections for simple cases of short- and open-circuited transmission lines, to a general case for a transmission line terminated by any complex load $Z_L = R_L + jX_L$. The circuit is shown in Fig.29. In this case, the reflection coefficient Γ_L would be a general complex number with magnitude less than or equal to unity, and can be expressed as function of Z_L using (4.62). The voltage and current phasors can be expressed in terms of the complex Γ_L

$$V(z) = V^+e^{-j\beta z} + V^-e^{+j\beta z} = V^+e^{-j\beta z}(1+\Gamma_L e^{+2j\beta z}) = V^+e^{-j\beta z}(1+\Gamma(z))$$
$$I(z) = \frac{V^+}{Z_c}e^{-j\beta z} - \frac{V^-}{Z_c}e^{+j\beta z} = \frac{V^+}{Z_c}e^{-j\beta z}(1-\Gamma_L e^{+2j\beta z}) = \frac{V^+}{Z_c}e^{-j\beta z}(1-\Gamma(z))$$
$$\Gamma(z) = \frac{V^-e^{+j\beta z}}{V^+e^{-j\beta z}} = \frac{V^-}{V^+}e^{-2j\beta z} = \Gamma_L e^{-2j\beta z},\ \Gamma(z=0) = \Gamma_L = \frac{V^-}{V^+}$$
$$\Gamma_L = \frac{Z_L - Z_c}{Z_L + Z_c},\ |\Gamma_L| = |\frac{R_L + jX_L - Z_c}{R_L + jX_L + Z_c}| = \sqrt{\frac{(R_L - Z_c)^2 + X_L^2}{(R_L + Z_c)^2 + X_L^2}} \leq 1,\ R_L, Z_c \geq 0 \quad (4.92)$$

Γ_L is the ratio of the backward to the forward voltage waves, measured at the load point. The ratio, measured at any general location z, is $\Gamma(z)$. The magnitudes of the voltage and current phasors can be expressed in terms of $\Gamma(z)$ as follows.

$$|V(z)| = |V^+||e^{-j\beta z}||(1+\Gamma_L e^{+2j\beta ig.z})| = |V^+||(1+\Gamma(z))|$$
$$|I(z)| = |\frac{V^+}{Z_c}||e^{-j\beta z}||(1-\Gamma_L e^{+2j\beta z})| = |\frac{V^+}{Z_c}||(1-\Gamma(z))|$$
$$\frac{|V(z)|}{|V^+|} = |(1+\Gamma(z))|,\quad \frac{|I(z)|Z_c}{|V^+|} = |(1-\Gamma(z))|$$
$$|\Gamma(z)| = |\Gamma_L e^{2j\beta z}| = |\Gamma_L||e^{2j\beta z}| = |\Gamma_L| \leq 1$$
$$\arg(\Gamma(z)) = \arg(\Gamma_L e^{2j\beta z}) = \arg(\Gamma_L) + \arg(e^{2j\beta z}) = \arg(\Gamma_L) + 2\beta z \quad (4.93)$$

We will first study the z dependence of the total voltage phasor $V(z)$. The analysis can then be similarly extended to study the z dependence of the current phasor $I(z)$. The z dependence of the magnitude $|V(z)|$ is determined by the magnitude of the factor $(1+\Gamma(z))$, which is equal to the magnitude $|V(z)|$ normalized with respect to $|V^+|$.

9.5.1. *Phasor Diagram.* The magnitude of the factor $(1+\Gamma(z))$, which determines the z dependence of $|V(z)|$, maybe graphically evaluated using a phasor diagram, shown in Fig.29. The above factor is sum of the unit-phasor, which is independent of z, and a variable phasor $\Gamma(z)$. Further, the magnitude of the variable phasor $\Gamma(z)$ is equal to $|\Gamma_L|$, which is independent of z. Whereas, the phase of the of the variable phasor $\Gamma(z)$ is dependent on z. Therefore, the variable phasor $\Gamma(z)$, and accordingly the total factor $(1+\Gamma(z))$, trace a circular locus for different observation points z.

As shown in (4.93), the phase of of $\Gamma(z)$ is a linear function of z. As the observation point moves away from the load end ($z=0$) towards the source, we have z increasing negatively, in which case the variable phase $\arg(\Gamma(z))$ would increase negatively. Therefore, the circular locus of $(1+\Gamma(z))$ in Fig.29 is shown to trace clockwise, as the observation point moves away from the load.

9.5.2. *Periodicity of Standing Wave.* The locus in Fig.29 traces a full circle when $2\beta z$ changes by 2π. Accordingly, the pattern for $|V(z)|$ would repeat over a distance

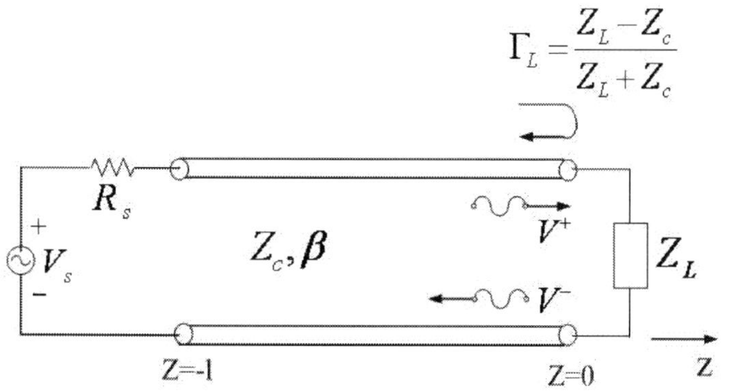

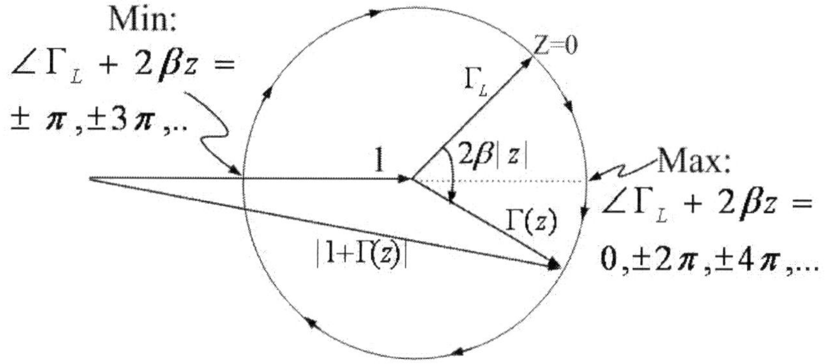

FIGURE 29

$\Delta z = \pi/\beta$, which is $\lambda/2$. This is same as the periodicity of the standing waves for a short-circuit or an open-circuit load, studied in the last sections.

$$2\beta\Delta = 2\pi, \quad \Delta = \frac{\pi}{\beta} = \frac{\pi}{(2\pi/\lambda)} = \frac{\lambda}{2} \tag{4.94}$$

9.5.3. *Standing Wave Ratio.* The magnitude of the total factor $(1+\Gamma(z))$ periodically oscillates between a maximum value equal to $(1+|\Gamma_L|)$ and a minimum value equal to $(1-|\Gamma_L|)$. The resulting voltage phasor magnitude $|V(z)|$ is sketched in Fig.30 as a function of z, showing the oscillatory nature. The ratio of the maximum and minimum magnitude of the voltage phasor is determined by $|\Gamma_L|$. This ratio is referred as the standing wave ratio (SWR). SWR is always greater than or equal to 1. SWR is one when $\Gamma_L = 0$, which occurs when $Z_L = Z_c$. SWR is infinity when $|\Gamma_L| = 1$, which occurs when the load is a short or an open circuit.

$$\frac{|V(z)|_{max}}{|V(z)|_{min}} = \frac{|V^+||(1+|\Gamma_L|)|}{|V^+||(1-|\Gamma_L|)|} = \frac{|(1+|\Gamma_L|)|}{|(1-|\Gamma_L|)|} = SWR \geq 1$$
$$SWR = 1, \ \Gamma_L = 0, \ Z_L = Z_c; \ SWR = +\infty, \ |\Gamma_L| = 1, \ |Z_L| = 0, \ \infty \tag{4.95}$$

The lowest value of $SWR = 1$ represents a wave with no backward wave ($V^- = 0$). Therefore, the total signal in this case consists of only a forward wave, which is a purely propagating wave with no standing-wave behavior. The highest value of $SWR = \infty$, on the other hand, represents a wave with both the forward and backward waves of equal magnitudes. This special case occurs when the load termination is either a short circuit or an open circuit. The wave behavior in these specific cases have been studied in the last sections. They exhibit a perfect standing-wave behavior, with signal "nulls" which are frozen or fixed in time at periodic locations. Now, the lowest value of $SWR = 1$ represents a perfect propagating wave, and the highest value of $SWR = \infty$ represents a perfect standing wave. Accordingly, a general wave with SWR between these two limits may be interpreted as a "partial standing wave" or a "partial propagating wave." Larger is the value of SWR, closer is the wave behavior to a perfect standing wave with $|\Gamma_L| = 1$. On the other hand, smaller is the value of SWR, closer is the wave behavior to a pure propagating wave, with $\Gamma_L = 0$.

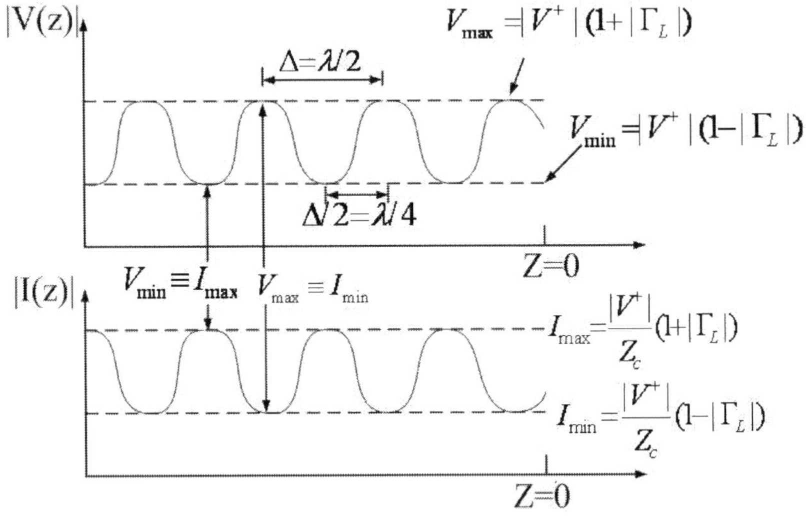

FIGURE 30

9.5.4. *Standing-Wave Maximas and Minimas.* The magnitude is maximum when the phase of the variable phasor $\Gamma(z)$ is $0, \pm 2\pi, \pm 4\pi, ...$, which corresponds to the right-most point on the real axis of the circular locus in Fig.29. Similarly, the magnitude is minimum when the phase of $\Gamma(z)$ is $\pm\pi, \pm 3\pi, \pm 5\pi, ...$, which corresponds to the left-most point on the real axis of the circular locus of Fig.29. The phase of $\Gamma(z)$ at the load point ($z = 0$), is equal to the phase angle $\arg(\Gamma_L)$, which is defined in the range $-\pi < \arg(\Gamma_L) < +\pi$. If $\arg(\Gamma_L) > 0$, the load point in the phasor diagram of Fig.29 would be in the upper half of the circular locus. In this case, as the locus moves clockwise, with observation point moving away from load, the first maximum point on the circular locus would be reached before the first minimum point. Therefore, the first voltage maximum would occur closer to load than first minimum. The distance $z = -d_{max}$ of the first voltage maximum may be related to $\arg(\Gamma_L)$

$$\begin{aligned} \arg(\Gamma(z)) &= \arg(\Gamma_L) + 2\beta z = \arg(\Gamma_L) - 2\beta d_{max} = 0 \\ \arg(\Gamma_L) &= 2\beta d_{max} = 4\pi\frac{d_{max}}{\lambda}, \ \frac{d_{max}}{\lambda} = \frac{\arg(\Gamma_L)}{4\pi} \end{aligned} \tag{4.96}$$

Similarly, one may see that when $\arg(\Gamma_L) < 0$, the first voltage minimum would occur closer to load than first maximum. This distance $z = -d_{min}$ of the first voltage minima may be related to $\arg(\Gamma_L)$.

$$\begin{aligned} \arg(\Gamma(z)) &= \arg(\Gamma_L) + 2\beta z = \arg(\Gamma_L) - 2\beta d_{min} = -\pi \\ \arg(\Gamma_L) &= 2\beta d_{min} - \pi = 4\pi\frac{d_{min}}{\lambda} - \pi, \ \frac{d_{min}}{\lambda} = \frac{\arg(\Gamma_L)}{4\pi} + \frac{1}{4} \end{aligned} \tag{4.97}$$

The standing-wave pattern of $|V(z)|$ repeats with periodicity $\Delta = \lambda/2$, as per (4.94). Therefore, once the first voltage maximum and minimum points are located, other maximum and minimum points may be located periodically, every half-wavelength apart. Further, as seen from the phasor diagram of Fig.29, consecutive maximum and minimum points are half-cycle apart, which corresponds to a distance of $\lambda/4$. Therefore, we need to locate only one maximum (or one minimum) point. The neighboring minimum (or maximum) point can be located $\lambda/4$ from the initial reference point. All other maximum and minimum points can then be periodically located.

9.5.5. *Current Versus Voltage Standing Wave.* The pattern for the current magnitude $|I(z)|$, as expressed in (4.93), may be evaluated using a procedure similar to that used for $|V(z)|$. As seen from (4.93), the z dependent factor for $|I(z)|$ is $|(1 - \Gamma(z))|$. This is in contrast with that for $|V(z)|$, which is $|(1 + \Gamma(z))|$. The difference between the z dependent factors for $|V(z)|$ and $|I(z)|$ is simply to have $-\Gamma(z)$ instead of $+\Gamma(z)$. $-\Gamma(z)$ and $+\Gamma(z)$ are diametrically opposite of each other on a complex phasor plane. Therefore, $|I(z)|$ can be evaluated using the same phasor diagram of Fig.29. Except, the locus of the z variation of $|I(z)|$ would be diametrically opposite, or half-cycle away, from that of $|V(z)|$ on the phasor circle of Fig.29. As we have discussed earlier, half of the circular locus is equivalent to a distance of $\lambda/4$. Accordingly, the z dependence for $I(z)$ would be the same as that for $|V(z)|$, but shifted by a distance $\lambda/4$. When the voltage magnitude $|V(z)|$ is maximum, the corresponding current magnitude $|I(z)|$ would be minimum, and vice versa. The pattern for $|I(z)|$ is sketched in Fig.30, as compared with that for $|V(z)|$, showing the relative positions of their maximum and minimum locations. The maximum and minimum values of $|I(z)|$ differ from those of $|V(z)|$ by a factor of Z_c. However, the ratio of the maximum and minimum values of $|I(z)|$ is the same as that for $|V(z)|$, which is the standing wave ratio SWR.

9.6. Impedance Transformation. The voltage and current along a transmission line, terminated by a load impedance Z_L at $z = 0$ (Fig.29) change with location z, as expressed in (4.92). The ratio of the total voltage $V(z)$ and current $I(z)$ phasors at a location z is an equivalent impedance $Z(z)$ seen across the two conductors of the transmission line at the location z.

$$Z(z) = \frac{V(z)}{I(z)} = \frac{V^+e^{-j\beta z} + V^-e^{-j\beta z})}{(\frac{V^+}{Z_c}e^{-j\beta z} - \frac{V^-}{Z_c}e^{-j\beta z})} = \frac{V^+(e^{-j\beta z} + \Gamma_L e^{-j\beta z})}{(V^+/Z_c)(e^{-j\beta z} - \Gamma_L e^{-j\beta z})}$$
$$= Z_c\frac{(e^{-j\beta z} + \Gamma_L e^{-j\beta z})}{(e^{-j\beta z} - \Gamma_L e^{-j\beta z})} = Z_c\frac{(1+\Gamma(z))}{(1-\Gamma(z))},\ \Gamma_L = \frac{V^-}{V^+},\ \Gamma(z) = \Gamma_L e^{-2j\beta z} \quad (4.98)$$

Using the relationship in (4.92) between the reflection coefficient Γ_L and the load impedance Z_L in the above equation, one may express $Z(z)$ in terms of the load impedance Z_L.

$$Z(z) = Z_c\frac{(e^{-j\beta z} + \frac{Z_L - Z_c}{Z_L + Z_c}e^{+j\beta z})}{(e^{-j\beta z} - \frac{Z_L - Z_c}{Z_L + Z_c}e^{+j\beta z})} = Z_c\frac{Z_L(e^{-j\beta z} + e^{+j\beta z}) + Z_c(e^{-j\beta z} - e^{+j\beta z})}{Z_L(e^{-j\beta z} - e^{+j\beta z}) + Z_c(e^{-j\beta z} + e^{+j\beta z})}$$
$$= Z_c\frac{Z_L\cos(\beta z) - jZ_c\sin(\beta z)}{Z_c\cos(\beta z) - jZ_L\sin(\beta z)} = Z_c\frac{Z_L - jZ_c\tan(\beta z)}{Z_c - jZ_L\tan(\beta z)},\quad \Gamma_L = \frac{Z_L - Z_c}{Z_L + Z_c} \quad (4.99)$$

$Z(z)$ computed at $z = -l$ is referred to as Z_in. This is the input impedance seen across one end of a transmission line of length l, when terminated by a load impedance Z_L across the other end.

$$Z_{in} = Z(z = -l) = Z_c\frac{Z_L\cos(\beta l) + jZ_c\sin(\beta l)}{Z_c\cos(\beta l) + jZ_L\sin(\beta l)} = Z_c\frac{Z_L + jZ_c\tan(\beta l)}{Z_c + jZ_L\tan(\beta l)} \quad (4.100)$$

Z_{in} is in general different from the load impedance Z_L. Z_{in} is a periodic function of l. repeating for every incremental length of $l = \Delta = \lambda/2$, or equivalently for every increment of $\beta l = \beta\Delta = \pi/2$. This is because, the tangent function in the relation (4.100) repeats for increments of its argument βl by an angle $\pi/2$, which corresponds to the the above increment in the length.

As per the relationship (4.100), any given load impedance Z_L can be modified or transformed into a different new impedance by simply connecting a transmission line segment of length l. This principle carries significant practical importance, and is referred to as the impedance-transformation principle. The above relationship (4.100) between Z_{in} and Z_L is referred to as the impedance-transformation relationship.

We will verify simple specific cases for the general transformation relationship (4.100). Substituting $l = 0$ in the general relation of (4.100) we will get $Z_{in} = Z_L$. This should be expected, because in this case the input is directly connected across the load Z_L. As per the periodic nature of Z_{in}, discussed earlier, Z_{in} would also be equal to Z_L when l is an integral multiple of half wave-length.

$$Z_{in} = Z_c\frac{Z_L + jZ_c\tan(\beta l)}{Z_c + jZ_L\tan(\beta l)} = Z_L,\ \tan(\beta l) = 0$$
$$\beta l = \frac{2\pi}{\lambda}l = n\pi,\ l = n\frac{\lambda}{2},\ n = 0, 1, 2... \quad (4.101)$$

By substituting $Z_L = Z_c$ in the general relation of (4.100), we would obtain the input impedance Z_{in} of a transmission line which is terminated by a load impedance equal to the characteristic impedance of the particular transmission line.

$$Z_{in} = Z_c \frac{Z_L + jZ_c \tan(\beta l)}{Z_c + jZ_L \tan(\beta l)} = Z_c, \; Z_L = Z_c \tag{4.102}$$

The reflection coefficient Γ_L in this case is zero, and the transmission line behaves as if it is an infinite long transmission line with no reflection. Accordingly, the input impedance is expected to be equal to that of an infinite-long transmission line, which is equal to the characteristic impedance of the line. This is consistent the above result (4.102).

By substituting $Z_L = 0$ or $Z_L = \infty$ in the general relation of (4.100), we would obtain the input impedance Z_{in} for a short- or open-circuited transmission line, respectively. The expressions for Z_{in} in these cases may be verified with those derived in (4.87) and (4.90), respectively,

$$\begin{aligned} Z_{in} = Z_c \frac{Z_L + jZ_c \tan(\beta l)}{Z_c + jZ_L \tan(\beta l)} &= jZ_c \tan(\beta l), \; Z_L = 0 \\ &= -jZ_c \cot(\beta l), \; Z_L = \infty \end{aligned} \tag{4.103}$$

9.6.1. *Quarter-Wave Transformer.* A specific case of the transmission line circuit of Fig.29, with length $l = \lambda/4$, is referred to as a quarter-wave transformer. The general relationship of (4.100) for this case may be expressed in a simple, useful form. For $l = \lambda/4$, we have $\beta l = (2\pi/\lambda)(\lambda/4) = \pi/2$, $\cos(\beta l) = 0$, and $\sin(\beta l) = 1$. Using these values in (4.100),

$$Z_{in} = Z_c \frac{Z_L \cos(\beta l) + jZ_c \sin(\beta l)}{Z_c \cos(\beta l) + jZ_L \sin(\beta l)} = Z_c \frac{jZ_c}{jZ_L} = \frac{Z_c^2}{Z_L}, \; l = \lambda/4 \tag{4.104}$$

Using the above relationship, the characteristic impedance Z_c of the transmission line of a quarter-wave transformer can be designed in order to transform any given real load $Z_L = R_L$ to any other desired real value $Z_{in} = R_{in}$.

$$Z_c = \sqrt{Z_L Z_{in}} = \sqrt{R_L R_{in}} \tag{4.105}$$

The length l for the quarter-wave transformer is designed to be $\lambda/4$. As per the periodic nature of Z_{in} we have discussed earlier, the same result would also be obtained by increasing the length by an integral multiple of $\lambda/2$. This corresponds to any design lengths of $l = \lambda/4 + n\lambda/2, n = 0, 1, 2....$

9.7. Time-Harmonic Solution for a Lossy Transmission Line. The equivalent circuit model for a lossy transmission line is shown in Fig.6. The voltage-current relations (4.28,4.29) for a lossy transmission line, which are derived from the circuit model of Fig.6 using the KCL and KVL equations, may be reproduced here for continuity.

$$\begin{aligned}\frac{\partial}{\partial z}V(z,t) &= -\underline{L}\frac{\partial}{\partial t}I(z,t) - \underline{R}I(z,t)\\ \frac{\partial}{\partial z}I(z,t) &= -\underline{C}\frac{\partial}{\partial t}V(z,t) - \underline{G}V(z,t)\end{aligned} \tag{4.106}$$

The parameters $\underline{L}$, $\underline{C}$ and $\underline{G}$ are expressed in (4.17), (4.21) and (4.22), respectively, which are derived from field relations based on a TEM condition.

The corresponding equations (4.13,4.14) for the electric and magnetic field, derived under the TEM condition, may also be reproduced.

$$\begin{aligned}\frac{\partial}{\partial z}E(z,t) &= -\mu\frac{\partial}{\partial t}H(z,t)\\ \frac{\partial}{\partial z}H(z,t) &= -\epsilon\frac{\partial}{\partial t}E(z,t) - \sigma E(z,t)\end{aligned} \tag{4.107}$$

Only the z and t variations of the fields are explicitly shown. The voltage-current equations in (4.106) and the electric-magnetic field equations in (4.107) may be recognized to be similar in form. The field equations (4.107) may be obtained from the voltage-current equations of (4.106) by replacing the voltage $V(z,t)$ and current $I(z,t)$ functions by the electric $E(z,t)$ and magnetic $H(z,t)$ fields, respectively, the inductance per unit length $\underline{L}$ and capacitance per unit length $\underline{C}$ by the material permeability μ and permittivity ϵ, respectively, and the parallel conductance per unit length $\underline{G}$ and the series resistance per unit length $\underline{R}$ by the material conductivity σ and conductor resistance equal to zero, respectively. We will develop in the following the solution only for the voltage-current equations (4.106), which can then be extended for the field equations (4.107), as per the above substitution.

Notice that the voltage-current equations (4.106), derived from the circuit model of a lossy transmission line in Fig.6, includes a non-zero resistance per unit length $\underline{R}$. Whereas, the equivalent parameter in the field equations of (4.107), which is the conductor resistance, is assumed to be zero. This is because the field analysis is based on the assumption of a TEM transmission line, which, as we have discussed in section 3.2.1, would not be valid unless the conductors are perfectly conducting with zero resistance. For the same reason, the derivation of the circuit parameters $\underline{L}$, $\underline{C}$, and $\underline{G}$ in section 2, which was also based on the TEM condition, was not applicable to the derivation of the series resistance per unit length $\underline{R}$. Accordingly, $\underline{R}$ remains so far undefined. However, assuming a small deviation from the ideal TEM condition when the conductors have finite but large conductivity, which is equivalent to having a small resistance per unit length $\underline{R}$, it may be possible to model $\underline{R}$ in an approximate manner for harmonic signals. This would be presented in a following section 9.7.9. We will now develop the harmonic wave function for a lossy line, in terms of the circuit parameters $\underline{L}$, $\underline{C}$, $\underline{R}$, $\underline{G}$, assuming that they are all available.

Using the basic phasor concept in Appendix B, the above voltage-current equations (4.106) for a lossy transmission line may be expressed in phasor forms. As followed in the analysis of a loss-less lines in pervious sections, we will use peak phasors as well for the present analysis. Also as followed in the analysis of loss-less lines, for the sake of notational economy we will use here the same variable to represent the phasor as well as its time-dependent function. The distinction between the variables is expected to be self evident from the context of the analysis and discussion.

$$\frac{\partial V(z)}{\partial z} = -j\omega \underline{L} I(z) - \underline{R} I(z), \ \frac{\partial I(z)}{\partial z} = -j\omega \underline{C} V(z) - \underline{G} V(z) \tag{4.108}$$

The above equations may be rearranged in the following forms.

$$\frac{\partial V(z)}{\partial z} = -j\omega(\underline{L} - j\frac{\underline{R}}{\omega})I(z), \ \frac{\partial I(z)}{\partial z} = -j\omega(\underline{C} - j\frac{\underline{G}}{\omega})V(z) \tag{4.109}$$

The phasor forms of the voltage-current equations for a loss-less transmission line are expressed in (4.75), and may be reproduced here.

$$\frac{\partial V(z)}{\partial z} = -j\omega \underline{L} I(z), \ \frac{\partial I(z)}{\partial z} = -j\omega \underline{C} V(z) \tag{4.110}$$

The above equations (4.110) for a loss-less transmission line may be obtained from the equations (4.108) or (4.109) for a lossy transmission line by substituting the loss parameters $\underline{R}$ and $\underline{G}$ equal to zero, as should be expected. Conversely, we may see that the phasor equations (4.109) for a lossy line may also be obtained from the equations (4.110) for a loss-less line, by following definition and substitution.

9.7.1. *Complex Inductance Per Unit Length.* Define a complex parameter $\underline{L}' = \underline{L} - j\underline{R}/\omega$. With this definition, we see that the first equations in (4.109) for a lossy line may be obtained from the first equation in (4.110) for a loss-less line by substituting the inductance per unit length $\underline{L}$, which is a real parameter, by the complex parameter $\underline{L}'$. Accordingly, $\underline{L}'$ may be referred to as the complex inductance per unit length, in contrast to the real value $\underline{L}$ for the inductance per unit length of a loss-less line.

$$\underline{L}' = \underline{L} - j\frac{\underline{R}}{\omega}, \ Im(\underline{L}') = -\frac{\underline{R}}{\omega} \leq 0, \ \underline{R}, \omega \geq 0 \tag{4.111}$$

Note that the imaginary part of $\underline{L}'$ is restricted to be negative assuming that $\underline{R}$ and the frequency ω are positive.

9.7.2. *Complex Capacitance Per Unit Length.* Define a complex parameter $\underline{C}' = \underline{C} - j\underline{G}/\omega$. With this definition, we see that the second equation in (4.109) for a lossy line may be obtained from the second equation in (4.110) for a loss-less line by substituting the capacitance per unit length $\underline{C}$, which is a real parameter, by the complex parameter $\underline{C}'$. Accordingly, $\underline{C}'$ may be referred to as the complex capacitance per unit length, in contrast to the real value $\underline{C}$ for the capacitance per unit length of a loss-less line.

$$\underline{C}' = \underline{C} - j\frac{\underline{G}}{\omega}, \ Im(\underline{C}') = -\frac{\underline{G}}{\omega} \leq 0, \ \underline{G}, \omega \geq 0 \tag{4.112}$$

Note that the imaginary part of $\underline{C}'$, like that for $\underline{L}'$ is also restricted to be negative assuming that $\underline{G}$ and the frequency ω are positive.

Harmonic solutions for a lossy transmission line may be obtained by independently solving the phasor equations in (4.108) or (4.109). However, we have already solved the phasor equations (4.110) and derived different useful parameters for a loss-less line in previous sections. One may also extend all solutions for a loss-less line to a lossy line by substituting $\underline{L}$ and $\underline{C}$ with the corresponding complex values $\underline{L}'$ and $\underline{C}'$, respectively.

9.7.3. *Complex Propagation Constant.* The wave function $f(z)$ in (4.77) for a loss-less transmission line may be extended for a lossy transmission line as per the above substitution.

$$\begin{aligned} f(z) = B^{\pm}e^{\mp j\beta' z} = A^{\pm}e^{j\phi^{\pm}}e^{\mp j\beta' z},\ \beta' = \omega\sqrt{\underline{L}'\underline{C}'},\ f \equiv V, I \\ f(z) = B^{\pm}e^{\mp\gamma z},\ \gamma = j\beta' = j\omega\sqrt{\underline{L}'\underline{C}'} \\ A^{\pm} = |B^{\pm}|,\ \arg(B^{\pm}) = \phi^{\pm} \end{aligned} \tag{4.113}$$

where the real propagation constant β for a loss-less transmission line is substituted with a complex parameter β'. β' is referred to as the complex propagation constant for a lossy transmission line. The variable $j\beta'$ is represented with a variable γ. which may also be referred to as the complex propagation constant with appropriate understanding.

$$\begin{aligned} \beta' = \omega\sqrt{\underline{L}'\underline{C}'} = \omega\sqrt{(\underline{L} - j\frac{\underline{R}}{\omega})(\underline{C} - j\frac{\underline{G}}{\omega})} = \beta - j\alpha,\ \gamma = j\beta = \alpha + j\beta \\ e^{\mp j\beta' z} = e^{\mp j(\beta - j\alpha)z} = e^{\mp\gamma z} = e^{\mp(\alpha + j\beta)} = e^{\mp j\beta z}e^{\mp\alpha z} \\ \beta > 0,\ \alpha \geq 0,\ \underline{L}, \underline{C}, \underline{R}, \underline{G} \geq 0 \end{aligned} \tag{4.114}$$

The square-root function in the above expression for β' is a double-value function. It is assumed that the value which has a positive real part is selected. This is consistent with the wave function in (4.77) for a loss-less line, where the real β expressed by the square-root expression $\beta = \omega\sqrt{\underline{LC}}$ was also selected to be positive (for $\omega > 0$). The double-sign notation used before β' in the propagation factor, as well as in the respective choices for related expressions, in (4.114) refer to a particular direction of propagation. They follow the same double-sign notation used in (4.40,4.78) for a loss-less case. As per a standard notation we have been following, the upper and lower sign of a double-sign notation refers to propagation in the $+z$ and $-z$ direction, respectively. The physical significance of a particular sign in the time-domain signal propagation in a lossy transmission line would be explained in the following section 9.7.4.

Both $\underline{L}'$ and $\underline{C}'$, as shown in (4.111,4.112), have negative imaginary and positive real parts. This means that they both are on the fourth quadrant of the complex plane. It may be shown that the resulting solution for $\beta' = \omega\sqrt{\underline{L}'\underline{C}'}$ would be either in the second or the fourth quadrant, with the two values negative of each other. If we must have the real part to be positive, as per the earlier assumption, out of the two possible values of β' the one in the fourth quadrant must be selected. This means that the imaginary part of β' would be negative, while the real part is selected to be positive. The magnitude of the imaginary part is represented by α. The real part of β' is represented by β. It should be clarified, this β for a lossy line would be in general different from the β for the corresponding loss-less line, and they would be equal to each other only when the loss parameters in the lossy line approach to zero. The β is simply used as a generic variable, which represents the real propagation constant of any general transmission line.

Small-Loss Approximation:

When the losses in the transmission line are sufficiently small, the expression for the complex propagation constant may be simplified in an approximate from.

$$
\begin{aligned}
\beta' &= \omega\sqrt{\underline{LC}} = \omega\sqrt{(\underline{L} - j\frac{R}{\omega})(\underline{C} - j\frac{G}{\omega})} = \omega\sqrt{\underline{L}(1 - j\frac{R}{\omega\underline{L}})\underline{C}(1 - j\frac{G}{\omega\underline{C}})} \\
&= \omega\sqrt{\underline{LC}}\sqrt{(1 - j\frac{R}{\omega\underline{L}})(1 - j\frac{G}{\omega\underline{C}})} \simeq \omega\sqrt{\underline{LC}}(1 - j\frac{R}{2\omega\underline{L}})(1 - j\frac{G}{2\omega\underline{C}}) \\
&\simeq \omega\sqrt{\underline{LC}}(1 - j(\frac{R}{2\omega\underline{L}} + \frac{G}{2\omega\underline{C}})) = \omega\sqrt{\underline{LC}} - j(\frac{R}{2\sqrt{\underline{L}/\underline{C}}} + \frac{G\sqrt{\underline{L}/\underline{C}}}{2}) \\
&= \omega\sqrt{\underline{LC}} - j(\frac{R}{2Z_c} + \frac{GZ_c}{2}) \simeq \beta - j\alpha,\ Z_c = \sqrt{\frac{\underline{L}}{\underline{C}}} \\
&\beta \simeq \omega\sqrt{\underline{LC}},\ \alpha \simeq \frac{R}{2Z_c} + \frac{GZ_c}{2} \\
&\frac{R}{\omega\underline{L}}, \frac{G}{\omega\underline{C}} \ll 1,\ \underline{R} \ll \omega\underline{L},\ \underline{G} \ll \omega\underline{C}
\end{aligned}
\tag{4.115}
$$

where Z_c is the characteristic impedance of the transmission line, when there is no loss. The above approximation is possible when $\underline{R}$ and $\underline{G}$ are sufficiently smaller than $\omega\underline{L}$ and $\omega\underline{C}$, respectively.

9.7.4. *Wave Function in Time-Dependent Form.* Now, the time-dependent form of the wave function $f(z,t)$ may be expressed using the complex phasor expression of $f(z)$ in (4.113) with β' given as (4.114).

$$
\begin{aligned}
f(z,t) &= Re(A^{\pm}e^{j\phi^{\pm}}e^{\mp j\beta' z}e^{j\omega t}) = Re(A^{\pm}e^{j\phi^{\pm}}e^{\mp j\beta z}e^{\mp\alpha z}e^{j\omega t}) \\
&= A^{\pm}e^{\mp\alpha z}Re(e^{j\phi^{\pm}}e^{\mp j\beta z}e^{j\omega t}) = A^{\pm}e^{\mp\alpha z}\cos(\omega t \mp \beta z + \phi^{\pm})
\end{aligned}
\tag{4.116}
$$

The wave function is illustrated in Figs.31,32 for a simple case with $A^{\pm} = 1$ and $\phi^{\pm} = 0$. Figs.31,32 show the snapshots of the wave function at times $t = 0$ and $t = \Delta t$, for the two cases, respectively when the upper or the lower sign is selected.

$$
\begin{aligned}
f(z,t=0) = A^{\pm}e^{\mp\alpha z}\cos(\mp\beta z+\phi^{\pm}) &= A^{\pm}e^{\mp\alpha z}\cos(\beta z\mp\phi^{\pm}) \\
f(z,t=\triangle t) &= A^{\pm}e^{\mp\alpha z}\cos(\omega\Delta t\mp\beta z+\phi^{\pm}) \\
= A^{\pm}e^{\mp\alpha z}\cos(\beta z\mp\omega\Delta t\mp\phi^{\pm}) &= A^{\pm}e^{\mp\alpha z}\cos(\beta(z\mp\frac{\omega}{\beta}\Delta t)\mp\phi^{\pm}) \\
&= A^{\pm}e^{\mp\alpha z}\cos(\beta(z\mp\triangle z)\pm\phi^{\pm}),\ \triangle z=\frac{\omega}{\beta}\Delta t \qquad (4.117)
\end{aligned}
$$

The above wave function in (4.116,4.117), illustrated in Figs.31,32 for a lossy transmission line, may be compared and contrasted with that in (4.78,4.80), illustrated in Fig.24 for a loss-less case. The peak amplitude for a loss-less case is independent of distance z. That is not the case for a lossy transmission line, where the peak amplitude has a $e^{-\alpha z}$ or $e^{+\alpha z}$ variation in the z direction, depending on if the $-$ or the $+$ sign is selected before β', respectively. The entire wave function in (4.78,4.80) for a loss-less transmission line shifts with time, uniformly along the $-z$ or $+z$ direction, as shown in Fig.24. Whereas, in reference to the wave function in (4.116,4.117) for a lossy transmission line, the amplitude envelope $e^{\mp\alpha z}$ remains unchanged in time, while only the oscillating cosine function shifts along the $+z$ or $-z$ direction under the constant envelope. This is shown in Figs.31,32.

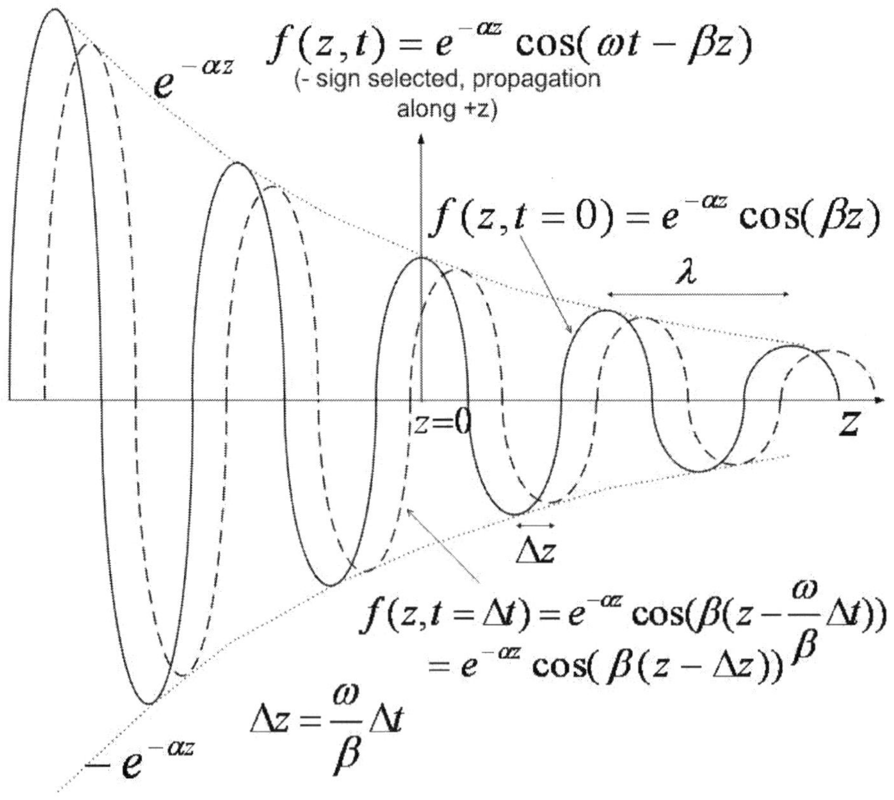

FIGURE 31

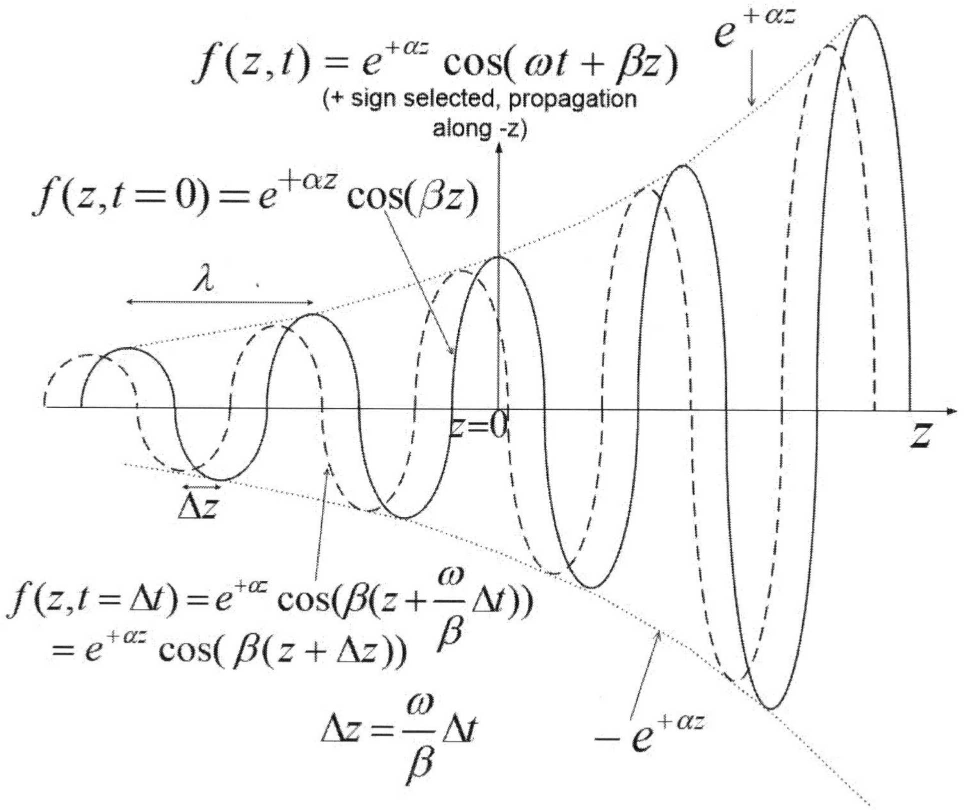

FIGURE 32

9.7.5. *Propagation Velocity and Wavelength.* The oscillating part of the wave function for a lossy line shifts in the $+z$ or $-z$ direction when $-$ or $+$ sign is selected, respectively, before the propagation constant β'. This is in consistency with the notation used in (4.78), and with related discussion and interpretation in sections 9.2,4.4. Except, in the present case of a lossy line, the shift refers to only the oscillating part, not the entire wave function. The parameter v_p is the ratio of distance Δz the oscillating part of the wave function shifts its relative position along the propagation direction, over a given time Δt.

$$v_p = \frac{\Delta z}{\Delta t} = \frac{\omega}{\beta} = \frac{\omega}{Re(\beta')} \tag{4.118}$$

The wave length λ may be defined as the periodicity of the oscillating part of the wave function. Clearly, in contrast with the wave function for a loss-less transmission line, the entire wave function for a lossy transmission line is no longer periodic,

$$\beta\lambda = 2\pi, \ \lambda = \frac{2\pi}{\beta} = \frac{2\pi}{Re(\beta')} = \frac{2\pi v_p}{\omega} = \frac{v_p}{f} \tag{4.119}$$

Note that the velocity for a loss-less line is $v_p = \omega/\beta = 1/\sqrt{\underline{L}\underline{C}}$. If we simply substitute $\underline{L}$ and $\underline{C}$ in the expression of v_p for a loss-less transmission line by the corresponding complex values $\underline{L}'$ and $\underline{C}'$ for a lossy transmission line, we will get a complex velocity v_p'.

$$v_p' = \frac{\omega}{\beta'} = \frac{1}{\sqrt{\underline{L}'\underline{C}'}} \tag{4.120}$$

Finding a complex velocity v_p' in this manner may be consistent with the substitution approach we have been following for the analysis of a lossy transmission line. However, the complex velocity so obtained may not be physically meaningful without proper interpretation. As we understand now, the complex velocity for a lossy transmission line may be interpreted as a real velocity of the oscillatory part as given in (4.118), confined under a fixed envelope $e^{\mp\alpha z}$. The real velocity v_p and the complex velocity v_p' may be related as follows.

$$v_p = \frac{\omega}{\beta} = \frac{\omega}{Re(\beta')},\ v_p' = \frac{\omega}{\beta'},\ \frac{1}{v_p} = Re(\frac{1}{v_p'}) \tag{4.121}$$

Note that real part of the complex velocity v_p' would not be equal to the real velocity v_p.

Small-Loss Approximation:

When the losses in the transmission line are sufficiently small, the expression for the complex propagation constant is approximated in (4.115). In this case, as per (4.115), the real part $\beta = Re(\beta')$ is approximately equal to $\omega\sqrt{\underline{L}\underline{C}}$, which is equal to the β for the loss-less case. Using this result in (4.118), we see that the velocity v_p for a transmission line with small loss is approximately equal to that without the loss.

$$v_p = \frac{\omega}{Re(\beta')} \simeq \frac{\omega}{\omega\sqrt{\underline{L}\underline{C}}} = \frac{1}{\sqrt{\underline{L}\underline{C}}},\ \underline{R} \ll \omega\underline{L},\ \underline{G} \ll \omega\underline{C} \tag{4.122}$$

9.7.6. *Attenuation Constant.* As discussed, the peak amplitude for a lossy transmission line has a $e^{-\alpha z}$ or $e^{+\alpha z}$ variation in the z direction, depending on if the $-$ or the $+$ sign is selected before β', in which case the velocity of propagation is along the $+z$ and $-z$ directions, respectively. The $e^{-\alpha z}$ variation decays along the $+z$ direction, which corresponds to the direction of propagation for this case. The $e^{+\alpha z}$ variation decays along the $-z$ direction, which also corresponds to the direction of propagation for this case. In other words, the peak signal amplitude decays along the direction of propagation in both cases. The signal decay is a result of power dissipation in the lossy filling material and the lossy conductors of the transmission line. In contrast, there is no power dissipation in a loss-less transmission line, and in consistence the signal peak amplitude for this case is uniform with distance z, without any decay. The decay factor α for a lossy transmission line, which is the magnitude of the imaginary part of β', is referred to as the attenuation constant.

$$\alpha = -Im(\beta') = -Im(\omega\sqrt{\underline{L}'\underline{C}'}) = |Im(\beta')| = |Im(\omega\sqrt{\underline{L}'\underline{C}'})| \tag{4.123}$$

Usable Length:

The distance δ, over which the signal decays in a particular transmission line by a factor of $e^{-1} = 0.37$, may be used as a general measure for the usable length of the transmission line.

$$e^{-\alpha\delta} = e^{-1} \simeq 0.37, \;\; \delta = \frac{1}{\alpha} = \frac{1}{-Im(\beta')} = \frac{1}{-Im(\omega\sqrt{\underline{L}'\underline{C}'})} \tag{4.124}$$

Small-Loss Approximation:

As derived in (4.115), the attenuation constant may be approximated when there is small loss.

$$\alpha = \frac{1}{\delta} = -Im(\beta') \simeq (\frac{R}{2Z_c} + \frac{GZ_c}{2}), \;\; \underline{R} << \omega\underline{L}, \;\; \underline{G} << \omega\underline{C} \tag{4.125}$$

The attenuation constant α depends on the ratio of $\underline{R}$ to Z_c, and the product of $\underline{G}$ and Z_c. Accordingly, for a given value of $\underline{R}$, which means a given quality of the conductors used in the transmission line, the α would be larger for a smaller value of the characteristic impedance Z_c. The effect of Z_c is opposite with respect to the $\underline{G}$ of the filling material. For a given value of $\underline{G}$, which means a given insulation quality of the filling material, the α would be smaller for a smaller value of the characteristic impedance Z_c.

9.7.7. *Complex Characteristic Impedance.* The general form of the harmonic wave function of (4.113) for a lossy transmission line applies to the voltage as well as the current signals. We will express the individual wave functions for the voltage or the current signal by substituting the coefficient $B^{\pm}$ in (4.113) by $V^{\pm}$ or $I^{\pm}$, respectively.

$$V(z) = V^{\pm}e^{\mp j\beta' z}, \;\; I(z) = I^{\pm}e^{\mp j\beta' z} \tag{4.126}$$

Using the above expressions in the first voltage-current relations of (4.109), and using the definitions for $\underline{L}'$ and $\underline{C}'$ in (4.111,4.112), one may relate the coefficients $V^{\pm}$ and $I^{\pm}$. This establishes an explicit relationship between the voltage $V(z)$ and $I(z)$.

$$\frac{\partial V(z)}{\partial z} = (\mp j\beta')V^{\pm}e^{\mp j\beta' z} = -j\omega\underline{L}'I(z) = -j\omega\underline{L}'I^{\pm}e^{\mp j\beta' z}$$

$$\frac{V^{\pm}}{I^{\pm}} = \frac{-j\omega\underline{L}'}{\mp j\beta'} = \pm\frac{\omega\underline{L}'}{\beta'} = \pm\frac{\omega\underline{L}'}{\omega\sqrt{\underline{L}'\underline{C}'}} = \pm\sqrt{\frac{\underline{L}'}{\underline{C}'}} = \pm Z_c'$$

$$\frac{V(z)}{I(z)} = \frac{V^{\pm}}{I^{\pm}} = \pm Z_c' \tag{4.127}$$

The same result may also be obtained by using the second equation of (4.109), in stead of the first equation used in the above derivation. As the above result shows, the ratio of voltage $V(z)$ and current $I(z)$ phasors is independent of the location z. The parameter Z_c' is in general a complex parameter, referred to as the complex characteristic impedance of the lossy transmission line. There is a $\pm$ sign before Z_c' in the above relationship between $V(z)$ and $I(z)$. This simply means that, for a given voltage $V(z)$ across the transmission line at any given location z, the current $I(z)$ would switch its sign or direction when the direction of propagation is switched.

The above derivation, and the expression of the complex characteristic impedance Z_c', are equivalent to those in section 4.6 for a loss-less line. The expression for the complex characteristic impedance Z_c' for a lossy transmission line could be obtained from the expression of the real characteristic impedance Z_c in (4.47) for a loss-less transmission line, by substituting the real parameters $\underline{L}$ and $\underline{C}$ for the loss-less case by the respective complex parameters $\underline{L}'$ and $\underline{C}'$ for the lossy case.

$$Z_c' = \sqrt{\frac{\underline{L}'}{\underline{C}'}},\ Z_c = \sqrt{\frac{\underline{L}}{\underline{C}}},\ \underline{L}' = \underline{L} - j\frac{R}{\omega},\ \underline{C}' = \underline{C} - j\frac{G}{\omega}$$
$$Z_c' = |Z_c'|e^{j\arg(Z_c')} = R_c + jX_c$$
$$R_c \geq 0,\ X_c \gtrless 0 \tag{4.128}$$

The complex value of Z_c' may be expressed in terms of its magnitude $|Z_c'|$ and phase angle $\arg(Z_c')$, or its real R_c and imaginary X_c parts. The square-root expression for Z_c', as shown in the above equation, would in general have two complex values, negative of each other. We need to select the value which has a positive real part R_c. This is consistent with the positive value for the real characteristic impedance Z_c for a loss-less transmission line. Having a negative real part would otherwise result in a negative power, which would be non-physical. A negative real part for the characteristic impedance would mean that, when a particular source is connected to excite a signal in a transmission line, power would be driven into a signal source from a transmission line, instead of the other way around as one would obviously expect. With the above choice of Z_c', having a positive real part, the corresponding imaginary part X_c may have a positive or negative value, depending on the relative levels of the imaginary parts of $\underline{L}'$ and $\underline{C}'$, or equivalently the relative phase angles of $\underline{L}'$ and $\underline{C}'$. When the phase angles for $\underline{L}'$ and $\underline{C}'$ are equal, the complex characteristic impedance Z_c', could be a real number, even in the presence of non-zero loss. This is in contrast to the complex value for the propagation constant β' of a lossy transmission line, which would always have a non-zero imaginary part under the non-zero loss condition.

The phase difference between the voltage $V(z)$ and current $I(z)$ phasors is determined by the phase angle of the complex characteristic impedance Z_c'. This is in general a non-zero number for a lossy transmission line, and may be a positive or negative angle as discussed above. In contrast, the voltage and current waveforms for a loss-less transmission line are in phase with each other, because the characteristic impedance in this case is a purely real number. The ratio between the magnitudes of the voltage and current phasors, which is the ratio of the peak amplitudes of the harmonic waveforms of the voltage and current signals, is determined by the magnitude $|Z_c'$ of the characteristic impedance.

Time-Domain Expressions:

The contribution from the magnitude and phase angle of Z_c' may be explicitly shown in the time-dependent expressions for the voltage and current, derived from the respective phasors in (4.126,4.127).

$$V(z,t) = Re(V(z)e^{j\omega t}) = Re(V^{\pm}e^{\mp j\beta' z}e^{j\omega t}) = Re(|V^{\pm}|e^{j\arg(V^{\pm})}e^{\mp j\beta z}e^{\mp\alpha z}e^{j\omega t})$$
$$= |V^{\pm}|e^{\mp\alpha z}Re(e^{j\arg(V^{\pm})}e^{\mp j\beta z}e^{j\omega t}) = |V^{\pm}|e^{\mp\alpha z}\cos(\omega t \mp \beta z + \arg(V^{\pm}))$$
$$I(z,t) = Re(I(z)e^{j\omega t}) = Re(I^{\pm}e^{\mp j\beta' z}e^{j\omega t}) = Re(\pm\frac{V^{\pm}}{Z'_c}e^{\mp j\beta' z}e^{j\omega t})$$
$$= Re(\pm\frac{|V^{\pm}|}{|Z'_c|}e^{j(\arg(V^{\pm}-\arg(Z'_c))}e^{\mp j\beta z}e^{\mp\alpha z}e^{j\omega t})$$
$$= \pm\frac{|V^{\pm}|}{|Z'_c|}e^{\mp\alpha z}Re(e^{j(\arg(V^{\pm})-\arg(Z'_c))}e^{\mp j\beta z}e^{j\omega t})$$
$$= \pm\frac{|V^{\pm}|}{|Z'_c|}e^{\mp\alpha z}\cos(\omega t \mp \beta z + \arg(V^{\pm}) - \arg(Z'_c))$$
$$\frac{|V^{\pm}|}{|I^{\pm}|} = |Z'_c|, \ \arg(V^{\pm}) - \arg(\pm I^{\pm}) = \arg(Z'_c) \quad (4.129)$$

Small-Loss Approximation:

The general expression for Z'_c may be simplified for a lossy transmission line when the loss parameters are sufficiently small. That is often the case in many practical transmission lines.

$$Z'_c = \sqrt{\frac{\underline{L}'}{\underline{C}'}} = \sqrt{\frac{\underline{L} - j\frac{\underline{R}}{\omega}}{\underline{C} - j\frac{\underline{G}}{\omega}}} = \sqrt{\frac{\underline{L}(1 - j\frac{\underline{R}}{\omega\underline{L}})}{\underline{C}(1 - j\frac{\underline{G}}{\omega\underline{C}})}} = \sqrt{\frac{\underline{L}}{\underline{C}}}\sqrt{\frac{(1 - j\frac{\underline{R}}{\omega\underline{L}})}{(1 - j\frac{\underline{G}}{\omega\underline{C}})}}$$
$$\simeq \sqrt{\frac{\underline{L}}{\underline{C}}}(1 - j\frac{\underline{R}}{2\omega\underline{L}})(1 + j\frac{\underline{G}}{2\omega\underline{C}}) \simeq \sqrt{\frac{\underline{L}}{\underline{C}}}(1 - j\frac{\underline{R}}{2\omega\underline{L}} + j\frac{\underline{G}}{2\omega\underline{C}})$$
$$= Z_c(1 - j\frac{\underline{R}}{2\omega\underline{L}} + j\frac{\underline{G}}{2\omega\underline{C}}); \ \frac{\underline{R}}{\omega\underline{L}}, \frac{\underline{G}}{\omega\underline{C}} << 1; \ \underline{R} << \omega\underline{L}, \ \ \underline{G} << \omega\underline{C} \quad (4.130)$$

The series resistance $\underline{R}$ due to the conductors contributes a negative imaginary part, whereas the parallel conductance $\underline{G}$ due to the filling material contributes a positive imaginary part for Z'_c.

9.7.8. *Field Solutions for a Lossy Transmission Line.* We will now examine the time-harmonic solution for the electric and magnetic fields of a lossy, TEM transmission line. The voltage-current relations for a general lossy transmission line, and the corresponding electric-magnetic field relations established under the TEM condition, are given in (4.106,4.107). Discussions following (4.106,4.107) indicate the functional similarities between the voltage-current and the electric-magnetic field relations, and prescribe how any solution for the voltage and current signals may be extended for the corresponding electric and magnetic fields of a lossy, TEM transmission line, through suitable substitution of variables. As per the prescription, the general form of the time-harmonic wave function in (4.113,4.114) would apply as well to the electric and magnetic fields of a lossy, TEM transmission line, when the circuit parameters $\underline{L}$, $\underline{C}$, $\underline{R}$ and $\underline{G}$ in (4.114) are substituted by μ, ϵ, zero, and σ, respectively.

Complex Propagation Constant and Complex Permittivity:

The complex propagation constant β' in the wave function of the lossy, TEM transmission line may be expressed by making the above substitutions in (4.114).

$$\beta' = \omega\sqrt{\mu(\epsilon - j\frac{\sigma}{\omega})} = \beta - j\alpha \tag{4.131}$$

When the material medium is loss-less, with the conductivity $\sigma = 0$, we have $\beta' = \beta = \sqrt{\mu\epsilon}$. Comparing the expressions under the lossy and the loss-less conditions, we see that he complex parameter $\epsilon' = \epsilon - j\sigma/\omega$ analytically replaces for the real permittivity ϵ when the material becomes lossy. Accordingly, ϵ' is referred to as the complex permittivity of a lossy material.

$$\epsilon' = \epsilon - j\frac{\sigma}{\omega}, \ \beta' = \omega\sqrt{\mu(\epsilon - j\frac{\sigma}{\omega})} = \omega\sqrt{\mu\epsilon'} = \beta - j\alpha \tag{4.132}$$

The electric and magnetic fields that exist in the filling material medium are directly associated with the voltage and current which appear at the conductor terminals of the transmission lines. Therefore, their variations along the z direction must be directly connected. The complex propagation constant β', associated with the z variation of the fields of lossy, TEM transmission line, will have to be the same as that for its voltage and current waves. Accordingly, the result for β' in (4.131) will have to be equal to that from (4.114).

$$\beta' = \omega\sqrt{\underline{L}'\underline{C}'} = \omega\sqrt{(\underline{L} - j\frac{\underline{R}}{\omega})(\underline{C} - j\frac{\underline{G}}{\omega})} = \omega\sqrt{\mu(\epsilon - j\frac{\sigma}{\omega})} = \omega\sqrt{\mu\epsilon'} \tag{4.133}$$

As we have discussed before, the conductors of a TEM transmission line will have to be perfectly conducting. Accordingly, the equivalent series resistance $\underline{R}$ for a TEM transmission line is zero. Using this result in (4.133) we get a relationship between the circuit and material parameters of a lossy, TEM transmission line.

$$\underline{L}\underline{C}' = \mu\epsilon'$$

$$\underline{L}(\underline{C} - j\frac{\underline{G}}{\omega}) = \mu(\epsilon - j\frac{\sigma}{\omega}), \ \underline{L}\underline{C} = \mu\epsilon, \ \frac{\underline{G}}{\underline{C}} = \frac{\sigma}{\epsilon} \tag{4.134}$$

The relation $\underline{L}\underline{C} = \mu\epsilon$ is the same as that we have already developed in (4.43) for a loss-less, TEM transmission line. We have an additional relation $\underline{G}/\underline{C} = \sigma/\epsilon$ when there is material loss, which we have also already developed in (4.23).

Material Medium and the TEM Condition:

Consider a general, lossy transmission line filled with an arbitrary material distribution, for which $\mu\epsilon$ and σ/ϵ are different at different locations over its cross section. The fields at different points on the cross section would propagate with a different complex propagation constant, β', as given by (4.131). As per the the z dependence $e^{\mp j\beta' z} = e^{\mp j\beta z}e^{\mp \alpha z}$ of the wave function, the field would appear at a distance z with a different phase shift (or time delay) and different attenuation for the different cross-sectional points. Accordingly, the cross-sectional field distribution would constantly change as the wave propagates over increasing distance z. Assuming that the cross-sectional fields at a given z satisfy Maxwell's equations, the changing fields may no longer satisfy Maxwell's equations at all distance $-\infty < z < +\infty$. That would be an invalid situation. Therefore, one may conclude that the basic TEM condition we have used to begin with our field analysis in section 1.1, leading to the field equations (4.13,4.14) or (4.107), and wave function (4.113), would have to be invalid when $\mu\epsilon$ and σ/ϵ are not unique or constant over the cross section of a transmission line.

Conversely, the basic TEM condition would be valid for a lossy transmission line, if, and only if, $\mu\epsilon$ and σ/ϵ are constants over the entire cross section. In this case, the field at each point on the cross-section of the transmission line would propagate with the the same phase shift and attenuation, thus maintaining the same relative field distribution over the entire cross section at a distance z. Validity of Maxwell's equations at one location z would, therefore, guarantee its validity at any other location.

The above argument is similar to that we made in section 4.5.1 to establish that $\mu\epsilon$ will have to be a constant for a loss-less, TEM transmission line. Here we extend the argument to a TEM transmission line with material loss, specifically under a time-harmonic excitation. The requirement of having constant values of $\mu\epsilon$ and σ/ϵ may be satisfied by having a single material uniformly filling the entire space between the two conductors of the transmission line. In general, however, one may have different materials covering different segments of the filling space, such that the product $\mu\epsilon$ and the ratio σ/ϵ of each material segment, evaluated for a harmonic signal of a given frequency, are the same.

Combining the above material condition with (4.134), we show that the circuit parameters $\underline{LC}$ and $\underline{G}/\underline{C}$ for a lossy, TEM transmission line are constants, which are dependent only on the parameters of the filling material, but are independent of the physical dimensions of the transmission line, This is a significant condition, which is not evident from the expressions of the individual parameters $\underline{L}$, $\underline{C}$ and $\underline{G}$ we derived in section 2.

Complex Wave Impedance:

The substitution approach used in the derivation of the complex propagation constant may also be used to extend the expression for the voltage-current ratio in (4.127) to express the ratio of the electric and magnetic field.

$$\frac{E}{H} = \sqrt{\frac{\mu}{\epsilon - j\frac{\sigma}{\omega}}} = \sqrt{\frac{\mu}{\epsilon'}} = \eta', \ \epsilon' = \epsilon - j\frac{\sigma}{\omega}$$

$$\eta' = \sqrt{\frac{\mu}{\epsilon}} = \eta, \ \epsilon' = \epsilon, \ \sigma = 0 \tag{4.135}$$

η' is a complex number, referred to as the complex wave impedance. This is in contrast to the real wave impedance $\eta = \sqrt{\mu/\epsilon}$, as given in (4.48) for a loss-less medium. The expression of η' in (4.135) may be obtained from that of η in (4.48), by replacing the real permittivity ϵ by the complex permittivity $\epsilon' = \epsilon - j\sigma/\omega$, which we defined in (4.132).

A complex value of η' means that the electric and magnetic fields are not in phase with each other for a general lossy medium. In contrast, the phase difference would be zero for a loss-less material for which the wave impedance η is a real parameter. Also, from (4.135) the phase angle for η' may be seen to be positive. or equivalently the imaginary part of η' is positive. In other words, the complex wave impedance η' for a conducting material with non-zero $\sigma > 0$ is inductive. Accordingly, the phase angle of the electric field E is less than that of the magnetic field H. This assumes that the unit vectors $\hat{e}$ and $\hat{h}$, in reference to which the harmonic functions $E(t)$ and $H(t)$ are respectively specified, are properly defined where $\hat{e} \times \hat{h}$ is along the direction of propagation. The reference vectors $\hat{e}$ and $\hat{h}$ were established in section 1.2.

Small-Loss Approximation:

When loss in the material medium is sufficiently small, so that the conductivity σ is much smaller than $\omega\epsilon$, the expressions of the complex propagation β' may be approximated in a simpler form.

$$\beta' = \beta - j\alpha = \omega\sqrt{\mu\epsilon'} = \omega\sqrt{\mu(\epsilon - j\frac{\sigma}{\omega})} = \omega\sqrt{\mu\epsilon(1 - j\frac{\sigma}{\omega\epsilon})}$$

$$= \omega\sqrt{\mu\epsilon}\sqrt{(1 - j\frac{\sigma}{\omega\epsilon})} \simeq \omega\sqrt{\mu\epsilon}(1 - j\frac{\sigma}{2\omega\epsilon}), \quad \frac{\sigma}{\omega\epsilon} << 1, \ \sigma << \omega\epsilon$$

$$\beta \simeq \omega\sqrt{\mu\epsilon}, \ \alpha \simeq \omega\sqrt{\mu\epsilon}(\frac{\sigma}{2\omega\epsilon}) = \frac{\sigma}{2}\sqrt{\frac{\mu}{\epsilon}} = \frac{\sigma\eta}{2} \tag{4.136}$$

where $\eta = \sqrt{\mu/\epsilon}$ is the real wave impedance of the transmission line when there is no loss. As (4.136) shows, the real part β of the complex propagation constant with the small loss is approximately equal to that without any loss. This is consistent with the approximate result in (4.115), established in terms of the circuit parameters. Using this approximate result in (4.118,4.119), the velocity v_p and wave length λ of the transmission line would also be approximately the same, with or without the small loss,

The complex wave impedance η' in (4.135) may also be similarly approximated for the small-loss condition.

$$\eta' = \eta_r + j\eta_i = \sqrt{\frac{\mu}{\epsilon'}} = \sqrt{\frac{\mu}{\epsilon - j\frac{\sigma}{\omega}}} = \sqrt{\frac{\mu}{\epsilon(1 - j\frac{\sigma}{2\omega})}} = \sqrt{\frac{\mu}{\epsilon}}\sqrt{\frac{1}{1 - j\frac{\sigma}{2\omega}}}$$

$$\simeq \sqrt{\frac{\mu}{\epsilon}}\omega(1 + j\frac{\sigma}{2\omega\epsilon}), \quad \frac{\sigma}{\omega\epsilon} << 1, \ \sigma << \omega\epsilon$$

$$\eta_r = Re(\eta') \simeq \sqrt{\frac{\mu}{\epsilon}}, \ \eta_i = Im(\eta') \simeq \sqrt{\frac{\mu}{\epsilon}}\frac{\sigma}{2\omega\epsilon} = \frac{\sigma\eta}{2\omega\epsilon} \tag{4.137}$$

As the above equation shows, the real part η_r of the wave impedance with the small loss is approximately equal to η, which is the wave impedance of the same medium without the loss. The imaginary part η_i of the wave impedance is positive, and is proportional to the conductivity σ, under the small loss condition.

9.7.9. *Resistance Per Unit Length.* The circuit and field analysis of a transmission line, which we have developed in this chapter, are strictly based on the TEM condition, This assumes that the conductors of the transmission line are ideally loss-less, with conductivity $\sigma_c = \infty$. Any loss in the conductors, where the conductivity σ_c is finite, the TEM condition is not satisfied. As discussed in sections 1.1,4.5.1, the voltage and current of the transmission line can not be properly defined, and the transmission line can not be modeled by any equivalent circuit. However, as discussed in section 3.2.1, for most practical cases with very good conductors, where the conductivity σ_c is very high, any deviation from the ideal TEM condition is expected to be small. The small deviation in this case may be approximately modeled by using a series resistance $\underline{R}$ per unit length in the equivalent circuit of Fig.6. In the following we will provide a reasonable basis for this model, and derive simple expressions for $\underline{R}$ under the low-loss condition.

With suitable assumptions, the fields inside the conductors may be modeled as a uniform plane plane-wave. The theory of a uniform plane wave will be presented in a following chapter. But, we will now borrow a few of the basic concepts from the plane-wave theory for the present development. The electric and magnetic fields for a uniform plane wave, propagating in an infinite medium, happen to satisfy the same relations (4.107) as the fields in a TEM transmission line. Accordingly, the solutions for the harmonic fields in section 9.7.8 for a general lossy medium, may also be applied for a uniform plane wave in a conducting medium. The resulting complex propagation constant β', as expressed in (4.131), may be approximated for a a good conductor with high conductivity σ_c. The imaginary part of β' is the attenuation constant α_c.

$$\beta' = \beta_c - j\alpha_c = \omega\sqrt{\mu_c(\epsilon_c - j\frac{\sigma_c}{\omega}))} \simeq \omega\sqrt{-j\frac{\mu_c\sigma_c}{\omega}} = \sqrt{\frac{\mu_c\sigma_c}{2\omega}}(1-j)$$

$$\alpha_c = -Im(\beta') \simeq \sqrt{\frac{\mu_c\sigma_c}{2\omega}}, \quad \frac{\sigma_c}{\omega} >> \epsilon_c \qquad (4.138)$$

The subscript c for different variables refer to a conducting medium. For a good conducting medium with a high conductivity σ_c, the attenuation constant α_c would be large, resulting in a strong decay of the fields with a $e^{-\alpha_c s}$ variation with distance s. The permeability μ_c of a conductor is often assumed to be equal to that ($= \mu_0$) for the free space.

The ratio of the orthogonal electric and magnetic fields, given by the complex wave impedance η' in (4.135) would also apply to a uniform plane wave in a conducting medium. The expression of η' in (4.135) may be approximated for a conducting medium with high conductivity σ_c. We are particularly interested in the real part η_{rc} of the wave impedance.

$$\frac{E}{H} = \eta' = \eta_{rc} + j\eta_{ic} = \sqrt{\frac{\mu_c}{\epsilon_c - j\frac{\sigma_c}{\omega}}} \simeq \sqrt{\frac{\mu_c\omega}{-j\sigma_c}} = \sqrt{\frac{\mu_c\omega}{2\sigma_c}}(1+j)$$

$$\eta_{rc} = Re(\eta') \simeq \sqrt{\frac{\mu_c\omega}{2\sigma_c}}, \quad \frac{\sigma_c}{\omega} >> \epsilon_c \qquad (4.139)$$

Any spatial variations of the fields in the filling medium just outside of a conductor surface S_1 or S_2 (see Fig.1), on a plane tangential to the surface, closely track those in the conducting medium just in the other side of the surface, as required by boundary conditions (Ch.2, section 4.4). The spatial variations of the fields in the filling medium just outside of the conductor surface, and accordingly those in the conducting medium just inside, are assumed to be significantly weaker than those associated with a strong attenuation of the fields in the conducting medium. Therefore, in relative terms, the variations tangential to the conductor surface may be ignored, and the fields in the conductor may be modeled to be locally uniform on a plane parallel to the surface S_1 or S_2. The field variation in the conductor along the remaining dimension, which is along a normal to the boundary surface S_1 or S_2, may now be approximated as a uniform plane wave, propagating normal to the boundary surface. This approximate model would apply in local vicinity of each and every point on the boundary surface S_1 or S_2. Further, due to the strong attenuation, the wave would decay to a negligible value, over a very thin layer close to the boundary surface. This effect is referred to as the skin effect, and the thickness of the layer is measured by a parameter called the skin depth. The wave will not be able to "see" beyond this very thin layer. Assuming that the thickness of the conductors is sufficiently larger than the skin depth, the conducting medium may be modeled as a semi-infinite medium in the normal direction away from the boundary surface S_1 or S_2.

In summary, the fields inside a conductor are approximated as a uniform plane-wave, propagating into the conducing medium along a normal to its boundary surface, S_1 or S_2. And, the conducting medium may be treated as a semi-infinite medium in the direction of propagation. This model would be valid at all locations on the boundary surface S_1 or S_2.

The power carried into a conductor, through a unit area on the surface S_1 or S_2, is equal to the magnitude of the Poynting vector $E(t)H(t)$, where $E(t)$ and $H(t)$ are the amplitudes of the orthogonal electric and magnetic fields, respectively, of the uniform plane wave. This power is dissipated inside the conductor, and the time average of the above power per unit area may be defined as the average power-loss density P_d. For time-harmonic waves, the above time-average can be expressed by phasor analysis (see Appendix B), using peak-phasors E and H for the electric and magnetic fields, respectively.

$$P_d = < E(t)H(t) > = \frac{1}{2}Re(EH^*) = \frac{1}{2}Re(\eta' HH^*) = \frac{1}{2}\eta_{rc}|H|^2 = \frac{1}{2}R_s|H|^2$$
$$\frac{E}{H} = \eta' = \eta_{rc} + j\eta_{ic},\ Re(\eta') = \eta_{rc} = R_s = \sqrt{\frac{\mu_c\omega}{2\sigma_c}} \qquad (4.140)$$

where the complex wave impedance η' and its real part η_{rc} for the conducting medium is expressed in (4.139). The real or resistive part η_{rc} of the complex wave impedance of a conducting medium is also referred to as the surface resistance R_s of the conductor.

Now, the average power loss $\underline{P}_{loss}$ over one unit length $\Delta z = 1$ of the transmission line may be expressed by integrating P_d in (4.140) over the two surfaces S_1 and S_2, covering the unit-length segment. This is equal to the sum of the line integrals of P_d over the cross-sectional boundary lines of the two surfaces S_1 and S_2 (see Fig.1). These boundary lines of the conductors are referred to as l_{c1} and l_{c2}, respectively, which align with specific magnetic field lines l_h at the boundaries

$$\underline{P}_{loss} = \int_{S_1,\Delta z=1} P_{d1} ds + \int_{S_2,\Delta z=1} P_{d2} ds = \int_{S_1,\, l_{c1}} P_{d1} dl + \int_{S_2,\, l_{c2}} P_{d2} dl$$
$$= \int_{l_{c1}} \frac{1}{2} R_s |H|^2 dl + \int_{l_{c2}} \frac{1}{2} R_s |H|^2 dl \qquad (4.141)$$

In reference to the circuit model of Fig.6, the parameter $\underline{R}$ is the equivalent series resistance produced due to the finite conductivity of the conductors. The average power loss in $\underline{R}$ may be expressed as $1/2|I|^2\underline{R}$, This power loss in $\underline{R}$ would be equal to the $\underline{P}_{loss}$ derived above in (4.141). Accordingly, the parameter $\underline{R}$ may be expressed in terms of integrals of magnetic field and surface resistances $R_s = \eta_{rc}$ of the conductors.

$$\underline{P}_{loss} = \frac{1}{2}|I|^2\underline{R} = \int_{l_{c1}} \frac{1}{2} R_s |H|^2 dl + \int_{l_{c2}} \frac{1}{2} R_s |H|^2 dl$$
$$\underline{R} = \frac{\int_{l_{c1}} R_s |H|^2 dl + \int_{l_{c2}} R_s |H|^2 dl}{|I|^2}, \quad R_s = \sqrt{\frac{\mu_c \omega}{2\sigma_c}} \qquad (4.142)$$

The conductivity σ_c, and consequently the surface resistance R_s, may be different for the two conductors, or may even be different at different points on the same conductor. This situation is accommodated in the above general expression of $\underline{R}$, where R_s is maintained as an integration variable. For a simple practical case with identical conducting materials, distributed uniformly over the two conducting bodies, the R_s would be a constant parameter which may be taken out of the integration in (4.142).

$$\underline{R} = R_s \frac{\int_{l_{c1}} |H|^2 dl + \int_{l_{c2}} |H|^2 dl}{|I|^2} = \sqrt{\frac{\mu_c \omega}{2\sigma_c}} \frac{\int_{l_{c1}} |H|^2 dl + \int_{l_{c2}} |H|^2 dl}{|I|^2} \qquad (4.143)$$

10. Specific Transmission Line Geometries

We have modeled signal propagation in terms of equivalent parameters $\underline{L}$, $\underline{C}$, $\underline{G}$ and $\underline{R}$. These parameters are defined in the equations (4.17), (4.21), (4.22) and (4.143), respectively, for a general transmission line. In this section, we will derive these parameters for a few specific transmission lines of common interest. We will analyze the electric and magnetic fields of the specific transmission lines under loss-less conditions. The loss-less condition may allow simplification in the analysis and understanding. The capacitance per unit length $\underline{C}$ and the inductance per unit length $\underline{L}$ are derived from the fields using their standard definitions in (4.17) and (4.21), respectively. The expression for the equivalent capacitance per unit length $\underline{C}$, derived under the loss-less condition, is then related to the equivalent conductance per unit length $\underline{G}$ using the simple relationship (4.23) between the two parameters. The derivation for $\underline{L}$, $\underline{C}$ and $\underline{G}$ confirm to the TEM condition in a rigorous sense. This assumes a uniform material medium and ideal conductors with infinite conductivity. In contrast, when a finite conductivity is introduced in the conductors, the associated series resistance per unit length $\underline{R}$, as expressed in (4.143), is derived in an approximate sense using the approach outlined in section 9.7.9. Introduction of a non-zero $\underline{R}$ violates the basic TEM condition, and therefore the circuit model of Fig.6 in this case would not be valid in a rigorous sense, as discussed in section 9.7.9. For conductors with very high conductivity, the deviation from the ideal TEM condition is expected to be relatively small, which may be modeled in an approximate sense by introducing the small resistance parameter $\underline{R}$ in the equivalent circuit of Fig.6. Under the high-conductivity condition, it is possible to derive the simple expression for the $\underline{R}$, as given in (4.143), using a time-harmonic analysis.

The transmission line is assumed to be infinite long along the z direction, and the signal is assumed to propagate along $+z$ direction for reference. As suggested earlier, we will perform the field analysis of the specific transmission lines, under the loss-less condition. We assume a uniform material medium over the cross section of the transmission line, and the parameters μ and ϵ are assumed to be constants which are ideally valid for all time variations. As we have discussed in section 4.5.1, this means that the transmission line would have TEM fields, with equal velocity for all time variations. The circuit model of Fig.5 would be rigorously valid, with constant parameters $\underline{L}$ and $\underline{C}$ that are valid for all time variations. In this case, the derivation of $\underline{L}$ and $\underline{C}$ should be independent of any specific time-dependence of the waveform. Therefore, for simplicity of analysis and understanding, we may choose for the present study a waveform with its time-derivative approaching zero ($\partial/\partial t \to 0$). We know from section 4.4 that the shape of the waveform would remain unchanged as the signal propagates along a given direction. The waveform at any location z is simply a shifted version of the the waveform at location $z = 0$ with a delay equal to z/v_p. Accordingly, the time derivative of the waveform at all locations would be the same, which approaches zero as per our choice in the present study. Now, using the voltage-current relations (4.18,4.25), we show that the derivative in z would approach zero at all locations z, because we have the time derivative approaching zero everywhere. Accordingly, the voltage and current signals, and their related fields, may be assumed to be independent of z for the present study. This is a significant simplification. The results, so obtained, would however apply for general waveforms.

10.1. Parallel Strip transmission Line. Figure 33 shows the geometry of a parallel-plate transmission line, of width W and plate separation d. The permeability and permittivity of the medium between the two plates are μ and ϵ, respectively. The fields are mostly confined in the space between the two plates. Some of the fields fringe outside the region, and the fringing effect is expected to be reduced for larger values of W or smaller values of d. The fringing fields may be neglected, and accordingly the fields outside the limited region ($W \times d$) between the plates may be assumed to be zero, when $W >> d$. In this case, any structure or material outside the parallel-plate region would ideally have no effect on the performance of the transmission line. That would be the case, if the material medium outside the parallel-plate region is different from that inside the parallel-plate region. Similarly, if the width of one of the plates is extended, while keeping the width of the other plate W, the characteristics of the transmission line would remain unchanged. As per the above discussion, the transmission line configuration of Fig.34 may be considered equivalent to a parallel-plate transmission line of Fig.33, when $W >> d$. The configuration of Fig.34 may be recognized as a transmission line commonly used on a printed circuit board, and is commonly referred to as a printed-circuit transmission line, or also as a microstrip line. The results derived for a parallel-plate transmission line under the condition $W >> d$ would apply as well to a microstrip line. We will limit our derivation under this condition $W >> d$.

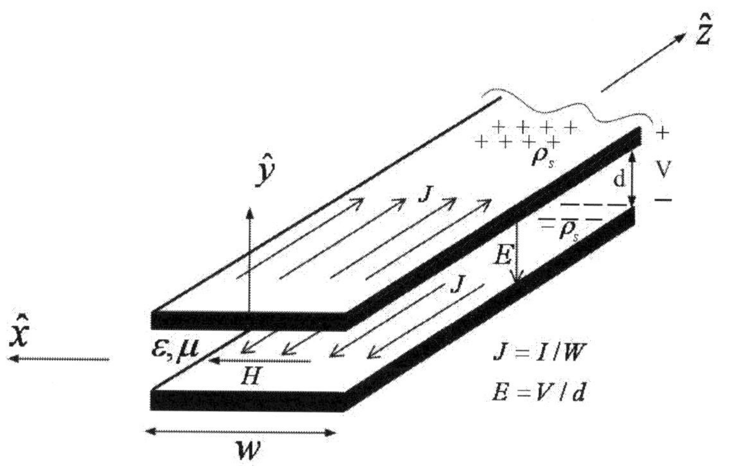

FIGURE 33

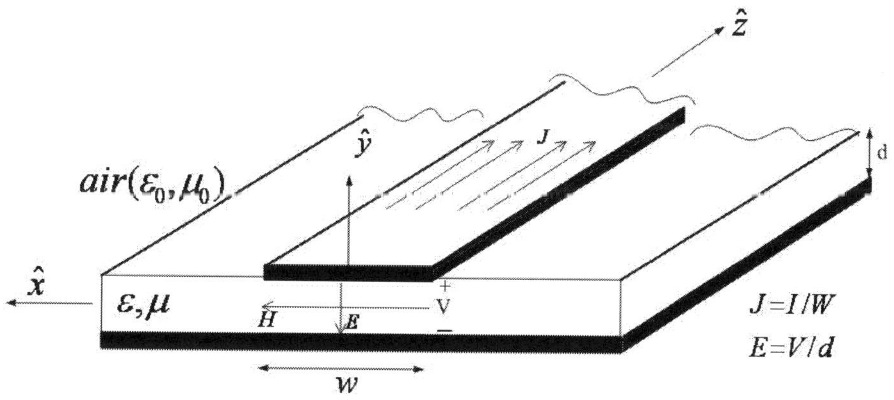

FIGURE 34

The fields of the parallel-plate transmission line of Fig.33 may be analyzed as a superposition of the individual fields produced by the charges and currents on each of the plates. We will derive first the fields $\overline{E_1}$ and $\overline{H_1}$ due to the top plate, as shown in Fig.35-37. The procedure can be repeated to derive the fields due to the bottom plate. The simple difference in this case is that the location of the bottom plate is different from that of the top plate, and the charges and current flow on the bottom plate are negative of those on the top plate. As explained above, for simplicity in this derivation, the charge distribution ρ_s and current distribution J, and the related electric and magnetic fields, are assumed to be independent of z. The results would, however, apply to general waveforms. Further, as discussed before, we would also assume that the width W of the transmission line is very large, approaching infinity in an ideal sense. In this case, the charge density ρ_s and current density J may be assumed to be approximately uniform in the x direction. The results for $\underline{L}$ and $\underline{G}$ derived using this assumption are expected to apply closely for practical cases with $W >> d$

10.1.1. *Electric Field and Capacitance Per Unit Length.* Fig.35 shows the electric field produced by the top plate only. The plate is is considered to be of infinite size in z and x dimensions, with a uniform charge density ρ_s, as we discussed. The resulting electric field would be uniform in x and z. Under this situation, we may show by simple symmetry argument that any electric field components in the z and x directions, which are parallel to the plate, would be zero. This may be seen by rotating the structure by 180^0 about the y axis. Let the x component of the electric field at at a given location be E_x. After the 180^0 rotation of the plate, the field would rotate to $-E_x$. A infinite sized charge plate, with a uniform distribution of ρ_s is invariant to the rotation. Therefore, we expect that the resulting fields to be invariant to the rotation. Accordingly, E_x and $-E_x$ should be equal. This can happen only if E_x is zero. Similarly, the z component of the electric field may also be shown to be zero. Having the z component of the electric field to be zero is consistent with the TEM nature of the transmission line.

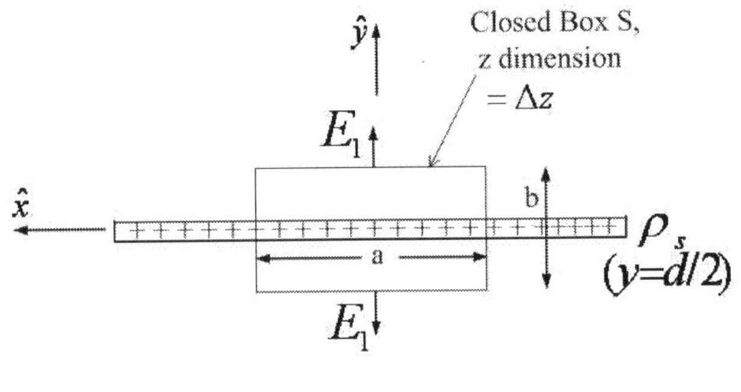

FIGURE 35

With both the x and z components of the electric field shown to be zero, we are left with the analysis of only the y component of the electric field. By symmetry, the y component of the electric field above the plate should be negative of that below the plate, as shown in Fig.35. With this, and other considerations discussed earlier, apply Gauss' law to the closed surface S in Fig.35. The electric flux through all sides, other than the top and bottom of S, would be zero. This is because, we have $E_x = E_z = 0$ Also, the contribution from the top and bottom sides would be equal to each other.

$$2a\triangle z\epsilon E_1 = a\triangle z\rho_s, \quad E_1 = \frac{\rho_s}{2\epsilon} \tag{4.144}$$

$$\begin{aligned}
\overline{E}_1 &= E_1\hat{y} = \frac{\rho_s}{2\epsilon}\,\hat{y}, \quad y > d/2 \\
&= -E_1\hat{y} = -\frac{\rho_s}{2\epsilon}\,\hat{y}, \quad y < d/2 \\
E_{x1} &= E_{z1} \;=\; 0
\end{aligned} \tag{4.145}$$

Similarly, the electric field $\overline{E}_2$ due to the lower plate may be expressed as follows. Note that the charge density on the lower plate is $-\rho_s$

$$\begin{aligned}
\overline{E}_2 &= E_2\hat{y} = -\frac{\rho_s}{2\epsilon}\,\hat{y}, \quad y > -d/2 \\
&= -E_2\hat{y} = \frac{\rho_s}{2\epsilon}\,\hat{y}, \quad y < -d/2 \\
E_{x2} &= E_{z2} \;=\; 0
\end{aligned} \tag{4.146}$$

The total field $\overline{E}$ is obtained by adding $\overline{E}_1$ in (4.145) and $\overline{E}_2$ in (4.146).

$$\begin{aligned}
\overline{E} = -E\hat{y} &= \overline{E}_1 + \overline{E}_2 = (E_1 + E_2)\hat{y} = 0, \quad y > d/2 \\
&= (-E_1 + E_2)\hat{y} = -\frac{\rho_s}{\epsilon}\,\hat{y}, \quad -d/2 < y < d/2 \\
&= (-E_1 - E_2)\hat{y} = 0, \quad y < -d/2 \\
E = \frac{\rho_s}{\epsilon}, &\quad E_x = E_z \;=\; 0
\end{aligned} \tag{4.147}$$

Note that the field uniform between the plates in the region $-d/2 < y < +d/2$, and is zero outside the region. Now, using the above field in (4.21), we get the expression for the capacitance per unit length $\underline{C}$ as the ratio of the charge per unit length $\Delta z = 1$ to the voltage V across the plates.

$$\underline{C} = \frac{\triangle Q}{V} = \frac{\rho_s W\triangle z}{Ed} = \frac{\epsilon W}{d}, \quad \triangle z = 1 \tag{4.148}$$

10.1.2. *Magnetic Field and Inductance Per Unit Length.* We first consider the magnetic field produced by the top plate at $y = +d/2$. It is considered to be of infinite size in z and x dimensions, with a uniform current density J flowing along the z direction. The resulting magnetic field would be uniform in x and z.

We may apply Gauss' law for the magnetic field to a closed box S, symmetrically placed around the top plate, as shown in Fig.36. This would show that the y component of the magnetic field is zero. The procedure would be very similar to the application of Gauss' law for the electric field in Fig.35, except that the magnetic charge density, unlike

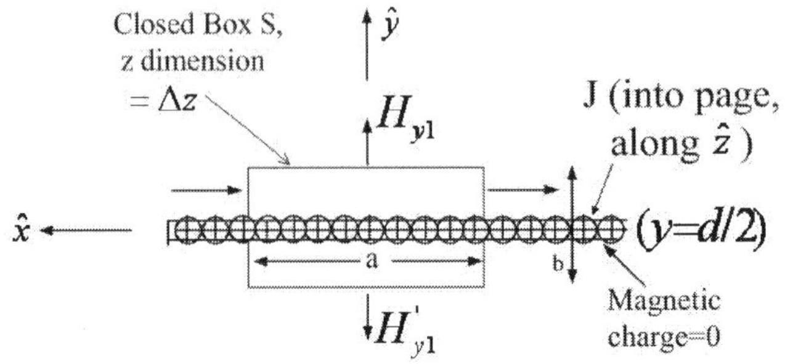

FIGURE 36

the electric charge density, is zero. We need to calculate the magnetic flux through the box S in Fig.36. Any magnetic flux from the left or front side of the box S would cancel with those from th right or back sides, respectively. This is because any x and z directed magnetic field is uniform in x and z. Accordingly, if the magnetic flux enters into the box through the left or front side, an equal amount of magnetic flux would exit through the right or back sides, respectively. By symmetry consideration, any y component of the magnetic field H_{y1} at the top face of S would be equal in magnitude, but opposite in sign, as compared to that $(= H'_{y1})$ on the bottom face of S. Therefore, the magnetic flux contributions from the top side of the box S in Fig.36 would be $a\Delta z\mu H_{y1}$, which would be equal that for the bottom side of S. Adding the contributions from all sides of the box S, the total magnetic flux through S would be $2a\Delta z\mu H_{y1}$. This is equal to the total magnetic charge inside S, which is zero. Accordingly, the y component of the magnetic field H_{y1} will have to be zero.

$$2a\triangle z\mu H_{y1} = 0, \quad H_{y1} = 0 \tag{4.149}$$

Next, we will derive the x component of the magnetic field due to the upper plate, under the conditions we have already established. Let the x component of the magnetic field below the plate be H_1. If the plate is rotated 180^0 about the z axis, the x component of the magnetic field above the rotated plate would be $-H_1$. The uniform current distribution on the plate is invariant with respect to the above rotation. Accordingly, $-H_1$ would also be the x component of the magnetic field above the original current distribution, as shown in Fig.37. Apply Ampere's law to the closed loop l_c in Fig.37.

$$2aH_1 = aJ, \quad H_1 = \frac{J}{2} = \frac{I}{2W} \tag{4.150}$$

The contribution from the top segment of l_c in Fig.37 is seen to be equal to that from the bottom segment, each of which is equal to aH_1. The contributions from the two side segments of l_c would be zero. This is because the y component of the magnetic field zero, as shown in (4.149). Accordingly, the magnetic field integral over the entire close loop l_c is $2aH_1$, as expressed in (4.150). As per Ampere's law, this is equal to Ja, which is the total current inside the loop l_c.

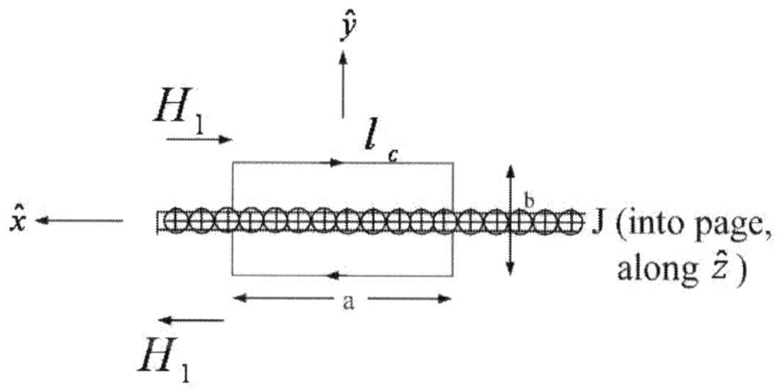

FIGURE 37

We have derived the x component of the magnetic field in (4.150), expressed in relation to the current density J. The z component of the magnetic field may be similarly derived, and shown to be equal to zero. This is accomplished by repeating the above procedure, by applying Ampere's law in Fig.37 to a rectangular loop, placed around the plate current on the yz plane. In this case the component of the current J (which is along the z direction) normal to the loop would be zero, which would lead to having the z component of the magnetic field equal to zero. This is in contrast to a non-zero current through the loop in the xy plane, shown in Fig.37, which resulted in a non-zero value for the x component of the magnetic field in (4.150).

The assumption of the TEM nature of a transmission line, as discussed in section 1.1, requires that there is no field component along the direction of propagation, which is the z direction. Having the z component of the magnetic field equal to zero is consistent with this TEM condition.

Now, the magnetic field $\overline{H}_1$ due to the top plate may be expressed as follows.

$$
\begin{aligned}
\overline{H}_1 &= -H_1\hat{x} = -\frac{J}{2}\,C\hat{x}, \quad y > d/2 \\
&= +H_1\hat{x} = +\frac{J}{2}\,\hat{x}, \quad y < d/2 \\
H_{y1} &= H_{z1} \;=\; 0
\end{aligned}
\tag{4.151}
$$

The magnetic field $\overline{H}_2$ due to the lower plate may be similarly expressed. Note that the current density on the lower plate is of equal magnitude ($= J$) as that on the top plate, but is directed in an opposite direction. The switch in the current direction would have to be properly adjusted in the expression for $\overline{H}_2$.

$$
\begin{aligned}
\overline{H}_2 &= H_2\hat{x} = \frac{J}{2}\,\hat{x}, \quad y > -d/2 \\
&= -H_2\hat{x} = -\frac{J}{2}\,\hat{x}, \quad y < -d/2 \\
H_{y2} &= H_{z2} \;=\; 0
\end{aligned}
\tag{4.152}
$$

The total field $\overline{H}$ is obtained by adding $\overline{H}_1$ in (4.151) and $\overline{H}_2$ in (4.152).

$$
\begin{aligned}
\overline{H} = H\hat{x} &= (-H_1 - H_2)\hat{x} = 0, \quad y > d/2 \\
&= (H_1 - H_2)\hat{y} = J\;\hat{x}, \quad -d/2 < y < d/2 \\
&= (H_1 + H_2)\hat{x} = 0, \quad y < -d/2 \\
H \;=\; J, \quad H_y &= H_z \;=\; 0
\end{aligned}
\tag{4.153}
$$

Note that the field is uniform between the plates in the region $-d/2 < y < +d/2$, and is zero outside the region. Now, using the above field in (4.17), we get the expression for the inductance per unit length $\underline{L}$ as the ratio of the magnetic flux in the x direction, through the yz cross-section of the transmission line over unit length $\Delta z = 1$, to the total current I. The yz cross-section of the transmission line, used for the flux calculation, is shown in Fig.38.

$$
\underline{L} = \frac{\Delta\Psi_m}{I} = \frac{\mu H d \Delta z}{JW} = \frac{\mu J d}{JW} = \frac{\mu d}{W}, \quad \Delta z = 1
\tag{4.154}
$$

10.1.3. *Conductance Per Unit Length.* We will express the conductance per unit length $\underline{G}$ using the expression for the capacitance per unit length $\underline{C}$, which we have already derived. This uses the basic relationship (4.23).

$$
\underline{G} = \frac{W\sigma}{d}, \quad \frac{\underline{G}}{\underline{C}} = \frac{\sigma}{\epsilon}, \quad \underline{C} = \frac{W\epsilon}{d}
\tag{4.155}
$$

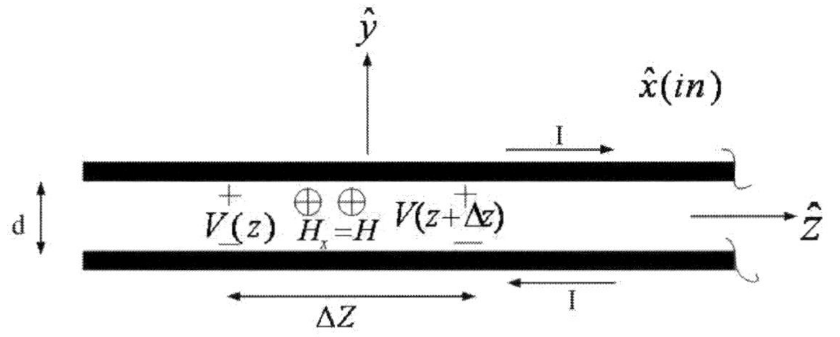

FIGURE 38

$\underline{G}$ is the conductance of the material body across the plate distance d, with cross-sectional area $W\Delta z$, which is equal to W for $\Delta z = 1$.

10.1.4. *Resistance Per Unit Length.* The resistance per unit length $\underline{R}$ will be expressed using the model discussed in section 9.7.9. This assumes that the current and fields in the two conducting plates are concentrated very close to their interfaces with the material medium between the plates. The confinement region, called the skin depth δ, is assumed to be much smaller than the thickness Δ_d of the conductors. In this case, $\underline{R}$ is expressed using (4.143), and the magnetic field of (4.153)

$$\underline{R} = R_s \frac{\int\limits_{l_{c1}} |H|^2 dl + \int\limits_{l_{c2}} |H|^2 dl}{|I|^2} = R_s \frac{\int\limits_{l_{c1}} |J|^2 dl + \int\limits_{l_{c2}} |J|^2 dl}{|JW|^2} = \frac{2R_s}{W} = \frac{1}{W}\sqrt{\frac{2\mu_c\omega}{2\sigma_c}}$$
$$l_{c1} : y = +\frac{d}{2},\ l_{c2} : y = -\frac{d}{2},\ H_{l_{c1}} = H_{l_{c2}} = J,\ I = JW,\ R_s = \sqrt{\frac{\mu_c\omega}{2\sigma_c}} \tag{4.156}$$

where R_s is the real part of the wave impedance of the conducting material used in the two conductors, referred to as the surface resistance. The two conductors are assumed to be made of identical materials. with permeability μ_c and conductivity σ_c.

10.1.5. *Propagation Velocity.* Let us calculate the propagation velocity v_p for a parallel-plate line in a loss-less case, using the expressions for $\underline{L}$ and $\underline{C}$ we have derived in this section.

$$v_p = \frac{1}{\sqrt{\underline{LC}}} = \frac{1}{\sqrt{\mu\epsilon}},\ \underline{L} = \frac{\mu d}{W},\ \underline{C} = \frac{\epsilon W}{d} \tag{4.157}$$

The result may be verified with (4.43). This confirms that the propagation velocity of a parallel-plate line, like any other TEM transmission line, is independent of the dimensions the parallel-plate structure. The velocity is dependent only on the parameters of the material medium filling between the two conductors. .

10.1.6. *Characteristic Impedance.* The characteristic impedance for a parallel-plate transmission line, under loss-less condition, may be derived using the expressions for $\underline{L}$ and $\underline{C}$ in (4.47).

$$Z_c = \sqrt{\frac{\underline{L}}{\underline{C}}} = \sqrt{\frac{\mu}{\epsilon}}\frac{d}{W} = \frac{\eta d}{W}, \ \underline{L} = \frac{\mu d}{W}, \ \underline{C} = \frac{W\epsilon}{d}, \ \eta = \sqrt{\frac{\mu}{\epsilon}} \tag{4.158}$$

The characteristic impedance, Z_c, increases with increasing distance, d, or reducing width W. Accordingly, a parallel-plate line designed with either a very large distance d, or a very narrow plate width W, would approach an open circuit (infinity impedance). This is often undesirable in practical designs. On the other hand, when the distance d is very small (when the two plates are very close to each other) or the width W is very large, the characteristic impedance Z_c would be close to zero or a short circuit. Such a design with a small characteristic impedance, like the above case with a large impedance, is also practically undesirable.

The characteristic impedance is proportional to the wave impedance η of the material medium. This is a typical behavior for all TEM transmission lines.

10.2. Coaxial Transmission Line. Fig.39 shows the geometry of a coaxial line. We will analyze the electric and magnetic fields for the coaxial line, from which the capacitance per unit length $\underline{C}$ and inductance per unit length $\underline{L}$ can be derived. As in the case of a parallel-plate line in the last section, the coaxial line will be assumed to be lossless for the field analysis. The time derivative of waveform would be assumes to approach zero $\partial/\partial t \to 0$ in the present study. Consequently, as discussed before, the voltage and current signals, and the related fields and charge distribution, would be independent of z in the present study. The values of $\underline{C}$ and $\underline{L}$ obtained with this simplification would, however, be valid for general signals.

The equivalent conductance per unit length $\underline{G}$ is related to $\underline{C}$ using (4.23). The equivalent series resistance $\underline{R}$ is derived in an approximate sense using (4.143).

10.2.1. *Electric Field and Capacitance Per Unit Length.* As mentioned, we have a uniform charge distribution along z, and the transmission line is infinite long. The resulting electric field would be uniform in z. One may show that the $\hat{z}$ component of the electric field, E_z, produced by the uniform charge distribution is zero. This may be accomplished by a simple symmetry argument. Consider turning the transmission line by 180^0, such that the $+z$ and $-z$ axes of the transmission line are interchanged. The rotation would turn the $+E_z$ field into the $-E_z$ field . However, the uniform charge distribution of the infinite-long transmission line is invariant to the above rotation. In other words, the transmission line would look physically indistinguishable after the rotation. Accordingly, all the fields should also be invariant with the rotation. Therefore, E_z should be equal to $-E_z$. This can happen only if E_z is zero, as we expected. Having E_z equal to zero is also consistent with the TEM condition of the transmission line.

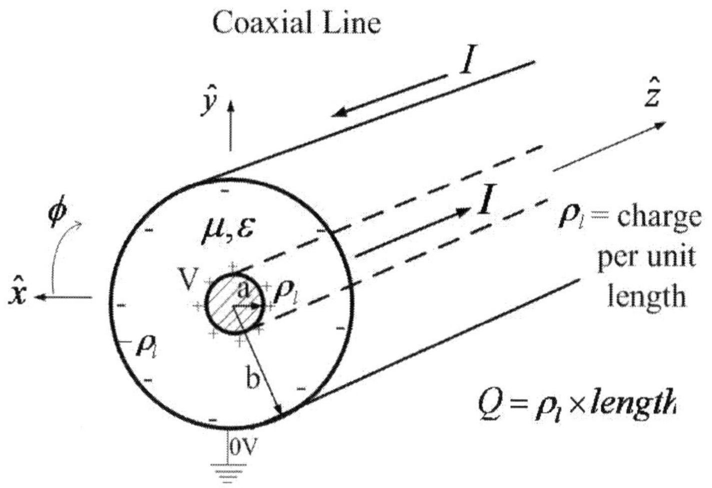

FIGURE 39

The $\hat{\phi}$ component of the electric field E_ϕ may also be shown to be equal to zero. The transmission line is symmetric in ϕ. Therefore, any E_ϕ we may have should also be independent of ϕ. Now consider line integration of the electric field over a circle of radius ρ. It would be equal to $2\pi\rho E_\phi$, obtained by multiplying E_ϕ at the distance ρ from the center by $2\pi\rho$ which is the line length around the circle. As per Faraday's law, the line integral would be zero when the magnetic field component H_z is zero. This is the case, as shown from the magnetic field analysis in the next section, and as also expected from the TEM condition. Accordingly, the line integral $2\pi\rho E_\phi$ and therefore E_ϕ should be zero.

Both the $\hat{z}$ and $\hat{\phi}$ components of the electric field are zero. This leaves the only remaining component E_ρ to be analyzed. Apply Gauss' law to a closed cylindrical surface S, as shown in Fig.40. The electric flux through the two sides of S are zero, because E_z is zero. Therefore, the total flux through S is contributed only from the curved surface S_1.

$$2\pi\rho\epsilon E_\rho = \triangle Q = \rho_l, \; E_\rho = \frac{\rho_l}{2\pi\epsilon\rho}$$
$$\overline{E} = E\hat{\rho} = E_\rho\hat{\rho} = \frac{\rho_l}{2\pi\epsilon\rho}\hat{\rho}, \; E_z = E_\phi = 0 \tag{4.159}$$

The E_ρ is assumed to be uniform over the curved surface S_1 due to the ϕ symmetry. Now, using the above field in (4.21), we get the expression for the capacitance per unit length $\underline{C}$ as the ratio of the charge ΔQ per unit length $\Delta z = 1$ to the voltage V across the plates.

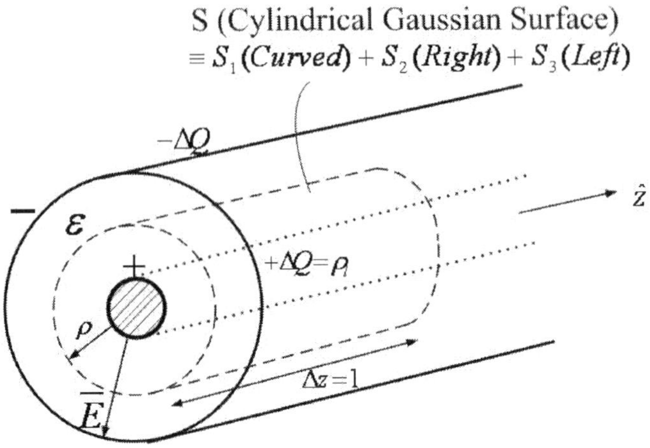

FIGURE 40

$$V = \int_a^b E_\rho d\rho = \frac{\rho_l}{2\pi\epsilon}\int_a^b \frac{d\rho}{\rho} = \frac{\rho_l}{2\pi\epsilon}\ln\frac{b}{a}$$

$$\underline{C} = \frac{\triangle Q}{V} = \frac{\rho_{l\triangle z}}{V} = \frac{2\pi\epsilon}{\ln\frac{b}{a}}, \ \triangle z = 1 \quad (4.160)$$

10.2.2. *Magnetic Field and Inductance Per Unit Length.* We assume the current flow in the z direction. For the present study, we also assume the time-derivative to approach zero. Accordingly, we may use Biot-Savart's law to relate the magnetic field to the current I. As per Biot-Savart's law, the magnetic field is perpendicular to the current direction. Accordingly, the z directed magnetic field H_z would be zero. This is consistent with the basic TEM condition.

With $H_z = 0$, we may apply Gauss's law for the magnetic field to the surface S in Fig.40. This procedure is similar to that in applying Gauss' law for the electric field, used in the derivation of the electric field in the last section 10.2.1. Except, we have magnetic charge equal to zero. Accordingly, we would have $H_\rho = 0$.

$$2\pi\rho\mu H_\rho = 0, \ H_\rho = 0 \quad (4.161)$$

Having $H_\rho = 0$ is also consistent with the condition of orthogonality between the electric and magnetic field directions, shown in section 1.2. We have already shown in (4.159) that the electric field is in the $\hat{\rho}$ direction. According to the orthogonality condition, the the magnetic field will have to be directed perpendicular to the $\hat{\rho}$ direction. This means the $\hat{\rho}$ component of the magnetic field will have to be zero, consistent with (4.161).

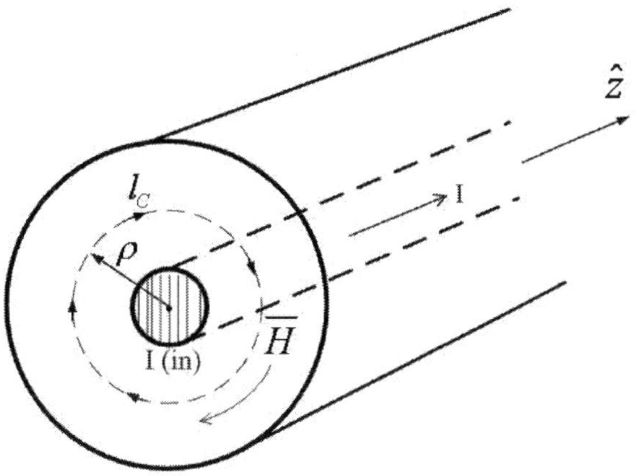

FIGURE 41

Both the $\hat{z}$ and $\hat{\rho}$ components of the magnetic field are zero. This leaves the only remaining component H_ϕ to be analyzed. Apply Ampere's law to a circle of radius ρ as shown in Fig.41.

$$2\pi\rho H_\phi = I,\ H_\phi = \frac{I}{2\pi\rho}$$

$$\overline{H} = H\hat{\phi} = H_\phi\hat{\phi} = \frac{I}{2\pi\rho}\hat{\phi},\ H_z = H_\rho = 0 \tag{4.162}$$

The H_ϕ is assumed to be uniform over the circle due to the ϕ symmetry. Now, using the above field in (4.17), we will get the expression for the inductance per unit length $\underline{L}$. $\underline{L}$ is the ratio of the magnetic flux $\Delta\psi_m$ in the $\hat{\phi}$ direction, passing through the ρz cross-section of the transmission line, over unit length $\Delta z = 1$. to the total current I. The ρz cross-section of the transmission line, used for the flux calculation, is shown in Fig.42.

$$\triangle\Psi_m = \int_a^b \mu H_\phi d\rho\triangle z = \frac{\mu I}{2\pi}\int_a^b \frac{d\rho}{\rho} = \frac{\mu I}{2\pi}\ln\frac{b}{a},\ \triangle z = 1$$

$$\underline{L} = \frac{\triangle\Psi_m}{I} = \frac{\mu}{2\pi}\ln\frac{b}{a} \tag{4.163}$$

10.2.3. *Conductance Per Unit Length.* We will express the conductance per unit length $\underline{G}$ using the formula for the capacitance per unit length $\underline{C}$, which we have already derived. This uses the basic relationship (4.23).

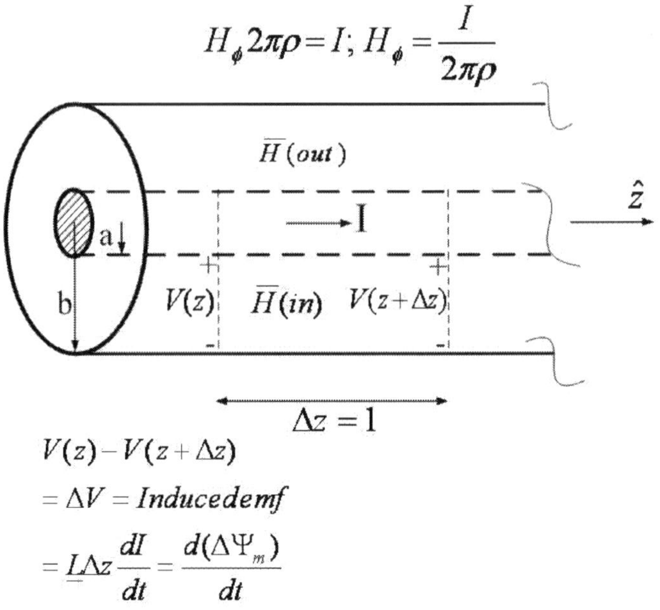

FIGURE 42

$$\underline{G} = \frac{2\pi\sigma}{\ln\frac{b}{a}},\ \frac{\underline{G}}{\underline{C}} = \frac{\sigma}{\epsilon},\ \underline{C} = \frac{2\pi\epsilon}{\ln\frac{b}{a}} \tag{4.164}$$

$\underline{G}$ is the conductance measured across the two concentric, cylindrical conductors, when the conductivity of the material between the two conductors is σ, and the length of the cylinder is $\Delta z = 1$. .

10.2.4. *Resistance Per Unit Length.* The resistance per unit length $\underline{R}$ will be expressed using the model discussed in section 9.7.9. This assumes that the current and fields in the two conducts are concentrated very close to their surfaces in contact with the inner material medium. The confinement thickness or the skin depth δ is assumed to be much smaller than the thickness Δ_ρ of the outer conductor and the radius a of the inner conductor. In this case, $\underline{R}$ is expressed using (4.143), and the magnetic field of (4.162)

$$\underline{R} = R_s \frac{\int_{l_{c1}} |H|^2 dl + \int_{l_{c2}} |H|^2 dl}{|I|^2} = R_s \frac{\int_{l_{c1}} |\frac{I}{2\pi b}|^2 dl + \int_{l_{c2}} |\frac{I}{2\pi a}|^2 dl}{|I|^2}$$
$$= R_s(\frac{1}{2\pi b} + \frac{1}{2\pi a}) = \sqrt{\frac{\mu_c\omega}{2\sigma_c}}(\frac{1}{2\pi b} + \frac{1}{2\pi a})$$
$$l_{c1} : \rho = b,\ l_{c2} : \rho = a,\ H_{l_{c1}} = \frac{I}{2\pi b},\ H_{l_{c2}} = \frac{I}{2\pi a},\ R_s = \sqrt{\frac{\mu_c\omega}{2\sigma_c}} \tag{4.165}$$

where R_s is the real part of the wave impedance of the conducting material used in the two conductors, referred to as the surface resistance. The two conductors are assumed to be made of identical materials. with permeability μ_c and conductivity σ_c.

10.2.5. *Propagation Velocity.* Let us calculate the propagation velocity v_p for a coaxial line in a loss-less case, using the expressions for $\underline{L}$ and $\underline{C}$ we have derived in this section.

$$v_p = \frac{1}{\sqrt{\underline{L}\underline{C}}} = \frac{1}{\sqrt{\mu\epsilon}}, \; \underline{L} = \frac{\mu}{2\pi}\ln\frac{b}{a}, \; \underline{C} = \frac{2\pi\epsilon}{\ln\frac{b}{a}} \tag{4.166}$$

The result may be verified with (4.43). This confirms that the propagation velocity of a coaxial line, like any TEM transmission line, is independent of the dimensions of a coaxial transmission line. The velocity is dependent only on the parameters of the material medium filling between the inner and outer conductors. .

10.2.6. *Characteristic Impedance.* The characteristic impedance for a coaxial transmission line, under loss-less condition, may be derived using the expressions for $\underline{L}$ and $\underline{C}$ in (4.47).

$$Z_c = \sqrt{\frac{\underline{L}}{\underline{C}}} = \frac{1}{2\pi}\sqrt{\frac{\mu}{\epsilon}}\ln\frac{b}{a} = \frac{\eta}{2\pi}\ln\frac{b}{a}, \; \underline{L} = \frac{\mu}{2\pi}\ln\frac{b}{a}, \; \underline{C} = \frac{2\pi\epsilon}{\ln\frac{b}{a}}, \; \eta = \sqrt{\frac{\mu}{\epsilon}} \tag{4.167}$$

The characteristic impedance, Z_c, increases with the radius of the outer conductor, b, and approaches infinity as b approaches an infinitely large value. The characteristic impedance Z_c increases for smaller value of a, and approaches infinity as a approaches zero. In other words, a coaxial line designed with either a very large outer conductor, or a very thin inner conductor, would approach an open circuit (infinity impedance). Both the above situations are often undesirable in practical designs.

When the radii of the outer and inner conductors approach equal values, that is $b \to a$, the characteristic impedance would approach zero. In other words, in this limit the transmission line would behave close to a short circuit. Such designs with a small characteristic impedance, like the above cases with a large impedance, are also practically undesirable.

The characteristic impedance is proportional to the wave impedance η of the material medium. This is a typical behavior for all TEM transmission lines.

CHAPTER 5

INTRODUCTION TO ELECTROMAGNETIC WAVES II: Plane-Wave Propagation in Free Space and Uniform Media

A plane wave is the simplest form of wave propagation in an open medium. The basic relationship for such a wave, propagating in a specific direction in the free space or any other uniform, isotropic and lossless medium, are derived. Ampere's law and Faraday's law, together constituting Maxwell's equations, are the fundamental basis of the wave relations. The results are then generalized for propagation in any arbitrary direction. The concept of vector potential functions are introduced from which field components can be conveniently derived using differential relations. Propagation of plane wave across planar interface between uniform media are then modeled by enforcing appropriate boundary conditions for electric and magnetic fields.

1. Basic Plane-Wave Relations

A uniform plane wave is the most basic form of an electromagnetic wave in an empty space, or any other isotropic, uniform medium. We will consider a uniform medium with dielectric and magnetic constants equal to ϵ and μ, respectively. These parameters would be equal to ϵ_0 and μ_0, respectively, for an empty space. We also assume that the medium is loss less, with conductivity $\sigma = 0$. The wave propagates along a particular direction, for example along z. The amplitude of the plane wave is uniform on a plane (xy plane) perpendicular to direction of propagation.

$$\frac{\partial}{\partial x} = 0 = \frac{\partial}{\partial y}, \frac{\partial}{\partial z} \neq 0. \tag{5.1}$$

Under the above conditions we will derive basic relationships that satisfy Ampere's and Faraday's laws everywhere. We will assume a time-harmonic wave with general time variation of the form $Re[Ae^{j\omega t}] = |A|\cos(\omega t + \angle A)$. The A is the complex phasor representing a general cosinusoidal function with a given angular frequency ω, having amplitude $|A|$ and phase $\angle A$. Using a phasor representation, instead of using the explicit cosinusoidal time function, often allows significant mathematical simplification, and is therefore commonly used in signal analysis (appendix B). The phasor for the time derivative of an original function is a simple multiplication of $j\omega$ to the original phasor. This property would allow treatment of time derivatives as equivalent algebraic operations.

1.1. Application of Ampere's and Faraday's Laws. In Fig.1 we select closed loops C_{xy}, C_{yz}, C_{zx}, on three orthogonal planes, placed around an arbitrary point (x, y, z) in space, for application of Ampere's and Faraday's laws. Consider first a closed loop C_{xy} on the xy plane, and apply Ampere's law.

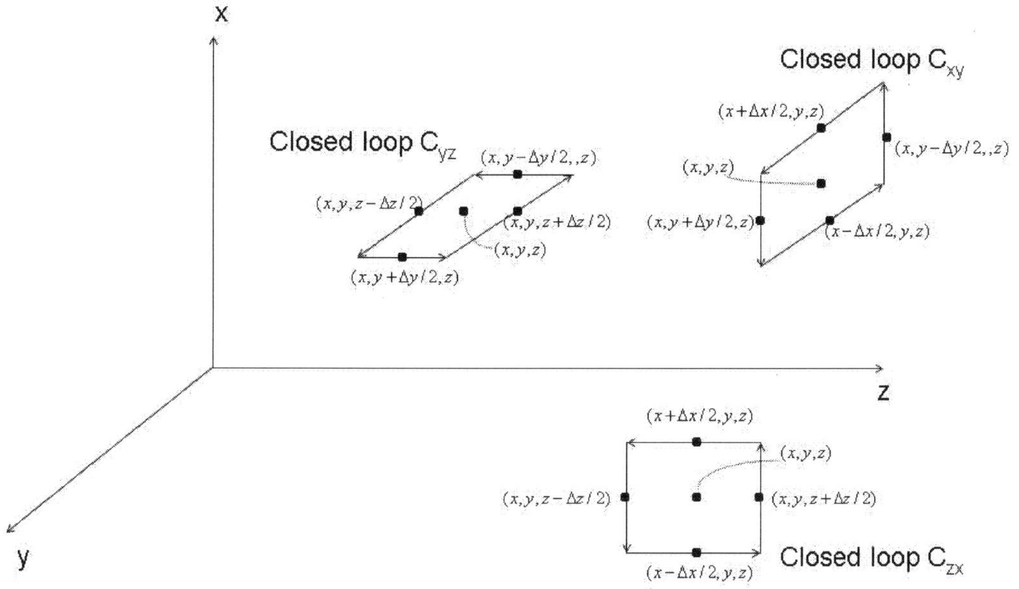

FIGURE 1. Closed loop paths C_{xy}, C_{yz}, C_{zx} considered on the xy, yz, zx planes, respectively, for treatment of uniform plane-wave relationships. The loops are rectangular in shape, with its center located at an arbitrary point with coordinates (x, y, z).

$$\sum_{C_{xy}} H_{ti}\Delta l_i = J_z \Delta x \Delta y, \tag{5.2}$$

$$J_z\Delta x\Delta y = (J_{cz} + J_{dz})\Delta x\Delta y = (\sigma + \epsilon\frac{\partial}{\partial t})E_z\Delta x\Delta y = \epsilon\frac{\partial E_z}{\partial t}\Delta x\Delta y;\ \sigma = 0, \tag{5.3}$$

$$\sum_{C_{xy}} H_{ti}\Delta l_i = H_x(x, y - \frac{\Delta y}{2}, z)\Delta x + H_y(x + \frac{\Delta x}{2}, y, z)\Delta y$$
$$-H_x(x, y + \frac{\Delta y}{2}, z)\Delta x - H_y(x - \frac{\Delta x}{2}, y, z)\Delta y = 0. \tag{5.4}$$

The subscript $'i'$ refers to a length segment along the loop C_{xy}. The subscript $'t'$ refers to the tangential field component along the path direction of the loop C_{xy} over the ith length segment Δl_i. Because the conductivity σ of the free space in (5.3) is zero, the conduction current density J_{cz} is zero. The total z-directed current density J_z in (5.3) is, therefore, contributed only by the displacement current $J_{dz} = \epsilon\frac{\partial E_z}{\partial t}$. As assumed in (5.1), there is no variation of amplitudes in x and y. Accordingly, the first and third terms, and the second and the last terms in (5.4) are equal to each other, resulting in the total expression to be zero. Now, using (5.3,5.4) in (5.2) we get,

$$(\epsilon \frac{\partial E_z}{\partial t})(\Delta x \Delta y) = 0,\ E_z = 0;\ (\frac{\partial}{\partial t} \neq 0). \tag{5.5}$$

Similarly, applying Faraday's law to a C_{xy} loop we will get $H_z = 0$.

$$\begin{aligned} \sum_{C_{xy}} E_{ti}\Delta l_i &= -\mu \frac{\partial H_z}{\partial t} \Delta x \Delta y \\ = E_x(x, y - \frac{\Delta y}{2}, z)\Delta x &+ E_y(x + \frac{\Delta x}{2}, y, z)\Delta y \\ -E_x(x, y + \frac{\Delta y}{2}, z)\Delta x &- E_y(x - \frac{\Delta x}{2}, y, z)\Delta y = 0, \\ -\mu \frac{\partial H_z}{\partial t} \Delta x \Delta y &= 0,\ H_z = 0;\ (\frac{\partial}{\partial t} \neq 0). \end{aligned} \tag{5.6}$$

Because $E_z = 0 = H_z$, both electric and magnetic fields are transverse to the direction of propagation (z). Now $\overline{E}$ can be directed along any direction on the xy plane. Because the choice of coordinate directions are arbitrary, let us choose coordinates such that

$$\overline{E} = \hat{x}E_x,\ E_y = 0. \tag{5.7}$$

Now repeat the above procedure for the closed loop C_{yz} on the yz plane (Fig.1). Apply first Ampere's law.

$$\begin{aligned} \sum_{C_{yz}} H_{ti}\Delta l_i = H_y(x, y, z - \frac{\Delta z}{2})\Delta y &+ H_z(x, y + \frac{\Delta y}{2}, z)\Delta z \\ -H_y(x, y, z + \frac{\Delta z}{2})\Delta y &- H_z(x, y - \frac{\Delta y}{2}, z)\Delta z \\ = -\frac{\partial H_y}{\partial z}\Delta y \Delta z &= \epsilon \frac{\partial}{\partial t} E_x \Delta y \Delta z, \\ \frac{\partial H_y}{\partial z} &= -j\omega\epsilon E_x\ (\frac{\partial}{\partial t} \equiv j\omega). \end{aligned} \tag{5.8}$$

Similarly, applying Faraday's law to the loop C_{yz}, we get,

$$\sum_{C_{yz}} E_{ti}\Delta l_i = E_y(x, y, z - \frac{\Delta z}{2})\Delta y + E_z(x, y + \frac{\Delta y}{2}, z)\Delta z$$
$$-E_y(x, y, z + \frac{\Delta z}{2})\Delta y - E_z(x, y - \frac{\Delta y}{2}, z)\Delta z$$
$$= -\mu\frac{\partial H_x}{\partial t}\Delta y\Delta z = -\frac{\partial E_y}{\partial z}\Delta y\Delta z = 0,$$
$$H_x = 0 \ (\frac{\partial}{\partial t} \equiv j\omega \neq 0). \tag{5.9}$$

Here we use E_y to be zero as assumed in (5.7). Finally, repeat the above procedure for the closed loop C_{zx} on the zx plane (Fig.1). We will give only the final results. Ampere's and Faraday's laws will result in the following two equations, respectively.

$$\frac{\partial H_x}{\partial z} = j\omega\epsilon E_y = 0, H_x = 0 \ (\frac{\partial}{\partial z} \neq 0), \tag{5.10}$$
$$\frac{\partial E_x}{\partial z} = -j\omega\mu H_y. \tag{5.11}$$

1.2. One-Dimensional Wave Equation and Wave Function. Equation (5.10) is consistent with (5.9), and is not a new result. This leaves us with five governing conditions (5.5,5.6,5.8,5.9,5.11), of which (5.5,5.6,5.9) are trivial equations for zero field components. The remaining two equations (5.8,5.11) need to be solved for the only two non-zero field components E_x and H_y. Differentiating (5.8) with respect to z, and then using (5.11), we get,

$$\frac{\partial^2 H_y}{\partial z^2} = -j\omega\epsilon\frac{\partial E_x}{\partial z} = (-j\omega\epsilon)(-j\omega\mu)H_y = -\omega^2\mu\epsilon H_y. \tag{5.12}$$

This is an equation involving only H_y. Similarly, differentiate (5.11) with respect to z, and then use (5.8). This would lead to an equation for only E_x.

$$\frac{\partial^2 E_x}{\partial z^2} = -\omega^2\mu\epsilon E_x. \tag{5.13}$$

As can be seen comparing (5.12,5.13), E_x and H_y satisfy the same general differential equation.

$$(\frac{\partial^2}{\partial z^2} + \omega^2\mu\epsilon)A(z) = 0; \ A = E_x, \ H_z. \tag{5.14}$$

Equation (5.14) is a wave equation in one dimension, the general solution for which maybe expressed in the following form.

$$A(z) = A_0 e^{\mp jkz}; \ A_0 = E_{x0}, \ H_{y0}; k = \omega\sqrt{\mu\epsilon}. \tag{5.15}$$

The above solution is referred to as the phasor wave function in one dimension, The wave function $A_0 e^{\mp jkz}$ may be expressed in its time-dependent form as follows.

$$A_0 e^{\mp jkz} \equiv |A_0| \cos(\omega t \mp kz + \angle A_0) = Re(A_0 e^{-jkz} e^{j\omega t}). \tag{5.16}$$

1.3. Wave Length and Wave Number. At a given time, the wave function (5.16) varies cosinusoidally in z, with its spatial periodicity defined as the wave length λ. The wave length is related to k, which is called the wave number or the propagation constant in the given medium.

$$k\lambda = 2\pi, \ \lambda = \frac{2\pi}{k} = \frac{2\pi}{\omega\sqrt{\mu\epsilon}} = \frac{1}{f\sqrt{\mu\epsilon}}. \tag{5.17}$$

1.4. Propagation Direction and Velocity. The wave nature may be visualized by examining the argument of the wave function (5.16). After a small positive time interval Δt the same function would be reproduced at a positive shift $\Delta z = +\Delta t\frac{\omega}{k}$ in position, if (-) sign is selected in (5.15,5.16). In this case, the function "moves forward" in positive z direction as time progresses, and is therefore referred to as a forward ($+z$) propagating harmonic wave. On the other hand, the same function is reproduced at a negative shift $\Delta z = -\Delta t\frac{\omega}{k}$ in position if (+) sign is selected in (5.15,5.16). In this case, the function "moves backwards" in negative z direction as time progresses, and is therefore called a backward ($-z$) propagating harmonic wave. The propagation velocity of the wave is defined from the time rate of forward or backward shift in position.

$$v_p \stackrel{\triangle}{=} \frac{|\Delta z|}{\Delta t} = \frac{\omega}{k} = \frac{1}{\sqrt{\mu\epsilon}}. \tag{5.18}$$

1.5. Wave Impedance. Equations (5.14,5.15) provides the same functional forms of solution for both E_x and H_y, having only different amplitudes. The solutions for E_x and H_y from (5.14,5.15) may now be used in either (5.8) or (5.11) to relate their relative amplitudes.

$$\frac{E_x}{H_y} = \frac{k}{\omega\epsilon} = \frac{\omega\mu}{k} = \sqrt{\frac{\mu}{\epsilon}} = \eta, \ \text{for } +z\,\text{propagation}$$
$$\frac{E_x}{H_y} = -\eta, \ \text{for } -z\,\text{propagation}. \tag{5.19}$$

The parameter η is dependent on the material property of the propagation medium, and is called the wave impedance of the medium.

2. General Propagation Along Arbitrary Direction

Results of (5.15) and (5.19) may be generalized for propagation in an arbitrary direction $\hat{s}$ in the following forms, independent of any specific choice of a coordinate system.

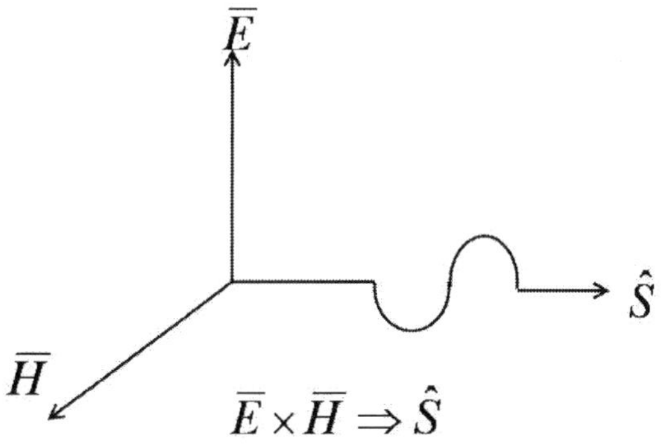

FIGURE 2. Electric field vector $\overline{E}$, magnetic field vector $\overline{H}$, and the direction of propagation $\hat{s}$ constitute three orthogonal directions.

$$\overline{E}, \overline{H} = (\overline{E}_0, \overline{H}_0)e^{-jks} \tag{5.20}$$

$$\overline{H} = \frac{\hat{s} \times \overline{E}}{\eta}, \ \overline{E} = (\overline{H} \times \hat{s})\eta, \tag{5.21}$$

where $\hat{s}$ is the unit vector along the direction of propagation, and s is the distance measured along $\hat{s}$.

Equation (5.21) may be expressed in words, in simple conceptual terms: (a) $\hat{s}$, $\overline{E}$, $\overline{H}$ are three mutual orthogonal directions, such that the cross-product $\overline{E} \times \overline{H}$ of the electric and magnetic field vectors is directed along $\hat{s}$ (see Fig.2). (b) The ratio $\frac{|\overline{E}|}{|\overline{H}|}$ of the magnitudes of the electric and magnetic fields is η. The condition (a) provides the relative orientations of the field vectors and the direction of propagation, whereas (b) provides the relative field amplitudes. Accordingly, if one field (electric or magnetic) vector and the direction of the wave propagation is specified, the other field vector (direction as well as magnitude) is completely known, given material parameters.

The general propagation factor in (5.20), and the relative vector directions in (5.21) are expressed with respect to the direction of propagation $\hat{s}$, where s is the distance along the direction of propagation. It maybe useful to express them with respect to a reference rectangular coordinate system (x, y, z) for common analytical treatment.

2.1. Propagation factor in Rectangular Coordinates. Let us first transform the propagation factor (5.20) to a common rectangular coordinate system. The propagation direction $\hat{s}$ is oriented at angles (θ', ϕ') with respect to the common coordinates, as shown in Fig.3.

$$s = \overline{r} \cdot \hat{s}, \ e^{-jks} = e^{-jk\hat{s}\cdot\overline{r}} = e^{-j\bar{k}\cdot\overline{r}}, \tag{5.22}$$

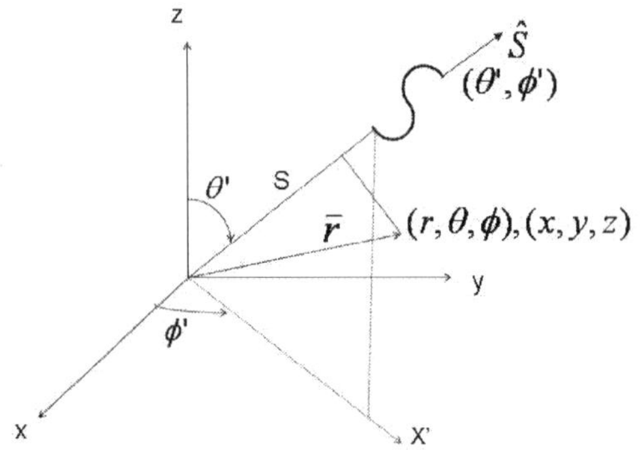

FIGURE 3. The propagation direction $\hat{s}$ is shown with an arbitrary orientation with respect to a common coordinate system. $\hat{s}$ is referenced with primed angles (θ', ϕ'), whereas the position vector $\hat{r}$ for a general point of observation is referenced with unprimed angles (θ, ϕ).

where $\bar{r}$ is the position vector at the point of observation, and $\bar{k}$ is defined as the wave-number vector, with its components k_x, k_y and k_z along the x, y and z directions, respectively.

$$\bar{r} = x\hat{x} + y\hat{y} + z\hat{z}, \tag{5.23}$$

$$\bar{k} \stackrel{\triangle}{=} k\hat{s} = k_x\hat{x} + k_y\hat{y} + k_z\hat{z}; \tag{5.24}$$

$$k_x = k\sin\theta'\cos\phi',\ k_y = k\sin\theta'\sin\phi',\ k_z = k\cos\theta' \tag{5.25}$$

$$k_x^2 + k_y^2 + k_z^2 = k^2. \tag{5.26}$$

Using (5.23,5.24) in (5.22) we get the general propagation factor in the rectangular coordinates.

$$e^{-jks} = e^{-j\bar{k}\cdot\bar{r}} = e^{-jk_x x}e^{-jk_y y}e^{-jk_z z} \tag{5.27}$$

The parameters k_x, k_y and k_z are referred to as the wave numbers in x, y and z directions, respectively, and are related in (5.25) to the angle parameters θ' and ϕ' of the direction of propagation.

2.2. Relationships Between Rectangular Field Components. Now consider the vector relation in (5.21), and express equivalent relationships between the rectangular components of the fields. Note that fields can be oriented along any arbitrary direction on a plane perpendicular to $\hat{s}$. However, they can be decomposed into two independent orthogonal directions. The field relations for the two orthogonal cases maybe derived separately, and then superposed to obtain relations for a general wave. In principle any two orthogonal directions may be considered for decomposition. We will consider two orthogonal directions defined in reference to the z axis. Consider a plane passing through the $\hat{s}$ and $\hat{z}$ directions. The two orthogonal cases are defined as follows. (I) The electric field $\overline{E}$ lies on the zs plane. Equivalently, the magnetic field lies orthogonal to the zs plane, meaning perpendicular or transverse to the $\hat{z}$ axis. Accordingly, this case is referred to as the transverse magnetic to z wave, or the TM_z wave. (II) When the electric field $\overline{E}$ is perpendicular to the the zs plane, which is perpendicular or transverse to the $\hat{z}$ axis. This case is referred to as the transverse electric to z wave, or the TE_z wave. Equivalently, the magnetic field in this case lies on the zs plane.

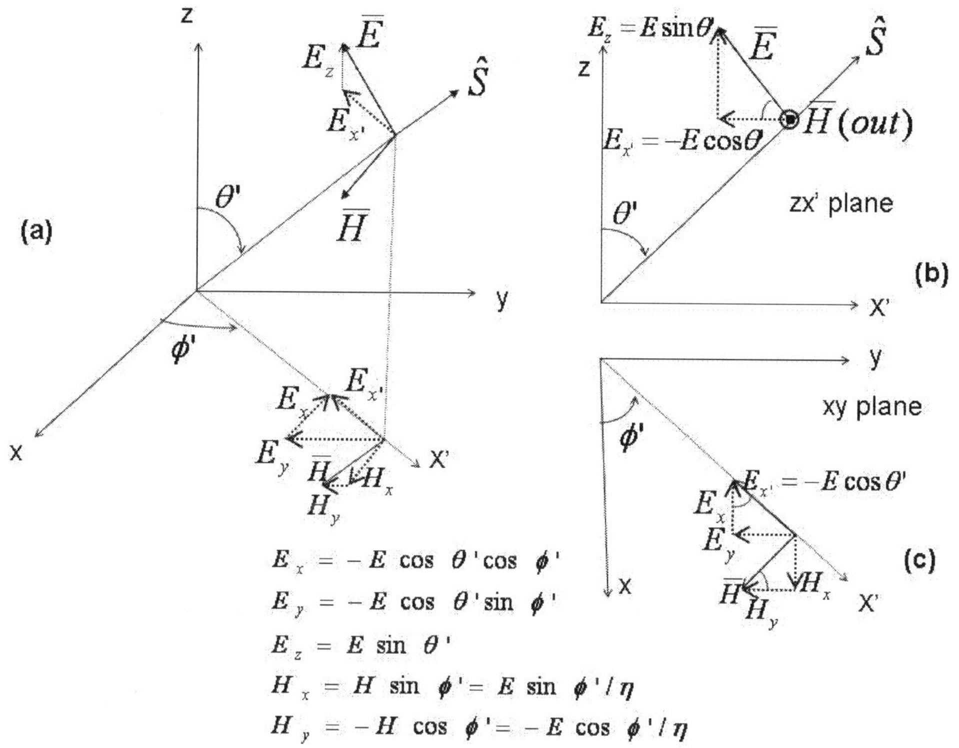

FIGURE 4. Propagation in arbitrary direction $\hat{s}$, showing different fields for a TM_z wave, with $\overline{E}$ on the zs plane and $\overline{H}$ perpendicular to the z direction ($H_z = 0$). The field orientation in 3-D coordinates are shown in (a). The field components on the (b) zs or zx' plane and (c) on the xy plane are also shown for clarity.

TM_z Wave

As discussed, in this case $\overline{H}$ is perpendicular to the zs plane. This means $H_z = 0$, and therefore $\overline{H}$ is on the xy plane. On the other hand, $\overline{E}$ is on the zs plane, and therefore has a non-zero z component ($E_z \neq 0$). The orientation of the propagation direction and field components are shown in Fig.4. A new coordinate x' is introduced for convenience, directed along the intersection of the zs plane and the xy plane. For clarity, field orientations on the zs (or zx') and the xy planes are also shown in Fig.4(b) and (c), respectively, in addition to the 3-D orientations in Fig.4(a). It maybe be useful to express all fields in relations to E_z. The derivation maybe done geometrically from Fig.4, in terms of the angle θ' and ϕ', using the basic plane-wave relations of (5.20,5.21). Then, the θ' and ϕ' dependence may be expressed in terms of the wave numbers k_x, k_y and k_z using the relations of (5.25)

$$\begin{aligned}
E_{x'} &= -E_z \cot\theta' = -E_z \frac{k_z}{k_\rho}, \\
E_x &= E_{x'} \cos\phi' = E_{x'} \frac{k_x}{k_\rho} = -E_z \frac{k_z k_x}{k_\rho^2}, \\
E_y &= E_{x'} \sin\phi' = E_{x'} \frac{k_y}{k_\rho} = -E_z \frac{k_y k_z}{k_\rho^2}, \\
H_x &= +\sin\phi' H = +\frac{k_y}{k_\rho}\frac{E}{\eta} = +\frac{k_y}{\eta k_\rho}\frac{E_z}{\sin\theta'} = +\frac{k_y k}{k_\rho^2 \eta} E_z, \\
H_y &= -\cos\phi' H = -\frac{k_x}{k_\rho}\frac{E}{\eta} = -\frac{k_x}{k_\rho \eta}\frac{E_z}{\sin\theta'} = -\frac{k_x k}{k_\rho^2 \eta} E_z, \\
H_z &= 0; \\
k_\rho &= \sqrt{k_x^2 + k_y^2}.
\end{aligned} \tag{5.28}$$

TE_z Wave

In this case $\overline{E}$ is perpendicular to the zs plane. This means $E_z = 0$, and therefore $\overline{E}$ is on the xy plane. On the other hand, $\overline{H}$ is on the zs plane, and therefore has a non-zero z component ($H_z \neq 0$). The propagation direction and field orientations in this case are shown in Fig.5, which may be compared with the TM_z case in Fig.4.

In contrast to the TM_z case, here it may be useful to express all fields in relations to H_z. The relations maybe derived geometrically from Fig.5, as it was done in the TM_z case from Fig.4, using basic plane-wave conditions of (5.20,5.21). An alternate approach may also be possible, such that the field relations in the TE_z case can be directly obtained from those in (5.28) for the TM_z case, by simple switching or substitution of variables. With this objective, the orientation of the $\overline{E}$ and $\overline{H}$ fields in the present TE_z case maybe seen in comparison to those in the TM_z case. The direction of $\overline{E}$ in the TM_z case maybe substituted with $\overline{H}$ in the TE_z case, and then the direction of $\overline{H}$ in the TM_z case is to be substituted with $-\overline{E}$ in the TE_z case. Such relative orientations of $\overline{E}$ and $\overline{H}$ are such that the direction of $\overline{E} \times \overline{H}$ remains unchanged along the specific direction of propagation $\hat{S}$, as per the conditions (5.20,5.21). Accordingly, the field relationship in the TE_z case may also be directly obtained from (5.28) by replacing components of $\overline{E}$ with those for $\overline{H}$, but replacing components of $\overline{H}$ with negative of those for $\overline{E}$. In addition, η needs to be replaced with $\frac{1}{\eta}$ to adjust for the impedance ratio, which is also equivalent to switching μ with ϵ.

$$
\begin{aligned}
H_x &= -H_z \frac{k_z k_x}{k_\rho^2}, \\
H_y &= -H_z \frac{k_y k_z}{k_\rho^2}, \\
E_x &= -\frac{k_y k \eta}{k_\rho^2} H_z, \\
E_y &= +\frac{k_x k \eta}{k_\rho^2} H_z, \\
E_z &= 0.
\end{aligned}
\tag{5.29}
$$

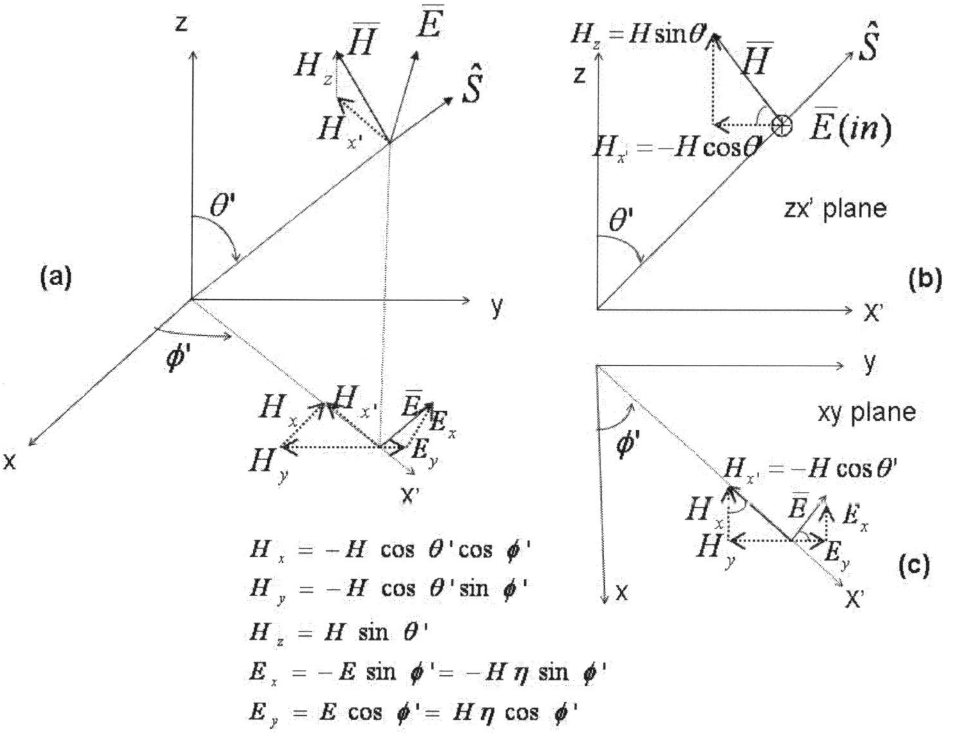

FIGURE 5. Propagation in arbitrary direction $\hat{s}$, showing different fields for a TE_z wave, with $\overline{H}$ on the zs plane and $\overline{E}$ perpendicular to the z direction ($E_z = 0$). The field orientation in 3-D coordinates are shown in (a). The field components on the (b) zs or zx' plane and (c) on the xy plane are also shown for clarity.

2.3. Evanescent Wave. The general propagation function (5.27) and the field relationships (5.28,5.29) would satisfy Ampere's and Faraday's laws for all real values of k_x, k_y and k_z, as long as (5.26) is valid. When each of the wave numbers k_x, k_y, k_z are less than k, one can see from (5.25) that the propagation direction is physically along a real (θ', ϕ') direction. When that is not the case, consider for example,

$$\begin{aligned} k_x^2 + k_y^2 > k^2,\ k_z^2 &= k^2 - k_x^2 - k_y^2 < 0, \\ k_z = \mp j\sqrt{k_x^2 + k_y^2 - k_0^2} &= \mp j\alpha_z,\ \alpha_z > 0, \\ e^{-jk_z z} = e^{\mp\alpha_z z},\ e^{+jk_z z} &= e^{\pm\alpha_z z}\ . \end{aligned} \tag{5.30}$$

In this case the wave number k_z is imaginary. Accordingly, a propagation factor in z would be exponentially decaying or growing along z, depending if k_z is positive or negative imaginary, and if the wave is propagating in $+z$ or $-z$ direction. Consider first a wave propagating along $+z$ away from its source, in which case the required propagation function is $e^{-jk_z z}$ (see (5.15) and section 1.4). The choice of a negative imaginary value of $k_z = -j\alpha$ would lead to having the propagation function $e^{-jk_z z} = e^{-\alpha_z z}$. This is physically valid, because it represents a decaying amplitude along the propagation direction $+z$. The opposite choice of a positive imaginary $k_z = j\alpha$ would lead to a propagation function $e^{+\alpha_z z}$, exponentially growing along the direction of propagation $+z$, and would therefore be invalid. Similarly, for a wave propagating away from source along $-z$, the required propagation function $e^{+jk_z z}$ would be physically valid for a negative imaginary $k_z = -j\alpha$, because it would exponentially decay as $e^{+jk_z z} = e^{+\alpha_z z}$ along the propagation direction $-z$, The other possibility with a positive imaginary $k_z = j\alpha$, resulting in a $e^{-\alpha_z z}$ growth along the propagation direction $-z$, would be invalid.

In either case of propagation in positive or negative z, a choice of an negative imaginary $k_z = -j\alpha$ is seen to be physically valid, representing an exponentially decaying wave along the direction of propagation. Such waves that decay exponentially away from source are called evanescent waves. Because in this case k_z in (5.30) is imaginary, it does not correspond to any real propagation angles (θ', ϕ'). Mathematically, θ', ϕ' become in general complex angles. In fact, all the above results (5.28,5.29), which could be obtained through simple geometric visualization from Figs.3,4,5 for real (θ', ϕ') and (k_x, k_y, k_z), can be analytically applicable for any real or complex angles (θ', ϕ'), or equivalent real or complex (k_x, k_y, k_z), as long as (5.26) is satisfied.

3. Potential Functions

It may be useful to define a function $\psi_{zm} = j\omega\epsilon E_z/k_\rho^2$ for a TM_z wave and $\psi_{ze} = j\omega\mu H_z/k_\rho^2$ for a TE_z wave. Then, the respective field components can be expressed in terms of these new functions as follows.

TM_z Wave

For a TM_z wave, from (5.28) we have

$$E_x = \frac{-k_z k_x}{j\omega\epsilon}\psi_{zm},$$
$$E_y = \frac{-k_y k_z}{j\omega\epsilon}\psi_{zm},$$
$$E_z = \frac{k_\rho^2}{j\omega\epsilon}\psi_{zm},$$
$$H_x = -jk_y\psi_{zm},$$
$$H_y = jk_x\psi_{zm},$$
$$H_z = 0. \tag{5.31}$$

TE_z Wave

Similarly, For a TE_z wave, from (5.29) we have

$$H_x = \frac{-k_z k_x}{j\omega\mu}\psi_{ze},$$
$$H_y = \frac{-k_y k_z}{j\omega\mu}\psi_{ze},$$
$$H_z = \frac{k_\rho^2}{j\omega\mu}\psi_{ze},$$
$$E_x = jk_y\psi_{ze},$$
$$E_y = -jk_x\psi_{ze},$$
$$E_z = 0. \tag{5.32}$$

These new functions ψ_{zm} and ψ_{ze} are referred to as the TM_z and TE_z potentials, respectively.

3.1. Differential Relations to Fields in Rectangular Coordinates. The spatial dependence of the wave is given by the propagation function of (5.27), from which different spatial derivatives can be derived. A spatial derivative operation maybe seen as equivalent to an algebraic multiplication. This observation maybe used to express different field components, directly in terms of the above potential functions ψ_{zm} and ψ_{ze}. These relations would be valid for any value of the wave numbers k_x, k_y and k_z

$$\frac{\partial}{\partial x} \equiv -jk_x, \ \frac{\partial}{\partial y} \equiv -jk_y, \ \frac{\partial}{\partial z} \equiv -jk_z,$$
$$\frac{\partial^2}{\partial x^2} + \frac{\partial^2}{\partial y^2} \equiv -k_\rho^2 = -(k_x^2 + k_y^2), \ \frac{\partial^2}{\partial x^2} + \frac{\partial^2}{\partial y^2} + \frac{\partial^2}{\partial z^2} \equiv -(k_x^2 + k_y^2 + k_z^2) = -k^2. \tag{5.33}$$

TM_z Wave

For a TM_z wave, we get from (5.31,5.33),

$$E_x = \frac{1}{j\omega\epsilon}\frac{\partial^2 \psi_{zm}}{\partial x \partial z},$$
$$E_y = \frac{1}{j\omega\epsilon}\frac{\partial^2 \psi_{zm}}{\partial y \partial z},$$
$$E_z = -\frac{1}{j\omega\epsilon}(\frac{\partial^2}{\partial x^2} + \frac{\partial^2}{\partial y^2})\psi_{zm} = \frac{1}{j\omega\epsilon}(k^2 + \frac{\partial^2}{\partial z^2})\psi_{zm},$$
$$H_x = \frac{\partial \psi_{zm}}{\partial y},$$
$$H_y = -\frac{\partial \psi_{zm}}{\partial x},$$
$$H_z = 0. \tag{5.34}$$

TE_z **Wave**

Similarly, For a TE_z wave, from (5.32,5.33) we have

$$H_x = \frac{1}{j\omega\mu}\frac{\partial^2 \psi_{ze}}{\partial x \partial z},$$
$$H_y = \frac{1}{j\omega\mu}\frac{\partial^2 \psi_{ze}}{\partial y \partial z},$$
$$H_z = -\frac{1}{j\omega\mu}(\frac{\partial^2}{\partial x^2} + \frac{\partial^2}{\partial y^2})\psi_{ze} = \frac{1}{j\omega\mu}(k^2 + \frac{\partial^2}{\partial z^2})\psi_{ze},$$
$$E_x = -\frac{\partial \psi_{ze}}{\partial y},$$
$$E_y = \frac{\partial \psi_{ze}}{\partial x},$$
$$E_z = 0. \tag{5.35}$$

Relating ψ_{zm} and ψ_{ze} to different field components in (5.34,5.35) is similar in principle to relating electric fields to a scalar potential in static electric field. Once ψ_{zm} and ψ_{ze} are known, all the field components in the TM_z and TE_z waves can be derived. Thus, they maybe treated as two primary variables for a complete description of the total wave. This is analogous to a potential function for a static field from which all electric field components are derived through a gradient derivative function. Accordingly, ψ_{zm} and ψ_{ze} are commonly referred to as the potential functions for the TM_z or TE_z waves. However, unlike a static potential which is a scalar function, the ψ_{zm} and ψ_{ze} have inherent directionality through the reference direction of choice (in the present case z) for the TM and TE decomposition in (5.28,5.29). Therefore, vector functions defined as $\overline{A} = \psi_{zm}\hat{z}$ and $\overline{F} = \psi_{ze}\hat{z}$ maybe more specifically referred to as the vector TM_z and TE_z potentials, respectively.

$$\overline{A} = \psi_{zm}\hat{z}, \ \overline{F} = \psi_{ze}\hat{z} \tag{5.36}$$

3.1.1. *Potential-Field Relations Using Differential Operators.* The above field relations (5.34,5.35) maybe expressed in terms of differential operators (appendix A) applied upon the vector potentials $\overline{A}$ and $\overline{F}$ of (5.36), respectively for the TM_z and TE_z waves.

$$\begin{aligned} \overline{E} &= \frac{1}{j\omega\epsilon}(\overline{\nabla}(\overline{\nabla}\cdot\overline{A}) + k^2\overline{A}), \ \overline{H} = \overline{\nabla}\times\overline{A}, \\ \overline{H} &= \frac{1}{j\omega\mu}(\overline{\nabla}(\overline{\nabla}\cdot\overline{F}) + k^2\overline{F}), \ \overline{E} = -\overline{\nabla}\times\overline{F}. \end{aligned} \tag{5.37}$$

The above concept of the TM_z anf TE_z waves may be extended for TM_p and TE_p waves with respect to any general direction $\hat{p}$ in space, by using respective vector potentials $\overline{A}$ and $\overline{F}$ directed along the $\hat{p}$. Unlike the potential-field relations (5.34,5.35) which are applicable specifically for the TM_z TE_z waves, expressed in the rectangular coordinates, the relations (5.37) would work for any general orientation of the $\overline{A}$ and $\overline{F}$ vectors, and for any general coordinate system. This may be verified in the following for spherical and cylindrical coordinates.

3.2. Differential Relations to Fields in Spherical Coordinates. Substituting $z = r\cos\theta$, $x = r\sin\theta\cos\phi$, $y = r\sin\theta\sin\phi$ in the propagation factor of (5.27) we get the expression for the propagation factor in terms of the spherical coordinates r, θ and ϕ. This can then be differentiated to obtain the expressions for the spatial derivatives with respect to r, θ and ϕ.

$$\frac{\partial}{\partial r} \equiv -j[k_z\cos\theta + \sin\theta(k_x\cos\phi + k_y\sin\phi)], \tag{5.38}$$

$$\frac{1}{r}\frac{\partial}{\partial\theta} \equiv -j[-k_z\sin\theta + \cos\theta(k_x\cos\phi + k_y\sin\phi)], \tag{5.39}$$

$$\frac{1}{r}\frac{\partial}{\partial\phi} \equiv -j\sin\theta(-k_x\sin\phi + k_y\cos\phi). \tag{5.40}$$

We have related in (5.28,5.29) the rectangular field components to the potentials ψ_{zm} and ψ_{ze}. Now, the field components in the spherical coordinates may also be related to ψ_{zm} and ψ_{ze} by trigonometric transformations from the rectangular to spherical field components.

<u>TM_z **Wave**</u>

For a TM_z wave, using (5.28,5.31) and geometric transformation from rectangular to spherical components we get,

.

$$E_r = E_z \cos\theta + E_x \sin\theta\cos\phi + E_y \sin\theta\sin\phi$$
$$= \frac{\psi_{zm}}{j\omega\varepsilon}[+k_\rho^2 \cos\theta - k_z \sin\theta(k_x\cos\phi + k_y\sin\phi)], \quad (5.41)$$
$$E_\theta = -E_z \sin\theta + E_x \cos\theta\cos\phi + E_y \cos\theta\sin\phi$$
$$= \frac{\psi_{zm}}{j\omega\epsilon}[-k_\rho^2 \sin\theta - k_z \cos\theta(k_x\cos\phi + k_y\sin\phi)], \quad (5.42)$$
$$E_\phi = -E_x \sin\phi + E_y \cos\phi$$
$$= \frac{\psi_{zm}}{j\omega\epsilon}k_z(k_x\sin\phi - k_y\cos\phi), \quad (5.43)$$
$$H_r = H_z \cos\theta + H_x \sin\theta\cos\phi + H_y \sin\theta\sin\phi$$
$$= \psi_{zm}(-jk_y\cos\phi + jk_x\sin\phi)\sin\theta, \quad (5.44)$$
$$H_\theta = -H_z \sin\theta + H_x \cos\theta\cos\phi + H_y \cos\theta\sin\phi$$
$$= j\psi_{zm}(-k_y\cos\phi + k_x\sin\phi)\cos\theta, \quad (5.45)$$
$$H_\phi = -H_x \sin\phi + H_y \cos\phi$$
$$= j\psi_{zm}(+k_y\sin\phi + k_x\cos\phi). \quad (5.46)$$

By observation and superposition, we can write the final differential relations from ψ_{zm} to different spherical TM_z field components. Using (5.41,5.38) and (5.33),

$$E_r = \frac{1}{j\omega\epsilon}(k^2\cos\theta + \frac{\partial}{\partial r}\frac{\partial}{\partial z})\psi_{zm}. \quad (5.47)$$

Using (5.42,5.39) and (5.33),

$$E_\theta = \frac{1}{j\omega\epsilon}(-k^2\sin\theta + \frac{1}{r}\frac{\partial}{\partial\theta}\frac{\partial}{\partial z})\psi_{zm}. \quad (5.48)$$

Using (5.43,5.40)and (5.33),

$$E_\phi = \frac{1}{j\omega\epsilon}(\frac{1}{r\sin\theta}\frac{\partial}{\partial\phi}\frac{\partial}{\partial z}\psi_{zm}). \quad (5.49)$$

Comparing (5.44) and (5.40),

$$H_r = \frac{1}{r}\frac{\partial\psi_{zm}}{\partial\phi}. \quad (5.50)$$

Comparing (5.45) and (5.40),

$$H_\theta = \frac{\cot\theta}{r}\frac{\partial\psi_{zm}}{\partial\phi} = H_r\cot\theta. \quad (5.51)$$

Using (5.38,5.39) in (5.46),

$$H_\phi = -(\sin\theta\frac{\partial}{\partial r} + \frac{\cos\theta}{r}\frac{\partial}{\partial\theta})\psi_{zm}. \tag{5.52}$$

It may be verified that the above expressions may also be established using the differential-operator relations (5.37,5.36) for $\overline{A}$, with the gradient and curl operators expressed in the spherical, but the divergence operator in the rectangular coordinates (appendix A). Note that in the differential relations (5.47,5.48) we have, for analytical simplicity, mixed z-derivative (in the rectangular coordinate) with $r-$ and $\theta-$ derivatives (in the spherical coordinates). One may replace the $z-$derivative by its equivalent differential form in spherical coordinates, as needed.

$$\frac{\partial}{\partial z} = \frac{\partial r}{\partial z}\frac{\partial}{\partial r} + \frac{\partial\theta}{\partial z}\frac{\partial}{\partial\theta} + \frac{\partial\phi}{\partial z}\frac{\partial}{\partial\phi} = \cos\theta\frac{\partial}{\partial r} - \frac{\sin\theta}{r}\frac{\partial}{\partial\theta}. \tag{5.53}$$

TE_z Wave

Similar steps (5.41)-(5.52) can be repeated for a TE_z wave by transforming rectangular components in (5.29) to spherical components, and then using spherical differential operations (5.38-5.40) to simplify. We will give the final relations from ψ_{ze} to spherical TE_z field components. Alternatively, final results for the TE_z case may be obtained from those for the TM_z case in (5.47-5.52) by suitable switching of variables as was done to obtain the TE_z field relationships (5.29) from the TM_z field relationships of (5.28).

$$H_r = \frac{1}{j\omega\mu}(k^2\cos\theta + \frac{\partial}{\partial r}\frac{\partial}{\partial z})\psi_{ze}, \tag{5.54}$$

$$H_\theta = \frac{1}{j\omega\mu}(-k^2\sin\theta + \frac{1}{r}\frac{\partial}{\partial\theta}\frac{\partial}{\partial z})\psi_{ze}, \tag{5.55}$$

$$H_\phi = \frac{1}{j\omega\mu}(\frac{1}{r\sin\theta}\frac{\partial}{\partial\phi}\frac{\partial}{\partial z}\psi_{ze}), \tag{5.56}$$

$$E_r = -\frac{1}{r}\frac{\partial\psi_{ze}}{\partial\phi}, \tag{5.57}$$

$$E_\theta = -\frac{\cot\theta}{r}\frac{\partial\psi_{ze}}{\partial\phi} = E_r\cot\theta, \tag{5.58}$$

$$E_\phi = +(\sin\theta\frac{\partial}{\partial r} + \frac{\cos\theta}{r}\frac{\partial}{\partial\theta})\psi_{ze}. \tag{5.59}$$

It may be verified that the above expressions may also be established using the differential-operator relations (5.37,5.36) for $\overline{F}$, with the gradient and curl operators expressed in the spherical, but the divergence operator in the rectangular coordinates (appendix A). As for the TM_z fields, in the differential relations (5.54,5.55) for the TE_z

fields we have also, for analytical simplicity, mixed z-derivative (in the rectangular coordinate) with $r-$ and $\theta-$ derivatives (in the spherical coordinates). One may replace the $z-$derivative by its equivalent differential form (5.53) in spherical coordinates, as needed.

3.3. Differential Relations to Fields in Cylindrical Coordinates. We follow the same steps as in the last section for deriving relations to field components in the cylindrical coordinates. Substituting $x = \rho\cos\phi$, $y = \rho\sin\phi$ in (5.27) we get the expression for the propagation factor in terms of the cylindrical coordinates ρ, ϕ and z. The propagation factor can then be differentiated to obtain the expressions for the spatial derivatives with respect to ρ, ϕ and z.

$$\frac{\partial}{\partial\rho} \equiv -j(k_x\cos\phi + k_y\sin\phi), \tag{5.60}$$

$$\frac{1}{\rho}\frac{\partial}{\partial\phi} \equiv -j(k_y\cos\phi - k_x\sin\phi), \tag{5.61}$$

$$\frac{\partial}{\partial z} \equiv -jk_z. \tag{5.62}$$

We have related in (5.28,5.29) the rectangular field components to the potentials ψ_{zm} and ψ_{ze}. Now, the field components in the cylindrical coordinates may be related to ψ_{zm} and ψ_{ze} by trigonometric transformations from the rectangular to cylindrical field components. The z component of the field is identical in both the rectangular and the cylindrical coordinates.

TM_z **Wave**

For a TM_z wave, using (5.28) and necessary transformation from rectangular to cylindrical components we get,

$$
\begin{aligned}
E_\rho &= E_x \cos\phi + E_y \sin\phi \\
&= \frac{\psi_{zm}}{j\omega\varepsilon}[-k_z(k_x \cos\phi + k_y \sin\phi)], \qquad (5.63) \\
E_\phi &= -E_x \sin\phi + E_y \cos\phi \\
&= \frac{\psi_{zm}}{j\omega\epsilon}[k_z(k_x \sin\phi - k_y \cos\phi)], \qquad (5.64) \\
E_z &= \frac{\psi_{zm}}{j\omega\epsilon}k_\rho^2 \\
&= \frac{\psi_{zm}}{j\omega\epsilon}(k^2 - k_z^2), \qquad (5.65) \\
H_\rho &= H_x \cos\phi + H_y \sin\phi \\
&= -j\psi_{zm}(k_y \cos\phi - k_x \sin\phi), \qquad (5.66) \\
H_\phi &= -H_x \sin\phi + H_y \cos\phi \\
&= j\psi_{zm}(k_y \sin\phi + k_x \cos\phi), \qquad (5.67) \\
H_z &= 0. \qquad (5.68)
\end{aligned}
$$

The final differential relations from ψ_{zm} to different cylindrical TM_z field components can now be expressed by observation, combining the relations (5.60-5.62) and (5.63-5.68). Using (5.63) and (5.60,5.62),

$$E_\rho = \frac{1}{j\omega\epsilon}(\frac{\partial}{\partial\rho}\frac{\partial}{\partial z})\psi_{zm}. \qquad (5.69)$$

Using (5.64) and (5.61,5.62),

$$E_\phi = \frac{1}{j\omega\epsilon}(\frac{1}{\rho}\frac{\partial}{\partial\phi}\frac{\partial}{\partial z})\psi_{zm}. \qquad (5.70)$$

Using (5.65) and (5.62),

$$E_z = \frac{1}{j\omega\epsilon}(k^2 - \frac{\partial^2}{\partial z^2})\psi_{zm}. \qquad (5.71)$$

Comparing (5.66) and (5.61),

$$H_\rho = \frac{1}{\rho}\frac{\partial\psi_{zm}}{\partial\phi}. \qquad (5.72)$$

Comparing (5.67) and (5.60),

$$H_\phi = -\frac{\partial}{\partial\rho}\psi_{zm}. \qquad (5.73)$$

We also have,

$$H_z = 0, \tag{5.74}$$

which is the same component as in the rectangular coordinates.

It may be verified that the above expressions confirms to the differential-operator relations (5.37,5.36) for $\overline{A}$, with the operators expressed in the cylindrical coordinates (appendix A).

<u>TE_z **Wave**</u>

Similar steps (5.60)-(5.74) can be repeated for a TE_z wave by transforming rectangular components in (5.29) to cylindrical components, and then using cylindrical differential operations (5.60-5.62) to simplify. We will give the final relations from ψ_{ze} to cylindrical TE_z field components. Alternatively, final results for the TE_z case may be obtained from those for the TM_z case in (5.69-5.74) by suitable switching of variables as was done to obtain the TE_z field relationships (5.29) from the TM_z field relationships of (5.28).

$$H_\rho = \frac{1}{j\omega\mu}(\frac{\partial}{\partial\rho}\frac{\partial}{\partial z})\psi_{ze}, \tag{5.75}$$

$$H_\phi = \frac{1}{j\omega\mu}(\frac{1}{\rho}\frac{\partial}{\partial\phi}\frac{\partial}{\partial z})\psi_{ze}, \tag{5.76}$$

$$H_z = \frac{1}{j\omega\mu}(k^2 - \frac{\partial^2}{\partial z^2})\psi_{ze}, \tag{5.77}$$

$$E_\rho = -\frac{1}{\rho}\frac{\partial}{\partial\phi}\psi_{ze}, \tag{5.78}$$

$$E_\phi = \frac{\partial}{\partial\rho}\psi_{ze}, \tag{5.79}$$

$$E_z = 0. \tag{5.80}$$

It may be verified that the above expressions confirms to the differential-operator relations (5.37,5.36) for $\overline{F}$, with the operators expressed in the cylindrical coordinates (appendix A).

4. Plane Wave Incident Normal to a Planar Interface

Consider a plane wave is incident upon a planar interface between two media, propagating along the normal axis to the interface. This is shown in Fig.6. The electric field of the incident wave is assumed to be of unit amplitude. The material media on the two sides of the interface at $z = 0$ are uniformly distributed. The material properties are independent of the vector direction of any applied field (isotropic material). We expect that a reflected wave is generated, propagating back into the first medium from where the incident wave approached the interface. The amplitude of the electric field of this reflected wave is referred to as the reflection coefficient Γ. We also expect to have another wave, with the amplitude of its electric field T, to propagate in the second medium, on the other side of the interface. T is referred to as the transmission coefficient. The amplitude distribution of the incident wave, which is propagating normal to the interface, is uniform in the transverse (xy) plane. The material environment is also uniform in the xy plane. Therefore, the resulting reflected and transmitted waves are also expected to have amplitude distribution which is uniform on the xy plane, with variation only in the z direction. This means, the reflected and transmitted waves would propagate along the z direction, which is normal to the interface.

In reference to the basic characteristics of a plane wave, we know that the electric field vector will have to be perpendicular to the direction of propagation. Accordingly, the electric field vector of the incident wave would be directed parallel to the plane of the interface, which is on the xy plane in Fig.6. However, there is no specific preference for any particular choice for the vector direction on the xy plane. This is because, the material media are assumed to be isotropic in nature, meaning the material characteristics are independent of any specific direction of the field vector. For the sake of reference, we assume the electric field vector of the incident wave to be along the $\hat{x}$ direction. Any other choice would also lead to the same final results for the reflection and transmission coefficients.

We have specified the reference direction for the electric field vector of the incident wave. Now, we need to specify the directions for the electric field vectors for the reflected and transmitted waves. As explained, all the three waves propagate normal to the interface, along the z direction. Accordingly, as per the basic plane-wave characteristics, the reflected and transmitted wave, just like the incident wave, should have the respective electric field vectors oriented on the xy plane. We assume that electric field vectors for the transmit and reflected waves are aligned with that of the incident wave, along the the reference x direction. This may be justified based on a simple symmetry consideration. Let us consider that the electric field vector of the reflected or the transmitted wave is oriented at a general angle ϕ to that of the incident wave (x axis). Due to physical symmetry of the problem on both sides of the x axis, it is as well possible that the orientation angle be $-\phi$, instead of $+\phi$. These two orientations can not both be valid, except when ϕ is zero or $\pm 180^0$. Accordingly, the electric field of the transmit and the reflected wave will have to be parallel to the x axis, and for reference are assumed to be along the $+x$ direction.

In Fig.6 we have shown the relative directions and amplitudes of the electric and magnetic fields, associated with the incident, reflected and transmitted waves. These are based on the considerations we just discussed, and are consistent with the basic characteristics of a uniform plane wave. The incident, reflected, and transmitted fields are referred with subscripts (i), (r) and (t), respectively. The incident wave is propagating in the first medium, along the $+z$ direction, with the z-dependence of its phasor expressed as $e^{-jk_1 z}$. Similarly, the transmitted wave is propagating in the second medium, also along in the $+z$ direction, with its phasor dependence of $e^{-jk_2 z}$. The reflected wave propagates along the $-z$ direction, which is opposite to that of the incident and transmitted waves. It propagates in the first medium, with a phasor dependence of $e^{+jk_1 z}$. The magnetic field shares the same z dependence as the electric field, with only different vector directions and relative amplitudes. The amplitudes and vector directions of the magnetic fields for the incident, reflected and transmitted waves are to be properly determined, in relations to those of the respective electric fields. As per the basic plane-wave relations in sections 1,2. the vector direction and amplitude of the magnetic field in Fig.6 are expressed as

$+\hat{y}1/\eta_1$ for the incident wave, as $+\hat{y}T/\eta_2$ for the transmitted wave, both with a positive reference sign, but as $-\hat{y}\Gamma/\eta_1$ with a negative sign for the reflected wave.

$$\overline{E}_i = 1\hat{x}e^{-jk_1z},\ \overline{E}_r = \Gamma\hat{x}e^{+jk_1z},\ \overline{E}_t = T\hat{x}e^{-jk_2z}$$
$$\overline{H}_i = \frac{1}{\eta_1}\hat{y}e^{-jk_1z},\ \overline{H}_r = -\frac{\Gamma}{\eta_1}\hat{y}e^{+jk_1z},\ \overline{H}_t = \frac{T}{\eta_2}\hat{y}e^{-jk_2z} \tag{5.81}$$

The total fields in the medium #1 is the sum of the incident and reflected fields, and those in the medium #2 is described by only the transmitted wave.

$$\begin{aligned}
\overline{E} &= \overline{E}_i + \overline{E}_r = \hat{x}(e^{-jk_1z} + \Gamma e^{+jk_1z}),\ z < 0 \\
&= \overline{E}_t = T\hat{x}e^{-jk_2z},\ z > 0 \\
\overline{H} &= \overline{H}_i + \overline{H}_r = \hat{y}(\frac{1}{\eta_1}e^{-jk_1z} - \frac{\Gamma}{\eta_1}e^{+jk_1z}),\ z < 0 \\
&= \overline{H}_t = \frac{T}{\eta_2}\hat{y}e^{-jk_2z},\ z > 0
\end{aligned} \tag{5.82}$$

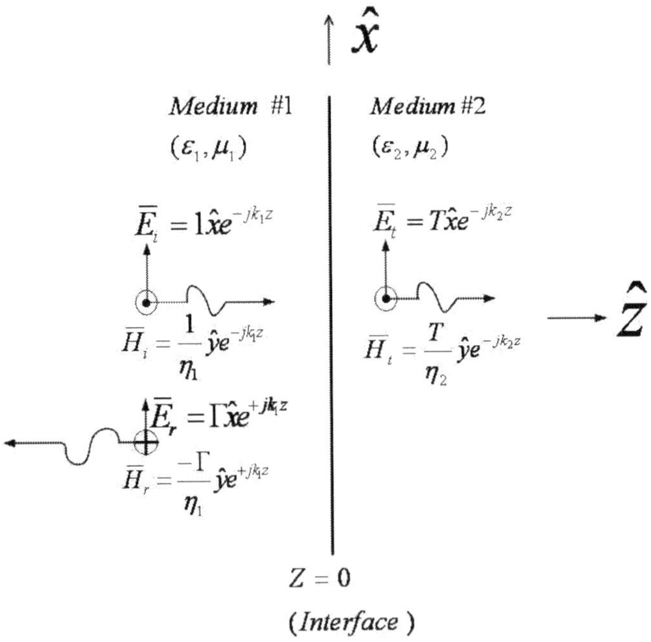

FIGURE 6

Proper boundary conditions, developed in chapter 2, section 4.4, will have to be enforced across the interface between the two media in Fig.6. The components of the total electric as well as magnetic fields in the medium #1, which are parallel to the interface, each evaluated at the interface at $z = 0$, will have to be equal to those in medium #2. These boundary conditions may be established using the field expressions of (5.81,5.82).

$$E_x = E_{xi} + E_{xr} = E_t, z = 0$$
$$1 + \Gamma = T \tag{5.83}$$
$$H_y = H_{yi} + H_{yr} = H_{xt}, z = 0$$
$$\frac{1}{\eta_1} - \frac{\Gamma}{\eta_1} = \frac{T}{\eta_2}, \quad (1 - \Gamma) = T\frac{\eta_1}{\eta_2} \tag{5.84}$$

The above two equations (5.83,5.84) may be solved for the reflection coefficient Γ and transmission coefficient T.

$$T = \frac{2\eta_2}{\eta_1 + \eta_2}, \quad \Gamma = \frac{\eta_2 - \eta_1}{\eta_2 + \eta_1} \tag{5.85}$$

It may be noted, for simplicity we presumed that the electric field vectors of the reflected and transmitted waves were directed along $\hat{x}$, which is the same as the vector direction of the incident electric field. This was justified based on a symmetry consideration. Alternately, for analytical generality one may like to explicitly include additional electric field components along the $\hat{y}$ direction, each for the reflected and transmitted waves with different unknown amplitudes. With these unknown components, one may consistently proceed through the similar analysis procedure we have presented above, based on basic plane-wave principles and required boundary conditions. The analysis would then lead to prove that the above unknown amplitudes for the y directed electric fields would have to be zero, as we expected. If this simple condition can be obvious and assumed to begin with, based on simple symmetry argument, the additional analytical effort would clearly be avoided. Different kinds of symmetry principles are routinely used in electromagnetic modeling for similar analytical simplification.

4.1. Equivalent Transmission Line Model. The above expressions for the reflection coefficient Γ and transmission coefficient T are similar to those derived in chapter 4, section 7 for a voltage wave incident at the junction of two transmission lines. It may be useful to recognize the equivalence between the field modeling at a material interface to that of the voltage-current modeling at the junction between two transmission lines. The voltage V across a transmission line is equivalent to the electric field E_x of a plane wave in Fig.6. And, the current I along the transmission line is equivalent to the magnetic field H_y of a plane wave in Fig.6. The characteristic impedance Z_c of a transmission line is defined as the positive or negative ratio of the V and I, for a forward or a backward propagating wave, respectively. Similarly, in Fig.6 the equivalent impedance of the first medium $Z_1 = \eta_1$ is equal to the positive ratio of the E_{xi} and H_{yi} fields of the incident wave propagating in the forward $+z$ direction, or the negative ratio of the E_{xr} and H_{yr} fields of the reflected wave propagating in the backward $-z$ direction. The equivalent impedance of the second medium in Fig.6 is $Z_2 = \eta_2$, which is similarly defined as the positive ratio of the E_{xt} and H_{yt} fields of the transmit wave, propagating in the forward $+z$ direction. There is no backward propagating wave in the second medium in Fig.6. In case there was a backward propagating wave in the second medium (if there was an incident wave in second medium), the equivalent impedance $Z_2 = \eta_2$ would be equal to the negative ratio of the respective E_x and H_y fields.

$$V \equiv E_x,\ I \equiv H_y$$
$$Z_1 = \eta_1 = \frac{E_{xi}}{H_{yi}} = -\frac{E_{xr}}{H_{yr}}$$
$$Z_2 = \eta_2 = \frac{E_{xt}}{H_{yt}} \tag{5.86}$$

Note that the total voltage and current are continuous across the junction of two transmission lines in chapter 4, section 7, Fig.14. This continuity condition led to the solution for the reflection and transmission coefficient derived in chapter 4, section 7. This is equivalent to solving for the reflection and transmission coefficient in Fig.6 by satisfying the continuity of the equivalent fields - that is the continuity of the total E_x and H_y fields - on the two sides of the interface.

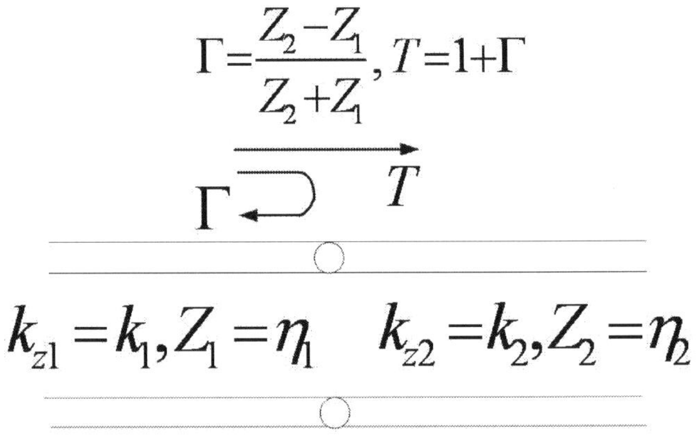

FIGURE 7

Based on the above equivalence principles, the reflection and transmission coefficients in (5.85) for the plane-wave problem of Fig.6 could also be obtained from an equivalent transmission model shown in Fig.7. The results for the transmission-line analysis in chapter 4, section 7 can be directly used, by substituting for the appropriate equivalent impedance parameters shown in Fig.7 .

$$\Gamma = \frac{Z_2 - Z_1}{Z_2 + Z_1} = \frac{\eta_2 - \eta_1}{\eta_2 + \eta_1},\ T = 1 + \Gamma = \frac{2Z_2}{Z_2 + Z_1} = \frac{2\eta_2}{\eta_2 + \eta_1} \tag{5.87}$$

The above result may be verified with (5.85), obtained from field analysis.

The transmission line model may also be similarly extended to a general problem of wave incidence at an arbitrary angle to an interface, to be covered in the following section. Further, the transmission line model can also be extended to the analysis of wave propagation in multiple layers of material media, which would be addressed in section 6. Accordingly, all general theory and physical concepts for the propagation of voltage-current signals in transmission lines can be conveniently shared with those of the propagation of plane waves in material layers. A general transmission-line modeling, with suitable definition of different transmission-line parameters, would be a valuable tool for the analysis and understanding of wave propagation through single or multiple material interfaces.

5. Plane Wave Incident at an Arbitrary Angle with a Planar Interface

Consider a plane wave is incident upon a planar interface, with its propagation direction inclined at an arbitrary angle with respect to the interface. This is a more general situation, compared to the specific case of normal incidence presented in section 4. The basic physical principles and analysis procedure applied in section 4 need to be properly extended to model the general problem considered here.

As per the basic plane-wave principles, the electric and magnetic field vectors are to be directed on a plane (referred to as the field-vector plane) perpendicular to the direction of propagation. However, there are different possible vector orientations on the particular field-vector plane. Fig.8 shows the geometry of wave incidence at a general inclination to the interface, with a specific vector orientation. A plane which is passing through the direction of propagation, and is perpendicular to the interface plane at $z = 0$ is referred to as the plane of incidence. We choose the xz plane as the plane of incidence, as shown in Fig.8. The direction of propagation of the incident wave is inclined to the z axis (which is normal to the interface) at a general angle θ_i, referred to as the incidence angle.

For a general incidence angle θ_i, other than zero (normal incidence), different possible vector orientations on the field-vector plane (Fig.8) would have different relative orientation with respect to the interface. Accordingly, the incident wave would "see" the material interface differently for different specific vector orientations. Therefore, the resulting reflected and transmitted waves are expected be dependent upon the specific vector orientation. This is in contrast with the special case of normal incidence we studied in the last section 4, where all possible vector orientations happened to be parallel to the interface, and accordingly the results were independent of any specific vector orientation.

Any particular vector direction on a field-vector plane can always be decomposed into two orthogonal components. Accordingly, we would need to separately analyze for only two orthogonal vector orientations. The results for the two independent cases can then be superposed to obtain results for any arbitrary vector orientation. Choices for the two orthogonal field directions are commonly specified in the following alternate references:

With respect to the planes of incidence and interface: It may be convenient to select the orthogonal directions in reference to the plane of incidence (xz plane) and the plane of interface ($z = 0$ plane). Consider the electric field vector of the incident wave. Let us choose one of the orthogonal directions for the electric field to be aligned on the plane of incidence (xz plane), and the other perpendicular to it (y direction). The second orthogonal component of the electric field, which is in the y direction, would be parallel to the interface. The corresponding magnetic field vectors for the above two electric field orientations would be parallel to the interface (y direction), or on the plane of incidence (xz plane), respectively. The specific vector orientation we have shown in Fig.8 refers to the former case, with its electric field in the y direction and the magnetic field on the xz plane. The other orthogonal orientation is shown in Fig.10, with the electric field on the xz plane and the magnetic field parallel to the y direction.

With respect to the normal to interface: It may also be convenient to describe the two orthogonal orientations, discussed above, in reference to the z direction, which is normal to the interface. For the first case shown in Fig.8, the y directed electric field is orthogonal or *transverse* to the z direction. Whereas, the corresponding magnetic field on the xz plane in general has a component along the z axis. In other words, the magnetic field here is, in general, not orthogonal or *not transverse* to the z axis. Accordingly, the incident wave in this case is commonly referred to as a Transverse-Electric-to-z, or a TE_z wave. For the second case represented in Fig.10, the y directed magnetic field is orthogonal or *transverse* to the z direction. Whereas, the corresponding electric field on the xz plane in general has a component along the z axis. In other words, the electric field here is, in general, not orthogonal or *not transverse* to the z axis. Accordingly, the incident wave in this case is commonly referred to as a Transverse-Magnetic-to-z, or a TM_z wave. A similar classification of TE_z and TM_z waves was also used in section 2.2 in modeling field relations for a general plane wave, propagating in any arbitrary direction in the three-dimensional (x, y, z) space.

It may be noted, for a special situation of normal incidence ($\theta_i = 90^0$), both the electric and the magnetic fields in either case of a TE_z (Fig.8) or a TM_z (Fig.10) wave would be *transverse* to the z direction. Accordingly, the specific case when the incident wave is propagating normal to the interface, may be referred to as a Transverse-Electro-Magnetic-to-z, or a TEM_z wave.

The two cases of orthogonal field orientations, the TE_z and TM_z waves, would be separately treated in the following sections 5.1, and 5.2, respectively, to model reflection from a planar interface.

5.1. Electric Field Vector Parallel to Interface, TE_z Wave. This case is shown in Fig.8, with the electric field vector in the y direction. The basic procedure to model this problem is similar to the case of normal incidence, analyzed in section 4. The specific differences here are in proper description of (1) the orientations of various field vectors, (2) orientation of the propagation directions and the respective wave-number vectors, and (3) of the wave propagation phasors. The relative orientations of the electric and magnetic fields are shown in Fig,8, based on the basic principles of a plane wave derived in sections 1,2. They may be expressed in the common rectangular (xyz) coordinates, through a simple trigonometric decomposition. Similarly, the propagation directions for the incident, reflected and transmitted waves, and accordingly the respective wave-number vectors, may also be trigonometrically decomposed into their components in the common xyz coordinates. The propagation phasors are then expressed in the rectangular coordinates using the rectangular components of the respective wave number vectors. Basic symmetry considerations, similar to those used for the case of normal incidence in section 4, may also be used here. The incident wave, which is propagating on the xz plane, is uniform in the y coordinate (see Fig.8). The characteristics of the material media are also uniform in y. Accordingly, based on basic symmetry consideration, we expect that the reflected and transmitted waves would also be uniform in y, with variation only in the xz plane. This would require that the propagation directions of the reflected and transmitted waves would have to be on the xz plane, as shown in Fig.8.

The incident wave is selected to be TE_z, having its magnetic field vector aligned on the xz plane, as shown in Fig.8. The vector field directions for the reflected and transmitted waves may be assumed also to be TE_z. Like for the case of a normal incidence in section 4, this assumption may be justified based on a simple symmetry consideration. Let us consider that the magnetic field vectors of the reflected or the transmitted wave is

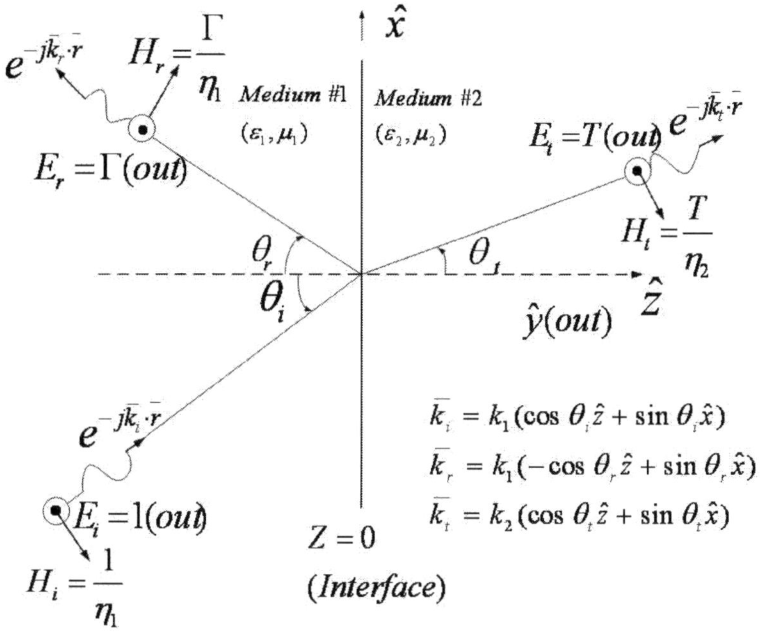

FIGURE 8

tilted at a general angle ϕ with respect to the xz plane, having a non-zero component in the $\hat{y}$ direction. Due to physical symmetry of the problem in Fig.8 about the xz plane, it may be argued that the tilt angle could as well be $-\phi$, instead of $+\phi$ (tilted toward negative y axis instead of the positive y axis). These two orientations can not both be valid, except when ϕ is zero or $\pm 180^0$. Accordingly, the magnetic fields of the transmitted and the reflected wave will have to be aligned with the xz plane, with no y component. The respective electric fields would have to be parallel to tbe y direction, and therefore arc transverse to the z axis. As per the definition of a TE_z wave, this means that the reflected and transmitted waves, like the incident wave, are also TE_z

Based on the basic principles discussed above, the electric and magnetic fields for the incident, reflected and transmitted waves are expressed in the following.

$$\begin{aligned}
\overline{E}_i &= 1\hat{y}e^{-jk_1(\cos\theta_i z+\sin\theta_i x)} \\
\overline{H}_i &= \tfrac{1}{\eta_1}(-\hat{x}\cos\theta_i + \hat{z}\sin\theta_i)e^{-jk_1(\cos\theta_i z+\sin\theta_i x)} \\
\overline{E}_r &= \Gamma\hat{y}e^{-jk_1(-\cos\theta_r z+\sin\theta_r x)} \\
\overline{H}_r &= \tfrac{\Gamma}{\eta_1}(+\hat{x}\cos\theta_r + \hat{z}\sin\theta_r)e^{-jk_1(-\cos\theta_r z+\sin\theta_r x)} \\
\overline{E}_t &= T\hat{y}e^{-jk_2(\cos\theta_t z+\sin\theta_t x)} \\
\overline{H}_t &= \tfrac{T}{\eta_2}(-\hat{x}\cos\theta_t + \hat{z}\sin\theta_t)e^{-jk_2(\cos\theta_t z+\sin\theta_t x)}
\end{aligned} \quad (5.88)$$

The total fields in the medium #1 is the sum of the incident and reflected fields, and those in the medium #2 are described by only the transmitted wave.

$$\begin{aligned}
\overline{E} &= \overline{E}_i + \overline{E}_r,\ z < 0 \\
&= \overline{E}_t,\ z > 0 \\
\overline{H} &= \overline{H}_i + \overline{H}_r,\ z < 0 \\
&= \overline{H}_t,\ z > 0 \qquad (5.89)
\end{aligned}$$

Proper boundary conditions, developed in chapter 2, section 4.4 will have to be enforced across the interface between the two media in Fig.8. We will find the component of the total electric field in the medium #1, which is parallel to the interface, evaluated at the interface by substituting $z = 0_-$, which will have to be equal to that in medium #2, evaluated at $z = 0_+$. In reference to Fig.8, the total electric field in this case is along the $\hat{y}$ direction, which is parallel to the interface. Accordingly, the required boundary condition for the electric field may be established using (5.88,5.89).

$$\begin{aligned}
E_y = E_{yi} + E_{yr} &= E_{yt}, z = 0 \\
1e^{-jk_1 \sin\theta_i x} + \Gamma e^{-jk_1 \sin\theta_r x} &= Te^{-jk_2 \sin\theta_r x} \qquad (5.90)
\end{aligned}$$

Similarly, we find the component of the total magnetic field in the medium #1, which is parallel to the interface, evaluated at the interface by substituting $z = 0_-$, which will have to be equal to that in medium #2, evaluated at $z = 0_+$. In reference to Fig.8, the magnetic field has both $\hat{x}$ and $\hat{z}$ components, out of which only the $\hat{x}$ component is parallel to the interface. Accordingly, the required boundary condition for the magnetic field may be established in terms of the x component of the magnetic field using (5.88,5.89).

$$\begin{aligned}
H_x = H_{xi} + H_{xr} &= H_{xt}, z = 0 \\
\frac{-\cos\theta_i}{\eta_1} e^{-jk_1 \sin\theta_i x} + \frac{\Gamma \cos\theta_r}{\eta_1} e^{-jk_1 \sin\theta_r x} &= \frac{-T\cos\theta_t}{\eta_2} e^{-jk_2 \sin\theta_t x} \qquad (5.91)
\end{aligned}$$

The different terms in equations (5.90,5.91) consist of a phase function and an amplitude factor. In order to satisfy (5.90,5.91) at all locations of x, we need to have each phase function to be equal to each other.

$$k_1 \sin\theta_i = k_1 \sin\theta_r = k_2 \sin\theta_t \qquad (5.92)$$

With the above phase relationships established, the amplitude parts in (5.90,5.91) may be expressed as follows.

$$1 + \Gamma = T \qquad (5.93)$$

$$(1-\Gamma)\frac{\cos\theta_i}{\eta_1} = \frac{T\cos\theta_t}{\eta_2},\ 1 - \Gamma = T\left(\frac{\eta_1 \cos\theta_t}{\eta_2 \cos\theta_i}\right) \qquad (5.94)$$

By adding (5.93,5.94) we may solve for the unknown transmission coefficient T, which can then be used in (5.93) to solve for the reflection coefficient Γ.

$$
\begin{aligned}
2 &= T[1 + \tfrac{\eta_1 \cos\theta_t}{\eta_2 \cos\theta_i}] \\
T &= \tfrac{2\eta_2 \cos\theta_i}{\eta_1 \cos\theta_t + \eta_2 \cos\theta_i} = \tfrac{2\eta_2 / \cos\theta_t}{\eta_1/\cos\theta_i + \eta_2/\cos\theta_t} \\
\Gamma &= T - 1 = \tfrac{\eta_2/\cos\theta_t - \eta_1/\cos\theta_i}{\eta_2/\cos\theta_t + \eta_1/\cos\theta_i}
\end{aligned}
\tag{5.95}
$$

The above expressions for the reflection and transmission coefficients may be compared with those for the case of normal incidence in (5.85). For the case of normal incidence, we have $\theta_i = 0$, which means $\theta_r = 0 = \theta_t$, as per the relationship (5.92). This means the reflected and transmitted waves would also propagate normal to the interface, as we expected. Under this condition, the expressions for the reflection and transmission coefficients in (5.95) would lead to the same results we derived in (5.85) specifically for the normal incidence.

5.1.1. *Equivalent Transmission Line Model.* The above expressions for the reflection coefficient Γ and transmission coefficient T, for the inclined incidence of the TE_z wave may be directly obtained using an equivalent transmission line model. The basic principles of an equivalent transmission line model is similar to those we discussed in section 4.1 for the case of normal incidence. Like the impedance definitions used in (5.86), equivalent impedances for the two media in Fig.8 can be defined using the transverse field expressions of (5.88). The impedances are represented with a subscript TE, in order to specifically refer to the TE_z wave in the present case.

$$
\begin{aligned}
V \equiv E_y,\ & I \equiv -H_x \\
Z_{TE1} = -\frac{E_{yi}}{H_{xi}} = \frac{E_{yr}}{H_{xr}} &= \frac{\eta_1}{\cos\theta_i} \\
Z_{TE2} = -\frac{E_{yt}}{H_{xt}} &= \frac{\eta_2}{\cos\theta_t}
\end{aligned}
\tag{5.96}
$$

The E_x and H_y fields in the impedance definitions of section 4.1 are replaced here by E_y and $-H_x$, respectively. The new reference field directions, as compared to those in section 4.1, may be seen simply by rotating the x, y directions in Fig.6 to $y, -x$ directions in Fig.8, respectively. Accordingly, the equivalent impedance Z_{TE1} in (5.96) for the first medium of Fig.8 is defined as the ratio of E_{yi} and $-H_{xi}$ fields of the incident wave. Like the impedance definition of section 4.1, the Z_{TE1} in the present case may also be defined using the reflected fields, but the expression would have a negative sign compared to that using the incident fields. Accordingly, Z_{TE1} is defined as the ratio of the E_{yr} and H_{xr} fields of the reflected wave. The equivalent impedance Z_{TE2} of the second medium in Fig.8 is similarly defined as the ratio of the E_{yt} and $-H_{xt}$ fields of the transmit wave.

Note that the total voltage and current are continuous across the junction of two transmission lines in Fig.9. Equivalently in Fig.8, only the components of the total electric and magnetic field vectors, that are parallel to the interface, are continuous on the two sides of the interface. Keeping this equivalence in mind, the equivalent voltage and current for Fig.8 are defined in (5.96) using *only* the *parallel* components of field vectors, *not* the total field vectors including both the parallel and normal components to interface. Particularly for the magnetic field in Fig.8, only the H_x component, which is parallel to the boundary, is used in the definition of (5.96), not the total magnetic field including also the H_z component. However, no such distinction is necessary for the electric field in Fig.8, because the total electric field vector in Fig.8 happens to be parallel to the interface. For the case of normal incidence in Fig.6, all field vectors happened to be parallel to the interface, and accordingly no such distinction was necessary for either the electric or the magnetic fields.

It may be useful to express the impedances in (5.96) using the wave number in the respective media in the normal (z) direction.

$$Z_{TE1} = \frac{\eta_1}{\cos\theta_i} = \frac{\eta_1 k_1}{k_1 \cos\theta_i} = \eta_1(\frac{k_1}{k_{z1}}), \; Z_{TE2} = \frac{\eta_2}{\cos\theta_t} = \frac{\eta_2 k_2}{k_2 \cos\theta_t} = \eta_2(\frac{k_2}{k_{z2}}) \tag{5.97}$$

Consistent with the above impedance expressions for the two media, we may use a general definition for the equivalent impedance for a TE_z wave, applicable to any material medium.

$$Z_{TE} = \eta\frac{k}{k_z}, \; k_z = \sqrt{k^2 - k_x^2}, \; k_x = k_1 \sin\theta_i \tag{5.98}$$

The wave number k_z in the z direction may be related to the wave number k_x in the x direction. Further, the wave number k_x is the same in both the media as per the phase relations (5.92), which can be expressed in terms of the wave number in the incidence medium and the incidence angle.

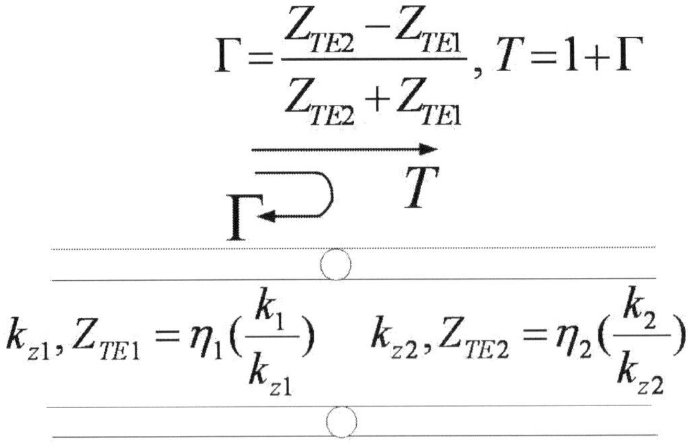

FIGURE 9

Based on the above equivalence principles, the reflection and transmission coefficients for the plane-wave problem in Fig.8 could be obtained using an equivalent transmission model shown in Fig.9. The expressions in (5.95) could be directly obtained using transmission-line equations in chapter 4, section 7, and substituting the appropriate equivalent impedance parameters as shown in Fig.9.

$$\Gamma = \frac{Z_{TE2}-Z_{TE1}}{Z_{TE2}+Z_{TE1}} = \frac{\eta_2/\cos\theta_t-\eta_1/\cos\theta_i}{\eta_2/\cos\theta_t+\eta_1/\cos\theta_i} = \frac{\eta_2(k_2/k_{z2})-\eta_1(k_1/k_{z1})}{\eta_2(k_2/k_{z2})+\eta_1(k_1/k_{z1})}$$
$$T = 1+\Gamma = \frac{2Z_{TE2}}{Z_{TE2}+Z_{TE1}} \tag{5.99}$$

The reflection and transmission coefficients at a transmission line junction, as derived in the chapter 4, section 7, refer to the relative amplitudes of the respective voltage signal. As per the definitions in (5.97), the voltage is equivalent to the E_y field. Therefore, the reflection and transmission coefficients of (5.99), obtained from an equivalent transmission line model, refer to the relative amplitudes of the respective E_y fields. In the present problem, the total electric field vector consists of only the E_y component. Therefore, the results in (5.99) as well refer to the reflection and transmission coefficients of the total electric field.

The equivalent transmission line modeling may be conveniently extended to model the propagation of the TE_z wave in multiple layers of material media. This will be addressed in section 6.

5.2. Magnetic Field Vector Parallel to Interface, TM_z Wave. The different fields for this case are shown in Fig.10, with its magnetic field vector parallel to the y direction, and the electric field vector on the xz plane which is the plane of incidence. As discussed before, the incidence wave in this case is referred to as a Transverse-Magnetic-to-z or a TM_z wave. This is because the y directed magnetic field in this case is *transverse* to the z direction. This is in contrast to the case covered in section 5.1, which was referred to as the TE_z wave.

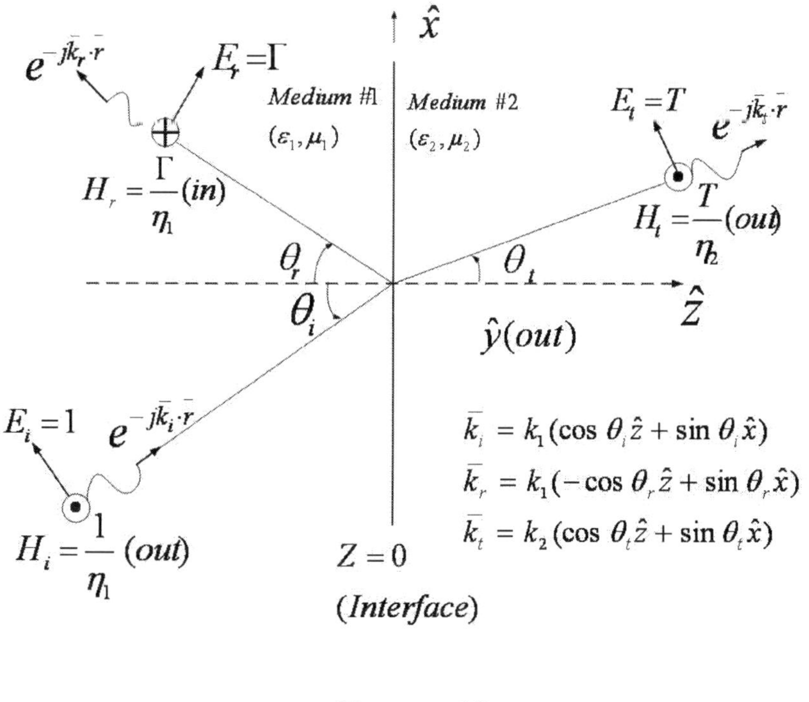

FIGURE 10

Here we will follow the similar analysis steps used for the TE_z wave in section 5.1. Basic plane-wave principles and symmetry considerations are used to determine relative orientations of the propagation directions and field vectors. The electric and magnetic fields in Fig.10, for the incident, reflected and transmitted waves may be expressed as follows.

$$\begin{aligned}
\overline{E}_i &= (\cos\theta_i\hat{x} - \sin\theta_i\hat{z})e^{-jk_1(\cos\theta_i z+\sin\theta_i x)} \\
\overline{H}_i &= \tfrac{1}{\eta_1}\hat{y}e^{-jk_1(\cos\theta_i z+\sin\theta_i x)} \\
\overline{E}_r &= \Gamma(\cos\theta_r\hat{x} + \sin\theta_r\hat{z})e^{-jk_1(-\cos\theta_r z+\sin\theta_r x)} \\
\overline{H}_r &= -\tfrac{\Gamma}{\eta_1}\hat{y}e^{-jk_1(-\cos\theta_r z+\sin\theta_r x)} \\
\overline{E}_t &= T(\cos\theta_t\hat{x} - \sin\theta_t\hat{z})e^{-jk_2(\cos\theta_t z+\sin\theta_t x)} \\
\overline{H}_t &= \tfrac{T}{\eta_2}\hat{y}e^{-jk_2(\cos\theta_t z+\sin\theta_t x)}
\end{aligned} \quad (5.100)$$

The required boundary conditions for the electric and magnetic fields are established at the interface, $z = 0$, using the above field expressions.

$$\begin{aligned}
E_x = E_{xi} + E_{xr} &= E_{xt},\ z = 0 \\
\cos\theta_i e^{-jk_1\sin\theta_i x} + \Gamma\cos\theta_r e^{-jk_1\sin\theta_r x} &= T\cos\theta_t e^{-jk_2\sin\theta_t x}
\end{aligned} \quad (5.101)$$

$$\begin{aligned}
H_y = H_{yi} + H_{yr} &= H_{yt},\ z = 0 \\
\tfrac{1}{\eta_1}e^{-jk_1\sin\theta_i x} - \tfrac{\Gamma}{\eta_1}e^{-jk_1\sin\theta_r x} &= \tfrac{T}{\eta_2}e^{-jk_1\sin\theta_t x}
\end{aligned} \quad (5.102)$$

Equating the phase factors in (5.101,5.102) results in the same relation we derived in (5.92) for the TE_z wave.

$$k_1\sin\theta_i = k_1\sin\theta_r = k_2\sin\theta_t \quad (5.103)$$

With the above phase relationships established, the amplitude parts in (5.101,5.102) may be related.

$$(1+\Gamma)\cos\theta_i = T\cos\theta_t,\ 1+\Gamma = T\tfrac{\cos\theta_t}{\cos\theta_i} \quad (5.104)$$

$$\tfrac{1}{\eta_1} - \tfrac{\Gamma}{\eta_1} = \tfrac{T}{\eta_2},\ 1-\Gamma = T\tfrac{\eta_1}{\eta_2} \quad (5.105)$$

Equations (5.104) and (5.105) may be added to solve for the transmission coefficient T, which can then be used in (5.105) to solve for the reflection coefficient Γ.

$$\begin{aligned}
2 &= T(\tfrac{\eta_1}{\eta_2} + \tfrac{\cos\theta_t}{\cos\theta_i}) \\
T &= \tfrac{2\eta_2\cos\theta_i}{\eta_1\cos\theta_i+\eta_2\cos\theta_t} \\
\Gamma = 1 - T\tfrac{\eta_1}{\eta_2} &= \tfrac{\eta_2\cos\theta_t-\eta_1\cos\theta_i}{\eta_2\cos\theta_t+\eta_1\cos\theta_i}
\end{aligned} \quad (5.106)$$

The above expressions for the reflection and transmission coefficients may be compared with those for the case of normal incidence in (5.85). For the case of normal incidence, we have $\theta_i = 0$, which means $\theta_r = 0 = \theta_t$, as per the relationship (5.103). This means the reflected and transmitted waves would also propagate normal to the interface, as we expected. Under this condition, the expressions for the reflection and transmission coefficients in (5.106) would lead to the same results we derived in (5.85), which we specifically derived for the normal incidence.

5.2.1. *Equivalent Transmission Line Model.* The above expressions for the reflection coefficient Γ and transmission coefficient T, for the case of inclined incidence of the TM_z wave, may be directly obtained using an equivalent transmission line model. The basic principles of an equivalent transmission line model is similar to those we discussed in section 4.1 for the case of normal incidence. Like the impedance definitions in (5.86), equivalent impedances for the two media in Fig.10 can be defined using the transverse field expressions of (5.100). The impedances are represented with a subscript TM, in order to specifically refer to the TM_z wave in the present case.

$$
\begin{aligned}
V \equiv E_x,\ I \equiv H_y \\
Z_{TM1} = \frac{E_{xi}}{H_{yi}} = -\frac{E_{xr}}{H_{yr}} = \eta_1 \cos\theta_i \\
Z_{TM2} = \frac{E_{xt}}{H_{yt}} = \eta_2 \cos\theta_t
\end{aligned}
\tag{5.107}
$$

As in the definitions of section 4.1, the equivalent impedance Z_{TM1} of the first medium in Fig.10 is defined as the positive ratio of E_{xi} and H_{yi} fields of the incident wave, or the negative ratio of E_{xr} and H_{yr} fields of the reflected wave. The equivalent impedance Z_{TM2} of the second medium in Fig.10 is similarly defined as the ratio of the E_{xt} and H_{yt} fields of the transmit wave.

As in section 5.1 for a TE_z wave, for the present case of a TM_z wave the equivalent voltage and current are defined in (5.107) using the components of field vectors, which are parallel to the interface. Particularly for the electric field in Fig.10, only the E_x component, which is parallel to the boundary, is used in the definition of (5.107), not the total electric field including both the E_x and the E_z components. However, no such distinction is necessary for the magnetic field in Fig.10, because the total magnetic field vector in Fig.10 happens to be parallel to the interface.

It may be useful to express the impedances in (5.107) using the wave number k_z in the respective media.

$$
\begin{aligned}
Z_{TM1} = \eta_1 \cos\theta_i = \frac{\eta_1 k_1 \cos\theta_i}{k_1} = \eta_1(\frac{k_{z1}}{k_1}) \\
Z_{TM2} = \eta_2 \cos\theta_t = \frac{\eta_2 k_2 \cos\theta_t}{k_2} = \eta_2(\frac{k_{z2}}{k_2})
\end{aligned}
\tag{5.108}
$$

Consistent with the above impedance expressions for the two media, we may use a general definition for the equivalent impedance for a TM_z wave, applicable to any material medium.

$$Z_{TM} = \eta\frac{k_z}{k},\ k_z = \sqrt{k^2 - k_x^2},\ k_x = k_1 \sin\theta_i \tag{5.109}$$

The wave number k_z may be related to the wave number k_x. Further, the wave number k_x is the same in both the media as per the phase relations (5.103), which may be expressed in terms of the wave number in the incidence medium and the incidence angle.

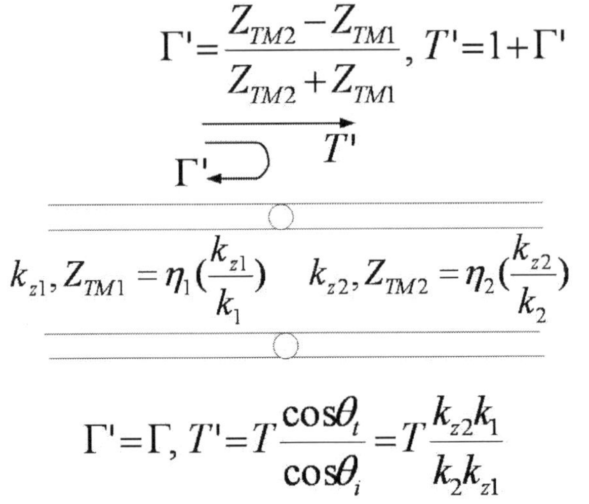

FIGURE 11

Based on the equivalence parameters in (5.108), the reflection and transmission coefficients for the plane-wave problem in Fig.10 could be obtained using an equivalent transmission model shown in Fig.11. The transmission-line formulas in section 7 may be used, with required impedance parameters for the equivalent transmission lines defined in (5.108).

$$\Gamma' = \frac{Z_{TM2} - Z_{TM1}}{Z_{TM2} + Z_{TM1}} = \frac{\eta_2 \cos\theta_t - \eta_1 \cos\theta_i}{\eta_2 \cos\theta_t + \eta_1 \cos\theta_i} = \frac{\eta_2(k_{z2}/k_2) - \eta_1(k_{z1}/k_1)}{\eta_2(k_{z2}/k_2) + \eta_1(k_{z1}/k_1)}$$
$$T' = 1 + \Gamma' = \frac{2Z_{TM2}}{Z_{TM2} + Z_{TM1}} = \frac{2\eta_2 \cos\theta_t}{\eta_2 \cos\theta_t + \eta_1 \cos\theta_i} \tag{5.110}$$

The reflection and transmission coefficients at a transmission line junction, as derived in section 7, refer to the respective relative amplitudes of the voltage signal. As per the definitions in (5.108), the voltage is equivalent to the E_x field. Therefore, the reflection and transmission coefficients of (5.110), obtained from an equivalent transmission line model, refer to the relative amplitudes of the respective E_x fields. They are represented in (5.110) with primed parameters, Γ' and T'. The reflection and transmission coefficients for the total electric field in Fig.10 are represented with unprimed parameters Γ and T, respectively, which may be related with the primed parameters using the field expressions (5.100).

$$\Gamma' = \frac{E_{xr}}{E_{xi}}\Big|_{z=0} = \frac{\Gamma\cos\theta_r}{1\cos\theta_i} = \Gamma,\ T' = \frac{E_{xt}}{E_{xi}}\Big|_{z=0} = \frac{T\cos\theta_t}{1\cos\theta_i}$$
$$\Gamma = \Gamma' = \frac{\eta_2\cos\theta_t - \eta_1\cos\theta_i}{\eta_2\cos\theta_t + \eta_1\cos\theta_i} = \frac{Z_{TM2} - Z_{TM1}}{Z_{TM2} + Z_{TM1}}$$
$$T = T'\frac{\cos\theta_i}{\cos\theta_t} = \frac{2\eta_2\cos\theta_i}{\eta_2\cos\theta_t + \eta_1\cos\theta_i} = (\frac{2Z_{TM2}}{Z_{TM2} + Z_{TM1}})(\frac{\cos\theta_i}{\cos\theta_t}) \tag{5.111}$$

These expressions for Γ and T may be verified with the results of (5.106), which we independently derived based on the field analysis and suitable boundary conditions. The field analysis may be increasingly difficult to implement when we have multiple layers of interface. However, the equivalent transmission line modeling may be conveniently extended to study the propagation of the TM_z wave in multiple layers of material media. We will discuss the modeling of multilayered media in section 6.

5.3. Snell's Law. The equation (5.92) was derived by equating the phase factors for the TE_z waves at the interface. The same relation (5.103) is also obtained by equating the phase factors for the TM_z wave in section 5.2. (5.92,5.103) establish a relationship between the propagation directions for the incident, reflected and transmitted waves. The angle of reflection θ_r is equal to the angle of incidence θ_i. Whereas, the angle of transmission θ_t is in general different from the angle of incidence θ_i. The ratio of the sine of the transmit angle θ_t to that of the incident angle θ_i is equal to the ratio of the wave numbers in the second to that in the first medium. This ratio is equal to the ratio of the wave length in the first to that in the second medium, which is also the ratio of the propagation velocities in the first to that in the second medium.

$$\sin\theta_i = \sin\theta_r,\ \theta_i = \theta_r$$
$$k_1\sin\theta_i = k_2\sin\theta_t,\ \tfrac{\sin\theta_i}{\sin\theta_t} = \tfrac{k_2}{k_1} = \tfrac{\lambda_1}{\lambda_2} = \tfrac{v_{p1}}{v_{p2}} \tag{5.112}$$

The above relationship is referred to as Snell's law of reflection and transmission.

5.4. Total Internal Reflection. As per Snell's law in (5.112), if the velocity in the first medium or medium of incidence is larger than that in the second medium or the medium of transmission, then the incidence angle θ_i would be larger than the transmit angle θ_t. In other words, the transmitted wave bends closer to the normal axis to the interface, when it enters into a medium with a smaller propagation velocity, compared to the medium of incidence. Whereas, if the wave enters into a medium with a larger velocity, compared to the medium of incidence, then the transmitted wave would bend away from the normal axis to the interface.

Now, consider particularly the situation, when the medium #1, which is the medium of incidence, to be denser than the medium #2, which is the medium of transmission. That is $v_{p2} > v_{p1}$. As explained, in this case, we have $\theta_t > \theta_i$, which means the wave bends away from the normal axis to the interface.

$$\frac{\sin\theta_i}{\sin\theta_t} = \frac{v_{p1}}{v_{p2}} < 1$$
$$\sin\theta_t > \sin\theta_i, \;\; \theta_t > \theta_i \tag{5.113}$$

Accordingly, there should be a certain angle incidence θ_i, when the transmitted wave would propagate in a direction just parallel to the interface, with $\theta_t = 90^0$. This angle of incidence is called the critical angle of incidence θ_c. This value of incidence angle $\theta_i = \theta_c$ may be found from Snell's law of (5.112), by substituting $\theta_t = 90^0$

$$\frac{\sin\theta_c}{\sin 90^0} = \frac{v_{p1}}{v_{p2}} = \sin\theta_c, \theta_c = \arcsin(\frac{v_{p1}}{v_{p2}}) \tag{5.114}$$

Now, it would be interesting to see what would happen when the incidence angle is larger than the critical angle, that is $\theta_i > \theta_c$, From (5.113,5.114), we get

$$\sin\theta_t = \sin\theta_i\frac{v_{p2}}{v_{p1}} > \sin\theta_c\frac{v_{p2}}{v_{p1}} = 1$$
$$\sin\theta_t > 1, \;\; \theta_i > \theta_c \tag{5.115}$$

There is no real angle θ_t, for which the above relation would be satisfied, because we have $\sin\theta_t > 1$. When the sine of the transmit angle is greater than unity, which means the corresponding sine-squared value is also greater than unity or the cosine-squared value is negative: $\cos^2\theta_t = 1 - \sin^2\theta_t < 0$. Equivalently, the corresponding cosine value $\cos\theta_t$ would be purely imaginary, positively or negatively.

Let us see the situation in terms of the wave numbers in the transmit medium. When $\cos\theta_t$ is a purely imaginary, the wave number k_{zt} in the transmit medium would be purely imaginary, positively or negatively.

$$k_{xt} = k_2\sin\theta_t > k_2$$
$$k_{zt} = k_2\cos\theta_t = \pm j\alpha_{zt}, \;\; R_e(k_{zt}) = 0, \;\; Im(k_{zt}) = \pm\alpha_{zt} \tag{5.116}$$

The propagation function in the z direction determines the variation of the wave along the normal axis. When k_{zt} is purely imaginary, the propagation function would be an exponentially growing or decaying function in the transmit medium. This would mean that the transmit wave would either grow or decay exponentially along the z axis. The possibility of an exponentially growing function would mean that the wave would keep growing as one moves far away from the interface in the transmit medium, approaching infinity as one approaches $z \to \infty$. This is clearly a non physical situation, Accordingly, the exponential decay of the transmit wave is the only physical possibility. This corresponds to having the $\cos\theta_t$ and the wave number k_{zt} in the transmit medium to be negatively imaginary

$$e^{-jk_2\cos\theta_t z} = e^{-jk_{z2}z} = e^{-\alpha_{z2}z}, \;\; Im(k_{zt}) = -\alpha_{zt} < 0 \tag{5.117}$$

We may also view the situation in terms of the equivalent impedance of the transmit medium for the TE_z and TM_z cases. These impedances were defined in (5.96,5.107) in terms of the $\cos\theta_t$, and would be purely imaginary when $\cos\theta_t$ is purely imaginary.

$$\begin{aligned} Z_{TM2} &= \eta_2 \cos\theta_t, Re(Z_{TM2}) = 0,\ \ Im(Z_{TM2}) < 0 \\ Z_{TE2} &= \eta_2/\cos\theta_t, Re(Z_{TE2}) = 0,\ \ Im(Z_{TE2}) > 0 \end{aligned} \tag{5.118}$$

Z_{2TM} is negatively imaginary, whereas Z_{TE2} is positively imaginary. The reflection coefficient Γ is expressed in (5.99,5.111) in terms of these equivalent impedances. Let us see the magnitude of the reflection coefficients. Now we have the equivalent impedance in the second medium purely imaginary. Whereas the equivalent impedance, Z_{TE1} or Z_{TM2}, in the first medium is still real, because $\cos\theta_i$ is real, or k_{z1} is real. Under this situation, the magnitude of the reflection coefficient in (5.99,5.111) can be shown to be unity.

$$\begin{aligned} |\Gamma_{TM}|^2 &= \frac{|Z_{TM2}|^2 + |Z_{TM1}|^2}{|Z_{TM2}|^2 + |Z_{TM1}|^2} = 1 \\ |\Gamma_{TE}|^2 &= \frac{|Z_{TE2}|^2 + |Z_{TE1}|^2}{|Z_{TE2}|^2 + |Z_{TE2}|^2} = 1 \end{aligned} \tag{5.119}$$

When the magnitude of reflection in both TM_z and TE_z cases is unity, this means that the signal amplitude of reflection is equal to that of the incident wave. Equivalently, all the signal power associated with the incident wave is reflected back to the reflected wave. Therefore, by the principle of power conservation, no power is transmitted into the second medium. Accordingly, this situation is referred to as total internal reflection. The dielectric interface behaves like a perfect reflector (mirror) in the sense that the magnitude of reflection is unity, or that all the incident power is reflected back.

However, unlike a short or open circuit where $\Gamma = -1$ or $+1$ with phases 180° and 0°, respectively, for a total internal reflection the phases of Γ_{TE} or Γ_{TE} is not 180° or 0°. Because we have Z_{TE2} positively imaginary and Z_{TE1} is real, as discussed earlier, the phase of the reflection coefficient $\angle\Gamma_{TE}$ for the TE_z wave, as expressed in (5.99), can be shown to be positive. This refers to the principal value of the phase $(0 < \angle\Gamma_{TE} < \pi)$. Therefore, the total reflection in this case may be interpreted as a perfect reflector with $\Gamma = 1$, placed at a virtual location displaced from the actual, physical interface, toward the negative z direction (displaced into the incident medium).

Similarly, the phase of the reflection coefficient $\angle\Gamma_{TM}$ for the TM_z wave, as expressed in (5.111), can be shown to be negative $(-\pi < \angle\Gamma_{TM} < 0)$. Therefore, the total reflection in this case may be interpreted as a perfect reflector with $\Gamma = 1$, placed at a virtual location displaced from the actual, physical interface, toward the positive z direction (displaced into the transmit medium).

5.5. Brewster Angle and Total Power Transmission. It may be possible that, for a specific set of material parameters for the two media on the two sides of an interface, there exists a specific angle of incidence θ_i, at which the reflection coefficient Γ is zero. In this case, no signal power would be reflected back to the incident medium. Equivalently, using the principle of power conservation, all signal power will pass through the interface into the transmit medium. The angle of incidence θ_i for this condition of total power transmission is referred as the Brewster angle of incidence θ_{BW}. We will analyse such a possibility, and find the required Brewster angle, for the two separate cases of TE_z and TM_z waves.

5.5.1. *Transverse Electric (TE_z) Wave.* In this case the expression of the reflection coefficient Γ is given in (5.99). $\Gamma = 0$ in (5.99) would be zero, when the numerator in (5.99) is zero.

$$\begin{aligned}\Gamma_{TE} = 0,\ \eta_2/\cos\theta_t = \eta_1/\cos\theta_i,\ \theta_i = \theta_{BW}^{TE} \\ \tfrac{\eta_2^2}{\cos^2\theta_t} = \tfrac{\eta_1^2}{\cos^2\theta_i},\ \eta_2^2\cos^2\theta_i = \eta_1^2\cos^2\theta_t \\ \eta_1^2(1-\sin^2\theta_t) = \eta_2^2(1-\sin^2\theta_i)\end{aligned} \tag{5.120}$$

Using Snell's law (5.112) to relate the transmit and incident angles, we can solve (5.120) for the required incidence angle $\theta_i = \theta_{BW}^{TE}$.

$$\sin^2\theta_t = \sin^2\theta_i \times \tfrac{v_{p2}^2}{v_{p1}^2} = \sin^2\theta_i \tfrac{\mu_1\epsilon_1}{\mu_2\epsilon_2}$$

$$\eta_1^2(1-\sin^2\theta_i \tfrac{\mu_1\epsilon_1}{\mu_2\epsilon_2}) = \eta_2^2(1-\sin^2\theta_i)$$

$$\eta_1^2 - \eta_2^2 = \sin^2\theta_i[\eta_1^2\tfrac{\mu_1\epsilon_1}{\mu_2\epsilon_2} - \eta_2^2] \tag{5.121}$$

$$\sin^2\theta_{BW}^{TE} = \sin^2\theta_i = \tfrac{(\eta_1^2-\eta_2^2)\mu_2\epsilon_2}{\eta_1^2\mu_1\epsilon_1-\eta_2^2\mu_2\epsilon_2} \tag{5.122}$$

Consider a more specific situation, with $\mu_2 = \mu_1$. This is often the case when the two materials are dominantly dielectric in nature, with negligible magnetic property. Such materials are characterized by the same magnetic permeability, $\mu_1 = \mu_2 = \mu_0$, which is the permeability of the free space medium. Under such conditions, (5.122) for the Brewster angle may be simplified.

$$\sin^2\theta_{BW}^{TE} = \tfrac{(\frac{1}{\epsilon_1}-\frac{1}{\epsilon_2})\epsilon_2}{0} = \infty \tag{5.123}$$

Clearly, under the above condition no solution would exist for θ_{BW}^{TE}. Accordingly, when the two materials have the same magnetic property, it would not be possible to achieve total transmission of signal power through the interface using a TE_z wave .

Consider another possibility when $\epsilon_2 = \epsilon_1$, but $\mu_2 \neq \mu_1$, Under such condition, the expression (5.122) for the Brewster angle may be simplified.

$$\sin^2\theta_{BW}^{TE} = \tfrac{(\mu_1-\mu_2)\mu_2}{\mu_1^2-\mu_2^2} = \tfrac{\mu_2}{\mu_1+\mu_2} < 1$$

$$\theta_{BW}^{TE} = \sin^{-1}\sqrt{\tfrac{\mu_2}{\mu_1+\mu_2}} \tag{5.124}$$

Solution exists in this case for θ_{BW}^{TE}, for any values of μ_1 and μ_2. This is because, the sine of the Brewster angle, as given in (5.124), is always less than unity, irrespective of the relative values of the material permeability μ_1 and μ_2.

5.5.2. *Transverse Magnetic (TM_z) Wave.* In this case the expression of the reflection coefficient Γ is given in (5.111). Γ in (5.111) would be zero, when the numerator in (5.111) is zero.

$$\Gamma_{TM} = 0, \ \eta_2 \cos\theta_t = \eta_1 \cos\theta_i, \ \theta_i = \theta_{BW}^{TM}$$
$$\eta_2^2 \cos^2\theta_t = \eta_2^2(1 - \sin^2\theta_t) = \eta_1^2 \cos^2\theta_i = \eta_1^2(1 - sin^2\theta_i) \tag{5.125}$$

Using Snell's law (5.112) to relate the transmit and incident angles, we can solve (5.125) for the required incidence angle $\theta_i = \theta_{BW}^{TM}$.

$$\sin^2\theta_t = \sin^2\theta_i \times \frac{v_{p2}^2}{v_{p1}^2} = \sin^2\theta_i \frac{\mu_1\epsilon_1}{\mu_2\epsilon_2}$$
$$\eta_1^2(1 - \sin^2\theta_i \frac{\mu_1\epsilon_1}{\mu_2\epsilon_2}) = \eta_2^2(1 - \sin^2\theta_i)$$
$$\eta_2^2 - \eta_1^2 = \eta_2^2 \sin^2\theta_t - \eta_1^2 \sin^2\theta_i = \sin^2\theta_i[\eta_2^2 \frac{\mu_1\epsilon_1}{\mu_2\epsilon_2} - \eta_1^2]$$
$$\sin^2\theta_{BW}^{TM} = \sin^2\theta_i = \frac{(\eta_2^2 - \eta_1^2)\mu_2\epsilon_2}{\eta_2^2\mu_1\epsilon_1 - \eta_1^2\mu_2\epsilon_2} \tag{5.126}$$

Consider a more specific situation, with $\mu_2 = \mu_1$. This is often the case when the two materials are dominantly dielectric in nature, with negligible magnetic property. Such materials are characterized by the same magnetic permeability, $\mu_1 = \mu_2 = \mu_0$, which is the permeability of the free space medium. Under such conditions, (5.126) for the Brewster angle may be simplified.

$$\sin^2\theta_{BW}^{TM} = \frac{(\eta_2^2 - \eta_1^2)\epsilon_2}{\eta_2^2\epsilon_1 - \eta_1^2\epsilon_2} = \frac{(\frac{1}{\epsilon_2} - \frac{1}{\epsilon_1})\epsilon_2}{\frac{\epsilon_1}{\epsilon_2} - \frac{\epsilon_2}{\epsilon_1}}$$
$$= \frac{\epsilon_2(\epsilon_1 - \epsilon_2)}{\epsilon_1^2 - \epsilon_2^2} = \frac{\epsilon_2}{\epsilon_1 + \epsilon_2} < 1$$
$$\theta_{BW}^{TM} = \sin^{-1}\sqrt{\frac{\epsilon_2}{\epsilon_1 + \epsilon_2}} \tag{5.127}$$

Solution exists in this case for the Brewster angle θ_{BW}^{TM}, with any arbitrary values of ϵ_1 and ϵ_2. This is because, the sine of the Brewster angle, as given in (5.127) is always less than unity, irrespective of the relative values of the material permittivities ϵ_1 and ϵ_2.

Consider another possibility when $\epsilon_2 = \epsilon_1$, but $\mu_2 \neq \mu_1$, Under such condition, the expression (5.126) for the Brewster angle may be simplified.

$$\sin^2\theta_{BW}^{TM} = \frac{(\mu_2 - \mu_1)\mu_2}{\mu_1\mu_2 - \mu_1\mu_2} = \infty \tag{5.128}$$

Clearly, under the above condition no solution would exist for θ_{BW}^{TM}. Accordingly, when the two materials have the same dielectric property, it would not be possible to achieve total transmission of signal power through the interface using a TM_z wave.

6. Wave Propagation Through Multiple, Layered Interfaces

Wave propagation in multiple layers of material media, as shown in Fig.12, can be modeled using similar procedure used for a single material interface, presented in sections 4 for normal incidence, 5.1 and 5.2 for inclined incidence. The electric and magnetic fields in the different layers may be expressed in terms of unknown wave amplitudes, using basic plane-wave principles, The different unknown parameters can then be solved by enforcing boundary conditions at each interface. Such analytical process can be increasingly complex for larger number of layers. However as a simpler alternative, the equivalent transmission line models which we developed for the case of a single interface, may be extended to a multilayer problem.

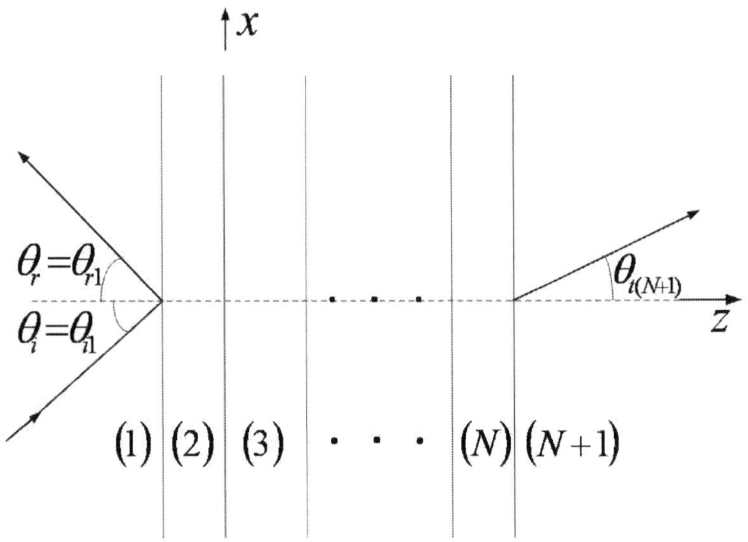

FIGURE 12

The equivalent transmission line model for the multilayer structure of Fig.12 is shown in Fig.13. Using equivalent transmission-line parameters for different layers, one may use transmission line theory to solve the equivalent circuit of Fig.13. Depending on the TE_z or the TM_z nature of the incident wave, the equivalent impedances would be different as expressed in (5.97) and (5.108), respectively. However, the propagation constant in different equivalent transmission lines is the wave number k_z in the corresponding medium, expressed independent of the TE_z or TM_z nature of the incident wave. k_z in a particular medium is a function of the wave numbers k_x and k of the medium. The wave number k_x is the same in all media, as per the boundary condition requirement (5.92,5.103), and may be expressed in terms of the wave number and incidence angle in the first medium. k_z is to be positive when it real, so that a propagation phasor of the form $e^{-jk_z z}$ refers to a forward propagation in the $+z$ direction, as per the standard assumption. However, when k_z is imaginary, it has to be negatively imaginary so that it would refer to an exponentially decaying wave, as discussed in section 5.4.

The impedance and propagation constant for the equivalent transmission line of a given medium is expressed in terms of the wave impedance η, and the wave numbers $k = \omega\sqrt{\mu\epsilon}$ and k_z in the particular medium.

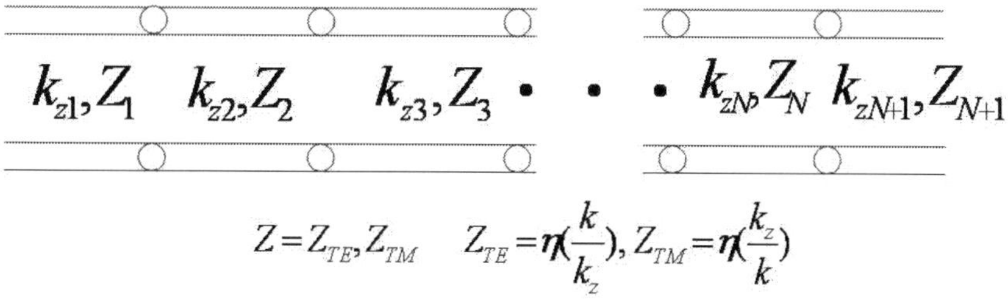

FIGURE 13

$$
\begin{aligned}
Z_{TE} &= \frac{\eta}{\cos\theta} = \eta(\frac{k}{k\cos\theta}) = \eta(\frac{k}{k_z}) \\
Z_{TM} &= \eta\cos\theta = \eta(\frac{k\cos\theta}{k}) = \eta(\frac{k_z}{k}) \\
k_z = \sqrt{k^2 - k_x^2} &= \sqrt{k^2 - k_{x1}^2} = \sqrt{k^2 - k_1^2 \sin^2\theta_i} \\
&Re(k_z) > 0, \; Im(k_z) < 0
\end{aligned} \tag{5.129}
$$

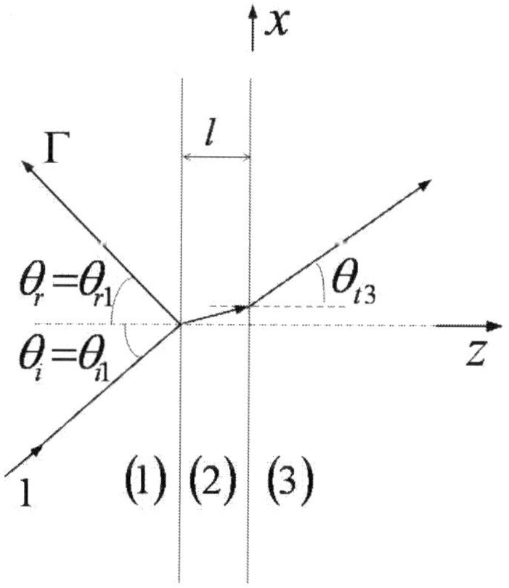

FIGURE 14

6.1. Quarter-Wave Matching Layer. As a simple example, consider a three-layer problem shown in Fig.14, with its equivalent transmission line model shown in Fig.15. In reference to the transmission line theory of section 9.6.1, when the second layer is properly designed to be a quarter-wave long and with a suitable equivalent impedance, the input reflection can be zero. The second layer, so designed, can be considered as a "matching" layer between two dissimilar media.

$$l = \frac{\lambda_{z2}}{4} = \frac{\pi}{2k_{z2}} = \frac{\pi}{2\sqrt{k_2^2 - k_1^2 \sin^2\theta_i}}, \quad Z_2 = \sqrt{Z_1 Z_3} \tag{5.130}$$

The above conditions may be solved for the required material parameter η_2 and length l of the matching layer. The impedance condition needs to be solved first for the material parameter of the second medium, which would be different depending upon the TE_z or the TM_z nature of the incident wave and the incident angle θ_i. Then, based on this material parameter, the length of the second layer will have to be designed equal to $\lambda_{z2}/2$. Accordingly, the design length l would also be different depending upon the TE_z or the TM_z nature of the incident wave and the incident angle θ_i. The design length $l = \lambda_{z2}/4$ is the smallest required value, and may be increased by any integral multiples of $\lambda_{z2}/2$, as per basic transmission line theory.

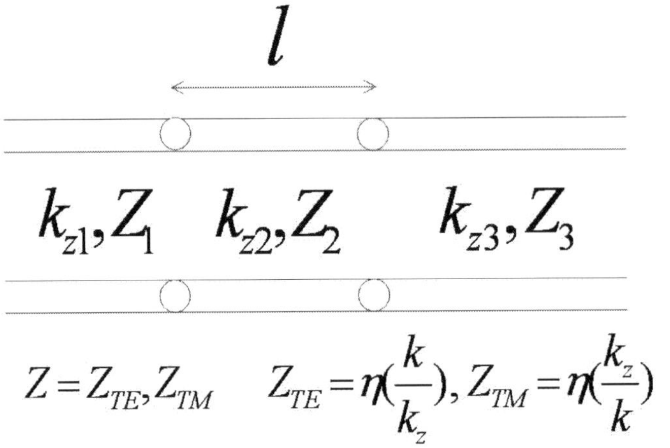

FIGURE 15

6.2. Half-Wave Matching Layer. The quarter-wave design we presented above is meant for matching between two dissimilar layers 1 and 3. In reference to Figs.14,15, consider another situation when the layers 1 and 3 are identical. The layer 2 may be considered as a space or a gap between the layers 1 and 3, filled by a suitable material. In general, the filling material between the two identical layers would produce undesirable reflection at the filling interface. However, consider that the thickness l of the filling layer is designed to be an integral multiple of $\lambda_{z2}/2$. With this design length, based on transmission line theory, the equivalent impedance seen at the interface of the layers 1 and 2 (Fig.14) can be shown to be equal to the characteristic impedance of the third medium Z_3, which is in this case also equal to that of the first medium Z_1. Accordingly, the incident wave would "see" as if there is a continuation of the first medium, resulting in zero reflection.

$$l = n\frac{\lambda_{z2}}{2} = n\frac{\pi}{k_{z2}}, \quad n = 1, 2, \cdots \tag{5.131}$$

There is no particular impedance requirement for the second layer in this design, unlike the quarter-wave design we presented earlier for matching between two dissimilar layers. This means that there is no specific requirement for the material parameters of the second layer, irrespective of the TE_z or TM_z nature of the incident wave. The thickness of the second layer l is designed based on k_{z2}, which is dependent on a choice of the material parameters of the second layer and the incidence angle θ_{i1}. The wave number k_{z2}, and accordingly the design thickness l, are independent of the TE_z or the TM_z nature of the incident signal. As per (5.131), a suitable thickness l can be designed, only if k_{z2} is real. This requires k_2^2 to be greater than $k_1^2 \sin\theta_{i1}$.

CHAPTER 6

RADIATION FROM ELEMENTARY SOURCES

A given distribution of time-varying electric current or charge constitutes a basic source for electromagnetic fields and radiation. Electromagnetic fields generated from some most elementary source configurations are studied in this chapter. More complex source configurations, that are used in practical antennas for radio communication, would be studied separately using the elementary sources as the building blocks. The elementary sources studied here include: (1) planar current distributions which are of infinite extent in space; (2) a current element of length l much less than the wavelength, localized at the origin, which may also be referred to as a charge dipole element in reference to the two charges $\pm Q(t)$ accumulated at its two ends (poles); (3) an infinite-length line current; (4) a current loop of radius a much less than the wavelength; (5) a small current or dipole element placed on a dielectric substrate; and (6) a dynamic charge Q, with a general accelerating motion. The planar current distribution (1) of infinite extent is not practical, but its simple fields are useful for simple conceptual understanding. Ideal planar sources (1) of infinite extent, with general spatially-sinusoidal current distribution, can also be used to model waves generated by any practical finite-sized current distribution, through superposition or Fourier integration. The waves generated by the small current or dipole element (2) can also be used to model waves generated by any arbitrary current distribution, by discretizing the distribution into such small elements, and then spatially superposing or integrating the contributions from the discretized small elements. In this regard, the sources (1) or (2) are alternate forms of basic source elements, using which any other source configurations can be synthesized.

The other elementary sources (3-6), although could be mathematically modeled building upon the basic sources (1) or (2), their fields exhibit certain fundamental characteristics distinct from those of (1) and (2). The line current (3) produces cylindrical waves, in contrast to the plane waves generated by the source (1) or spherical waves generated by the source (2), and therefore maybe considered as an elementary building block of cylindrical waves. The small-loop current (4) on the xy plane produces Transverse-Electric to z, or TE_z, waves having z-directed electric field $E_z = 0$ but z-directed magnetic field $H_z \neq 0$, in contrast with the Transverse-Magnetic to z, or TM_z, waves having the $H_z = 0$ but the $E_z \neq 0$, produced by a z-directed dipole element (2). In other words, the loop on the xy plane is an elementary source of the TE_z radiation, and in contrast the z-directed dipole is an elementary source of the TM_z radiation. Accordingly, the loop element (4) on the xy plane maybe considered a "dual source" as compared to the z-directed dipole element (2), having their respective fields orthogonal to each other. The source (5) is made of the same small-dipole element of (2), but it operates in the presence of a planar dielectric substrate. Modeling of the source (5) would be similar to that of the source (2), except reflections from the substrate structure need to be added. Accordingly, the two sources (2) and (5) are physically and theoretically not that different. However, the presence of the dielectric substrate in (5) introduces a new wave mechanism, where a part of the signal radiates in the form of a "surface wave." The physical concept and mathematical analysis of the surface wave associated with the source (5) are distinct from those of the free-space waves radiated from all the other basic sources studied here, thus justifying the study of the source (5) as a special, elementary source. The source (6) is a moving charge under a general dynamic condition in a free-space medium. The study of the source (6) introduces fundamental dynamics of a moving charge, which is the most basic element of all electromagnetic fields and radiation. A charge moving with a uniform velocity produces an electric field which is different from that of a static charge, with an added magnetic field produced due to the relativistic effects of its motion. Like a static charge, the charge moving with a uniform velocity does not experience a net force, referred to as the "self-force," applied on itself, and does not radiate any power. However, radiation would result when the charge would accelerate. In this case, there would be a non-zero net self-force with different contributions due to a first (first time-derivative of velocity) or a second (second time-derivative of velocity) order acceleration.

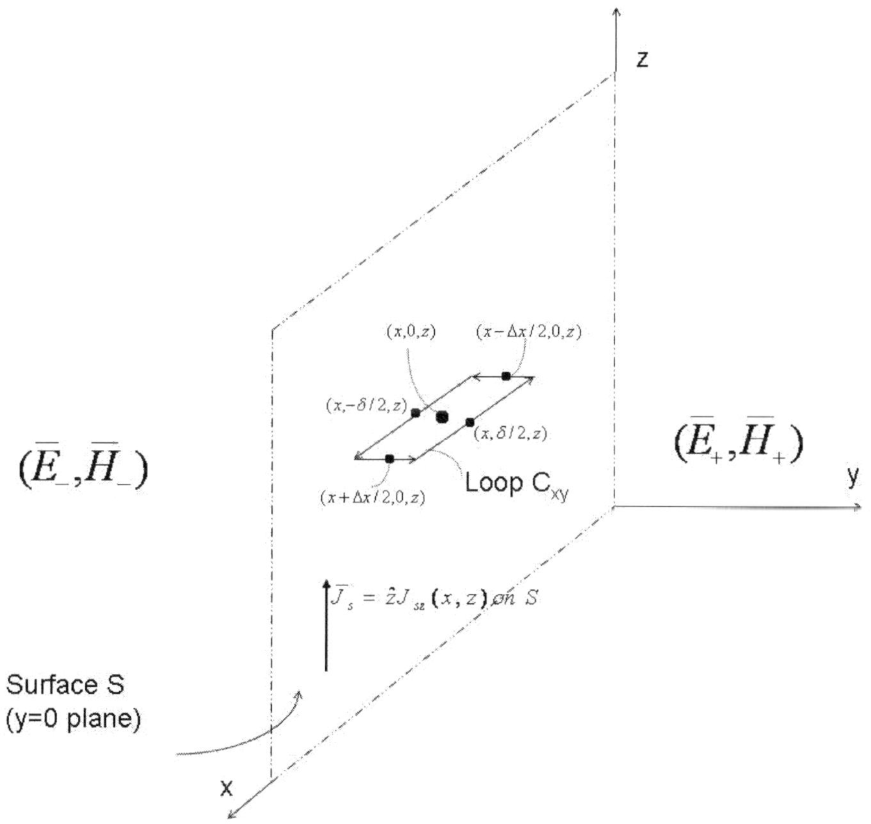

FIGURE 1. Current sheet source $\overline{J}_s = \hat{z}J_{sz}(x,z)$ A/m placed on a surface at $y=0$. A loop C_{xy} parallel to the xy plane is considered around the source at a general location $(x,0,z)$, for applying Ampere's and Faraday's laws, in order to establish boundary conditions across the source $\overline{J}_s$. A similar loop C_{yz} may also be considered parallel to the yz plane.

1. Boundary Condition Across a Surface Current

Consider a $\hat{z}$ directed surface current $\overline{J}_s = \hat{z}J_{sz}(x,z)$ placed on a plane surface $y=0$. The surrounding media on the two sides of the source are general in nature, and could be different from each other. Consider a small loop C_{xy} placed across the source surface S, as shown in Fig.1. Apply Ampere's law to the loop C_{xy}, with $\delta \simeq 0$.

$$\sum_{C_{xy}} H_{ti}\triangle l_i = H_{x-}(x,0,z)\triangle x - H_{x+}(x,0,z)\triangle x = J_{sz}(x,z)\triangle x. \tag{6.1}$$

In the above derivation, the contributions from the two sides of C_{xy} with length δ are zero. This is because the path lengths δ are assumed to be zero, thus placing the other two sides of C_{xy} just next to the current sheet on its opposite sides. Further, the area of the loop $\Delta x\delta$ is also infinitesimally small because δ is zero. Therefore, any additional current (displacement or conduction) through C_{xy} contributed from the surrounding media would be zero. Accordingly, the total current through C_{xy} is produced only due to the source current J_{sz}. Now, the equation (6.1) would be valid for any arbitrary differential element Δx, which may be cancelled from two sides of the equation.

$$H_{x-}(x,0,z) - H_{x+}(x,0,z) = J_{sz}(x,z). \tag{6.2}$$

Similarly, applying Faraday's law to the loop C_{xy} in Fig.1 we would have a condition for the electric field.

$$E_{x-}(x,0,z) - E_{x+}(x,0,z) = 0. \tag{6.3}$$

Notice that, in contrast with (6.2), the right side of the equation (6.3) is zero. This is because, unlike Ampere's law, Faraday's law does not have a source contribution. The induced voltage along the loop C_{xy} is contributed only by the total magnetic flux through C_{xy}, which is zero because δ is zero (same situation as in the derivation of (6.2).) Hypothetically, if there was a "magnetic current" in nature, and a magnetic sheet current was placed on the surface S, then the right side of (6.1) would have been non-zero. However, because there is no "magnetic charge" and consequently no "magnetic current" in nature, any source contribution in the application of of Faraday's law is zero for all physical situations.

One may consider another loop C_{yz} on the yz plane in Fig.1, with y dimension $\delta \simeq 0$, and z dimension Δz. The current $\overline{J}_s$ flows parallel to this loop, and therefore the current component normal to the loop is zero. Applying Ampere's and Faraday's laws, using similar steps as applied to the loop C_{xy}, we would get the following relations.

$$H_{z-}(x,0,z) - H_{z+}(x,0,z) = 0, \tag{6.4}$$

$$E_{z-}(x,0,z) - E_{z+}(x,0,z) = 0. \tag{6.5}$$

The conditions (6.2-6.5) are referred to as the necessary boundary conditions across a z-directed current source sheet J_{sz}. These conditions establish relationships with only the x and z components of the electric and magnetic fields, which are seen to be directed tangential to the surface S on which the current $\overline{J}_s$ flows. These field components may be referred to simply as tangential components, with a subscript t, and the necessary boundary conditions may often be referred to as tangential field boundary conditions. The conditions (6.3) and (6.5) simply state that both components of the electric field tangential to the source surface S are continuous across the source.

$$\overline{E}_{t-} - \overline{E}_{t+} = 0, \ \overline{E}_{t-} = \overline{E}_{t+}. \tag{6.6}$$

On the other hand, according to the conditions (6.2) and (6.4), only the tangential component of $\overline{H}$ which is directed along the current direction (z direction) is continuous across the boundary, whereas the other component (x component) is discontinuous. The discontinuity condition (6.2) and the continuity condition (6.4) maybe together expressed in a general vector form.

$$\hat{y} \times (\overline{H}_{t+} - \overline{H}_{t-}) = \hat{y} \times (\overline{H}_{+} - \overline{H}_{-}) = \overline{J}_s, \tag{6.7}$$

where the field components are to be computed close to the source surface. The general vector condition (6.7) would also work if $\overline{J}_s$ is directed along the x direction, instead of the z direction assumed in the above derivation, and consequently would work for any direction of current flow on the surface S. Further, in the above derivation the source surface was arbitrarily assumed to be on the xz plane, with normal along $\hat{y}$. The condition (6.7) maybe modified for a more general situation where the surface S is arbitrarily oriented with $\hat{n}_+$ as the normal unit vector towards the side referred to with a subscript +.

$$\hat{n}_+ \times (\overline{H}_{+} - \overline{H}_{-}) = \overline{J}_s. \tag{6.8}$$

Further more, although the above relations were explicitly deduced considering a planar source sheet on S for simplicity, the basic concept and consequently the final boundary conditions in the general forms (6.6) and (6.8) would apply to an arbitrary curved source surface, with arbitrary media on its two sides. The equations (6.6) and (6.8) are referred to as the general source boundary conditions. These boundary conditions are used as the theoretical basis for relating the amplitude of a source current distribution to that of the external electromagnetic field.

2. Current Source for Radiation of a Plane Wave

2.1. Propagation Perpendicular to the Source Plane. As shown in Fig.2, $\overline{J}_s$ is a current sheet (A/m) placed at $z = 0$ plane ($-\infty < x, y < +\infty$). It produces a plane wave ($\overline{E}_+, \overline{H}_+$) for $z > 0$, and ($\overline{E}_-, \overline{H}_-$) for $z < 0$, as per the relations (5.20,5.21).

$$\begin{aligned}
\overline{E}_+ &= \hat{x} A e^{-jkz}, \\
\overline{H}_+ &= \hat{y} \frac{A}{\eta} e^{-jkz}, \\
\overline{E}_- &= \hat{x} B e^{+jkz}, \\
\overline{H}_- &= -\hat{y} \frac{B}{\eta} e^{+jkz}.
\end{aligned} \tag{6.9}$$

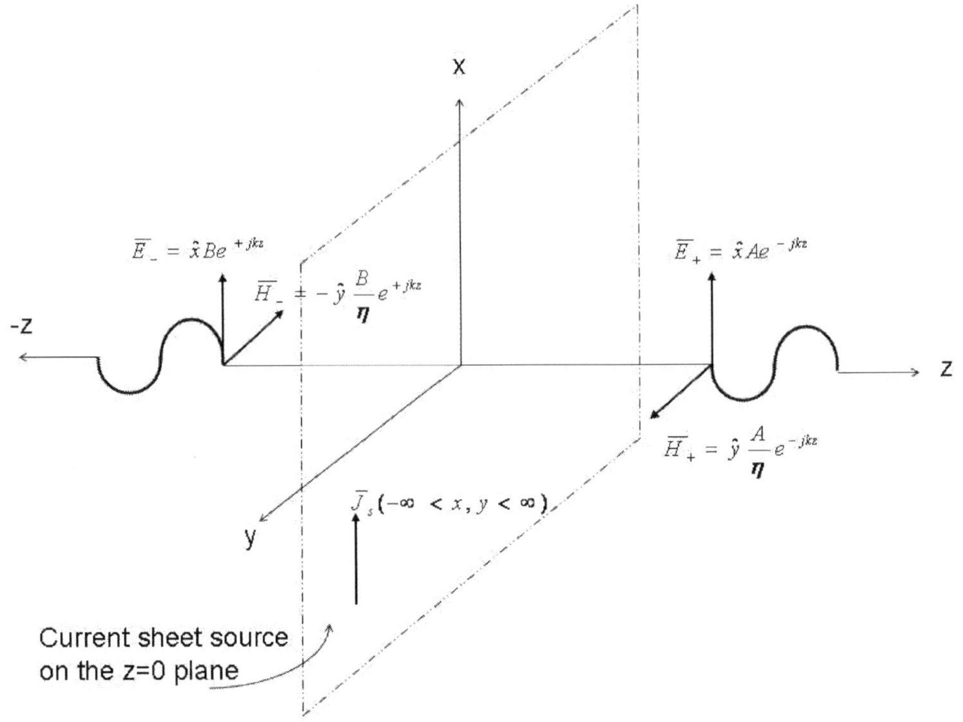

FIGURE 2. Current sheet source $\overline{J}_s$ placed at z = 0 to produce plane waves radiating along the $\pm z$ directions.

We want to find $\overline{J}_s$ to produce the above radiation. Faraday's law needs to be satisfied across the source. This can be achieved by applying the general boundary condition (6.6) for the electric field, developed in the last section. Accordingly, the electric field component which is tangential to the source plane should be continuous on both sides of the source.

$$E_{x+}(z=0) = E_{x-}(z=0), \; A = B. \tag{6.10}$$

Equation (6.10) means that the amplitudes of $\overline{E}_+$ and $\overline{E}_-$ are the same, as should be expected from symmetry.

Similarly, Ampere's law needs to be applied across the source. This is achieved by applying the general boundary condition (6.8) for the magnetic field, developed in the last section. Accordingly, the magnetic field component which is tangential to the source plane would be discontinuous, with the difference related to the source current.

$$\hat{n}_+ \times [\overline{H}_+(z=0) - \overline{H}_-(z=0)] = \overline{J}_s, \; \hat{n}_+ = \hat{z} \; ,$$
$$-\hat{x}(\frac{A}{\eta} + \frac{B}{\eta}) = \overline{J}_s \; . \tag{6.11}$$

Combining (6.10) and (6.11) we get

$$\overline{J}_s = \hat{x} J_{sx} = -\frac{2A}{\eta}\hat{x} = -\frac{2B}{\eta}\hat{x}, \; A = B = -\frac{\eta}{2} J_{sx}. \tag{6.12}$$

2.2. Power Flow. Consider a section $(\Delta x \times \Delta y)$ of the source sheet derived above. The voltage across this section of source J_{sx} is equal to $-E_x \Delta x = -A\Delta x = +\Delta V_x$, where the source voltage is defined such that it is increasing along the current direction. And, the total current flowing through this section is equal to $J_{sx}\Delta y = \Delta I_x$. This results in the following complex power ΔP_c delivered from the above source section, which is twice the powers delivered to each side(represented with subscripts + and −) . The average powers $\Delta P_\pm$ is the real part of the respective complex powers. We will assume RMS complex phasors for the voltage and current (appendix B).

$$\begin{aligned} \Delta P_c &= \Delta V_x \Delta I_x^* = -A J_{sx}^* \Delta x \Delta y \\ &= \frac{2|A|^2}{\eta}\Delta x \Delta y \; = 2\Delta P_{c+} = 2\Delta P_{c-}, \\ \Delta P = Re(\triangle P_c) &= Re(\Delta V_x \Delta I_x^*) = Re(-A J_{sx}^* \Delta x \Delta y) \\ &= \frac{2|A|^2}{\eta}\Delta x \Delta y \; = 2\Delta P_+ = 2\Delta P_-. \end{aligned} \tag{6.13}$$

The complex and average powers are same here, because the wave impedance η is a real number, when the electric and magnetic fields are in phase with each other. Now define a complex power density $P_{dc\pm}$ as the power per unit area of the source, delivered to a particular side represented by a subscript + or −. The average power density $P_{d\pm}$ is the real part of the corresponding complex power density.

$$\begin{aligned} P_{dc+} = P_{dc-} &= \frac{\Delta P_{c+}}{\Delta x \Delta y} = \frac{|A|^2}{\eta} \\ &= E_{x+} H_{y+}^* (z > 0); \\ &= E_{x-}(-H_{y-}^*)\,(z < 0), \\ P_{d+} = Re(P_{dc+}) &= P_{d-} = Re(P_{dc+}). \end{aligned} \tag{6.14}$$

Above we considered uniform plane waves radiating along $\pm z$ directions, produced by a current sheet on the xy plane. Similarly, radiation along any other direction (x or y) can be produced by a current sheet on a plane orthogonal to it (yz or xz planes). The same derivation discussed above may be repeated, except with different co-ordinates, or the results of (6.9-6.14) be modified by proper rotation of co-ordinates.

2.3. Poynting Vector. It would be useful to express the power flow in a general form, applicable independent of any specific choice of coordinates. With this objective in mind, a general complex Poynting vector $\overline{S}$ is defined.

$$\overline{S} \triangleq \overline{E} \times \overline{H}^*. \tag{6.15}$$

The direction of $\overline{S}$ as defined by a unit vector $\hat{s} = \overline{S}/|\overline{S}|$ is the direction of the wave propagation. Using the field expressions of (6.10) in the present case, this would be along $+z$ for forward and $-z$ for backward propagation, as expected. The complex power density P_{dc}, which is the complex power flowing through an unit area, with surface normal along $\hat{n}$, is defined using the Poynting vector as follows.

$$P_{dc} = \overline{S} \cdot \hat{n}. \tag{6.16}$$

In the present case, the complex power density flowing normal to the source ($\hat{n} = \hat{z}$) is $(\overline{E} \times \overline{H}^*) \cdot \hat{z}$. This can be seen to be equal to $E_x H_y^*$ for $(z > 0)$, and $-E_x H_y^*$ for $(z < 0)$, using the field expressions of (6.9), which is consistent with the specific results in (6.14).

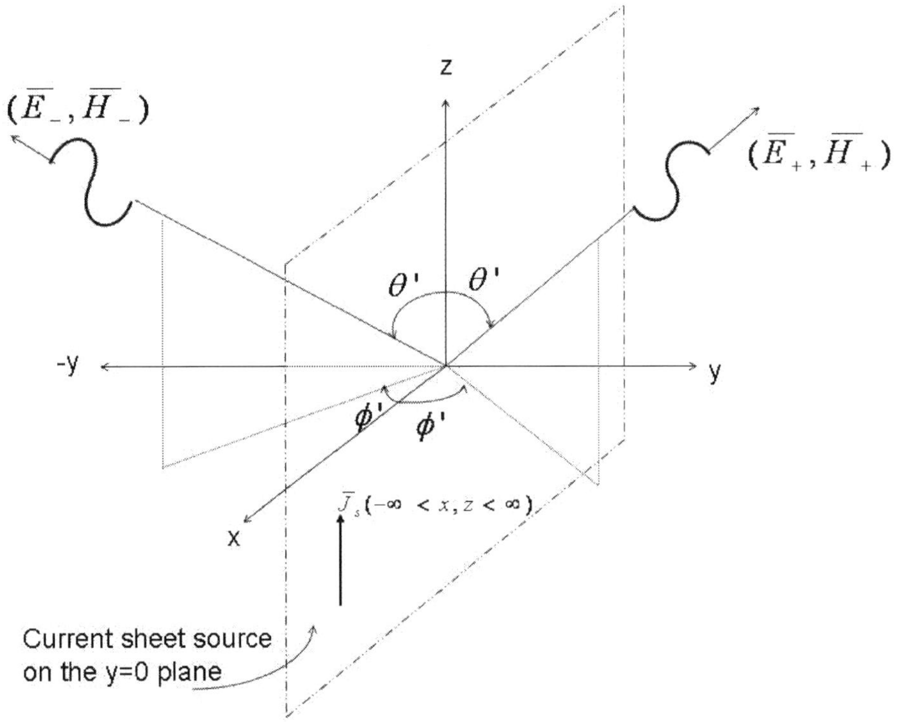

FIGURE 3. Current sheet source $\overline{J}_s$ placed at the $y = 0$ plane to produce plane waves radiating along the (θ', ϕ') and $(\theta', -\phi')$ directions.

2.4. Propagation at an Arbitrary Angle with the Source Plane. The source analysis in the last section can be extended to other general situations when the wave propagates at arbitrary angle with respect to the source plane. Consider, for example, a current source on the xz plane which generates a general plane wave $(\overline{E}_+, \overline{H}_+)$ propagating in a certain direction for $y > 0$, and a similar wave $(\overline{E}_-, \overline{H}_-)$ for $y < 0$ with proper symmetry in their field and propagation directions. This is shown in Fig.3. Because the wave is propagating in an arbitrary direction with respect to the reference coordinates, it may be transverse magnetic TM_z or transverse electric TE_z type, as discussed in relation to general plane wave propagation. Consider TM_z waves for now. Similar procedure can be followed for TE_z waves. We will express all field components of the wave in terms of a vector magnetic potential function ψ_{zm+} with a coefficient A for $y > 0$, and ψ_{zm-} with a coefficient B for $y < 0$. Consider first the region $y > 0$.

$$\psi_{zm+} = Ae^{-jk_x x}e^{-jk_z z}e^{-jk_y y}. \tag{6.17}$$

Various field components can be completely expressed using potential-field relations of (5.34).

$$\begin{aligned}
E_{x+} &= -\frac{\psi_{zm+}k_z k_x}{j\omega\epsilon} = -\frac{Ak_z k_x}{j\omega\epsilon}e^{-jk_x x}e^{-jk_z z}e^{-jk_y y},\\
E_{y+} &= -\frac{\psi_{zm+}k_z k_y}{j\omega\epsilon} = -\frac{Ak_z k_y}{j\omega\epsilon}e^{-jk_x x}e^{-jk_z z}e^{-jk_y y},\\
E_{z+} &= \frac{\psi_{zm+}k_\rho^2}{j\omega\epsilon} = \frac{Ak_\rho^2}{j\omega\epsilon}e^{-jk_x x}e^{-jk_z z}e^{-jk_y y},\\
H_{x+} &= -jk_y\psi_{zm+} = -jk_y Ae^{-jk_x x}e^{-jk_z z}e^{-jk_y y},\\
H_{y+} &= jk_x\psi_{zm+} = jk_x Ae^{-jk_x x}e^{-jk_z z}e^{-jk_y y},\\
H_{z+} &= 0.
\end{aligned} \tag{6.18}$$

It may be noted that the waves are propagating along real $(\theta', \pm\phi')$ directions, or are evanescent, depending on the values of k_x and k_z, given the constraint $k_x^2 + k_y^2 + k_z^2 = k_\rho^2 + k_z^2 = k^2$. Similar expressions for $y < 0$ can be written using vector potential function ψ_{zm-} with a coefficient B. The potential-field relations (5.34) are used, with k_y switched to -k_y. This is because the y-propagation will now be $e^{+jk_y y}$ instead of $e^{-jk_y y}$ for $y > 0$.

$$\psi_{zm-} = Be^{-jk_x x}e^{-jk_z z}e^{+jk_y y}. \tag{6.19}$$

$$\begin{aligned}
E_{x-} &= -\frac{\psi_{zm-}k_z k_x}{j\omega\epsilon} = -\frac{Bk_z k_x}{j\omega\epsilon}e^{-jk_x x}e^{-jk_z z}e^{+jk_y y},\\
E_{y-} &= \frac{\psi_{zm-}k_z k_y}{j\omega\epsilon} = \frac{Bk_z k_y}{j\omega\epsilon}e^{-jk_x x}e^{-jk_z z}e^{+jk_y y},\\
E_{z-} &= \frac{\psi_{zm-}k_\rho^2}{j\omega\epsilon} = \frac{Bk_\rho^2}{j\omega\epsilon}e^{-jk_x x}e^{-jk_z z}e^{+jk_y y},\\
H_{x-} &= jk_y\psi_{zm-} = jk_y Be^{-jk_x x}e^{-jk_z z}e^{+jk_y y},\\
H_{y-} &= jk_x\psi_{zm-} = jk_x Be^{-jk_x x}e^{-jk_z z}e^{+jk_y y},\\
H_{z+} &= 0.
\end{aligned} \tag{6.20}$$

We need to find $\overline{J}_s$ to produce the above TM_z radiation. Faraday's and Ampere's laws will have to be satisfied across the source plane by applying necessary boundary conditions. First apply boundary condition (6.6) for the electric field component tangential to the source plane at $y = 0$.

$$\begin{aligned} E_{x+}(y=0_+) &= E_{x-}(y=0_-), \\ E_{z+}(y=0_+) &= E_{z-}(y=0_-), \\ A &= B. \end{aligned} \tag{6.21}$$

The above result represents expected symmetry of the wave amplitudes on opposite sides of the source. Similarly apply the boundary condition (6.8) for the magnetic field at $y = 0$.

$$\begin{aligned} \hat{n}_+ \times (\overline{H}_+(y=0_+) - \overline{H}_-(y=0_-) = \overline{J}_s = \hat{x}J_{sx} + \hat{z}J_{sz},\ \hat{n}_+ = \hat{y}\ , \\ H_{z+}(y=0_+) - H_{z-}(y=0_-) = J_{sx} = 0\ , \\ -H_{x+}(y=0_+) + H_{x-}(y=0_-) = J_{sz} = jk_y(A+B)e^{-jk_xx}e^{-jk_zz}\ . \end{aligned} \tag{6.22}$$

By solving the conditions from (6.21,6.22) we can relate the potential function to the current amplitude.

$$\begin{aligned} \overline{J}_s = \hat{z}J_{sz} = \hat{z}J_{szo}e^{-jk_xx}e^{-jk_zz},\ J_{szo} = 2jAk_y,\ A = B = \frac{J_{szo}}{2jk_y}, \\ \psi_{zm} = \psi_{zm\pm} = Ae^{-jk_xx}e^{-jk_zz}e^{\mp jk_yy} = \frac{J_{szo}}{2jk_y}e^{-jk_xx}e^{-jk_zz}e^{\mp jk_yy}. \end{aligned} \tag{6.23}$$

Once the potential function ψ_{zm} is determined, any component of the electric or magnetic field can be obtained from (6.18,6.20). In fact, the field components in any required coordinate system can be derived from the potential using suitable differential relations (chapter 5, section 3), which are independent of the particular values of k_x, k_y or k_z. Accordingly, now that the potential is known, the problem is understood to be completely solved.

2.4.1. *TE_z Radiation.* The above process maybe similarly repeated for a TE_z radiation, using potential-field relations of chapter 5, section 3, for the TE_z wave. The final expressions for the TE_z potential functions $\psi_{ze\pm}$ in the sides $y \gtrless 0$, and the required surface current distribution $\overline{J}_s$ on the $y = 0$ plane, are provided here.

$$\begin{aligned} \overline{J}_s = \hat{x}J_{sx} + \hat{z}J_{sz} = (\hat{x}\frac{k_\rho^2}{k_xk_z}J_{sz0} + \hat{z}J_{szo})e^{-jk_xx}e^{-jk_zz}, \\ \psi_{ze} = \psi_{ze\pm} = \pm\frac{j\omega\mu J_{szo}}{2k_xk_z}e^{-jk_xx}e^{-jk_zz}e^{\mp jk_yy}. \end{aligned} \tag{6.24}$$

Unlike the TM_z radiation for which the required current source we derived earlier is directed along the z-direction ($J_{sx} = 0, J_{sz} \neq 0$), for the TE_z radiation both the $x-$ and $z-$ components of current source would be required.

2.4.2. *Radiation from a Current Source on the xy plane.* For the above derivations, the current source was placed on the xz plane at $y = 0$, radiating into the two sides $y \gtrless 0$ of the source. It may also be useful to follow similar derivations for current sources placed on the xy plane at $z = 0$, required to radiate TM_z or TE_z waves into the two sides $z \gtrless 0$ of the source.

The final expressions for the TM_z potential functions $\psi_{zm\pm}$ in the sides $z \gtrless 0$, and the required surface current distribution $\overline{J}_s$ on the $z = 0$ plane, are provided here.

$$\overline{J}_s = \hat{x}J_{sx} + \hat{y}J_{sy} = (\hat{x}\frac{k_x}{k_\rho} + \hat{y}\frac{k_y}{k_\rho})J_{s0}e^{-jk_xx}e^{-jk_yy},$$
$$\psi_{zm} = \psi_{zm\pm} = \pm\frac{jJ_{so}}{2k_\rho}e^{-jk_xx}e^{-jk_yy}e^{\mp jk_zz}. \quad (6.25)$$

Similarly, the final expressions for the TE_z potential functions $\psi_{ze\pm}$ in the sides $z \gtrless 0$, and the required surface current distribution $\overline{J}_s$ on the $z = 0$ plane, are provided here.

$$\overline{J}_s = \hat{x}J_{sx} + \hat{y}J_{sy} = (\hat{x}\frac{k_y}{k_\rho} - \hat{y}\frac{k_x}{k_\rho})J_{s0}e^{-jk_xx}e^{-jk_yy},$$
$$\psi_{ze} = \psi_{ze\pm} = \frac{j\omega\mu J_{so}}{2k_zk_\rho}e^{-jk_xx}e^{-jk_yy}e^{\mp jk_zz}. \quad (6.26)$$

Note that, in order to produce either the TM_z or the TE_z radiations, both the $x-$ and $y-$ components of the current source are required to be placed on the xy plane at $z = 0$. This is unlike the TM_z radiation produced by a current source placed on the xz plane at $y = 0$, in which case only one current component ($J_{sx} = 0, J_{sz} \neq 0$) was needed.

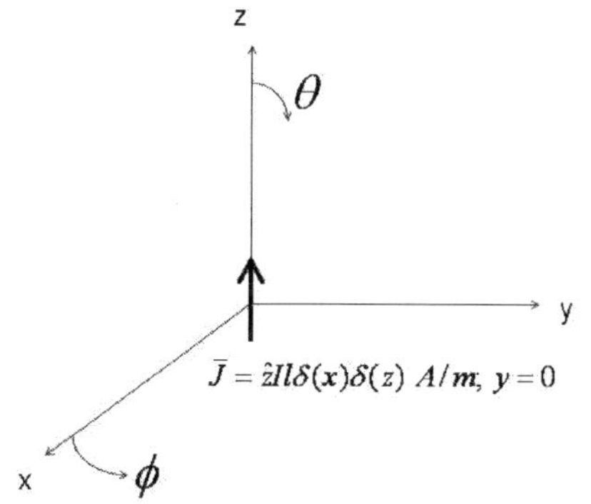

FIGURE 4. A current element of amplitude I and length $l << \lambda$, radiating into the surrounding medium.

3. Radiation from an Infinitesimal Current Source:

So far, the sources of radiation we have considered are of infinite size (infinite current sheets). Such sources, although are useful for theoretical understanding and conceptual development, practical sources are to be of finite physical size. In the following we will analyze radiation from a basic current element of small transverse and longitudinal dimensions.

Consider a current element of length $l << \lambda$, with a z-directed current I, shown in Fig.4. It may be expressed as a surface current $\overline{J}_s$ on the xz plane in the following form.

$$\overline{J}_s = \hat{z} I l \delta(x)\delta(z) = \hat{z} J_{sz}, \\ \iint J_{sz} dx dz = Il. \tag{6.27}$$

Define a two-dimensional Fourier transform $\Im_{xz}$ of this source on the xz plane.

$$\Im_{xz}\{\overline{J}_s\} = \tilde{\tilde{J}}_s = \int_{z=-\infty}^{\infty} \int_{x=-\infty}^{\infty} \overline{J}_s e^{-jk_x x} e^{-jk_z z} dx dz = \hat{z} Il. \tag{6.28}$$

Using inverse transform, $\overline{J}_s$ can be expressed as integration or superposition of current sheets of the form $Il\hat{z}e^{jk_x x}e^{jk_z z}$ over all values of k_x and k_z.

$$\overline{J}_s = \Im_{xz}^{-1}\{\tilde{\tilde{J}}_s\} = \frac{1}{4\pi^2} \iint_{-\infty}^{\infty} \hat{z} I l e^{jk_x x} e^{jk_z z} dk_x dk_z. \tag{6.29}$$

Accordingly, all potential or field components produced by this source can be expressed as inverse Fourier integration over k_x, k_z of the respective components produced due to a source current sheet of the form $\overline{J}_s = \hat{z} J_{sz} = \hat{z} I l e^{jk_x x} e^{jk_z z}$. Because the current sheet is z directed, the resulting fields would be of TM_z type, and its potential function ψ_{zm} can be expressed using (6.23) with $J_{sz0} = Il$. The total ψ_{zm} can then be determined by integrating the potential function for a current sheet with a given k_x and k_z over all possible values of $-\infty < (k_x, k_z) < +\infty$.

$$\psi_{zm} = \frac{1}{4\pi^2} \iint_{-\infty}^{\infty} \frac{Il}{2jk_y} e^{+jk_x x} e^{+jk_z z} e^{\mp jk_y y} dk_x dk_z. \tag{6.30}$$

Notice that we have modified the signs of k_x amd k_z of a current sheet in the present treatment, switched from that in the derivations of the last section. This is to be compatible with a standard inverse integral expression of (6.29). The integral in (6.30) can be reduced using standard Fourier transform techniques, as shown in the Appendix D, resulting in a basic wave function for the potential ψ_{zm}.

$$\psi_{zm} = \frac{Il}{4\pi r} e^{-jkr}. \tag{6.31}$$

The infinitesimal source considered here has spherical symmetry, and therefore is expected to produce a spherical wave. Accordingly it would be useful to express the spherical components of the resulting total fields using differential operations of (5.47-5.52) on (6.31) with respect to spherical variables. Significant simplification in the differential operations of (5.47-5.52) would be possible because the potential function in (6.31) is independent of θ and ϕ.

$$\begin{aligned} E_r &= \frac{\eta}{jk}(k^2\cos\theta + \frac{\partial}{\partial r}\cos\theta\frac{\partial}{\partial r})\psi_{zm} = \frac{\eta}{jk}(k^2 + \frac{\partial^2}{\partial r^2})\psi_{zm}\cos\theta \\ &= \frac{Il\eta}{2\pi jk}e^{-jkr}(\frac{1}{r^3} + \frac{jk}{r^2})\cos\theta, \\ E_\theta &= \frac{\eta}{jk}(-k^2\sin\theta + \frac{1}{r}\frac{\partial}{\partial\theta}\cos\theta\frac{\partial}{\partial r})\psi_{zm} = -\frac{\eta}{jk}(k^2 + \frac{1}{r}\frac{\partial}{\partial r})\psi_{zm}\sin\theta \\ &= +\frac{Il\eta}{jk4\pi}e^{-jkr}(-\frac{k^2}{r} + \frac{jk}{r^2} + \frac{1}{r^3})\sin\theta, \\ E_\phi &= 0, \\ H_r &= 0, \\ H_\theta &= 0, \\ H_\phi &= -\frac{\partial\psi_{zm}}{\partial r}\sin\theta = \frac{Il}{4\pi}e^{-jkr}(\frac{jk}{r} + \frac{1}{r^2})\sin\theta. \end{aligned} \quad (6.32)$$

3.1. Far-Field Radiation,($r \to \infty$). Equation (6.32) provides rigorous expressions for all field components, valid for all distance r from the source. Fields at large distances ($r >> \lambda$) are of particular interest for communication with receivers placed far away from the source. As $r \to \infty$, the terms in (6.32) with $\frac{1}{r}$ dependence will only contribute significantly compared to $\frac{1}{r^2}$ and $\frac{1}{r^3}$ terms. Accordingly, for $r >> \lambda$, the results in (6.32) can be simplified in the following simple forms.

$$\begin{aligned} E_r = 0 &= E_\phi = H_r = H_\theta, \\ E_\theta &= j\frac{Il\eta k}{4\pi r}e^{-jkr}\sin\theta, \\ H_\phi &= j\frac{Ilk}{4\pi r}e^{-jkr}\sin\theta. \end{aligned} \quad (6.33)$$

Notice that $H_\phi = \frac{E_\theta}{\eta}$ in the far field. This is because the spherical wave behaves approximately like a uniform plane wave in the far-field region. Also, the radial field components along the propagation direction $E_r = H_r = 0$ in the far field, which is also consistent with the uniform plane wave approximation.

3.2. Low Frequency Approximation. As $\omega \to 0$, $k \to 0$, the results in (6.32) can be approximated in the following useful forms.

$$
\begin{aligned}
E_r &= \frac{Il\eta\cos\theta}{jk2\pi r^3}, \\
E_\theta &= +\frac{Il\eta\sin\theta}{jk4\pi r^3}, \\
H_\phi &= \frac{Il}{4\pi r^2}\sin\theta, \\
E_\phi &= H_r = H_\theta = 0.
\end{aligned}
\tag{6.34}
$$

Observe that the magnetic fields in the above equation are the same as those from Biot-Savarts's law of magneto-statistics (2.71), but have been independently derived here from Ampere's and Faraday's laws, as static ($\omega \to 0$) limit from a general time-invariant condition. Further, the electric fields in (6.34) can be shown to be the same as those produced due to electro-static charges $\pm Q$ accumulated at the two ends (poles) of the current element, separated by a distance l, with $\frac{\partial Q}{\partial t} \equiv j\omega Q = I$ (see (2.69)). In reference to the equivalent di-polar charges, a current element may also be called a dipole element. When the static current changes into a general time-variant current, the corresponding dipole charges turn into general time-variant dipoles, and any resulting radiation from the time-variant current element is referred to as dipole radiation.

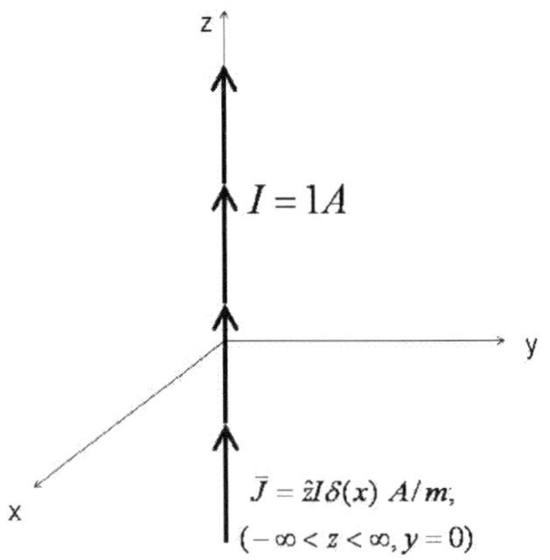

FIGURE 5. An infinite-length current I along the $\hat{z}$ direction, radiating a cylindrical wave into the surrounding medium.

4. Infinite-Length Line Current Source

Fig.5 shows an infinite-length line current source, oriented along the z-axis. The field produced by the line current maybe modeled through superposition or Fourier integration of planar current sources presented in section 2. Alternately, it can also be modeled as a superposition of infinitesimal current elements presented in section 3, through descretization of the line current and spatial integration over the entire length $(-\infty < z < +\infty)$ of the line current. We will now use the first approach. The line current of Fig.5 may be expressed as a surface current distribution on the $y = 0$ plane as follows. The current I is assumed to be one unit.

$$\bar{J}(x,z) = \hat{z}\ \delta(x),\ -\infty < z < +\infty, \tag{6.35}$$

$$\tilde{\tilde{J}}\ (k_x, z) = \int_{-\infty}^{\infty} \bar{J}(x,z)e^{-jk_x x}dx = \hat{z}, \tag{6.36}$$

$$\bar{J}(x,z) = \frac{1}{2\pi}\int_{\infty}^{\infty} \tilde{\tilde{J}}\ (k_x, z)e^{+jk_x x}dk_x$$

$$= \hat{z}\frac{1}{2\pi}\int_{-\infty}^{+\infty} e^{+jk_x x}dk_x. \tag{6.37}$$

Consider a current source of the following form, placed at $y = 0$:

$$\bar{J}(x,z) = \hat{z}e^{+jk_x x}, -\infty < z < +\infty. \tag{6.38}$$

The TM_z potential ψ_{zm} due to the above current source can be obtained from (6.23) by substituting $k_z = 0$, and replacing k_x *by* $-k_x$.

$$\psi_{zm} = \frac{1}{2jk_y}e^{+jk_x x}e^{\mp jk_y y},$$

$$k_y = \sqrt{k^2 - k_x^2},\ Re(k_y) > 0, Im(k_y) < 0. \tag{6.39}$$

The total current in (6.35,6.37) is the Fourier integration of the current distribution in (6.38). Accordingly, the TM_z potential for the current in (6.35,6.37) can be obtained by Fourier integration of (6.39) over all k_x,

$$\psi_{zm} = \frac{1}{2\pi}\int_{-\infty}^{+\infty} \frac{1}{2j\sqrt{k^2 - k_x^2}}e^{+jk_x x}e^{\mp jy\sqrt{k^2-k_x^2}}dk_x. \tag{6.40}$$

Because the line source of (6.35) has a uniform cylindrical symmetry with respect to azimuth angle ϕ, we expect that the resulting ψ_{zm} will be independent of ϕ. So it may be sufficient to evaluate (6.40) for $\phi = 0$ plane, when $y = 0$ and $\rho = x$.

$$\psi_{zm}(\rho) = \psi_{zm}(x = \rho, y = 0) = \frac{1}{2\pi}\int_{-\infty}^{+\infty} \frac{1}{2j\sqrt{k^2 - k_x^2}}e^{jk_x\rho}dk_x. \tag{6.41}$$

We will evaluate (6.41) by integrating over two separate regions of k_x,

$$\psi_{zm_1}(\rho) = \frac{1}{2\pi}\int_{-k}^{+k} \frac{1}{2j\sqrt{k^2 - k_x^2}} e^{jk_x\rho} dk_x; \; |k_x| \leq k,$$

$$\psi_{zm_2}(\rho) = \frac{1}{2\pi}\left(\int_{-\infty}^{-k} + \int_{k}^{\infty}\right)\frac{1}{2\sqrt{k_x^2 - k^2}} e^{+jk_x\rho} dk_x; \; k < |k_x| < \infty,$$

$$\psi_{zm}(\rho) = \psi_{zm_1}(\rho) + \psi_{zm_2}(\rho). \tag{6.42}$$

ψ_{zm1} may be evaluated by substituting $k_x = k\cos t$.

$$\psi_{zm1} = \frac{1}{2\pi}\int_0^{\pi} \frac{1}{2j} e^{jk\rho\cos t} dt = \frac{1}{4j} J_0(k\rho), \tag{6.43}$$

where $J_0(\cdot)$ is the Bessel Function of the first kind of order zero. We use the following standard integral identity of $J_0(\cdot)$ for the result of (6.43).

$$J_0(s) = \frac{1}{\pi}\int_0^{\pi} e^{js\cos t} dt = \frac{1}{\pi}\int_0^{\pi} \cos(s\cos t) dt. \tag{6.44}$$

ψ_{zm2} may be evaluated by substituting $k_x = k\cosh t$.

$$\psi_{zm2}(\rho) = \frac{1}{2\pi}\int_{+k}^{+\infty} \frac{1}{\sqrt{k_x^2 - k^2}} \cos(k_x\rho) dk_x = \frac{1}{2\pi}\int_{t=0}^{+\infty} \frac{\cos(k\rho\cosh t)}{k\sinh t}(k\sinh t) dt = -\frac{1}{4} Y_0(k\rho), \tag{6.45}$$

where $Y_0(\cdot)$ is the Bessel Function of the second kind of order zero. We use the following integral identity of $Y_0(\cdot)$ for the result of (6.45)

$$Y_0(s) = \frac{-2}{\pi}\int_0^{\infty} \cos(s\cosh t) dt. \tag{6.46}$$

Now, by combining (6.42,6.43,6.45) we get the expression of the total potential function ψ_{zm} in terms of Bessel and Hankel Functions.

$$\psi_{zm}(\rho) = \frac{1}{4j} J_0(k\rho) - \frac{1}{4} Y_0(k\rho)$$

$$= \frac{1}{4j} H_0^{(2)}(k\rho), \tag{6.47}$$

$$H_0^{(2)}(s) \triangleq J_0(s) - jY_0(s). \tag{6.48}$$

$H_0^{(2)}(\cdot)$ is the Hankel Function of the second kind, of order zero, which is related to $J_0(\cdot)$ and $Y_0(\cdot)$ as defined in (6.48). Basic characteristics and useful relationships for Bessel and Hankel functions of integer orders are presented in some details in the Appendix E.

Once the potential function ψ_{zm} is known, various field components in the cylindrical coordinates can be derived from (6.48) using the field-potential relations (5.69-5.74). Because we have $\frac{\partial}{\partial z} = 0$ and $\frac{\partial}{\partial \phi} = 0$ in this case, many field components in (5.69-5.74) will be zero.

$$E_z = \frac{\omega\mu}{j}\psi_{zm} = -\frac{\omega\mu}{4} H_0^{(2)}(k\rho) = -\frac{k\eta}{4} H_0^{(2)}(k\rho),$$

$$H_\phi = -\frac{\partial}{\partial\rho}\psi_{zm} = -\frac{k}{4j} H_0^{\prime(2)}(k\rho),$$

$$E_\rho = E_\phi = H_\rho = H_z = 0. \tag{6.49}$$

4.1. Source Impedance. Evaluate E_z at $\rho = 0$, using (6.49,6.41).

$$E_z(\rho = 0) = \frac{\omega\mu}{j}\psi_{zm}(\rho = 0), \tag{6.50}$$

$$\psi_{zm}(\rho = 0) = \frac{1}{2\pi}\int_{-\infty}^{+\infty} \frac{1}{2j\sqrt{k^2 - k_x^2}} dk_x. \tag{6.51}$$

The voltage seen across the source over length Δl is $\Delta V = -E_z(\rho = 0)\Delta l$, using which the average power ΔP radiated from length Δl can be expressed.

$$\Delta P = \frac{1}{2}Re(\Delta V I^*) = \frac{-\Delta l}{2}Re(E_z(\rho=0)\times 1) = \frac{-\Delta l}{2}Re(E_z(\rho=0)),$$
$$\frac{\Delta P}{\Delta l} = \frac{-1}{2}Re(E_z(\rho=0)) = \frac{-1}{2}Re(\frac{\omega\mu}{j}\psi_{zm}(\rho=0))$$
$$= \frac{-1}{2}Re[-\frac{\omega\mu}{2\pi}\int_{-\infty}^{+\infty}\frac{1}{2\sqrt{k^2-k_x^2}}dk_x]$$
$$= \frac{\omega\mu}{8\pi}\int_{-k}^{+k}\frac{1}{\sqrt{k^2-k_x^2}}dk_x = \frac{\omega\mu}{8\pi}\int_{t=-\frac{\pi}{2}}^{t=+\frac{\pi}{2}}\frac{1}{k\cos t}k\cos t dt \; (k_x = k\sin t)$$
$$= \frac{\omega\mu}{8} = \frac{k\eta}{8}. \qquad (6.52)$$

Define R_l as the radiation resistance per unit length,

$$\Delta P = \frac{1}{2}|I|^2 R_l = \frac{1}{2}R_l; I = 1,$$
$$R_l = \frac{k\eta}{4}\frac{\Omega}{m}. \qquad (6.53)$$

Observe that as $\omega \to 0$, or equivalently $k \to 0$, we would have $R_l \to 0$. This means that the line source does not radiate any power as zero frequency, as one would expect. Also, R_l is independent of radius of wire. Let us now examine the reactive part of the field E_z.

$$Im[E_z(\rho=0)] = Im[\frac{\omega\mu}{j}\psi_{zm}(\rho=0)]$$
$$= Im[-\frac{\omega\mu}{2\pi}\int_{-\infty}^{+\infty}\frac{1}{2\sqrt{k^2-k_x^2}}dk_x$$
$$= -\frac{\omega\mu}{2\pi}\int_{+k}^{+\infty}\frac{1}{\sqrt{k_x^2-k^2}}dk_x. \qquad (6.54)$$

It may be seen that the integrand of (6.54) has a $1/k_x$ trend as $k_x \to \infty$. The integral of $(1/k_x)dk_x$ diverges to $+\infty$ as the upper limit of integration goes to $+\infty$. Therefore, the expression (6.54) would be $-\infty$.

$$Im[E_z(\rho=0)] = -\infty, \;\; Im(\frac{\Delta P}{\Delta l}) = Im[-E_z(\rho=0)] = +\infty = X_l, \qquad (6.55)$$

where X_l is defined as the reactance per unit length. $X_l = +\infty$, which means that the reactance would be infinitely inductive. This would be the case if the source is an ideal line source with zero radius. Practical sources use metal conductors with a non-zero radius a, with currents flowing on the surface of the thin cylindrical conductor. This situation may be treated by having the source current still placed at the center of the conductor, but computing the electric field on the surface of the conductor ($\rho = a$) to determine the reactance seen by the current source. Following this concept, the inductance per unit length X_l for a practical line current of radius $a << \lambda$ may be derived from (6.47-6.49) as follows.

$$
\begin{aligned}
X_l = Im[-E_z(\rho = a)] &= Im[-\frac{\omega\mu}{j}\psi_{zm}(\rho = a; a << \lambda)] \\
= Im[-\frac{\omega\mu}{j}\frac{1}{4j}H_0^{(2)}(ka)] &= Im[\frac{\omega\mu}{4}H_0^{(2)}(ka)] \simeq -\frac{\omega\mu}{4}Y_0(ka) \\
&\simeq -\frac{\omega\mu}{4}\frac{2}{\pi}\ln(0.89 \times ka) = -\frac{\omega\mu}{2\pi}\ln(0.89ka). \qquad (6.56)
\end{aligned}
$$

Because the radius $a << \lambda$, small-argument approximation for the Hankel Function is used in the above derivation (see Appendix E). Note that $X_l \sim -\frac{\omega\mu}{2\pi}\ln(0.89ka)$ approaches $+\infty$ as a approaches 0, which is consistent with the result (6.55). Also, note that X_l approaches 0 as w approaches 0. The limiting relation $\omega \ln(\omega) \sim 0$, as $\omega \to 0$ is used for this result. Physically this means that a line current source produces no electric field in the limit of zero frequency or DC, and accordingly results in zero source voltage or source impedance.

4.2. Far-Field Approximation ($\rho \to \infty$). The field expressions in (6.49), together with approximations for the Bessel and Hankel functions for large arguments (see Appendix E) can lead to far-field ($\rho \to \infty$) expressions for a line current.

$$H_0^{(2)}(s)\,\Big|_{s\to\infty} = \sqrt{\frac{2j}{s\pi}}e^{-js}, \tag{6.57}$$

$$\begin{aligned} E_z\,\Big|_{\rho\to\infty} &= -\frac{k\eta}{4}\sqrt{\frac{2j}{\pi k\rho}}e^{-jk\rho} \\ &= -\frac{\eta}{2\sqrt{2}}\sqrt{\frac{jk}{\pi}}\frac{e^{-jk\rho}}{\sqrt{\rho}}, \end{aligned} \tag{6.58}$$

$$\begin{aligned} H_0'^{(2)}(s)\,\Big|_{s\to\infty} &= -jH_0^{(2)}(s)\,\Big|_{s\to\infty} \\ &= -j\sqrt{\frac{2j}{\pi s}}e^{-js}, \end{aligned} \tag{6.59}$$

$$\begin{aligned} H_\phi\Big|_{\rho\to\infty} &= +\frac{k}{4}\sqrt{\frac{2j}{\pi k\rho}}e^{-jk\rho} \\ &= +\frac{1}{2\sqrt{2}}\sqrt{\frac{jk}{\pi}}\frac{e^{-jk\rho}}{\sqrt{\rho}}. \end{aligned} \tag{6.60}$$

Notice that as $\rho \to \infty$, we have $\frac{E_z}{H_\phi} = -\eta$. This is because the cylindrical wave at large distances behave as a plane wave. Power radiated per unit length of the wire may be derived by integrating the Poynting vector in the far-field region.

$$\frac{\Delta P}{\Delta l} = \frac{1}{2}(-E_z H_\phi)_{\rho\to\infty} \times 2\pi\rho = \frac{1}{2}\frac{\eta}{8}\frac{k}{\pi\rho} \times 2\pi\rho = \frac{1}{2}\frac{\eta k}{4} = \frac{1}{2}R_l,\ R_l = \frac{\eta k}{4}. \tag{6.61}$$

The above result for the radiation resistance per unit length is consistent with that obtained in (6.53) from the near-field.

4.3. Low Frequency Approximation ($\omega \to 0$). When $\omega \to 0$, or equivalently $k \to 0$, the field expressions in (6.49) can be approximated using small-argument approximations for the Bessel and Hankel functions (see Appendix E). This is similar to the approximations made for small distances from the source in section 4.1.

$$\begin{aligned} H_0^{(2)}(s)\Big|_{z\to 0} \simeq -jY_0(s)\Big|_{s\to 0} &= \frac{-2j}{\pi}\ln(0.89s), \\ H_0'^{(2)}(s)\Big|_{s\to 0} &\simeq -\frac{2j}{\pi s}. \end{aligned} \tag{6.62}$$

Using the above approximation (6.62) in the field expressions of (6.49), we get,

$$E_z\Big|_{\omega\to 0} = -\frac{k\eta}{4}\frac{-2j}{\pi}\ln(0.89k\rho)\,\Big|_{k\to 0} = 0, \tag{6.63}$$

$$H_\phi\big|_{\omega\to 0} = -\frac{k}{4j} \times (\frac{-2j}{\pi k\rho}) = \frac{1}{2\pi\rho}. \tag{6.64}$$

In (6.63) we use the limiting relation $s \ln s\big|_{s\to 0} = 0$. Equations (6.63) and (6.64) are familiar DC conditions for a line current when the electric field is known to be zero, and the magnetic field is $\frac{1}{2\pi\rho}$ which can be derived from a DC-version of Ampere's law.

The low-frequency approximations (6.63), (6.64) for E_z and H_ϕ maybe derived without having to look up Bessel Function definitions and approximate formulas, using the following steps. First, derive the limiting expression, as $\omega \to 0$, for the ψ_{zm} integral in (6.41) and its derivative with respect to ρ, using suitable calculus and Fourier techniques. Then, using these limiting expressions and the field-potential relations of (5.69-5.74), derive the limiting expressions (6.63), (6.64) for E_z *and* H_ϕ.

The potential function (6.41) and its derivative with ρ may be simplified in the limit as ω approaches zero, using the following Fourier techniques.

$$\psi_{zm} = \frac{1}{2\pi}\int_{-\infty}^{+\infty} \frac{1}{2j\sqrt{k^2 - k_x^2}} e^{+jk_x\rho} dk_x$$

$$= \frac{1}{2\pi}\int_{-\infty}^{+\infty} \frac{1}{2j\sqrt{1-t^2}} e^{+jtk\rho} dt = \psi_{zm}(k\rho);\ t = k_x/k, \tag{6.65}$$

$$\psi'_{zm} = \frac{\partial}{\partial\rho}\psi_{zm} = \frac{1}{2\pi}\int_{-\infty}^{+\infty} \frac{1}{2j\sqrt{k^2 - k_x^2}} e^{+jk_x\rho}(jk_x) dk_x, \tag{6.66}$$

$$\psi'_{zm}|_{\omega,k\to 0} = \frac{1}{2\pi}\int_{-\infty}^{+\infty} \frac{jk_x}{2|k_x|} e^{+jk_x\rho} dk_x = \frac{1}{2\pi}\int_{-\infty}^{+\infty} \frac{j}{2} Sgn(k_x) e^{+jk_x\rho} dk_x, \tag{6.67}$$

$$\begin{aligned} Sgn(k_x) &= +1,\ k_x > 0; \\ &= -1,\ k_x < 0; \\ \frac{dSgn(k_x)}{dk_x} &= 2\delta(k_x). \end{aligned} \tag{6.68}$$

Equation (6.67) maybe equivalently expressed as,

$$\frac{j}{2} Sgn(k_x) = \int_{-\infty}^{+\infty} \psi'_{zm}(\rho) e^{-jk_x\rho} d\rho. \tag{6.69}$$

Differentiate both sides by k_x, and use delta-function relations.

$$j\delta(k_x) = \int_{-\infty}^{+\infty} \psi'_{zm}(\rho) e^{-jk_x\rho}(-j\rho)d\rho,$$

$$\frac{1}{2\pi}\int_{-\infty}^{+\infty} \delta(k_x)e^{+jk_x x}dk_x = -\rho\psi'_{zm}(\rho) = \frac{1}{2\pi},$$

$$\psi'_{zm}(\rho) = -\frac{1}{2\pi\rho},\ \psi_{zm}(\rho) = -\frac{\ln(k\rho)}{2\pi} + A. \tag{6.70}$$

In the last step of (6.70) we adjust the constant of integration A so that ψ_{zm} is a function of $(k\rho)$ (as expected from (6.65)). Now use field-potential relations of (5.69-5.74) and the results of (6.70) to obtain the field expressions in the zero-frequency limit.

$$E_\rho = E_\phi = H\rho = H_z = 0,$$

$$E_z|_{\omega,k\to 0} = \frac{\omega\mu}{j}\psi_{zm}|_{\omega\to 0} = \frac{j\omega\mu\ln(k\rho)}{2\pi}\Big|_{\omega,k\to 0} = 0, \tag{6.71}$$

$$H_\phi = -\frac{\partial}{\partial\rho}\psi_{zm} = \frac{1}{2\pi\rho}. \tag{6.72}$$

These are the same results (6.63),(6.64), but derived without having to use definitions and approximations for Bessel Functions.

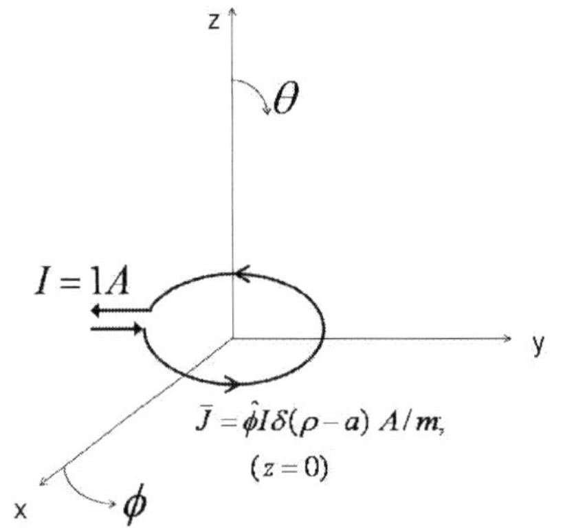

FIGURE 6. A current loop of amplitude I and radius $a << \lambda$, on the xy plane, radiating a TE_z wave into the surrounding medium.

5. Small Current Loop

Fig.6 shows a loop of uniform current $I = 1A$. The radius $(= a)$ of the loop is assumed to be much less than the wavelength λ. The loop maybe modeled as a surface current distribution on the xy plane as follows.

$$\bar{J} = \hat{\phi}\delta(\rho - a) = (-\sin\phi\hat{x} + \cos\phi\hat{y})\delta(\rho - a)$$

$$\tilde{\tilde{J}} = \int\int \bar{J} e^{-jk_x x} e^{-jk_y y} dx dy$$

$$= \int_{\phi=0}^{2\pi}\int_{\rho=0}^{\infty} \bar{J} e^{-jk_x\rho\cos\phi} e^{-jk_y\rho\sin\phi} \rho d\rho d\phi$$

$$= \int_{\phi=0}^{2\pi}\int_{\rho=0}^{\infty} (-\sin\phi\hat{x} + \cos\phi\hat{y})\delta(\rho - a) e^{-jk_x\rho\cos\phi} e^{-jk_y\rho\sin\phi} \rho d\rho d\phi$$

$$= \int_{\phi=0}^{2\pi} (-\sin\phi\hat{x} + \cos\phi\hat{y}) e^{-jk_x a\cos\phi} e^{-jk_y a\sin\phi} a d\phi$$

$$\simeq \int_{\phi=0}^{2\pi} a(-\sin\phi\hat{x} + \cos\phi\hat{y})(1 - jk_x a\cos\phi - jk_y a\sin\phi) d\phi$$

$$= +jk_y a^2\pi\hat{x} - jk_x a^2\pi\hat{y}. \tag{6.73}$$

$$\bar{J} = \frac{1}{4\pi^2}\int\int_{-\infty}^{\infty} \tilde{\tilde{J}}\, e^{jk_x x} e^{jk_y y} dk_x dk_y$$

$$= \frac{j\pi a^2}{4\pi^2}\int\int_{-\infty}^{\infty} (k_y\hat{x} - k_x\hat{y}) e^{jk_x x} e^{jk_y y} dk_x dk_y. \tag{6.74}$$

Comparing results of (6.26) in section 2.4, the present source would produce a TE_z wave with

$$\psi_{ze} = \frac{-j\pi a^2}{4\pi^2}\int\int_{-\infty}^{\infty} \frac{j\omega\mu\pi}{2k_z} e^{jk_x x} e^{jk_y y} e^{\mp jk_z z} dk_x dk_y$$

$$= \frac{1}{4\pi^2}\int\int_{-\infty}^{\infty} \frac{+ja^2\omega\mu\pi}{2jk_z} e^{jk_x x} e^{jk_y y} e^{\mp jk_z z} dk_x dk_y. \tag{6.75}$$

This inverse Fourier Transform can be simplified using the results from Appendix D.

$$\psi_{ze} = +j\omega\mu\pi a^2 \frac{e^{-jkr}}{4\pi r}. \tag{6.76}$$

Now, using the potential-field relations of (5.54-5.59) we get,

$$\begin{aligned} H_r &= \frac{1}{j\omega\mu}(k^2 \cos\theta + \frac{\partial}{\partial r}\cos\theta\frac{\partial}{\partial r})\psi_{ze} \\ &= \pi a^2 \frac{e^{-jkr}}{2\pi}(\frac{1}{r^3} + \frac{jk}{r^2})\cos\theta, \end{aligned} \tag{6.77}$$

$$\begin{aligned} H_\theta &= \frac{1}{j\omega\mu}(-k^2 \sin\theta + \frac{1}{r}\frac{\partial}{\partial\theta}\cos\theta\frac{\partial}{\partial r})\psi_{ze} \\ &= \frac{+\sin\theta}{j\omega\mu}(-k^2 - \frac{1}{r}\frac{\partial}{\partial r})\psi_{ze} \\ &= \pi a^2 \frac{e^{-jkr}}{4\pi}(\frac{-k^2}{r} + \frac{jk}{r^2} + \frac{1}{r^3})\sin\theta, \end{aligned} \tag{6.78}$$

$$E_\phi = \frac{\partial\psi_{ze}}{\partial r}\sin\theta = (+j\omega\mu\pi a^2)\frac{e^{-jkr}}{4\pi}(-\frac{jk}{r} - \frac{1}{r^2})\sin\theta, \tag{6.79}$$

$$H_\phi = E_r = E_\theta = 0. \tag{6.80}$$

5.1. Far-Field Approximation ($r \to \infty$)**.** The field expressions (6.77-6.80) maybe approximated in the far-field region ($r >> \lambda$) by retaining only the dominant terms with $1/r$ dependence.

$$H_r = 0, \tag{6.81}$$

$$\begin{aligned} H_\theta &= -\pi a^2 k^2 \frac{e^{-jkr}}{4\pi r}\sin\theta \\ &= -Sk^2 \frac{e^{-jkr}}{4\pi r}\sin\theta, \end{aligned} \tag{6.82}$$

$$E_\phi = +\omega\mu\pi a^2 k \frac{e^{-jkr}}{4\pi r}\sin\theta = +Sk^2\eta\frac{e^{-jkr}}{4\pi r}\sin\theta, \tag{6.83}$$

$$\frac{E_\phi}{H_\theta} = -\frac{\omega\mu}{k} = -\eta, \tag{6.84}$$

$$H_\phi = E_r = E_\theta = 0, \tag{6.85}$$

where $S = \pi a^2$ is the area of the loop. The radiated power P_{rad} maybe derived by integrating the Poynting vector in the far-field region.

$$2P_{rad} = \int_{\phi=0}^{2\pi} \int_{\theta=0}^{\pi} (-E_\phi H_\theta^*) r^2 \sin\theta d\theta d\phi$$

$$= \int_{\phi=0}^{2\pi} \int_{\theta=0}^{\pi} \frac{S^2 k^4 \eta}{16\pi^2} \sin^3\theta d\theta d\phi$$

$$= \frac{S^2 k^4 \eta}{16\pi^2} \times 2\pi \times \frac{4}{3} = 20 S^2 k^4 = R_{rad}; I = 1A. \quad (6.86)$$

5.2. Low Frequency Approximation ($\omega \to 0$)**.** The field expressions (6.77-6.80) maybe approximated in the low frequency limit, $\omega \to 0,\ k \to 0$, as follows.

$$H_r = \cos\theta \frac{(\pi a^2)}{2\pi r^3}, \quad (6.87)$$

$$H_\theta = \sin\theta \frac{(\pi a^2)}{4\pi r^3}, \quad (6.88)$$

$$H_\phi = E_r = E_\theta = 0,\ E_\phi \simeq 0. \quad (6.89)$$

5.3. Radiation Resistance R_{rad} from the Near Field. Using the potential expression of (6.75), and the field-potential relations (5.34), we would get the expression for the electric field. The complex power and the related input impedance can then be derived by integrating $-\bar{E} \cdot \bar{J}^*$ over the source surface.

$$\bar{E} = -\hat{x}\frac{\partial}{\partial y}\psi_{ze} + \hat{y}\frac{\partial}{\partial x}\psi_{ze}$$

$$= \frac{+1}{4\pi^2} \int\int \frac{a^2 \omega\mu\pi}{2k_z} (-jk_y \hat{x} + jk_x \hat{y}) e^{jk_x x} e^{jk_y y} e^{\mp jk_z z} dk_x dk_y, \quad (6.90)$$

$$2P = -\int\int \bar{E} \cdot \bar{J}^* ds = Z_{in} \quad (6.91)$$

$$= -\frac{1}{4\pi^2} \int\int_{-\infty}^{\infty} \frac{a^2 \omega\mu\pi}{2k_z} (-jk_y \hat{x} + jk_x \hat{y}) \cdot (\tilde{\tilde{J}}\ (k_x, k_y))^* dk_x dk_y. \quad (6.92)$$

Using the transform $\tilde{\tilde{J}}\ (k_x, k_y)$ from (6.73) in (6.92) we get,

$$Z_{in} = \frac{1}{4\pi^2} \int\int_{-\infty}^{\infty} \frac{a^2 \omega\mu\pi}{2k_z} (k_y^2 + k_x^2) a^2 \pi dk_x dk_y$$

$$= \frac{a^4}{8} k\eta \int_{k_\rho=0}^{\infty} \int_{\phi'=0}^{2\pi} \frac{k_\rho^2}{\sqrt{k^2 - k_\rho^2}} k_\rho dk_\rho d\phi' = R_{in} + jX_{in}. \quad (6.93)$$

$$R_{in} = R_{rad} = \frac{a^4}{8} k\eta \times 2\pi \int_{k_\rho=0}^{k} \frac{k_\rho^3}{\sqrt{k^2 - k_\rho^2}} dk_\rho$$

$$= \frac{a^4}{8} k\eta \times 2\pi \times k^3 \int_{t=o}^{\frac{\pi}{2}} \frac{\sin^3 t \times \cos t}{\cos t} dt \;(k_\rho = k \sin t)$$

$$= \frac{a^4}{8} k^4 \eta \times 2\pi \times \frac{4}{6} = S k_0^4 \times \frac{\eta}{6\pi} = 20 S^2 k_0^4 = R_{rad}\;. \tag{6.94}$$

The above result for the radiation resistance R_{in} is consistent with that of (6.86) obtained from the far-field. The input reactance X_{in} would be $+\infty$ (inductive). This is because the imaginary part of the integrand in (6.93) diverges to $+j\infty$ with a $+jk_\rho^2$-asymptotic trend for large values of the integration parameter k_ρ.

6. A Dipole Element on a Dielectric Substrate

In section 3 we modeled radiation from a dipole element placed in a free-space medium. Radiation from the dipole element, but in the presence of a dielectric slab structure, as shown in Fig.7, will be studied in this section. The new radiation source of Fig.7 is of practical significance as an antenna element, which can be etched or "printed" on a dielectric substrate using a printed-circuit fabrication process. The dielectric substrate provides a convenient mechanical support for the antenna, and the etching process allows accurate fabrication of the antenna using an economic printed-circuit technology. In reference to its modern printed-antenna applications, the dipole element of Fig.7 is commonly referred to as a printed-dipole element. The theory of the printed-dipole element of Fig.7 would serve as the fundamental basis for modeling any general antenna or electrical circuitry, which is printed on a conventional circuit board made of a dielectric substrate with a conducting ground plane. The theory of the simple printed-dipole element of Fig.7, with only a single layer of substrate material backed by a conducting ground plane, can as well be extended to more general layered media, consisting of any number of material layers, with or without conducting planes.

The fields of the printed-dipole element of Fig.7 can be derived following the same procedural steps used for the modeling of a free-space dipole in section 3, except additional fields reflected from the dielectric slab structure need to be included. Accordingly, modeling of the printed-dipole element of Fig.7 may not be formulationally that different from the modeling of the free-space dipole element presented in section 3. However, the resulting fields generated from a printed-dipole element can be fundamentally distinct from those of a free-space dipole element studied in section 3, and as well from other elementary sources studied in this chapter that operate in a free-space environment. The modeling of a printed-dipole element would also require some special mathematical treatment, in order to account for its new physical characteristics. Considering the distinct analytical and physical specialities, the printed dipole element of Fig.7 may be viewed as a fundamental elementary source of radiation, warranting special attention.

The dipole element of Fig.7 with current I and a small length l can be expressed as a surface current distribution $\bar{J}_s$ on the xy plane ($z = 0$), in the form of delta functions.

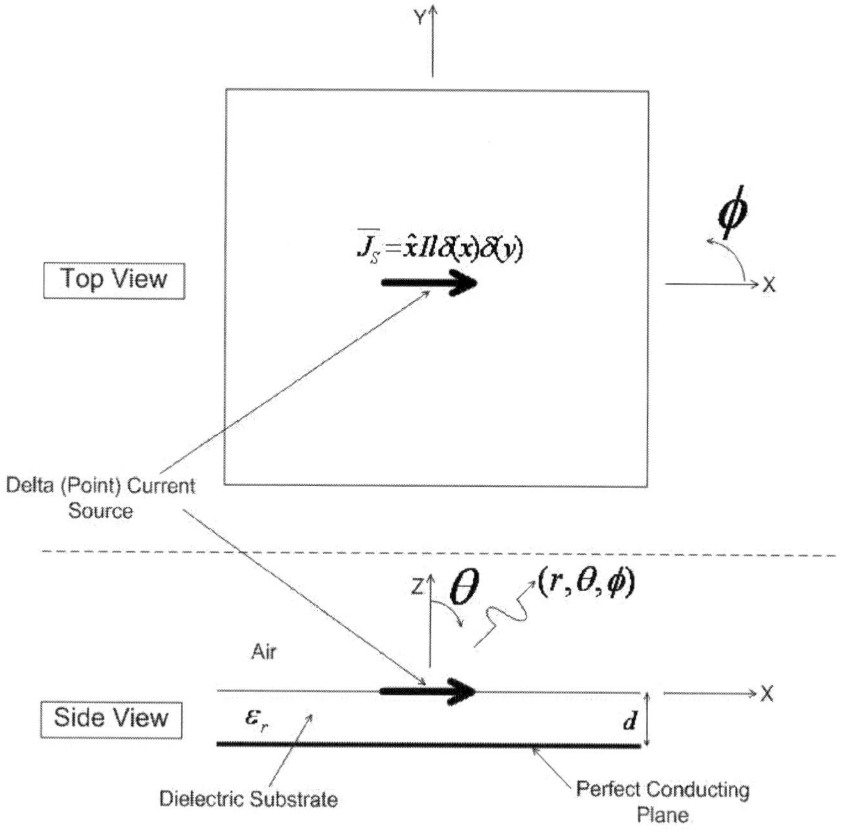

FIGURE 7. A planar current element of amplitude I and length $l << \lambda$, placed on a grounded dielectric substrate, radiating into the surrounding free-space (air) and the dielectric substrate.

The delta-function distribution can also be expressed as a Fourier integral, which may be interpreted as a sum or superposition of Fourier current sheets on the xy plane.

$$\bar{J}_s = \hat{x}\ Il\ \delta(x)\delta(y) = \hat{x}\frac{Il}{4\pi^2}\int_{-\infty}^{\infty}\int_{-\infty}^{\infty} e^{jk_x x} e^{jk_y y} dk_x dk_y. \tag{6.95}$$

Any potential or field component produced by the above dipole source can be expressed as a superposition or integration of the corresponding component produced by a basic Fourier current sheet of unit amplitude of the following form.

$$\bar{J}_s = \hat{x} e^{jk_x x} e^{jk_y y},\ z = 0. \tag{6.96}$$

We will start with modeling of the fields or potentials due to the current sheet of (6.96), which would then lead to that due to the printed-dipole current of (6.95), through a Fourier integration.

It may be noted that the free-space dipole element of Fig.4 was placed on the xz plane ($y = 0$). Whereas, the printed dipole element of Fig.7 we are now modeling is placed on the xy plane ($z = 0$). For convenience of presentation, we choose the normal to the dielectric slab structure, and accordingly to the plane of the dipole surface current distribution $\bar{J}_s$, in the z direction. This geometric orientation is consistent with that in chapter 5, used in modeling reflection of TM_z and TE_z waves from layered interfaces. The choice would allow convenient use of results from the reflection analysis of chapter 5, in modeling the interactions of the printed-dipole element of Fig.7 with its dielectric slab structure.

Modeling of the fields or potentials due to the current sheet of (6.96) is separated into two parts, as suggested in Fig.8. In the first part, analysis is performed without the dielectric slab, with only the free space medium both above and below the source. As in the modeling of a free-space dipole in section 3, the Fourier current sheet of (6.96) would generate plane waves propagating in two sides of the source, respectively in the $+z$ and $-z$ directions in the present configuration. We will express the plane waves as superposition of TM_z and TE_z waves, so that the reflection results from chapter 5 can be utilized, as we suggested above. In the second part of the analysis, the theory avaialable in chapter 5 is used to model reflections of the TM_z and TE_z waves specifically from the slab structure of Figs.7, 8. Results from the two parts of the modeling can then be added to constitute the total solution for the current sheet of (6.96). The two parts of the modeling are separately presented in the following sections. It may be mentioned, that an alternate formulation in the first part of the analysis, using only TM_x plane waves having magnetic field orthogonal to the reference source direction (x), could also be possible. However, this choice is not preferred because it would complicate modeling of the resulting reflected fields in the second part of the analysis, where the reflection theory available in chapter 5 can not be readily used.

6.1. TM_z and TE_z Waves Generated from a Current Sheet on the xy plane, Placed in a Free-Space Medium. The expressions for the current sheets on the xy plane ($z = 0$), required to generate only TM_z or TE_z waves in a free-space medium, are established in (6.25,6.26) of section 2.4. We will use those results here, except substitute k_x and k_y by $-k_x$ and $-k_y$, respectively, in order to match with the functional dependence of the Fourier sheet of (6.96). In the derivation of section 2.4, the angle ϕ' referred to the physical direction on the source plane, with respect to the reference $\hat{x}$ axis, of propagation of the wave generated by a current sheet. And, the cosine and sine of the angle were related to the wave numbers k_x, k_y and k_ρ as: $\cos\phi' = k_x/k_\rho$ and $\sin\phi' = k_y/k_\rho$. We will maintain the same relationship $\cos\phi' = k_x/k_\rho$ and $\sin\phi' = k_y/k_\rho$ as well in the present study. However, according to the change in the signs for k_x and k_y, as prescribed above, we need to substitute $\cos\phi'$ and $\sin\phi'$ in the results of section 2.4 by their respective negative values. Effectively, this amounts to having ϕ' in the present study equal to 180 degrees plus the actual angle, with respect to the $\hat{x}$ axis, of wave propagation on the source plane. Following the results of (6.25) in section 2.4, with the above considerations, a current source distribution $\bar{J}_{s1}$ on the xy plane, and the corresponding potential ψ_{TM1} of the TM_z wave produced by the source, are expressed as follows.

$$\bar{J}_{s1} = \frac{2j}{\eta_0}(\hat{x}\cos\phi' + \hat{y}\sin\phi')e^{+jk_x x}e^{+jk_y y}, \tag{6.97}$$

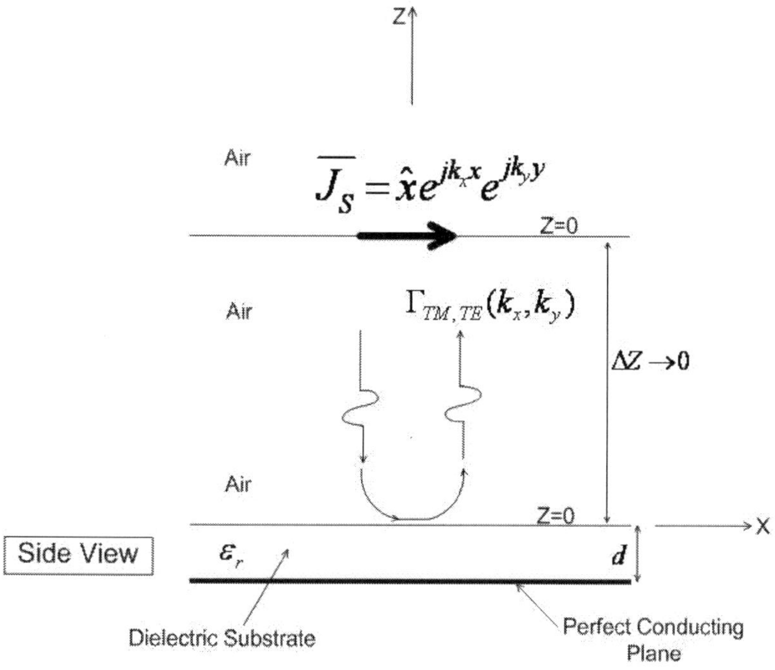

FIGURE 8

$$\begin{aligned}
\psi_{ZM1} &= \pm\frac{\omega\epsilon_0}{k_\rho k_0}e^{+jk_x x}e^{+jk_y y}e^{\mp jk_{z0}z}; \\
k_{z0} &= \sqrt{k_0^2 - k_\rho^2},\ k_\rho < k_0; \\
&= -j\sqrt{k_\rho^2 - k_0^2},\ k_\rho > k_0\ .
\end{aligned} \tag{6.98}$$

The upper and lower signs in a double-sign notation $\pm$ or $\mp$ refer to the medium above ($z > 0$) and below ($z < 0$) the source, respectively. Any field component can be derived from the above potential function using the field-potential relations of section 3 in chapter 5. We will specifically express here the electric field component E_{x1}.

$$E_{x1} = \frac{1}{j\omega\epsilon_0}\frac{\partial^2\psi_{ZM}}{\partial x\partial z} = \pm\frac{k_x k_{z0}}{k_\rho k_0 j}e^{+jk_x x}e^{+jk_y y}e^{\mp jk_{z0}z}. \tag{6.99}$$

Similarly, the current source distribution $\bar{J}_{s2}$ on the xy plane, and the corresponding potential ψ_{TE}, following the results of (6.26) in section 2.4, and consequently the electric field E_{x2} of the TE_z wave produced by the source using the field-potential relations of section 3 in chapter 5, are expressed as follows:

$$\bar{J}_{s2} = -\frac{2jk_z}{\omega\mu_0}(\hat{y}\cos\phi' - \hat{x}\sin\phi')e^{+jk_x x}e^{+jk_y y}, \tag{6.100}$$

$$\psi_{ZE2} = \frac{1}{k_\rho}e^{+jk_x x}e^{+jk_y y}e^{\mp jk_{z0}z}, \tag{6.101}$$

$$E_{x2} = -\frac{\partial\psi_{ZE}}{\partial y} = -\frac{jk_y}{k_\rho}e^{+jk_x x}e^{+jk_y y}e^{\mp jk_{z0}z}. \tag{6.102}$$

The current sheet $\bar{J}_s$ of (6.96) can be expressed as a superposition of the above two currents $\bar{J}_{s1}$ and $\bar{J}_{s2}$, and accordingly would produce both TM_z and TE_z waves. The respective fields from the TM_z and TE_z parts can be added to obtain the fields produced by the total source $\bar{J}_s$.

$$\bar{J}_s = \frac{\eta_0}{2j}\bar{J}_{s1}\cos\phi' + \frac{\omega\mu_0}{2jk_{z0}}\bar{J}_{s2}\sin\phi' = \hat{x}e^{+jk_x x}e^{+jk_y y}, \tag{6.103}$$

$$\begin{aligned}
\psi_{TM} &= \frac{\eta_0}{2j}\cos\phi'\psi_{TM1} = \pm\frac{1}{2jk_\rho}\cos\phi' e^{+jk_x x}e^{+jk_y y}e^{\mp jk_{z0}z},\\
\psi_{TE} &= \frac{\omega\mu_0}{2jk_{z0}}\sin\phi'\psi_{TE2} = \frac{\eta_0 k_0}{2jk_{z0}k_\rho}\sin\phi' e^{+jk_x x}e^{+jk_y y}e^{\mp jk_{z0}z},\\
E_x &= \frac{\eta_0}{2j}\cos\phi' E_{x1} + \frac{\omega\mu_0}{2jk_{z0}}\sin\phi' E_{x2}\\
&= (-\frac{\eta_0}{2}\frac{k_{z0}}{k_0}\cos^2\phi' - \frac{\eta_0}{2}\frac{k_0}{k_{z0}}\sin^2\phi')e^{+jk_x x}e^{+jk_y y}e^{\mp jk_{z0}z}\\
&= (-\frac{Z_{TM0}}{2}\cos^2\phi' - \frac{Z_{TE0}}{2}\sin^2\phi')e^{+jk_x x}e^{+jk_y y}e^{\mp jk_{z0}z}
\end{aligned} \tag{6.104}$$

$$Z_{TM0} \stackrel{\triangle}{=} \eta_0\frac{k_{z0}}{k_0} = \eta_0\cos\theta' = \frac{k_{z0}}{\omega\epsilon_0}, \tag{6.105}$$

$$Z_{TE0} \stackrel{\triangle}{=} \eta_0\frac{k_0}{k_{z0}} = \frac{\eta_0}{\cos\theta'} = \frac{\omega\mu_0}{k_{z0}}. \tag{6.106}$$

6.2. Reflection from the Dielectric Slab. The reflection coefficients Γ_{TM} and Γ_{TE}, respectively for a TM_z and TE_z wave, from the slab structure of Figs.7, 8 can be expressed using the theory of chapter 5 (see section 6). This is possible using an equivalent transmission line model with suitable equivalent characteristic impedance for the free-space and dielectric media.

$$\begin{aligned}
Z_{TM} &= jZ_{TM1}\tan(k_{z1}d), \\
Z_{TE} &= jZ_{TE1}\tan(k_{z1}d), \\
Z_{TM1} &= \eta_1\frac{k_{z1}}{k_1} = \frac{k_{z1}}{\omega\epsilon_1} = \frac{k_{z1}}{\omega\epsilon_0\epsilon_r}, \\
Z_{TE1} &= \eta_1\frac{k_1}{k_{z1}} = \frac{\omega\mu_0}{k_{z1}} = \frac{\omega\mu_1}{k_{z1}}, \\
\Gamma_{TM} &= \frac{Z_{TM} - Z_{TM0}}{Z_{TM} + Z_{TM0}}, \\
\Gamma_{TE} &= \frac{Z_{TE} - Z_{TE0}}{Z_{TE} + Z_{TE0}}, \\
k_{z1} &= \sqrt{k_1^2 - k_\rho^2},\ k_\rho < k_1; \\
&= -j\sqrt{k_\rho^2 - k_1^2},\ k_\rho > k_1; \\
k_1 &= k_0\sqrt{\epsilon_r}. \qquad (6.107)
\end{aligned}$$

6.3. Total Solution for a Current Sheet: Free-Space Waves Plus Slab Reflection. The above reflections Γ_{TM} and Γ_{TE}, added to the respective potential or fields generated by the current-sheet of (6.96) in a free-space medium, derived earlier, would result in the total potential or fields generated by the current sheet in the presence of the slab environment of Figs.7, 8.

$$\begin{aligned}
E_x = [-\frac{Z_{TM0}}{2}\cos^2\phi'(1+\Gamma_{TM}) - \frac{Z_{TE0}}{2}\sin^2\phi'(1+\Gamma_{TE})]e^{+jk_x x}e^{+jk_y y}e^{-jk_{z0}z} \\
= (-Z^t_{TM}\cos^2\phi' - Z^t_{TE}\sin^2\phi')e^{+jk_x x}e^{+jk_y y}e^{-jk_{z0}z},\ z > 0, \\
Z^t_{TM} = \frac{Z_{TM0}}{2}(1+\Gamma_{TM}) = \frac{Z_{TM0}}{2} \times \frac{2Z_{TM}}{Z_{TM}+Z_{TM0}} = Z_{TM0} \parallel Z_{TM}, \\
Z^t_{TE} = \frac{Z_{TE0}}{2}(1+\Gamma_{TE}) = \frac{Z_{TE0}}{2} \times \frac{2Z_{TE}}{Z_{TE}+Z_{TE0}} = Z_{TE0} \parallel Z_{TE}. \qquad (6.108)
\end{aligned}$$

It may be noted, that the above reflection coefficients, Γ_{TM} and Γ_{TE}, are defined in reference to the electric field components, E_x or E_y, parallel to the dielectric surface ($xy-$ *plane*), which is consistent with a standard notational practice. Reflection coefficients for the potentials or other field components may require a + or − sign before the above reflection coefficients, in order to account for proper relative signs between the respective reflected and incident parts. For example, the relative signs between the incident and reflected parts of the potential ψ_{TM} for a TM_z wave can be shown to be opposite of those of the corresponding E_x or E_y field (see chapter 5). Accordingly, the reflection coefficient for the TM_z potential is negative of that expressed above for the E_x or E_y. On the other hand, the relative signs between the incident and reflected parts of the potential ψ_{TE} for a TE_z wave would be same as those of the corresponding E_x or E_y field, and therefore they would share the same reflection coefficient. The reflection coefficient for the H_x or

H_y component of the magnetic fields can be shown to be negative of that for the E_x or E_y electric fields, as expressed above, for both the TM_z as well as the TE_z waves.

6.4. Surface-Wave Singularity. In contrast with the field expression of (6.104) for a free-space medium, the field expression of (6.108) for a slab environment can be shown to be singular at specific values of $k_\rho = k_{\rho p}$. That means, when the wave number of the current sheet of (6.96) is equal to the singular value, the field generated would be infinite for a finite amplitude of the current-sheet source. Equivalently, a finite, non-zero amplitude of the field, with a wave number equal to the singular value, can be sustained with a zero amplitude of the current sheet. The fields associated with the singular wave number may, therefore, be referred to as a source-free mode of the slab structure. In the following analysis, a real solution for the wave number $k_\rho = k_{\rho p}$ on the xy plane is seen to be grater than the wave number k_0 in the free-space medium. From (6.98), the corresponding wave number $k_z = k_{z0} = -j\alpha_{z0}$ in the free-space along the normal (z) direction would be negatively imaginary. Therefore, the z-variation $e^{-jk_{z0}z} = e^{-\alpha_{z0}z}$ of the associated source-free mode would exponentially decay away from the surface, to a negligible value beyond a sufficient distance above the surface of the slab. In other words, the fields of the source-free mode in the region above the slab is confined only over a limited distance above the surface of the slab interface. Accordingly, the associated wave may also be referred to as a "surface wave." In the following we will derive the expressions of a real wave number $k_{\rho p}$ for such a source-free, surface wave mode by equating the field expression in (6.108), when evaluated at at the real $k_\rho = k_{\rho p}$, to infinity. Two sets of solutions can be formulated by independently equating the TM_z or the TE_z part of the field in (6.108) to infinity. The respective solution would be the real wave number of a TM_z or a TE_z surface wave, supported by the slab structure.

6.4.1. *TM_z Surface Wave.* The wave number $k_\rho = k_{\rho p}$ for a TM_z surface wave is obtained by equating the impedance Z^t_{TM} in (6.108) to infinity. Once the solution for $k_{\rho p}$ is obtained, we will then approximate the impedance function Z^t_{TM} in the vicinity of $k_\rho \simeq k_{\rho p}$ for use in following analyses.

<u>Surface Wave Number</u>

$$Z^t_{TM} = \frac{Z_{TM0}Z_{TM}}{Z_{TM0}+Z_{TM}} = \frac{Z_{TM0}jZ_{TM1}\tan(k_{z1}d)}{Z_{TM0}+jZ_{TM1}\tan(k_{z1}d)}$$

$$= \frac{\frac{k_{z0}}{\omega\epsilon_0}j\frac{k_{z1}}{\omega\epsilon_0\epsilon_r}\tan(k_{z1}d)}{\frac{k_{z0}}{\omega\epsilon_0}+j\frac{k_{z1}}{\omega\epsilon_0\epsilon_r}\tan(k_{z1}d)} = \frac{k_{z0}}{\omega\epsilon_0}\times\frac{jk_{z1}\tan(k_{z1}d)}{k_{z0}\epsilon_r+jk_{z1}\tan(k_{z1}d)}. \qquad (6.109)$$

A singularity or pole for the impedance function Z^t_{TM} would occur at $k_\rho = k_{\rho p}$, where the denominator in the above expression would be zero. A solution would be possible for a real value of $k_{\rho p}$ in the range $k_0 < k_{\rho p} < k_1$.

$$
\begin{aligned}
jk_{z1}\tan(k_{z1}d) + k_{z0}\epsilon_r &= 0, \\
k_x^2 + k_y^2 = k_{\rho p}^2,\ k_0 < k_{\rho p} < k_1,\ k_{\rho p} &= k_0\sqrt{\epsilon_e}; 1 < \epsilon_e < \epsilon_r, \\
j\sqrt{k_1^2 - k_{\rho p}^2}\tan(d\sqrt{k_1^2 - k_{\rho p}^2}) - j\epsilon_r\sqrt{k_{\rho p}^2 - k_0^2} &= 0, \\
\sqrt{k_1^2 - k_{\rho p}^2}\tan(d\sqrt{k_1^2 - k_{\rho p}^2}) &= \epsilon_r\sqrt{k_{\rho p}^2 - k_0^2}, \\
\sqrt{\epsilon_r - \epsilon_e}\tan(dk_0\sqrt{\epsilon_r - \epsilon_e}) &= \epsilon_r\sqrt{\epsilon_e - 1}, \\
\tan(dk_0\sqrt{\epsilon_r - \epsilon_e}) &= \frac{\epsilon_r\sqrt{\epsilon_e - 1}}{\sqrt{\epsilon_r - \epsilon_e}}. \qquad (6.110)
\end{aligned}
$$

It can be useful to rearrange the above equation in terms of suitable normalized variables $\bar{\epsilon}_e$ and $\bar{d}$.

$$
\tan(\bar{d}\sqrt{1 - \bar{\epsilon}_e}) = \epsilon_r\sqrt{\frac{\bar{\epsilon}_e}{1 - \bar{\epsilon}_e}},\ \bar{d} = k_0 d\sqrt{\epsilon_r - 1},\ \bar{\epsilon}_e = \frac{\epsilon_e - 1}{\epsilon_r - 1}. \qquad (6.111)
$$

The real propagation constants $k_{\rho p} = \beta_{sw}$ for different TM_z surface waves are plotted in Fig.9 as a function of the substrate thickness d, in the normalized form derived in (6.111). The results are plotted together with similar data for different TE_z surface waves, derived in the following section.

In the above derivation, the slab is assumed to be ideally lossless. In practice, however, there is always a non-zero amount of loss in every physical material. The loss would approach an infinitesimally small, but non-zero value for a realistic slab with practically very little loss. Such a realistic small loss may be introduced by replacing the dielectric constant ϵ_r with a complex number $\epsilon_r' = \epsilon_r - j\delta_\epsilon$, having a small non-zero, negative imaginary part $-j\delta_\epsilon$ (see chapter 4, section 9.7.8). The corresponding wave number k_1' for the dielectric medium would also be a complex number, $k_1' = k_1 - j\delta_1$, with an additional small negative imaginary part $-j\delta_1$, which would represent signal attenuation in a uniform medium made of the particular dielectric material. The above solution for the wave number of a surface wave may be modified for such a lossy dielectric material, by replacing the parameter ϵ_r with its complex value ϵ_r'. Like the wave number k_1' for the uniform dielectric medium, we should expect the wave number for the surface wave to also be a complex number $k_{\rho p}' = k_{\rho p} - j\delta_p$ with an added small negative imaginary part $-j\delta_p$, representing attenauation of the resulting surface wave.

$$
k_{\rho p}' = k_{\rho p} - j\delta_p,\ \delta_p > 0,\ \delta_p \to 0. \qquad (6.112)
$$

Residue of the Impedance Function

We will approximate the total impedance function Z_{TM}^t, for wave numbers in the vicinity of $k_\rho \simeq k_{\rho p}$. The relative magnitude of the impedance Z_{TM}^t for $k_\rho \simeq k_{\rho p}$ would represent the strength of the surface wave that can be excited by the current source.

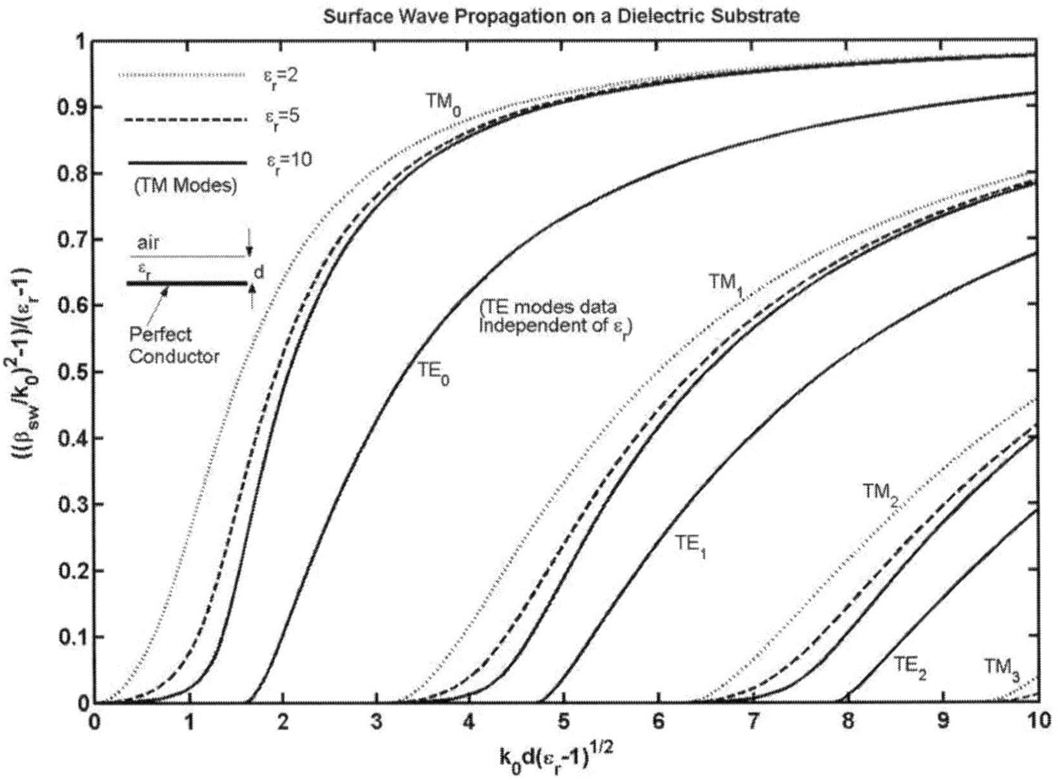

FIGURE 9

$$Z^t_{TM} = \frac{k_{z0}}{\omega\epsilon_0} \times \frac{jk_{z1}\tan(k_{z1}d)}{jk_{z1}\tan(k_{z1}d) + k_{z0}\epsilon_r}$$

$$= \frac{k_{z0}}{\omega\epsilon_0} \times \frac{jk_{z1}\sin(k_{z1}d)}{jk_{z1}\sin(k_{z1}d) + k_{z0}\epsilon_r\cos(k_{z1}d)} = \frac{jk_{z1}k_{z0}}{\omega\epsilon_0} \times \frac{\sin(k_{z1}d)}{D_m},$$

$$D_m = jk_{z1}\sin(k_{z1}d) + k_{z0}\epsilon_r\cos(k_{z1}d)$$

$$\simeq a(k_\rho^2 - k_{\rho p}^2),\ k_\rho^2 \simeq k_{\rho p}^2, \tag{6.113}$$

$$Z^t_{TM} \simeq \frac{a'}{(k_\rho^2 - k_{\rho p}^2)},\ k_\rho \simeq k_{\rho p};\ a' = \frac{jk_{z1}k_{z0}}{\omega\epsilon_0}\sin(k_{z1}d)\Big|_{k_\rho = k_{\rho p}} \times (\frac{1}{a}). \tag{6.114}$$

The denominator D_m of the impedance expression Z^t_{TM} is zero at the wave number $k_\rho = k_{\rho p}$. Further, the wave-numbers k_{z0} and k_{z1} in (6.98, 6.107) are functions of k_ρ^2, which result also in a k_ρ^2 dependence for the denominator as well as for the total impedance function. The above functional requirements justifies the basic functional forms used in (6.113, 6.114) for approximation of the denominator D_m and the total impedance Z^t_{TM}, in the vicinity of the pole at $k_\rho = k_{\rho p}$, or equivalently at $k_\rho^2 = k_{\rho p}^2$. The coefficient a can be derived by differentiating both sides of (6.113) and evaluating them at $k_\rho = k_{\rho p}$. The coefficient a' of the impedance function Z^t_{TM} is then related to a. The coefficient a' is referred to as the residue of the impedance function Z^t_{TM} with respect to k^2_ρ, at the pole $k^2_\rho = k^2_{\rho p}$.

$$
\begin{aligned}
D'_m = \frac{dD_m}{dk_\rho} &= -jk_{z1}\cos(k_{z1}d)\frac{k_{\rho p}d}{k_{z1}} - j\sin(k_{z1}d)\frac{k_{\rho p}}{k_{z1}} + \\
&\qquad k_{z0}\epsilon_r \sin(k_{z1}d)\frac{k_{\rho p}d}{k_{z1}} - \epsilon_r\cos(k_{z1}d)\frac{k_{\rho p}}{k_{z0}} \\
&= -\frac{k_{\rho p}}{k_{z1}k_{z0}}[\cos(k_{z1}d)(jk_{z1}k_{z0}d + \epsilon_r k_{z1}) + \sin(k_{z1}d)(jk_{z0} - \epsilon_r k^2_{z0}d)] \\
&= -\frac{k_{\rho p}A}{k_{z1}k_{z0}},\ A = [\cos(k_{z1}d)k_{z1}(\epsilon_r + \alpha_{z0}d) + \sin(k_{z1}d)(1+\epsilon_r\alpha_{z0}d)\alpha_{z0}], \\
&\qquad k_{z0} = -j\alpha_{z0},\ \alpha_{z0} = k_0\sqrt{\epsilon_e - 1},
\end{aligned}
\tag{6.115}
$$

$$
\begin{aligned}
D'_m &= \frac{dD_m}{dk_\rho} \simeq 2ak_{\rho p} = -\frac{k_{\rho p}A}{k_{z1}k_{z0}};\ k_\rho \simeq k_{\rho p}, \\
a &= -\frac{1}{2k_{z1}k_{z0}}A\Big|_{k_\rho = k_{\rho p}},\ a' = -\frac{j2(k_{z1}k_{z0})^2}{\omega\epsilon_0 A}\sin(k_{z1}d)\,\Big|_{k_\rho=k_{\rho p}} \\
&= -\frac{j2(k_{z1}k_{z0})^2}{\omega\epsilon_0[\cot(k_{z1}d)k_{z1}(\epsilon_r + \alpha_{z0}d) + (1+\epsilon_r\alpha_{z0}d)\alpha_{z0}]}\,\Big|_{k_\rho=k_{\rho p}}.
\end{aligned}
\tag{6.116}
$$

The above expression for the residue a' can be simplified further using (6.110), and the expressions of k_{z0} and k_{z1} in (6.98, 6.107).

$$
\begin{aligned}
a' &= -\frac{j2(k_{z1}k_{z0})^2}{\omega\epsilon_0[\frac{k_{z1}}{\alpha_{z0}}k_{z1}(\epsilon_r + \alpha_{z0}d) + (1+\epsilon_r\alpha_{z0}d)\alpha_{z0}]}\,\Big|_{k_\rho=k_{\rho p}} \\
&= \frac{j2k^2_{z1}\alpha^3_{z0}}{\omega\epsilon_0[k^2_{z1}(\epsilon_r + \alpha_{z0}d) + (1+\epsilon_r\alpha_{z0}d)\alpha^2_{z0}]}\,\Big|_{k_\rho=k_{\rho p}} \\
&= \frac{j2k^2_0\eta_0(\epsilon_r - \epsilon_e)(\epsilon_e - 1)^{3/2}}{[(\epsilon_r - \epsilon_e)(\epsilon_r + k_0d(\epsilon_e-1)^{1/2}) + (1+\epsilon_r k_0 d(\epsilon_e - 1)^{1/2})(\epsilon_e - 1)]}.
\end{aligned}
\tag{6.117}
$$

As defined in (6.110), $\epsilon_e = (k_{\rho p}/k_0)^2$ is the effective dielectric constant associated with the singular wave number $k_{\rho p}$.

As noted before, the real singularity $k_{\rho p}$ would change to a complex value $k'_{\rho p}$ of (6.112) for a realistic material with a small but non-zero loss. The residue of the impedance functions at a surface-wave singularity, as derived above, need to be re-evaluated with any material loss. It may be shown that the residue will essentially remain unchanged when a small material loss is introduced. However, the total singular function of (6.114), which consists of the residue as its coefficient and the remaining basic singular part, may significantly change in the vicinity of the singularity, with the substitution of (6.112) for a small loss. The significant change occurs due to the strong variation of the basic singular part of (6.114), while the residue remaining essentially unchanged for the small loss. These considerations for the location and residue for a surface-wave singularity with a realistic slab material would be important in analysis of the surface-wave radiation in section 6.5.2. This applies as well to TE_z surface waves studied in the following.

6.4.2. TE_z *Surface Wave.* The wave number $k_{\rho p}$ and the residue of a TE_z surface wave would be derived by following the same procedural steps used above for a TM_z wave. In the following we will present important derivational steps and significant results.

Surface Wave Number

$$\begin{aligned} Z^t_{TE} &= \frac{Z_{TE0}Z_{TE}}{Z_{TE0}+Z_{TE}} = \frac{Z_{TE0}jZ_{TE1}\tan(k_{z1}d)}{Z_{TE0}+jZ_{TE1}\tan(k_{z1}d)} \\ &= \frac{\frac{\omega\mu_0}{k_{z0}}j\frac{\omega\mu_0}{k_{z1}}\tan(k_{z1}d)}{\frac{\omega\mu_0}{k_{z0}}+j\frac{\omega\mu_0}{k_{z1}}\tan(k_{z1}d)} = \frac{j\omega\mu_0\tan(k_{z1}d)}{k_{z1}+jk_{z0}\tan(k_{z1}d)}. \end{aligned} \tag{6.118}$$

A singularity of the above impedance function would occur for a real value of $k_\rho = k_{\rho p} = k_0\sqrt{\epsilon_e}$, in the range of $k_0 < k_{\rho p} < k_1$, or equivalently in the range $1 < \epsilon_e < \epsilon_r$.

$$\begin{aligned} & jk_{z0}\tan(k_{z1}d) + k_{z1} = 0, \\ k_0 < k_{\rho p} < k_1,\ & dk_0\sqrt{\epsilon_r - 1} > \frac{\pi}{2},\ \frac{d}{\lambda_0} > \frac{1}{4\sqrt{\epsilon_r-1}}, \\ \tan(dk_0\sqrt{\epsilon_r - \epsilon_e}) &= \frac{-k_{z1}}{jk_{z0}} = \frac{-\sqrt{\epsilon_r-\epsilon_e}}{\sqrt{\epsilon_e - 1}}. \end{aligned} \tag{6.119}$$

We will rearrange the above equation in terms of normalized variables $\bar{\epsilon}_e$ and $\bar{d}$, that can be useful for evaluation independent of the actual frequency, dielectric constant or thickness of choice.

$$\tan(\bar{d}\sqrt{1-\bar{\epsilon}_e}) = -\sqrt{\frac{1}{\bar{\epsilon}_e} - 1},\ \bar{d} = dk_0\sqrt{\epsilon_r - 1},\ \bar{\epsilon}_e = \frac{\epsilon_e - 1}{\epsilon_r - 1}. \tag{6.120}$$

The propagation constants $\beta_{sw} = k_{\rho p}$ for different TE_z surface waves are plotted in Fig.9 as a function of the substrate thickness d, in the normalized form derived in (6.120). The results are plotted together with similar results for different TM_z surface waves, derived in (6.111).

The normalized relationship in (6.111) for the effective dielectric constant of a TM_z wave may be compared with that in (6.120). Unlike the normalized relationship (6.120) for the TE_z wave, the relationship in (6.111) for the TM_z wave is still a function of the actual dielectric constant ϵ_r. Accordingly, the TM_z relationship of (6.111) needs to be re-evaluated for a different choice of the dielectric constant ϵ_r, whereas the normalized relationship (6.120) for the TE_z wave does not need such re-evaluation. This is an important observation, which is reflected in the computed plots for the normalized effective dielectric constants in Fig.9.

Residue of the Impedance Function

$$Z^t_{TE} = \frac{j\omega\mu_0 \tan(k_{z1}d)}{jk_{z0}\tan(k_{z1}d) + k_{z1}} = \frac{j\omega\mu_0 \sin(k_{z1}d)}{jk_{z0}\sin(k_{z1}d) + k_{z1}\cos(k_{z1}d)} = \frac{j\omega\mu_0 \sin(k_{z1}d)}{D_e},$$

$$D_e = jk_{z0}\sin(k_{z1}d) + k_{z1}\cos(k_{z1}d) \simeq b(k_\rho^2 - k_{\rho p}^2),\ k_\rho \simeq k_{\rho p};$$

$$Z^t_{TE} = \frac{b'}{(k_\rho^2 - k_{\rho p}^2)},\ k_\rho \simeq k_{\rho p};\ b' = j\omega\mu_0 \sin(k_{z1}d)\,\Big|_{k_\rho = k_{\rho p}} \times (\frac{1}{b}),$$

$$\begin{aligned} D'_e = \frac{dD_e}{dk_\rho} &= -jk_{z0}\cos(k_{z1}d)\frac{k_{\rho p}d}{k_{z1}} - j\sin(k_{z1}d)\frac{k_{\rho p}}{k_{z0}} \\ &\quad + k_{z1}\sin(k_{z1}d)\frac{k_{\rho p}d}{k_{z1}} - \frac{k_{\rho p}}{k_{z1}}\cos(k_{z1}d) \\ &= -\frac{k_{\rho p}}{k_{z1}k_{z0}}[\cos(k_{z1}d)(jk_{z0}^2 d + k_{z0}) + \sin(k_{z1}d)(jk_{z1} - dk_{z0}k_{z1})] \\ &= -\frac{k_{\rho p}}{k_{z1}k_{z0}}[\cos(k_{z1}d)k_{z0}(1+jdk_{z0}) + \sin(k_{z1}d)jk_{z1}(1+jdk_{z0})] \\ &= \frac{k_{\rho p}j}{-jk_{z1}\alpha_{z0}}[\cos(k_{z1}d)\alpha_{z0}(1+d\alpha_{z0}) - \sin(k_{z1}d)k_{z1}(1+d\alpha_{z0})] \\ &= -\frac{k_{\rho p}}{k_{z1}\alpha_{z0}}B;\ B = [\cos(k_{z1}d)\alpha_{z0}(1+d\alpha_{z0}) - \sin(k_{z1}d)k_{z1}(1+d\alpha_{z0})] \\ &= 2bk_{\rho p},\ k_\rho \simeq k_{\rho p};\ b = -\frac{B}{2k_{z1}\alpha_{z0}}\,\Big|_{k_\rho = k_{\rho p}}, \end{aligned}$$

$$\begin{aligned} b' &= \frac{-j2\omega\mu_0 k_{z1}\alpha_{z0}\sin(k_{z1}d)}{B}\,\Big|_{k_\rho = k_{\rho p}} \\ &= \frac{-j2\omega\mu_0 k_{z1}\alpha_{z0}}{[\cot(k_{z1}d)\alpha_{z0}(1+d\alpha_{z0}) - k_{z1}(1+d\alpha_{z0})]}\,\Big|_{k_\rho = k_{\rho p}}. \end{aligned} \tag{6.121}$$

The above expression for the residue b' is simplified further using (6.119), and the expression of k_{z0} and k_{z1} in (6.98, 6.107).

$$\begin{aligned} b' &= \frac{-j2\omega\mu_0 k_{z1}\alpha_{z0}}{[(\frac{-jk_{z0}}{k_{z1}})\alpha_{z0}(1+d\alpha_{z0}) - k_{z1}(1+d\alpha_{z0})]}\,\Big|_{k_\rho = k_{\rho p}} \\ &= \frac{j2k_0^2\eta_0(\epsilon_r - \epsilon_e)(\epsilon_e - 1)^{1/2}}{(1+dk_0(\epsilon_e - 1)^{1/2})(\epsilon_r - 1)}. \end{aligned} \tag{6.122}$$

6.5. Fields and Input Impedance of a Printed-Dipole Element. We derived the fields and potentials produced by the current sheet of (6.96, 6.103) using the steps (6.106, 6.107, 6.108), as superposition of TM_z and TE_z parts. Now, the fields or potentials produced by the printed-dipole current of (6.95) can be expressed by a Fourier integral of the results of (6.106, 6.107, 6.108). However, unlike in the similar derivation in section 3 for a free-space dipole, the Fourier integral in the present study of a printed-dipole element may not be analytically evaluated into a simple form, which would be valid at all locations in rigorous terms. The integral may need to be evaluated computationally. We will express the electric field component E_x for the printed dipole element of (6.95), in a Fourier integral form. The equivalent voltage V_{in} seen at the input of the printed dipole is equal to the negative E_x field evaluated at the center of the source $(x, y, z) = (0, 0, 0)$ multiplied by the length l of dipole.

$$E_x(x, y, z > 0) =$$

$$\frac{Il}{4\pi^2} \int_{-\infty}^{\infty} \int_{-\infty}^{+\infty} (-Z^t_{TM} \cos^2 \phi' - Z^t_{TE} \sin^2 \phi') e^{+jk_x x} e^{+jk_y y} e^{-jk_{z0} z} dk_x dk_y, \qquad (6.123)$$

$$V_{in} = -E_x(0,0,0)l = \frac{Il^2}{4\pi^2} \int_{-\infty}^{\infty} \int_{-\infty}^{+\infty} (Z^t_{TM} \cos^2 \phi' + Z^t_{TE} \sin^2 \phi') dk_x dk_y,$$

$$Z_{in} = \frac{V_{in}}{I} = \frac{l^2}{4\pi^2} \int_{-\infty}^{\infty} \int_{-\infty}^{+\infty} (Z^t_{TM} \cos^2 \phi' + Z^t_{TE} \sin^2 \phi') dk_x dk_y = R_{in} + jX_{in}$$

$$= \frac{l^2}{4\pi^2} \int_{k_\rho=0}^{\infty} \int_{\phi'=0}^{2\pi} (Z^t_{TM} \cos^2 \phi' + Z^t_{TE} \sin^2 \phi') k_\rho d\phi' dk_\rho. \qquad (6.124)$$

6.5.1. *Power Radiated into the Free-Space Medium.* Like for a small free-space dipole element, any field or potential for a printed dipole element is expected to approach infinite magnitude, as one approaches the central point of the source. Accordingly, the magnitudes of V_{in} and Z_{in} are also expected to be infinity. These quantities are expected to be of finite magnitude only for a practical printed dipole element with a non-zero length and width, with a realistic current distribution different from the ideal delta-function distribution of (6.95). Input impedance for such practical printed dipole elements would be covered in chapter 9. The magnitude of Z_{in} in (6.124) for the printed-dipole element of (6.95) would approach infinity, due to an infinite value only for the imaginary part of the Z_{in}, which represents infinite reactive fields or stored energy near the source. The real part of Z_{in} may still be finite, representing a finite value for the time-averaged power radiated from the printed-dipole element of (6.95) with a finite current I. In the following, only the real part R_{in} of the integral (6.124), as related to the time-averaged input power $P_{in} = \frac{1}{2}|I|^2 R_{in}$, will be derived. This assumes that $|I|$ is the peak current amplitude. The integral for the R_{in} would be evaluated computationally for specific sets of physical parameters.

It may be shown using (6.106, 6.107) that the individual impedance terms Z_{TM0} and Z_{TM}, and therefore the total impedance Z^t_{TM} of (6.108, 6.124), would be purely

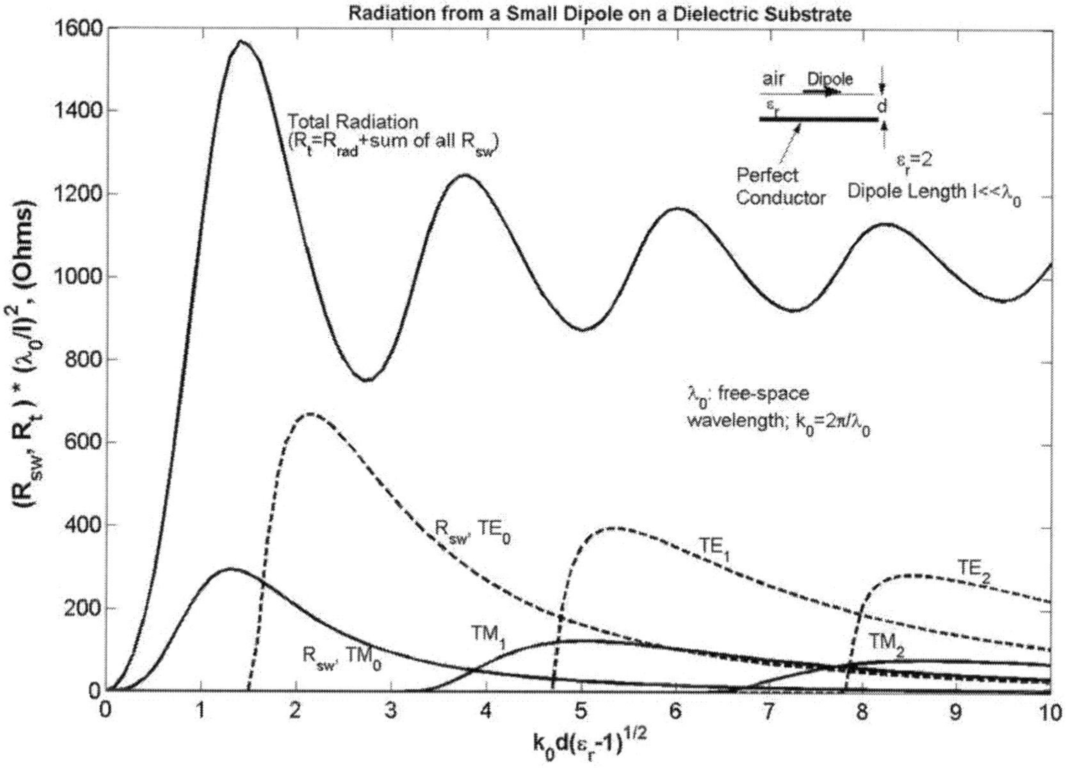

FIGURE 10

imaginary for wave numbers $k_\rho > k_0$, for a slab structure without any material loss (real k_1). Similarly, the corresponding impedance terms Z_{TE0} and Z_{TE} of (6.106, 6.107), and therefore the total impedance Z^t_{TE} of (6.108, 6.124), would also be purely imaginary, in the range $k_\rho > k_0$. Accordingly, the integration of (6.124) in the range $k_\rho > k_0$ would contribute only to the input reactance X_{in}. The range $k_\rho > k_0$ represents evanescent waves (see chapter 5, section 2.3). On the other hand, in the range $0 < k_\rho < k_0$, the Z_{TM} and Z_{TE} would still be purely imaginary for a loss-less slab, but the Z_{TM0} and Z_{TE0} would be purely real, leading to a non-zero real part for the Z^t_{TM} and Z^t_{TE}. This would contribute to the real part of the input impedance of (6.124) in the integration range $k_\rho < k_0$. The range $k_\rho < k_0$ represents propagating or radiating waves into the free-space, and the real part of the impedance of (6.124), evaluated in the integration range $k_\rho < k_0$, is referred to as the free-space radiation resistance R_{rad} of the printed-dipole element. The R_{rad} is related to the corresponding power P_{rad} radiated into the free-space region.

$$R_{rad} = Real[\frac{l^2}{4\pi^2} \int_{k_\rho=0}^{k_0} \int_{\phi'=0}^{2\pi} (Z^t_{TM} \cos^2 \phi' + Z^t_{TE} \sin^2 \phi') k_\rho d\phi' dk_\rho]$$

$$= \frac{l^2}{4\pi} \int_{k_\rho=0}^{k_0} Real(Z^t_{TM} + Z^t_{TE}) k_\rho dk_\rho,$$

$$P_{rad} = \frac{1}{2} |I|^2 R_{rad}. \tag{6.125}$$

Because Z^t_{TM} and Z^t_{TE} are functions of only k_ρ, independent of the angle ϕ', the ϕ' integration is analytically evaluated. The R_{rad} may be decomposed into R^{TM}_{rad} and R^{TE}_{rad}, as contributed from the TM_z and TE_z parts of the above integration.

$$R^{TM}_{rad} = \frac{l^2}{4\pi} \int_{k_\rho=0}^{k_0} Real(Z^t_{TM}) k_\rho dk_\rho, \; R^{TE}_{rad} = \frac{l^2}{4\pi} \int_{k_\rho=0}^{k_0} Real(Z^t_{TE}) k_\rho dk_\rho,$$

$$P^{TM}_{rad} = \frac{1}{2} |I|^2 R^{TM}_{rad}, \; P^{TE}_{rad} = \frac{1}{2} |I|^2 R^{TE}_{rad} \; . \tag{6.126}$$

The radiation resistance R_{rad} as computed using (6.125), together with all surface-wave resistances R_{sw} associated with surface-wave radiation to be derived in the following section, are plotted in Figs.10-12 for three different substrate dielectric constants $\epsilon_r = 2$, 5 and 10.

6.5.2. *Power Radiated as a Surface Wave.* As discussed above, the real part of the input impedance in (6.124) is contributed only from the integration range $k_\rho < k_0$. This is true, strictly when there is no material loss. For a practical situation with a small but non-zero material loss in the slab medium, the real part of (6.124) from the integration range $k_\rho < k_0$ would essentially remain unchanged, with only a small additional real value attributed to the material loss. On the other hand, with the material loss in the slab, the integrand in the range $k_\rho > k_0$ would no longer be purely imaginary, and the additional real part in this integration range would also contribute to the real part of the input impedance. However, the additional real part in the range $k_\rho > k_0$, introduced due to the small material loss, may not be as insignificant as that in the range $k_\rho < k_0$. This special situation may be attributed particularly to the singular nature of the integrand at the surface-wave poles $k_\rho = k_{\rho p}$. We will examine this special effect in the following.

Let us approximate the Z_{in} integrand of (6.124) in the vicinity of a surface-wave singularity $k_\rho = k_{\rho p}$. The integrand of (6.124) with respect to k_ρ consists of the impedance functions Z^t_{TM} and Z^t_{TE}, and an additional integration factor k_ρ resulting from converting the original integral of (6.124) from rectangular (k_x and k_y) to polar (k_ρ and ϕ') coordinates. The impedance functions Z^t_{TM} and Z^t_{TE} may be substituted with the respective approximate expressions derived in section 6.4, and the additional k_ρ factor may be substituted with the singular wave number $k_{\rho p}$, in order to approximate the integrand of (6.124) in the vicinity of $k_\rho = k_{\rho p}$. With the above substitutions, contribution to the

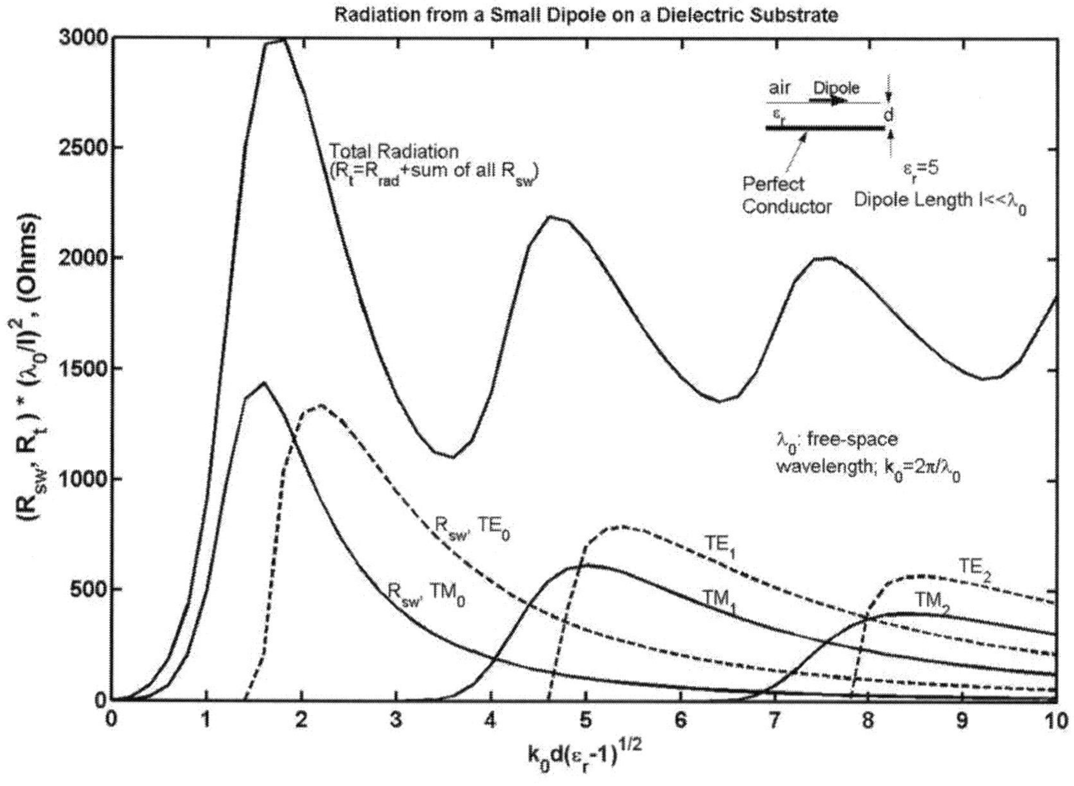

FIGURE 11

input impedance Z_{in} of (6.124) from a surface-wave singularity may be extracted in a simple form. The contribution from a TM_z or a TE_z singularity is represented by Z_{sw}^{TM} or Z_{sw}^{TE}, respectively.

$$Z_{in} = \frac{l^2}{4\pi^2} \int_{k_\rho=0}^{\infty} \int_{\phi'=0}^{2\pi} (Z_{TM}^t \cos^2\phi' + Z_{TE}^t \sin^2\phi') k_\rho d\phi' dk_\rho, \quad (6.127)$$

$$Z_{sw}^{TM} = \frac{l^2}{4\pi^2} \int_{k_\rho=0}^{\infty} \int_{\phi'=0}^{2\pi} \frac{a'}{(k_\rho^2 - k_{\rho p}^2)} \cos^2\phi' k_{\rho p} d\phi' dk_\rho = \frac{l^2}{4\pi} \int_{k_\rho=0}^{\infty} \frac{a'}{(k_\rho^2 - k_{\rho p}^2)} k_{\rho p} dk_\rho,$$

$$Z_{sw}^{TE} = \frac{l^2}{4\pi^2} \int_{k_\rho=0}^{\infty} \int_{\phi'=0}^{2\pi} \frac{b'}{(k_\rho^2 - k_{\rho p}^2)} \sin^2\phi' k_{\rho p} d\phi' dk_\rho = \frac{l^2}{4\pi} \int_{k_\rho=0}^{\infty} \frac{b'}{(k_\rho^2 - k_{\rho p}^2)} k_{\rho p} dk_\rho,$$

$$Z_{TM}^t \simeq \frac{a'}{(k_\rho^2 - k_{\rho p}^2)}, \quad Z_{TM}^t k_\rho \simeq \frac{a' k_{\rho p}}{(k_\rho^2 - k_{\rho p}^2)},$$

$$Z_{TE}^t \simeq \frac{b'}{(k_\rho^2 - k_{\rho p}^2)}, \quad Z_{TE}^t k_\rho \simeq \frac{b' k_{\rho p}}{(k_\rho^2 - k_{\rho p}^2)}. \quad (6.128)$$

It may be recognized, the singular function used in the above integral with coefficient a' is meant to capture the behavior of the integrand in the vicinity of the singularity. The behavior at or near the singularity carries information for the associated surface-wave, propagating away to large distances from the source. This concept is explained in later discussions. Variation of the singular function away form the singularity is not of critical physical significance. In principle, any other singular function with the same behavior near the singularity, but a different variation away from it, could be used. All such equivalent functions would lead to the same results for the surface-wave power derived here, as well as for evaluation of the surface-wave fields at large distances to be studied in the following section 6.6.2. The particular choice of the singular function we are using is mathematically simple, resulting in convenient Fourier analysis.

The integrands in (6.128) are simple functions of ϕ', which were analytically integrated over the range $0 < \phi' < 2\pi$. In the following, we will present mathematical treatment for the remaining k_ρ-integral, for only the TM_z part Z_{sw}^{TM}. The procedure can be repeated as well for the TE_z part Z_{sw}^{TE}.

The k_ρ integrand in (6.128) is an even function with respect to the positive and negative values of k_ρ. Therefore, for analytical convenience, the integral (6.128) over the range $0 < k_\rho < +\infty$ may be expressed as half of that over $-\infty < k_\rho < +\infty$. Further, the singular form of the integrand may be decomposed as a sum of two functions, one with a singularity at $k_\rho = k_{\rho p}$ and the other at $k_\rho = -k_{\rho p}$.

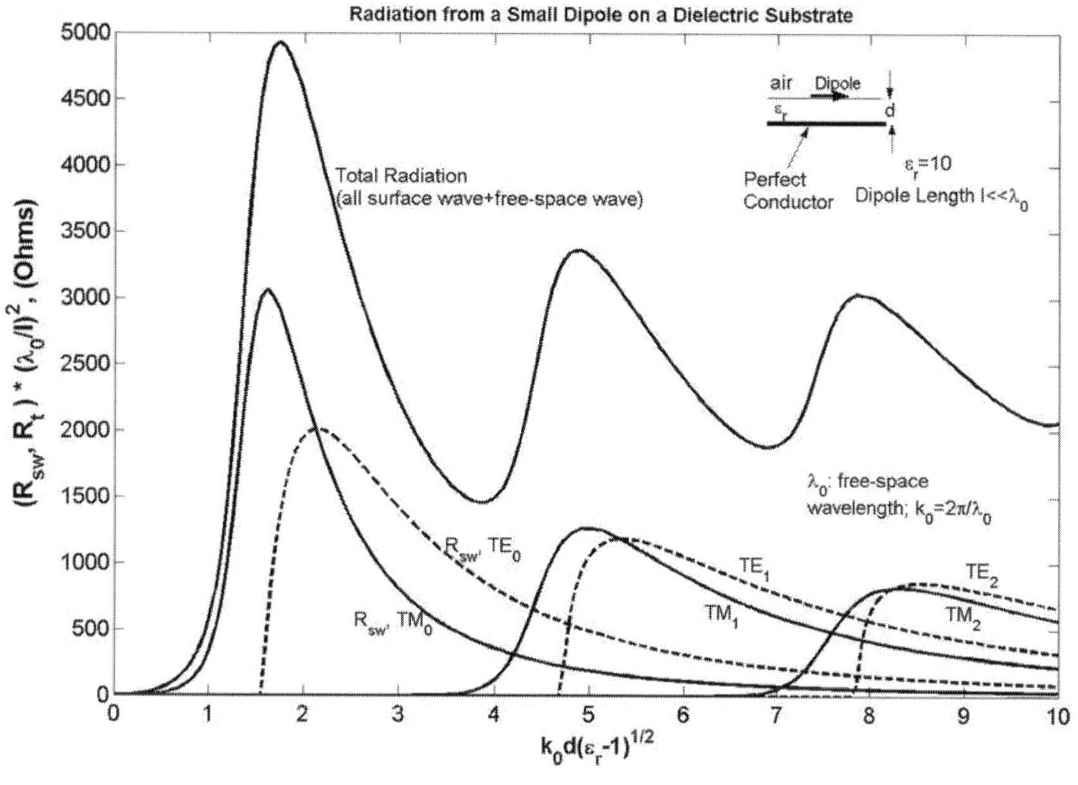

FIGURE 12

$$Z_{sw}^{TM} = \frac{l^2}{4\pi}\int_{k_\rho=0}^{\infty}\frac{a'k_{\rho p}}{(k_\rho^2-k_{\rho p}^2)}dk_\rho = \frac{l^2}{8\pi}\int_{k_\rho=-\infty}^{\infty}\frac{a'k_{\rho p}}{(k_\rho^2-k_{\rho p}^2)}dk_\rho$$

$$= \frac{l^2}{8\pi}\int_{k_\rho=-\infty}^{\infty}[\frac{1}{k_\rho-k_{\rho p}}-\frac{1}{k_\rho+k_{\rho p}}]\frac{a'}{2}dk_\rho = \frac{l^2a'}{16\pi}\int_{k_\rho=-\infty}^{\infty}[\frac{1}{k_\rho-k_{\rho p}}-\frac{1}{k_\rho+k_{\rho p}}]dk_\rho. \quad (6.129)$$

As discussed in section 6.4, the pole $k_{\rho p}$ need to be substituted by $k'_{\rho p} = k_{\rho p} - j\delta_p$ to model a realistic dielectric slab with a small but non-zero loss, where δ_p is a small positive number approaching zero.

$$Z_{sw}^{TM} = \frac{l^2 a'}{16\pi} \int_{k_\rho=-\infty}^{\infty} [\frac{1}{k_\rho - (k_{\rho p} - j\delta_p)} - \frac{1}{k_\rho + (k_{\rho p} - j\delta_p)}] dk_\rho$$

$$= \frac{l^2 a'}{16\pi} \int_{k_\rho=-\infty}^{\infty} [\frac{1}{(k_\rho - k_{\rho p}) + j\delta_p} - \frac{1}{(k_\rho + k_{\rho p}) - j\delta_p}] dk_\rho. \quad (6.130)$$

Now, consider a function $f(k_\rho)$ of the following basic form, and examine its characteristic as its parameter ζ approaches zero.

$$f(k_\rho;\zeta) = \frac{1}{k_\rho \pm j\zeta} = \frac{k_\rho \mp j\zeta}{k_\rho^2 + \zeta^2} = \frac{k_\rho}{k_\rho^2 + \zeta^2} \mp j\frac{\zeta}{k_\rho^2 + \zeta^2} = f_1(k_\rho;\zeta) \mp j f_2(k_\rho;\zeta),$$

$$f_1(k_\rho;\zeta) = \frac{k_\rho}{k_\rho^2 + \zeta^2};\ f_1(k_\rho;\zeta \to 0) = \frac{1}{k_\rho},\ k_\rho \neq 0,$$

$$f_2(k_\rho;\zeta) = \frac{\zeta}{k_\rho^2 + \zeta^2};\ f_2(k_\rho;\zeta \to 0) = 0,\ k_\rho \neq 0;\ = \infty,\ k_\rho = 0,$$

$$\int_{k_\rho=-\infty}^{\infty} f_2(k_\rho;\zeta) dk_\rho = \int_{k_\rho=-\infty}^{\infty} \frac{\zeta}{k_\rho^2 + \zeta^2} dk_\rho = \int_{t=-\pi/2}^{\pi/2} \frac{\zeta}{\zeta^2 \sec^2 t} \zeta \sec^2 t dt = \pi\ ,$$

$$k_\rho = \zeta \tan t;\ dk_\rho = (\zeta \sec^2 t) dt;\ t = \pm\pi/2,\ k_\rho = \pm\infty,$$

$$f_2(k_\rho;\zeta \to 0) = \pi\delta(k_\rho);\ f(k_\rho;\zeta \to 0) = \frac{1}{k_\rho} \mp j\pi\delta(k_\rho),$$

$$\int_{k_\rho=-\infty}^{\infty} f(k_\rho;\zeta \to 0) dk_\rho = \int_{k_\rho=-\infty}^{\infty} \big(\frac{1}{k_\rho} \mp j\pi\delta(k_\rho)\big) dk_\rho = \mp j\pi. \quad (6.131)$$

As ζ approaches zero, the first part f_1 of the function $f(k_\rho)$ would approach $1/k_\rho$ for all non-zero values of k_ρ with its integration over $-\infty < k_\rho < +\infty$ equal to zero, whereas the second part f_2 would exhibit a special functional behavior. As ζ approaches zero, the second part f_2 can be seen to approach zero for all non-zero values of k_ρ, but an infinite value for $k_\rho = 0$. Further, the integral of f_2 can be shown to be a finite value equal to π. Accordingly, the f_2 is represented by a $\delta(k_\rho)$ functional of amplitude π, leading to an equivalent representation of $f(k_\rho)$ as equal to $1/k_\rho \mp \pi\delta(k_\rho)$, under the above limiting condition $\zeta \to 0$.

Using the above functional behavior of $f(k_\rho)$, the integrand of (6.130) may also be expressed in terms of δ-functions.

$$Z_{sw}^{TM} == \frac{l^2a'}{16\pi} \int_{k_\rho=-\infty}^{\infty} [\frac{1}{(k_\rho - k_{\rho p}) + j\delta_p} - \frac{1}{(k_\rho + k_{\rho p}) - j\delta_p}] dk_\rho$$

$$= \frac{l^2a'}{16\pi} \int_{k_\rho=-\infty}^{\infty} [(\frac{1}{(k_\rho - k_{\rho p})} - j\pi\delta(k_\rho - k_{\rho p})) - (\frac{1}{(k_\rho + k_{\rho p})} + j\pi\delta(k_\rho - k_{\rho p}))] dk_\rho . \quad (6.132)$$

We may substitute $k'_\rho = k_\rho - k_{\rho p}$ in evaluating the first part of the integral, but substitute $k'_\rho = k_\rho + k_{\rho p}$ for the second part. This would result in having the pole location at $k'_\rho = 0 - j\delta_p$ and $k'_\rho = 0 + j\delta_p$, respectively for the first and second parts of the integral, such that the real part of the pole for each case is equal to zero,

$$Z_{sw}^{TM} = \frac{l^2a'}{16\pi} \int_{k'_\rho=-\infty}^{\infty} [\frac{1}{k'_\rho + j\delta_p} - \frac{1}{k'_\rho - j\delta_p}] dk'_\rho$$

$$= \frac{l^2a'}{16\pi} \int_{k'_\rho=-\infty}^{\infty} [(\frac{1}{k'_\rho} - j\pi\delta(k'_\rho)) - (\frac{1}{k'_\rho} + j\pi\delta(k'_\rho))] dk'_\rho = \frac{-jl^2a'}{8} = R_{sw}^{TM}. \quad (6.133)$$

The coefficient a' is a positive imaginary number, as derived in (6.117), leading to a real positive value for the $Z_{sw}^{TM} = R_{sw}^{TM}$. Note that the purely imaginary value for a' results in a purely imaginary integrand in (6.128) for Z_{sw}^{TM}, when evaluated along the real axis of k_ρ, with a real value of $k_{\rho p}$ associated with an ideal lossless material. However, under a realistic material condition, where $k_{\rho p}$ is substituted with $k'_{\rho p} = k_{\rho p} - j\delta_p$ having a small negative imaginary part, the purely imaginary original integrand of Z_{sw}^{TM} results in a purely real positive value for the $Z_{sw}^{TM} = R_{sw}^{TM}$. This is an important result of fundamental mathematical and physical significance. The resulting resistance value R_{sw}^{TM} is referred to as the TM_z surface-wave resistance, and the power $P_{sw}^{TM} = |I|^2 R_{sw}^{TM}/2$ is referred to as the surface-wave power, associated with the TM_z surface-wave mode with a propagation constant $k_\rho = k_{\rho p}$.

Note that the above non-zero resistance R_{sw}^{TM}, and the associated power P_{sw}^{TM}, result from integration of the delta-function part in (6.132, 6.133). This is contributed only from the immediate vicinity of the integrand at $k'_\rho = 0$, or $k_\rho = \pm k_{\rho p}$, which is equivalent to the vicinity of $k_\rho^2 = k_{\rho p}^2$. Because $k_{\rho p}$ is greater than k_0, the wave numbers associated with the above power generate only evanescent fields in the free-space region above the slab, that do not radiate into the free-space medium. The power may, however, be interpreted as a different form of radiation.

The resistance R_{sw}^{TM} in (6.133), and the resulting power P_{sw}^{TM}, is independent of any small value of δ_p, associated with any infinitesimally small, but non-zero, material loss in the slab. The non-zero amount of power P_{sw}^{TM} is to be dissipated in the entire slab medium, distributed in the form of very small, non-zero amount of material loss over finite local regions. This is possible by having the dissipation spread over a large volume, covering significant distances away from the source. The associated fields are also spread covering the large distances. The mechanism for the power P_{sw}^{TM} can be interpreted in the following way. The power first originates from the source in the form of a surface wave with the propagation constant $k_\rho = k_{\rho p} > k_0$. The surface-wave then spreads over the large distances from the source, and all the power P_{sw} eventually dissipates in the slab medium over the entire expanse of the wave, assuming that the slab medium realistically has a small, non-zero amount of material loss.

A similar basic interpretation would apply as well to a regular radiation process into a free-space medium. For the free-space radiation, the radiation power, originating from the source, spreads in the form of propagating waves into the entire free-space region. All the radiation power in this case would also dissipate in the form of material loss, distributed in small quantities over the large free-space region, if there is any small but non-zero loss mechanism introduced in the free-space medium. However, unlike in the above derivation of the surface-wave radiation power, we did not need to explicitly introduce any small material loss in the derivation of the free-space radiation power in the last section 6.5.1. This is a distinct difference between the analyses of the surface-wave and free-space radiation powers.

The above interpretation of radiation in terms of material loss is a fundamental concept. It views radiation power equivalently in the form of material loss distributed in the propagating medium, in the limit of its material-loss paramter approaching zero. As discussed above, the basic interpretation applies to the surface-wave as well as the free-space radiation, although there are distinct theoretical and physical differences between the two processes. The P_{sw}^{TM} is referred to as the surface-wave power, and the associated mechanism the surface-wave radiation process, in distinction from the regular radiation into a free-space medium. The free-space radiation propagates away from the source in all spatial directions (θ, ϕ) into the free-space medium above the slab. Whereas, the surface-wave radiation propagates only in lateral directions $(\theta = 90^0, \phi)$ from the source, parallel to the plane of the slab, but is evanescent in the free-space region.

The P_{sw}^{TM} and R_{sw}^{TM} specifically refer to a TM_z type of surface-wave. Derivation for a TE_z surface wave can be similarly repeated substituting the coefficient a' by the corresponding coefficient b' for a TE_z wave, with the following result.

$$Z_{sw}^{TE} = \frac{-jl^2 b'}{8} = R_{sw}^{TE}, \quad P_{sw}^{TE} = \frac{1}{2}|I|^2 R_{sw}^{TE}. \tag{6.134}$$

The surface-wave resistance R_{sw} for individual TM_z and TE_z modes, as computed using (6.133, 6.134), are plotted in Figs.10-12 for three different substrate dielectric constants $\epsilon_r = 2$, 5 and 10.

The large spatial spread of the surface-wave radiation fields, on a plane parallel to the slab surface, as we suggested earlier, can be analytically anticipated. We may examine the Fourier integral for a field or potential component of the form (6.123), which is an inverse transform from the transform variables (k_x, k_y) to the spatial variables (x, y). Any function that would spread over a larger spatial region in (x, y), on a plane parallel to the slab interface, must have relatively sharp variation of its transform with respect to (k_x, k_y), and vice-versa. This is a basic characteristic of Fourier transformation. Based on this basic characteristic, the singular behavior of the integrands Z_{TM}^t or Z_{TE}^t of (6.123) at

$k_\rho = (k_x^2+k_y^2)^{1/2} = k_{\rho p}$ is expected to generate a field that would spread over large distances $\rho = (x^2+y^2)^{1/2}$ in space. Accordingly, the Fourier integral of a field or potential component can be analytically approximated for large distances from the source, by approximating the integrand in the vicinity of a surface-wave singularity $k_\rho = k_{\rho p}$. In the following section, we will derive suitable approximate expressions for the surface-wave fields, using the above principle.

6.6. Fields at Large Distance from Source. For a free-space dipole in section 3, rigorous expressions for the potential and field components could be analytically derived in simple forms, which are valid for all distances. In contrast, rigorous expressions for a printed-dipole element may not be possible in such simple analytical forms. Only approximate expressions for the potential and field components of a printed-dipole can be derived in simple forms, that are valid for sufficiently large distances from the source. Such approximations are studied in the following.

Only the propagating radiation fields, which include both the free-space and surface-wave radiation parts, are expected to spread over large distances away from the source. The surface-wave fields propagate along a plane parallel to the slab surface, as a function of $\rho = (x^2 + y^2)^{1/2}$, but exponentially decay away from the slab surface as a function of z. Whereas the free-space radiation fields propagate away from the source in all radial directions in spherical coordinates, as a function of r. The rest of the fields are tightly confined near the source as non-propagating or non-radiating parts, in the form of reactive or stored energy. Therefore, in order to model the fields at a sufficiently large distance from the source one needs only the radiation fields, expressed in suitable approximate forms that are valid at the large distances. The free-space radiation fields consist of all propagating waves with wave numbers $k_\rho < k_0$. Therefore, we need to evaluate the Fourier integral of a field component of the form (6.123) only over the limited integration range $k_\rho < k_0$, approximated for large distances $(r \to \infty)$. On the other hand, the surface-wave contribution comes from the associated singularity at $k_{\rho p} > k_0$, and a Fourier analysis of the singular behavior of the integrand would lead to the desired approximation at a large distance ρ. We will analyze the free-space and surface-wave radiation parts separately.

6.6.1. *Free-Space Radiation Fields.* Let us evaluate, for example the electric field E_x of (6.123), at large distance $r \to \infty$ contributed by the free-space radiation. In the process, we will retain the separate contributions from its TM_z and TE_z parts of the field. Any other field or potential component can be similarly evaluated, by replacing the integrand for the E_x in (6.123), with that for the respective field or potential component. The integrand would be the required field or potential component, produced by a Fourier current sheet of (6.96), and is to be derived following the analysis of sections 6.1-6.3. Alternately, at a large distance $r \to \infty$ all other field components may be directly related to the TM_z and TE_z parts of the specific field E_x. This is possible because the fields in the far distances can be approximately treated as plane waves, which allow simple relationships between their field components as studied in the plane-waves chapter 5. We will follow the later approach to derive the field components E_θ and E_ϕ by relating them to the TE and TM parts of the field E_x.

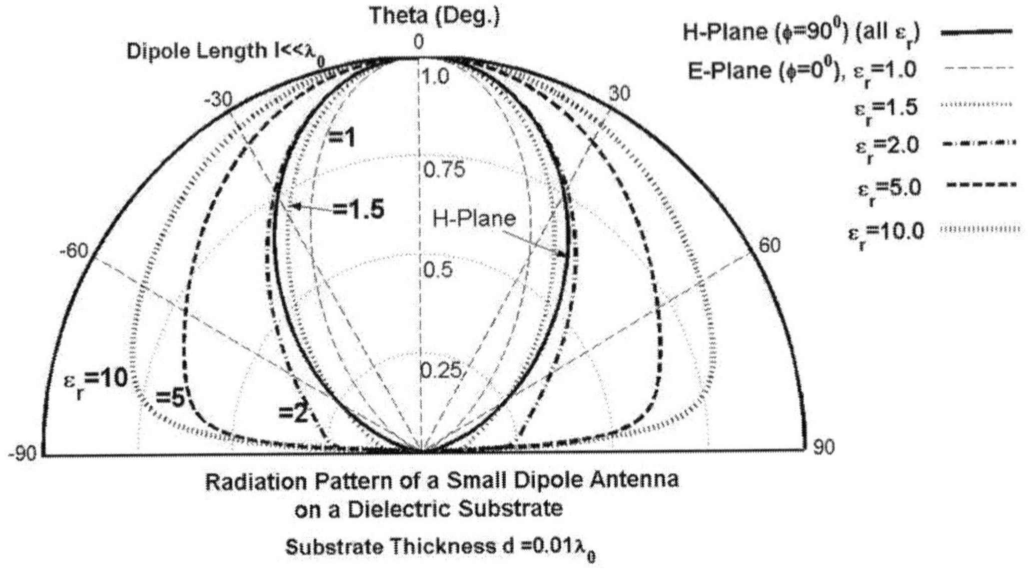

FIGURE 13

$$E_x(x, y, z > 0) =$$

$$\frac{Il}{4\pi^2} \int_{-\infty}^{\infty} \int_{-\infty}^{+\infty} (-Z_{TM}^t \cos^2 \phi' - Z_{TE}^t \sin^2 \phi') e^{+jk_x x} e^{+jk_y y} e^{-jk_{z0} z} dk_x dk_y, \tag{6.135}$$

$$E_x(r \to \infty) \simeq$$

$$\frac{Il}{4\pi^2} \int \int_{(k_x^2 + k_y^2 < k_0^2)} (-Z_{TM}^t \cos^2 \phi' - Z_{TE}^t \sin^2 \phi') e^{+jk_x x} e^{+jk_y y} e^{-jk_{z0} z} dk_x dk_y$$

$$= \frac{Il}{4\pi^2} \int \int_{(k_x^2 + k_y^2 < k_0^2)} (-Z_{TM}^t \cos^2 \phi' - Z_{TE}^t \sin^2 \phi')$$

$$\times e^{+jk_0 r (\sin\theta' \cos\phi' \sin\theta \cos\phi + \sin\theta' \sin\phi' \sin\theta \sin\phi - \cos\theta' \cos\theta)} k_0^2 \sin\theta' \cos\theta' d\theta' d\phi'$$

$$= \frac{Il}{4\pi^2} \int_{\phi'=0}^{2\pi} \int_{\theta'=0}^{\pi/2} (-Z_{TM}^t \cos^2 \phi' - Z_{TE}^t \sin^2 \phi')$$

$$\times e^{+jk_0 r (\sin\theta' \sin\theta \cos(\phi - \phi') - \cos\theta' \cos\theta)} k_0^2 \sin\theta' \cos\theta' d\theta' d\phi' \ , \tag{6.136}$$

$$(k_x = k_0 \sin\theta' \cos\phi',\ k_y = k_0 \sin\theta' \sin\phi',\ k_{z0} = \sqrt{k_0^2 - k_x^2 - k_y^2} = k_0 \cos\theta',$$
$$dk_x dk_y = k_0^2 \sin\theta' \cos\theta' d\theta' d\phi',\ Re(k_{z0}) > 0,\ Im(k_{z0}) < 0,\ 0 < \theta < \pi/2)\ .$$

We will use the stationary-phase method of integration in Appendix F, section 2.3, with the following substitutions. Note that the integration variables in the above integral are θ' and ϕ', with θ and ϕ appearing as specific parameters, whereas in Appendix F, section 2.3, the integration variables are θ and ϕ.

$$f(\theta', \phi') = -\frac{Il}{4\pi^2}[Z_{TM}^t(k_\rho = k_0 \sin\theta') \cos^2\phi'$$
$$+Z_{TE}^t(k_\rho = k_0 \sin\theta') \sin^2\phi'] k_0^2 \sin\theta' \cos\theta',\ \alpha = k_0 r,$$
$$g(\theta', \phi') = -(\sin\theta' \sin\theta \cos(\phi' - \phi) - \cos\theta' \cos\theta),$$

$$\frac{\partial g}{\partial \theta'} = -(\cos\theta' \sin\theta \cos(\phi' - \phi) + \sin\theta' \cos\theta) = 0,$$
$$\frac{\partial g}{\partial \phi'} = \sin\theta' \sin\theta \sin(\phi' - \phi) = 0,$$
$$\theta_0' = \theta,\ \phi_0' = \phi + \pi\ (0 < \theta', \theta < \pi/2),$$

$$g(\theta_0', \phi_0') = -(\sin\theta_0' \sin\theta \cos(\phi_0' - \phi) - \cos\theta_0' \cos\theta) = +1,$$
$$\frac{\partial^2 g(\theta_0', \phi_0')}{\partial \theta'^2} = -(-\sin\theta_0' \sin\theta \cos(\phi_0' - \phi) + \cos\theta_0' \cos\theta) = -1,$$
$$\frac{\partial^2 g(\theta_0', \phi_0')}{\partial \phi'^2} = \sin\theta_0' \sin\theta \cos(\phi_0' - \phi) = -\sin^2\theta,$$
$$\frac{\partial^2 g(\theta_0', \phi_0')}{\partial \phi' \partial \theta'} = \cos\theta_0' \sin\theta \sin(\phi_0' - \phi) = 0\ . \tag{6.137}$$

The above parameters allow the Fourier integral of (6.136) to be expressed in the general two-dimensional integral form of Appendix F, section (2.3. Accordingly, the integral of (6.136) may be approximated by substituting the above parameters in the general result of the Appendix F, section 2.3, as expressed in the following.

Note that the actual limits of integration in (6.136) really do not matter for performing the stationary-phase integration, as discussed in the Appendix F. The general integration form in Appendix F, section 2.3, generically shows integration limits covering variables from $-\infty$ to $+\infty$, but the results would be the same for any integration limits that covers the stationary-phase point (θ_0', ϕ_0'). The integration limits $0 < \theta' < \pi/2$ and $0 < \phi' < 2\pi$ for (6.136) would cover any stationary point $\theta' = \theta_0' = \theta$ and $\phi' = \phi_0' = \phi$ as derived above, for all spatial directions of observation $0 < \theta < \pi/2$, $0 < \phi < 2\pi$. Therefore, the results of the integration in (6.136), when evaluated for $r \to \infty$, are expected to be the same as those with integration limits covering $-\infty$ to $+\infty$.

$$E_x(r,\theta,\phi)_{r\to\infty} \simeq -\frac{Il}{4\pi^2}[Z^t_{TM}(k_\rho = k_0\sin\theta)\cos^2\phi + Z^t_{TE}(k_\rho = k_0\sin\theta)\sin^2\phi]$$
$$\times k_0^2\sin\theta\cos\theta\times e^{-jk_0r}\times\sqrt{\frac{2\pi}{-jk_0r}}\times\sqrt{\frac{2\pi}{-jk_0r\sin^2\theta}}$$
$$= -jI(\frac{l}{\lambda_0})\frac{e^{-jk_0r}}{r}\cos\theta[Z^t_{TM}(k_\rho = k_0\sin\theta)\cos^2\phi + Z^t_{TE}(k_\rho = k_0\sin\theta)\sin^2\phi],$$
$$E_x^{TM}(r,\theta,\phi)_{r\to\infty} \simeq -jI(\frac{l}{\lambda_0})\frac{e^{-jk_0r}}{r}\cos\theta[Z^t_{TM}(k_\rho = k_0\sin\theta)\cos^2\phi],$$
$$E_x^{TE}(r,\theta,\phi)_{r\to\infty} \simeq -jI(\frac{l}{\lambda_0})\frac{e^{-jk_0r}}{r}\cos\theta[Z^t_{TE}(k_\rho = k_0\sin\theta)\sin^2\phi]. \quad (6.138)$$

We have retained the independent parts of E_x contributed from the TM_z and TE_z waves. The plane-wave relations from chapter 5 can now be used to express the E_θ and E_ϕ components. Note that the E_θ component is generated only from the TM_z wave, whereas the E_ϕ component is generated only by the TE_z wave.

$$E_\theta(r,\theta,\phi)_{r\to\infty} =$$
$$\frac{E_x^{TM}(r,\theta,\phi)_{r\to\infty}}{\cos\theta\cos\phi} \simeq -jI(\frac{l}{\lambda_0})\frac{e^{-jk_0r}}{r}[Z^t_{TM}(k_\rho = k_0\sin\theta)\cos\phi],$$
$$E_\phi(r,\theta,\phi)_{r\to\infty} =$$
$$-\frac{E_x^{TE}(r,\theta,\phi)_{r\to\infty}}{\sin\phi} \simeq jI(\frac{l}{\lambda_0})\frac{e^{-jk_0r}}{r}\cos\theta[Z^t_{TE}(k_\rho = k_0\sin\theta)\sin\phi]. \quad (6.139)$$

The magnetic fields $\bar{H} = (\hat{r}\times\bar{E})/\eta_0$, and the power density $P_d = [|E_\theta|^2+|E_\phi|^2]/(2\eta_0)$ which is the power flow per unit area in the radial direction, may also be expressed using the above electric fields. The radiated power P_{rad} can then be derived by integrating the power density over a large hemi-spherical surface above the printed dipole. This would provide an alternate method of derivation for the P_{rad} we have already expressed in (6.125). The result of (6.125) was derived using the electric field E_x evaluated at the center of the source ($r = 0$), whereas the alternate approach would use fields of (6.139) evaluated far from the source ($r \to \infty$). However, both contrasting approaches are expected to provide the same result. In other words, the free-space radiation power originating from the source is expected to be equal to the power captured away from the source over the closed hemi-sphere. This is based on the principle of power conservation, which assumes that the free-space medium is loss-less, as is the case in the derivation of (6.139).

$$H_\phi(r,\theta,\phi)_{r\to\infty} = \frac{1}{\eta_0}E_\theta(r,\theta,\phi)_{r\to\infty},\ H_\theta(r,\theta,\phi)_{r\to\infty} = -\frac{1}{\eta_0}E_\phi(r,\theta,\phi)_{r\to\infty},$$

$$P_d(r,\theta,\phi) = \frac{1}{2\eta_0}|E(r,\theta,\phi)|^2{}_{r\to\infty} = \frac{|I|^2}{2\eta_0 r^2}(\frac{l}{\lambda_0})^2\big[\cos^2\phi|Z^t_{TM}(k_\rho = k_0\sin\theta)|^2$$
$$+\cos^2\theta\sin^2\phi|Z^t_{TE}(k_\rho = k_0\sin\theta)|^2\big]. \quad (6.140)$$

The power radiation pattern, which is the angular dependence of the above power density function P_d, normalized to a unity maximum value, is plotted in Figs.13, 14 for different dielectric constants ϵ_r and two practical substrate thicknesses $d = 0.01\lambda_0, 0.05\lambda_0$. For a practical design with a small substrate thickness d, the pattern in the $\phi = 90^0$ plane is essentially independent of the substrate dielectric constant ϵ_r. On the other hand, in the plane $\phi = 0^0$, the pattern tends to have a wider angular spread for a larger ϵ_r and a smaller substrate thickness d.

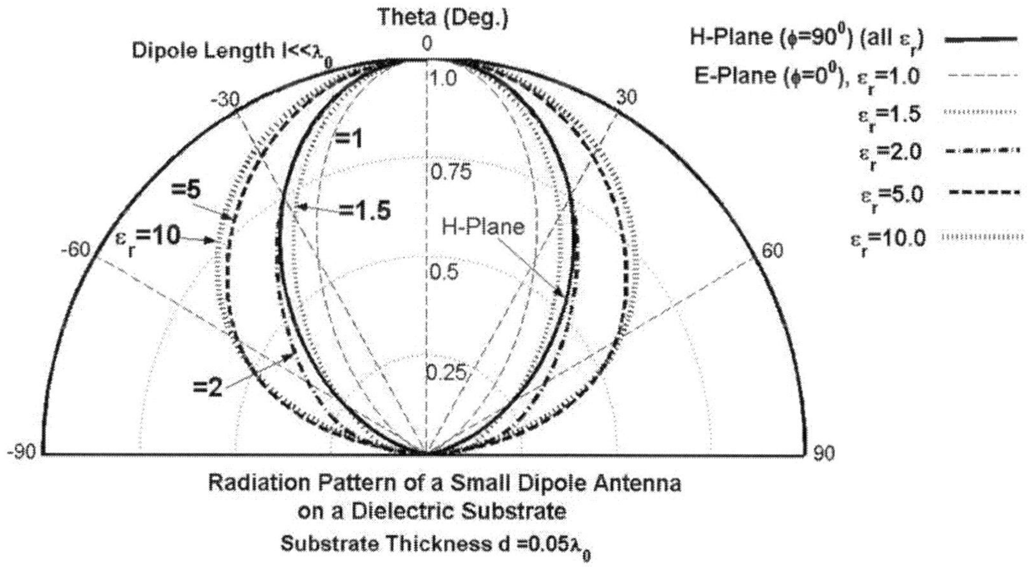

FIGURE 14

6.6.2. *Surface-Wave Radiation Fields.* As discussed before, approximation of a field or potential component at a large distance ρ from the source, contributed by the surface-wave radiation process, would be based on the singular behavior of the associated Fourier integrand. It would require evaluation of the Fourier integral by properly extracting the singular part(s) of the integral at one or more surface-wave pole(s). It would be convenient to evaluate the Fourier integral in polar variables k_ρ and ϕ'. The singularity occurs at a given value of the radial wave number $k_\rho = (k_x^2 + k_y^2)^{1/2} = k_{\rho p}$, independent of the angular variable ϕ'. Integration with respect to k_ρ can be simplified using Fourier analysis of singular functions, following related analysis for surface-wave power developed in section 6.5.2. On the other hand, the ϕ' integration can be evaluated using a stationary-phase method, which is valid for a large distance ρ from the source.

In the stationary-phase evaluation of the free-space radiation fields in last section 6.6.1, the Fourier integral was approximated for $r \to \infty$. In this case, all the three coordinates (x, y, z) simultaneously approached infinite magnitudes along a given direction (θ, ϕ) of observation. In contrast, the surface-wave fields would be evaluated for a fixed z coordinate, as the other two coordinates (x, y) on a plane parallel to the slab surface simultaneously approach infinite magnitudes along a given observation angle ϕ. In other words, the cylindrical distance $\rho = (x^2+y^2)^{1/2}$ approaches infinity in this case, for a given z. Evaluation at a different value of z would require the associated Fourier integrand to be calculated at the particular z.

In the following we will analyze the electric field E_x observed on the plane of the current source ($z = 0$), as $\rho \to \infty$. The Fourier integral of (6.123) for the field E_x is the integral of interest. The integrand in this case is the electric field produced by a Fourier current sheet of (6.96), evaluated at the $z = 0$ plane. This was derived in sections 6.1-6.3, using a transmission-line model of chapter 5 in order to account for the layering structure of the slab. Evaluation of the field at any other z, which may be inside ($z < 0$) or outside ($z > 0$) the slab medium, can be accomplished from the transmission-line analysis, by sampling the equivalent voltage across the transmission line circuit at the required plane. Any other field or potential component can be similarly evaluated, by selecting the required component in the current-sheet analysis of sections 6.1-6.3. We will specifically model contribution from a TM_z singularity in the impedance function Z^t_{TM}. We will assume that there is only one TM_z singularity. Any more TM_z singularity, or one or more TE_z singularity in the impedance function Z^t_{TE}, can be similarly treated.

$$E_x(x, y, z > 0) =$$

$$-\frac{Il}{4\pi^2} \int_{-\infty}^{\infty} \int_{-\infty}^{+\infty} (Z^t_{TM} \cos^2 \phi' + Z^t_{TE} \sin^2 \phi') e^{+jk_x x} e^{+jk_y y} e^{-jk_{z0} z} dk_x dk_y, \tag{6.141}$$

$$E_x(\rho, \phi; z = 0) = -\frac{Il}{4\pi^2} \int_{k_\rho=0}^{\infty} \int_{\phi'=0}^{2\pi} (Z^t_{TM}(k_\rho) \cos^2 \phi' + Z^t_{TE}(k_\rho) \sin^2 \phi')$$

$$\times e^{+jk_\rho \rho(\cos\phi' \cos\phi + \sin\phi' \sin\phi)} k_\rho d\phi' dk_\rho, \quad (k_x = k_\rho \cos\phi',\ k_y = k_\rho \sin\phi') \ ,$$

$$E_x(\rho, \phi; z = 0)_{\rho\to\infty} \simeq$$
$$-\frac{Il}{4\pi^2} \int_{k_\rho=0}^{\infty} \int_{\phi'=0}^{2\pi} (\frac{a'}{k_\rho^2 - k_{\rho p}^2} \cos^2\phi') e^{+jk_\rho \rho \cos(\phi'-\phi)} k_\rho d\phi' dk_\rho \Big|_{\rho\to\infty}$$
$$= -\frac{Il}{4\pi^2} \int_{k_\rho=0}^{\infty} \frac{a'}{k_\rho^2 - k_{\rho p}^2} [\int_{\phi'=0}^{2\pi} \cos^2\phi' e^{+jk_\rho \rho \cos(\phi'-\phi)} d\phi']_{\rho\to\infty} k_\rho dk_\rho ,$$

$$E_x(\rho, \phi; z = 0)_{\rho\to\infty} \simeq -\frac{Il}{4\pi^2} \int_{k_\rho=0}^{\infty} \frac{a'}{k_\rho^2 - k_{\rho p}^2} \zeta(k_\rho; \rho) k_\rho dk_\rho,$$
$$\zeta(k_\rho; \rho) = [\int_{\phi'=0}^{2\pi} \cos^2\phi' e^{+jk_\rho \rho \cos(\phi'-\phi)} d\phi']_{\rho\to\infty} . \tag{6.142}$$

Let us evaluate the ϕ' integral $\zeta(k_\rho; \rho)$ using the stationary-phase integration method of Appendix F. We will use the general single-dimensional form of integration in Appendix F, section 2.2, with the following substitutions. Proper references between the variables used in the Appendix F, section 2.2, and in the above integral ζ need to be made. The integration variable in the above integral is ϕ', with ϕ as a specific parameter, whereas in Appendix F, section 2.2, the integration variable is θ. The parameter α in the phase function of Appendix F, section 2.2, is substituted by $k_\rho \rho$ in the ζ integral.

$$f(\phi') = \cos^2\phi', \; g(\phi') = -\cos(\phi' - \phi), \; \alpha = k_\rho \rho,$$
$$g'(\phi') = \frac{dg}{d\phi'} = \sin(\phi' - \phi), \; g'(\phi'_0) = 0, \; \phi'_0 = \phi, \; \phi + \pi,$$
$$g''(\phi'_0) = \cos(\phi'_0 - \phi) = +1, \; -1; \; g(\phi'_0) = -1, \; +1 \; ,$$

$$\zeta(k_\rho; \rho) = \cos^2\phi e^{+jk_\rho \rho} \sqrt{\frac{2\pi}{jk_\rho \rho}} + \cos^2\phi e^{-jk_\rho \rho} \sqrt{\frac{2\pi}{-jk_\rho \rho}}$$
$$= \cos^2\phi \sqrt{\frac{2\pi}{k_\rho \rho}} (e^{+jk_\rho \rho} e^{-j\pi/4} + e^{-jk_\rho \rho} e^{+j\pi/4}) . \tag{6.143}$$

There are two stationary points at $\phi'_0 = \phi, \; \phi + \pi$ for the phase function $g(\phi')$. Therefore, the results of Appendix F, section 2.2 with the above substitution for the two stationary points, need to be added. Using the above result for ζ in (6.142), we get

$$
\begin{aligned}
&E_x(\rho,\phi;z=0)_{\rho\to\infty} \simeq \\
&-\frac{Il}{4\pi^2}\int\limits_{k_\rho=0}^{\infty}\frac{a'}{k_\rho^2-k_{\rho p}^2}\cos^2\phi\sqrt{\frac{2\pi}{k_\rho\rho}}(e^{+jk_\rho\rho}e^{-j\pi/4}+e^{-jk_\rho\rho}e^{+j\pi/4})k_\rho dk_\rho \\
&=-\frac{Il}{4\pi^2}\sqrt{\frac{2\pi}{\rho}}\cos^2\phi\int\limits_{k_\rho=0}^{\infty}\frac{a'\sqrt{k_\rho}}{k_\rho^2-k_{\rho p}^2}(e^{+jk_\rho\rho}e^{-j\pi/4}+e^{-jk_\rho\rho}e^{+j\pi/4})dk_\rho\ .
\end{aligned}
\tag{6.144}
$$

The above single-sided integral in the range $0 < k_\rho < +\infty$ can be expressed in a double-sided form covering $-\infty < k_\rho < +\infty$. This would be possible using suitable symmetry of the two parts of the above integrand for positive and negative values of k_ρ, and by defining the square-root term $\sqrt{k_\rho}$ to be $-j\sqrt{|k_\rho|}$ for a negative value of k_ρ. Then, the singular expression in terms of k_ρ^2, with singularity at $k_\rho^2 = k_{\rho p}^2$, can be decomposed into two parts with individual singularity at $k_\rho = +k_{\rho p}$ and $k_\rho = -k_{\rho p}$, respectively.

$$
\begin{aligned}
E_x(\rho,\phi;z=0)_{\rho\to\infty} &\simeq -\frac{Il}{4\pi^2}\sqrt{\frac{2\pi}{\rho}}\cos^2\phi\int\limits_{k_\rho=-\infty}^{\infty}\frac{a'\sqrt{k_\rho}}{k_\rho^2-k_{\rho p}^2}e^{-jk_\rho\rho}e^{+j\pi/4}dk_\rho \\
&=-\frac{Il}{4\pi^2}\sqrt{\frac{2\pi}{\rho}}\cos^2\phi e^{+j\pi/4}\int\limits_{k_\rho=-\infty}^{\infty}\frac{a'\sqrt{k_\rho}}{2k_{\rho p}}\left(\frac{1}{k_\rho-k_{\rho p}}-\frac{1}{k_\rho+k_{\rho p}}\right)e^{-jk_\rho\rho}dk_\rho \\
&\simeq -\frac{Il}{4\pi^2}\sqrt{\frac{2\pi}{\rho}}\cos^2\phi e^{+j\pi/4}\int\limits_{k_\rho=-\infty}^{\infty}\frac{a'}{2k_{\rho p}}\left(\frac{\sqrt{k_{\rho p}}}{k_\rho-k_{\rho p}}-\frac{-j\sqrt{k_{\rho p}}}{k_\rho+k_{\rho p}}\right)e^{-jk_\rho\rho}dk_\rho \\
&=-\frac{Il}{4\pi^2}\sqrt{\frac{\pi}{2\rho k_{\rho p}}}\cos^2\phi e^{+j\pi/4}a'\int\limits_{k_\rho=-\infty}^{\infty}\left(\frac{1}{k_\rho-k_{\rho p}}+j\frac{1}{k_\rho+k_{\rho p}}\right)e^{-jk_\rho\rho}dk_\rho\ .
\end{aligned}
\tag{6.145}
$$

The coefficients for the two singular parts were adjusted by substituting the square-root function $\sqrt{k_\rho}$, as we defined earlier, evaluated at the respective singular points $k_\rho = +k_{\rho p}$ and $k_\rho = -k_{\rho p}$. As discussed in section 6.4, the wave number $k_{\rho p}$ in the denominators of the two singular parts needs to be substituted by $k'_{\rho p} = k_{\rho p} - j\delta_p$ for a realistic material medium with a small but non-zero loss, where δ_p is a positive real number approaching zero. Similar treatment was also needed in section 6.5.2 for evaluation of the surface-wave radiation power, under the physically realistic material condition.

$$E_x(\rho, \phi; z = 0)_{\rho \to \infty} \simeq$$

$$-\frac{Il}{4\pi^2}\sqrt{\frac{\pi}{2\rho k_{\rho p}}}\cos^2\phi e^{+j\pi/4}a' \int_{k_\rho=-\infty}^{\infty} \left(\frac{1}{k_\rho - k_{\rho p} + j\delta_p} + j\frac{1}{k_\rho + k_{\rho p} - j\delta_p}\right)e^{-jk_\rho \rho}dk_\rho$$

$$= -\frac{Il}{4\pi^2}\sqrt{\frac{\pi}{2\rho k_{\rho p}}}\cos^2\phi e^{+j\pi/4}a' \int_{k'_\rho=-\infty}^{\infty} \left(\frac{e^{-jk_{\rho p}\rho}}{k'_\rho + j\delta_p} + j\frac{e^{+jk_{\rho p}\rho}}{k'_\rho - j\delta_p}\right)e^{-jk'_\rho \rho}dk'_\rho \, . \quad (6.146)$$

The first and second parts of the above integral were reduced by substituting the variable $k'_\rho = k_\rho - k_{\rho p}$ and $k'_\rho = k_\rho + k_{\rho p}$, respectively. Now, the k'_ρ integral in the above equation can be simplified using the results of 6.131, where the contribution due to the small loss factor $\delta_p \to 0$ is represented by a delta function.

$$\gamma = \int_{k'_\rho=-\infty}^{\infty} \left(\frac{e^{-jk_{\rho p}\rho}}{k'_\rho + j\delta_p} + j\frac{e^{+jk_{\rho p}\rho}}{k'_\rho - j\delta_p}\right)_{\delta_p \to 0} e^{-jk'_\rho \rho}dk'_\rho$$

$$= e^{-jk_{\rho p}\rho} \int_{k'_\rho=-\infty}^{\infty} \left(\frac{1}{k'_\rho} - j\pi\delta(k'_\rho)\right)e^{-jk'_\rho \rho}dk'_\rho +$$

$$je^{+jk_{\rho p}\rho} \int_{k'_\rho=-\infty}^{\infty} \left(\frac{1}{k'_\rho} + j\pi\delta(k'_\rho)\right)e^{-jk'_\rho \rho}dk'_\rho$$

$$= e^{-jk_{\rho p}\rho}(-j\pi sgn(\rho) - j\pi) + je^{+jk_{\rho p}\rho}(-j\pi sgn(\rho) + j\pi)$$

$$= -j2\pi e^{-jk_{\rho p}\rho}u(\rho) - 2\pi e^{+jk_{\rho p}\rho}u(-\rho) = -j2\pi e^{-jk_{\rho p}\rho}, \; \rho > 0 \, , \quad (6.147)$$

$$u(\rho) = \frac{1}{2}(1 + sgn(\rho)) = +1, \; \rho > 0; \; = 0, \; \rho < 0 \, ,$$

$$u(-\rho) = \frac{1}{2}(1 - sgn(\rho)) = 0, \; \rho > 0; \; = 1, \; \rho < 0 \, .$$

The second part in the above equation 6.147, expressed in the form of the step function $u(-\rho)$, would be zero for all physical values of $\rho > 0$. The radial distance ρ in cylindrical coordinates is defined as a positive distance, making a negative value of ρ physically non-existent or meaningless. We have used basic delta-function mathematics, and the following Fourier relation for the *sgn* function, in the above integration.

$$\int_{k'_\rho=-\infty}^{\infty} \frac{1}{k'_\rho}e^{-jk'_\rho \rho}dk'_\rho = -j\pi sgn(\rho); \; sgn(\rho) = +1, \; \rho > 0; \; = -1, \; \rho < 0 \, . \quad (6.148)$$

In order to establish the above relation, it would be sufficient to verify the corresponding derivative relation obtained by differentiating both sides with respect to ρ. In this process, the derivative of the *sgn* function with a unit amplitude is recognized to be a delta function $2\delta(\rho)$ with two units of amplitude.

$$-j\int_{k'_\rho=-\infty}^{\infty} e^{-jk'_\rho\rho}dk'_\rho = -j2\pi\delta(\rho), \quad \frac{1}{2\pi}\int_{k'_\rho=-\infty}^{\infty} e^{-jk'_\rho\rho}dk'_\rho = \delta(\rho) \ . \tag{6.149}$$

The above relation may be verified as a basic inverse-Fourier integral for the delta function $\delta(\rho)$.

The result for the k'_ρ integral of (6.147) can be used in (6.146) to obtain the expression for the field E_x at a large distance, contributed by a TM_z surface-wave singularity. Contribution from a TE_z singularity can be similarly derived, where the coefficient a' would be substituted with b', and the factor $\cos^2\phi$ with $\sin^2\phi$. A superscript TM or TE with the field component would represent contribution from the TM_z or TE_z surface waves, respectively.

$$E_x^{TM}(\rho,\phi;z=0)_{\rho\to\infty} \simeq Il\frac{a'}{2}\sqrt{\frac{1}{2\pi\rho k_{\rho p}}}\cos^2\phi e^{+j3\pi/4}e^{-jk_{\rho p}\rho},$$
$$E_x^{TE}(\rho,\phi;z=0)_{\rho\to\infty} \simeq Il\frac{b'}{2}\sqrt{\frac{1}{2\pi\rho k_{\rho p}}}\sin^2\phi e^{+j3\pi/4}e^{-jk_{\rho p}\rho}. \tag{6.150}$$

Note the $1/\sqrt{\rho}$ dependence of the above field magnitude, contributed from a surface-wave radiation. The corresponding power density, or power flow per unit area, in the radial direction $\hat{\rho}$, would have a $1/\rho$ dependence, which would result in a constant amount of power flow when integrated over a large cylindrical surface of radius ρ, centered around the source. This is in contrast with a $1/r$ dependence in (6.139) for the field, contributed from a free-space radiation. This would result in a $1/r^2$ dependence for the power density of the free-space radiation, resulting in a constant amount of radiation power over a large spherical surface of radius r. In reference to the above basic characteristics, the surface wave radiation is recognized as a cylindrical wave, whereas the free-space radiation is a spherical wave.

7. Fields of a Moving Charge, and Electrodynamics of Radiation

A moving electric charge is physically the most basic source of time-varying electromagnetic fields, which would lead to radiation under a non-uniform motion of the charge. The field produced by a moving charge is fundamentally different from that of a static charge. Further, the field of a moving charge with a uniform velocity is also different from that of a charge in general motion. In the next section, we will analyze the fields of a moving charge with a uniform velocity. The fields of a charge in general motion will be modeled in the following sections. The analyses and results would serve as an introduction to basic electrodynamics and the origin or radiation.

7.1. Moving Charge with a Uniform Velocity v. A moving electric charge is physically the most basic source of time-varying electromagnetic fields and radiation. The field produced by a moving charge is fundamentally different from that of a static charge. Further, the field of a moving charge with a uniform velocity is also different from that of an accelerating charge. We will analyze in the following the fields of a moving charge with a uniform velocity, as shown in Fig.15. The analysis and results is to be an introduction to electrodynamic problems.

We will model the problem of Fig.15 as a surface charge distribution $\rho_s(x, z, t)$ on the xz plane.

$$\rho_s(x, z, t) = Q\delta(z - vt)\delta(x),$$

$$\widetilde{\rho}_s(k_x, k_z, t) = \int\int \rho_s(x, z, t)e^{-jk_z z}e^{-jk_x x}dxdz = Qe^{-jk_z vt}$$

$$\rho_s(x, z, t) = \frac{1}{4\pi^2}\int\int_{-\infty}^{+\infty} \widetilde{\rho}_s(k_x, k_z, t)e^{jk_z z}e^{jk_x x}dk_x dk_z$$

$$= \frac{1}{4\pi^2}\int\int_{-\infty}^{+\infty} Qe^{-jk_z vt}e^{jk_z z}e^{jk_x x}dk_x dk_z. \tag{6.151}$$

Consider first the fields due to a surface charge distribution of the form:

$$\rho_s(x, z, t) = Qe^{j\omega t}e^{-jk_z z}e^{-jk_x x}. \tag{6.152}$$

The total field due to the charge distribution (6.151) can be expressed as the Fourier integration of the field due to the distribution of (6.152) over all $-\infty < k_x, k_y < +\infty$ after replacing ω by $+k_z v$, and then substituting k_z and k_x by $-k_z$ and $-k_x$, respectively.

Let us relate the surface charge density of (6.152) to an equivalent surface current density.

$$\bar{J}(x, z) = \hat{z}J_z(x, z) = \hat{z}J_{z0}e^{j\omega t}e^{-jk_x x}e^{-jk_z z}, \ \omega = k_z v. \tag{6.153}$$

When the surface current density $\bar{J}$ varies along its direction of current flow, the derivative of $\bar{J}$ along the current flow would relate to the surface charge accumulation. This principle is governed by conservation or continuity of charge.

$$[J_z(x, z + \triangle z) - J_z(x, z)]\Delta x = -\frac{\partial \rho_s(x,z)}{\partial t}\Delta x\Delta z = \frac{\partial J_z(x,z)}{\partial z}\Delta x\Delta z, \tag{6.154}$$

$$\frac{\partial J_z(x,z)}{\partial z} = -\frac{\partial \rho_s(x,z)}{\partial t}. \tag{6.155}$$

Now, using (6.152,6.153) in (6.155), we get,

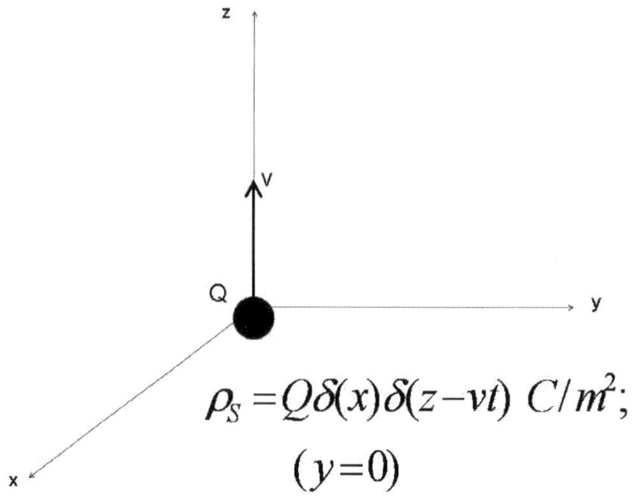

FIGURE 15. A charge Q, moving along the z axis with a uniform velocity v. It can be modeled as a surface charge distribution on the $y = 0$ plane, with surface charge density $\rho_s = Q\delta(x)\delta(z - vt)\ C/m^2$. When $v \neq 0$, it produces a magnetic field $\overline{H}$ in addition to the electric field $\overline{E}$. The electric field $\overline{E}$ for $v \neq 0$ is different from that for a static charge with $v = 0$.

$$-jk_z J_z = -j\omega\rho_s = -j\omega Q e^{j\omega t} e^{-jk_z z} e^{-jk_x x},$$
$$J_z = \frac{Q\omega}{k_z} e^{j\omega t} e^{-jk_z z} e^{-jk_x x} = Qv e^{j\omega t} e^{-jk_z z} e^{-jk_x x}. \tag{6.156}$$

Now, the $z-$directed current distribution in (6.153,6.156) will result in a TM_z field, with its potential function ψ_{zm} expressed using (6.23).

$$\psi_{zm} = \frac{Qv}{2jk_y} e^{j\omega t} e^{-jk_z z} e^{-jk_x x} e^{\mp jk_y y}, \tag{6.157}$$
$$k = \frac{\omega}{c} = k_z \frac{v}{c} < k_z, \tag{6.158}$$

Note that the wave number k is less than k_z for $v < c$. This means the field would be evanescent, without any power radiating away from the source. Accordingly, k_y would be negative imaginary.

$$k_y = \sqrt{k^2 - k_x^2 - k_z^2} = \sqrt{-k_x^2 - k_z^2(1 - \frac{v^2}{c^2})} = -j\sqrt{k_x^2 + k_z^2(1 - \frac{v^2}{c^2})}. \tag{6.159}$$

It maybe convenient to define new variables k_z' and z'.

$$k'_z = k_z\alpha,\ z' = z/\alpha,\ k_z z = k'_z z',\ \alpha = \sqrt{1 - \frac{v^2}{c^2}}, \tag{6.160}$$

$$\psi_{zm} = \frac{Qv}{2\sqrt{k_x^2 + k_z'^2}} e^{jk_z vt} e^{-jk_x x} e^{-jk_z z} e^{\mp\sqrt{k_x^2+k_z'^2}y}. \tag{6.161}$$

Now, as discussed before, the total ψ_{zm} can be derived by replacing k_z, k'_z and k_x by $-k_z$, $-k'_z$ and $-k_x$, respectively, in the equation (6.161), and integrating over all $-\infty < k_z, k_x < \infty$.

$$\psi_{zm} = \frac{1}{4\pi^2} \int\int_{-\infty}^{+\infty} \frac{Qv}{2\sqrt{k_x^2 + k_z'^2}} e^{-jk_z vt} e^{+jk_x x} e^{+jk'_z z'} e^{\mp\sqrt{k_x^2+k_z'^2}y} dk_z dk_x. \tag{6.162}$$

Let us evaluate the potential ψ_{zm} at $t = 0$. The $e^{-jk_z vt}$ factor in the inverse Fourier integral of (6.162) implies that at any other time $t \neq 0$ the potential function, and accordingly all field functions, would be shifted along z by distance vt.

$$\psi_{zm}(z,t) = \psi_{zm}(z - vt),$$
$$\psi_{zm}(z,t=0) = \frac{1}{4\pi^2} \int\int_{-\infty}^{+\infty} \frac{Qv}{2\sqrt{k_x^2+k_z'^2}} e^{+jk_x x} e^{+jk'_z z'} e^{\mp\sqrt{k_x^2+k_z'^2}y} dk'_z \frac{1}{\alpha} dk_x. \tag{6.163}$$

From Appendix D, we have the following Fourier inverse relationship.

$$\frac{1}{4\pi r} e^{-jkr} = \frac{1}{4\pi^2} \int\int_{-\infty}^{+\infty} \frac{Qv}{2jk_y} e^{+jk_x x} e^{+jk_z z} e^{\mp jk_y y} dk_z dk_x. \tag{6.164}$$

Taking the limits for (6.164) as $\omega \to 0, k \to 0$, when $k_y = \sqrt{k^2 - k_x^2 - k_z^2} = -j\sqrt{k_x^2 + k_z^2}$, we would have

$$\frac{1}{4\pi r} = \frac{1}{4\pi^2} \int\int_{-\infty}^{+\infty} \frac{Qv}{2\sqrt{k_x^2 + k_z^2}} e^{+jk_x x} e^{+jk_z z} e^{\mp\sqrt{k_x^2+k_z^2}} dk_z dk_x. \tag{6.165}$$

Comparing this expression to that of ψ_{zm} in (6.163) we get

$$\psi_{zm}(z,t=0) = \frac{1}{\alpha}\frac{Qv}{4\pi r}\bigg|_{z \to z' = \frac{z}{\alpha}} = \frac{1}{\alpha}\frac{Qv}{4\pi r'},\ r' = \sqrt{x^2 + y^2 + z'^2}. \tag{6.166}$$

Now that the potential ψ_{zm} is known, we can derive the field expressions using field-potential relations. For a wave of the form $e^{j\omega t}e^{-jk_x x}e^{-jk_y y}e^{-jk_z z}$, the field-potential relations (5.34) are expressed as follows.

$$\bar{E} = \frac{1}{j\omega\epsilon}[-k_z k_x\hat{x} - k_z k_y\hat{y} + (k^2 - k_z^2)\hat{z}]\psi_{zm},$$
$$\bar{H} = [-jk_y\hat{x} + jk_x\hat{y}]\psi_{zm}. \tag{6.167}$$

Because we have $\omega = k_z v,\ k = \omega/c = k_z v/c$, (6.167) maybe revised in the following form.

$$\bar{E} = \frac{1}{\epsilon v}[+jk_x\hat{x} + jk_y\hat{y} + j\alpha^2 k_z]\psi_{zm},$$
$$\bar{E} = \frac{1}{\epsilon v}[-\frac{\partial}{\partial x}\hat{x} - \frac{\partial}{\partial y}\hat{y} - \alpha\frac{\partial}{\partial z'}\hat{z}]\psi_{zm}, \tag{6.168}$$
$$\bar{H} = [\hat{x}\frac{\partial}{\partial y} - \hat{y}\frac{\partial}{\partial x}]\psi_{zm}. \tag{6.169}$$

The differential operations (6.168,6.169) are applicable for all values of k_x,k_y, k_z, and so would be vallid for the total ψ_{zm} of (6.162,6.163,6.166) as well. Using the operations (6.168,6.169) on the ψ_{zm} expression of (6.166) we get the electric and magnetic fields at $t = 0$.

$$\bar{E}(z, t = 0) = \frac{1}{\alpha\epsilon} \times \frac{Q}{4\pi r'^3}[x\hat{x} + y\hat{y} + \alpha z'\hat{z}] = \frac{Q}{4\pi\epsilon\alpha r'^3}[x\hat{x} + y\hat{y} + z\hat{z}], \tag{6.170}$$
$$\bar{H}(z, t = 0) = \frac{Qv}{\alpha}\frac{1}{4\pi r'^3}[-y\hat{x} + x\hat{y}] = \frac{Qv}{4\pi\alpha r'^3}\rho\hat{\phi}. \tag{6.171}$$

As discussed earlier, at any other time $t \neq 0$ the fields would be shifted versions of the above fields. Let us examine the electric field on the xy plane, at $t = 0, z = 0$. In this case, $\bar{E} = \frac{1}{\alpha} \times \frac{Q}{4\pi\epsilon\rho^2}\hat{\rho}$, which is larger compared to that of a static charge by factor $1/\alpha$. Along the z axis, $x = y = 0$, we would have $\bar{E} = \frac{Q}{4\pi\epsilon z'^2}\hat{z} = \frac{Q\alpha^2}{4\pi\epsilon z^2}\hat{z}$, which is same as that of a static charge with z replaced by $z' = z/\alpha$. In this case, the field is smaller compared to that of a static charge by a factor of α^2. The field lines on the xz plane at $t = 0$ are sketched in Fig.16, showing the asymmetric behavior. In the limit as the velocity v approaches the speed of light c, the fields along the z direction and its orthogonal component in the (xy) plane would approach zero and infinity, respectively.

Note the scaling factor α, which accounts for the velocity of the charge. This factor is commonly encountered in relativistic transformation. This is derived using the electromagnetic theory directly through Maxwell's equations, without explicit need for the relativistic effects. This should clearly suggest direct physical founding of the electrodynamics in the relativity theory, and vice versa.

Let us examine the divergence operation $\nabla \cdot (\epsilon\bar{E}) = \nabla \cdot (\bar{D})$ on the electric field expression in (6.168).

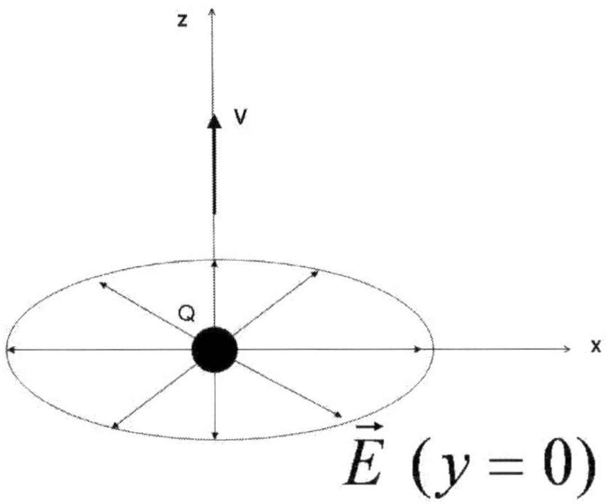

FIGURE 16. Sketch of the electric field lines on the ($y = 0$) plane of a charge moving along the z axis with a uniform velocity. The field is smaller along the direction of motion, and is larger along directions transverse to the direction of motion. In the limit as the velocity v approaches the speed of light c, the fields along the above orthogonal directions approach zero and infinity, respectively.

$$
\begin{aligned}
\nabla \cdot (\epsilon \bar{E}) &= (\hat{x}\frac{\partial}{\partial x} + \hat{y}\frac{\partial}{\partial y} + \hat{z}\frac{\partial}{\partial z}) \cdot (\epsilon \bar{E}) \\
&= \frac{1}{v}[-\frac{\partial^2}{\partial x^2} - \frac{\partial^2}{\partial y^2} - \frac{\partial}{\partial z}\alpha\frac{\partial}{\partial z'}]\psi_{zm} \\
&= \frac{1}{v}[-\frac{\partial^2}{\partial x^2} - \frac{\partial^2}{\partial y^2} - \frac{\partial^2}{\partial z'^2}]\frac{Qv}{\alpha 4\pi r'},
\end{aligned}
\tag{6.172}
$$

which maybe shown to be zero, independent of velocity v. This means Gauss's law is valid independent of the velocity of motion. However, from the results of (6.170) we know that Coulomb's law of static electricity does not apply when the velocity is non-zero. In other words, Gauss' law is more general and universal compared to Coulomb's law. This is a significant observation.

Now let us approximate the magnetic field $\bar{H}$ from (6.171) for small v, in which case $\alpha \simeq 1$.

$$
\bar{H} \simeq \frac{Qv\rho}{4\pi r^3}\hat{\phi}, \ v << c. \tag{6.173}
$$

This is same as the magnetic field expression for a DC current element with $I\triangle l = Qv$.

As discussed, for the present problem of a charge with uniform velocity, we have $k_z = \frac{\omega}{v} = k\frac{c}{v}$, which is greater than k for all velocity $v < c$. Therefore, the field is evanescent, which means the charge would not radiate power away from source. However, if the motion does not maintain a uniform velocity, from the Fourier analysis one would see that a given wave number k_z will no longer be associated with only a single frequency $\omega = k_z v$. Instead, there would be a spectrum covering in general all frequencies $-\infty < \omega < +\infty$, part of which having $k > k_z$ would result in power radiating away from the source.

7.2. Stored Energy of a Charge Moving with a Uniform Velocity. We have derived in section (7.1) the fields of a point charge Q moving with a uniform velocity v. As per the expressions (6.170,6.171), the field magnitudes approach infinity at the charge location, as $x, y, z \to 0$, or $r \to 0$. Having such infinite fields would be physically unrealistic. However, unlike the ideal point charge we have assumed in the derivation of (6.170,6.171), a realistic charge would have a non-zero size with certain charge distribution. Let us assume a spherically shaped charge of radius r_0, with a uniform distribution over the spherical body. We need to be a bit more specific, because size and shape depends on the velocity of observer in reference to the charge. The charge would look to be an ideal spherical distribution, with a radius r_0, only when seen in a moving reference frame (primed reference frame) at rest with respect to the charge. The observer in this primed reference frame is moving at the same velocity v as the charge. In this reference frame, the charge is limited in the spherical region defined by $r' = (x'^2 + y'^2 + z'^2)^{\frac{1}{2}} \leq r_0$. r_0 may be referred to as the "rest-radius." Now, when the charge is moving along the z axis, with a uniform velocity v with respect to a stationary reference frame (unprimed coordinates), the z-dimension would look compressed by a factor $\alpha = (1 - v^2/c^2)^{1/2}$, as per the relativistic transformation (6.160), with the x and y dimensions remaining unaffected. Accordingly, the charge distribution in the unprimed frame would be limited over the region $r' = (x^2 + y^2 + \frac{z^2}{\alpha^2})^{\frac{1}{2}} \leq r_0$. This is an ellipsoid region in the unprimed frame, with its z-dimension smaller by a factor $\alpha = (1 - v^2/c^2)^{1/2}$, as compared with the x and y dimensions.

With the above consideration, we may express the fields of the realistic charge distribution, moving with a velocity v along the z axis. Like the fields of a point charge, moving with the same velocity v along the z axis, the fields of the spherical charge distribution at any arbitrary time t can be obtained from those at time $t = 0$, by simply shifting the field pattern by a distance of vt along the z direction. So, it is necessary to describe the fields only at time $t = 0$. The field expressions of (6.170,6.171) for time $t = 0$, which we have derived for a point charge, would also be valid for the spherical charge distribution, when observed external to the charged region. The spherical charge would look like a point charge, as seen outside the charge distribution, $r' > r_0$.

$$\begin{aligned} \overline{E}(t=0) &= \tfrac{Q}{4\pi\epsilon\alpha r'^3}(x\hat{x} + y\hat{y} + z\hat{z}) = \tfrac{Q\overline{r}}{4\pi\epsilon\alpha r'^3}, \\ \overline{H}(t=0) &= \tfrac{Qv}{4\pi\alpha r'^3}(-y\hat{x} + y\hat{y}) = \tfrac{Qv}{4\pi\alpha r'^2}\sin\theta'\hat{\phi}; \quad r' > r_0 \end{aligned} \tag{6.174}$$

For an observation point $r' < r_0$, only a part $(= Q')$ of the total charge Q contained inside a smaller sphere of radius r' would contribute to the observed field. Considering a uniform density ρ_v of charge distribution, Q' may be expressed as:

$$\rho_v = \frac{Q}{\frac{4}{3}\pi r_0^3}, \quad Q' = \rho_v \frac{4}{3}\pi r'^3 = Q\Big(\frac{r'}{r_0}\Big)^3 \tag{6.175}$$

The field expressions in the region $r' < r_0$ can be obtained by replacing Q with Q' in the field expressions of (6.174).

$$\overline{E}(t=0) = \frac{Q\overline{r}}{4\pi\epsilon\alpha r_0^3}, \quad \overline{H}(t=0) = \frac{Qvr'}{4\pi\alpha r_0^3}\sin\theta'\hat{\phi}; \quad r' < r_0 \tag{6.176}$$

The above field distribution would approach zero as $r, r' \to 0$, as should be realistically expected.

It would be useful to derive the total stored energy W of the moving charge, resulting from the above field distribution. W consists of electric field energy W_e and magnetic field energy W_m, each partly contributed from the external $(r' > r_0)$ and the internal $(r' > r_0)$ regions of the spherical charge. As discussed, we know that the charge distribution is spherically symmetric in terms of the primed variables, not the unprimed variables. However, the energy of the moving charge, as observed in the unprimed reference frame, is to be expressed as an integration of the energy density functions with respect to the unprimed variables. Therefore, by proper substitution of variables, it would be convenient to transform the integration variables from the unprimed coordinates to the primed coordinates $(x' = x, y' = y, z' = \frac{z}{\alpha})$, or $(r', \theta', \phi' = \phi)$. This would allow integration in the spherical coordinates in the primed frame, by making use of the spherical symmetry of the charged and external regions.

Because the fields at $t \neq 0$ are simply shifted versions of those at $t = 0$, the total energy would be independent of time t of observation. One may compute the constant energy by integrating the energy density at $t = 0$. Let us first derive the electric stored energy W_e, which consists of two parts, W_{e1} contributed from outside the charged region $r' > r_0$ and W_{e2} from inside the charged region $r' < r_0$.

$$\begin{aligned} W_e &= \int\int\int \tfrac{1}{2}\epsilon\left|\overline{E}(t=0)\right|^2 dv = \int\int\int \tfrac{1}{2}\epsilon\left|\overline{E}(t=0)\right|^2 dxdydz \\ &= \int\int\int \tfrac{1}{2}\epsilon\left|\overline{E}(t=0)\right|^2 dxdy(\alpha\, dz') = \int\int\int \tfrac{1}{2}\epsilon\left|\overline{E}(t=0)\right|^2 \alpha dv' \\ &= W_{e1}(r' > r_0) + W_{e2}(r' < r_0) \end{aligned} \tag{6.177}$$

$$\begin{aligned} W_{e1} &= \int_{\theta'=0}^{\pi}\int_{\phi'=0}^{2\pi}\int_{r'=r_0}^{\infty} \tfrac{1}{2}\epsilon\left|\overline{E}(t=0)\right|^2 \alpha dv' = \int_{\theta'=0}^{\pi}\int_{\phi'=0}^{2\pi}\int_{r'=r_0}^{\infty} \tfrac{1}{2}\epsilon\left(\tfrac{Q}{4\pi\epsilon\alpha r'^3}\right)^2 r^2\alpha r'^2 \sin\theta' dr'd\phi'd\theta' \\ &= \frac{Q^2}{32\pi^2\epsilon\alpha}\int_0^{\pi}\int_0^{2\pi}\int_{r_0}^{\infty} \frac{(x^2+y^2+\alpha^2 z'^2)}{r'^4}\sin\theta' dr'd\phi'd\theta' \\ &= \frac{Q^2}{32\pi^2\epsilon\alpha}\int_0^{\pi}\int_0^{2\pi}\int_{r_0}^{\infty} \frac{(r'^2\sin^2\theta' + \alpha^2 r'^2\cos^2\theta')}{r'^4}\sin\theta' dr'd\phi'd\theta' \\ &= \frac{Q^2}{32\pi^2\epsilon\alpha}\cdot\frac{2\pi}{r_0}\int_0^{\pi}\left(1 - \tfrac{v^2}{c^2}\cos^2\theta'\right)\sin\theta' d\theta' = \frac{Q^2}{8\pi\epsilon\alpha r_0}\left[1 - \tfrac{v^2}{3c^2}\right] \end{aligned} \tag{6.178}$$

$$W_{e2} = \int_{\theta'=0}^{\pi} \int_{\phi'=0}^{2\pi} \int_{r'=0}^{r_0} \tfrac{1}{2}\epsilon \left|\overline{E}(t=0)\right|^2 \alpha dv' = \int_{\theta'=0}^{\pi} \int_{\phi'=0}^{2\pi} \int_{r'=0}^{r_0} \tfrac{1}{2}\epsilon \left(\tfrac{Q}{4\pi\epsilon\alpha r_0^3}\right)^2 r^2 \alpha r'^2 \sin\theta' dr' d\phi' d\theta'$$

$$= \tfrac{Q^2}{32\pi^2\epsilon\alpha r_0^6} \int_0^{\pi}\int_0^{2\pi}\int_0^{r_0} (x^2 + y^2 + z'^2\alpha^2) r'^2 \sin\theta' \alpha'^2 dr' d\phi' d\theta'$$

$$= \tfrac{Q^2}{32\pi^2\epsilon\alpha r_0^6} \int_0^{\pi}\int_0^{2\pi}\int_0^{r_0} (r'^2 \sin^2\theta' + \alpha^2 r'^2 \cos^2\theta') r'^2 \sin\theta' \alpha'^2 dr' d\phi' d\theta'$$

$$= \tfrac{Q^2}{32\pi^2\epsilon\alpha r_0^6} \cdot \tfrac{2\pi r_0^5}{5} \int_0^{\pi} \left(1 - \tfrac{v^2}{c^2}\cos^2\theta'\right) \sin\theta' d\theta' = \tfrac{Q^2}{16\pi\epsilon\alpha r_0} \tfrac{1}{5}\left(2 - \tfrac{2}{3}\tfrac{v^2}{c^2}\right) = \tfrac{1}{5} W_{e1} \tag{6.179}$$

$$W_e = W_{e1} + W_{e2} = \frac{6}{5} W_{e1} \tag{6.180}$$

Similarly, we derive the total magnetic stored energy W_m, which consists of two parts, W_{m1} contributed from outside the charged region $r' > r_0$ and W_{m2} from inside the charged region $r' < r_0$.

$$W_m = \int\int\int \tfrac{1}{2}\mu \left|\overline{H}(t=0)\right|^2 dv = \int\int\int \tfrac{1}{2}\mu \left|\overline{H}(t=0)\right|^2 dxdydz$$
$$= \int\int\int \tfrac{1}{2}\mu \left|\overline{H}(t=0)\right|^2 dxdy\alpha dz' = \int\int\int \tfrac{1}{2}\mu \left|\overline{H}(t=0)\right|^2 \alpha dv'$$
$$= W_{m1}(r' > r_0) + W_{m2}(r' < r_0) \tag{6.181}$$

$$W_{m1} = \int_{\theta'=0}^{\pi} \int_{\phi'=0}^{2\pi} \int_{r'=r_0}^{\infty} \tfrac{1}{2}\mu \left|\overline{H}(t=0)\right|^2 \alpha dv'$$

$$= \int_{\theta'=0}^{\pi} \int_{\phi-=0}^{2\pi} \int_{r'=r_0}^{\infty} \tfrac{1}{2}\mu \left(\tfrac{Qv}{4\pi\alpha r'^2}\right)^2 \sin^2\theta' \alpha r'^2 \sin\theta' dr' d\phi' d\theta'$$

$$= \tfrac{Q^2 v^2 \mu}{32\pi^2\alpha} \int_0^{\pi}\int_0^{2\pi}\int_{r_0}^{\infty} \tfrac{1}{r'^2} \sin^3\theta' dr' d\phi' d\theta'$$

$$= \tfrac{Q^2 v^2 \mu}{32\pi^2\alpha} \tfrac{2\pi}{r_0} \int_0^{\pi} \sin^3\theta' d\theta' = \tfrac{Q^2 v^2 \mu}{16\pi^2\alpha r_0} \cdot \tfrac{4}{3} = \tfrac{Q^2}{8\pi\epsilon\alpha r_0} \cdot \tfrac{2v^2}{3c^2} \tag{6.182}$$

$$W_{m2} = \int_{\theta'=0}^{\pi} \int_{\phi'=0}^{2\pi} \int_{r'=0}^{r_0} \tfrac{1}{2}\mu \left|\overline{H}(t=0)\right|^2 \alpha dv'$$

$$= \int_{\theta'=0}^{\pi} \int_{\phi'=0}^{2\pi} \int_{r'=0}^{r_0} \tfrac{1}{2}\mu \left(\tfrac{Qvr'}{4\pi\alpha r_0^3}\right)^2 \sin^2\theta' \alpha r'^2 \sin\theta' dr' d\phi' d\theta'$$

$$= \tfrac{Q^2 v^2 \mu}{32\pi^2 \alpha r_0^6} \int_0^{\pi}\int_0^{2\pi}\int_0^{r_0} r'^4 \sin^3\theta' dr' d\phi' d\theta' = \tfrac{Q^2 v^2 \mu}{32\pi^2 \alpha r_0^6} \times \tfrac{2\pi r_0^5}{5} \int_0^{\pi} \sin^3\theta' d\theta'$$

$$= \tfrac{Q^2 v^2 \mu}{16\pi\alpha r_0} \times \tfrac{1}{5} \times \tfrac{4}{3} = \tfrac{Q^2 v^2 \mu}{8\pi\alpha r_0} \times \tfrac{1}{5} \times \tfrac{2}{3} = \tfrac{Q^2}{8\pi\alpha r_0 \epsilon} \times \tfrac{1}{5} \times \tfrac{2}{3}\tfrac{v^2}{c^2} = \tfrac{1}{5} W_{m1} \tag{6.183}$$

$$W_m = W_{m1} + W_{m2} = \frac{6}{5} W_{m1} \tag{6.184}$$

The total energy W stored in the fields of the moving charge Q is obtained by adding the total electric energy W_e and the total magnetic energy W_m.

$$W = W_e + W_m = \frac{6}{5}(W_{e1} + W_{m1})$$

$$= \frac{6}{5} \times \frac{Q^2}{8\pi\epsilon\alpha r_0}(1 + \frac{1}{3}\frac{v^2}{c^2}) = \frac{3}{20}\frac{Q^2}{\pi\epsilon\alpha r_0}(1 + \frac{1}{3}\frac{v^2}{c^2}) \tag{6.185}$$

The total energy W is a function of the velocity v. As mentioned, the energy is independent of the time t of observation, when the charge is moving with a constant velocity v. This independence of the stored energy W with time may also be viewed in terms of applied self force and energy conservation. One may observe that the vector electric field $\overline{E}$ at each pair of points on the body of the charge, which are diametrically opposite with respect to the charge center, are negative of each other. That is, $\overline{E}(-x, -y, -z) = -\overline{E}(x, y, z)$. Accordingly, the corresponding force-pairs acting on the charge, produced due to the electric fields of the charge itself (called the electric self-force), would cancel with each other. This would result in a zero total self-force due to the electric field. The total self force due to the magnetic field may also be shown to be zero, due to similar symmetry consideration. Therefore, the total "self-force," due to both the electric and the magnetic fields, is zero. No external force would be needed to counter this zero self-force, or equivalently no external energy be required, in order to maintain the uniform motion of the charge. As per the principle of conservation of energy, this would mean that the total energy of the charge should remain unchanged or constant over the motion. This is consistent with our earlier observation of having the stored energy independent of the time. In other words, having a constant stored energy, associated with the electric and magnetic fields of the charge in uniform motion, is also equivalent to having zero total "self-force" applied on the charge.

The above analysis and conclusion, that the self force is zero for a uniform motion, is based on having the uniform motion maintained for all time $-\infty < t < +\infty$. However, the same condition of zero self force may also be extended for a locally uniform motion, having no change of velocity only at a given reference time $t = 0$, which means that the time-derivative of the velocity vector is zero only at $t = 0$. This assumes that future

($t > 0$) motion of the charge does not affect the present fields, which is a basic physical principle of causality. And, any electromagnetic disturbance produced due to change in the charge's past ($t < 0$) motion has propagated farther away from the charge at $r \to 0$. This is because the charge's own physical speed during the past motion is assumed to be less than the propagation speed (speed of light) of the electromagnetic disturbance.

7.3. Self Force on an Accelerating Charge, Associated with Change in Stored Energy. The above situation would be fundamentally different for a charge moving with a non-uniform velocity. The stored energy would now change in time, associated with a changing velocity. This would require external energy, or equivalently non-zero external force $\overline{F}_{as}$, to be applied along the velocity to compensate for the change in the stored energy. The applied external force must be negative of the force $\overline{F}_s$ produced by the field on the charge. In other words, there would be a non-zero force $\overline{F}_s$ applied on the charge, produced by itself. This "self-force" $\overline{F}_s$, contributed by the change in the stored field energy, may be referred to as the "reactive" self force. The is because the stored energy (as opposed to dissipated or lost energy), which is responsible for the force, is commonly referred to as the reactive energy. In addition to the incremental charge in stored or reactive energy, there may be additional energy escaping away from a charge, in non-uniform motion, in the form of radiation. Any such radiated power would also contribute to an additional "self-force" $\overline{F}_r$, which may be referred to as "radiative" self-force. The radiative self-force on a charge in non-uniform motion will be examined later in section 7.5. We would now evaluate the reactive self-force $\overline{F}_s$ from the variation of the stored energy with the charge velocity.

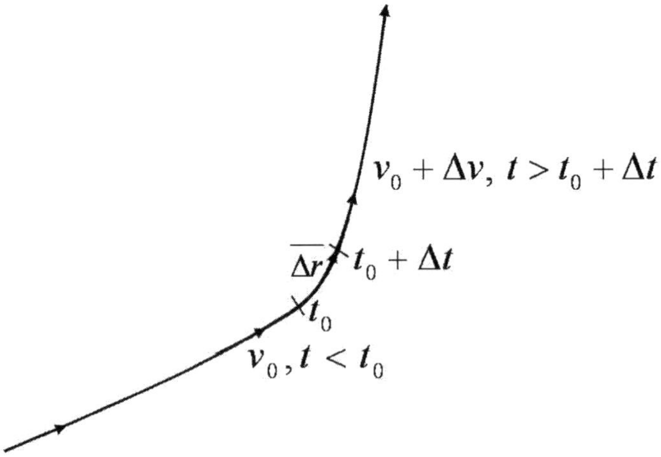

FIGURE 17

Let a charge be moving with a constant velocity with magnitude v_0 for $t < t_0$, where the direction of the velocity may be in any fixed direction (Fig.17). The total stored energy W_0 for $t < t_0$ is expressed as

$$W_0 = \frac{3Q^2}{20\pi\epsilon r_0} \frac{\left(1 + \frac{1}{3}\frac{v_0^2}{c^2}\right)}{\sqrt{1 - \frac{v_0^2}{c^2}}} \tag{6.186}$$

Let the magnitude of the velocity change to $v_0 + \triangle v$, and remain constant for all $t > t_0 + \triangle t_0$ (Fig.17). The direction of the new uniform velocity may or may not be the same as the prior velocity. Now, let us wait long enough for the transients of the change in motion to settle, when we have the fields of the charge with new uniform velocity be established everywhere in space. The change in stored energy $\triangle W$ can be expressed by differentiating W_0 with v_0.

$$W_0 + \triangle W = W_0 + \frac{\partial W_0}{\partial v_0}\triangle v$$

$$\triangle W = \frac{\partial W_0}{\partial v_0}\triangle v = \frac{3Q^2\triangle v}{20\pi\epsilon r_0}\left[\frac{\frac{2}{3}\frac{v_0}{c^2}}{\sqrt{1-\frac{v_0^2}{c^2}}} + \frac{\left(1+\frac{1}{3}\frac{v_0^2}{c^2}\right)}{\left(1-\frac{v_0^2}{c^2}\right)^{\frac{3}{2}}}\cdot\frac{v_0^2}{c^2}\right]$$

$$= \frac{3Q^2\triangle v}{20\pi\epsilon r_0}\frac{\frac{v_0}{c^2}\left[\frac{1}{3}\left(1-\frac{v_0^2}{c^2}\right)+\left(1+\frac{1}{3}\frac{v_0^2}{c^2}\right)\right]}{\left(1-\frac{v_0^2}{c^2}\right)^{\frac{3}{2}}} = \frac{3Q^2\triangle v}{20\pi\epsilon r_0}\frac{\frac{v_0}{c^2}\left(\frac{5}{3}-\frac{1}{3}\frac{v_0^2}{c^2}\right)}{\left(1-\frac{v_0^2}{c^2}\right)^{\frac{3}{2}}} \tag{6.187}$$

Let us express the change, $\triangle v$, in the magnitude of the velocity to corresponding changes in their vector components $\triangle v_x, \triangle v_y$ and $\triangle v_z$ along three orthogonal directions.

$$v_0 = \sqrt{v_{ox}^2 + v_{oy}^2 + v_{oz}^2}$$

$$\triangle v = \frac{\partial v_0}{\partial v_{ox}}\triangle v_x + \frac{\partial v_0}{\partial v_{oy}}\triangle v_y + \frac{\partial v_o}{\partial v_{oz}}\triangle v_z$$

$$= \frac{v_{ox}\triangle v_x + v_{oy}\triangle v_y + v_{oz}\triangle v_z}{v_o}$$

$$\triangle W = \frac{3Q^2}{20\pi\epsilon r_0}\frac{\left(\frac{5}{3}-\frac{1}{3}\frac{v_0^2}{c^2}\right)}{\left(1-\frac{v_0^2}{c^2}\right)^{\frac{3}{2}}} \times \left(\frac{v_{0x}}{c^2}\triangle v_x + \frac{v_{0y}}{c^2}\triangle v_y + \frac{v_{0z}}{c^2}\triangle v_z\right) \tag{6.188}$$

This change in stored energy is to be supplied by an external force $\overline{F}_{as} = -\overline{F}_s$. The external force $\overline{F}_{as}$ and the corresponding self force $\overline{F}_s$ can only exist during the transition period $\triangle t$. This is became, as we have explained before, there would be no self force on the charge when in uniform motion over the time periods $t > t_o + \triangle t$ and $t < t_0$.

$$\triangle W = \overline{F}_{as}\cdot\overline{\triangle r} = \overline{F}_{as}\cdot(\triangle x\hat{x} + \triangle y\hat{y} + \triangle z\hat{z}) = -\overline{F}_s\cdot\overline{\triangle r}, \tag{6.189}$$

where $\overline{\triangle r}$ is vector displacement over time $\overline{\triangle t}$, with $\triangle x$, $\triangle y$ and $\triangle z$ the corresponding displacements along orthogonal rectangular co-ordinates.

$$
\begin{aligned}
\frac{\triangle W}{\triangle t} &= \overline{F}_{as} \cdot \left(\frac{\triangle x}{\triangle t}\hat{x} + \frac{\triangle y}{\triangle t}\hat{y} + \frac{\triangle z}{\triangle t}\hat{z}\right) \\
&= \overline{F}_{as} \cdot \left(v_{ox}\hat{x} + v_{oy}\hat{y} + v_{oz}\hat{z}\right)
\end{aligned}
$$

$$
= \frac{3Q^2}{20\pi\epsilon r_0} \frac{\left(\frac{5}{3} - \frac{1}{3}\frac{v_0^2}{c^2}\right)}{\left(1 - \frac{v_0^2}{c^2}\right)^{\frac{3}{2}}} \times \left(\frac{v_{ox}}{c^2}\frac{\triangle v_x}{\triangle t} + \frac{v_{oy}}{c^2}\frac{\triangle v_y}{\triangle t} + \frac{v_{oz}}{C^2}\frac{\triangle v_z}{\triangle t}\right)
$$

$$
\overline{a}_0 = a_{0x}\hat{x} + a_{0y}\hat{y} + a_{0z}\hat{z} = \frac{\triangle v_x}{\triangle t}\hat{x} + \frac{\triangle v_y}{\triangle t}\hat{y} + \frac{\triangle v_z}{\triangle t}\hat{z}
$$

$$
\begin{aligned}
\frac{\triangle W}{\triangle t} &= \frac{3Q^2}{20\pi\epsilon r_0} \frac{\left(\frac{5}{3} - \frac{1}{3}\frac{v_0^2}{c^2}\right)}{\left(1 - \frac{v_0^2}{c^2}\right)^{\frac{3}{2}}} \times \frac{\overline{a}_0}{c^2} \cdot \left(v_{0x}\hat{x} + v_{0y}\hat{y} + v_{0z}\hat{z}\right) \\
&= \overline{F}_{as} \cdot \left(v_{0x}\hat{x} + v_{0y}\hat{y} + v_{0z}\hat{z}\right) \\
\overline{F}_{as} &= \frac{3Q^2}{20\pi\epsilon r_0} \frac{\left(\frac{5}{3} - \frac{1}{3}\frac{v_0^2}{c^2}\right)}{\left(1 - \frac{v_0^2}{c^2}\right)^{\frac{3}{2}}} \frac{\overline{a}_0}{c^2} = -\overline{F}_s = -\overline{E}_s Q \\
\overline{E}_s &= \frac{-3Q}{20\pi\epsilon r_0} \frac{\left(\frac{5}{3} - \frac{1}{3}\frac{v_0^2}{c^2}\right)}{\left(1 - \frac{v_0^2}{c^2}\right)^{\frac{3}{2}}} \frac{\overline{a}_0}{c^2}
\end{aligned}
\tag{6.190}
$$

$\overline{E}_s$ is the reactive self-field due to change in the stored energy. This self-field at a given time t_0 depends on the instantaneous values of the velocity magnitude v_0 and acceleration $\overline{a}_0$ at the particular time t_0, We have selected t_0 to be any arbitrary instant of time. Therefore, the formula may be extended to any general time instant t, by substituting v_0 and $\overline{a}_0$ respectively with the general velocity magnitude $v(t)$ and acceleration $\overline{a}(t)$ at the given time t.

$$
\begin{aligned}
\overline{E}_s(t) &= \frac{-3Q}{20\pi\epsilon r_0} \frac{\left(\frac{5}{3} - \frac{1}{3}\frac{v(t)^2}{c^2}\right)}{\left(1 - \frac{v(t)^2}{c^2}\right)^{\frac{3}{2}}} \frac{\overline{a}(t)}{c^2} \\
\overline{F}_s(t) &= \overline{E}_s(t)Q = -\overline{F}_{as}(t)
\end{aligned}
\tag{6.191}
$$

7.4. Electromagnetic Mass. The applied force $\overline{F}_{as}$ and the self force $\overline{F}_s$ are proportional to the acceleration $\overline{a}$. Consider particularly a classical limit, with a small velocity compared to the speed of light, that is $v << c$. In this case, the ratio of the applied force to the acceleration maybe considered an equivalent inertial mass m_s, as per Newton's law of motion. This equivalent mass $m_s = (\overline{F}_{as}/\overline{a})$, $v << c$, may be referred to as the "electromagnetic mass" of the charge, which is contributed due to the velocity-dependent stored energy in its own electromagnetic field.

$$\frac{\overline{F}_{as}}{\overline{a}} = (\frac{3Q^2}{20\pi\epsilon r_0 c^2})\frac{\left(\frac{5}{3} - \frac{1}{3}\frac{v^2}{c^2}\right)}{\left(1 - \frac{v^2}{c^2}\right)^{\frac{3}{2}}}$$

$$\frac{\overline{F}_{as}}{\overline{a}}\Big|_{v\to 0} = \frac{Q^2}{4\pi\epsilon r_0 c^2} = \frac{\mu Q^2}{4\pi r_0} = m_s \tag{6.192}$$

There is this interesting possibility that the mass of a particular elementary charge, such as an electron or a proton, may be entirely due to its electromagnetic effects. In other words, the origin of the inertial mass could possibly be fundamentally electromagnetic in nature. If that was the case, the expression for the total electromagnetic energy W should have had the velocity dependence as per the Einstein relation:

$$W = mc^2 = \frac{m_s c^2}{\sqrt{1 - (\frac{v}{c})^2}} = \frac{Q^2}{4\pi\epsilon r_0\sqrt{1 - (\frac{v}{c})^2}} \tag{6.193}$$

Equivalently, if the above expression (6.193) for W was used in place of (6.185), it may be shown that the derivation steps (6.186) through (6.192) would have led to the following expression for the ratio of $\overline{F}_{as}$ and $\overline{a}$

$$\frac{\overline{F}_{as}}{\overline{a}} = \frac{1}{v}\frac{dW}{dv} = \frac{1}{v}\frac{d}{dv}(\frac{m_s c^2}{\sqrt{1 - (\frac{v}{c})^2}}) = \frac{m_s}{(1 - (\frac{v}{c})^2)^{3/2}} \tag{6.194}$$

Clearly, the energy expression of (6.185) neither has the same velocity dependence, nor has the same rest-value as $v \to 0$, as the desired relation (6.193) we expect to have if the inertial mass was purely electromagnetic in nature. Consequently, the $\overline{F}_{as}/\overline{a}$ expression of (6.192) also does not have the desired velocity dependence of (6.194). Accordingly, we may conclude that there be additional physical effects in order to possibly explain the origin of the total mass of a charged element. Such effects may either be in the form of new forces and fields, fundamentally different from the electromagnetic fields we have considered here, or the electromagnetic theory used in the modeling may not be accurate and needs to be modified. The new effects would be such that the resulting total mass of the charge exhibits characteristics consistent with the known principles of classical and relativistic mechanics. Accordingly, the new contributions should compensate for the improper velocity dependence of the energy W and $\overline{F}_{as}/\overline{a}$ expressions in (6.185) and (6.192), respectively. The total new mass m_0 may be interpreted as the sum of the electromagnetic mass m_s of (6.192) and additional mass contribution due to the new effects. The balancing new contributions should be such that the total new energy W and the new ratio of the applied force $\overline{F}_{as}$ and acceleration $\overline{a}$, would be expressed in the desired forms of (6.193) and (6.194), respectively, with the total new mass m_0 now replacing for the electromagnetic mass m_s.

Another closely related issue may be addressed here. Consider the simple situation of a stationary charge. We assume a suitable distribution of the charge over a spherical region of non-zero radius r_0. The forces due to the electric field produced by the charge, measured at two diametrically opposite points on the body of the charge itself, can be seen to be negative of each other. Accordingly, all force elements add up to a zero total force, as we have assumed. However, force at each point of the charge body is clearly not zero. This would mean that each piece could pull away from the center of the charge, leading to a mechanically unstable body. In other words, the basic field model of the charge we now have is not self consistent. The additional new effects, which we have presumed could lead to realistic velocity dependence for its total energy and equivalent mass, may also produce zero force at each point of the charge distribution. The charge distribution could very well be different from a uniform distribution we may have assumed for simplicity. The stability condition may also have to be extended under a general motion of the charge body. Such a possible solution, which is mechanically realistic and self-consistent from the different basic considerations, could be fundamentally satisfactory and philosophically significant.

7.5. Self Force on a Charge Produced Due to Radiative Fields. The self-force $\overline{F}_s$ derived in the last section is contributed due to change in the stored or reactive energy. This self-force is referred to as the reactive self-force. We will now analyze a charge in general motion, and examine any self-force produced due to radiation from the charge. For simplicity, we will consider a charge moving along the z axis, with no motion in the x or y directions. The derivation of the self-force under this one-dimensional motion may be extended to a general three-dimensional motion. The functional dependence of the dominant force to the charge and motion parameters is expected to remain valid for a general motion of the charge.

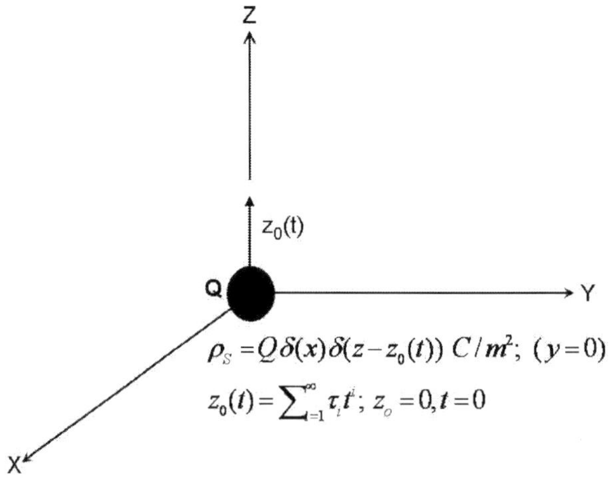

FIGURE 18

Fig.18 shows the geometry of a charge Q moving in one dimension along the z axis. The location of the charge is defined by a general time function $z = z_0(t)$. The charge Q may be modeled as a surface-charge distribution ρ_s, placed on the $y = 0$ plane, expressed in the following form.

$$\rho_s(x,z) = Q\delta(x)\delta(z - z_0(t))\ C/m^2,\ y = 0$$
$$z_0(t) = \tau_1 t + \tau_2 t^2 + ... = \textstyle\sum_{i=1}^{\infty} \tau_i t^i;\ z_0 = 0, t = 0$$
$$\tau_1 = \frac{dz_0(t)}{dt}\Big|_{t=0} = v_0,\ 2\tau_2 = \frac{d^2 z_0(t)}{dt^2}\Big|_{t=0} = a_0,\ 6\tau_3 = \frac{d^3 z_0(t)}{dt^3}\Big|_{t=0} = \dot{a}_0 \qquad (6.195)$$

$\rho_s(x, z)$ may also be expressed as a two-dimensional Fourier integral of its transform function $\tilde{\rho}_s(k_x, k_z)$, where k_x and k_z are the Fourier-transform variables in the x and z dimensions.

$$\tilde{\rho}_s = \int\int \rho_s e^{-jk_x x} e^{-jk_z z} dxdz = Qe^{-jk_z z_0(t)}$$

$$\rho_s = \frac{1}{4\pi^2}\int\int_{-\infty}^{\infty} \tilde{\rho}_s e^{jk_x x} e^{jk_z z} dk_x dk_z$$

$$= \frac{Q}{4\pi^2}\int_{k_z}\int_{k_x} e^{-jk_z z_0(t)} e^{jk_x x} e^{jk_z z} dk_x dk_z \qquad (6.196)$$

$$e^{-jk_z z_0(t)} = \textstyle\sum_{n=0}^{\infty} \frac{1}{n!}(-jk_z z_0(t))^n = \sum_{n=0}^{\infty}\sum_{m=0}^{\infty} b_{mn} k_z^n t^{n+m} \frac{(-j)^n}{n!}$$

$$z_0^n(t) = \textstyle\sum_{m=0}^{\infty} b_{mn} t^{n+m} = (\sum_{i=1}^{\infty} \tau_i t^i)^n$$

$$b_{0n} = \tau_1^n, n \geq 0$$

$$b_{1n} = 0, n = 0; \ = n\tau_1^{n-1}\tau_2, n \geq 1$$

$$b_{2n} = 0, n = 0; \ = \tau_3, n = 1;$$

$$= \tau_1^{n-2}\tau_2^2 n_{C_{n-2}} + \tau_1^{n-1}\tau_3 n_{C_{n-1}} = \tau_1^{n-2}\tau_2^2\tfrac{n(n-1)}{2} + \tau_1^{n-1}\tau_3 n, n \geq 2 \qquad (6.197)$$

The time-dependent part $e^{-jk_z z_0(t)}$ of (6.196), which is decomposed in the summation form of (6.197), maybe expressed as a Fourier integral of its frequency-transform $F(k_z, w)$. This can be used to express $\rho_s(x, z)$ in (6.196) as a three-dimensional integral of k_x k_z and w.

$$F(k_z, w) = \int_{-\infty}^{\infty} e^{-jk_z z_0(t)} e^{-jwt} dt = \Im_t(e^{-jk_z z_0(t)})$$

$$= \sum_{n=0}^{\infty}\sum_{m=0}^{\infty} b_{mn} k_z^n \frac{(-j)^n}{n!} 2\pi(n+m)! \frac{\delta(w)}{(jw)^{m+n}} = \sum_{n=0}^{\infty}\sum_{m=0}^{\infty} f_{mn} \frac{\delta(w)}{w^{m+n}} k_z^n$$

$$\Im_t(t^n) = 2\pi n! \frac{\delta(w)}{(jw)^n}$$

$$\Im_w^{-1}\frac{\delta(w)}{w^n} = \frac{1}{2\pi}\int_{-\infty}^{\infty} \frac{\delta(w)}{w^n} e^{jwt} dw = \frac{t^n j^n}{2\pi n!}, \ n \geq 0$$

$$f_{mn} = b_{mn}\frac{(-j)^n}{n!} 2\pi \frac{(n+m)!}{j^{m+n}}; f_{mn} = 0, \ n < 0$$

$$e^{-jk_z z_0(t)} = \Im_w^{-1} F(k_z, w) = \frac{1}{2\pi}\int_{-\infty}^{\infty} F(k_z, w) e^{jwt} dw$$

$$\rho_s = \frac{Q}{8\pi^3}\int_w\int_{k_z}\int_{k_x} F(k_z, w) e^{jk_x x} e^{jk_z z} e^{jwt} dk_x dk_z dw \qquad (6.198)$$

A time-dependent surface-charge distribution $\rho_s(x,z)$ may be modeled as an equivalent surface-current distribution $J_z(z,t)$, using the relationship in (6.155). The z derivative of the z directed surface current J_z is the negative of the time derivative of the surface charge ρ_z.

$$\frac{\partial J_z}{\partial z} = \frac{-\partial \rho_s}{\partial t},\ \bar{J} = \hat{z}J_z,\ J_z = \frac{Q}{8\pi^3}\int_w\int_{k_z}\int_{k_x}\frac{-w}{k_z}F(k_z,w)e^{jk_x x}e^{jk_z z}e^{jwt}dk_x dk_z dw \quad (6.199)$$

The vector-magnetic potential ψ_{zm} produced due to the surface current J_z can now be expressed, using which different fields can be derived. The current-potential-field relations of chapters 5 and 6 can be used for the derivation.

$$\psi_{zm}(x,y\gtrless 0,z,t) =$$
$$\frac{Q}{8\pi^3}\int_w\int_{k_z}\int_{k_x}(\frac{-w}{k_z})\frac{1}{2jk_y}F(k_z,w)e^{jk_x x}e^{jk_z z}e^{\mp jk_y y}e^{jwt}dk_x dk_z dw$$
$$k_y = \sqrt{k^2-k_z^2-k_x^2} = \sqrt{k_\rho^2-k_x^2},\ R_e(k_y)\gtrless 0, Im(k_y)\lesseqgtr 0, w\gtrless 0.$$
$$E_z(x,y\gtrless 0,z,t) =$$
$$\frac{Q}{8\pi^3}\int_w\int_{k_z}\int_{k_x}\frac{(k^2-k_z^2)}{jw\epsilon}(\frac{-w}{k_z})\frac{1}{2jk_y}F(k_z,w)e^{jk_x x}e^{jk_z z}e^{\mp jk_y y}e^{jwt}dk_x dk_z dw$$
$$= \frac{Q}{16\pi^3\epsilon}\int_w\int_{k_z}\int_{k_x}\frac{(k^2-k_z^2)}{k_z k_y}F(k_z,w)e^{jk_x x}e^{jk_z z}e^{\mp jk_y y}e^{jwt}dk_x dk_z dw$$
$$E_x(x,y\gtrless 0,z,t) =$$
$$\frac{Q}{8\pi^3}\int_w\int_{k_z}\int_{k_x}\frac{-k_x k_z}{jw\epsilon}(\frac{-w}{k_z})\frac{1}{2jk_y}F(k_z,w)e^{jk_x x}e^{jk_z z}e^{\mp jk_y y}e^{jwt}dk_x dk_z dw$$
$$= \frac{Q}{16\pi^3\epsilon}\int_w\int_{k_z}\int_{k_x}-\frac{kx}{k_y}F(k_z,w)e^{jk_x x}e^{jk_z z}e^{\mp jk_y y}e^{jwt}dk_x dk_z dw$$
$$E_y(x,y\gtrless 0,z,t) =$$
$$\frac{Q}{8\pi^3}\int_w\int_{k_z}\int_{k_x}\mp\frac{k_y k_z}{jw\epsilon}(\frac{-w}{k_z})\frac{1}{2jk_y}F(k_z,w)e^{jk_x x}e^{jk_z z}e^{\mp jk_y y}e^{jwt}dk_x dk_z dw$$
$$= \frac{Q}{16\pi^3\epsilon}\int_w\int_{k_z}\int_{k_x}\mp F(k_z,w)e^{jk_x x}e^{jk_z z}e^{\mp jk_y y}e^{jwt}dk_x dk_z dw \quad (6.200)$$

The fields E_x, E_y and E_z we have expressed include contributions from both the radiation as well as evanescent fields. We will examine only parts of the fields, E_{xr}, E_{yr}, E_{zr}, produced due to the mechanism of radiation, where signal power propagates away from the charge. For a given w, the radiated fields are described by all k_x, k_z, such that $k_x^2+k_z^2 \le k^2$, $k_x^2 \le k^2-k_z^2 = k_\rho^2$. Under this condition, the wave number k_y would be a real number, which results in signal propagation away from the source, in the $\pm y$ directions. In contrast, when the above condition for the wave number spectrum k_x and k_z is not satisfied, the wave number k_y would be imaginary. This is associated with an evanescent field, producing reactive or stored energy around the source. This reactive fields are accounted for in the last section to calculate the total stored energy and the associated self force $\overline{F}_s$ and equivalent mass m_s. We are interested now in the propagating or radiating part of the fields. Further, we are particularly interested in the self-field due to radiation, as measured at the center of the charge. To this end, we first derive the radiation field along the path of the charge motion ($x = y = 0$), and then obtain the self-field by having $z, t \to 0$, which specifies the instantaneous location of the source charge at $t = 0$.

It may be seen that the integrand in (6.200) for the total field component $E_x(x = y = 0, z, t)$, evaluated along the path of the charge, is an odd function of k_x. This would result in the total integral for $E_x(x = y = 0, z, t)$ to be zero. Accordingly, the total self-field $E_x(x = y = 0, z \to 0, t \to 0)$ and its radiative part $E_{xr}(x = y = 0, z \to 0, t \to 0)$ would also be zero. The y dependence of total field component $E_y(x = y = 0, z, t)$, as expressed in (6.200), has an odd symmetry in y, about the charge plane at $y = 0$. This means that the E_y field, measured in one side of the charge with $y > 0$ is negative of that observed at symmetric points on the opposite side with $y < 0$. Accordingly, the total self-field component $E_y(x = y = 0, z \to 0, t \to o)$, and hence its radiative part $E_{yr}(x = y = 0, z \to 0, t \to 0)$, would also be zero.

$$\begin{aligned} E_{xr}(x = 0, y = 0, z, t) &= 0 \\ E_{yr}(x = 0, y = 0, z, t) &= 0 \end{aligned} \tag{6.201}$$

This leaves us to evaluate only the z component of the radiative field $E_{zr}(x = 0, y = 0, z, t)$, as expressed in (6.200), from which the self field can be obtained by having $z = 0$ and $t \to 0$. . Let us first perform the k_x integration.part of the radiative field $E_{zr}(x = y = 0)$ in (6.200).

$$\begin{aligned} \int_{k_x=-k_\rho}^{k_\rho} \frac{e^{jk_x x}}{\sqrt{k_\rho^2 - k_x^2}} dk_x &= \int_{\phi'=\pi}^{0} \frac{e^{jk_\rho \cos\phi' x}}{k_\rho \sin\phi'} (-k_\rho \sin\phi') d\phi'; \;\; k_x = k_\rho \cos\phi' \\ &= \int_{\phi'=0}^{\pi} e^{jk_\rho x \cos\phi'} d\phi' = \pi J_0(k_\rho x) = \pi (x = 0) \end{aligned} \tag{6.202}$$

$J_0(.)$ is the Bessel function of the first kind of order zero (see Appendix E). Using the above result in (6.200), we get,

$$\begin{aligned} E_{zr}(x = 0, y = 0, z, t) &= \frac{Q}{16\pi^2\epsilon} \int_w \int_{k_z=-k}^{k} \frac{k^2 - k_z^2}{k_z} F(k_z, w) e^{jk_z z} e^{jwt} dk_z dw \\ &= \frac{Q}{16\pi^2\epsilon} \int_{w=-\infty}^{\infty} \int_{k_z=-k}^{k} \frac{k^2 - k_z^2}{k_z} (\sum_{n=0}^{\infty} \sum_{m=0}^{\infty} f_{mn} \frac{\delta(w)}{w^{m+n}} k_z^n) e^{jk_z z} e^{jwt} dk_z dw \end{aligned} \tag{6.203}$$

As discussed, we need to find any "self-force" or "self-field" due to the radiative effects, by evaluating $E_{zr}(x = 0, y = 0, z, t)$ with $z \to 0$ and $t \to 0$. Let us now perform the k_z integration of $E_{zr}(x = 0, y = 0, z, t)$, when $z \to 0$

$$\int_{-k}^{k} \frac{k^2 - k_z^2}{k_z} k_z^n dk_z = (k^2 \frac{k_z^n}{n} - \frac{k_z^{n+2}}{n+2})\Big|_{-k}^{k}$$
$$= 2(\frac{k^{n+2}}{n} - \frac{k^{n+2}}{n+2}) = k^{n+2}\frac{4}{n(n+2)}, n\ odd; \ = 0, n\ even$$
$$E_{zr}(x=0, y=0, z=0, t)$$
$$= \frac{Q}{16\pi^2\epsilon}\int_{w=-\infty}^{\infty}\sum_{n,odd}\sum_{m=0}^{\infty} f_{mn}\frac{\delta(w)}{w^{m+n}}\frac{4}{n(n+2)}\frac{w^{n+2}}{c^{n+2}}e^{jwt}dw$$
$$= \frac{Q}{16\pi^2\epsilon}\int_{w=-\infty}^{\infty}\sum_{n,odd}\sum_{m=0}^{\infty} f_{mn}\frac{\delta(w)}{w^{m-2}}\frac{4}{n(n+2)}\frac{1}{c^{n+2}}e^{jwt}dw \qquad (6.204)$$

We are now left with the $w-$integration for $E_{zr}(x=0, y=0, z=0, t)$. Examine the w-integration of each term in (6.204)

$$\textstyle\int_{-\infty}^{\infty}\frac{\delta(w)}{w^{m-2}}e^{jwt}dw = 0, m < 2; \ = 1, m = 2;$$
$$= \tfrac{t^{m-2}j^{m-2}}{(m-2)!}, m \geq 2 \qquad (6.205)$$

As $t \rightarrow 0$, only $m = 2$ would produce a non-zero value $(= 1)$ for the above integration.

$$E_{zr}(x=0, y=0, z=0, t\rightarrow 0) = \sum_{n,odd} E_{zrn}$$
$$= \frac{Q}{16\pi^2\epsilon}\sum_{n,odd} f_{2n}\frac{4}{n(n+2)}\frac{1}{c^{n+2}}$$
$$= \frac{Q}{16\pi^2\epsilon}\sum_{n,odd} b_{2n}\frac{(-j)^n}{n!}2\pi\frac{(n+2)!}{j^{n+2}}\frac{4}{n(n+2)}\frac{1}{c^{n+2}}$$
$$= \frac{Q}{8\pi\epsilon}\sum_{n,odd} b_{2n}\frac{(n+1)\times 4}{j^{2n+2}\times n\times c^{n+2}} = \frac{Q}{2\pi\epsilon}\sum_{n,odd}\frac{(n+1)b_{2n}}{nc^{n+2}} \qquad (6.206)$$

We have expressed the radiative "self-field" rigorously using a relatively simple series, the first few terms of which may be more explicitly expressed using (6.195,6.197) in (6.206):

$$
\begin{aligned}
n = 1, E_{zr1} &= \frac{Q}{2\pi\epsilon}\frac{2b_{21}}{c^3} = \frac{Q}{2\pi\epsilon} \times \frac{2\tau_3}{c^3} = \frac{Q}{6\pi\epsilon}\frac{\dot{a}_0}{c^3} \\
n = 3, E_{zr3} &= \frac{Q}{2\pi\epsilon} \times \frac{4}{3}\frac{b_{23}}{c^5} = \frac{Q}{2\pi\epsilon}[\frac{4\tau_1\tau_2^2}{c^5} + \frac{4\tau_1^2\tau_3}{c^5}] \\
&= \frac{Q}{2\pi\epsilon}[\frac{v_0}{c}(\frac{a_0}{c^2})^2 + (\frac{v_0}{c})^2\frac{2}{3}\frac{\dot{a}_0}{c^3}] \\
n \geq 3, E_{zrn} &= \frac{Q}{2\pi\epsilon}\frac{1}{c^{n+2}}[\tau_1^{n-2}\tau_2^2\frac{(n^2-1)}{2} + \tau_1^{n-1}\tau_3(n+1)] \\
&= \frac{Q}{2\pi\epsilon}[(\frac{\tau_1}{c})^{n-2}(\frac{\tau_2}{c^2})^2\frac{(n^2-1)}{2} + (\frac{\tau_1}{c})^{n-1}(\frac{\tau^3}{c^3})(n+1)] \\
&= \frac{Q}{2\pi\epsilon}[(\frac{v_0}{c})^{n-2}(\frac{a_0}{c^2})^2\frac{(n^2-1)}{8} + (\frac{v_0}{c})^{n-1}(\frac{\dot{a}_0}{c^3})\frac{n+1}{6}]
\end{aligned} \tag{6.207}
$$

where v_0, a_0 and $\dot{a}_0$ are the instantaneous velocity, acceleration and derivative of acceleration of the charge, respectively. Expression in terms of the instantaneous parameters v_0, a_0, and $\dot{a}_0$ is more useful than that using the parameters τ_1, τ_2, τ_3. We have derived now the field at $t \to 0$, and the expression can be used to compute the field at any other reference time t, if the instantaneous parameters v_0, a_0 and $\dot{a}_0$ are substituted by the corresponding instantaneous values at the particular time. In other words, we have the formula to compute the time-dependent radiation self field, $E_{zr}(t)$, once we have the time dependent expressions for $v_0(t)$, $a_0(t)$, $\dot{a}_0(t)$.

Notice that the "self field" is zero when the charge is moving with a constant velocity. As we have discussed, this result should be expected, because the total field energy remain unchanged in time. Further, the radiative "self-field" is seen to be zero when $\dot{a}_0 = 0$, $v_0 = 0$, even if it might be accelerating with non-zero a_0. The only situation when a charge would experience a radiative self field with $v_0 = 0$, is when its $\dot{a}_0$ is non-zero. Interestingly, one may also observe that the radiative self-field only depends on v_0, a_0 and $\dot{a}_0$, and is independent of all other higher order time derivative of motion of the charge. This assumes that the charge is a point charge, and accordingly the self-field is measured at the point location of the charge. For a charge with a non-zero size, the self-field would be evaluated as a sum or integration over the non-zero volume of the charge. The fields away from the center of the charge, and consequently the total self-field over the entire charge distribution, would then depend on higher-order motion parameters. These higher-order contributions would go to zero as the radius of the charge r_0 go to zero. Accordingly, any such higher-order terms would be proportional to positive powers of r_0. This is in contrast to the reactive self-force, which is inversely proportional to the radius r_0

The radiative "self-force" F_{zr} produced by the radiative "self-field" E_{zr} is expressed as $F_{zr} = E_{zr}Q$. The corresponding external force F_{zre} necessary to maintain the motion of the charge would be negative of F_{zr}. Accordingly, the external instantaneous power P_r utilized for the radiative motion may be expressed as

$$
\begin{aligned}
P_r = F_{zre}v_0 = -F_{zr}v_0 = -E_{zr}Qv_0 = \textstyle\sum_{n=1,3\ldots}^{\infty} P_{rn} \\
P_{rn} = -E_{zrn}Qv_0
\end{aligned} \tag{6.208}
$$

Using the principle of power conservation, we may equate P_r to the corresponding radiated power propagating away from the charge.

7.5.1. *Radiation from an Oscillatory Charge.* The above results may be applied, for illustration, to an oscillatory charge Q, whose motion is described by its time-dependent position $z_0(t) = \Delta_0 \sin(\omega t)$ along the z-axis, where Δ_0 is the maximum amplitude of displacement from the center. We will first find (a) the expression of the radiative self-field E_{zr}, assuming $v_0(t) = \Delta_0\omega \cos(\omega t) << c$, by keeping only the most significant term in (6.206,6.207), and (b) the instantaneous and average radiated power using the self-field of (a) in (6.208).

$$\begin{aligned}
(a)\ v_0(t) &= \frac{dz_0(t)}{dt} = \Delta_0\omega \cos(\omega t), \\
a_0(t) &= \frac{dv_0(t)}{dt} = -\Delta_0\omega^2 \sin(\omega t), \\
\dot{a}_0(t) &= \frac{da_0(t)}{dt} = -\Delta_0\omega^3 \cos(\omega t), \\
E_{zr}(t) &= \sum_{n=1,3}^{\infty} E_{zrn}(t) \approx E_{zr1}(t) \\
&= \frac{Q}{6\pi\epsilon}\frac{\dot{a}_0(t)}{c^3} = -\frac{Q}{6\pi\epsilon}\frac{\Delta_0\omega^3}{c^3}\cos(\omega t), \\
(b)\ P_r(t) &= -E_{zr}(t)Qv_0(t) \\
&\approx -E_{zr1}(t)Qv_0(t) \\
&= \frac{Q^2}{6\pi\epsilon}\frac{{\Delta_0}^2\omega^4}{c^3}\cos^2(\omega t), \\
< P_r > &= \frac{1}{T}\int_0^T P_r(t)dt = \frac{Q^2}{6\pi\epsilon}\frac{{\Delta_0}^2\omega^4}{c^3}\frac{1}{2}.
\end{aligned} \tag{6.209}$$

Now, the moving charge maybe expressed in the form of a line charge distribution $\rho_l(z,t)$.

$$\begin{aligned}
\rho_l(z,t) &= Q\delta(z - z_0(t)) = Q\delta(z - \Delta_0 \sin(\omega t)), \\
\frac{\partial}{\partial z}I(z,t) &= -\frac{\partial\rho_l(z,t)}{\partial t} \\
&= Q\delta'(z - z_0(t))\frac{dz_0(t)}{dt} \\
&= Q\Delta_0\omega \cos(\omega t)\delta'(z - \Delta \sin(\omega t)), \\
I(z,t) &= Q\Delta_0\omega \cos(\omega t)\delta(z - \Delta_0 \sin(\omega t)) \\
&= I_0 l \cos(\omega t)\delta(z - \Delta_0 \sin(\omega t)) \\
&\approx I_0 l \cos(\omega t)\delta(z);\ I_0 l = Q\Delta_0\omega, \Delta_0 \to 0.
\end{aligned} \tag{6.210}$$

This is essentially equivalent to a current element $I_0 \cos(\omega t)$, distributed over a small length l, placed at the origin $z = 0$. We may need a stationary charge Q, in addition to the harmonic current element of amplitude $(I_0 l)$ which carries no net charge, in order to equivalently model all fields of the oscillatory charge Q. We may ignore this extra stationary charge, which does not radiate any power, in the calculation of power radiated from the oscillatory charge. We assumed Δ_0 to be sufficiently small ($\Delta_0 \to 0$), such that the charge velocity v_0 in (6.209) would also be sufficiently small compared to the speed of light c, as we originally assumed for the problem. This would justify using only the most significant term from (6.206,6.207) for calculation of the radiative self-force, and the resulting radiated power in (6.209).

Now, the average radiated power from (6.209) maybe expressed in terms of the equivalent current parameter $(I_0 l)$ of the moving charge from (6.210).

$$
\begin{aligned}
< P_r > &= \frac{Q^2 {\Delta_0}^2 \omega^2}{12\pi\epsilon c}\frac{\omega^2}{c^2} = \frac{{I_0}^2 l^2}{12\pi} \times \sqrt{\frac{\mu}{\epsilon}}\frac{4\pi^2}{\lambda^2} \\
&= I_0^2 (\frac{l}{\lambda})^2 \frac{\eta}{12\pi} 4\pi^2 = I_0^2 (\frac{l}{\lambda})^2 \times 40\pi^2.
\end{aligned}
\tag{6.211}
$$

One may define radiation resistance R_{rad} as:

$$
\begin{aligned}
\frac{1}{2} I_0^2 R_{rad} &=< P_r >= I_0^2 (\frac{l}{\lambda})^2 40\pi^2, \\
R_{rad} &= 80\pi^2 (\frac{l}{\lambda})^2.
\end{aligned}
\tag{6.212}
$$

Significantly, this is the same expression of the R_{rad} we have derived in (9.8), using far-fields of a time-harmonic current element. In contrast to the derivation of (9.8), the present derivation calculates the radiation power and the R_{rad} from an electrodynamic near field of the charge, and is more general in scope that would account for the oscillatory or any other arbitrary time-dependence of the charge motion.

It may be recognized, that the power radiated from the oscillatory charge in (6.209) is dominantly generated by the parameter $\dot{a}_0$, which is the derivative of the acceleration, not the basic acceleration a_0 of the charge.

CHAPTER 7

ANTENNA FUNDAMENTALS

An *antenna* is a device used for transmission or reception of *electromagnetic waves,* allowing *radio-frequency* communication between distant locations across empty space or other material media. It is fabricated using conducting and/or dielectric materials, designed with desirable radiation/reception characteristics, and employs suitable coupling arrangements to an external port for input/output connection. The *input/output port* may be used for exciting the antenna from a radio-frequency *signal source,* or for coupling any radio-frequency signal received by the antenna to an output load, respectively when operated as a *transmitter* or a *receiver.* Depending on the particular application and technical requirements, an antenna may take a very simple form, like a piece of metal wire or rod used in a *car or home radio,* or can be much more complex in shape and size for more sophisticated applications in *wireless communication* and *radar.* In the following sections we will study the fundamental operation of a general antenna structure, independent of its specific configuration and application, and deduce basic performance parameters commonly used in the antenna technology.

We will analyze basic antenna parameters under four operating conditions: (1) antenna as a transmitter, (2) antenna as a receiver, (3) antenna in a transmit-receive communication link, and (4) antenna as a scatterer.

1. Antenna as a Transmitter

1.1. Basic Far-Field Relations. Fig.1 shows the geometry of an arbitrary antenna, placed at the origin of a spherical co-ordinate system (r,θ,ϕ). The electric field $\overline{E}$, and the magnetic field $\overline{H}$ of any arbitrary antenna, as observed in the far-field region, can be expressed in a general form as follows. We assume that the antenna is excited from a time-harmonic source, and the field expressions represent the corresponding complex phasors.

$$\overline{E}(r,\theta,\phi) = \hat{\theta}E_\theta(r,\theta,\phi) + \hat{\phi}E_\phi(r,\theta,\phi), \tag{7.1}$$

$$\overline{H}(r,\theta,\phi) = \hat{\theta}H_\theta(r,\theta,\phi) + \hat{\phi}H_\phi(r,\theta,\phi), \tag{7.2}$$

$$|\overline{E}(r,\theta,\phi)| = \sqrt{|E_\theta(r,\theta,\phi)|^2 + |E_\phi(r,\theta,\phi)|^2}, \tag{7.3}$$

$$\overline{H} = \frac{\hat{r}\times\overline{E}}{\eta_0}, \ \frac{|\overline{E}|}{|\overline{H}|} = \frac{E_\theta}{H_\phi} = -\frac{E_\phi}{H_\theta} = \eta_0 = \sqrt{\frac{\mu_0}{\epsilon_0}} = 120\pi. \tag{7.4}$$

The far-field region may be loosely defined to be sufficiently far away from the antenna, i.e. $r \to \infty$. In this region, the radiating fields behave like a plane wave, having no field component along the radial ($\hat{r}$) direction. Using basic principles of a plane wave in chapter 5 (see (5.19), (5.21)), one can relate the magnetic field $\overline{H}$ in the far-field region to the corresponding electric field $\overline{E}$ in a quite simple manner, as expressed in (7.4). Therefore, it may be convenient and sufficient to use only the electric field $\overline{E}$ to model various far-field antenna parameters.

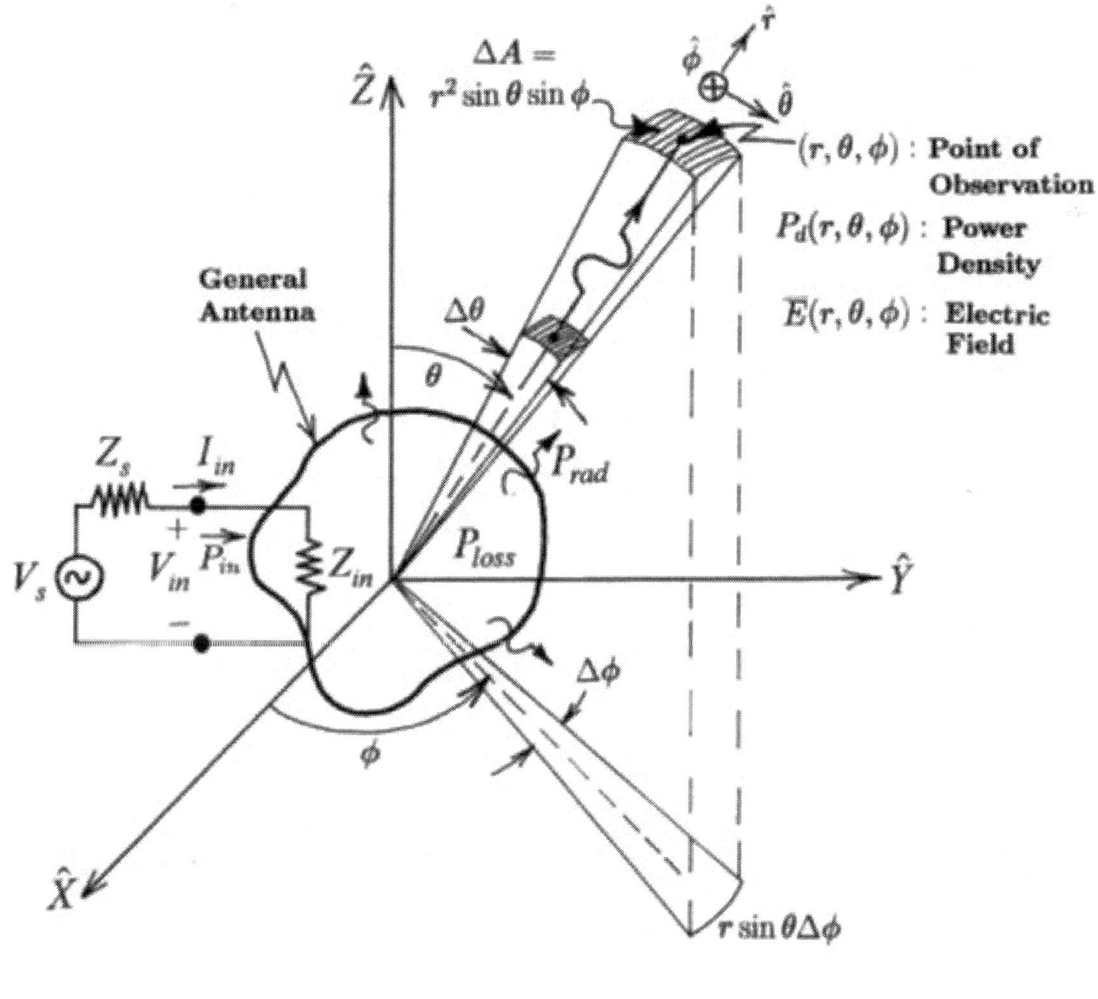

FIGURE 1. A general radiating antenna in a spherical coordinate system, showing its radiation fields and input excitation.

1.2. Radiation Power Density. The electromagnetic power flows in radial directions $\hat{r}$. The power density of radiation, P_d, is defined as the power flow per unit area, ΔA, perpendicular to the power flow.

$$\triangle P = \overline{S}(r,\theta,\phi)\cdot\hat{n}\Delta A = P_d(r,\theta,\phi)\hat{r}\cdot\hat{n}\triangle A, \tag{7.5}$$

where the vector $\bar{S}(r,\theta,\phi) = \hat{r}P_d(r,\theta,\phi)$ is called the power-flow vector or the Poynting vector, and $\hat{n}$ is normal to the surface ΔA. The power density, P_d, is related to the electric-field magnitude.

$$P_d(r,\theta,\phi) = P_{d\theta}(r,\theta,\phi) + P_{d\phi}(r,\theta,\phi) = \frac{|\overline{E}|^2}{2\eta_0}, \tag{7.6}$$

$$P_{d\theta} = \frac{|E_\theta|^2}{2\eta_0}, \quad P_{d\phi} = \frac{|E_\phi|^2}{2\eta_0}. \tag{7.7}$$

1.3. Power Per Solid Angle. The *solid angle,* $\triangle\Omega$, sometimes referred to as the *cone angle,* produced by a surface element $\overline{\triangle A} = \triangle A\hat{n}$, is defined as follows:

$$\triangle\Omega = \frac{\overline{\triangle A}\cdot\hat{r}}{r^2} = \frac{(\Delta A\hat{n})\cdot\hat{r}}{r^2}. \tag{7.8}$$

The unit of a solid angle is *Steradian,* or *Sr* for short. The solid angle formed by the angle segments $(\triangle\theta,\triangle\phi)$, shown in Fig.1, can be expressed as:

$$\triangle\Omega = \frac{\triangle A}{r^2} = \frac{r^2\sin\theta\triangle\theta\triangle\phi}{r^2} = \sin\theta\triangle\theta\triangle\phi. \tag{7.9}$$

Using (7.8,7.9) in (7.5), one may rewrite the power flow $\triangle P$ through the cone angle $\Delta\Omega$, formed by the angle segments $(\triangle\theta,\triangle\phi)$ (see Fig.1.)

$$\triangle P = P_d(r,\theta,\phi)r^2\triangle\Omega = P_d(r,\theta,\phi)r^2\sin\theta\triangle\theta\triangle\phi, \tag{7.10}$$

$$\frac{\triangle P}{\triangle\Omega} = P_d(r,\theta,\phi)r^2 = [P_{d\theta}(r,\theta,\phi) + P_{d\phi}(r,\theta,\phi)]r^2. \tag{7.11}$$

The power ΔP is conserved while propagating through the cone angle $\Delta\Omega$. Therefore, from (7.10), $P_d(r,\theta,\phi)r^2$ should be independent of r. In other words, $P_d(r,\theta,\phi)$, and its constituent parts $P_{d\theta}(r,\theta,\phi)$ and $P_{d\phi}(r,\theta,\phi)$, should be inversely proportional to r^2.

$$P_{d\theta}(r,\theta,\phi) = \frac{p_{d\theta}(\theta,\phi)}{r^2}, \quad P_{d\phi}(r,\theta,\phi) = \frac{p_{d\phi}(\theta,\phi)}{r^2}, \quad P_d(r,\theta,\phi) = \frac{p_d(\theta,\phi)}{r^2}, \tag{7.12}$$

$$p_d(\theta,\phi) = p_{d\theta}(\theta,\phi) + p_{d\phi}(\theta,\phi), \tag{7.13}$$

$$\frac{\triangle P}{\triangle \Omega} = p_d(\theta, \phi). \tag{7.14}$$

$p_d(\theta, \phi)$ is called the *power per solid angle*, with W/Sr as its unit. $p_d(\theta, \phi)$ consists of two parts, $p_{d\theta}(\theta, \phi)$ and $p_{d\phi}(\theta, \phi)$, which correspond to the $\hat{\theta}$ and $\hat{\phi}$ components of the radiated electric field, respectively.

Now, from (7.12,7.6,7.7), the magnitudes of the total electric field $\bar{E}$, and its components E_θ and E_ϕ, should be inversely proportional to r.

$$|E_\theta(r, \theta, \phi)| = \frac{f_\theta(\theta, \phi)}{r}, \quad |E_\phi(r, \theta, \phi)| = \frac{f_\phi(\theta, \phi)}{r}, \quad |\overline{E}(r, \theta, \phi)| = \frac{f(\theta, \phi)}{r}, \tag{7.15}$$

$$p_{d\theta} = \frac{f_\theta^2}{2\eta_0}, \quad p_{d\phi} = \frac{f_\phi^2}{2\eta_0}, \quad p_d = \frac{f^2}{2\eta_0}; \; f^2 = f_\theta^2 + f_\phi^2. \tag{7.16}$$

1.4. Radiation Pattern. p_d, $p_{d\theta}$, $p_{d\phi}$ are the r-independent parts of the corresponding power density functions, and f, f_θ, f_ϕ are the r-independent parts of the corresponding electric field functions. These power density and field functions may be normalized with respect to their maximum total values p_{dmax} and f_{max}, respectively .

$$\underline{p}_d(\theta, \phi) = \frac{p_d(\theta, \phi)}{p_{dmax}}, \quad \underline{p}_{d\theta}(\theta, \phi) = \frac{p_{d\theta}(\theta, \phi)}{p_{dmax}}, \quad \underline{p}_{d\phi}(\theta, \phi) = \frac{p_{d\phi}(\theta, \phi)}{p_{dmax}}, \tag{7.17}$$

$$\underline{f}(\theta, \phi) = \frac{f(\theta, \phi)}{f_{max}}, \quad \underline{f}_\theta(\theta, \phi) = \frac{f_\theta(\theta, \phi)}{f_{max}}, \quad \underline{f}_\phi(\theta, \phi) = \frac{f_\phi(\theta, \phi)}{f_{max}}, \tag{7.18}$$

$$\underline{p}_d = \underline{p}_{d\theta} + \underline{p}_{d\phi} \le 1; \; \underline{f} = \sqrt{\underline{f}_\theta^2 + \underline{f}_\phi^2} \le 1; \; \underline{p}_d = \underline{f}^2, \; \underline{p}_{d\theta} = \underline{f}_\theta^2, \; \underline{p}_{d\phi} = \underline{f}_\phi^2. \tag{7.19}$$

The above normalized functions $\underline{p}_d, \underline{p}_{d\theta}$ and $\underline{p}_{d\phi}$, are called the total-, theta-, and phi- power radiation patterns, respectively. Similarly, $\underline{f}, \underline{f}_\theta$ and $\underline{f}_\phi$, are called the total-, theta-, and phi- field radiation patterns, respectively. An example of a total power-radiation pattern, $\underline{p}_d(\theta, \phi = \phi_0)$, is shown in Fig.2, in both rectangular and polar forms. Radiation patterns are normally plotted with respect to one angle variable, θ or ϕ, keeping the other variable fixed. Alternatively, the radiation pattern may be plotted using a three-dimensional graph, including both θ and ϕ variables.

1.5. Beam Width and Side Lobe Level. As shown in Fig.2, an antenna has a direction of maximum radiation, such that $\underline{p}_d(\theta_{max}, \phi = \phi_0) = 1$ (in Fig.2 θ_{max} is assumed 0^0). The lobe surrounding the maximum radiation is called the *main lobe*, or the *main beam*. The width of the main beam, defined using the angles of first nulls in both sides of θ_{max}, is called the *first-null beam width* or FNBW. Sometimes it is more useful to define the beam width using the angles of half-power radiation, relative to the maximum radiation. The beam width defined in such manner is called the *half-power beam width* or HPBW. The lobes of the radiation pattern in both sides of the main lobe are called the *side lobes*, and the maximum level of radiation in the side lobes is called the *side lobe level* or SLL. In many applications it is desirable to have a sufficiently low SLL, in order to minimize unwanted interference away from the main lobe of communication. This also helps in efficient use of radiated power, by utilizing most of the radiated power in the desired main lobe of communication. Often, antennas also radiate to the direction opposite to the main beam, and the corresponding lobe is called the *back lobe*. The level of maximum radiation in the back lobe is called the *back lobe level* or BLL. A low BLL may be desirable to minimize interference to the back side, or also for efficient use of the radiated power.

It may be noted, depending on the specific application in mind, the desired radiation pattern may take many different forms. Also, in addition to the total power radiation pattern, one may be interested in the theta- or phi- power patterns in order to evaluate the polarization properties of radiation. The polarization properties are discussed separately in the following section.

1.6. Polarization Unit Vector. It is seen from (7.15) that the magnitude of the electric field is inversely proportional to r. Assuming linearity, the electric field should be proportional to the antenna input current I_{in}. In addition, the phase of the electric field must vary as $e^{-jk_0 r}$ in order to account for the wave propagation in the $\hat{r}$ direction, where $k_0 = \frac{2\pi}{\lambda_0} = \frac{2\pi f}{c}$ is called the wave number in free space, λ_0 is the free-space wave length, $c = 3 \times 10^8\ m/s$ is the velocity of the electromagnetic wave in free space, and f is the frequency of operation. Accordingly, the complete phasor of the E-field vector may be expressed in the following form.

$$\overline{E}(r,\theta,\phi) = I_{in}\frac{e^{-jk_0 r}}{r}\bar{e}(\theta,\phi) = I_{in}\frac{e^{-jk_0 r}}{r}[\hat{\theta}e_\theta(\theta,\phi) + \hat{\phi}e_\phi(\theta,\phi)], \tag{7.20}$$

$$E_\theta(r,\theta,\phi) = I_{in}\frac{e^{-jk_0 r}}{r}e_\theta(\theta,\phi), \quad E_\phi(r,\theta,\phi) = I_{in}\frac{e^{-jk_0 r}}{r}e_\phi(\theta,\phi), \tag{7.21}$$

$$p_d = \frac{|I_{in}|^2|\bar{e}|^2}{2\eta_0}, \quad p_{d\theta} = \frac{|I_{in}|^2|\bar{e}_\theta|^2}{2\eta_0}, \quad p_{d\phi} = \frac{|I_{in}|^2|\bar{e}_\phi|^2}{2\eta_0}, \tag{7.22}$$

$$k_0 = \frac{2\pi}{\lambda_0}, \quad \lambda_0 = \frac{c}{f}, \quad c = 3 \times 10^8\ m/s. \tag{7.23}$$

The (θ,ϕ)-dependent part, $\bar{e}(\theta,\phi)$, of the E-field is a complex vector, called the *transmit vector*, which is separated into its magnitude and a complex unit vector $\hat{\rho}$.

$$\bar{e}(\theta,\phi) = |\bar{e}(\theta,\phi)|\hat{\rho}(\theta,\phi), \tag{7.24}$$

$$|\bar{e}(\theta,\phi)| = \sqrt{[|e_\theta(\theta,\phi)|^2 + |e_\phi(\theta,\phi)|^2]}. \tag{7.25}$$

The complex unit vector $\hat{\rho}$ is called the *transmit polarization unit vector*, or simply the *polarization unit vector*. $\hat{\rho}$ contains information about the phase and direction of the transmit radiation. $\hat{\rho}$ has $\hat{\theta}$ and $\hat{\phi}$ components, and has a unit magnitude.

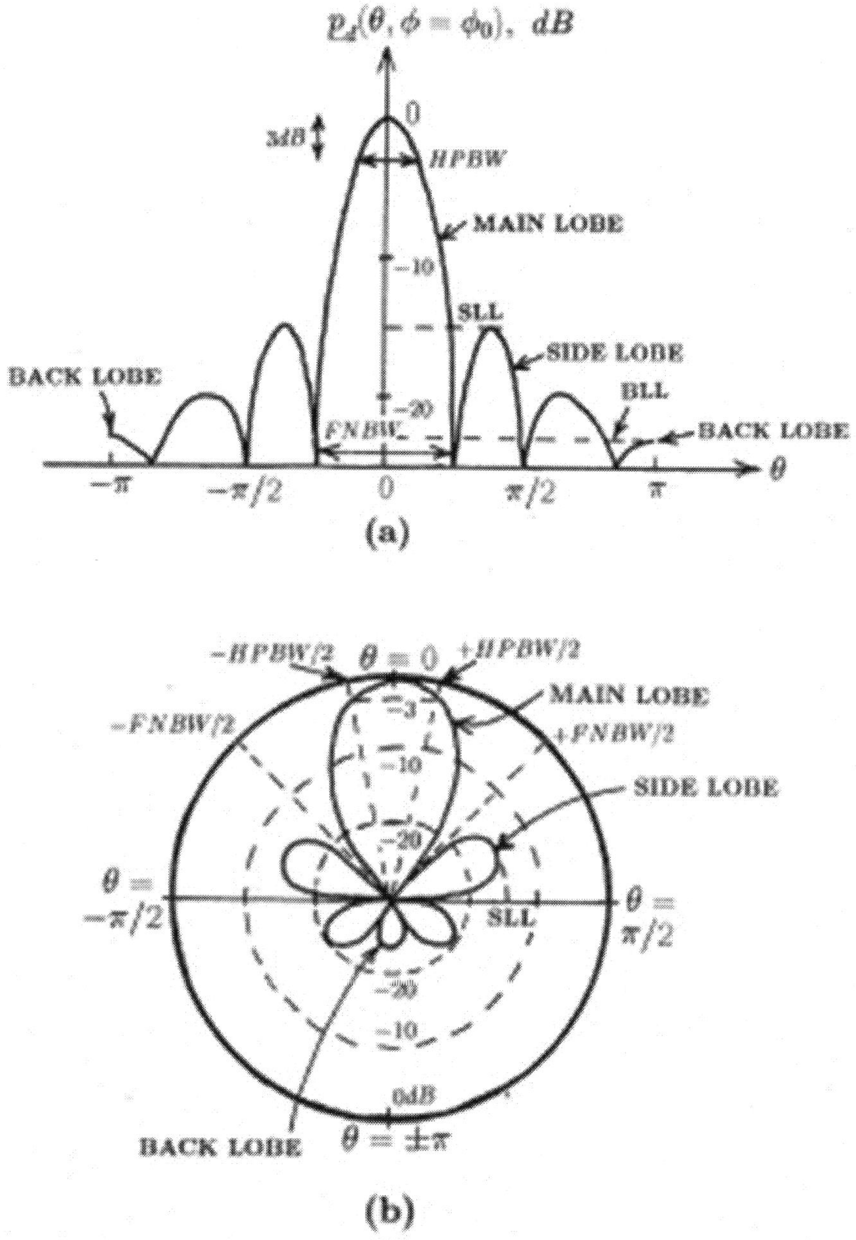

FIGURE 2. An example of a power radiation pattern $\underline{p}_d$, plotted in (a) rectangular, and (b) polar forms, as a function of θ, for a fixed value of $\phi = \phi_0$. HPBW(Half Power Beam Width), FNBW(First-Null Beam Width), SLL(Side Lobe Level), and BLL(Back Lobe Level) are shown, indicating the main lobe, side lobe and back lobe of radiation.

$$\hat{\rho}(\theta, \phi) = \hat{\theta}\rho_\theta(\theta, \phi) + \hat{\phi}\rho_\phi(\theta, \phi); \; |\hat{\rho}(\theta, \phi)| = 1, \tag{7.26}$$

$$\rho_\theta = \frac{e_\theta}{|\bar{e}|} = \frac{E_\theta}{|\bar{E}|},\ \rho_\phi = \frac{e_\phi}{|\bar{e}|} = \frac{E_\phi}{|\bar{E}|}, \tag{7.27}$$

$$\arg(\rho_\theta) = \arg(e_\theta) = \alpha_\theta,\ \arg(\rho_\phi) = \arg(e_\phi) = \alpha_\phi,\ \alpha_\phi - \alpha_\theta = \triangle\phi. \tag{7.28}$$

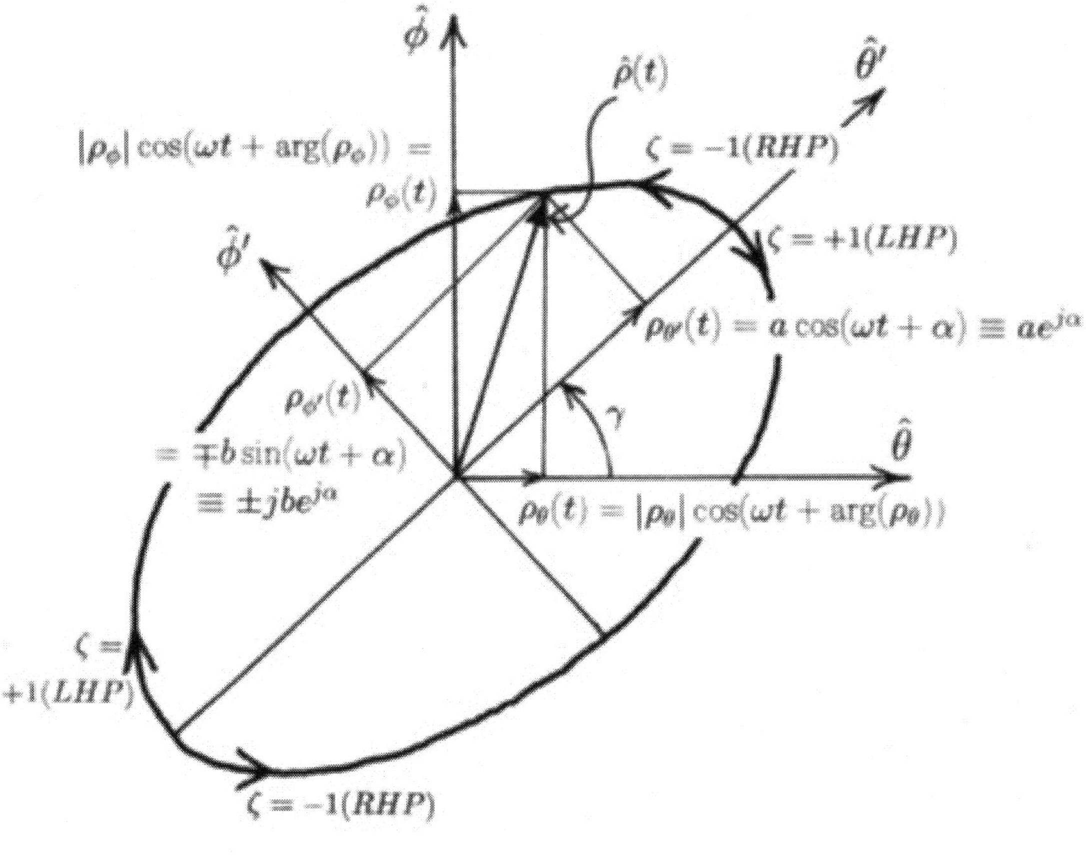

FIGURE 3. Polarization unit vector $\hat{\rho}$ is shown in two coordinate systems: the original coordinate system (θ, ϕ), and a rotated coordinate system (θ', ϕ'). The locus of the unit vector $\hat{\rho}(t)$ is in general an ellipse, with its major axis aligned along $\hat{\theta}'$, which is in general tilted by an angle $\hat{\gamma}$ with respect to the original axis $\hat{\theta}$.

1.7. Polarization Ellipse and Axial Ratio. The polarization unit vector, $\hat{\rho}$, is originally expressed in the (θ, ϕ) coordinates. The locus of the vector, with time, is in general an ellipse, with its major and minor axes oriented along $\hat{\theta}'$ and $\hat{\phi}'$ respectively, as shown in Fig.3. The polarization vector $\hat{\rho}(\theta, \phi)$ may be transformed into the new coordinate system (θ', ϕ'):

$$\begin{aligned} \hat{\rho}(\theta,\phi) = e^{j\alpha(\theta,\phi)}\hat{\rho}'(\theta,\phi) = e^{j\alpha(\theta,\phi)}[\hat{\theta}'a(\theta,\phi) + \zeta(\theta,\phi)\hat{\phi}'jb(\theta,\phi)]; \\ a, b \geq 0, \;\; a \geq b, \;\; \hat{\theta}' \times \hat{\phi}' = \hat{r}, \end{aligned} \tag{7.29}$$

$$\begin{aligned} \zeta = +1, \;\; for\ CW\ wave; \\ \zeta = -1, \;\; for\ CCW\ wave. \end{aligned} \tag{7.30}$$

ζ is the rotational parameter of the wave, indicating in which manner the polarization ellipse rotates with time. In Fig.3, $\zeta = +1$, , when the polarization ellipse of the particular wave turns clockwise (CW), and $\zeta = -1$, , when the polarization ellipse turns counter-clockwise (CCW). The CW and CCW waves are also referred to, respectively, as left-handed polarized (LHP) and right-handed polarized (RHP), when the wave propagates in the $\hat{r} = \hat{\theta}' \times \hat{\phi}'$ direction (which is the case here). A more standardized convention may be followed in order to describe the rotation of the polarization vector, independent of the coordinate system one chooses to use. A wave is called a RHP (Right-Hand Polarized) wave, when the direction of propagation of the wave corresponds to the thumb of the right hand, as the wave polarization vector spins around in the directions of the other fingers of the right hand (with fingers curved in a rotational orientation). Alternately, the wave is called a LHP (Left-Hand Polarized) wave, when the left hand matches the above directions and orientations.

Using simple trigonometry for rotating the coordinates by an angle γ (see Fig.3), we can write the following transformation equations.

$$e^{j\alpha}a = \rho_\theta \cos\gamma + \rho_\phi \sin\gamma, \tag{7.31}$$

$$e^{j\alpha}jb\zeta = -\rho_\theta \sin\gamma + \rho_\phi \cos\gamma. \tag{7.32}$$

The unknown parameters a, b, γ and α, as well as the rotational parameter ζ, can be solved in terms of ρ_θ and ρ_ϕ, using the pair of complex equations (7.31,7.32). γ is called the tilt angle, a the major axis, b the minor axis, and α the phase of the ellipse. The ratio of the major and minor axes is called the Axial Ratio (AR).

$$AR = \frac{a}{b} \equiv 20\log\frac{a}{b}(dB). \tag{7.33}$$

The above parameters of the polarization ellipse are expressed in the following, conditional upon the phase difference between ρ_θ and ρ_ϕ ($\arg(\rho_\phi) - \arg(\rho_\theta) = \triangle\phi$) . In the following it is assumed that we choose $-\pi/2 \leq \gamma \leq \pi/2$, and $a \geq b$. It is also assumed that one represents the phase difference, $\triangle\phi$, in the domain $-\pi < \triangle\phi \leq \pi$.

- **Case A, Linearly Polarized Wave**. In this case $\triangle\phi = 0\ or\ \pi,\ or\ Im(\rho_\phi\rho_\theta^*) = 0$.

$$a = \sqrt{|\rho_\theta|^2 + |\rho_\phi|^2} = 1,\ b = 0,\ AR = \infty,\ \alpha = \alpha_\theta,$$
$$\gamma = \arctan(\frac{|\rho_\phi|}{|\rho_\theta|})\ if\ \triangle\phi = 0,\ \gamma = -\arctan(\frac{|\rho_\phi|}{|\rho_\theta|})\ if\ \triangle\phi = \pi,\ \zeta = undefined. \quad (7.34)$$

In this case, the polarization ellipse becomes a straight line, with $b = 0$, $AR = \infty$, and the rotational parameter ζ becomes irrelevant.

- **Case B, Circularly Polarized Wave.** In this case $\triangle\phi = \pm\pi/2$ (or equivalently $Re(\rho_\phi\rho_\theta^*) = 0,$) and $|\rho_\theta| = |\rho_\phi|$.

$$a = b = |\rho_\theta| = |\rho_\phi|,\ \gamma = 0,\ \alpha = \alpha_\theta,\ \zeta = Im(\frac{\rho_\phi}{\rho_\theta}),\ AR = 1 \equiv 0dB. \quad (7.35)$$

In this case the polarization ellipse becomes a circle, with $a = b$, and axial ratio $AR = 1 (\equiv 0dB)$. The wave is called *circularly polarized* wave, or CP for short. The tilt angle is arbitrary, and is chosen to be zero. When the wave is RHP ($\zeta = -1$), it is called *right-hand circularly polarized* or RHCP, and on the other hand, when the wave is LHP ($\zeta = +1$), it is called *left-hand circularly polarized* or LHCP.

- **Case C, Elliptically Polarized Wave With $\gamma = 0$ or $\pi/2$.** In this case, $\triangle\phi = \pm\pi/2$ (or equivalently $Re(\rho_\phi\rho_\theta^*) = 0,$) but $|\rho_\theta| \neq |\rho_\phi|$.

$$if\ |\rho_\theta| > |\rho_\phi|, \begin{cases} a = |\rho_\theta|,\ b = |\rho_\phi|,\ \gamma = 0,\ \alpha = \alpha_\theta, \\ \zeta = +1,\ if\ \triangle\phi = +\pi/2;\ \zeta = -1,\ if\ \triangle\phi = -\pi/2; \end{cases} \quad (7.36)$$

$$if\ |\rho_\theta| < |\rho_\phi|, \begin{cases} a = |\rho_\phi|,\ b = |\rho_\theta|,\ \gamma = +\pi/2,\ \alpha = \alpha_\phi, \\ \zeta = +1,\ if\ \triangle\phi = +\pi/2;\ \zeta = -1,\ if\ \triangle\phi = -\pi/2. \end{cases} \quad (7.37)$$

This is the special situation of an elliptical wave with the major and minor axes coinciding with the original coordinates. Except, depending on the relative magnitudes of ρ_θ and ρ_ϕ, one needs to choose the tilt angle 0 or $\pi/2$ degrees, as indicated.

- **Case D, General Elliptically Polarized Wave.** This includes all cases other than A, B, and C, when $\triangle\phi \neq 0, \pm\pi/2,\ \pi$, (irrespective of $|\rho_\theta|$ and $|\rho_\phi|$).

$$\begin{aligned} \zeta &= +1\ (LHP),\ if\ Im(\rho_\phi\rho_\theta^*) > 0,\ (or\ \sin\triangle\phi > 0); \\ \zeta &= -1\ (RHP),\ if\ Im(\rho_\phi\rho_\theta^*) < 0,\ (or\ \sin\triangle\phi < 0). \end{aligned} \quad (7.38)$$

When $\zeta = +1$, the wave is called *left-hand elliptically polarized*, or LHEP, and when $\zeta = -1$, the wave is called *right-hand elliptically polarized*, or RHEP.

$$\tan 2\gamma = \frac{2Re(\rho_\phi\rho_\theta^*)}{|\rho_\theta|^2 - |\rho_\phi|^2} = \frac{2|\rho_\theta||\rho_\phi|\cos\triangle\phi}{|\rho_\theta|^2 - |\rho_\phi|^2};\ -\frac{\pi}{2} \leq \gamma \leq +\frac{\pi}{2}. \quad (7.39)$$

Note that there can be two values of $-\pi/2 \leq \gamma \leq \pi/2$ satisfying the above equation. One needs to eliminate one of them as follows:

$$\begin{aligned} 0 < \gamma < \pi/2, \; if \; Re(\rho_\phi \rho_\theta^*) > 0 \; (or \; \cos\triangle\phi > 0); \\ -\pi/2 < \gamma < 0, \; if \; Re(\rho_\phi \rho_\theta^*) < 0 \; (or \; \cos\triangle\phi < 0), \end{aligned} \tag{7.40}$$

$$\begin{aligned} a &= \sqrt{|\rho_\theta|^2 \cos^2\gamma + |\rho_\phi|^2 \sin^2\gamma + 2|\rho_\theta||\rho_\phi| \sin\gamma \cos\gamma \cos\triangle\phi} \\ &= \sqrt{\frac{1}{2}[|\rho_\theta|^2 + |\rho_\phi|^2 + \{|\rho_\theta|^4 + |\rho_\phi|^4 + 2|\rho_\theta|^2|\rho_\phi|^2 \cos(2\triangle\phi)\}^{1/2}}, \end{aligned} \tag{7.41}$$

$$\begin{aligned} b &= \sqrt{|\rho_\theta|^2 \sin^2\gamma + |\rho_\phi|^2 \cos^2\gamma - 2|\rho_\theta||\rho_\phi| \sin\gamma \cos\gamma \cos\triangle\phi} \\ &= \sqrt{\frac{1}{2}[|\rho_\theta|^2 + |\rho_\phi|^2 - \{|\rho_\theta|^4 + |\rho_\phi|^4 + 2|\rho_\theta|^2|\rho_\phi|^2 \cos(2\triangle\phi)\}^{1/2}}, \end{aligned} \tag{7.42}$$

$$\begin{aligned} \alpha &= \arg(\rho_\theta \cos\gamma + \rho_\phi \sin\gamma) \\ &= \alpha_\theta + \arctan\left[\frac{|\rho_\phi| \sin\gamma \sin\triangle\phi}{|\rho_\phi| \sin\gamma \cos\triangle\phi + |\rho_\theta| \cos\gamma}\right]. \end{aligned} \tag{7.43}$$

It should be remembered that all the above parameters a, b, γ, α, ζ, AR, and the unit vectors $\hat{\theta}'$, $\hat{\phi}'$ are functions of (θ, ϕ). Fig.4 shows a four-quadrant chart for $\triangle\phi$ that may be used to determine the type of polarization (linear, circular or elliptical, left-handed or right-handed,) and the corresponding tilt angle γ. The other parameters a, b, AR, and α can then be determined from the above cases as appropriate.

1.8. Total Radiated Power and Radiation Efficiency. From equations (7.10,7.12), the total radiated power can be determined by integrating the power density function $P_d(r, \theta, \phi)$, over a closed spherical surface.

$$P_{rad} = \int_{\theta=0}^{\pi} \int_{\phi=0}^{2\pi} P_d(r, \theta, \phi) r^2 \sin\theta \, d\phi \, d\theta = \int_{\theta=0}^{\pi} \int_{\phi=0}^{2\pi} p_d(\theta, \phi) \sin\theta \, d\phi \, d\theta. \tag{7.44}$$

The total radiated power, P_{rad}, is less than or equal to the input power, P_{in}, supplied at the input of the antenna. A part of the input power, P_{Loss}, is dissipated as heat in the material (conductor or dielectric) of the antenna itself.

$$P_{rad} = P_{in} - P_{Loss} = P_{in}(1 - \frac{P_{Loss}}{P_{in}}) = \eta_r P_{in}, \tag{7.45}$$

$$\eta_r = 1 - \frac{P_{Loss}}{P_{in}}. \tag{7.46}$$

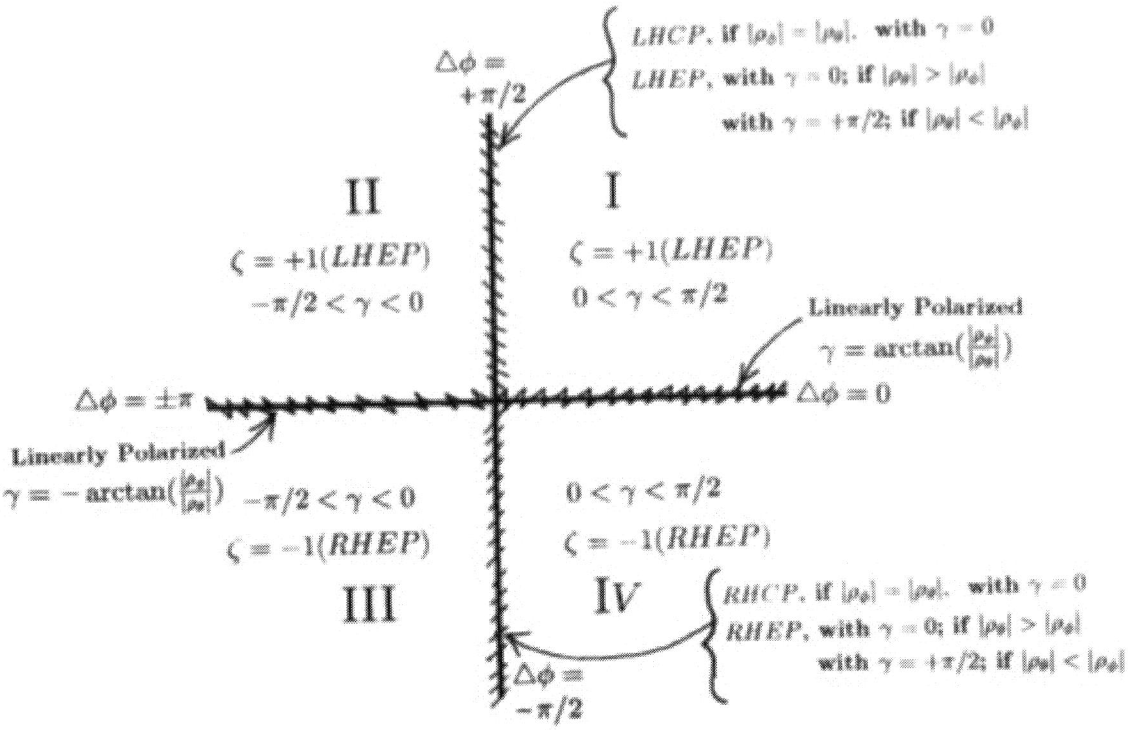

FIGURE 4. Different quadrants of the phase difference $\triangle\phi = \arg(\rho_\phi) - \arg(\rho_\theta)$, which determine the tilt angle γ and rotational parameter ζ.

η_r is called the radiation efficiency. A part of the P_{Loss} is dissipated as dielectric loss (P_{ld}), a part as conductor loss (P_{lc}), and a part is sometimes lost as unwanted surface wave in the antenna substrate material (P_{sw}), with efficiencies associated with the individual loss mechanisms as η_d, η_c and η_{sw}, respectively.

$$P_{Loss} = P_{ld} + P_{lc} + P_{sw}, \tag{7.47}$$

$$\eta_d = 1 - \frac{P_{ld}}{P_{in}}, \; \eta_c = 1 - \frac{P_{lc}}{P_{in}}, \; \eta_{sw} = 1 - \frac{P_{sw}}{P_{in}}. \tag{7.48}$$

If the individual losses are small, the total radiation efficiency can be approximated as the product of individual parts.

$$\eta_r = 1 - \frac{P_{ld} + P_{lc} + P_{sw}}{P_{in}} \simeq (1 - \frac{P_{ld}}{P_{in}})(1 - \frac{P_{lc}}{P_{in}})(1 - \frac{P_{sw}}{P_{in}}) = \eta_c\eta_d\eta_{sw}. \tag{7.49}$$

1.9. Antenna Input Impedance and Radiation Resistance. An antenna is seen by the input source as an equivalent impedance Z_{in} (see Fig.1), with real and imaginary parts R_{in} and X_{in}, respectively. The imaginary part is due to stored energy in the antenna field. The real part R_{in} consists of two parts, R_{rad} and R_{Loss}, respectively contributed due to the radiated power P_{rad} and the power loss P_{Loss} in the antenna.

$$Z_{in} = R_{in} + jX_{in} = R_{rad} + R_{Loss} + jX_{in}, \; R_{in} = R_{rad} + R_{Loss}, \tag{7.50}$$

$$P_{rad} = \frac{|I_{in}|^2 R_{rad}}{2}, \; P_{Loss} = \frac{|I_{in}|^2 R_{Loss}}{2}, \; P_{in} = \frac{|I_{in}|^2 (R_{rad} + R_{Loss})}{2}, \; \eta_r = \frac{R_{rad}}{R_{rad} + R_{Loss}}. \tag{7.51}$$

R_{rad} is called the *radiation resistance*, and R_{Loss} is called the *loss resistance* of the antenna.

1.10. Resonant Frequency and Bandwidth. The input impedance Z_{in} of the antenna is a function of the frequency of operation. Fig.5 shows the magnitude of the input impedance of an example antenna as a function of frequency. In this case, the antenna impedance looks like a parallel *RLC* resonant circuit. The frequency, f_r, for which the impedance magnitude is maximum, or equivalently the reactance is zero, is often defined as the *resonant frequency.* It is desirable to operate the antenna at this resonant frequency, so that it can be easily matched to an input transmission line with a real characteristic impedance. The frequency span $\triangle f_r$, beyond which the impedance magnitude falls below half of the resonant value, may be defined as the *impedance bandwidth* of the antenna. The antenna may be usable in this frequency band. Outside the antenna's bandwidth, the input power to the antenna would be significantly reflected due to impedance mismatch, resulting in poor radiation.

It may be noted that Fig.5 is only an example case. Depending on the type of the antenna and its application, the resonant frequency and bandwidth may be defined based on the performance parameter one desires for the particular application, such as the gain, directivity, axial ratio, etc. In such cases, the resonant frequency is defined as the frequency of best performance, and the bandwidth as the frequency span over which the performance can be tolerated, as dictated by the application. It is meaningful to call the resonant frequency and the bandwidth of an antenna with an appropriate parameter prefix, such as the impedance-resonance frequency, the impedance-bandwidth, gain-resonant frequency, the gain-bandwidth, the axial-ratio-resonant frequency, the axial-ratio-bandwidth, etc. Often, the definitions based on the different performance parameters have strong correlations with each other.

1.11. Directivity. How well an antenna directs the radiation in a particular angle is quantified by comparing with an ideal isotropic radiator. An isotropic radiator is defined as an antenna that radiates uniformly in all angles. Let the total radiated power from a given antenna, P_{rad}, also be radiated from the ideal isotropic radiator. The resulting power density, $P_{d(iso)}$, at a given distance r, is independent of θ and ϕ.

$$P_{d(iso)} = \frac{P_{rad}}{4\pi r^2}. \tag{7.52}$$

The ratio of the power density $P_d(r, \theta, \phi)$ from the actual antenna, in a given direction (θ, ϕ), with that from the isotropic radiator $P_{d(iso)}$, is called the *directive gain function* $D_g(\theta, \phi)$.

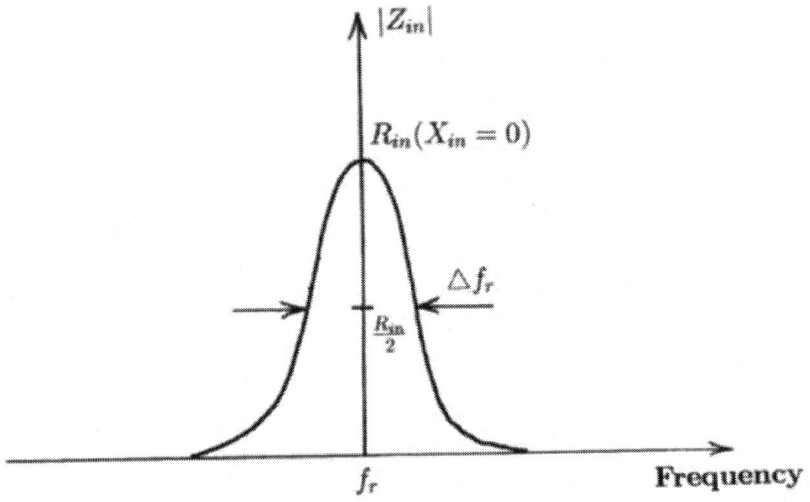

FIGURE 5. Frequency dependence of the impedance magnitude of an example antenna, showing the resonant frequency f_r and the bandwidth $\triangle f_r$ of the antenna.

$$D_g(\theta,\phi) = \frac{P_d(r,\theta,\phi)}{P_{d(iso)}} = \frac{4\pi r^2 P_d(r,\theta,\phi)}{P_{rad}} = \frac{4\pi p_d(\theta,\phi)}{P_{rad}}, \tag{7.53}$$

$$P_d(r,\theta,\phi) = \frac{P_{rad}}{4\pi r^2} D_g(\theta,\phi),\ p_d(\theta,\phi) = \frac{P_{rad}}{4\pi} D_g(\theta,\phi). \tag{7.54}$$

The maximum value of $D_g(\theta,\phi)$, occurring along the main beam direction $(\theta_{max},\phi_{max})$ is called the *directivity*, D.

$$Directivity = D = D_g(\theta,\phi)_{max} = D_g(\theta_{max},\phi_{max}). \tag{7.55}$$

Now, using (7.44) in (7.53), we can express $D_g(\theta,\phi)$ using either $p_d(\theta,\phi)$, or the power pattern $\underline{p}_d(\theta,\phi)$.

$$D_g(\theta,\phi) = \frac{4\pi p_d(\theta,\phi)}{\int_{\phi=0}^{2\pi}\int_{\theta=0}^{\pi} p_d(\theta,\phi)\sin\theta\ d\theta\ d\phi} = \frac{4\pi \underline{p}_d(\theta,\phi)}{\int_{\phi=0}^{2\pi}\int_{\theta=0}^{\pi} \underline{p}_d(\theta,\phi)\sin\theta\ d\theta\ d\phi}. \tag{7.56}$$

D_g, R_{rad} and $\overline{e}$ relationship.

Combining the equations (7.22, 7.51, 7.54), we can arrive at an useful relationship between the field vector $\overline{e}(\theta, \phi)$, directive gain function $D_g(\theta, \phi)$ and the radiation resistance R_{rad}.

$$\frac{|I_{in}|^2|\overline{e}(\theta,\phi)|^2}{2\eta_0} = \frac{|I_{in}|^2 R_{rad}}{8\pi} D_g(\theta,\phi), \ |\overline{e}(\theta,\phi)| = \sqrt{\frac{\eta_0 R_{rad}}{4\pi} D_g(\theta,\phi)}. \tag{7.57}$$

1.12. Gain. It may also be useful to find the ratio of the power density $P_d(r, \theta, \phi)$ from a given antenna, to that $(= P'_{d(iso)})$ from an ideal isotropic radiator, when supplied by the same input power P_{in} in both cases (unlike the same P_{rad} considered in the last section.) The ratio of $P_d(r, \theta, \phi)$ to $P'_{d(iso)}$ is called the *gain function* $G(\theta, \phi)$.

$$P'_{d(iso)} = \frac{P_{in}}{4\pi r^2}, \tag{7.58}$$

$$G(\theta,\phi) = \frac{P_d(r,\theta,\phi)}{P'_{d(iso)}} = \frac{4\pi r^2 P_d(r,\theta,\phi)}{P_{in}} = \frac{4\pi p_d(\theta,\phi)}{P_{in}}, \tag{7.59}$$

$$P_d(r,\theta,\phi) = \frac{P_{in}}{4\pi r^2} G(\theta,\phi) = \frac{P_{rad}}{4\pi r^2 \eta_r} G(\theta,\phi); \ p_d(\theta,\phi) = \frac{P_{in}}{4\pi} G(\theta,\phi). \tag{7.60}$$

Comparing the equations (7.59,7.53), and using the efficiency relationship (7.45), we obtain a simple relationship between $G(\theta, \phi)$ and $D_g(\theta, \phi)$.

$$G(\theta,\phi) = \eta_r D_g(\theta,\phi). \tag{7.61}$$

The maximum value of the gain function is referred to as the maximum gain, or simply the *gain* G

$$Gain = G = G(\theta,\phi)_{max} = G(\theta_{max}, \phi_{max}) = \eta_r D. \tag{7.62}$$

G, R_{in} and $\overline{e}$ relationship.

Combining the equations (7.22, 7.51, 7.60), we can arrive at an useful relationship between the field vector $\overline{e}(\theta, \phi)$, gain function $G(\theta, \phi)$ and the antenna input resistance R_{in}.

$$\frac{|I_{in}|^2|\overline{e}(\theta,\phi)|^2}{2\eta_0} = \frac{|I_{in}|^2 R_{in}}{8\pi} G(\theta,\phi), \ |\overline{e}(\theta,\phi)| = \sqrt{\frac{\eta_0 R_{in}}{4\pi} G(\theta,\phi)}. \tag{7.63}$$

1.13. Beam Efficiency. The *beam efficiency* η_{beam} of an antenna is defined as the ratio of the radiated power, $P_{mainbeam}$, confined in the antenna main lobe, to the total radiated power P_{rad}. This quantifies how well the antenna utilizes the total radiated power in the desired main beam.

$$\eta_{beam} = \frac{P_{mainbeam}}{P_{rad}} = \frac{\int\int_{mainbeam} p_d(\theta,\phi)\sin\theta d\theta d\phi}{\int_{\phi=0}^{2\pi}\int_{\theta=0}^{\pi} p_d(\theta,\phi)\sin\theta d\theta d\phi}. \tag{7.64}$$

2. Antenna as a Receiver

In this section we will view an antenna as a receiving element. Before we can model the receiving antenna, we will first model a uniform plane wave, incident on the antenna placed at the origin of a spherical coordinate system (see Fig.6).

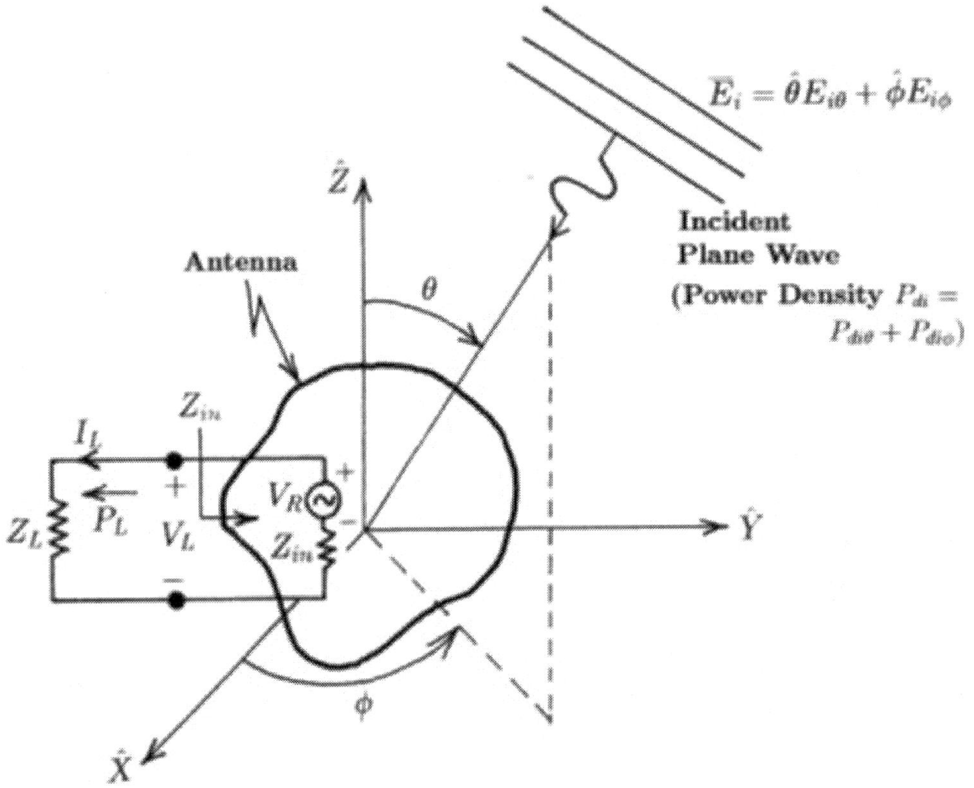

FIGURE 6. An arbitrary antenna in a receiving mode of operation, showing an incident plane wave and a Thevenin's equivalent model at the input terminals.

2.1. Incident Plane Wave. The incident plane wave has both $\hat{\theta}$ and $\hat{\phi}$ components, with arbitrary amplitudes and phases. Unlike a radiating wave propagating radially away from the origin, this plane wave is propagating toward the origin with a e^{+jk_0r} phase factor.

$$\bar{E}_i = \hat{\theta}E_{i\theta} + \hat{\phi}E_{i\phi} = \bar{e}_i e^{+jk_0r} = (\hat{\theta}e_{i\theta} + \hat{\phi}e_{i\phi})e^{+jk_0r}, \tag{7.65}$$

$$\bar{e}_i = \hat{\theta}e_{i\theta} + \hat{\phi}e_{i\phi} = |e_i|\hat{\rho}_i = |\bar{e}_i|(\hat{\theta}\rho_{i\theta} + \hat{\phi}\rho_{i\phi}), \tag{7.66}$$

$$|\bar{e}_i| = \sqrt{|e_{i\theta}|^2 + |e_{i\phi}|^2},\ \rho_{i\theta} = \frac{e_{i\theta}}{|\bar{e}_i|},\ \rho_{i\phi} = \frac{e_{i\phi}}{|\bar{e}_i|}, \tag{7.67}$$

$$|\hat{\rho}_i| = \sqrt{|\rho_{i\theta}|^2 + |\rho_{i\phi}|^2} = \frac{\sqrt{|e_{i\theta}|^2 + |e_{i\phi}|^2}}{|\bar{e}_i|} = 1, \tag{7.68}$$

$$P_{dinc} = \frac{|\overline{E}_i|^2}{2\eta_0} = \frac{|\bar{e}_i|^2}{2\eta_0}. \tag{7.69}$$

$\hat{\rho}_i$ is the polarization unit vector of the incident wave. Like the radiating polarization unit vector $\hat{\rho}$, the incident polarization unit vector $\hat{\rho}_i$ may be transformed to a new coordinate system $(\hat{\theta}'_i, \hat{\phi}'_i)$, aligned with the major and minor axes of the incident polarization ellipse, respectively. The angle between the $\hat{\theta}$ and $\hat{\theta}'_i$ axes is the tilt angle γ_i

$$\hat{\rho}_i = e^{j\alpha_i}\hat{\rho}'_i = e^{j\alpha_i}(a_i\hat{\theta}'_i - \zeta_i j b_i\hat{\phi}'_i);\ a_i, b_i \geq 0,\ a_i \geq b_i,\ \hat{\theta}'_i \times \hat{\phi}'_i = \hat{r}, \tag{7.70}$$

$$\begin{aligned} \zeta_i &= +1,\ for\ LHP\ wave; \\ \zeta_i &= -1,\ for\ RHP\ wave. \end{aligned} \tag{7.71}$$

The equations in section 1.7 may be used here in order to determine the various parameters of the incident wave. The major and minor axes a_i and b_i, axial ratio AR_i, tilt angle γ_i, the rotation parameter ζ_i, the phase angle α_i should be replaced for the corresponding variables in section 1.7: a, b, AR, γ, ζ and α, respectively. However, it may be noted that, in contrast with equation (7.29), the equation (7.70) has a '-' sign before ζ_i. This is to take into account the propagation of the incident wave in the $-\hat{r}$ direction, instead of the $+\hat{r}$ direction of propagation for the outgoing radiated field. The change in the direction of propagation, by definition, automatically switches the "handedness" (right- or left-handed) of the wave.

2.2. Receive Voltage Vector. As shown in Fig.6, the receiving antenna is modeled as a Thevenin's equivalent circuit at its output terminals. The open-circuit voltage V_R can be written as a linear combination of the $\theta-$ and the $\phi-$ components of the incident fields.

$$V_R(\theta,\phi) = V_{R\theta}(\theta,\phi) + V_{R\phi}(\theta,\phi) = e_{\theta i}v_{R\theta}(\theta,\phi) + e_{\phi i}v_{R\phi}(\theta,\phi) = \bar{e}_i \cdot \bar{v}_R(\theta,\phi), \tag{7.72}$$

$$\bar{v}_R(\theta,\phi) = \hat{\theta}v_{R\theta}(\theta,\phi) + \hat{\phi}v_{R\phi}(\theta,\phi) = |\bar{v}_R(\theta,\phi)|\hat{\rho}_R(\theta,\phi), \tag{7.73}$$

$$\hat{\rho}_R(\theta,\phi) = \hat{\theta}\rho_{R\theta}(\theta,\phi) + \hat{\phi}\rho_{R\phi}(\theta,\phi), \tag{7.74}$$

$$v_R = |\bar{v}_R| = \sqrt{|v_{R\theta}|^2 + |v_{R\phi}|^2},\ \rho_{R\theta} = \frac{v_{R\theta}}{|\bar{v}_R|},\ \rho_{R\phi} = \frac{v_{R\phi}}{|\bar{v}_R|}, \tag{7.75}$$

$$|\hat{\rho}_R| = \sqrt{|\rho_{R\theta}|^2 + |\rho_{R\phi}|^2} = \frac{\sqrt{|v_{R\theta}|^2 + |v_{R\phi}|^2}}{|\bar{v}_R|} = 1. \tag{7.76}$$

$\overline{v}_R(\theta,\phi)$ is called the *receive vector*, and the unit vector $\hat{\rho}_R(\theta,\phi)$ is called the *receive polarization unit vector*. Like the radiation and incident polarization unit vectors $\hat{\rho}$ and $\hat{\rho}_i$, the receive polarization unit vector may be transformed to a new co-ordinate system $(\hat{\theta}'_R, \hat{\phi}'_R)$, aligned with the major and minor axes of the receive polarization ellipse. The angle between the $\hat{\theta}$ and $\hat{\theta}'_R$ axes is the tilt angle γ_R.

$$\hat{\rho}_R = e^{j\alpha_R}\hat{\rho}'_R = e^{j\alpha_R}(a_R\hat{\theta}'_R + \zeta_R j b_R\hat{\phi}'_R);\ a_R, b_R \geq 0,\ a_R \geq b_R,\ \hat{\theta}'_R \times \hat{\phi}'_R = \hat{r}, \tag{7.77}$$

$$\begin{aligned} \zeta_R &= +1,\ for\ LHP\ receiver; \\ \zeta_R &= -1,\ for\ RHP\ receiver. \end{aligned} \tag{7.78}$$

Like the incident polarization unit vector in equation (7.70), the equations in the section 1.7 may also be used here in order to determine the various parameters for the receive polarization unit vector $\hat{\rho}_R$. It may be noted that, in contrast with equation (7.70), the equation (7.77) has a '+' sign before ζ_R. The rotational parameter ζ_R as a rule follows the same convention as for ζ, which is different from that for the incident wave ζ_i.

2.3. Received Power and Effective Area. Once the Thevenin's voltage V_R of the receiving antenna is known, the load power P_L delivered to a load impedance Z_L can be easily obtained (see Fig.6).

$$V_R = \overline{e}_i \cdot \overline{v}_R = |\overline{e}_i||\overline{v}_R|\hat{\rho}_i \cdot \hat{\rho}_R, \tag{7.79}$$

$$V_L = \frac{V_R Z_L}{Z_L + Z_{in}},\ I_L = \frac{V_R}{Z_L + Z_{in}},\ P_L = \frac{1}{2}\left|\frac{V_R}{Z_L + Z_{in}}\right|^2 R_L. \tag{7.80}$$

We define the effective area A_e of a receiving antenna as the ratio of the load power P_L and the incident power density P_{dinc}. The receive antenna is seen as an equivalent capture area for the incident wave. The equivalent area A_e of an antenna may be clearly distinguished from the physical area of an antenna. They can be significantly different from each other, in general. For example, a commonly-used wire antenna has a non-zero effective area allowing a fairly good reception, while its physical area is very small - ideally zero.

$$A_e = \frac{P_L}{P_{dinc}} = \frac{R_L\eta_0}{|\overline{e}_i|^2}\left|\frac{V_R}{Z_L + Z_{in}}\right|^2, \tag{7.81}$$

$$A_e = \frac{R_L\eta_0}{|\overline{e}_i|^2}\frac{|\overline{v}_R|^2|\overline{e}_i|^2|\hat{\rho}_i\cdot\hat{\rho}_R|^2}{|Z_L + Z_{in}|^2} = \frac{R_L\eta_0|\overline{v}_R|^2|\hat{\rho}_i\cdot\hat{\rho}_R|^2}{|Z_L + Z_{in}|^2}. \quad (7.82)$$

The effective area can be maximized in two ways. First, by matching the load impedance for maximum power transfer, which requires $Z_L = Z_{in}^*$. Second, by adjusting the incident and receive polarization vectors such that $|\hat{\rho}_i \cdot \hat{\rho}_R| = 1 = |\hat{\rho}'_i \cdot \hat{\rho}'_R|$, whenever possible. Under the above conditions, the maximum effective area A_{em} can be derived from (7.82).

$$A_{em}(\theta,\phi) = \frac{\eta_0|\overline{v}_R(\theta,\phi)|^2}{4R_{in}};\ Z_L = Z_{in}^*,\ |\hat{\rho}_i\cdot\hat{\rho}_R| = 1 = |\hat{\rho}'_i\cdot\hat{\rho}'_R|. \quad (7.83)$$

$|\hat{\rho}_i.\hat{\rho}_R|$ is called the *polarization mismatch factor* between the incident wave and the receive antenna. It can be shown that the polarization mismatch factor is maximum possible (unity) only when $\gamma_i = \gamma_r$, $\zeta_i = \zeta_r$, $AR_i = AR_r$ ($a_i = a_R$, $b_i = b_R$). From (7.70,7.77), one can see that the above condition is equivalent to having $\hat{\rho}'_i = \hat{\rho}'^*_R$. Other possible practical situations may be considered. (1) For a given receive antenna (given ζ_r and AR_r) and a given incident plane wave (given ζ_i and AR_i), and a given load impedance Z_L, the receive power P_L can be maximized by just turning the antenna about the $\hat{r}$ axis, such that the major axes of the incident and receive polarization vectors are aligned ($\gamma_i = \gamma_R$). (2) Every other parameter remaining the same, the received power is larger if the rotational parameters are matched ($\zeta_i = \zeta_r$), compared to when they are mismatched ($\zeta_i \neq \zeta_r$).

2.4. Received Noise Temperature. Fig.7a shows the noise-equivalent model of an antenna. Like an ohmic resistance, the input resistance of the antenna, R_{in}, is associated with a noise source v_{na}, with equivalent noise temperature T_{ea} **[25, 26]**. However, unlike an ohmic resistance, T_{ea} is not the physical temperature of the antenna. v_{na} consists of two parts, v_{nL} and v_{nr}, with noise temperatures T_a and T_{er}, respectively associated with the loss resistance R_{Loss} and radiation resistance R_{rad} of the antenna. v_{nL} is generated due to the thermal process in the antenna itself, and therefore T_a is equal to the physical temperature of the antenna. Whereas, T_{er} is associated with the incoming radiation noise, which depends on the temperature distribution of the surrounding objects that generate it.

$$\overline{v}_{nL}^2 = 4kT_aR_{Loss},\ \overline{v}_{nr}^2 = 4kT_{er}R_{rad},\ \overline{v}_{na}^2 = 4kT_{ea}R_{in};\ \overline{v}_{na}^2 = \overline{v}_{nL}^2 + \overline{v}_{nr}^2\ . \quad (7.84)$$

The over-bar represents root-mean-square (rms) values per unit bandwidth. Let the total noise-power, per unit bandwidth, delivered by the antenna to a conjugate-matched load $Z_L = Z_{in}^*$, be P_{na}. The respective constituent parts delivered by v_{nL} and v_{nR} are P_{nL} and P_{nr}. These noise powers can be expressed using the respective noise temperatures and antenna radiation efficiency η_r (from (7.51)), which would then provide a relationship between T_{ea}, T_a and T_{er}.

$$P_{nL} = \frac{\overline{v}_{nL}^2}{4R_{in}} = kT_a\frac{R_{Loss}}{R_{in}} = kT_a(1-\eta_r), \quad (7.85)$$

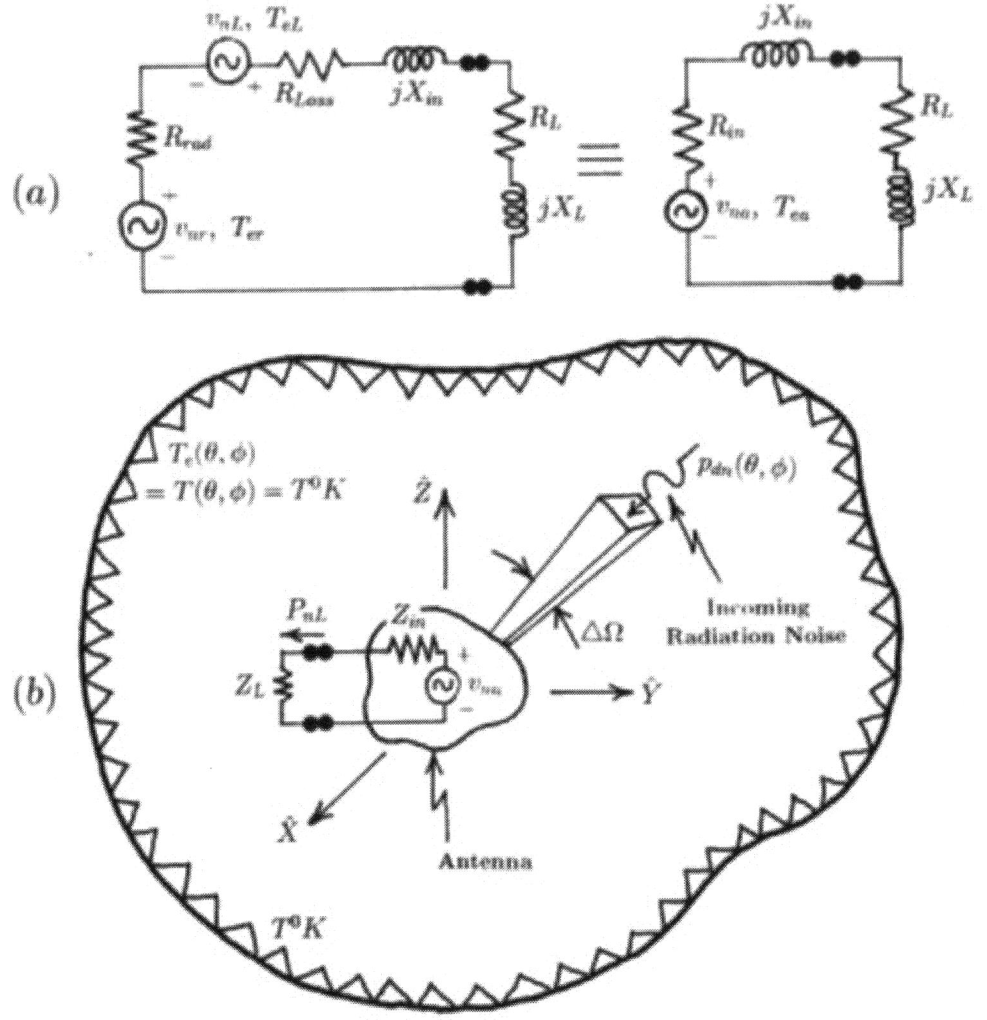

FIGURE 7. Noise-equivalent circuit of an antenna, showing noise voltage v_{nr} received through radiation, and v_{nl} produced by the antenna itself due to material-loss resistance R_{Loss}. (b) Antenna placed in a closed "perfectly absorbing" chamber at temperature T^0K. p_{dn} = noise power per unit solid angle, as seen by an ideal isotropic antenna.

$$P_{nr} = \frac{\overline{v}_{nr}^2}{4R_{in}} = kT_{er}\frac{R_{rad}}{R_{in}} = kT_{er}\eta_r, \tag{7.86}$$

$$P_{na} = \frac{\overline{v}_{na}^2}{4R_{in}} = kT_{ea} = P_{nL} + P_{nr};\ T_{ea} = \eta_r T_{er} + (1 - \eta_r)T_a. \qquad (7.87)$$

Relationship between T_{er} and radiation-noise density

Let $p_{dn}(\theta, \phi)$ be the noise-power density of the incoming radiation, per unit solid angle, per unit bandwidth, as seen by an ideal, matched "isotropic" antenna (gain function $G_{iso}(\theta, \phi) = 1$, matched effective area $= A_{em(iso)}$). We can then express the matched radiation noise P_{nr} received by an arbitrary antenna, in terms of $p_{dn}(\theta, \phi)$ and the antenna's effective area $A_{em}(\theta, \phi)$ relative to $A_{em(iso)}$.

$$P_{nr} = \int\limits_{\Omega} p_{dn}(\theta, \phi) \frac{A_{em}(\theta, \phi)}{A_{em(iso)}} d\Omega = \int\limits_{\Omega} p_{dn}(\theta, \phi) \frac{G(\theta, \phi)}{G_{iso}} d\Omega = \int\limits_{\Omega} p_{dn}(\theta, \phi) G(\theta, \phi) d\Omega. \quad (7.88)$$

It may be noted, in the above equation (7.88) we have used the $G(\theta, \phi) \sim A_{em}(\theta, \phi)$ relationship, to be developed in (7.110). Then, using (7.86,7.61) in (7.88) we can relate the radiation noise-temperature T_{er} with $p_{dn}(\theta, \phi)$ and the directive gain function $D_g(\theta, \phi)$

$$\eta_r k T_{er} = \int\limits_{\Omega} p_{dn}(\theta, \phi) G(\theta, \phi) d\Omega;\ T_{er} = \int\limits_{\Omega} \frac{p_{dn}(\theta, \phi)}{k} D_g(\theta, \phi) d\Omega. \qquad (7.89)$$

Relationship between T_{er} and spatial temperature distribution $T_e(\theta, \phi)$

Consider first a special situation, when the antenna is surrounded by perfect absorbers, and the entire system is in thermal equilibrium at the temperature $T = T_a$, as shown in Fig.7b. Under this condition, the noise temperature of the antenna should be equal to the surrounding temperature. Consequently, using (7.87), the radiation-noise temperature of the antenna can be shown to be the temperature of the surrounding objects T

$$T_{ea} = T = T_a = \eta_r T_{er} + (1 - \eta_r)T_a;\ T_{er} = T. \qquad (7.90)$$

Here, due to symmetry of the situation, $p_{dn}(\theta, \phi)$ is independent of the angle of arrival (θ, ϕ). With this assumption, and using (7.56,7.90) in (7.89), we get a relationship between p_{dn} and the absorber temperature T

$$T_{er} = T = \int\limits_{\Omega} \frac{p_{dn}(\theta, \phi)}{k} D_g(\theta, \phi) d\Omega = \frac{p_{dn}}{k} \int\limits_{\Omega} D_g(\theta, \phi) d\Omega = \frac{4\pi p_{dn}}{k};\ p_{dn} = \frac{kT}{4\pi}. \quad (7.91)$$

We can extend (7.91) to arbitrary situations, when the surrounding objects are at different physical temperatures $T(\theta, \phi)$, or can be modeled with noise-equivalent temperatures $T_e(\theta, \phi)$. This will relate $p_{dn}(\theta, \phi)$ to $T(\theta, \phi)$ or $T_e(\theta, \phi)$. Then, using (7.89), we can relate T_{er} to the surrounding temperature profile $T_e(\theta, \phi)$ via $D_g(\theta, \phi)$.

$$p_{dn}(\theta,\phi) = \frac{kT_e(\theta,\phi)}{4\pi}, \; T_{er} = \frac{1}{4\pi}\int_{\Omega} T_e(\theta,\phi) D_g(\theta,\phi) d\Omega. \tag{7.92}$$

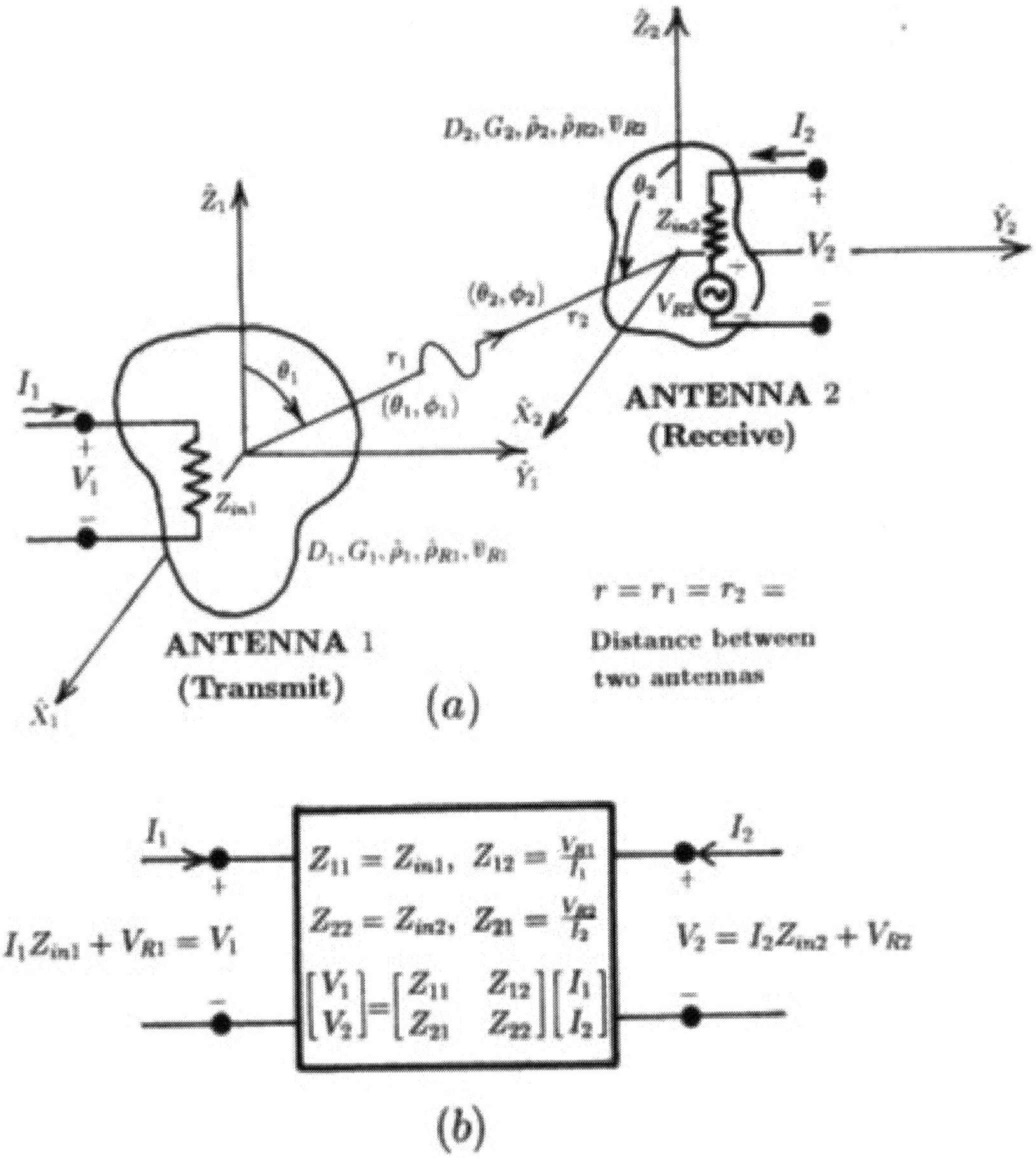

FIGURE 8. (a) A communication link between two antennas, which can be viewed as a two-port circuit as shown in (b).

3. Transmit-Receive Communication Link

Figure 8 shows a communication link between two antennas, with their antenna parameters indicated with a subscript 1 or 2. Each antenna is described with respect to its own spherical coordinate system, represented with subscripts 1 or 2, having origin at the center of the corresponding antenna. First let us treat antenna 1 as the transmit antenna. The transmit field, $\overline{E}_1$, radiated from antenna 1 is viewed as the incident wave, $\overline{e}_i$, for the receiving antenna 2.

$$\overline{E}_1(r_1, \theta_1, \phi_1) = I_1 \frac{e^{-jk_0 r_1}}{r_1} \overline{e}_1(\theta_1, \phi_1) = \overline{e}_i, \tag{7.93}$$

$$|\overline{e}_i| = \frac{|I_1||\overline{e}_1(\theta_1, \phi_1)|}{r_1}, \quad \hat{\rho}_i = e^{j(\arg(I_1) - k_0 r_1)} \hat{\rho}_1(\theta_1, \phi_1). \tag{7.94}$$

Note that the incident polarization unit vector $\hat{\rho}_i$, as seen by receive antenna 2, is equal to the transmit polarization unit vector $\hat{\rho}_1$ of antenna 1 multiplied by the phase factor $e^{j(\arg(I_1)-k_0 r_1)}$ due to propagation over distance r_1 and the phase of the input current I_1. The magnitude of the incident wave in (7.94) may be expressed in terms of the input impedance R_{in1} of antenna 1 (or, alternately the input power P_{in1} of antenna 1) using equation (7.63).

$$|\overline{e}_i| = |I_1| \sqrt{\frac{\eta_0 R_{in1}}{4\pi r_1^2} G(\theta_1, \phi_1)} = \sqrt{\frac{2\eta_0 P_{in1}}{4\pi r_1^2} G(\theta_1, \phi_1)}. \tag{7.95}$$

The Thevenin's voltage, or the open-circuit voltage V_{R2}, at the terminals of antenna 2 can now be written in different forms using equations (7.79, 7.93, 7.94, 7.95).

$$V_{R2} = I_1 \frac{e^{-jk_0 r_1}}{r_1} \overline{e}_1(\theta_1, \phi_1) \cdot \overline{v}_{R2}(\theta_2, \phi_2) \quad , \tag{7.96}$$

$$V_{R2} = \frac{|I_1||\overline{e}_1(\theta_1, \phi_1)||\overline{v}_{R2}(\theta_2, \phi_2)|}{r_1} e^{j(\arg(I_1)-k_0 r_1)} \hat{\rho}_1(\theta_1, \phi_1) \cdot \hat{\rho}_{R2}(\theta_2, \phi_2), \tag{7.97}$$

$$V_{R2} = \left[\sqrt{\frac{2\eta_0 P_{in1}}{4\pi r_1^2} G_1(\theta_1, \phi_1)} \right] |\overline{v}_{R2}(\theta_2, \phi_2)| e^{j(\arg(I_1)-k_0 r_1)} \hat{\rho}_1(\theta_1, \phi_1) . \hat{\rho}_{R2}(\theta_2, \phi_2). \tag{7.98}$$

3.1. Transmit-Receive Link as a Two-Port Circuit. If antenna 2 is also excited by an input current I_2, antenna 1 will receive a Thevenin's terminal voltage V_{R1}. Repeating the above procedure, V_{R1} may be expressed by interchanging the subscripts 1 and 2

$$V_{R1} = I_2 \frac{e^{-jk_0 r_2}}{r_2} \overline{e}_2(\theta_2, \phi_2) \cdot \overline{v}_{R1}(\theta_1, \phi_1). \tag{7.99}$$

The link between antenna 1 and 2 may be seen as a two-port circuit, as illustrated in Fig.8b. The total input voltages V_1 and V_2, respectively across the terminals of antenna 1 and 2, may be expressed as a superposition of the received voltage and the transmit voltage.

$$V_1 = I_1 Z_{in1} + V_{R1},\ V_2 = I_2 Z_{in2} + V_{R2}, \tag{7.100}$$

$$Z_{12} = \frac{V_{R1}}{I_2} = \frac{e^{-jk_0 r_2}}{r_2} \bar{e}_2(\theta_2, \phi_2) \cdot \bar{v}_{R1}(\theta_1, \phi_1), \tag{7.101}$$

$$Z_{21} = \frac{V_{R2}}{I_1} = \frac{e^{-jk_0 r_1}}{r_1} \bar{e}_1(\theta_1, \phi_1) \cdot \bar{v}_{R2}(\theta_2, \phi_2);\ r_1 = r_2. \tag{7.102}$$

Assuming the circuit to be reciprocal, we can equate $Z_{12} = Z_{21}$ (see 2.128, appendix G). Then,

$$\bar{e}_1(\theta_1, \phi_1) \cdot \bar{v}_{R2}(\theta_2, \phi_2) = \bar{e}_2(\theta_2, \phi_2) \cdot \bar{v}_{R1}(\theta_1, \phi_1). \tag{7.103}$$

The above equation (7.103) is an important relationship in antenna theory, called the antenna reciprocity relationship.

3.2. Relationship Between Transmit and Receive Vectors of an Arbitrary Antenna. Using (7.103) one can relate the transmit and receive vectors of any arbitrary antenna (replaced for antenna 1), once we have the transmit and receive vectors of one specific known antenna that can be substituted for antenna 2.

Let us use a small dipole antenna, with unit input current and length l, as the reference antenna 2. Let this dipole current be oriented in the $-\hat{\theta}_1$ direction, such that the radiation from the dipole antenna 2 is maximum towards the antenna 1. Alternately, one may visualize the current of the dipole antenna 2 to be oriented along the $\hat{z}_2$ axis (see Fig.8a). Then, let the $\hat{z}_2$ axis be turned to align in the $-\hat{\theta}_1$ direction, so that in the line of communication we have $\hat{\theta}_1 = \hat{\theta}_2$ and $\theta_2 = 90^0$. Under this situation, using the results from (6.33,9.4), we can find the transmit and receive vectors of the dipole antenna 2.

$$\bar{I} = I_2 \hat{z}_2 = -\hat{\theta}_1,\ I_2 = 1, \tag{7.104}$$

$$\bar{e}_2(\theta_2, \phi_2) = \frac{j\eta_0 l}{2\lambda_0} \hat{\theta}_2 = \frac{j\eta_0 l}{2\lambda_0} \hat{\theta}_1,\ \ \bar{v}_{R2}(\theta_2, \phi_2) = l\hat{\theta}_2 = l\hat{\theta}_1,\ \hat{\theta}_1 = \hat{\theta}_2. \tag{7.105}$$

Now, using (7.105) in (7.103), we get the relationship between the $\hat{\theta}$ components of the transmit and receive vectors of the arbitrary antenna 1.

$$\bar{e}_1(\theta_1, \phi_1) \cdot \hat{\theta}_1 l = \frac{j\eta_0 l}{2\lambda_0} \hat{\theta}_1 \cdot \bar{v}_{R1}(\theta_1, \phi_1),\ e_{1\theta}(\theta_1, \phi_1) = \frac{j\eta_0}{2\lambda_0} v_{R1\theta}(\theta_1, \phi_1). \tag{7.106}$$

One may repeat the above procedure by reorienting the dipole antenna 2 such that the current is in the $-\hat{\phi}_1$ direction. This would provide the relationship between the $\hat{\phi}$ components of the transmit and receive vectors.

$$e_{1\phi}(\theta_1, \phi_1) = \frac{j\eta_0}{2\lambda_0} v_{R1\phi}(\theta_1, \phi_1). \tag{7.107}$$

Combining (7.106, 7.107) we get the relationship between the complete transmit and receive vectors.

$$\bar{e}_1(\theta_1, \phi_1) = \frac{j\eta_0}{2\lambda_0} \bar{v}_{R1}(\theta_1, \phi_1). \tag{7.108}$$

Because we are considering an arbitrary antenna for antenna 1, and the direction of consideration (θ_1, ϕ_1) is arbitrary, we may generalize for any antenna and any direction of radiation/reception.

$$\bar{e}(\theta, \phi) = \frac{j\eta_0}{2\lambda_0} \bar{v}_R(\theta, \phi), \ |\bar{e}(\theta, \phi)| = \frac{\eta_0}{2\lambda_0} |\bar{v}_R(\theta, \phi)|, \ \hat{\rho}(\theta, \phi) = j\hat{\rho}_R(\theta, \phi). \tag{7.109}$$

Equation (7.109) is an useful universal relationship between the transmit and receive vectors of any antenna.

<u>$G \sim A_{em}$ **Relationship**</u>

Consider now the relationship between the magnitudes of the transmit and receive vectors in (7.109.) Then use equation (7.63) to relate $|\bar{e}|$ to G, and equation (7.83) to relate $|\bar{v}_R|$ to maximum effective area A_{em}. Thus, we get an equivalent universal relationship between $G(\theta, \phi)$ and $A_{em}(\theta, \phi)$.

$$\frac{|\bar{e}(\theta, \phi)|^2}{|\bar{v}_R(\theta, \phi)|^2} = (\frac{\eta_0}{2\lambda_0})^2 = \frac{G(\theta, \phi)}{A_{em}(\theta, \phi)} \frac{\eta_0^2}{16\pi}, \ \frac{G(\theta, \phi)}{A_{em}(\theta, \phi)} = \frac{4\pi}{\lambda_0^2}. \tag{7.110}$$

3.3. Friis Transmission Formula. Using the universal $\bar{e} \sim \bar{v}_R$ relationship of (7.109) in the equation (7.96), we can express the received Thevenin's voltage at the antenna 2 terminals, using transmit vectors of the two antennas.

$$V_{R2} = I_1 \frac{e^{-jk_0 r_1}}{r_1} \bar{e}_1(\theta_1, \phi_1) \cdot \bar{e}_2(\theta_2, \phi_2) \frac{2\lambda_0}{j\eta_0} \ , \tag{7.111}$$

$$|V_{R2}| = \frac{|I_1|}{r_1} |\bar{e}_1(\theta_1, \phi_1)||\bar{e}_2(\theta_2, \phi_2)| \frac{2\lambda_0}{\eta_0} |\hat{\rho}_1(\theta_1, \phi_1).\hat{\rho}_2(\theta_2, \phi_2)| \ . \tag{7.112}$$

Equation (7.63) may be used to relate $|\bar{e}|$'s of (7.112) to corresponding G's.

$$|V_{R2}|^2 = (\frac{\lambda_0}{4\pi r_1})^2 4|I_1|^2 R_{in1} R_{in2} G_1(\theta_1, \phi_1) G_2(\theta_2, \phi_2) |\hat{\rho}_1(\theta_1, \phi_1).\hat{\rho}_2(\theta_2, \phi_2)|^2 \ , \tag{7.113}$$

$$|V_{R2}|^2 = \left(\frac{\lambda_0}{4\pi r_1}\right)^2 8P_{in1}R_{in2}G_1(\theta_1,\phi_1)G_2(\theta_2,\phi_2)|\hat{\rho}_1(\theta_1,\phi_1).\hat{\rho}_2(\theta_2,\phi_2)|^2 \ . \tag{7.114}$$

Once V_{R2} is known, the load voltage V_{L2} and load power P_{L2}, when the antenna 2 is loaded by an arbitrary load impedance Z_{L2}, is found from (7.80).

$$P_{L2} = \frac{8R_{in2}R_{L2}}{2|Z_{L2}+Z_{in2}|^2}\left(\frac{\lambda_0}{4\pi r_1}\right)^2 P_{in1}G_1(\theta_1,\phi_1)G_2(\theta_2,\phi_2)|\hat{\rho}_1(\theta_1,\phi_1).\hat{\rho}_2(\theta_2,\phi_2)|^2 \ , \tag{7.115}$$

$$P_{L2max} = \left(\frac{\lambda_0}{4\pi r_1}\right)^2 P_{in1}G_1(\theta_1,\phi_1)G_2(\theta_2,\phi_2); \ Z_{L2} = Z^*_{in2}, \ |\hat{\rho}_1 \cdot \hat{\rho}_1| = 1 \ , \tag{7.116}$$

$$P_{L2} = P_{L2max}\left(\frac{8R_{in2}R_{L2}}{2|Z_{L2}+Z_{in2}|^2}\right)|\hat{\rho}_1(\theta_1,\phi_1).\hat{\rho}_2(\theta_2,\phi_2)|^2 \ . \tag{7.117}$$

The above equations (7.115, 7.116, 7.117) are called the *Friis transmission formula*, which relates the received load power to the input power through the antenna and load parameters.

3.4. Polarization Mismatch Factor. $\hat{\rho}_1(\theta_1,\phi_1).\hat{\rho}_2(\theta_2,\phi_2)$ in the above Friss formulas is the *polarization mismatch factor* M_p.

$$M_p = \hat{\rho}_1(\theta_1,\phi_1)\cdot\hat{\rho}_2(\theta_2,\phi_2). \tag{7.118}$$

Using (7.29) (also see Fig.3) M_p may be alternately represented in the coordinates of the respective polarization ellipses.

$$M_p = e^{j\alpha_1}\hat{\rho}'_1(\theta_1,\phi_1)\cdot\hat{\rho}'_2(\theta_2,\phi_2)e^{j\alpha_2}. \tag{7.119}$$

It should be remembered, however, that (7.118) applies for general conditions, where the polarization unit vectors are referenced with respect to the coordinate systems of the corresponding antennas (see Fig.8), that in general can have arbitrary relative orientation with respect to each other. It may be useful to express M_p for a simpler situation, when the axes $(\hat{x}_1,\hat{y}_1,\hat{z}_1)$ and $(\hat{x}_2,\hat{y}_2,\hat{z}_2)$ are respectively aligned with each other (with different origins). Under this situation, in the direction of communication between the antennas we have $\hat{\theta}_1\cdot\hat{\theta}_2 = 1$, but $\hat{\phi}_1\cdot\hat{\phi}_2 = -1$.

$$M_p = \hat{\rho}_1\cdot\hat{\rho}_2 = (\rho_{1\theta}\hat{\theta}_1 + \rho_{1\phi}\hat{\phi}_1)\cdot(\rho_{2\theta}\hat{\theta}_2 + \rho_{2\phi}\hat{\phi}_2) = \rho_{1\theta}\rho_{2\theta} - \rho_{1\phi}\rho_{2\phi}. \tag{7.120}$$

M_p Using $\hat{\rho}$ Vectors in Common Coordinates

Consider again the above situation of aligned coordinates for the two antennas of a communication link. However, consider that the polarization unit vectors of the two antennas, $\hat{\rho}_1$ and $\hat{\rho}_2$, are expressed while the two antennas are placed in a common coordinate system (r,θ,ϕ), with origin at the antenna center:

$$\hat{\rho}_1 = \rho_{1\theta}\hat{\theta} + \rho_{1\phi}\hat{\phi}, \; \hat{\rho}_2 = \rho_{2\theta}\hat{\theta} + \rho_{2\phi}\hat{\phi}. \tag{7.121}$$

Under this situation, one may see that the polarization mismatch factor, $M_p = \rho_{1\theta}\rho_{2\theta} - \rho_{1\phi}\rho_{2\phi}$, is no longer equal to $\hat{\rho}_1 \cdot \hat{\rho}_2$. The correct expression for M_p is as follows.

$$M_p = \rho_{1\theta}\rho_{2\theta} - \rho_{1\phi}\rho_{2\phi} = \hat{\rho}_{1s} \cdot \hat{\rho}_2 = \hat{\rho}_1 \cdot \hat{\rho}_{2s}; \; \hat{\rho}_{1s} = \hat{\rho}_1 \big|_{\hat{\phi} \to -\hat{\phi}}, \; \hat{\rho}_{2s} = \hat{\rho}_2 \big|_{\hat{\phi} \to -\hat{\phi}}, \tag{7.122}$$

$$|M_p| = |\hat{\rho}_{1s} \cdot \hat{\rho}_2| = |\hat{\rho}_1 \cdot \hat{\rho}_{2s}|, \tag{7.123}$$

where the unit vectors with subscript s refer to original vectors with $-\hat{\phi}$ switched for $\hat{\phi}$. As may be noted, this essentially switches the "handedness" of the coordinates.

4. Antenna as a Scatterer

Fig.9 shows a plane wave, $\overline{E}_i$, being scattered from an arbitrary antenna. $\overline{E}_i$ is incident from angle (θ_i, ϕ_i), in general having both $\hat{\theta}$ and $\hat{\phi}$ components of the electric field,

$$\overline{E}_i = e^{jk_0r}(e_{i\theta}\hat{\theta} + e_{i\phi}\hat{\phi}), \; P_{di} = P_{di}^{\theta} + P_{di}^{\phi}; \; P_{di}^{\theta} = \frac{|e_{i\theta}|^2}{2\eta_0}, \; P_{di}^{\phi} = \frac{|e_{i\phi}|^2}{2\eta_0}. \tag{7.124}$$

The scattered field $\overline{E}_s(r, \theta, \phi; \theta_i, \phi_i)$ in the far-field region can be decomposed into two parts. The first part, $\overline{E}_{so}(r, \theta, \phi; \theta_i, \phi_i)$, is the scattering with the antenna input terminals open-circuited ($I_L = 0$). The second part, $\overline{E}_r(r, \theta, \phi; \theta_i, \phi_i)$, is the re-radiated field due to the load current $I_L = -I_{in}$, when the antenna is loaded with a finite impedance.

$$\overline{E}_s = \overline{E}_{so} + \overline{E}_r = \frac{e^{-jk_0r}}{r}[e_{i\theta}\overline{e}_s^{\theta}(\theta, \phi; \theta_i, \phi_i) + e_{i\phi}\overline{e}_s^{\phi}(\theta, \phi; \theta_i, \phi_i)]. \tag{7.125}$$

The scattered field is a linear combination of the incident field components $e_{i\theta}$ and $e_{i\phi}$. As for a radiating antenna, in the far-field region the scattered field also has the $\frac{e^{-jk_0r}}{r}$ dependence, that account for the propagation phase and the free-space loss. $\overline{e}_s^{\theta}$ and $\overline{e}_s^{\phi}$ may respectively be referred to as the θ- and ϕ scattering vectors, and are the r-independent part of the scattered field due to one unit of $e_{i\theta}$ and $e_{i\phi}$, respectively.

The part due to the open-circuited antenna may be separately expressed in the following form, where the subscript o refers to the corresponding values in the open-circuited condition.

$$\overline{E}_{so} = \frac{e^{-jk_0r}}{r}[e_{i\theta}\overline{e}_{so}^{\theta}(\theta, \phi; \theta_i, \phi_i) + e_{i\phi}\overline{e}_{so}^{\phi}(\theta, \phi; \theta_i, \phi_i)]. \tag{7.126}$$

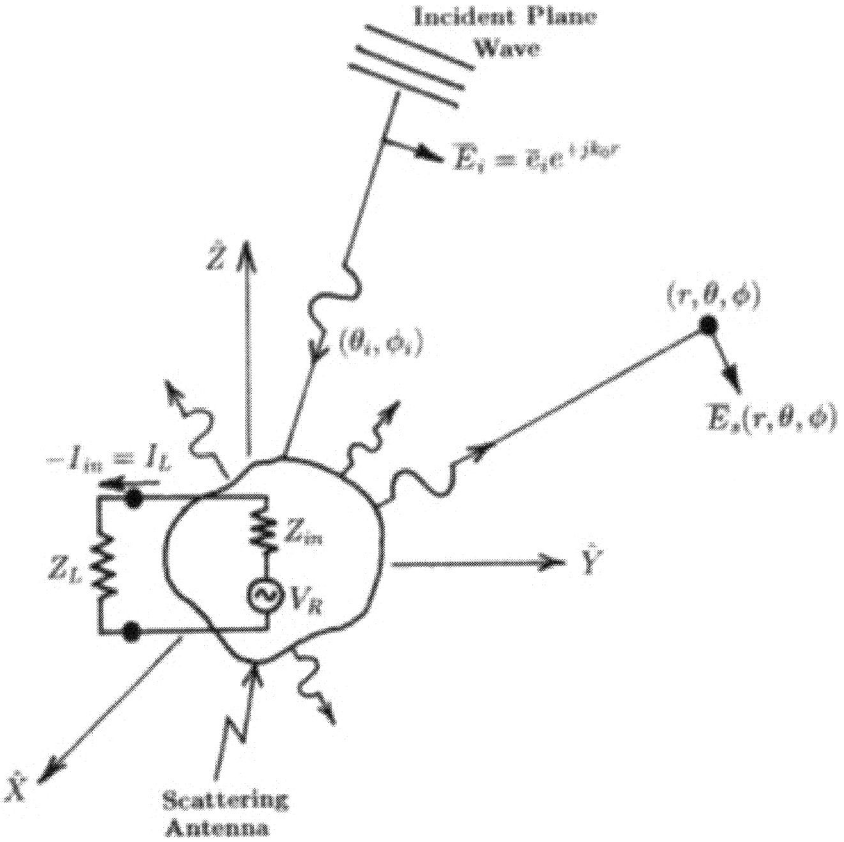

FIGURE 9. A plane wave incident from angle (θ_i, ϕ_i), is scattered from an antenna. The scattered signal $\overline{E}_s(r, \theta, \phi)$ is observed at a point (r, θ, ϕ). The scattered signal is affected by the load impedance Z_L.

Using (7.20,7.72, 7.80), the re-radiated part can be expressed in terms of the input current $I_{in} = -I_L$, the transmit vector $\overline{e}(\theta, \phi)$, and the receive vector $\overline{v}_R(\theta_i, \phi_i)$. The input/load current is induced by the incident field, related by the receive vector $\overline{v}_R$.

$$\overline{E}_r = \frac{e^{-jk_0 r}}{r} I_{in}(\theta_i, \phi_i)\overline{e}(\theta, \phi) = -\frac{e^{-jk_0 r}}{r}[\frac{\overline{e}_i \cdot \overline{v}_R(\theta_i, \phi_i)}{Z_{in} + Z_L}\overline{e}(\theta, \phi)]. \quad (7.127)$$

Now the individual parts in (7.126, 7.127) may be combined with (7.125) to obtain the following relationships.

$$\overline{e}_s^\theta = \overline{e}_{so}^\theta - \frac{v_{R\theta}(\theta_i, \phi_i)}{Z_{in} + Z_L}\overline{e}(\theta, \phi), \; \overline{e}_s^\phi = \overline{e}_{so}^\phi - \frac{v_{R\phi}(\theta_i, \phi_i)}{Z_{in} + Z_L}\overline{e}(\theta, \phi). \quad (7.128)$$

4.1. Intercepted Power and Radar Cross-Section. Let us consider first a θ-polarized incident field ($e_{i\phi} = 0$) . Under this condition, the power-density, P_{ds}^θ, of the scattered field at a given location (r, θ, ϕ) can be expressed using the resulting scattered E-field (from (7.125,7.124)). For brevity of presentation, we may consider the angle of incidence (θ_i, ϕ_i), and the observation point (r, θ, ϕ), as implicit variables, as appropriate.

$$P_{ds}^\theta(r, \theta, \phi; \theta_i, \phi_i) = \frac{|e_{i\theta}|^2 |\overline{e}_s^\theta|^2}{2\eta_0 r^2} = \frac{P_{di}^\theta |\overline{e}_s^\theta|^2}{r^2} = \frac{P_{di}^\theta [|e_{s\theta}^\theta|^2 + |e_{s\theta}^\theta|^2]}{r^2}. \quad (7.129)$$

We define an *effective intercept power*, P_{int}^θ, associated with a scatterer. P_{int}^θ is the equivalent power intercepted by the scattering antenna, which when re-radiated by an isotropic antenna provides the same power density at the observation location as the scattering wave. P_{int}^θ is a function of the angle of observation (θ, ϕ). We then define the corresponding *radar cross section*, σ^θ, as the ratio of the intercept power and the incident power density P_{di}^θ.

$$P_{ds}^\theta = \frac{P_{int}^\theta(\theta, \phi; \theta_i, \phi_i)}{4\pi r^2} = \frac{P_{di}^\theta \sigma^\theta}{4\pi r^2}; \; \sigma^\theta(\theta, \phi; \theta_i, \phi_i) = \frac{P_{int}^\theta}{P_{di}^\theta} = \frac{4\pi r^2 P_{ds}^\theta}{P_{di}^\theta}. \quad (7.130)$$

Using (7.129) in (7.130), we get the relationship between the θ-radar cross section σ^θ and the θ-scattering vector $\overline{e}_s^\theta$. σ^θ is further decomposed into σ_θ^θ and σ_ϕ^θ, that respectively correspond to the θ and ϕ components of $\overline{e}_s^\theta = \hat{\theta} e_{s\theta}^\theta + \hat{\phi} e_{s\phi}^\theta$.

$$\sigma^\theta(\theta, \phi; \theta_i, \phi_i) = 4\pi |\overline{e}_s^\theta|^2 = 4\pi [|e_{s\theta}^\theta|^2 + |e_{s\phi}^\theta|^2] = \sigma_\theta^\theta(\theta, \phi; \theta_i, \phi_i) + \sigma_\phi^\theta(\theta, \phi; \theta_i, \phi_i). \quad (7.131)$$

Similarly, the radar cross-section may be defined for the ϕ-polarized incident field, with the θ-superscripts exchanged for ϕ.

$$P_{ds}^\phi(r, \theta, \phi; \theta_i, \phi_i) = \frac{P_{int}^\phi(\theta, \phi; \theta_i, \phi_i)}{4\pi r^2} = \frac{P_{di}^\phi |\overline{e}_s^\phi|^2}{r^2} = \frac{P_{di}^\phi [|e_{s\theta}^\phi|^2 + |e_{s\theta}^\phi|^2]}{r^2}, \quad (7.132)$$

$$\sigma^\phi(\theta, \phi; \theta_i, \phi_i) = \frac{4\pi r^2 P_{ds}^\phi}{P_{di}^\phi} = 4\pi [|e_{s\theta}^\phi|^2 + |e_{s\theta}^\phi|^2] = \sigma_\theta^\phi(\theta, \phi; \theta_i, \phi_i) + \sigma_\phi^\phi(\theta, \phi; \theta_i, \phi_i). \quad (7.133)$$

5. Related Reading

For related studies on fundamental antenna parameters, or to gain exposure to more advanced concepts of antenna theory, one may refer to some of the standard antenna books **[27]**-**[32]**.

CHAPTER 8

ANTENNA PERFORMANCE IN A SCATTERING ENVIRONMENT

Antennas used in mobile communication systems often operate in an environment with surrounding scattering objects. Unlike in a open space communication with a single line-of-sight signal path, signals in these scattering environments connect with the antennas from multiple directions, that may be considered either as signal distributions or as signals from discrete angles. Antenna performance in such a scattering environment may require some different treatment and interpretation. Due to the complexity or randomness of the scattering structures, the signals may exhibit certain stochastic characteristics that may determine the overall communication performance. Antenna analysis under deterministic as well as stochastic scattering conditions will be presented here. Such analytical modeling will help in understanding and optimal design of communication systems in general scattering conditions.

1. Scattering Environment

It may be convenient to analyze antenna operation in a scattering environment in its receive mode. Operation in a transmit mode can be directly related to that in the receive mode through reciprocity theory. First we need to properly define the incoming fields approaching a given receive antenna, and then model the response to them by the antenna relating to its output terminal voltage and received power.

1.1. Discrete Scattered Fields. Consider a scattering environment with N incoming signal paths, as shown in Fig.1. Each signal path can be modeled as a plane wave propagating in the $-\hat{r}_k$ direction towards the antenna, with electric and magnetic fields $\bar{E}_i$ and $\bar{H}_i$, respectively. Each signal is incident from angle $\Omega_k \equiv (\theta_k, \phi_k)$.

$$\bar{E}_k(\Omega_k) = \bar{e}_k(\Omega_k)e^{+jk_0r_k}; k = 1, 2, ..N. \tag{8.1}$$

$$\overline{H}_k(\Omega_k) = \frac{(-\hat{r}_k) \times \overline{E}_k(\Omega_k)}{\eta_0} = \overline{e}_k(\Omega_k) \times \hat{r}_k \frac{e^{+jk_or_k}}{\eta_o} \tag{8.2}$$

$$\overline{e_i}(\Omega_k) = \hat{\theta}e_{i\theta}(\Omega_k) + \hat{\phi}e_{i\phi}(\Omega_k). \tag{8.3}$$

The associated total power density is $P_{di}(\Omega_k)$, which may be decomposed into its parts $P_{di\theta}(\Omega_k)$ and $P_{di\theta}(\Omega_k)$ contributed by the $\theta-$ and $\phi-$ polarizations.

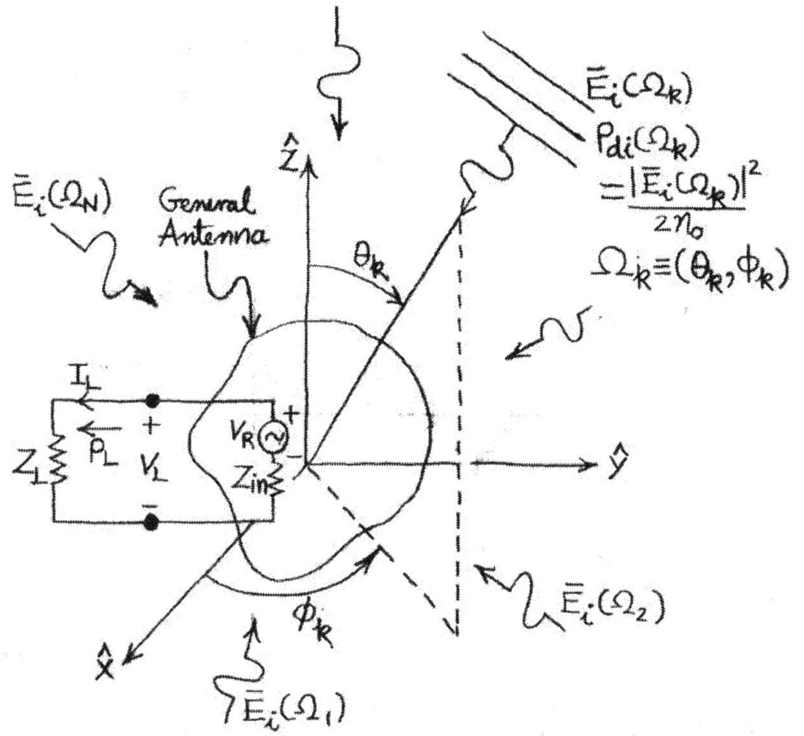

FIGURE 1. A receiving antenna operating in a general scattering environment. The incoming signals are shown here as incident from N discrete angular directions Ω_k, which may also be viewed as signal distributions for large N.

$$P_{di}(\Omega_k) = \frac{|\overline{E}_i(\Omega_k)|^2}{2\eta_o} = \frac{|\overline{e}_i(\Omega_k)|^2}{2\eta_o} = \frac{|e_{i\theta}(\Omega_k)|^2 + |e_{i\phi}(\Omega_k)|^2}{2\eta_o} = P_{di\theta}(\Omega_k) + P_{di\phi}(\Omega_k)$$

$$P_{di\theta}(\Omega_k) = \frac{|e_{i\theta}(\Omega_k)|^2}{2\eta_o}, P_{di\phi}(\Omega_k) = \frac{|e_{i\phi}(\Omega_k)|^2}{2\eta_o}. \quad (8.4)$$

1.2. Continuum Scattered Fields. Consider also a situation where incoming signals are modeled as a continuum over angles $\Omega \equiv (\theta, \phi)$. The plane wave coming between Ω - $\frac{\Delta\Omega}{2} < \Omega < \Omega + \frac{\Delta\Omega}{2}$ is expressed with electric field $\Delta\overline{E}_i(\Omega)$, magnetic field $\Delta\overline{H}_i(\Omega)$, and corresponding power density $\Delta P_{di}(\Omega)$.

$$\Delta\overline{E}_i(\Omega) = \Delta\overline{e}_i(\Omega)e^{+jk_or};\ \Omega - \frac{\Delta\Omega}{2} < (\theta,\phi) < \Omega + \frac{\Delta\Omega}{2} \tag{8.5}$$

$$\Delta\overline{H}_i(\Omega) = \Delta\overline{e}_i(\Omega) \times \hat{r}\frac{e^{+jk_or}}{\eta_0} \tag{8.6}$$

$$\Delta P_{di}(\Omega) = \frac{|\Delta\overline{e}_i(\Omega)|^2}{2\eta_o} = \Delta P_{di\theta}(\Omega) + \Delta P_{di\phi}(\Omega)$$

$$\Delta P_{di\theta}(\Omega) = \frac{|\Delta e_{i\theta}(\Omega)|^2}{2\eta_o}, \Delta P_{di\phi}(\Omega) = \frac{|\Delta e_{i\phi}(\Omega)|^2}{2\eta_o} \tag{8.7}$$

2. Received Signal Using a General Antenna

The total signal voltage and power received by a general antenna in a scattering environment can be expressed by superposition. Related fundamental antenna parameters, such as the receive voltage vector, $\overline{v}_R(\Omega)$, maximum effective area $A_{em}(\Omega)$, antenna gain $G(\Omega)$, transmit vector $\overline{e}(\Omega)$, which were devreloped in part-I can be used. In a discrete scattering environment, the total open-circuit receive voltage V_R can be expressed as:

$$V_R = \sum_k \overline{v}_R(\Omega_k)\cdot\overline{e}_i(\Omega_k)$$

$$= \sum_k |\overline{v}_R(\Omega_k)|\hat{\rho}_R(\Omega_k)\cdot\hat{\rho}_i(\Omega_k)|\overline{e}_i(\Omega_k)| = \sum_k |\overline{v}_R(\Omega_k)|\chi(\Omega_k)|\overline{e}_i(\Omega_k)|; \tag{8.8}$$

$$\chi(\Omega_k) = \hat{\rho}_i(\Omega_k)\cdot\hat{\rho}_R(\Omega_k). \tag{8.9}$$

Where $|\chi(\Omega)| = M_p(\Omega)$ is the polarization match factor in a given direction and $\angle\chi$ is the relative phase of the receive voltage, contributed both by the phases of the incoming signal as well as the antenna response. Both phase and magnitude of χ are important in a scattering environment. This is in contrast with antenna performance in a open-space environment with a single incoming signal path, where only the magnitude of χ, which accounts for polarization matching, contributes to the output power. For a continuum scattering environment, the equivalent expression is an integral over all solid angles.

$$V_R = \int_\Omega |\overline{v}_R(\Omega)|\chi(\Omega)|\Delta\overline{e}_i(\Omega)| \tag{8.10}$$

One may recall from antenna fundamentals, that for any arbitary antenna, the receive voltage vector $\overline{v}_R(\Omega)$ may be related to other antenna parameters in different ways

$$|\overline{v}_R|^2 = \frac{4R_{in}}{\eta_0}A_{em} = \frac{4R_{in}}{\eta_0}.\frac{\lambda_0^2 G}{4\pi} = \frac{R_{in}\lambda_0^2 G}{\pi\eta_0} \tag{8.11}$$

$$\overline{v}_R = \frac{-j2\lambda_0}{\eta_0}\overline{e},\ |\overline{v}_R| = \frac{2\lambda_0}{\eta_0}|\overline{e}|,\ \hat{\rho}_R = -j\hat{\rho} \tag{8.12}$$

The first equation relates the magnitude of $|\overline{v}_R|$ to the antenna Gain G when used as a transmitter. The second equation relates $\overline{v}_R$ to the electric field vector $\overline{e}$ when the antenna used as a transmitter with input current $I_{in} = 1$.

Using the above relations (8.11,8.12) in (8.8-8.10), the receive voltage V_R can be expressed in alternate forms for a discrete or a continuum scattering environment.

$$V_R = \sum_k \frac{-2j\lambda_0}{\eta_0}\overline{e}(\Omega_k)\cdot\overline{e}_i(\Omega_k) = \frac{-2j\lambda_0}{\eta_0}\sum_k \overline{e}(\Omega_k)\cdot\overline{e}_i(\Omega_k) \quad (8.13)$$

$$= \sum_k \sqrt{\frac{R_{in}\lambda_0^2}{\pi\eta_0}G(\Omega_k)}\chi(\Omega_k)|\overline{e}_i(\Omega_k)| = \sqrt{\frac{R_{in}\lambda_0^2}{\pi\eta_0}}\sum_k \sqrt{G(\Omega_k)}\chi(\Omega_k)|\overline{e}_i(\Omega_k)| \quad (8.14)$$

$$V_R = \frac{-2j\lambda_0}{\eta_0}\int_\Omega \overline{e}(\Omega)\cdot\Delta\overline{e}_i(\Omega) \quad (8.15)$$

$$= \sqrt{\frac{R_{in}\lambda_o^2}{\pi\eta_o}}\int_\Omega \sqrt{G(\Omega)}\chi(\Omega)|\Delta\overline{e}_i(\Omega)| \quad (8.16)$$

The total received power P can be expressed in terms of the receiver voltage V_R using circuit relations.

$$P_L = \frac{|V_R|^2 R_L}{2|Z_{in}+Z_L|^2} = \frac{R_{in}\lambda_0^2}{2\pi\eta_0}\frac{R_L}{|Z_{in}+Z_L|^2}\left|\sum_k \sqrt{G(\Omega_k)}\chi(\Omega_k)|\overline{e}_i(\Omega_k)|\right|^2; \quad (8.17)$$

$$P_L = \frac{R_{in}\lambda_o^2}{2\pi\eta_o}\frac{R_L}{|Z_{in}+Z_L|^2}\left|\int_\Omega \sqrt{G(\Omega)}\chi(\Omega)|\Delta\overline{e_i}.(\Omega)|\right|^2. \quad (8.18)$$

It should be clear that the summation and the $|.|^2$ operations in (8.17), and the integration and the $|.|^2$ operations in (8.18) can not be interchanged.

2.1. Reference Power by a Matched Isotropic Antenna. Consider a reference lossless, isotropic antenna for which gain $G = 1$. Consider also that the power P_L received by the reference antenna is ideally maximized in three ways: (a) impedance matching the load such that $Z_L = Z_{in}^*$, $R_L = R_{in}$, (b) polarization matching with the input signals at all angles, where $|\chi(\Omega)| = 1$, and (c) phase matching at all angles such that $\angle\chi(\Omega)$ is a constant, resulting in constructive interference of incoming signal at the antenna output. Except the impedance matching, other ideal matching conditions are often not achieved for a general practical antenna in an arbitrary scattering environment. However, the maximum possible power $P_{max}(iso)$ received by an hypothetical isotropic antenna under these ideal matched conditions would be meaningful as a reference available power at the location, with which performance of other antennas may be compared with. $P_{max}(iso)$ is obtained by simplifying (8.17,8.18) with the above three conditions.

$$P_{max}(iso) = \frac{\lambda_0^2}{8\pi\eta_0}\Big[\sum_k |\bar{e}_i(\Omega_k)|\Big]^2 = \frac{\lambda_0^2}{4\pi}\Big[\sum_k \sqrt{P_{di}(\Omega_k)}\Big]^2 \tag{8.19}$$

$$= \frac{\lambda_0^2}{8\pi\eta_0}\Big[\int_\Omega |\Delta\bar{e}_i(\Omega)|\Big]^2 = \frac{\lambda_0^2}{4\pi}\Big[\int_\Omega \sqrt{\Delta P_{di}(\Omega)}\Big]^2 \tag{8.20}$$

2.2. Effective Gain G_e. As discussed, it would be useful to measure the performance of a given antenna with respect to a reference matched, isotropic receiver. The ratio of signal power from a given antenna in a particular scattering environment to that from a matched isotropic antenna may be referred to as the effective gain G_e. This is expressed as the ratio of (8.17) for a discrete, or (8.18) for a distributed, scattering environment to (8.20).

$$G_e = \frac{P_L}{P_{max}(iso)} = \frac{4R_{in}R_L}{|Z_L+Z_{in}|^2}\frac{\Big|\sum_k \sqrt{G(\Omega_k)}\chi(\Omega_k)|\bar{e}_i(\Omega_k)|\Big|^2}{\Big[\sum_k |\bar{e}_i(\Omega_k)|\Big]^2}, \tag{8.21}$$

$$= \frac{4R_{in}R_L}{|Z_L+Z_{in}|^2}\frac{\Big|\int_\Omega \sqrt{G(\Omega)}\chi(\Omega)|\Delta\bar{e}_i(\Omega)|\Big|^2}{\Big[\int_\Omega |\Delta\bar{e}_i(\Omega)|\Big]^2}, \tag{8.22}$$

$$0 < G_e < \infty \tag{8.23}$$

The effective gain in general can be greater or less than unity. This is in contrast with the maximum gain G of any antenna, which is always greater than or equal to unity. The effective gain, and hence the received signal power, can be improved by designing the gain function $G(\Omega)$ larger towards angles from which stronger signals are incident and smaller towards angles from which relatively weaker signals are incident. As discussed in the context of an isotropic antenna, the phase response and polarization response for a general antenna would also have to be designed so that they are matched to the incoming signals in order to maximize the output power. In addition, the load impedance will have to be impedance matched with the antenna's input impedance for maximum power delivery to the load.

2.3. Maximum Effective Gain G_{em}. For a given antenna with a particular gain function $G(\Omega)$, and for a given scattered incoming signals, one may calculate a theoretical upper bound for the effective gain. This would correspond to calculating a theoretical maximum P_{max} for the received power of (8.17,8.18). P_{max} can be derived from (8.17,8.18) by using impedance matching condition ($Z_L = Z_{in}^*$, $R_L = R_{in}$) and assuming the polarization and phase distributions of the signal to be perfectly matched to the antenna ($|\chi(\Omega)| = 1$, $\angle\chi(\Omega) =$ a constant.)

$$P_L \leq P_{max} = \frac{\lambda_0^2}{8\pi\eta_0}\Big[\sum_k \sqrt{G(\Omega_k)}|\bar{e}_i(\Omega_k)|\Big]^2 = \frac{\lambda_0^2}{4\pi}\Big[\sum_k \sqrt{G(\Omega_k)P_{di}(\Omega_k)}\Big]^2 \tag{8.24}$$

$$= \frac{\lambda_0^2}{8\pi\eta_0}\Big[\int_\Omega \sqrt{G(\Omega)}|\Delta\bar{e}_i(\Omega)|\Big]^2 = \frac{\lambda_0^2}{4\pi}\Big[\int_\Omega \sqrt{G(\Omega)\Delta P_{di}(\Omega)}\Big]^2 \tag{8.25}$$

G_{em}, the corresponding upper bound for G_e, can then be determined by substituting $P_L = P_{max}$ in (8.23).

$$G_e \leq G_{em} = \frac{P_{max}}{P_{max}(iso)} = \frac{\Big[\sum_k \sqrt{G(\Omega_k)}|\bar{e}_i(\Omega_k)|\Big]^2}{\Big[\sum_k |\bar{e}_i(\Omega_k)|\Big]^2}, \tag{8.26}$$

$$= \frac{\Big[\int_\Omega \sqrt{G(\Omega)}|\Delta\bar{e}_i(\Omega)|\Big]^2}{\Big[\int_\Omega |\Delta\bar{e}_i(\Omega)|\Big]^2}, \tag{8.27}$$

$$0 \leq G_{em} \leq \infty \tag{8.28}$$

2.4. Performance at Different Antenna Locations. As mentioned, impedance matching can be easily implemented by adjusting the load impedance, which would improve the effective gain. Whereas, the ideal polarization and phase matching conditions are unlikely to be satisfied for a fixed, given antenna, in an arbitrary scattering environment, except for simple situations. Some improvement could be seen by turning the antenna around its origin so as to better fit the given polarization and phase distributions to the incoming signal. However, the location of the antenna can be used as an additional effective variable for optimization. This is in contrast to an antenna operating in a free-space environment with a single incoming signal path, where the received power is independent of local shift of the antenna position.

A shift of the antenna location can be modeled by effectively changing the phase of either the transmit or the receive unit vector.

$$\hat{\rho}_R(\Omega_k;\bar{r}_a) = \hat{\rho}_R(\Omega_k;\bar{r}_a = 0)e^{+jk_0\hat{r}_k\cdot\bar{r}_a} \tag{8.29}$$

$$\chi(\Omega_k;\bar{r}_a) = \hat{\rho}_i(\Omega_k)\cdot\hat{\rho}_R(\Omega_k;\bar{r}_a) = \chi(\Omega_k;0)e^{jk_0\Delta\varphi(\Omega_k;\bar{r}_a)} \tag{8.30}$$

$$\Delta\varphi(\Omega_k;\bar{r}_a) = k_0(\hat{r}_k\cdot\bar{r}_a). \tag{8.31}$$

Using the above expression of $\chi(\Omega)$ in (8.17,8.18), the received power level is expressed having explicit dependence with the antenna location $\bar{r}_a$. The resulting phase distribution of χ over different angles is desirable to be as close as possible to a constant. This would mean the phases of the received voltage at the antenna output terminal, contributed by different incident signals, would constructively add with each other to a larger total value, consequently increasing the received signal power. Essentially, by shifting its position the antenna would sense the interference or standing-wave pattern of the incoming signals in space, properly transformed by the antenna receive characteristics. The received power would vary as the antenna is moved around this interference field, reaching a local maximum when the antenna rides on a "peak" of the interference pattern. This may explain how the signal quality of a mobile phone or a radio often fades up and down as one physically moves around in a house or a office building.

3. Random Scattering Environment and Signal Statistics

The analysis in the last section would be applicable for operation at a particular location or its immediate vicinity. If one moves to a different region, the incoming scattered signal distribution may be different, and as a result the received voltage and power would be different. However, due to the randomness of the scattering environment, the incoming signals may show certain statistical trends, which we will examine in this section.

3.1. Incoming Field Statistics.

3.1.1. *Field Average.* The average of a particular signal at a given location, coming from a given direction Ω, is the average of the signal collected over a large number of samples, m, taken at different physical positions in the neighborhood of the location. For example, consider $e_{i\theta}$, the θ-component of the incident electric field. The average of $e_{i\theta}$ may be notationally referred to either as $< e_{i\theta}(\Omega) >$ or as $\underline{e}_{i\theta}(\Omega)$.

$$< e_{i\theta}(\Omega) > \equiv \underline{e}_{i\theta}(\Omega) = \lim_{M\to\infty} \frac{1}{M} \sum_{m=1}^{M} e_{i\theta m}(\Omega) \tag{8.32}$$

where $e_{i\theta}$ is a complex quantity, with amplitude and phase.

Let us examine the physical process involved when the signal travels from a transmit source to a given location of a receive antenna. The signal reaches the particular point after going through multiple reflections from random scattering objects in the region and travelling through various free-space paths between reflections. In this process the signal experiences different path delays through free space and phase shifts due to reflection at the scattering bodies. This results in random distribution of signal delays or phase shifts experienced by the signal. Therefore, the relative phase of the signal would be likely between 0 and 2π. Further, the magnitude of the incoming signal may be assumed to be independent of its phase.

On the above basis, it may be shown using standard probability modeling that the signal's field average is zero. This should be true at all angles of arrival Ω.

$$e_{i\theta}(\Omega) = xe^{j\alpha}, \; f_{x\alpha}(x,\alpha) = f_x(x)f_\alpha(\alpha),$$

$$< e_{i\theta} > = \int_0^{2\pi}\int_0^{\infty} xe^{j\alpha} f_{x\alpha}(x,\alpha)\,dx\,d\alpha = \int_0^{\infty} x f_x(x)\,dx \int_0^{2\pi} e^{j\alpha} f_\alpha(\alpha)\,d\alpha = 0, \tag{8.33}$$

$$f_\alpha(\alpha) = \frac{1}{2\pi} \tag{8.34}$$

where f_x, f_α and $f_{x\alpha}$ are the probability distribution functions for amplitude, phase, and their joint distributions, respectively. The joint distribution is equal to the product of the individual distributions, because the individual variables are assumed to be independent of each other. Because all values of the phase α $0 < \alpha < 2\pi$, are equally likely, on an average they add up to zero, resulting in zero average of the signal. This is true irrespective of its amplitude statistics f_x. The conclusion is valid for the ϕ component of the field as well.

$$\underline{e}_{i\phi}(\Omega) =< e_{i\phi}(\Omega) >= 0, \tag{8.35}$$

$$\underline{\bar{e}}_i(\Omega) =< \bar{e}_i(\Omega) >= \hat{\theta} < e_{i\theta}(\Omega) > +\hat{\phi} < e_{i\phi}(\Omega) >= 0 \tag{8.36}$$

3.1.2. *Average Incident Power Density.* The average for the incident power density functions $P_{di\theta}(\Omega)$, $P_{di\phi}(\Omega)$, and $P_{di}(\Omega)$, which are related to the incident field components in (8.4), can be defined using similar sample-averaging procedure used in (8.32). The average for $P_{di\theta}(\Omega)$, for example, is defined as follows as it relates to the average for $|e_{i\theta}(\Omega)|^2$.

$$< P_{di\theta}(\Omega) >= \underline{P}_{di\theta}(\Omega) =< \frac{|e_{i\theta}(\Omega)|^2}{2\eta_0} >=< \frac{e_{i\theta}(\Omega)e^*_{i\theta}(\Omega)}{2\eta_0} > \tag{8.37}$$

$$= \frac{1}{2\eta_0} \lim_{M\to\infty} \frac{1}{M} \sum_m e_{i\theta m}(\Omega)e^*_{i\theta m}(\Omega) \tag{8.38}$$

The averages for $P_{di\phi}(\Omega)$ and $P_{di}(\Omega)$ may be similarly defined, and are related to $|e_{i\phi}(\Omega)|^2$ and $|\bar{e}_i(\Omega)|^2$, respectively. .

$$< P_{di\phi}(\Omega) >= \underline{P}_{di\phi}(\Omega) =< \frac{|e_{i\phi}(\Omega)|^2}{2\eta_0} >= \frac{1}{2\eta_0} < e_{i\phi}(\Omega)e^*_{i\phi}(\Omega) > \tag{8.39}$$

$$< P_{di}(\Omega) >= \underline{P}_{di}(\Omega) =< \frac{|e_{i\theta}(\Omega)|^2 + |e_{i\phi}(\Omega)|^2}{2\eta_0} >= \underline{P}_{di\theta}(\Omega) + \underline{P}_{di\phi}(\Omega) \tag{8.40}$$

3.1.3. *Correlation Between Different Fields.* In addition to field and power-density averages in a given direction, it will be useful to evaluate the expected value of the complex-conjugate product of the same signal component at two different angular directions, or of different signal components at the same or different angular directions. Consider first the θ component of the electric field at two different angles Ω_l and Ω_k. The expected value of their conjugate product is its average over a large number of samples m collected at different physical positions in the vicinity of the given location.

$$< e_{i\theta}(\Omega_k)e^*_{i\theta}(\Omega_l) >= \frac{1}{M} \sum_M e_{i\theta m}(\Omega_k)e^*_{i\theta m}(\Omega_l) \tag{8.41}$$

If one would assume that the signals at different angles are independent, and we also know that they have zero mean, then we have

$$< e_{i\theta}(\Omega_k)e^*_{i\theta}(\Omega_l) >=< e_{i\theta}(\Omega_k) >< e^*_{i\theta}(\Omega_l) >= 0; k \neq l,$$
$$=< |e_{i\theta}(\Omega_k)|^2 >= 2\eta_o < P_{di\theta}(\Omega_k) >; \; k = l. \tag{8.42}$$

For a distributed scattering environment, the equivalent equations are expressed as,

$$< e_{i\theta}(\Omega)e^*_{i\theta}(\Omega') >=< |e_{i\theta}(\Omega)|^2 > \delta(\Omega - \Omega') = 2\eta_o < P_{di\theta}(\Omega) > \delta(\Omega - \Omega') \tag{8.43}$$

$$\delta(\Omega - \Omega') = 1, \; \Omega = \Omega' \tag{8.44}$$
$$= 0, \; \Omega \neq \Omega' \tag{8.45}$$

The expected value $< e_{i\theta}(\Omega_k)e^*_{i\theta}(\Omega_l) >$ reflects how similar or correlated the signals are at the particular angles k and l. A correlation coefficient γ_θ between the signals at different angles is defined in terms of the above expected value. This is essentially a normalized form of $< e_{i\theta}(\Omega_k)e^*_{i\theta}(\Omega_l) >$ such that the maximum is unity.

$$\gamma_\theta(\Omega_k, \Omega_l) = \frac{< e_{i\theta}(\Omega_k)e^*_{i\theta}(\Omega_l) >}{\sqrt{< |e_{i\theta}(\Omega_k)|^2 >< |e_{i\theta}(\Omega_l)|^2 >}} = 0; \; k \neq l$$
$$= 1; \; k = l \tag{8.46}$$

$$\gamma_\theta(\Omega, \Omega') = \delta(\Omega - \Omega') \tag{8.47}$$

When the correlation coefficient is unity the two signals have similar spatial variation. When it is zero the two signals are considered dissimilar in the sense that their conjugate product at different locations averages to zero.

For the ϕ component of the electric field $e_{i\phi}$ one would also have similar relations as $e_{i\theta}$.

$$< e_{i\phi}(\Omega_k)e^*_{i\phi}(\Omega_l) >= 0; k \neq l$$
$$= 2\eta_o < P_{di\phi}(\Omega_k) >=< |e_{i\phi}(\Omega_k)|^2 >; k = l \tag{8.48}$$
$$< e_{i\phi}(\Omega)e^*_{i\phi}(\Omega') >=< |e_{i\phi}(\Omega)|^2 > \delta(\Omega - \Omega')$$
$$= 2\eta_o < P_{di\phi}(\Omega) > \delta(\Omega - \Omega') \tag{8.49}$$
$$\gamma_\phi(\Omega, \Omega') = 1\delta(\Omega - \Omega') \tag{8.50}$$

Correlation between $e_{i\theta}$ and $e_{i\phi}$ may not be as simple to estimate. Correlation coefficient $\gamma_{\theta\phi}$ between $e_{i\theta}$ and $e_{i\phi}$ at two different angles would still be zero, following similar logic just discussed for $e_{i\theta}$ or e_ϕ at two different angles.

$$< e_{i\theta}(\Omega_k)e^*_{i\phi}(\Omega_l) >= 0 = \gamma_{\theta\phi}(\Omega_k,\Omega_l); \; k \neq l \tag{8.51}$$

But for the same angle of arrival $k = l$, the θ and ϕ signal components would follow the same propagation paths of a sequence of reflections from the transmitter to the receiver, except they experience different complex reflection coefficients and cross-coupling at individual reflections. They both, however, experience same phase shifts or delays due to path length of propagation. Consider a simple case when in a region of observation the signal is received from a particular direction only through a direct line-of-sight path from the transmitter to the receiver antenna. The ratio between the cross polar field components $\frac{e_{i\phi}(\Omega)}{e_{i\theta}(\Omega)} = ae^{j\alpha}$, observed at the receiver, would be the same as that originated from the transmitter, irrespective of the sample location m. For this case, correlation coefficient $\gamma_{\theta\phi}$ may be evaluated as follows.

$$< e_{i\theta}(\Omega)e^*_{i\phi}(\Omega) >= ae^{-j\alpha} < e_{i\theta}(\Omega)e^*_{i\theta}(\Omega) >$$

$$= ae^{-j\alpha} < |e_{i\theta}(\Omega)|^2 >= ae^{-j\alpha}2\eta_o < P_{di\theta}(\Omega) > \tag{8.52}$$

$$< |e_{i\phi}(\Omega)|^2 >= a^2 < |e_{i\theta}(\Omega)|^2 \tag{8.53}$$

$$\gamma_{\theta\phi} = \frac{< e_{i\theta}(\Omega)e^*_{i\phi}(\Omega) >}{\sqrt{< |e_{i\theta}(\Omega)|^2 >< |e_{i\phi}(\Omega)|^2 >}} = e^{-j\alpha} \tag{8.54}$$

Here $|\gamma_{\theta\phi}| = 1$, meaning $e_{i\theta}$ and $e_{i\phi}$ are perfectly correlated. In the other extreme situation, when there is no direct line-of sight, and the phase difference between $e_{i\theta}$ and $e_{i\phi}$ scattering are randomly distributed (due to the nature of the scatterers and scattering mechanism), we would have

$$< e_{i\theta}(\Omega)e^*_{i\phi}(\Omega) >= \gamma_{\theta\phi}(\Omega) = 0. \tag{8.55}$$

In general, the correlation coefficient $\gamma_{\theta\phi}$ would be a complex number with $|\gamma_{\theta\phi}| \leq 1$.

$$\gamma_{\theta\phi}(\Omega) = \frac{< e_{i\theta}(\Omega)e^*_{i\phi}(\Omega) >}{\sqrt{< |e_{i\theta}(\Omega)|^2 >< |e_{i\phi}(\Omega)|^2 >}}, \quad 0 < |\gamma_{\theta\phi}(\Omega)| < 1 \tag{8.56}$$

The correlation would depend on both the transmitter polarization as well as polarization transfer characteristics of the scattering structures.

3.2. Statistics of Received Signal by a Reference Antenna. Let us use for reference an ideal isotropic antenna. The receive voltage vector $\bar{v}_R(\Omega)$ for this antenna is independent of angle Ω. The antenna's directive gain $D_g(\Omega)$ and gain function $G(\Omega)$ are also independent of angle Ω, and are both equal to unity. Using the relationship (8.11) between $\bar{v}_R(\Omega)$ and $G(\Omega)$,

$$|\bar{v}_R|^2 = \frac{R_{in}\lambda_0^2}{\pi\eta_0} \tag{8.57}$$

Such reference antennas we also used in earlier section to compare received signal in a definitive scattering condition. The expressions of voltage and power received by a general antenna are given in section 2, where the gain G may be substituted by 1 for an ideal isotropic antenna. Here we will study statistical properties of their received voltage and power, when the incoming fields have certain statistical trends as discussed above in section 3.1.

We will consider situations with different antenna polarizations. In these cases the antenna's complex polarization unit vector $\bar{\rho}_R$, which is also independent of angle Ω, will have different relative $\hat{\theta}$ and $\hat{\phi}$ components.

3.2.1. *$\hat{\theta}$-Polarized Reference Antenna.* Here $\hat{\rho}_R = e^{i\alpha_R}\hat{\theta}$, where α_R is a constant phase angle independent of Ω. Examine the open-circuit voltage $V_R = V_{R\theta}$ across the antenna.

$$V_R = \sum_k \bar{v}_R(\Omega_k) \cdot \bar{e}_i(\Omega_k) = |\bar{v}_R|\hat{\rho}_R \cdot \sum_k \bar{e}_i(\Omega_k) = |\bar{v}_R| e^{j\alpha_R} \sum_k e_{i\theta}(\Omega_k) = V_{R\theta} \tag{8.58}$$

$$\begin{aligned} \underline{V}_{R\theta} &\equiv < V_{R\theta} >= |\bar{v}_R| e^{j\alpha_R} < \sum_k e_{i\theta}(\Omega_k) > \\ &= |\bar{v}_R| e^{j\alpha_R} \sum_k < e_{i\theta}(\Omega_k) >= 0; < e_{i\theta}(\Omega) >= 0 \end{aligned} \tag{8.59}$$

Because the average of incident fields is zero, the average of the received voltage $V_{R\theta}$ is also zero. Examine now the load power due to this received voltage $V_{R\theta}$. The received power by an ideal isotropic antenna, when the antenna is impedance matched to the load, is referred to by the variable P_θ. Using (8.57) and (8.58), and the impedance matching condition $Z_{in} = Z_L^*$,

$$P_\theta = \frac{|V_{R\theta}|^2 R_L}{2|Z_{in} + Z_L|^2} = \frac{|V_{R\theta}|^2}{8R_{in}} = \frac{|\bar{v}_R|^2}{8R_{in}} |\sum_k e_{i\theta}(\Omega_k)|^2 = \frac{\lambda_0^2}{8\pi\eta_0} |\sum_k e_{i\theta}(\Omega_k)|^2 \tag{8.60}$$

$$\underline{P}_\theta \equiv < P_\theta >= \frac{\lambda_o^2}{8\pi\eta_o} < |\sum_k e_{i\theta}(\Omega_k)|^2 > \tag{8.61}$$

$$\begin{aligned} &= \frac{\lambda_0^2}{8\pi\eta_0} < \left(\sum_k e_{i\theta}(\Omega_k)\right)\left(\sum_l e_{i\theta}^*(\Omega_l)\right) > \\ = \frac{\lambda_o^2}{8\pi\eta_o} < \sum_k \sum_l e_{i\theta}(\Omega_k) e_{i\theta}^*(\Omega_l) > &= \frac{\lambda_o^2}{8\pi\eta_o} \sum_k \sum_l < e_{i\theta}(\Omega_k) e_{i\theta}^*(\Omega_l) > \\ &= \frac{\lambda_o^2}{8\pi\eta_o} \sum_k < e_{i\theta}(\Omega_k) e_{i\theta}^*(\Omega_k) >= \frac{\lambda_o^2}{8\pi\eta_o} \sum_k < |e_{i\theta}(\Omega_k)|^2 > \end{aligned} \tag{8.62}$$

Equation (8.42) is used above in order to simplify the double-summation form to a single summation form. Now, using (8.4) one may express $\underline{P}_\theta$ in terms of power densities incident from different angles.

$$\underline{P}_\theta = \frac{\lambda_o^2}{8\pi\eta_o}\sum_k 2\eta_o < P_{di\theta}(\Omega_k) > = \frac{\lambda_o^2}{4\pi}\sum_k < P_{di\theta}(\Omega_k) > \tag{8.63}$$

$$= \sum_k < P_{i\theta}(\Omega_k) >; \; P_{i\theta} \triangleq \frac{\lambda_0^2}{4\pi} P_{di\theta} \tag{8.64}$$

$P_{i\theta}$ is referred to as the $\theta-$polarized incident reference power from a given angle with respect to an isotropic receiver.

$$\underline{P}_\theta = \sum_k \underline{P}_{i\theta}(\Omega_k) = \sum_k \frac{\lambda_o^2}{4\pi}\underline{P}_{di\theta}(\Omega_k) \tag{8.65}$$

The above relations may be extended it to a distributed scattering environment

$$\underline{V}_{R\theta} = 0 \tag{8.66}$$

$$\underline{P}_\theta = \int_\Omega \frac{\lambda_o^2}{4\pi}\Delta\underline{P}_{di\theta}(\Omega) = \int_\Omega \Delta\underline{P}_{i\theta} \tag{8.67}$$

Average $\theta-$ Power Intensity Function

Define $p_{i\theta}$, called the incoming average $\theta-$power intensity function, as the average reference power, received by an ideal, $\hat{\theta}-$polarized matched isotropic antenna, per unit solid angle.

$$p_{i\theta}(\Omega) = \frac{\Delta\underline{P}_{i\theta}(\Omega)}{\Delta\Omega} = \frac{\lambda_o^2}{4\pi}\frac{\Delta\underline{P}_{di\theta}(\Omega)}{\Delta\Omega} \tag{8.68}$$

$$\underline{P}_\theta = \int_\Omega \Delta\underline{P}_{i\theta} = \int_\Omega p_{i\theta}(\Omega)\Delta\Omega \tag{8.69}$$

All derivations for the $\theta-$ polarized case, both in a discrete as well as a distributed scattering environment, may be repeated for a $\phi-$ polarized reference isotropic antenna as well as a reference antenna with general polarization. The results for a $\phi-$ polarized case would be direct extensions from that for the $\theta-$ polarized case. Whereas, the case with a general polarization would require more special attention.

3.2.2. *$\hat{\phi}$-Polarized Reference Antenna.* All key results for a $\phi-$ polarized case are listed here, following similar steps used for a $\theta-$ polarized case in the last section.

$$\underline{V}_{R\phi} = 0; \underline{e}_{i\phi}(\Omega) = 0 \tag{8.70}$$

$$\underline{P}_\phi = \sum_k \underline{P}_{i\phi}(\Omega_k) = \sum_k \frac{\lambda_o^2}{4\pi}\,\underline{P}_{di\phi}(\Omega_k) = \sum_k \frac{\lambda_o^2}{8\pi\eta_o} < |e_{i\phi}(\Omega_k)|^2 > \tag{8.71}$$

$$\underline{P}_\phi = \int_\Omega \frac{\lambda_o^2}{4\pi}\Delta\underline{P}_{di\phi}(\Omega) = \int_\Omega \Delta\underline{P}_{i\phi} \tag{8.72}$$

$P_{i\phi}$ is referred to as the $\phi-$ polarized incident reference power from a given angle with respect to an isotropic receiver.

<u>Average $\phi-$ Power Intensity Function</u>

Like for the $\theta-$ polarized case, define also $p_{i\phi}$, called the incoming average $\phi-$ power intensity function, as the average reference power received by an ideal, $\hat{\phi}-$ polarized matched isotropic antenna, per unit solid angle. The average received power, $\underline{P}_\phi$, by the antenna can be expressed as the integration of the intensity function over the entire space.

$$p_{i\phi}(\Omega) = \frac{\Delta\underline{P}_{i\phi}}{\Delta\Omega} = \frac{\lambda_o^2}{4\pi}\frac{\Delta\underline{P}_{di\phi}(\Omega)}{\Delta\Omega} \tag{8.73}$$

$$\underline{P}_\phi = \int_\Omega p_{i\phi}(\Omega)\Delta\Omega \tag{8.74}$$

3.2.3. *Reference Antenna with General Polarization.*

$$\begin{aligned} V_R &= \sum_k V_R(\Omega_k) = \sum_k \overline{v}_R(\Omega_k)\cdot\overline{e}_i(\Omega_k) \\ &= \sum_k v_{R\theta}(\Omega_k)e_{i\theta}(\Omega_k) + v_{R\phi}(\Omega_k)e_{i\phi}(\Omega_k) \end{aligned} \tag{8.75}$$

$$\underline{V}_R = \sum_k v_{R\theta}(\Omega_k)\underline{e}_{i\theta}(\Omega_k) + v_{R\phi}(\Omega_k)\underline{e}_{i\phi}(\Omega_k) = 0 \tag{8.76}$$

P is the matched received power by an ideal, isotropic antenna having a general polarizion.

$$P = \frac{|V_R|^2}{8R_{in}} = \frac{1}{8R_{in}} |\sum_k \overline{v}_R(\Omega_k).\overline{e}_i(\Omega_k)|^2$$

$$= \frac{|\overline{v}_R|^2}{8R_{in}} |\sum_k \hat{\rho}_R(\Omega_k).\overline{e}_i(\Omega_k)|^2 \ (|\overline{v}_R| = constant)$$

$$= \frac{\lambda_0^2}{8\pi\eta_0} \left(\sum_k \rho_{R\theta}(\Omega_k) e_{i\theta}(\Omega_k) + \rho_{R\phi}(\Omega_k) e_{i\phi}(\Omega_k) \right)$$

$$\times \left(\sum_l \rho^*_{R\theta}(\Omega_l) e^*_{i\theta}(\Omega_l) + \rho^*_{R\phi}(\Omega_l) e^*_{i\phi}(\Omega_l) \right) \quad (8.77)$$

The average, $\underline{P}$, of the above power function may be expanded in the following double summation form.

$$\underline{P} \equiv < P > = \frac{\lambda_o^2}{8\pi\eta_o} \sum_k \sum_l < \rho_{R\theta}(\Omega_k) \rho^*_{R\theta}(\Omega_l) e_{i\theta}(\Omega_k) e^*_{i\theta}(\Omega_l) >$$

$$+ < \rho_{R\theta}(\Omega_k) \rho^*_{R\phi}(\Omega_l) e_{i\theta}(\Omega_k) e^*_{i\phi}(\Omega_l) >$$

$$+ < \rho_{R\phi}(\Omega_k) \rho^*_{R\theta}(\Omega_l) e_{i\phi}(\Omega_k) e^*_{i\theta}(\Omega_l) >$$

$$+ < \rho_{R\phi}(\Omega_k) \rho^*_{R\phi}(\Omega_l) e_{i\phi}(\Omega_k) e^*_{i\phi}(\Omega_l) > \quad (8.78)$$

The incoming field statistics described in section 3.1 may be used here for simplification. All the cross-average terms at different angles may be assumed to be zero. Accordingly, the double summation form can then be simplified to a single summation form,

$$\underline{P} - \frac{\lambda_o^2}{8\pi\eta_o} \sum_k |\rho_{R\theta}(\Omega_k)|^2 < c_{i\theta}(\Omega_k) c^*_{i\theta}(\Omega_k) >$$

$$+ \rho_{R\theta}(\Omega_k) \rho^*_{R\phi}(\Omega_k) < e_{i\theta}(\Omega_k) e^*_{i\phi}(\Omega_k) >$$

$$+ \rho^*_{R\theta}(\Omega_k) \rho_{R\phi}(\Omega_k) < e^*_{i\theta}(\Omega_k) e_{i\phi}(\Omega_k) >$$

$$+ |\rho_{R\phi}(\Omega_k)|^2 < e_{i\phi}(\Omega_k) e^*_{i\phi}(\Omega_k) > \quad (8.79)$$

$$\underline{P} = \frac{\lambda_o^2}{8\pi\eta_o} \sum_k |\rho_{R\theta}(\Omega_k)|^2 < |e_{i\theta}(\Omega_k)|^2 > + |\rho_{R\phi}(\Omega_k)|^2 < |e_{i\phi}(\Omega_k)|^2 >$$

$$+ 2Real \left(\rho_{R\theta}(\Omega_k) \rho^*_{R\phi}(\Omega_k) \gamma_{\theta\phi}(\Omega_k) 2\eta_o \sqrt{\underline{P}_{di\theta}(\Omega_k)} \sqrt{\underline{P}_{di\phi}(\Omega_k)} \right)$$

$$= \frac{\lambda_o^2}{4\pi} \sum_k |\rho_{R\theta}(\Omega_k)|^2 \underline{P}_{di\theta}(\Omega_k) + |\rho_{R\phi}(\Omega_k)|^2 \underline{P}_{di\phi}(\Omega_k)$$

$$+ 2\sqrt{\underline{P}_{di\theta}(\Omega_k) \underline{P}_{di\phi}(\Omega_k)} \times Real \left(\rho_{R\theta}(\Omega_k) \rho^*_{R\phi}(\Omega_k) \gamma_{\theta\phi}(\Omega_k) \right) \quad (8.80)$$

The complex correlation coefficient $\gamma_{\theta\phi}$, defined in (8.56), is used here. It can be shown that $\underline{P}$ would be maximum if the following conditions are possibly satisfied:

$$\begin{aligned}\angle\rho_{R\theta} - \angle\rho_{R\phi} &= -\angle\gamma_{\theta\phi}(\Omega)\\ \frac{|\rho_{R\theta}(\Omega)|}{|\rho_{R\phi}(\Omega)|} &= \sqrt{\frac{\underline{P}_{di\theta}(\Omega)}{\underline{P}_{di\phi}(\Omega)}}\\ |\gamma_{\theta\phi}| &= 1\end{aligned} \tag{8.81}$$

The above conditions essentially require that the antenna is polarization matched to the channel polarization statistics, and that the θ and ϕ components be perfectly correlated. These ideal conditions may not often be practically realizable, except in simple special situations. However, the maximum power levels may be useful as reference power levels with which performance of other antennas may be compared.

Under these above conditions, the maximum average power available may be simplified as follows.

$$\underline{P} = \frac{\lambda_o^2}{4\pi}\sum_k \left(|\rho_{R\theta}(\Omega_k)|\sqrt{\underline{P}_{di\theta}}(\Omega_k) + |\rho_{R\phi}(\Omega_k)|\sqrt{\underline{P}_{di\phi}(\Omega_k)}\right)^2 \tag{8.82}$$

$$|\rho_{R\theta}(\Omega)|^2 + |\rho_{R\phi}(\Omega)|^2 = 1 = |\rho_{R\theta}(\Omega)|^2\left(1 + \frac{\underline{P}_{di\phi}(\Omega)}{\underline{P}_{di\theta}(\Omega)}\right)$$

$$|\rho_{R\theta}(\Omega)|^2 = \frac{\underline{P}_{di\theta}(\Omega)}{\underline{P}_{di\phi}(\Omega) + \underline{P}_{di\theta}(\Omega)}, \quad |\rho_{R\phi}(\Omega)|^2 = \frac{\underline{P}_{di\phi}(\Omega)}{\underline{P}_{di\phi}(\Omega) + \underline{P}_{di\theta}(\Omega)} \tag{8.83}$$

$$\left(|\rho_{R\theta}(\Omega_k)|\sqrt{\underline{P}_{di\theta}}(\Omega_k) + |\rho_{R\phi}(\Omega_k)|\sqrt{\underline{P}_{di\phi}(\Omega_k)}\right)^2$$

$$= \left[\frac{\underline{P}_{di\theta}(\Omega_k)}{\sqrt{\underline{P}_{di\theta}(\Omega_k) + \underline{P}_{di\phi}(\Omega_k)}} + \frac{\underline{P}_{di\phi}(\Omega_k)}{\sqrt{\underline{P}_{di\theta}(\Omega_k) + \underline{P}_{di\phi}(\Omega_k)}}\right]^2 \tag{8.84}$$

$$= \underline{P}_{di\theta}(\Omega_k) + \underline{P}_{di\phi}(\Omega_k) \tag{8.85}$$

$$\underline{P} = \sum_k \frac{\lambda_o^2}{4\pi}\left(\underline{P}_{di\theta}(\Omega_k) + \underline{P}_{di\phi}(\Omega_k)\right) \tag{8.86}$$

$$= \sum_k \underline{P}_{i\theta}(\Omega_k) + \underline{P}_{i\phi}(\Omega_k) = \underline{P}_\theta + \underline{P}_\phi = \sum_k \underline{P}_i(\Omega_k) \tag{8.87}$$

P_i is referred to as the total reference power incident from a given angle, measured with respect to a matched (polarization and impedance) isotropic antenna. The maximum average available power under the above matched conditions is the sum of average reference powers for the θ and ϕ polarized cases.

The final results may be extended for distributed scattering as well.

$$\underline{V}_R = 0 \tag{8.88}$$

$$\underline{P} = \int\limits_{\Omega} \Delta \underline{P}_{i\theta} + \int\limits_{\Omega} \Delta \underline{P}_{i\phi} = \underline{P}_{\theta} + \underline{P}_{\phi} \tag{8.89}$$

Average Total Power Intensity Function

Like for the $\theta-$ and $\phi-$ polarized cases, define p_i, called the incoming average total power intensity function, as the average reference power, received by an ideal, polarization- and impedance= matched isotropic antenna, per unit solid angle. The average received power, $\underline{P}$, by the antenna can be expressed as the integration of the total intensity function over the entire space.

$$p_i(\Omega) = p_{i\theta}(\Omega) + p_{i\phi}(\Omega) = \frac{\Delta \underline{P}_i(\Omega)}{\Delta\Omega} \tag{8.90}$$

$$\underline{P} = \underline{P}_{\theta} + \underline{P}_{\phi} = \int\limits_{\Omega} p_{i\theta}(\Omega) d\Omega + \int\limits_{\Omega} p_{i\phi}(\Omega) d\Omega = \int\limits_{\Omega} p_i(\Omega) d\Omega \tag{8.91}$$

3.3. Average Power Distribution Functions. Based on the average incident power intensity functions defined in (8.68,8.73,8.91), for the $\theta-$, $\phi-$, and total powers, it is useful to define their respective distribution functions by deviding them by the respective total reference powers. The distribution functions, accordingly, would have unity integration over the entire space.

$$f_{\theta}(\Omega) \triangleq \frac{p_{i\theta}(\Omega)}{\underline{P}_{\theta}} = \frac{\lambda_o^2}{4\pi}.\frac{\Delta \underline{P}_{di\theta}(\Omega)}{\Delta\Omega} \times \frac{1}{\underline{P}_{\theta}} = \frac{\Delta P_{i\theta}(\Omega)}{\Delta\Omega} \times \frac{1}{\underline{P}_{\theta}} \tag{8.92}$$

$$\int\limits_{\Omega} f_{\theta}(\Omega) d\Omega = \frac{\int\limits_{\Omega} p_{i\theta}(\Omega) d\Omega}{\underline{P}_{\theta}} = 1 \tag{8.93}$$

$f_{\theta}(\Omega)$ is called $\hat{\theta}$ - polarized average power distribution function. Similarly, we may define $f_{\phi}(\Omega)$ as the $\hat{\phi}$ - polarized average power distribution function, $f(\Omega)$ as the total matched average power distribution function.

$$f_{\phi}(\Omega) = \frac{p_{i\phi}(\Omega)}{\underline{P}_{\phi}}, \int\limits_{\Omega} f_{\phi}(\Omega) d\Omega = 1 \tag{8.94}$$

$$f(\Omega) = \frac{p_i(\Omega)}{\underline{P}} = \frac{p_i(\Omega)}{\underline{P}_\theta + \underline{P}_\phi} = \frac{p_{i\theta}(\Omega) + p_{i\phi}(\Omega)}{\underline{P}_\theta + \underline{P}_\phi} = \frac{f_\theta(\Omega)\underline{P}_\theta + f_\phi(\Omega)\underline{P}_\phi}{\underline{P}_\theta + \underline{P}_\phi} \tag{8.95}$$

$$\int_\Omega f(\Omega) d\Omega = 1 \tag{8.96}$$

3.4. Average Cross-Polarization Level. The average cross polarization level ψ is the defined as the ratio of the average $\theta-$ and $\phi-$ polarized reference powers receeived by an ideal, isotropic antenna.

$$\frac{\underline{P}_\phi}{\underline{P}_\theta} = \psi \tag{8.97}$$

The cross-polarization level ψ may be used to relate the total power distribution function to the $\theta-$ and $\phi-$ polar distribution functions. Using (8.97) in (8.95), the following relationship is established.

$$f(\Omega) = \frac{1}{1+\psi} f_\theta(\Omega) + \frac{\psi}{1+\psi} f_\phi(\Omega) \tag{8.98}$$

$$\int_\Omega f(\Omega)\Delta\Omega = \frac{1}{1+\psi} \int_\Omega f_\theta(\Omega)\Delta\Omega + \frac{\psi}{1+\psi} \int_\Omega f_\phi(\Omega)\Delta\Omega = 1 \tag{8.99}$$

4. General Antenna in a Random Scattering Environment

Here, the derivation follows the same steps as in section 3.2.3 for isotropic antenna with general polarisation, except $\overline{v}_R$ is no more constant with Ω, and input impedance may not be in general matched to load.

$$\frac{|\overline{v}_R|^2}{|\overline{v}_R|^2_{isotropic}} = G(\Omega) = \frac{|\overline{v}_R|^2}{\frac{\lambda_o^2 R_{in}}{\pi\eta_o}} \tag{8.100}$$

Accordingly, the different expressions of received voltage or power in section 3.2.3 will have to be properly multiplied by $\sqrt{G(\Omega)}$ for receive voltage, or $G(\Omega)$ for receive power. The receive voltage V_R, expressed in the form of (8.75,8.76) are valid for general antennas as well. Thus the average receive voltage for a general antenna is zero.

$$\underline{V}_R = 0 \tag{8.101}$$

The derivation for the impedance-matched received power, P_{rec}, for a general antenna will follow the same steps for (8.80). The final result will have to be multiplied by $G(\Omega_k)$ inside the summation over angle k.

$$\underline{P}_{rec} = \frac{\lambda_o^2}{4\pi}\sum_k G(\Omega_k)\Big[|\rho_{R\theta}(\Omega_k)|^2\underline{P}_{di\theta}(\Omega_k) + |\rho_{R\phi}(\Omega_k)|^2\underline{P}_{di\phi}(\Omega_k)$$

$$+2\sqrt{\underline{P}_{di\theta}(\Omega_k)\underline{P}_{di\phi}(\Omega_k)}Real\big(\rho_{R\theta}(\Omega_k)\rho^*_{R\phi}(\Omega_k)\gamma_{\theta\phi}(\Omega_k)\big)\Big]$$

$$= \sum_k G(\Omega_k)\Big[|\rho_{R\theta}(\Omega_k)|^2\underline{P}_{i\theta}(\Omega_k) + |\rho_{R\phi}(\Omega_k)|^2\underline{P}_{i\phi}(\Omega_k)$$

$$+2\sqrt{\underline{P}_{i\theta}(\Omega_k)\underline{P}_{i\phi}(\Omega_k)}|\rho_{R\theta}(\Omega_k)||\rho_{R\phi}(\Omega_k)||\gamma_{\theta\phi}(\Omega_k)|Real(e^{j\alpha(\Omega_k)+j\alpha'(\Omega_k)})\Big] \quad (8.102)$$

$$\rho_{R\theta}(\Omega_k)\rho^*_{R\phi}(\Omega_k)) = |\rho_{R\theta}(\Omega_k)||\rho_{R\phi}(\Omega_k)|e^{j\alpha'(\Omega_k)}, \quad \gamma_{\theta\phi}(\Omega_k) = |\gamma_{\theta\phi}(\Omega_k)|e^{j\alpha(\Omega_k)} \quad (8.103)$$

The equivalent expression for a distributed scattering is as follows

$$\underline{P}_{rec} = \int_\Omega G(\Omega)\Big[p_{i\theta}(\Omega)|\rho_{R\theta}(\Omega)|^2 + p_{i\phi}|\rho_{R\phi}(\Omega)|^2$$

$$+2Real(\rho_{R\theta}(\Omega)\rho^*_{R\phi}(\Omega)\gamma_{\theta\phi}(\Omega)\sqrt{p_{i\theta}(\Omega)p_{i\phi}(\Omega)})\Big]\Delta\Omega \quad (8.104)$$

4.1. Different Performance Conditions:

Case I: Perfect Cross-Polar Correlation, and Antenna Polarization Matched

$$\angle\rho_{R\theta} - \angle\rho_{R\phi} = -\angle\gamma_{\theta\phi}, \ |\gamma_{\theta\phi}| = 1, \ \left|\frac{\rho_{R\theta}}{\rho_{R\phi}}\right| = \sqrt{\frac{\underline{P}_{i\theta}(\Omega)}{\underline{P}_{i\phi}(\Omega)}} = \sqrt{\frac{p_{i\theta}(\Omega)}{p_{i\phi}(\Omega)}} \quad (8.105)$$

This is the same matching condition as described in (8.81) for an isotropic antenna. The maximum received power, $\underline{P}_{max}$, can be derived from (8.102,8.104) using the above matching condition. The derivation will follow that for (8.87), except now the gain function $G(\Omega)$ needs to be multiplied.

$$\underline{P}_{max} = \sum_k G(\Omega_k)(\underline{P}_{i\theta}(\Omega_k) + \underline{P}_{i\phi}(\Omega_k)) = \sum_k G(\Omega_k)\underline{P}_i(\Omega_k) \quad (8.106)$$

$$\underline{P}_{max} = \int_\Omega G(\Omega)p_i(\Omega)\Delta\Omega = \underline{P}\int_\Omega G(\Omega)f(\Omega)\Delta\Omega \quad (8.107)$$

Here we use the distribution function $f(\Omega)$, defined in (8.95), for a distributed scattering environment.

Case II: Perfect Cross-Polar Correlation, but Antenna Polarization Arbitrary

In this case $|\gamma_{\theta\phi}| = 1$, which means perfect $\theta - \phi$ correlation. However, the antenna polarization is not matched to the scattering fields. The average received power $\underline{P}'_{max}$ may be obtained from (8.102,8.104) by substituting $|\gamma_{\theta\phi}|$ equal to unity.

$$\underline{P}'_{max} = \sum_k G(\Omega_k) \Big[\underline{P}_{i\theta}(\Omega)|\rho_{R\theta}(\Omega)|^2 + \underline{P}_{i\phi}(\Omega)|\rho_{R\phi}(\Omega)|^2$$
$$+2\sqrt{\underline{P}_{i\theta}(\Omega)\underline{P}_{i\phi}(\Omega)}|\rho_{R\theta}(\Omega)||\rho_{R\phi}(\Omega)|Real(e^{j(\alpha(\Omega)+\alpha'(\Omega))})\Big] \quad (8.108)$$

$$\underline{P}'_{max} = \int_\Omega G(\Omega) \Big[\underline{p}_{i\theta}(\Omega)|\rho_{R\theta}(\Omega)|^2 + \underline{p}_{i\phi}(\Omega)|\rho_{R\phi}(\Omega)|^2$$
$$+2\sqrt{\underline{p}_{i\theta}(\Omega)\underline{p}_{i\phi}(\Omega)}|\rho_{R\theta}(\Omega)||\rho_{R\phi}(\Omega)|Real(e^{j(\alpha(\Omega)+\alpha'(\Omega))})\Big] d\Omega \quad (8.109)$$

Case III: Antenna Polarization Matched, but Cross-Polar Correlation Arbitrary

$$\angle\rho_{R\theta} - \angle\rho_{R\phi} = -\angle\gamma_{\theta\phi}, \quad \left|\frac{\rho_{R\theta}}{\rho_{R\phi}}\right| = \sqrt{\frac{\underline{P}_{i\theta}(\Omega)}{\underline{P}_{i\phi}(\Omega)}} = \sqrt{\frac{p_{i\theta}(\Omega)}{p_{i\phi}(\Omega)}}, \; |\gamma_{\theta\phi}| < 1 \quad (8.110)$$

The maximum received power, $\underline{P}''_{max}$, can be derived from (8.102,8.104) using the above condition. The derivation will follow that for (8.87), except now the gain function $G(\Omega)$ needs to be multiplied and $|\gamma_{\theta\phi}| \neq 1$.

$$\underline{P}''_{max} = \sum_k G(\Omega_k) \Big[\underline{P}_i(\Omega_k) - 2\sqrt{\underline{P}_{i\theta}(\Omega_k)\underline{P}_{i\phi}(\Omega_k)}|\rho_{R\theta}(\Omega_k)||\rho_{R\phi}(\Omega_k)|(1 - |\gamma_{\theta\phi}(\Omega_k)|)\Big]$$
$$= \underline{P}_{max} - \sum_k G(\Omega_k) 2\sqrt{\underline{P}_{i\theta}(\Omega_k)\underline{P}_{i\phi}(\Omega_k)}|\rho_{R\theta}(\Omega_k)||\rho_{R\phi}(\Omega_k)|(1 - |\gamma_{\theta\phi}(\Omega_k)|)$$
$$= \underline{P}_{max} - \sum_k G(\Omega_k) \frac{2\underline{P}_{i\theta}(\Omega_k)\underline{P}_{i\phi}(\Omega_k)}{\underline{P}_i(\Omega_k)}|(1 - |\gamma_{\theta\phi}(\Omega_k)|) \quad (8.111)$$

$$\underline{P}''_{max} = \underline{P}_{max} - \int_\Omega G(\Omega) \frac{2p_{i\theta}(\Omega_k)p_{i\phi}(\Omega_k)}{p_i(\Omega_k)}|(1 - |\gamma_{\theta\phi}(\Omega_k)|)\Delta\Omega \quad (8.112)$$

4.1.1. *Average Power Loss Factors.* Comparing the above cases, $\underline{P}_{max}$ is the maximum possible average power that may be available under ideal matching conditions. $\underline{P}'_{max}$, $\underline{P}''_{max}$, as well as $\underline{P}_{rec}$ for a general situation are less than the ideal limit $\underline{P}_{max}$. $\eta_\gamma = \underline{P}''_{max}/\underline{P}_{max}$ may be referred to as the cross-polar correlation loss factor, $\eta_\rho = \underline{P}'_{max}/\underline{P}_{max}$

the polarization loss factor, and $\eta_{\gamma\rho} = \underline{P}_{rec}/\underline{P}_{max}$ the total loss factor due to polarization and cross-polar correlation.

In all the above cases we have assumed impedence matching condition, $Z_L = Z_{in}^*$. If that is not the case, the received power will have to be multiplied by $\frac{4R_L R_{in}}{|Z_{in}+Z_L|^2}$ (impedence match factor) for general impedance conditions.

$$\underline{P}_{rec} = \underline{P}_{rec}(Z_L = Z_{in}^*)\frac{4R_L R_{in}}{|Z_{in} + Z_L|^2} \tag{8.113}$$

$$\eta_Z = \frac{4R_L R_{in}}{|Z_{in} + Z_L|^2} \tag{8.114}$$

η_Z is the impedance mismatch loss factor. The total average power loss factor η_p (impedence, polarization and $\hat{\theta} - \hat{\phi}$ correlation loss) is the product of η_Z and $\eta_{\gamma\rho}$.

$$\eta_p = \eta_Z \eta_{\gamma\rho} = \eta_{\gamma\rho}\frac{4R_L R_{in}}{|Z_{in} + Z_L|^2} \tag{8.115}$$

4.2. Effective Average Antenna Gain G_{ea}. It would be useful to define an antenna parameter G_{ea}, called the effective average antenna gain, as the ratio of the average received power, $\underline{P}_{rec}$, of a given antenna to that, $\underline{P}$ of a reference isotropic antenna operating under ideal conditions of polarization matching as well as perfect cross-polar correlation.

The idea is to evaluate the antenna's coupling to the external incoming radiation, not the circuit performance at the antenna input terminals. Therefore, for the above evaluation the test antenna and the reference antenna are assumed to be operating under impedance matched conditions ($Z_L = Z_{in}^*$).

$$G_{ea} = \frac{\underline{P}_{rec}}{\underline{P}_{rec}\big|_{\substack{isotropic \\ matched}}} = \frac{\underline{P}_{rec}}{\underline{P}} = \frac{\underline{P}_{rec}}{\underline{P}_\theta + \underline{P}_\phi} \tag{8.116}$$

where $\underline{P}_{rec}$ is expressed in (8.102,8.104), and $\underline{P}$ is expressed in (8.87,8.89). $\underline{P}$ in the denominator is the total reference power level, and does not consider the power distribution over anlges Ω. It also does not depend upon the specific antenna used for reception. On the other hand, the receive power $\underline{P}_{rec}$ is dependent upon power distribution function, antenna gain function, as well as polarization matching and cross-polar correlation. We would consider useful practical situations where G_{ea} may be expressed in simple, meaningful forms.

<u>Case I: G_{ea} with Perfect Cross-Polar Correlation and Antenna Polarization Matching</u>

This is the same maximum power condition discussed above in section 4.1, case-I.

$$
\begin{aligned}
\underline{P}_{rec} = \underline{P}_{max} &= \underline{P} \int_{\Omega} G(\Omega) f(\Omega)\, d\Omega \\
&= (\underline{P}_{\theta} + \underline{P}_{\phi}) \int_{\Omega} G(\Omega) f(\Omega)\, d\Omega \\
G_{ea} = \frac{\underline{P}_{max}}{\underline{P}_{\theta} + \underline{P}_{\phi}} &= \int_{\Omega} G(\Omega) f(\Omega)\, d\Omega
\end{aligned}
\tag{8.117}
$$

which is essentially the average of the gain function $G(\Omega)$ weighted by the total power distribution function $f(\Omega)$.

As we know, the above ideal matching conditions to achieve $\underline{P}_{max}$ may not be satisfied in general practical situations with arbitrary polarization, although performances close to the ideal limit are possible. Situations where the scattering field is purely $\theta-$ or $\phi-$ polarised, and the antenna used is purely $\theta-$ or $\phi-$ polarized, respectively, are simple obvious cases where the above matching conditions and average gain formula would apply.

Under the above matched conditions, one may observe that when the power distribution function is uniform, $f(\Omega) = \frac{1}{4\pi}$, the average gain G_{ea} would be always unity independent of the particular antenna used. Practically, this means that all antennas under the above matching conditions would receive the same amount of average power for a uniformly distributed scattering environment.

$$
G_{ea} = \frac{1}{4\pi} \int_{\Omega} G(\Omega) d\Omega = 1; \ f(\Omega) = \frac{1}{4\pi} \tag{8.118}
$$

This is because the integration of the gain function $G(\Omega)$ over the entire space is 4π for all antennas. This assumes that the antenna is lossless. Otherwise the integration of $G(\Omega)$ over the entire space would be $4\pi\eta_r$, where η_r is the radiation efficiency. This would lead to $G_{ea} = \eta_r$.

A similar useful situation is when the power distribution is uniform in ϕ, but localized in equatorial plane $\theta = \pi/2$. Further, assume that the gain function $G(\theta, \phi)$ can be expressed in separable form as a product of θ and ϕ functions.

$$f(\Omega) = \delta(\theta - \pi/2)\frac{1}{2\pi}, \quad G(\theta, \phi) = h(\theta)g(\phi) \tag{8.119}$$

$$\int_{\Omega} G(\Omega)d\Omega = 4\pi = \int_{\phi=0}^{2\pi}\int_{\theta=0}^{\pi} G(\theta,\phi)\sin\theta d\theta d\phi = \int_{0}^{\pi} h(\theta)\sin\theta d\theta \int_{0}^{2\pi} g(\phi)d\phi \tag{8.120}$$

$$\begin{aligned} G_{ea} &= \int_{\Omega} G(\Omega)f(\Omega)d\Omega = \int_{\phi=0}^{2\pi}\int_{\theta=0}^{\pi} h(\theta)g(\phi)\delta(\theta-\pi/2)\frac{1}{2\pi}\sin\theta d\theta d\phi \\ &= \frac{h(\frac{\pi}{2})}{2\pi}\int_{0}^{2\pi} g(\phi)d\phi = \frac{h(\frac{\pi}{2})}{2\pi}\cdot\frac{4\pi}{\int_{0}^{\pi} h(\theta)\sin\theta d\theta} = \frac{2h(\frac{\pi}{2})}{\int_{0}^{\pi} h(\theta)\sin\theta d\theta} \end{aligned} \tag{8.121}$$

In this case, G_{ea} only depends on $h(\theta)$, independent of the specific $\phi-$ dependent part $g(\phi)$. As an example, consider a small vertical dipole with $G(\theta, \phi) = 1.5\sin^2\theta$. Using this gain function in above expression of average gain, we get $G_{ea} = 1.5$. Any other antenna with the same $\sin^2\theta$ dependence of the gain function but with a different $\phi-$ dependence will provide the same effective gain of 1.5 for the above scattering environment.

Case II: G_{ea} for Uncorrelated Cross-Polar Fields

In this case the correlation coefficient $\gamma_{\theta\phi} = 0$. $\underline{P}_{rec}$ can be expressed using (8.102,8.104), by substituting $\gamma_{\theta\phi} = 0$. The power distribution functions f_θ and f_ϕ, and the average cross-polar level ψ are used for simplification.

$$\begin{aligned} \underline{P}_{rec} &= \int G(\Omega)|\rho_{R\theta}(\Omega)|^2 p_{i\theta}(\Omega)d\Omega + \int G(\Omega)|\rho_{R\phi}(\Omega)|^2 p_{i\phi}(\Omega)d\Omega \\ &= \underline{P}_\theta \int G_\theta(\Omega)f_\theta(\Omega)d\Omega + \underline{P}_\phi \int G_\phi(\Omega)f_\phi(\Omega)d\Omega \\ &= \frac{\underline{P}}{1+\psi}\int G_\theta(\Omega)f_\theta(\Omega)d\Omega + \frac{\underline{P}\psi}{1+\psi}\int G_\phi(\Omega)f_\phi(\Omega)d\Omega \end{aligned} \tag{8.122}$$

where G_θ and G_ϕ are the parts of the total gain G contributed from the $\theta-$ and $\phi-$ polarization fields, respectively.

$$G_\theta(\Omega) = G(\Omega)|\rho_{R\theta}(\Omega)|^2, \; G_\phi(\Omega) = G(\Omega)|\rho_{R\phi}(\Omega)|^2, \; G(\Omega) = G_\theta(\Omega) + G_\phi(\Omega) \tag{8.123}$$

One may define $G_{ea\theta}$, $G_{ea\phi}$ as the effective gains for the $\theta-$ and $\phi-$ polarized fields.

$$G_{ea\theta} = \int G_\theta(\Omega)f_\theta(\Omega)d\Omega, \; G_{ea\phi} = \int G_\phi(\Omega)f_\phi(\Omega)d\Omega \tag{8.124}$$

Equations (8.122,8.124) can be used with (8.116) to obtain an useful expression for G_{ea}.

$$
\begin{aligned}
G_{ea} = \frac{P_{rec}}{P} &= \frac{1}{1+\psi}\int G_\theta(\Omega) f_\theta(\Omega) d\Omega + \frac{\psi}{1+\psi}\int G_\phi(\Omega) f_\phi(\Omega) d\Omega \\
&= \frac{1}{1+\psi} G_{ea\theta} + \frac{\psi}{1+\psi} G_{ea\phi}
\end{aligned}
\tag{8.125}
$$

As an illustrative simple case, consider an antenna which is polarized purely in the θ direction, where $G_\phi = 0 = G_{ea\phi}$, and $G = G_\theta$, and the power distribution function f_θ is uniform

$$
\begin{aligned}
G_{ea} &= \frac{1}{1+\psi}\int G_\theta(\Omega) f_\theta(\Omega) d\Omega = \frac{1}{1+\psi}\int G(\Omega)\frac{1}{4\pi} d\Omega \\
&= \frac{1}{1+\psi};\ \gamma_{\theta\phi} = 0,\ G_\phi = 0,\ f_\theta(\Omega) = \frac{1}{4\pi}
\end{aligned}
\tag{8.126}
$$

Note that here G_{ea} would be in general less than unity, which means that the antenna would receive less power than a matched, isotropic reference antenna.

4.3. Correlation between Signals from Two Arbitrary Antennas. In the last section, we studied statistics of signal voltage and power received by a particular antenna. It helped to understand and evaluate the average power received by given antennas under various scattering environments. In this section we will study how signals received from two different antennas operating at the same location, or two same or different antennas operating at two different locations, correlate with each other.

Consider antenna #1 located at the origin of a reference coordinate system. The open-circuit received voltage V_{R1} is expressed as in (8.9), which is sum of contributions from signals incident from different angles Ω_k. Each contribution is expressed as the dot product of the incident electric field and the receive voltage vector $\bar{v}_{R1}$ of the antenna, as defined by antenna fundamentals.

$$
\begin{aligned}
V_{R1} &= \sum_k v_{R1\theta}(\Omega_k) e_{i\theta}(\Omega_k) + v_{R1\phi}(\Omega_k) e_{i\phi}(\Omega_k) \\
&= \sum_k \bar{v}_{R1}(\Omega_k)\cdot\bar{e}_i(\Omega_k) = \frac{2j\lambda_o}{\eta_o}\sum_k \bar{e}_1(\Omega_k).\bar{e}_i(\Omega_k)
\end{aligned}
\tag{8.127}
$$

where the receive voltage vector $\bar{v}_{R1}$ of the antenna and its transmit electric field vector $\bar{e}_1$ are related by the fundamental antenna equation (8.12). Consider another antenna #2, which may be identical to or different from the antenna #1. Antenna #2 is physically located at a general position $\bar{r}_2$ with respect to the coordinate reference of antenna #1, as defined by its rectangular coordinates (x_2, y_2, z_2). The received voltage V_{R2} by the second antenna can be expressed by a similar expression as for V_{R1}, except the incident fields must be properly phase shifted in order to account for the path delay due to $\bar{r}_2$.

$$V_{R2} = \frac{2j\lambda_o}{\eta_o} \sum_k \bar{e}_2(\Omega_k) \cdot \bar{e}_i(\Omega_k) e^{+j\Delta\varphi(\Omega_k;\bar{r}_2)} \quad (8.128)$$

$$\bar{r}_2 = x_2\hat{x} + y_2\hat{y} + z_2\hat{z} \quad (8.129)$$

$$\Delta\varphi(\Omega_k;\bar{r}_2) = k_o(\hat{r}(\Omega_k).\bar{r}_2) = k_0(x_2 \sin\theta_k \cos\phi_k + y_2 \sin\theta_k \sin\phi_k + z_2 \cos\theta_k) \quad (8.130)$$

Using the above expressions of V_{R1} and V_{R2}, the conjugate product of them can be expressed as follows.

$$\begin{aligned} V_{R1}V_{R2}^* &= \frac{4\lambda_o^2}{\eta_o^2} \Big(\sum_k \bar{e}_1(\Omega_k).\bar{e}_i(\Omega_k) \Big) \Big(\sum_l \bar{e}_2^*(\Omega_l).\bar{e}_i^*(\Omega_l) e^{-jk_0\Delta\varphi(\Omega_l;\bar{r}_2)} \Big) \\ &= \frac{4\lambda_o^2}{\eta_o^2} \sum_k \sum_l \Big(e_{1\theta}(\Omega_k)e_{i\theta}(\Omega_k) + e_{1\phi}(\Omega_k)e_{i\phi}(\Omega_k) \Big) \\ &\quad \times \Big(e_{2\theta}^*(\Omega_l)e_{i\theta}^*(\Omega_l) + e_{2\phi}^*(\Omega_l)e_{i\phi}^*(\Omega_l) \Big) e^{-jk_0\Delta\varphi(\Omega_l;\bar{r}_2)} \end{aligned} \quad (8.131)$$

We are now ready to evaluate the statistical relations between V_{R1} and V_{R2}. We know from (8.101) that averages of V_{R1} and V_{R2} are zero. This may also be seen here by averaging of (8.127,8.128), and assuming that the incoming field averages $< \bar{e}_i(\Omega) >$ are zero for all angles Ω. Now, it would be useful to evaluate the expected value of $V_{R1}V_{R2}^*$ over different sample positions. Samples are collected at different positions, while maintaining the same relative coordinates between the two antennas. This expected value of $V_{R1}V_{R2}^*$ would provide a measure of the correlation between the two signals. The statistical relations (8.42,8.48,8.51) can be used for simplification.

$$\begin{aligned} < V_{R1}V_{R2}^* >= \frac{4\lambda_o^2}{\eta_o} \sum_k \Big[& e_{1\theta}(\Omega_k)e_{2\theta}^*(\Omega_k) < |e_{i\theta}(\Omega_k)|^2 > \\ + e_{1\phi}(\Omega_k)e_{2\phi}^*(\Omega_k) < |e_{i\phi}(\Omega_k)|^2 > & + e_{1\theta}(\Omega_k)e_{2\phi}^*(\Omega_k) < e_{i\theta}(\Omega_k)e_{i\phi}^*(\Omega_k) > \\ & + e_{1\phi}(\Omega_k)e_{2\theta}^*(\Omega_k) < e_{i\phi}(\Omega_k)e_{i\theta}^*(\Omega_k) > \Big] e^{-jk_0\Delta\varphi(\Omega_k;\bar{r}_2)} \end{aligned} \quad (8.132)$$

The averaging of the double summation expression in (8.131) results in a single summation form of (8.132). This is because the average of the cross terms between different angles $l \neq k$ are zero, as described in (8.42,8.48,8.51).

The open-circuit voltages relate to the received power P_{rec} at the load under impedance matching conditions.

$$P_{rec1} = \frac{|V_{R1}|^2}{8R_{in1}}, \; P_{rec2} = \frac{|V_{R2}|^2}{8R_{in2}} \quad (8.133)$$

The average received power under impedance matching conditions are given by the effective agerage gain G_{e1} of the antenna and the reference average power $\underline{P} = \underline{P}_\theta + \underline{P}_\phi$, as defined in (8.116).

$$< |V_{R1}|^2 >= 8R_{in1}\underline{P}_{rec1} = 8R_{in1}G_{ea1}\underline{P},$$
$$< |V_{R2}|^2 >= 8R_{in2}\underline{P}_{rec2} = 8R_{in2}G_{ea2}\underline{P} \quad (8.134)$$

A complex correlation coefficient γ_{12} between the two antennas is defined using (8.132,8.134), by normalizing $< V_{R1}V_{R2}^* >$ such that $|\gamma_{12}| \le 1$

$$\gamma_{12} = \frac{< V_{R1}V_{R2}^* >}{\sqrt{< |V_{R1}|^2 >< |V_{R2}|^2 >}} = \frac{< V_{R1}V_{R2}^* >}{2\underline{P}\sqrt{R_{in1}R_{in2}G_{ea1}G_{ea2}}}, \; |\gamma_{12}| \le 1 \quad (8.135)$$

We are particularly interested in the effect of the scattering environment and the antenna's external radiation characteristics on various signal statistics. With this objective in mind, it is meaningful to consider the signal statistics of the open-circuit voltages, which are independent of the specific load impedances used. The actual load voltages V_{L1} and V_{L2} under general impedance conditions can always be related to the open-circuit voltages using antenna and load impedances. Accordingly, various signal statistics for V_{R1} and V_{R2} need to be multiplied by proper impedance factors in order to obtain the respective statistics for the load signals.

4.4. Antenna Diversity. In a particular scattering environment with given incident reference power levels $\underline{P}$, $\underline{P}_\theta$, $\underline{P}_\phi$, the average power received by a particular antenna is determined by the effective antenna gain G_{ea}. This will determine the signal quality and quality of communication when the particular antenna is used. The signal quality or reliability may possibly be improved further by employing more than one antenna, and processing the multiple signals for additional information. This may be achieved, for example, by selecting the antenna signal having the best power level. Other processing techniques may also be used for optimum results. Such techniques are referred to as antenna diversity.

The antenna diversity technique fundamentally relies on the assumption that the signals from different antennas are significantly dissimilar or uncorrelated from each other. If the signal from one antenna is low at a given situation, one would hope that the signal from another antenna is relatively high at the same situation, so that selecting the stronger signal would make up for the weak signal from the first antenna. This would likely be the case if the different antenna signals are significantly dissimilar or uncorrelated from each other. The correlation coefficient γ_{12} as defined in (8.135) would provide a quantitative measure of similarity or dissimilarity between two antenna signals. The two antenna signals are completely dissimilar when the coefficient is zero, and are completely similar if the coefficient is unity. It is desirable that γ_{12} is ideally equal to zero, or practically as small as possible in order to get best results from a diversity technique. We will examine in the following specific choices for the two antennas under specific scattering conditions, so that γ_{12} can be ideally zero. Because the normalizing denominator of γ_{12} in (8.135) is always a positive real number, it would suffice to examine if the numerator $< V_{R1}V_{R2}^* >$, as expressed in (8.132,) is zero.

4.4.1. *Antenna Pattern or Beam Diversity.* Consider two antennas having orthogonal power patterns. Ideally, this means $|\bar{e}_1(\Omega)|^2 = 0$ where $|\bar{e}_2(\Omega)|^2 \neq 0$, or equivalently $\bar{e}_1(\Omega) = 0$ where $\bar{e}_2(\Omega) \neq 0$. Applying the above condition to (8.132), we will have

$$e_{1\theta}(\Omega)e^*_{2\theta}(\Omega) = e_{1\phi}(\Omega)e^*_{2\phi}(\Omega) = e_{1\theta}(\Omega)e^*_{2\phi}(\Omega) = 0$$
$$< V_{R1}V^*_{R2} >= 0; for\ all\ \bar{r}_2 \tag{8.136}$$

Signals from two antennas having such orthogonal patterns or beams would be uncorrelated from each other, independent of their relative locations $\bar{r}_2$ as well as independent of the incident power distributions. Two or more of such antennas with same or comparable average effective gain G_{ea}, or equivalently having comparable average received power levels (the average effective gain and average received power will depend on incident power distribution, antenna polarization etc.) can be used as diversity antennas. The technique is referred to as pattern or beam diversity technique.

Ideally, one would need strictly orthogonal beams to achieve zero correlation. In practice, two antennas are used where one antenna radiates strongly over a angle segment over which the other antenna radiates at a much lower level. In such cases the correlation coefficient γ_{12} would be small enough to be useful as effective diversity channels. Antennas with distinct orthogonal beams would usually require larger antenna sizes. This often limits practical use of such pattern/beam diversity techniques. Minimal diversity benefits are achieved by using smaller antennas with broader overlapping patterns, resulting in relatively higher correlations. However, a simple, useful way to produce orthogonal patterns using small antennas is to place a metal plate in between them. The radiation towards the metal plate from each antenna will be mostly reflected, leaving only some diffracted radiation going across the plate. This results in each antenna radiating to orthogonal half spaces. For example, two microstrip patch antennas printed on opposite sides of a common ground plane can be used as effective beam diversity antennas.

4.4.2. *Antenna Polarization Diversity.* Consider a scattering environment where $< e_{i\theta}e^*_{i\phi} > = 0$, which means $\hat{\theta} - \hat{\phi}$ fields are uncorrelated. In addition, consider that antenna #1 is $\theta-$polarized while the other is $\phi-$ polarized. In other words, $e_{1\phi}(\Omega) = e_{2\theta}(\Omega) = 0$, for all Ω. Applying the above condition to (8.132), we will have

$$< e_{i\theta}(\Omega)e^*_{i\phi}(\Omega) >= 0, e_{1\phi}(\Omega) = e_{2\theta}(\Omega) = 0,\ for\ all\ \Omega$$
$$< V_{R1}V^*_{R2} >= 0; for\ all\ \bar{r}_2 \tag{8.137}$$

If incoming fields are uncorrelated in two orthogonal polarizations, then signals from two antennas having polarization vectors along the respective orthogonal directions would be uncorrelated. This is independent of relative positions between the antennas given by $\bar{r}_2$, and independent of any incoming power distribution pattern. Two such antennas having comparable effective gains or comparable average powers (the average power received by the two antennas will depend on the cross-polar level ψ and pattern of the scattered fields, as well as the antenna gain pattern) can be used as effective diversity channels. One ideally needs uncorrelated orthogonal fields. In practice, the θ and ϕ fields may exhibit some correlations depending on the nature of the scattering environment. Also, the cross-polar incoming power levels are often not comparable ($\psi \neq 1$), which means two cross-polar antennas with comparable gain patterns would receive unequal powers.

These two factors may practically reduce the effectiveness of the polarization diversity technique.

We considered above the θ and ϕ directions as the principal orthogonal directions. Any other orthogonal directions, constructed as linear combinations of the θ and ϕ directions can be used, as long as the incoming scattered fields are uncorrelated in those particular orthogonal directions. It may be noted that if the $\theta-$ and $\phi-$ polar fields have same power levels ($\psi = 1$), and they are uncorrelated, then any other orthogonal directions constructed by linear combinations of the θ and ϕ directions would have same power levels and would be completely uncorrelated. It may be shown that this would not be the case in general, when $\psi \neq 1$.

4.4.3. *Antenna Field Diversity.* Under a given scattering environment, it may be possible to have zero correlation between signals from two co-located antennas, if the transmit field functions $\bar{e}_1(\Omega)$ and $\bar{e}_1(\Omega)$ are properly designed. Consider a scattering environment with a given polarization of the incoming fields. For example, $\bar{e}_i(\Omega) = \hat{\theta} e_{i\theta}(\Omega)$ for all angles Ω, and $e_{i\phi}(\Omega) = 0$. We also assume that the two antennas are co-located, $\bar{r}_2 = 0$. With these conditions, (8.132) can be re-written as:

$$\begin{aligned} < V_{R1} V_{R2}^* > &= \frac{4\lambda_o^2}{\eta_o} \sum_k e_{1\theta}(\Omega_k) e_{2\theta}^*(\Omega_k) 2\eta_o P_{di\theta}(\Omega_k) \\ &= 8\lambda_o^2 \sum_k e_{1\theta}(\Omega_k) e_{2\theta}^*(\Omega_k) P_{di\theta}(\Omega_k) \end{aligned} \tag{8.138}$$

It may be convenient here to express the above equation for distributed scattering fields. We will use the power distribution functions defined in section 3.3 for simplification.

$$\begin{aligned} < V_{R1} V_{R2}^* > &= 8\lambda_o^2 \int_\Omega e_{1\theta}(\Omega) e_{2\theta}^*(\Omega) \Delta P_{di\theta}(\Omega) \\ &= 8\lambda_o^2 \int_\Omega e_{1\theta}(\Omega) e_{2\theta}^*(\Omega) \frac{4\pi}{\lambda_o^2} p_{i\theta}(\Omega) \Delta\Omega \\ &= 32\pi \underline{P}_\theta \int_\Omega e_{1\theta}(\Omega) e_{2\theta}^*(\Omega) f_\theta(\Omega) d\Omega \end{aligned} \tag{8.139}$$

The above expected value could be zero if the transmit electric field functions of the two antennas $e_{1\theta}(\Omega)$ and $e_{2\theta}^*(\Omega)$, weighted by the power distribution function $f_\theta(\Omega)$, are orthogonal to each other. For example, consider two antennas with $e_{1\theta} = a \sin\theta$ (a vertical dipole antenna), $e_{2\theta} = b\cos\phi$ (a wire loop antenna with loop axis along y,) where a and b are some constants independent of the angle Ω, and a uniform power distribution function $f_\theta(\Omega) = \frac{1}{4\pi}$. Under these conditions,

$$< V_{R1}V_{R2}^* >= 32\pi \underline{P}_\theta \int\limits_\Omega (a\sin\theta)(b^*\cos\phi) f_\theta(\Omega) d\Omega$$

$$= 32\pi \underline{P}_\theta \int\limits_{\phi=0}^{2\pi}\int\limits_{\theta=0}^{2\pi} (a\sin\theta)(b^*\cos\phi)\frac{1}{4\pi}\sin\theta d\theta d\phi = 0 \tag{8.140}$$

Such two antennas, assuming that they have comparable effective average gain, or in other words comparable average receive powers, can be used as two diversity channels. This technique makes use of the orthogonality of the transmit field functions, $e_{1\theta}(\Omega)$ and $e_{2\theta}(\Omega)$ of the antennas. Also, upon closer examination of the physical mechanisms of the above two antennas which produce such orthogonal field functions $e_{1\theta}(\Omega)$ and $e_{2\theta}(\Omega)$, it would be seen that the first antenna fundamentally senses the vertical electric field at the location, whereas the other senses a horizontal (y) magnetic field. Accordingly, such diversity techniques are referred to as the "field diversity" techniques.

For a $\theta-$polarized incoming electric field, there are three basic orthogonal "field" sensors: (1) a z-directed electric dipole, which senses the E_z field, (2) a wire loop antenna with loop axis along x, which senses the H_x field, and (3) a wire loop antenna with loop axis along y, which senses the H_y field. All these have $\theta-$ polarization of radiation along the horizontal plane ($\theta = \pi/2$), and exhibit orthogonal $e_\theta(\Omega)$ patterns with respect to uniform scattering distribution functions. Similarly, for a $\phi-$polarized incoming electric field, there are three basic orthogonal "field" sensors: (1) a x-directed electric dipole, which senses the E_x field, (2) a y-directed electric dipole, which senses the E_y field, and (3) a wire loop antenna with loop axis along z, which senses the H_z field. All these have $\phi-$ polarization of radiation along the horizontal plane ($\theta = \pi/2$), and exhibit orthogonal $e_\phi(\Omega)$ patterns with respect to uniform scattering distribution functions.

4.4.4. *Antenna Location Diversity.* Under a particular scattering environment and two given antennas, it may be possible to achieve zero correlation between two antenna signals by simply locating them at different relative positions. This technique is accordingly called antenna location diversity. We will illustrate this possibility for useful antenna and scattering parameters. It may make practical sense to use two identical antennas for this technique. This means $e_{1\theta} = e_{2\theta} = e_\theta$, $e_{1\phi} = e_{2\phi} = e_\phi$. Because the two antennas are located at different positions, we have $\bar{r}_2 \neq 0$. Also let us assume that $< e_{i\theta}e_{i\phi}^* > = 0$. Under these conditions, (8.132) can be simplified as:

$$< V_{R1}V_{R2}^* >= \frac{4\lambda_o^2}{\eta_o}\sum_k \left[|e_\theta(\Omega)|^2 2\eta_o P_{di\theta}(\Omega_k) + |e_\phi(\Omega_k)|^2 2\eta_o P_{di\phi}(\Omega_k)\right]e^{-j\Delta\varphi(\Omega_k;\bar{r}_2)}$$

$$= 8\lambda_o^2 \sum_k \left[|e_{1\theta}(\Omega_k)|^2 P_{di\theta}(\Omega_k) + |e_{1\phi}(\Omega_k)|^2 P_{di\phi}(\Omega_k)\right]e^{-j\Delta\varphi(\Omega_k;\bar{r}_2)} \tag{8.141}$$

We now express the above expected value for a distributed scattering environment. We will use the power distribution functions defined in section 3.3 for simplification.

$$< V_{R1}V_{R2}^* >= 8\lambda_o^2 \int_\Omega \left[|e_\theta(\Omega)|^2 \times \frac{4\pi}{\lambda_o^2} p_{i\theta}(\Omega) + |e_\phi(\Omega)|^2 \times \frac{4\pi}{\lambda_o^2} p_{i\phi}(\Omega)\right] e^{-j\Delta\varphi(\Omega;\bar{r}_2)} d\Omega$$
$$= 32\pi \int_\Omega \left[|e_\theta(\Omega)|^2 \underline{P}_\theta f_\theta(\Omega) + |e_\phi(\Omega)|^2 \underline{P}_\phi f_\phi(\Omega)\right] e^{-j\Delta\varphi(\Omega;\bar{r}_2)} d\Omega \quad (8.142)$$

Consider for illustration antennas having $|e_\theta|$ and $|e_\phi|$ both uniform in ϕ. A special case is to have an antenna with only one field component (either the $|e_\theta|$ or the $|e_\phi|$ field,) uniformly distributed in ϕ, with the other field zero along all angles. A vertical dipole or a wire loop antenna placed on the horizontal plane would satisfy this requirement. Also consider a scattering environment with average power distributions to be uniform in ϕ but localized only to the horizontal plane ($\theta = \pi/2$).

$$|e_\theta(\Omega)| = |e_\theta|,\ |e_\phi(\Omega)| = |e_\phi|;\ f_\theta(\Omega) = f_\phi(\Omega) = \delta(\theta - \frac{\pi}{2}) \times \frac{1}{2\pi} \quad (8.143)$$

$$< V_{R1}V_{R2}^* >= \frac{32\pi}{2\pi}\left[|e_\theta|^2 \underline{P}_\theta + |e_\phi|^2 \underline{P}_\phi\right] \int_{\phi=0}^{2\pi} e^{-j\Delta\varphi(\theta=\frac{\pi}{2},\phi;\bar{r}_2)} d\phi$$
$$= 16\left[|e_\theta|^2 \underline{P}_\theta + |e_\phi|^2 \underline{P}_\phi\right] \times C$$
$$C = \int_0^{2\pi} e^{-j\Delta\varphi(\theta=\pi/2,\phi;\bar{r}_2)} d\phi \quad (8.144)$$

Now, let us see how the parameter C, and proportionately $< V_{R1}V_{R2}^* >$, varies with different relative positioning $\bar{r}_2$ of the antenna #2. Consider antenna displacement along the x axis. In this case $\bar{r}_2 = x_2\hat{x}$. Using the expression for $\Delta\varphi$ in (8.130), the parameter C in (8.144) may be expressed as follows.

$$\bar{r}_2 = x_2\hat{x},\ \Delta\varphi(\theta = \pi/2, \phi; \bar{r}_2) = k_0 x_2 \cos\phi \quad (8.145)$$
$$C(x_2) = \int_0^{2\pi} e^{-jk_o x_2 \cos\phi} d\phi = 2\pi J_0(k_0 x_2) \quad (8.146)$$
$$< V_{R1}V_{R2}^* >= 32\pi\left[|e_\theta|^2 \underline{P}_\theta + |e_\phi|^2 \underline{P}_\phi\right] \times J_0(k_0 x_2) \quad (8.147)$$

where $J_0(\cdot)$ is a standard known function, called the Bessel Function of first kind of order zero. This function is sketched in Fig.2 as a function of the variable of our interest $k_0 x_2$. It is an oscillating function, often associated with cylindrical waves. The function is zero, and hence $< V_{R1}V_{R2}^* >$ would also be zero, for a discrete series of values, the first three of which are $k_0 x_2 = 2.4, 5.5, 8.7$. This is equivalent to having $x_2/\lambda_0 = 0.38, 0.88, 1.38$. If one selectes the smallest vaue of x_2, we need a separation of about $0.38\lambda_0$ in order to result in $< V_{R1}V_{R2}^* >= 0$, or zero correlation between the antenna signals. It can be shown that, for the specific situation under consideration here, the same displacement between the antennas along any direction on the xy plane would provide zero correlation. Because the two antennas here are identical, they also receive the same average signal power. Therefore, the two antennas with such displacements can then be used as effective diversity channels. Other locations listed above for zero correlation may also be used, but usually the smallest distance of separation is practically preferred. It may be observed from the functional trend of $J_0(.)$ (see Fig.2) that the oscillation peaks reduces with larger arguments. This means that large displacements would result in relatively low value of correlation. In other words, when the displacement between the two antennas are sufficiently large compared to wavelength, the signals from the two antennas would be practically uncorrelated with each other.

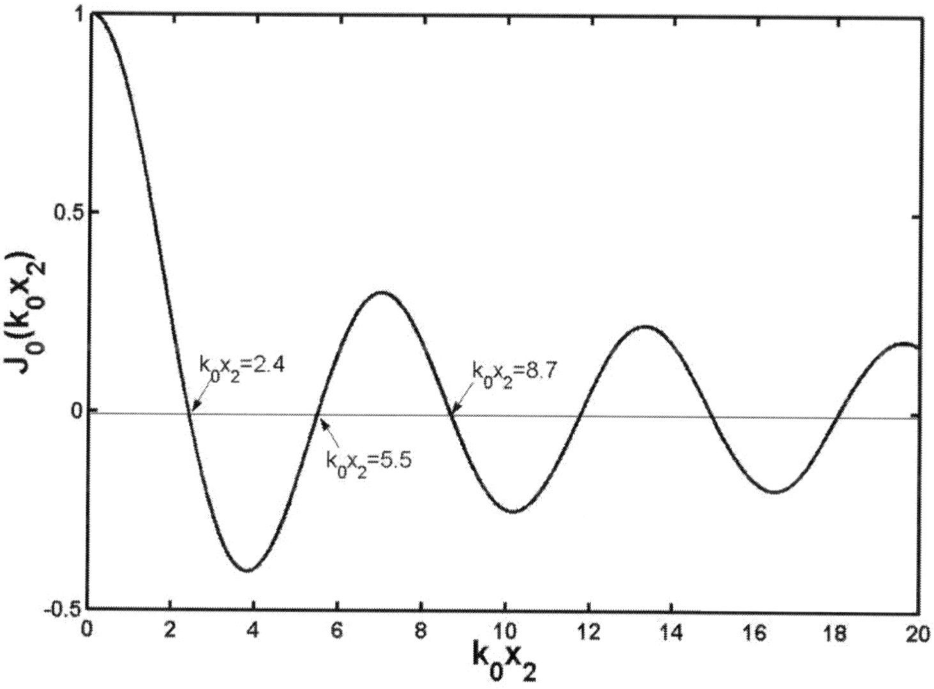

FIGURE 2. Bessel Function of the first kind, of order zero, sketched as a function of its argument kx_2. This governs how the signal correlation between two identical antennas depends upon their relative separation distance x_2, under given conditions. The antenna radiation patterns are uniform and they are operating in a scattering environment concentrated in the $\theta = \pi/2$ plane, with uniform power distribution in ϕ.

The above positioning for zero correlation is valid for the specific set of conditions we have considered. For other scattering distribution functions and antenna characteristics the analysis will have to be similarly performed to obtain the necessary relative positioning. In general situations analytical simplification may not always be possible, in which case numerical integration may be used to locate the proper positions.

CHAPTER 9

ANTENNA ELEMENTS

One can envision literally infinite possible antenna geometries, that can be designed in order to achieve a variety of performance features. Among these diverse configurations one may identify a few types of *basic antenna elements*, that are often used for general-purpose wireless applications. In the following sections we will analyze the radiation/reception mechanism in basic types of antenna elements, including (1) dipole and monopole antennas, (2) wire-loop antenna, (4) slot antenna, and (5) microstrip antenna. Selected other types of antennas in use today will also be introduced. Antenna behavior in the far-field region are used to derive power pattern, radiated power and radiation resistance, whereas field distribution close to the source (near-field region) particularly relate to the antenna reactance and resonance condition. The general definitions and notations from chapter 7 are used for this study. Useful parameters for some of the basic antenna elements covered in this section are summarized in Table I.

1. Small Dipole Antenna

A small dipole antenna may be visualized as a small length of current $\overline{I} = \hat{z}I_{in}$, excited by connecting a pair of small metal wires or rods to a radio-frequency source. This is shown in Fig.1. It is considered small when the length l is significantly smaller than the wave length ($l << \lambda_0$) of operation. This basic element is often used in many radio communication systems, particularly for communication in a relatively low frequency range (say $f < 10MHz$), where the wavelength of radiation is quite long ($\lambda_0 > 30m$). In this frequency range, a practical antenna of reasonable length $l < 3m$ can be approximated as a small dipole antenna, satisfying the required condition $l << \lambda_0$. The metal wire or rod is ideally made of a perfect conductor, when the ohmic loss in the conductor is zero. In practical situations when the wire is made from a very good conductor, unless the wire is too thin (which results in very high ohmic resistance,) only a small fraction of the input power is dissipated as ohmic loss. Most of the input power is thus radiated into the free space, resulting in a good radiation efficiency

Transmit Mode of Operation

The electric and magnetic fields of a small dipole antenna can be expressed using the theory of chapter 6, section 3. This is possible by modeling the antenna source current $I(z)$ as an impulse distribution $I_{in}l\delta(z)$ in z, so that the integration of the current over the antenna length is $I_{in}l$. The equivalent surface-current density $J_s(x, z)$ is a double-impulse distribution $I_{in}l\delta(x)\delta(z)$ in both x and z.

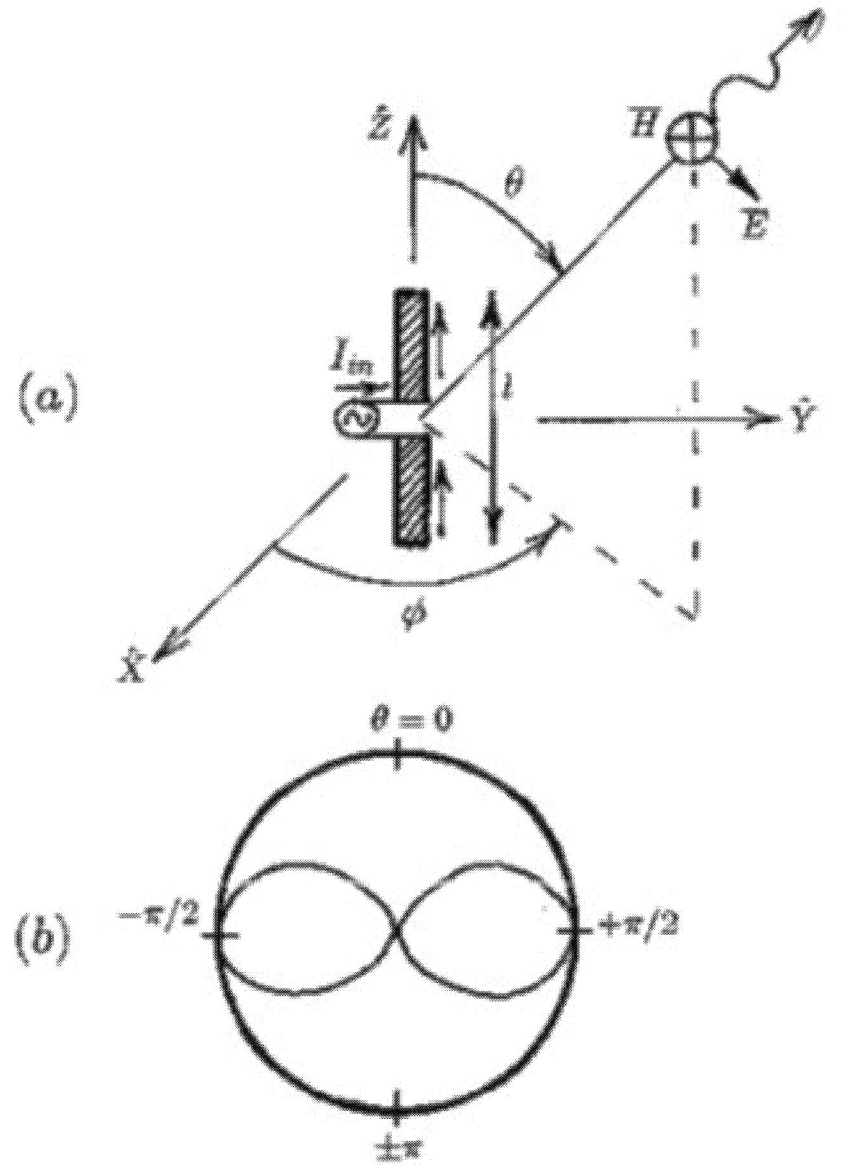

FIGURE 1

$$\overline{I}(z) = \hat{z}I(z) = \hat{z}I_{in}l\delta(z),\ \ x, y = 0;\ \int I(z)dz = I_{in}l,$$

$$\overline{J}_s(x, z) = \hat{z}J_s(x, z) = \hat{z}I_{in}l\delta(z)\delta(x),\ \ y = 0;\ \int J_s(x, z)dx = I(z). \tag{9.1}$$

The above impulse distribution theoretically represents a uniform current I_{in} over the entire antenna length $l << \lambda_0$. For a practical small dipole antenna, where the current distribution may be non-uniform over its length, the above impulse model (9.1) can be applicable by substituting an equivalent length l_e in place of its physical length l. l_e is to be defined such that the integration of the current over the length l is $I_{in}l_e$. Practical current distribution on dipole antennas, specifically in the limit of small length, would be discussed in section 2.

With the above impulse-distribution model for the antenna current, the analysis of chapter 6, section 3 can now be applied. We will particularly use here the field expressions far from the antenna as $r \to \infty$ (equation 6.33,) from which basic antenna parameters such as radiated power, radiation resistance and directivity would be derived.

$$\overline{E}(r,\theta,\phi) = \hat{\theta} I_{in} \frac{e^{-jk_0 r}}{r} \frac{j\eta_0 l}{2\lambda_0} \sin\theta = I_{in} \frac{e^{-jk_0 r}}{r} \overline{e}(\theta,\phi), \tag{9.2}$$

$$\overline{H}(r,\theta,\phi) = \frac{\hat{r} \times \overline{E}(r,\theta,\phi)}{\eta_0} = \hat{\phi} I_{in} \frac{e^{-jk_0 r}}{r} \frac{jl}{2\lambda_0} \sin\theta = I_{in} \frac{e^{-jk_0 r}}{r} \frac{jl}{2\lambda_0} (\hat{z} \times \hat{r}), \tag{9.3}$$

$$\overline{e}(\theta,\phi) = \hat{\theta} \frac{j\eta_0 l}{2\lambda_0} \sin\theta, \;\; \hat{\rho}(\theta,\phi) = j\hat{\theta}. \tag{9.4}$$

The dipole does not radiate in the vertical directions ($\theta = 0, \; \pi$), produces maximum radiation perpendicular to the current ($\theta = \pi/2$), with uniform variation in the azimuth. There is a $\sin\theta$ factor for variation of the radiation fields in the elevation, in proportion to the projection of the input current in the particular direction. The magnetic field $\overline{H}$ is in the $\hat{\phi}$ direction, and thus revolves around the wire in analogy to a magneto-static field due to a linear current. The electric field $\overline{E}$, and accordingly the polarization vector of the dipole antenna is in the $\hat{\theta}$ direction. Once the far-field expressions are known, then the power density P_d and power per unit solid angle p_d of the radiation can be derived using the theory of chapter 7.

$$P_d(r,\theta,\phi) = \frac{|\overline{E}|^2}{2\eta_0} = \frac{|I_{in}|^2\eta_0}{8r^2} (\frac{l}{\lambda_0})^2 \sin^2\theta, \;\; p_d(\theta,\phi) = r^2 P_d(r,\theta,\phi) = \frac{|I_{in}|^2\eta_0}{8} (\frac{l}{\lambda_0})^2 \sin^2\theta. \tag{9.5}$$

The power radiation pattern $\underline{p}_d(\theta,\phi)$ of the dipole antenna is a $\sin^2\theta$ function in elevation, and uniform in the azimuth.

$$\underline{p}_d(\theta,\phi) = \frac{p_d(\theta,\phi)}{p_{dmax}} = \sin^2\theta. \tag{9.6}$$

The total radiated power from the dipole antenna, for a given input current I_{in} can be obtained by integrating $p_d(\theta,\phi)$ over the entire sphere.

$$P_{rad} = \int_{\phi=0}^{2\pi}\int_{\theta=0}^{\pi} p_d(\theta,\phi)\sin\theta\, d\theta\, d\phi = \frac{|I_{in}|^2\eta_0\pi}{3}(\frac{l}{\lambda_0})^2 = \frac{|I_{in}|^2 R_{rad}}{2}, \tag{9.7}$$

$$R_{rad} = \frac{2\eta_0\pi}{3}(\frac{l}{\lambda_0})^2 = 80\pi^2(\frac{l}{\lambda_0})^2,\ R_{in} = R_{rad};\ \eta_r = 1. \tag{9.8}$$

The directive gain and directivity of the dipole antenna is obtained from $p_d(\theta,\phi)$ or $\underline{p}_d(\theta,\phi)$ using the theory of chapter 7.

$$D_g(\theta,\phi) = \frac{4\pi p_d(\theta,\phi)}{\int_{\phi=0}^{2\pi}\int_{\theta=0}^{\pi} p_d(\theta,\phi)\sin\theta\, d\theta\, d\phi} = \frac{4\pi \underline{p}_d(\theta,\phi)}{\int_{\phi=0}^{2\pi}\int_{\theta=0}^{\pi} \underline{p}_d(\theta,\phi)\sin\theta\, d\theta\, d\phi}, \tag{9.9}$$

$$D_g(\theta,\phi) = \frac{4\pi\sin^2\theta}{\int_{\phi=0}^{2\pi}\int_{\theta=0}^{\pi}\sin^3\theta\, d\theta\, d\phi} = \frac{3\sin^2\theta}{2};\ D = 1.5. \tag{9.10}$$

In an ideal situation when the ohmic loss is zero, or radiation efficiency is 1, the antenna gain is equal to the directivity. In practical cases using good conductors, the ohmic loss can be calculated using the metal conductivity, shape of the wire and the current distribution.

$$G(\theta,\phi) = \eta_r\frac{3\sin^2\theta}{2} \simeq \frac{3\sin^2\theta}{2},\ G \simeq 1.5;\ \eta_r \simeq 1. \tag{9.11}$$

Receive Mode of Operation

Fig.2 shows the dipole antenna in the receive mode of operation. The Thevenin's voltage excited at the antenna terminals can be obtained by taking the tangential component of the incident electric field and multiplying with the antenna length.

$$V_R = -lE_{zinc} = le_{i\theta}\sin\theta = \overline{v}_R(\theta,\phi)\cdot\overline{e}_i = v_{R\theta}(\theta,\phi)e_{i\theta} + v_{R\phi}(\theta,\phi)e_{i\phi}, \tag{9.12}$$

$$v_{R\theta}(\theta,\phi) = l\sin\theta,\ v_{R\phi}(\theta,\phi) = 0;\ \overline{v}_R(\theta,\phi) = \hat{\theta}l\sin\theta,\ \hat{\rho}_R(\theta,\phi) = \hat{\theta}. \tag{9.13}$$

The receive vector $\overline{v}_R$ is linearly polarized, with the receive polarization unit vector in the $\hat{\theta}$ direction. From the magnitude of the receive vector we can find the maximum effective area A_{em} using the theory of chapter 7.

$$A_{em}(\theta,\phi) = \frac{\eta_0|\overline{v}_R(\theta,\phi)|^2}{4R_{in}} = \frac{3\lambda_0^2\sin^2\theta}{8\pi}. \tag{9.14}$$

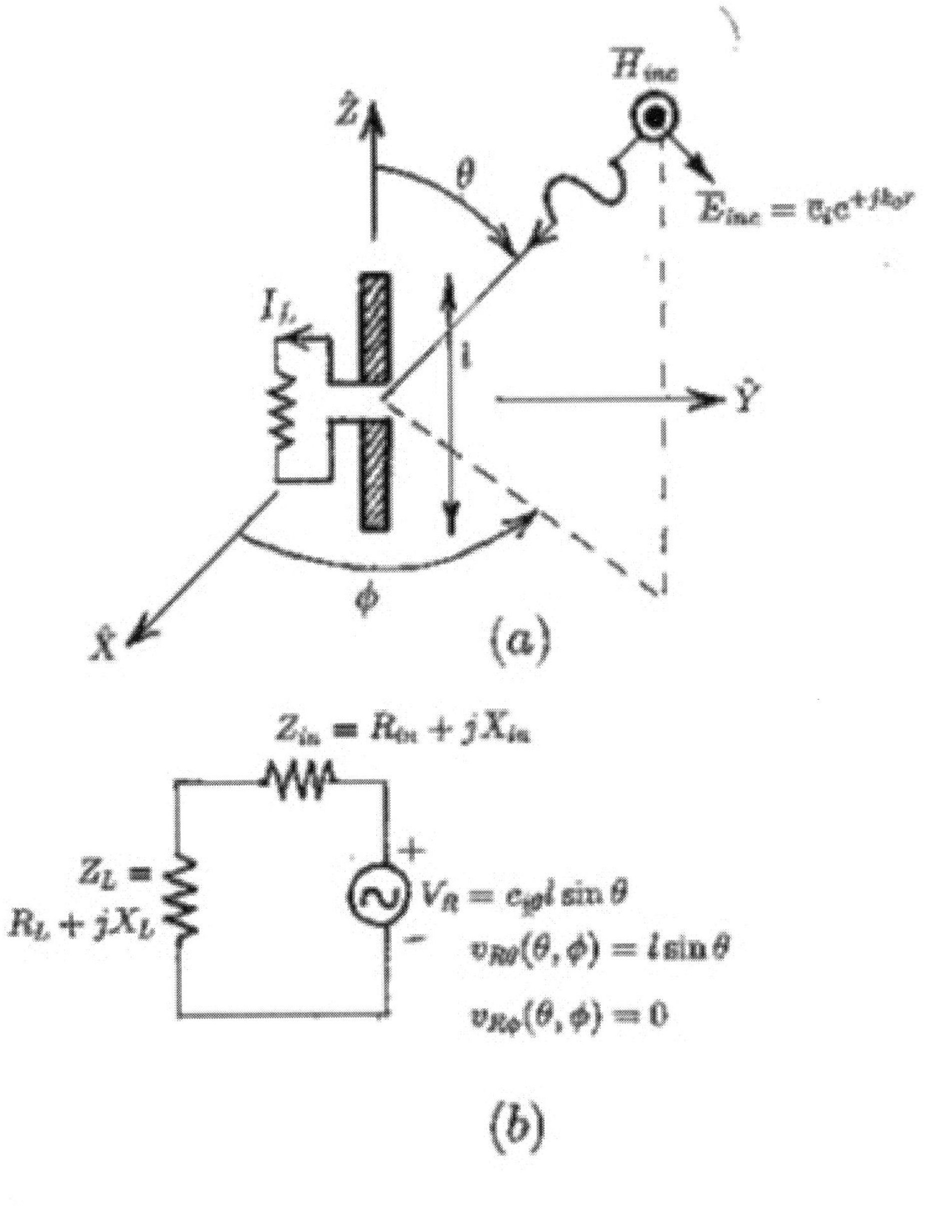

FIGURE 2

Notice that the ratio of $A_{em}(\theta, \phi)$ and $G(\theta, \phi)$ is equal to $\frac{\lambda_0^2}{4\pi}$, which is a constant independent of (θ, ϕ). This is a universal relationship applicable for any arbitrary antenna, as derived in chapter 7.

$$\frac{A_{em}(\theta,\phi)}{G(\theta,\phi)} = \frac{3\lambda_0^2 \sin^2\theta}{8\pi(1.5\sin^2\theta)} = \frac{\lambda_0^2}{4\pi}. \tag{9.15}$$

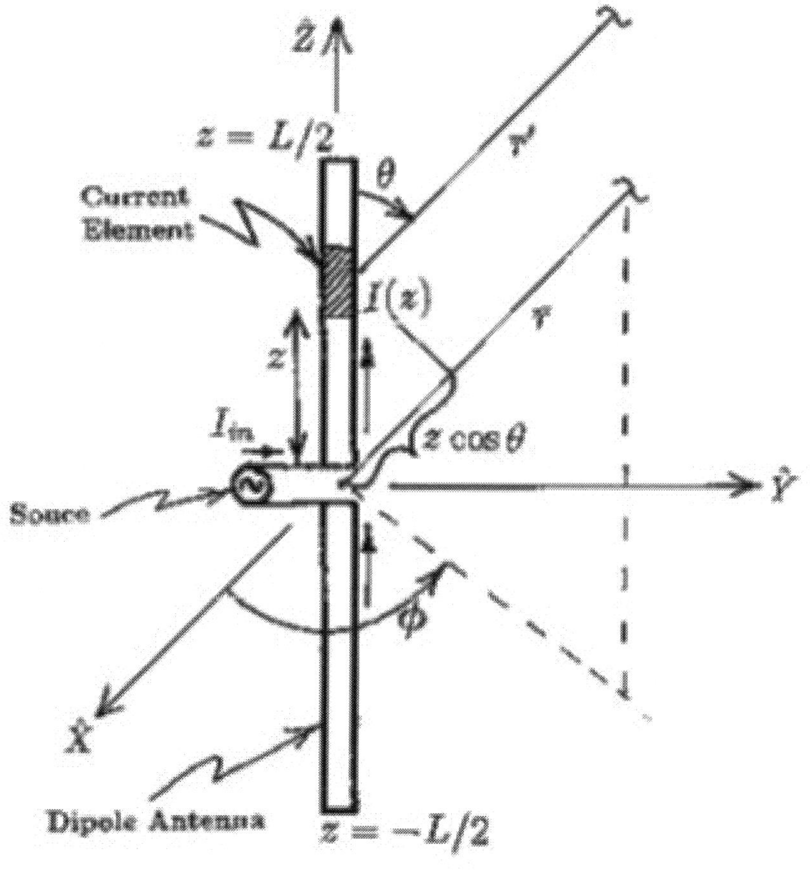

FIGURE 3

2. Dipole Antenna of Arbitrary Length: Far-Field Analysis

If the length of the dipole antenna is not electrically small, then one can derive the radiation fields by slicing the antenna into small pieces, and then superposing (integrating) the individual fields due to each small element. This is shown in Fig.3. The current distribution over the antenna length is assumed to be known. Each piece in Fig.3 can be modeled as a small dipole, with a different current of excitation and a different location.

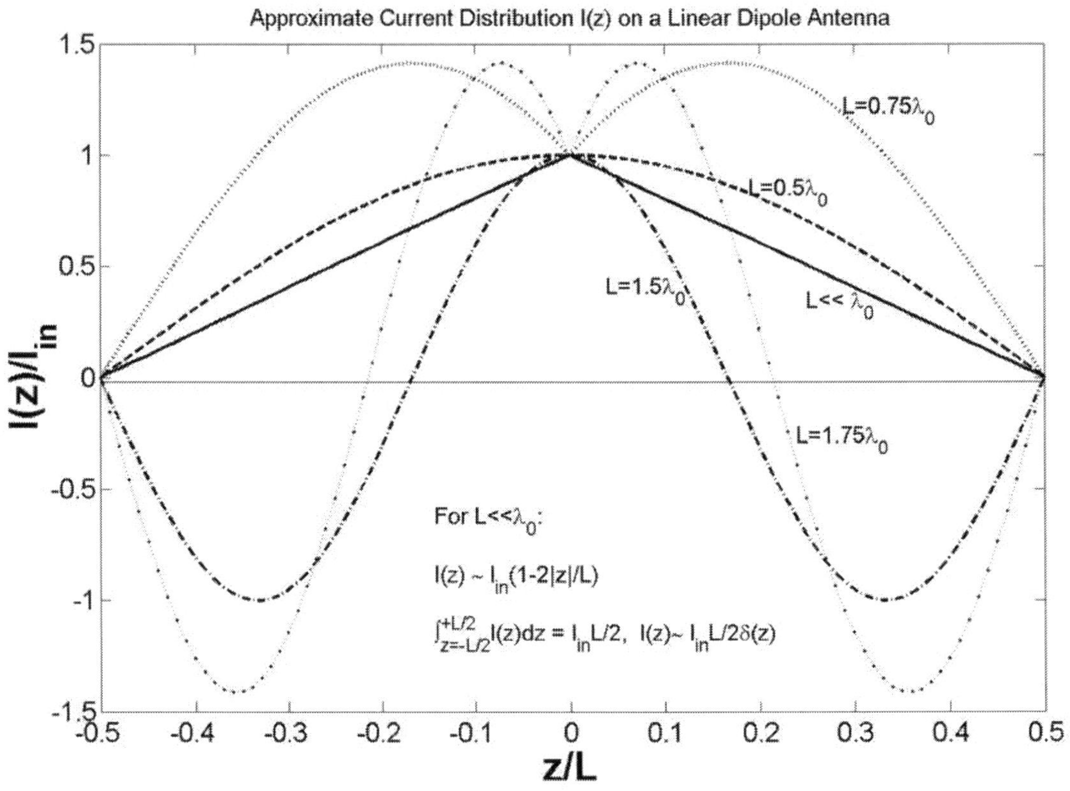

FIGURE 4

The elemental field $\triangle\overline{E}$ produced by a current element of length $\triangle z$, located at distance z from the center, with a current I_z can be written using (9.2, 9.3, 9.4).

$$\triangle\overline{E} = \hat{\theta} I(z) \frac{e^{-jk_0 r'}}{r'} \frac{j\eta_0}{2\lambda_0} \sin\theta \triangle z; \; r' = r - z\cos\theta. \tag{9.16}$$

Assuming that $z << r$ for the far-field calculation, we can approximate the above equation by approximating the r' in the denominator as r.

$$\triangle\overline{E} \simeq \hat{\theta} I(z) \frac{e^{-jk_0 r} e^{jk_0 z\cos\theta}}{r} \frac{j\eta_0}{2\lambda_0} \sin\theta \triangle z. \tag{9.17}$$

The total field can be obtained by integrating (9.17) with respect to z, from $z = -L/2$ to $z = +L/2$.

$$\overline{E} = \hat{\theta}\frac{e^{-jk_0 r}}{r}\frac{j\eta_0}{2\lambda_0}\sin\theta \int_{z=-L/2}^{L/2} I(z)e^{jk_0 z\cos\theta}dz. \tag{9.18}$$

If the current distribution $I(z)$ is known, or can be approximated by a suitable function, then the far field of the total antenna can be obtained using (9.18). One would expect the current distribution to exhibit a standing-wave pattern, with zero current at the the ends of the line ($z = \pm L/2$), and $I = I_{in}$ at the input terminal ($z = 0$). This behavior is similar to the current distribution along an open-circuited transmission line, having propagation constant k_0.

$$I(z) = I_{in}\frac{\sin\left[k_0(L/2 - |z|)\right]}{\sin(k_0 L/2)};\ I(z=0) = I_{in}. \tag{9.19}$$

The above current distribution is shown in Fig.4 for selected antenna lengths L. Using the current distribution of (9.19) in (9.18), and then performing the integration over z, one can obtain a close-form expression for the far-field distribution.

$$\begin{aligned}\overline{E}(r,\theta,\phi) &= \hat{\theta}\frac{e^{-jk_0 r}}{r}\frac{j\eta_0}{2\lambda_0}\sin\theta\int_{z=-L/2}^{L/2} I_{in}\frac{\sin[k_0(L/2-|z|)}{\sin(k_0 L/2)}e^{jk_0 z\cos\theta}dz \\ &= \hat{\theta}\frac{e^{-jk_0 r}}{r}\frac{j\eta_0 I_{in}}{2\pi}\left(\frac{\cos\left(k_0\cos\theta L/2\right) - \cos\left(k_0 L/2\right)}{\sin(k_0 L/2)\sin\theta}\right).\end{aligned} \tag{9.20}$$

Once the electric field is obtained, then basic antenna parameters can be derived using steps similar to those used for a small dipole antenna in section 1, using the definitions in chapter 7.

The radiated power P_{rad} is obtained by integrating the power-density function $P_d = |\overline{E}|^2/(2\eta_0)$ over a sphere, from which the radiation resistance R_{rad} can be derived. The necessary spherical integration involving the field expression of (9.20) may not be easily expressed in closed form, but could be evaluated using numerical integration.

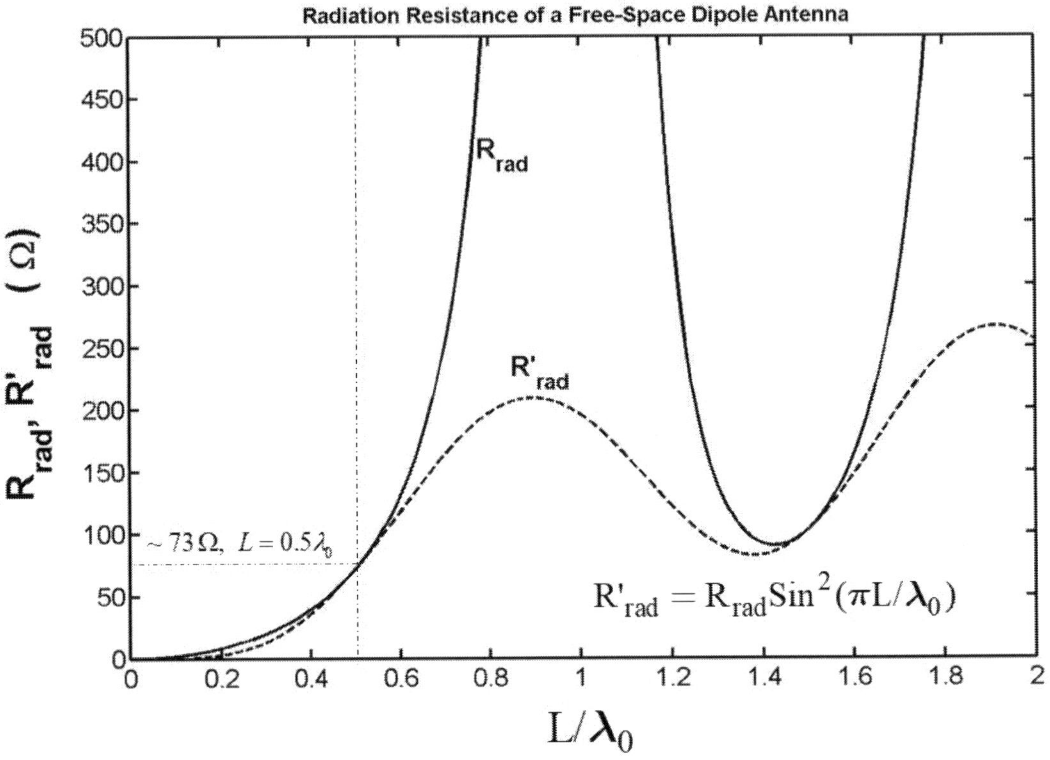

FIGURE 5

$$P_d(r,\theta,\phi) = \frac{p_d(\theta,\phi)}{r^2} = \frac{|\overline{E}|^2}{2\eta_0} = \frac{\eta_0|I_{in}|^2}{8\pi^2r^2}\Big(\frac{\cos(k_0\cos\theta L/2) - \cos(k_0L/2)}{\sin(k_0L/2)\sin\theta}\Big)^2,$$

$$P_{rad} = \int_{\phi=0}^{2\pi}\int_{\theta=0}^{\pi} P_d(r,\theta,\phi)r^2\sin\theta d\theta d\phi = \int_{\phi=0}^{2\pi}\int_{\theta=0}^{\pi} p_d(r,\theta,\phi)\sin\theta d\theta d\phi$$

$$= \frac{\eta_0|I_{in}|^2}{4\pi\sin^2(k_0L/2)}\int_{\theta=0}^{\pi}\frac{\Big(\cos(k_0\cos\theta L/2) - \cos(k_0L/2)\Big)^2}{\sin\theta}d\theta,$$

$$R_{rad} = \frac{2P_{rad}}{|I_{in}|^2} = \frac{\eta_0}{2\pi\sin^2(k_0L/2)}\int_{\theta=0}^{\pi}\frac{\Big(\cos(k_0\cos\theta L/2) - \cos(k_0L/2)\Big)^2}{\sin\theta}d\theta. \quad (9.21)$$

The radiation resistance R_{rad} evaluated using (9.21) is plotted in Fig.5 as a function of L/λ_0. Notice that R_{rad} approaches infinity when the antenna length L is an integral multiple of wavelength λ_0. This is because, in these cases the term $\sin(k_0L/2)$ in the denominator of the R_{rad} expression of (9.21) is zero. Accordingly, it may also be useful to evaluate $R'_{rad} = R_{rad}\sin^2(k_0L/2)$, which is plotted together with R_{rad} in Fig.5. Unlike R_{rad}, the new impedance parameter R'_{rad} is always finite. R'_{rad} may be expressed in terms of the radiated power P_{rad} and the maximum value of the current distribution $I(z)$ in (9.19).

$$R'_{rad} = \frac{2P_{rad}}{|I(z)_{max}|^2} = \frac{2P_{rad}}{|I_{in}/\sin(k_0L/2)|^2} = \sin^2(k_0L/2)R_{rad}. \tag{9.22}$$

The directive gain function $D_g(\theta, \phi)$ can be expressed using the power intensity distribution $p_d = r^2P_d$ and the radiated power P_{rad}, or the equivalent radiation resistance R_{rad}.

$$D_g(\theta,\phi) = \frac{4\pi p_d(\theta,\phi)}{P_{rad}} = \frac{4\pi p_d(\theta,\phi)}{\int_{\phi=0}^{2\pi}\int_{\theta=0}^{\pi} p_d(\theta,\phi)\sin\theta\, d\theta\, d\phi}$$
$$= \frac{8\pi p_d(\theta,\phi)}{|I_{in}|^2R_{rad}} = \frac{\eta_0}{\pi R_{rad}}\Big(\frac{\cos(k_0\cos\theta L/2) - \cos(k_0L/2)}{\sin(k_0L/2)\sin\theta}\Big)^2 \tag{9.23}$$

The directive gain function D_g is plotted in Fig.6 as a function of the elevation angle θ (D_g is independent of ϕ), for selected values of antenna length L. The directivity D, which is the maximum value of the $D_g(\theta,\phi)|_{max}$, and the corresponding value of $\theta = \theta_{max}$ at which this maximum value occurs, are plotted in Fig.7. As seen from Fig.7, the directivity starts with 1.5 for a small dipole ($L/\lambda_0 \to 0$), consistent with the results in section 1. The directivity then increases with lengh, and peaks at about 3.3 for a length of about $L = 1.3\lambda_0$, before it starts to reduce with length. The direction of maximum radiation θ_{max} continues to remain along 90^0 for antenna length less than about $L = 1.45\lambda_0$, beyond which it shifts to a different variable angle.

$$D = D_g(\theta,\phi)_{max} = D_g(\theta_{max},\phi) = \frac{\eta_0}{\pi R_{rad}}\Big(\frac{\cos(k_0\cos\theta_{max}L/2) - \cos(k_0L/2)}{\sin(k_0L/2)\sin\theta_{max}}\Big)^2 \tag{9.24}$$

Small-Length Limit ($L << \lambda_0$)

When the antenna length approaches zero ($L << \lambda_0$), the general current distribution $I(z)$ in (9.19) may be approximated as a "triangular-shaped" distribution (linear variation with z), as shown in Fig.4. Therefore, the integration of the current $I(z)$ over the antenna length would equal $I_{in}L/2$. Accordingly, the antenna current in this case maybe described as an impulse distribution $I_{in}L/2\delta(z)$.

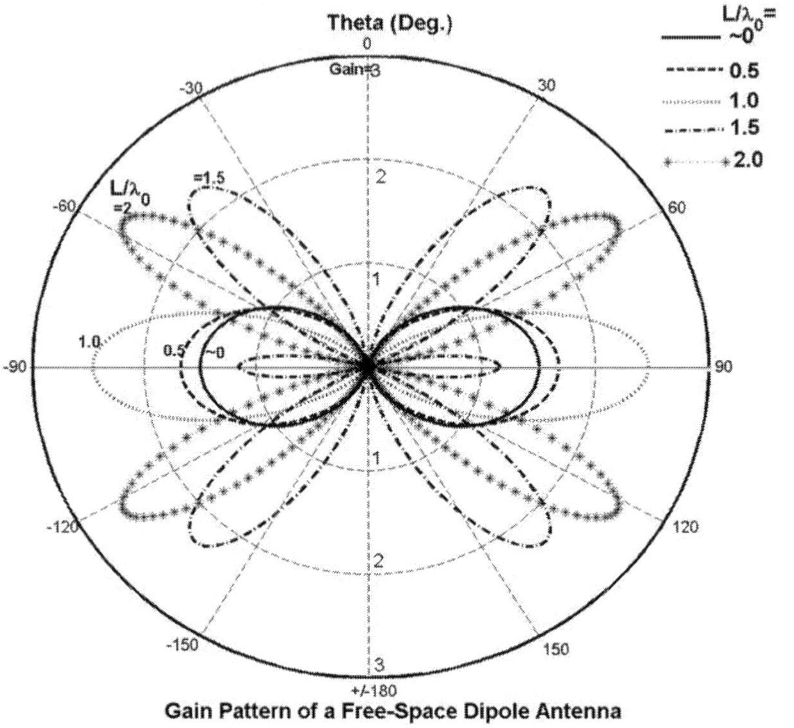

FIGURE 6

$$I(z) \simeq I_{in}\frac{L/2-|z|}{L/2} = I_{in}(1-2|z|/L);\ L << \lambda_0$$

$$\int\limits_{z=-L/2}^{+L/2} I(z)dz \simeq I_{in}L/2, \quad I(z) \simeq I_{in}L/2\delta(z). \tag{9.25}$$

The above "triangular" current distribution is practically realistic for a dipole antenna of small length. It starts with a zero current at the open ends ($z = \pm L$), and gradually increases to the maximum value I_{in} at the input point ($z = 0$).

The above distribution of (9.25) maybe compared with the distribution of (9.1) used for modeling a small-dipole element in section 1. The triangular current distribution of (9.25) (see Fig.4) is equivalent to an impulse distribution of (9.1), with a uniform current I_{in} over an effective length $l_e = L/2$. Accordingly, all derivations in section 1 can be reused for the realistic current distribution of (9.25). This is accomplished by simply substituting the antenna length l in the results of section 1 by an effective length $l_e = L/2$. For example,

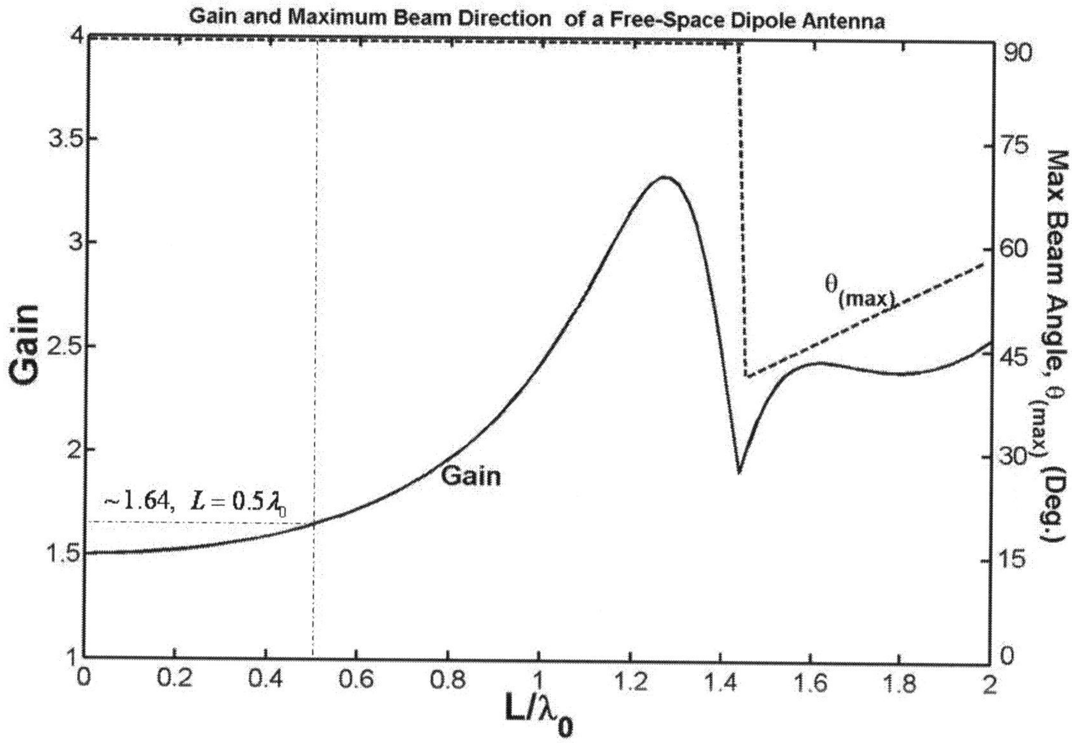

FIGURE 7

the radiated power P_{rad} and the radiation resistance R_{rad} may be expressed with the above substitution in (9.7) and (9.8), respectively.

$$R_{rad} = 20\pi^2(L/\lambda_0)^2, \; P_{rad} = 1/2|I_{in}|^2 R_{rad} = 10\pi^2(L/\lambda_0)^2|I_{in}|^2; \; L << \lambda_0. \tag{9.26}$$

These results could also be verified by approaching $L << \lambda_0$ in the general expressions of (9.21). The expressions for the directive gain function $D_g(\theta, \phi)$ and directivity D of section 1, which do not depend on the antenna length, would remain unchanged under substitution of the effective length.

Half-Wave Dipole Antenna

A specific design of a dipole antenna with $L = \lambda_0/2$, called a half-wave dipole, is often used because it operates close to a resonant condition of its input impedance. This results in an almost real value of the input impedance, which can be easily matched to an input source having a real source impedance for maximum power transfer. For this case, the above general expressions (9.19,9.20) can be simplified.

$$I(z) = I_{in}\cos(k_0 z) = I_{in}\cos\frac{2\pi z}{\lambda_0}, \tag{9.27}$$

$$\overline{E}(r,\theta,\phi) = \hat{\theta}\frac{e^{-jk_0r}}{r}\frac{j\eta_0 I_{in}}{2\pi}\Big(\frac{\cos(\frac{\pi}{2}\cos\theta)}{\sin\theta}\Big) = \hat{\theta}\frac{e^{-jk_0r}}{r}j60I_{in}\Big(\frac{\cos(\frac{\pi}{2}\cos\theta)}{\sin\theta}\Big). \tag{9.28}$$

The polarization vector is in the $\hat{\theta}$ direction, which is the same as that for a small dipole. The power density P_d and power per unit solid angle p_d can be expressed using (9.28).

$$P_d(r,\theta,\phi) = \frac{|E_\theta|^2}{2\eta_0} = \frac{15|I_{in}|^2}{\pi r^2}\Big(\frac{\cos(\frac{\pi}{2}\cos\theta)}{\sin\theta}\Big)^2,\ p_d(\theta,\phi) = \frac{15|I_{in}|^2}{\pi}\Big(\frac{\cos(\frac{\pi}{2}\cos\theta)}{\sin\theta}\Big)^2. \tag{9.29}$$

The direction of maximum radiation is along $\theta = \frac{\pi}{2}$, which is perpendicular to the direction of the antenna current. The power radiation pattern $\underline{p}_d(\theta,\phi)$ has the following form with respect to θ, having an uniform variation in the azimuth.

$$\underline{p}_d(\theta,\phi) = \frac{p_d(\theta,\phi)}{p_{dmax}} = \Big(\frac{\cos(\frac{\pi}{2}\cos\theta)}{\sin\theta}\Big)^2. \tag{9.30}$$

Now, the general expressions for radiated power P_{rad}, radiation resistance R_{rad}, directive gain $D_g(\theta,\phi)$, and directivity D for a dipole antenna of arbitrary length maybe simplified for a half-wave dipole antenna. This will require integrations involving the pattern function of (9.30).

$$P_{rad} = \int_{\phi=0}^{2\pi}\int_{\theta=0}^{\pi} p_d(\theta,\phi)\sin\theta\ d\theta\ d\phi = 30|I_{in}|^2\int_{\theta=0}^{\pi}\frac{\cos^2(\frac{\pi}{2}\cos\theta)}{\sin\theta}d\theta, \tag{9.31}$$

$$R_{rad} = \frac{2P_{rad}}{|I_{in}|^2} = 60\int_{\theta=0}^{\pi}\frac{\cos^2(\frac{\pi}{2}\cos\theta)}{\sin\theta}d\theta, \tag{9.32}$$

$$D_g(\theta,\phi) = \frac{4\pi\underline{p}_d(\theta,\phi)}{\int_{\phi=0}^{2\pi}\int_{\theta=0}^{\pi}\underline{p}_d(\theta,\phi)\sin\theta\ d\theta\ d\phi} = \frac{2\left(\frac{\cos(\frac{\pi}{2}\cos\theta)}{\sin\theta}\right)^2}{\int_{\theta=0}^{\pi}\frac{\cos^2(\frac{\pi}{2}\cos\theta)}{\sin\theta}\ d\theta}, \tag{9.33}$$

$$D = D_g(\theta,\phi)_{max} = \frac{2}{\int_{\theta=0}^{\pi}\frac{\cos^2(\frac{\pi}{2}\cos\theta)}{\sin\theta}\ d\theta}. \tag{9.34}$$

The θ-integral that identically appears in the above equations (9.31,9.32,9.33,9.34) may be computed numerically, which has an approximate value of 1.22. Using this value the above antenna parameters can be approximated as follows.

$$P_{rad} = 36.6|I_{in}|^2,\ D_g(\theta,\phi) = 1.64\left(\frac{\cos(\frac{\pi}{2}\cos\theta)}{\sin\theta}\right)^2,\ D = 1.64, R_{rad} = 73.2\Omega \tag{9.35}$$

The above results for the specific case of a half-wave dipole antenna maybe verified from the general data of Figs.5,7 with $L/\lambda_0 = 0.5$. Notice that the directivity D of a half-wave dipole ($= 1.64$) is not significantly different from that of a small dipole ($= 1.5$).

3. Dipole Antenna of Arbitrary Length: Rigorous Field Analysis

The modeling in the last section used radiation fields far from the antenna source. The far-field distributions only partially model the antenna characteristics, but are analytically much simpler compared to the rigorous field expressions that carry complete information about the antenna. Some antenna characteristics, such as radiation resistance, radiated power, pattern and directivity , could be derived from the far-field distributions, through relatively simple analysis. In contrast, antenna input reactance and resonance conditions, which are associated with the stored energy in the fields mostly confined in close vicinity of the antenna, can not be derived from a far-field analysis.

In this section, we will use a rigorous field model for a dipole antenna, and particularly use the field distribution near the antenna to compute the antenna impedance and resonance characteristics. The near-field analysis would provide a complete model of the antenna impedance, including both the real and impaginary parts. The real part may also be verified with the simple, far-field results from the last section. Whereas, the additional imaginary part of the impedance would be studied for its variation with frequency and antenna dimensions. Resonance conditions can be located where the antenna reactance would be zero, resulting in a resistive input impedance. The resonant conditions are particularly useful for practical designs, so that the antenna can be conveniently matched to a practical input source with a real source impedance, resulting in maximum power delivered to antenna radiation.

3.1. Strip-Dipole Model. We will model a thin dipole antenna of Fig.3 as a current strip of length L and an equivalent width w, as shown in Fig.8. We assume that the current on a narrow strip-dipole is along its length in the $\hat{z}$ direction, and would produce only TM_z fields. The TM_z potential and associated fields of a strip dipole can be modeled by following a similar analysis for a small-dipole element in chapter 6, section 3. In the derivation of chapter 6, section 3, we need to substitute the surface-current transform Il for a small dipole element by a new transform $\tilde{J}_z(k_x, k_z)$ for the strip dipole. Accordingly, the TM_z potential ψ_{zm} for a strip-dipole is expressed using (6.30), by substituting Il with $\tilde{J}_z(k_x, k_z)$.

$$\psi_{zm} = \frac{1}{4\pi^2} \int_{k_z=-\infty}^{+\infty} \int_{k_x=-\infty}^{+\infty} \frac{\tilde{J}_z(k_x, k_z)}{2jk_y} e^{jk_x x} e^{jk_z z} e^{\mp jk_y y} dk_x dk_z. \tag{9.36}$$

The $-$ or $+$ in the double-sign notation $\mp$ refers to $+$ ($y > 0$) or $-$ ($y < 0$) values of the observation point y, respectively. When k_y is real, which would be the case for $k_x^2 + k_z^2 < k_0^2$, it is assumed to be possitively real in order that the associated wave function $e^{\mp jk_y y}$ is propagating away from the source in the $\pm y$ directions. On the other hand, when k_y is an imaginary number, which would be the case for $k_x^2 + k_z^2 > k_0^2$, it needs to be negatively imaginary in order that the wave function $e^{\mp jk_y y}$ is decaying away from the source in the $\pm y$ directions (chapter 5, section 2.3).

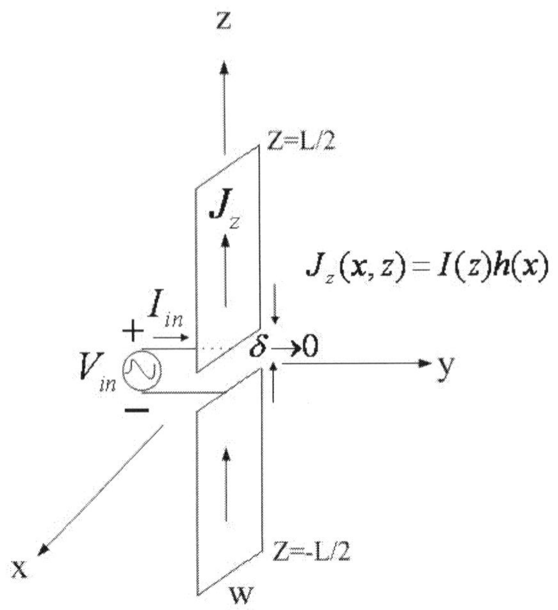

$$h(x,w) = \frac{1}{w}, \ |x| \leq w/2$$
$$= 0, \ |x| > w/2$$

FIGURE 8

$$\begin{aligned} k_y &= \sqrt{k_0^2 - k_x^2 - k_z^2}, \ Re(k_y) > 0, \ Im(k_y) < 0, \\ &= \sqrt{k_0^2 - k_x^2 - k_z^2}, \ k_x^2 + k_z^2 < k_0^2; \\ &= -j\sqrt{k_x^2 + k_z^2 - k_0^2}, \quad k_x^2 + k_z^2 > k_0^2. \end{aligned} \tag{9.37}$$

The z-directed electric field E_z is expressed using the potential function ψ_{zm} of (9.36) through the field-potential relation (5.34).

$$E_z = \frac{1}{j\omega\epsilon}(k_0^2 + \frac{\partial^2}{\partial z^2})\psi_{zm}$$
$$= \frac{1}{j\omega\epsilon} \times \frac{1}{4\pi^2} \int_{k_z} \int_{k_x} \frac{\tilde{J}_z(k_x, k_z)}{2jk_y}(k_0^2 - k_z^2) e^{jk_x x} e^{jk_z z} e^{\mp jk_y y} dk_x dk_z. \tag{9.38}$$

Assume that the current variation in the x direction is uniform over the width w of the strip. This is a reasonable assumption for a narrow strip ($w << \lambda_0$). The current variation along the z direction of a strip dipole would be same as that for a thin wire dipole, as expressed in (9.19). With these functional variations in x and z, we can derive the current transform function $\tilde{J}_z(k_x, k_z)$.

$$J_z(x,z) = I(z)h(x),\ I(z) = \int_{x=-w/2}^{+w/2} J_z(x,z)dx,\ \int_{x=-w/2}^{+w/2} h(x)dx = 1$$

$$h(x) = \frac{1}{w},\ |x| < \frac{w}{2};\ = 0,\ |x| > \frac{w}{2};$$

$$I(z) = I_{in}\frac{\sin[k(\frac{L}{2} - |z|)]}{\sin(\frac{k_0 L}{2})},\ |z| < L/2;\ = 0,\ |z| > L/2;$$

$$I(z) = I_{in}f(z),\ I(z=0) = I_{in},\ f(z=0) = 1. \tag{9.39}$$

$$\tilde{J}_z(k_x,k_z) = \int_{z=-L/2}^{L/2}\int_{x=-w/2}^{w/2} J_z(x,z)e^{-jk_x x}e^{-jk_z z}dxdz = I_{in}F(k_z)H(k_x), \tag{9.40}$$

$$H(k_x) = \int_{x=-w/2}^{w/2} h(x)e^{-jk_x x}dx = \frac{1}{w}(\frac{e^{-jk_x\frac{w}{2}} - e^{+jk_x\frac{w}{2}}}{-jk_x}) = \frac{\sin(k_x\frac{w}{2})}{(k_x\frac{w}{2})}, \tag{9.41}$$

$$F(k_z) = \int_{z=-L/2}^{L/2} f(z)e^{-jk_z z}dz = 2\int_0^{\frac{L}{2}} f(z)\cos(k_z z)dz$$

$$= \frac{2}{\sin(\frac{k_0 L}{2})}\int_0^{\frac{L}{2}} \sin[k_0(\frac{L}{2} - z)]\cos(k_z z)dz$$

$$= \frac{1}{\sin(\frac{k_0 L}{2})}\int_0^{\frac{L}{2}} [\sin(\frac{k_0 L}{2} - k_0 z + k_z z) + \sin(\frac{k_0 L}{2} - k_0 z - k_z z)]dz$$

$$= \frac{1}{\sin(\frac{k_0 L}{2})}\Big[\frac{\cos(\frac{k_0 L}{2} - (k_0 - k_z)z)}{(k_0 - k_z)} + \frac{\cos(\frac{k_0 L}{2} - (k_0 + k_z)z)}{(k_0 + k_z)}\Big]_0^{\frac{L}{2}}$$

$$= \frac{1}{\sin(\frac{k_0 L}{2})}[\frac{\cos(\frac{k_z L}{2}) - \cos(\frac{k_0 L}{2})}{(k_0 - k_z)} + \frac{\cos(\frac{k_z L}{2}) - \cos(\frac{k_0 L}{2})}{(k_0 + k_z)}]$$

$$= \frac{2k_0}{\sin(\frac{k_0 L}{2})}[\frac{\cos(\frac{k_z L}{2}) - \cos(\frac{k_0 L}{2})}{(k_0^2 - k_z^2)}]. \tag{9.42}$$

Now, the complex power P drawn from the antenna current maybe expressed by integrating the elemental power $dP = (1/2)\triangle V \triangle I^* = -(1/2)E_z(y=0)J_z^* dxdz$ over the strip surface. Using the E_z expression of (9.38) and the current expresions of (9.39-9.41) the above power integral in the space variables x and z can be simpified as an integral over the transform variables k_x and k_z.

$$\begin{aligned} dP &= \tfrac{1}{2}\triangle V \triangle I^* = \tfrac{1}{2}(-E_z(y=0)dz)(J_z^* dx) \\ &= -\tfrac{1}{2}E_z(y=0)J_z^* dxdz, \end{aligned} \tag{9.43}$$

$$\begin{aligned} P &= \int_z\int_x dP = -\frac{1}{2}\int_z\int_x E_z(x,y=0,z)J_z^*(x,z)dxdz \\ &= \frac{1}{16\pi^2\omega\epsilon}\int_{k_z}\int_{k_x} \frac{I_{in}F(k_z)H(k_x)}{k_y}(k_0^2-k_z^2) \\ &\times\Big(\int_z\int_x I_{in}^* f(z)h(x)e^{+jk_x x}e^{+jk_z z}dxdz\Big)dk_x dk_z \\ &= \frac{1}{16\pi^2\omega\epsilon}\int_{k_z}\int_{k_x} \frac{I_{in}F(k_z)H(k_x)}{k_y}(k_0^2-k_z^2)I_{in}^* F(-k_z)H(-k_x)dk_x dk_z \\ &= \frac{|I_{in}|^2}{16\pi^2\omega\epsilon}\int_{k_z}\int_{k_x} \frac{(k_0^2-k_z^2)}{k_y}F^2(k_z)H^2(k_x)dk_x dk_z. \end{aligned} \tag{9.44}$$

The source power is equated with the input power at the antenna feed terminals, expressed in terms of input current I_{in} and input impedance Z_{in}. Accordingly, the input impedance of the antenna can be expressed using the power integral of (9.44) and the current transform of (9.40-9.42).

$$\begin{aligned} P &= \frac{1}{2}V_{in}I_{in}^* = \frac{1}{2}Z_{in}\left|I_{in}\right|^2, \; Z_{in} = 2P/|I_{in}|^2, \\ Z_{in} &= \frac{1}{8\pi^2\omega\epsilon}\int_{k_z}\int_{k_x} \frac{(k_0^2-k_z^2)}{k_y}F^2(k_z)H^2(k_x)dk_x dk_z, \end{aligned} \tag{9.45}$$

where k_y is expressed by the square-root expression of (9.37).

It may be convenient to implement the above integration in polar coordinates, by defining suitable polar transform variables k_α and α.

$$Z_{in} == R_{in} + jX_{in} = R_{rad} + jX_{in}$$

$$= \frac{1}{8\pi^2\omega\epsilon} \int\limits_{k_\alpha=0}^{\infty} \int\limits_{\alpha=0}^{2\pi} \frac{(k_0^2 - k_\alpha^2 \sin^2\alpha)}{\sqrt{k_0^2 - k_\alpha^2}} F^2(k_\alpha \sin\alpha) H^2(k_\alpha \cos\alpha) k_\alpha d\alpha dk_\alpha,$$

$$k_x = k_\alpha \cos\alpha,\ k_z = k_\alpha \sin\alpha,\ k_\alpha^2 = k_x^2 + k_z^2. \tag{9.46}$$

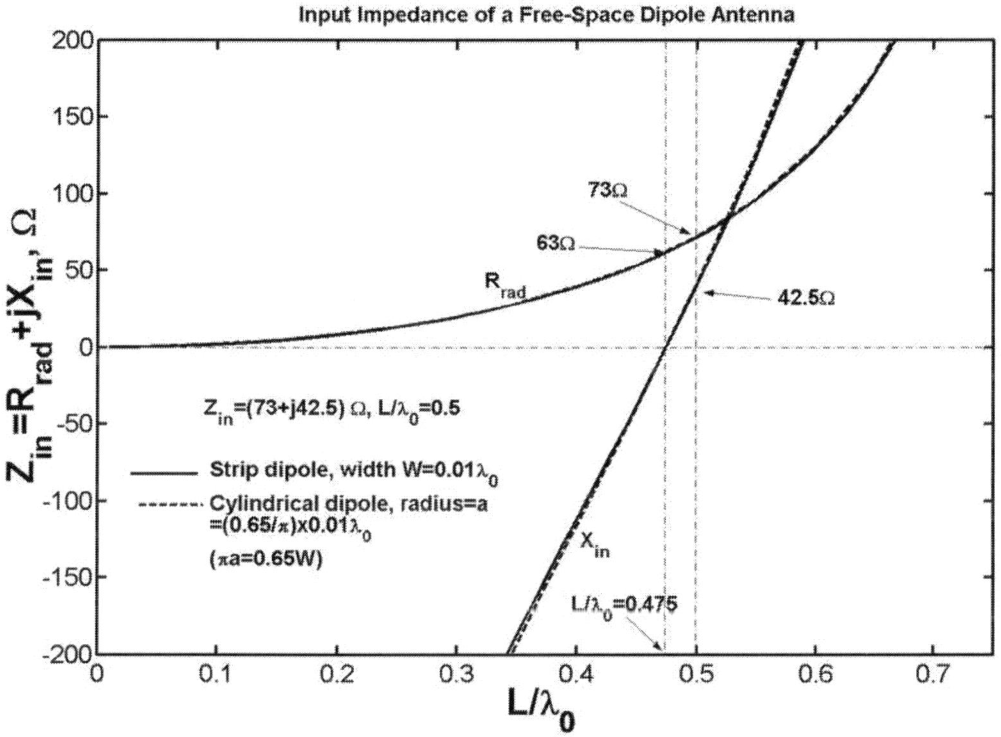

FIGURE 9

The real part ($= R_{in}$) of the input impedance would come only from integration range $0 < k_\alpha < k_0$, where $k_y = \sqrt{k_0^2 - k_\alpha^2}$ is real. On the other hand, the rest of the integration (9.46) for $k_\alpha > k_0$, where $k_y = \sqrt{k_0^2 - k_\alpha^2}$ is imaginary, would contribute only to the imaginary part of the impedance ($= X_{in}$). Proper choice of the sign for k_y is defined in (9.37). We assume that there is no material loss in the antenna, in which case R_{in} is contributed only by the power radiated to the outside space, and is therefore equal to the radiation resistance R_{rad} of the antenna.

R_{rad} and X_{in}, as computed using (9.46) for a strip diple, are plotted in Fig.9 for $w = 0.01\lambda_0$. R_{rad} maybe verified with resuts in Fig.5 derived from a far-field analysis in section 2. R_{rad} is essentially independent of the width w for narrow strips. When $L = 0.5\lambda_0$, where R_{rad} in Fig.9 is about 73Ω, the input reactance X_{in} is about 42.5Ω.

The antenna in Fig.9, with $w = 0.01\lambda_0$, is seen to resonate with $X_{in} = 0$ when the length is about $0.475\lambda_0$. This is somewhat less than an ideal resonant length of $0.5\lambda_0$, one might expect based on an ideal transmission line model of the antenna. Each half of the antenna maybe seen at the input as an open-ended, transmission-line stub, resulting in a zero input reactance when each half is $0.25\lambda_0$ (see chapter 4), or equivalently when the total antenna length is $0.5\lambda_0$. Deviation from the ideal condition maybe viewed in terms of any fringing fields at the open ends of the antenna, which would effectively extend the antenna length somewhat larger than the actual physical length. Equivalently, this would result in having a resonant physical length of the antnna somewhat smaller than the expected ideal value of $0.5\lambda_0$, as observed in Fig.9.

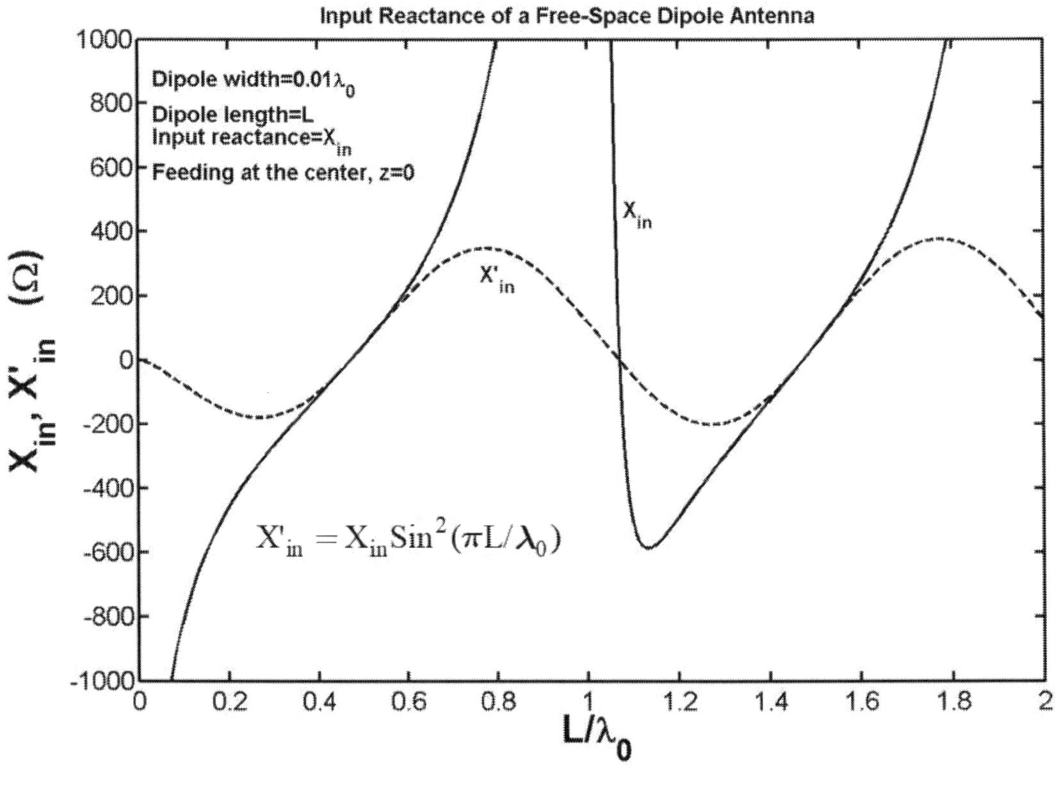

Figure 10

X_{in}, and a modified parameter $X'_{in} = X_{in} \sin^2(k_0 L/2)$, are plotted over a broader range of $0 < L/\lambda_0 < 2$ in Fig.10, showing additional resonance around $L \sim 1.5\lambda_0$. X_{in} approaches infinity when the length is an integral multiple of λ_0, whereas X'_{in} is finite at these locations. X'_{in} is the reactive counterpart to the modified resistance R'_{rad}, discussed in section 2 and plotted in Fig.5. Unlike R_{rad}, the antenna reactance X_{in} or X'_{in} is strongly dependent on the strip width w. X'_{in} for different values of w are plotted in Fig.11, showing stronger variation of the reactance with L/λ_0 for narrower strip width.

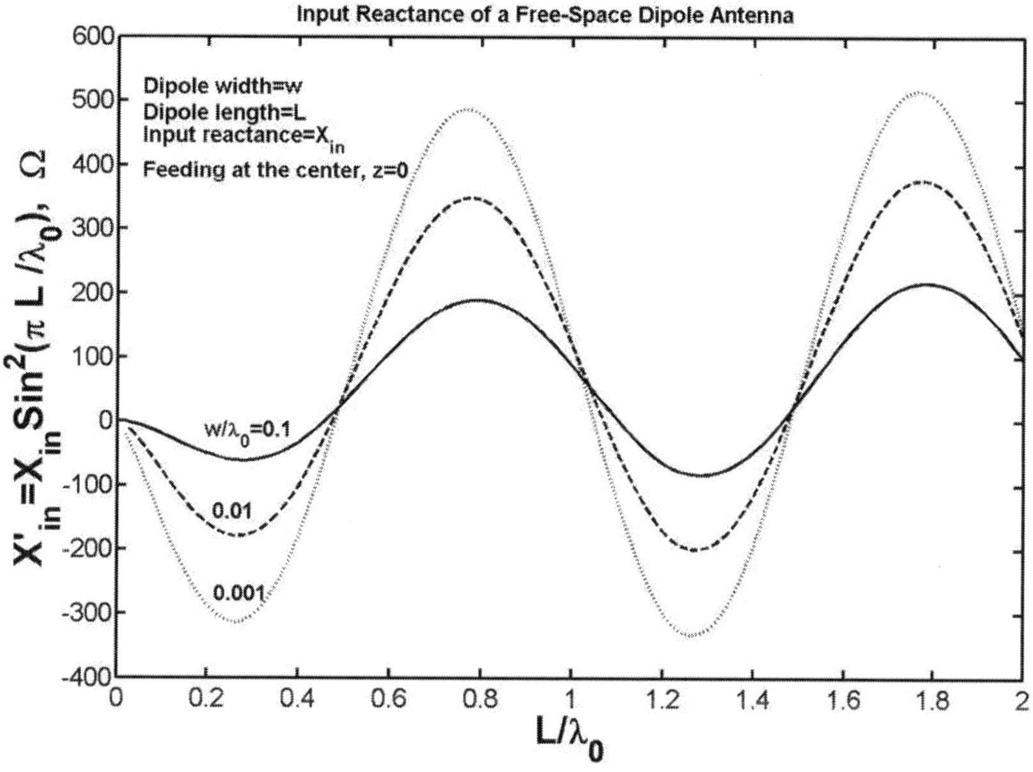

FIGURE 11

3.2. Cylindrical Dipole Model. In this section we will model a thin dipole antenna of Fig.3 by a cylindrical current distribution of radius a and length L, as shown in Fig.12..

$$J_z(z, \phi) = \frac{I(z)}{2\pi a}; \ \rho = a, \ 0 < \phi < 2\pi. \tag{9.47}$$

where $I(z)$ is the z variation of the current, as expressed in (9.19). When the dipole antenna is sufficiently thin, with $a << \lambda_0$, the above cylindrical current distribution may be replaced by a line current I_z at $\rho = 0$, for analytical convenience. The field produced by this equivalent line source is expected to accurately model the actual field produced by the original cylindrical current, over the region $\rho \geq a$. The equivalent line source maybe modeled as a surface-current distribution $J'_z(x, z)$, as expressed in (9.39) for a strip dipole but with its width w approaching zero. Accordingly, the corresponding current transform $\tilde{J}'(k_x, k_z)$ can be expressed by (9.40) with $w \to 0$. This is equivalent to replacing $H(k_x)$ in (9.41) with its limit for $w \to 0$, which is equal to 1.

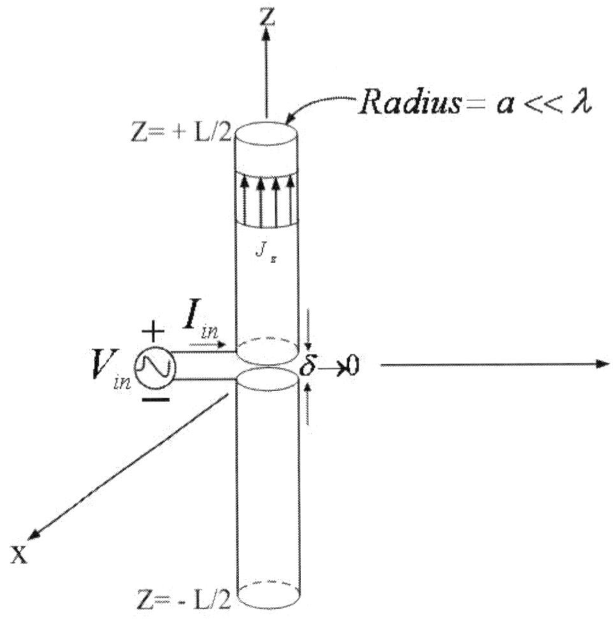

$$J_z = I(z)\frac{1}{2\pi a};\ \rho = a,\ 0 < \phi < 2\pi$$

FIGURE 12

$$\begin{aligned} J'_z(x,z) = I(z)h(x;w\to 0) \sim I(z)\delta(x) = I_{in}f(z)\delta(x),\ y=0, \\ \tilde{J}'_z(k_x,k_z) = \tilde{I}(k_z)H(k_x;w\to 0) = I_{in}F(k_z), \\ H(k_x;w\to 0) = \left.\frac{\sin(k_x w/2)}{(k_x w/2)}\right|_{w\to 0} = 1. \end{aligned} \quad (9.48)$$

We will now derive the electric field E_z produced by the equivelent source $J'_z(x,z)$, which would be used as a good approximation for the actual field, as explained earlier. The field expression of (9.38) can be used, by substituting the transform $\tilde{J}_z(k_x,k_z)$ with its limiting form of (9.48).

$$E_z = \frac{1}{j\omega\epsilon} \times \frac{1}{4\pi^2} \int\limits_{k_z} \int\limits_{k_x} \frac{I_{in}F(k_z)}{2jk_y}(k_0^2 - k_z^2)e^{jk_x x}e^{jk_z z}e^{\mp jk_y y}dk_x dk_z. \quad (9.49)$$

The above field is produced by the line current source of (9.48), which is cylindrically symmetric. The field should share the cylindrical symmetry of the current source, and therefore be independent of ϕ. Accordingly, the field evaluated at a general location $(\rho = a, \phi, z)$, on the surface of the cylindrical antenna, maybe equated to that at $(x = a, y = 0, z)$.

$$\begin{aligned} E_z(\rho = a, \phi, z) &= E_z(x = a, y = 0, z) \\ &= \frac{-1}{8\pi^2\omega\epsilon} \int\limits_{k_z}\int\limits_{k_x} \frac{I_{in}F(k_z)}{k_y}(k_0^2 - k_z^2)e^{jk_x a}e^{jk_z z}dk_x dk_z. \end{aligned} \tag{9.50}$$

The complex power delivered from the cylindrical source of (9.47) can be derived using a formulation similar to (9.43,9.44) used for a strip dipole. The above condition (9.50) of cylindrical symmetry for E_z maybe used for simplification.

$$\begin{aligned} dP &= 1/2\triangle V\triangle I^* = 1/2(-E_z(\rho = a)dz)(J_z^* a d\phi) \\ &= -1/2E_z(\rho = a)J_z^* a d\phi dz, \end{aligned} \tag{9.51}$$

$$P = \int\limits_{z=-\frac{L}{2}}^{\frac{L}{2}} \int\limits_{\phi=0}^{2\pi} -\frac{1}{2}E_z(\rho = a, \phi, z)J_z^* a d\phi dz \quad (J_z = \frac{I(z)}{2\pi a})$$

$$= -\frac{1}{2}\int\limits_{z=-\frac{L}{2}}^{\frac{L}{2}} E_z(\rho = a, \phi = 0, z)I^*(z)dz = -\frac{1}{2}\int\limits_{z=-L/2}^{L/2} E_z(x = a, y = 0, z)I^*(z)dz$$

$$= -\frac{1}{2}\int\limits_{z=-L/2}^{L/2} E_z(x = a, y = 0, z)I_{in}^* f(z)dz, \quad I^*(z) = I_{in}^* f(z). \tag{9.52}$$

Using the E_z expression of (9.50), the above power integral in space variable z can be simpified as an integral over transform variables k_x and k_z.

$$\begin{aligned} P &= -\frac{1}{2}\int\limits_{z=-\frac{L}{2}}^{+\frac{L}{2}} \frac{-1}{8\pi^2\omega\epsilon}\int\limits_{k_z}\int\limits_{k_x} \frac{I_{in}F(k_z)}{k_y}(k_0^2 - k_z^2)e^{jk_x a}e^{jk_z z}dk_x dk_z I_{in}^* f(z)dz \\ &= \frac{|I_{in}|^2}{16\pi^2\omega\epsilon}\int\limits_{k_z}\int\limits_{k_x} \frac{F(k_z)}{k_y}(k_0^2 - k_z^2)e^{jk_x a}\Big(\int\limits_{z=-L/2}^{L/2} f(z)e^{jk_z z}dz\Big)\, dk_x dk_z \\ &= \frac{|I_{in}|^2}{16\pi^2\omega\epsilon}\int\limits_{k_z}\int\limits_{k_x} \frac{F(k_z)}{k_y}(k_0^2 - k_z^2)e^{jk_x a}F(-k_z)dk_x dk_z. \end{aligned} \tag{9.53}$$

As in (9.45) for a strip dipole, the source power is equated with the input power at the antenna feed terminals, expressed in terms of input current I_{in} and input impedance Z_{in}. The input impedance is then expressed using the power integral of (9.53). The current transform $F(k_z)$ is the same transform function (9.42) used also for a strip dipole.

$$P = \frac{1}{2}V_{in}I_{in}^* = \frac{1}{2}|I_{in}|^2 Z_{in}, \; Z_{in} = \frac{2P}{|I_{in}|^2}, \tag{9.54}$$

$$\begin{aligned} Z_{in} = \frac{2P}{|I_{in}|^2} &= \frac{1}{8\pi^2\omega\epsilon}\int\limits_{k_z}\int\limits_{k_x}\frac{(k_0^2-k_z^2)}{k_y}F^2(k_z)e^{jk_xa}dk_xdk_z \\ &= \frac{1}{8\pi^2\omega\epsilon}\int\limits_{k_z}\int\limits_{k_x}\frac{(k_0^2-k_z^2)}{\sqrt{k_0^2-k_x^2-k_z^2}}F^2(k_z)e^{jk_xa}dk_xdk_z. \end{aligned} \tag{9.55}$$

The term e^{jk_xa} maybe decomposed to cosine and sine parts, $e^{jk_xa} = \cos(k_xa) + j\sin(k_xa)$. Considering the even symmetry of the rest of the integrand for k_x, the the odd symmetry of the sine part would not contribute to the total integration. Accordingly, the above integration (9.55) maybe expressed by replacing e^{jk_xa} with $\cos(k_xa)$.

$$Z_{in} = \frac{1}{8\pi^2\omega\epsilon}\int\limits_{k_z}\int\limits_{k_x}\frac{(k_0^2-k_z^2)}{\sqrt{k_0^2-k_x^2-k_z^2}}F^2(k_z)\cos(k_xa)dk_xdk_z. \tag{9.56}$$

The above integration may also be expressed in polar coordinates, by defining suitable polar transform variables k_α and α. Similar implementation was also used in (9.46) for a strip dipole.

$$\begin{aligned} Z_{in} &== R_{in} + jX_{in} = R_{rad} + jX_{in} \\ = \frac{1}{8\pi^2\omega\epsilon}\int\limits_{k_\alpha=0}^{\infty}\int\limits_{\alpha=0}^{2\pi} &\frac{(k_0^2-k_\alpha^2\sin^2\alpha)}{\sqrt{k_0^2-k_\alpha^2}}F^2(k_\alpha\sin\alpha)\cos(k_\alpha\cos\alpha a)k_\alpha d\alpha dk_\alpha, \\ k_x &= k_\alpha\cos\alpha, \; k_z = k_\alpha\sin\alpha, \; k_\alpha^2 = k_x^2+k_z^2. \end{aligned} \tag{9.57}$$

As in (9.46) for a strip dipole, the real part ($= R_{in}$) of the input impedance in (9.57) would come only from integration range $0 < k_\alpha < k_0$, where $k_y = \sqrt{k_0^2 - k_\alpha^2}$ is real. On the other hand, the rest of the integration (9.57) for $k_\alpha > k_0$, where $k_y = \sqrt{k_0^2 - k_\alpha^2}$ is imaginary, would contribute only to the imaginary part of the impedance ($= X_{in}$). Proper choice of the sign for k_y is defined in (9.37). Assuming that there is no material loss in the antenna, R_{in} is contributed only by the power radiated to the outside space, and therefore is equal to the radiation resistance R_{rad} of the antenna.

R_{rad} and X_{in}, as computed using (9.57) for a cylindrical dipole with radius $a = 0.065\lambda_0$, are compared in Fig.9 with data for a strip dipole with $w = 0.01\lambda_0$. R_{rad} may also be verified with the results in Fig.5, derived in section 2 from a far-field analysis

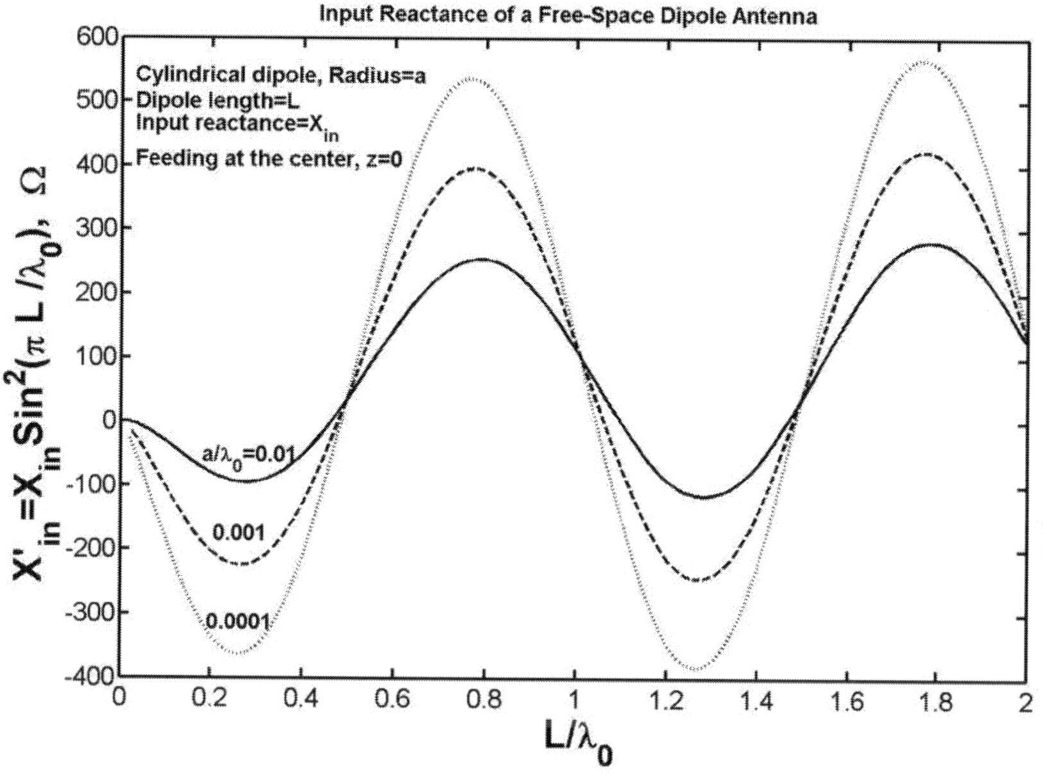

FIGURE 13

of a thin dipole. R_{rad} is found to be essentially independent of the radius a for a thin cylindrical dipole, which is equivalent to independence with width w for a strip dipole.

The resonant behavior of the reactance X_{in} for a cylindrical dipole in Fig.9, around $L = 0.5\lambda_0$, is similar to that for a strip dipole. Unlike the resistance R_{rad}, the reactance X_{in} is found to be strongly dependent on the radius a. $X'_{in} = X_{in} \sin^2(k_0 L/2)$ for a cylindrical dipole are plotted in Fig.13 for different radius a, showing stronger variation of the reactance with L/λ_0 for smaller radius a. This behavior maybe compared with Fig.11 for a strip dipole with different width w. The variation of X_{in} as a function of radius a is plotted in Fig.14 for $L = 0.25\lambda_0$. This is also compared with those for a strip dipole with $w = \pi a/0.65$ and $w = \pi a/0.70$. The results in Fig.14 suggest that a strip dipole of width w maybe considered equivalent to a cylindrical dipole of radius a, when $\pi a/w$ is about 0.65 or 0.7.

3.3. Mutual Coupling Between Dipole Antennas. Consider a situation, where two antenna elements are operating in the presence of each other, as shown in Fig.15. We consider here a strip-dipole model for each element. A cylindrical element may be modeled as a strip element with an equivalent width, as discussed in the last section. The first element in Fig.15 is centered at the origin of the coordinate system, while the second element is placed on the plane $y = y_0$, centered at $x = x_0$ and $z = z_0$. The currents J_{z1}

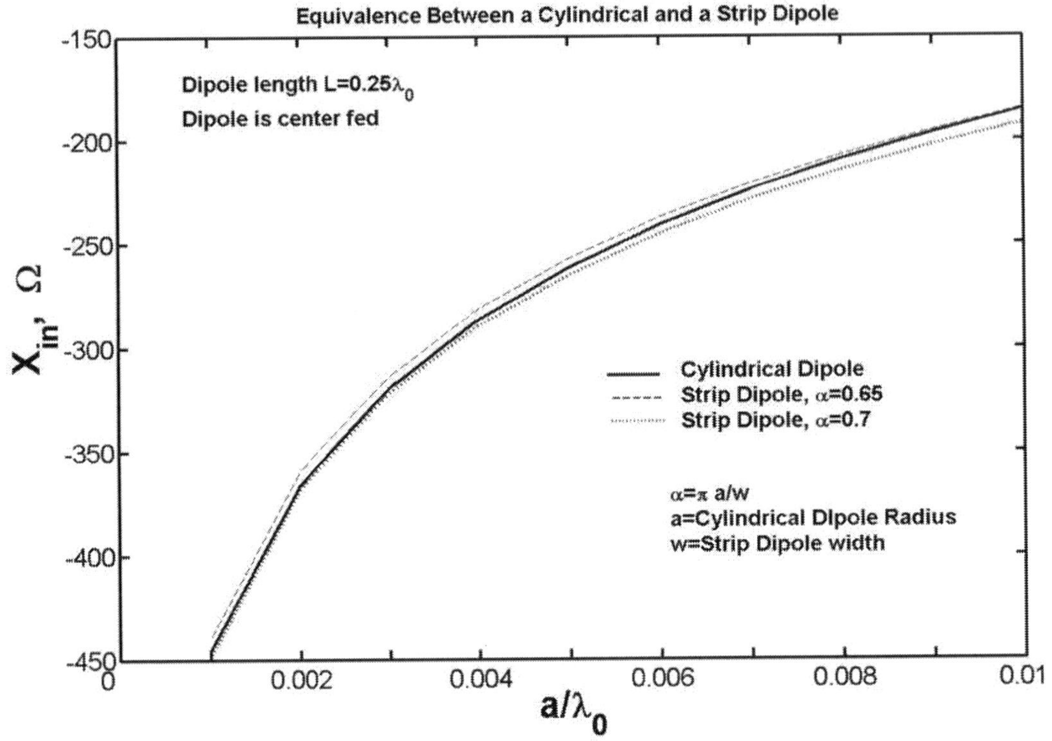

FIGURE 14

or J_{z2} on the two elements can be expressed using (9.39), by properly substituting the input current I_{in} by I_1 or I_2, L and w by the respective values for the particular element, and shifting the center coordinate for the second element to (x_0, z_0). Accordingly, the respective transforms $\tilde{J}_{z1}$ and $\tilde{J}_{z1}$ can then be expressed in terms of $\tilde{J}_z$ in (9.40) using suitable Fourier-transform relations.

$$\begin{aligned} J_{z1}(x,z) &= I_1 f(z;L_1)h(x;w_1),\ y=0, \\ J_{z2}(x,z) &= I_2 f(z-z_0;L_2)h(x-x_0;w_2),\ y=y_0. \\ \tilde{J}_{z1}(k_x,k_z) &= I_1 F(k_z;L_1)H(k_x;w_1), \\ \tilde{J}_{z2}(k_x,k_z) &= I_2 F(k_z;L_2)e^{-jk_z z_0}H(k_x;w_2)e^{-jk_x x_0}. \end{aligned} \tag{9.58}$$

If each antenna was operating as an isolated, single element, its input voltage would be equal to the product of its input current and input impedance, which we derived in the last two sections. When there are two elements, as in Fig.15, there would be an additional voltage at the input of each element, contributed by the fields generated from the other antenna. The additional voltage would be proportional to the input current at the other antenna, with the factor of proportionality defined as the mutual impedance Z_m. It can be shown by reciprocity theorem (chapter 2, section 4.3) that the mutual impedance

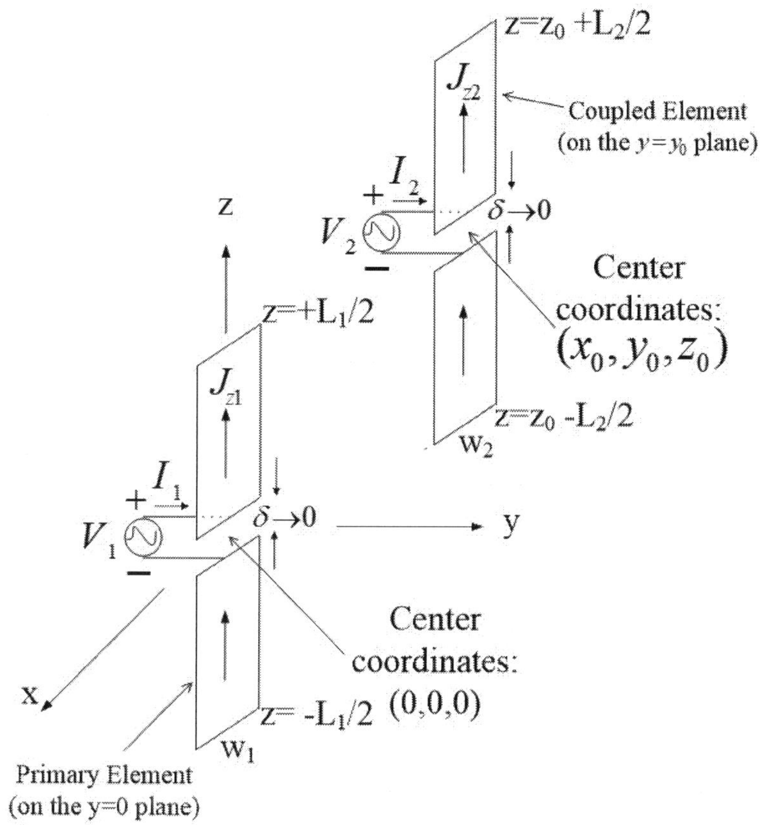

FIGURE 15

Z_{21} for coupling from element #1 to element #2 is same as Z_{12} from element #2 to #1. Accordingly, both the mutual impedances are represented by a common variable $Z_m = Z_{12} = Z_{21}$. Consider the voltage V_2 seen at the input of the second antenna element in Fig.15.

$$V_2 = I_2 Z_{in2} + I_1 Z_m. \tag{9.59}$$

The complex power at the input of the second antenna would also be different from that $(= 1/2|I_2|^2 Z_{in2})$ when the second antenna is operating alone. The additional complex power $\triangle P = P_{21}$ can be related to the mutual impedance Z_m.

$$P_2 = 1/2 V_2 I_2^* = 1/2|I_2|^2 Z_{in2} + 1/2 I_1 I_2^* Z_m = 1/2|I_2|^2 Z_{in2} + 1/2 P_{21},$$
$$P_{21} = \tfrac{1}{2} I_1 I_2^* Z_m, \; Z_m = \tfrac{2P_{21}}{I_1 I_2^*}. \tag{9.60}$$

P_{21} is the additional complex power generated from the element #2, contributed by the field due to the element #1. P_{21} can be derived by a similar procedure used in section 3.1. The equation (9.44) can be used to model P_{21} by properly substituting the electric field E_z with E_{z1} produced by the element #1, and the current J_z with J_{z2} on element #2.

$$P_{21} = \frac{1}{2}\int_z\int_x -E_{z1}(x, y = y_0, z)J^*_{z2}(x, z)dxdz$$
$$= \frac{1}{2}\int_z\int_x -E_{z1}(x, y = y_0, z)I^*_2 f(z; L_2)h(x; w_2)dxdz. \quad (9.61)$$

The field E_{z1} can be expressed using (9.38), by substituting the current J_z with J_{z1} on element #1.

$$E_{z1}(x, y, z) = \frac{-1}{8\pi^2\omega\epsilon}\int_{k_z}\int_{k_x} \frac{I_1F(k_z; L_1)H(k_x; w_1)}{k_y}(k_0^2 - k_z^2)$$
$$e^{jk_xx}e^{jk_zz}e^{\mp jk_yy}dk_xdk_z. \quad (9.62)$$

As in (9.38), the $(-)$ or $(+)$ sign in $e^{\mp jk_yy}$ refer to positive or negative value of the observation point y. The above E_{z1} in (9.62) can be used to simplify the spatial integral of (9.61) in the form of an integral in transform variables k_x and k_z.

$$P_{21} = \frac{1}{16\pi^2\omega\epsilon}\int_{k_x}\int_{k_z} \frac{I_1F(k_z; L_1)H(k_x; w_1)}{k_y}(k_0^2 - k_z^2)e^{\mp jk_yy_0}$$
$$\Big(\int_z\int_x I^*_2 f(z - z_0; L_2)h(x - x_0; w_2)e^{+jk_xx}e^{+jk_zz}dxdz\Big)dk_xdk_z$$
$$= \frac{1}{16\pi^2\omega\epsilon}\int_{k_z}\int_{k_x} \frac{I_1F(k_z, L_1)H(k_x; w_1)}{k_y}(k_0^2 - k_z^2)e^{\mp jk_yy_0}$$
$$I^*_2F(-k_z; L_1)H(-k_x; w_1)e^{+jk_xx_0}e^{+jk_zz_0}dk_xdk_z. \quad (9.63)$$

Transform definitions in (9.41,9.42), with suitable Fourier-transform relations, are used in the above simplification.

The above expression for P_{12} can now be used in (9.60) to obtain the mutual impedance Z_m.

$$Z_m = \frac{2P_{21}}{I_2^* I_1}$$

$$= \frac{1}{8\pi^2 \omega\epsilon} \int\limits_{k_z} \int\limits_{k_x} \frac{(k_0^2 - k_z^2)}{k_y} F(k_z; L_1) H(k_x; w_1) F(k_z; L_2) H(k_x; w_2)$$

$$e^{\mp jk_y y_0} e^{+jk_z z_0} e^{+jk_x x_0} dk_x dk_z$$

$$= \frac{1}{8\pi^2 \omega\epsilon} \int\limits_{k_z} \int\limits_{k_x} \frac{(k_0^2 - k_z^2)}{k_y} F(k_z; L_1) H(k_x; w_1) F(k_z; L_2) H(k_x; w_2)$$

$$e^{\mp jk_y y_0} \cos(k_z z_0) \cos(k_x x_0) dk_x dk_z. \quad (9.64)$$

The transform functions $H(k_x)$ and $F(k_z)$ are even-symmetric for positive and negative values of k_x and k_z, respectively. The terms $e^{jk_x x_0}$ and $e^{jk_z z_0}$ maybe decomposed to cosine and sine parts, such as $e^{jk_x x_0} = \cos(k_x x_0) + j\sin(k_x x_0)$, of which the sine parts exhibit odd symmetry for positive and negative values of k_x or k_z. Considering even symmetry of the rest of the integrand for k_x and k_z, the the odd symmetry of the sine parts would not contribute to the total integration. Accordingly, the above expression (9.64) is expressed by replacing $e^{jk_x x_0}$ and $e^{jk_z z_0}$ with $\cos(k_x x_0)$ and $\cos(k_z z_0)$, respectively.

Consistent with (9.62), the choice of $(-)$ or $(+)$ sign in the $e^{\mp jk_y y_0}$ term refers to positive or negative value for the coordinate y_0. k_y used in the above integral (9.64) is expressed in (9.37), with proper choice of its sign when it is real or imaginary. The above choices represent physically valid fields, with signal propagation or exponential decay away from the source.

For a simple case with $L_1 = L_2 = L$ and $w_1 = w_2 = w$, the integral for Z_m would appear in a form similar to Z_{in} in (9.45), except with additional factors $e^{\mp jk_0 y_0} \cos(k_z z_0) \cos(k_x x_0)$.

$$Z_m = \frac{1}{8\pi^2 \omega\epsilon} \int\limits_{k_x} \int\limits_{k_x} \frac{(k_0^2 - k_z^2)}{k_y} F^2(k_z; L) H^2(k_x; w)$$

$$e^{\mp jk_y y_0} \cos(k_z z_0) \cos(k_x x_0) dk_x dk_z. \quad (9.65)$$

In this case the mutual impedance Z_m would be equal to Z_{in}, if the distance between the elements is zero, or equivalently $x_0 = y_0 = z_0 = 0$.

For convenience, the integrals in (9.64,9.65) may also be expressed in polar form, with suitable choice of polar variables (k_α, α). This is similar to the implementation in (9.46).

$$Z_m = \frac{1}{8\pi^2\omega\epsilon} \int_{k_\alpha=0}^{\infty} \int_{\alpha=0}^{2\pi} \frac{(k_0^2 - k_z^2)}{k_y} F(k_z; L_1) H(k_x; w_1) F(k_z; L_2) H(k_x; w_2)$$

$$e^{\mp jk_y y_0} \cos(k_z z_0) \cos(k_x x_0) k_\alpha d\alpha dk_\alpha,$$

$$Z_m = \frac{1}{8\pi^2\omega\epsilon} \int_{k_\alpha=0}^{\infty} \int_{\alpha=0}^{2\pi} \frac{(k_0^2 - k_z^2)}{k_y} F^2(k_z; L) H^2(k_x; w)$$

$$e^{\mp jk_y y_0} \cos(k_z z_0) \cos(k_x x_0) k_\alpha d\alpha dk_\alpha; \; L_1 = L_2 = L, \; w_1 = w_2 = w,$$

$$k_x = k_\alpha \cos\alpha, \; k_z = k_\alpha \sin\alpha, \; k_y = \sqrt{k_0^2 - k_\alpha^2}. \qquad (9.66)$$

The above integration in the range $0 < k_\alpha < k_0$, where k_y is real, maybe seen to to contribute both real and imaginary parts for Z_m. On the other hand, integration outside the above range with $k_\alpha > k_0$, where k_y is imaginary, would contribute only imaginary part for Z_m. Further, in the range $k_\alpha > k_0$ the wave would be evanescent, with the term $e^{\mp jk_y y_0}$ exponentially decaying with $|y_0|$. Accordingly, integration in this evanescent range would result in a negligible contribution to the mutual coupling Z_m, for a sufficiently large distance y_0. Therefore, when y_0 is sufficiently large, one may need to implement the above integration only over the range $0 < k_\alpha < k_0$, where the wave is propagating. In other words, the coupling over large distances occur only through propagating waves, with negligible or no coupling through evanescent waves.

Figs.16,17 show the mutual impedance $Z_m = Z_{12} = Z_{21}$ between two identical, half-wave strip dipoles, computed from numerical integration of (9.66). Fig.16 shows the mutual impedance as a function of distance $d = x_0$, with $z_0 = y_0 = 0$, whereas Fig.17 is the corresponding data as a function of end-to-end distance $\delta = z_0 - L$, with $x_0 = y_0 = 0$. For a narrow dipole, the mutual impedance as a function of $d = y_0$, with $x_0 = z_0 = 0$, would be essentially identical to Fig.16. The coupling along the z direction in Fig.17 is relatively weaker compared to that in Fig.16 along x or y directions. There would be no coupling along the z direction in the far-field region, because the dipole does not radiate in the z direction ($\theta = 0$). As the separation distance $d = x_0$ approaches zero, the mutual coupling in Fig.16 approaches $Z_m \sim (73 + j42.5)\Omega$, which maybe checked with Fig.9 as the input impedance Z_{in} for a half-wave dipole antenna.

3.4. Method of Moments. The analyses in the last sections 3.1-3.3 assumed that the antenna current distribution $I(z)$ was given as in (9.39). Although (9.39) happens to closely describe the actual current for a linear antenna of Fig.8, one may desire a more accurate solution. In general configurations of wire antennas, a good description of the current distribution may not always be available, or, when available, the distribution might not be accurate enough for a particular application. Under such situations, a suitable method to accurately model antenna currents would be critical. In this section, we will present one possible approach. It is particularly formulated to find the antenna current distribution $I(z)$ for a linear dipole antenna of Fig.8, with arbitrary length L. The method may be similarly extended for general variations of the antenna shape or feeding location, or for configurations with multiple feeding terminals.

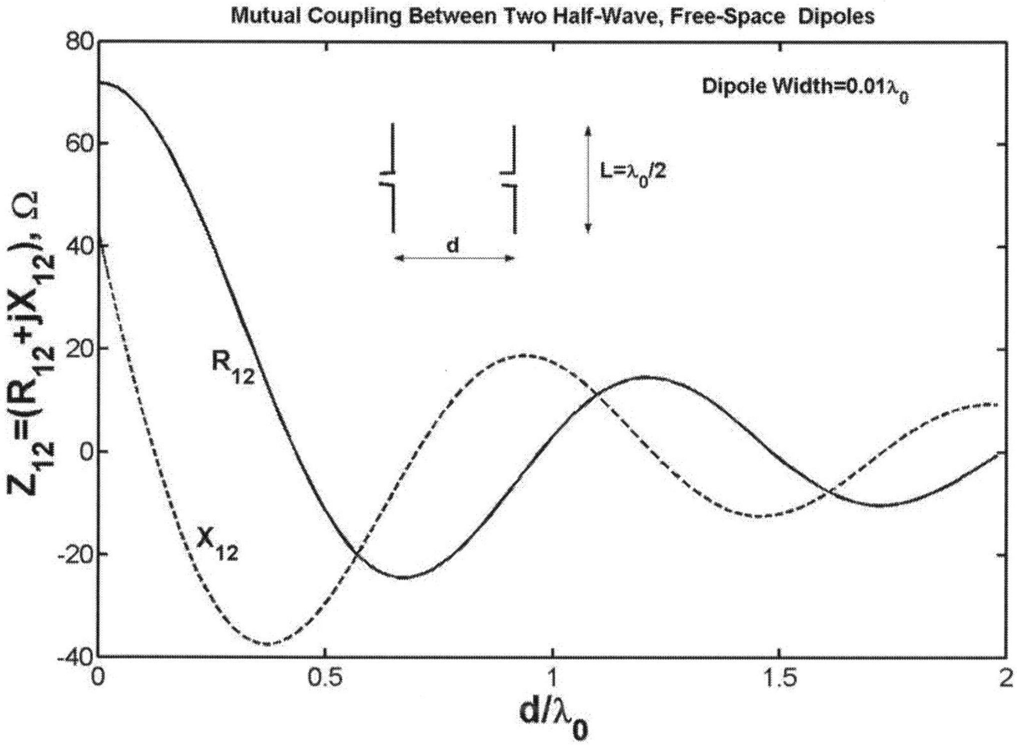

FIGURE 16

For the geometry of Fig.8, the x dependence of the antenna surface current $J_z(x,z)$ is expressed by a uniform distribution $h(x)$, as used in (9.39). Whereas, unlike (9.39), the z dependence $I(z)$ is now modeled as a superposition of N number of overlaping segments, each of length $\Delta L = L/(N+1)$. The physical lay-out of the segments is shown in Fig.18.

$$J_z(x,z) = I(z)h(x),$$
$$I(z) = \sum_{n=1}^{N} I_n f_n(z) = \sum_{n=1}^{N} I_n f(z - z_{on}; \triangle L),$$
$$\Delta L = \frac{2L}{N+1}, \; z_{on} = -\frac{L}{2} + \frac{\Delta L}{2} n,$$
$$f(z;L) = \frac{\sin[k_0(\frac{L}{2} - |z|)]}{\sin(k_0 \frac{L}{2})}, \; |z| < L/2; \; = 0, \; |z| > L/2,,$$
$$h(x) = \frac{1}{w}, \; |x| < w/2; \; = 0, \; |x| > w/2. \tag{9.67}$$

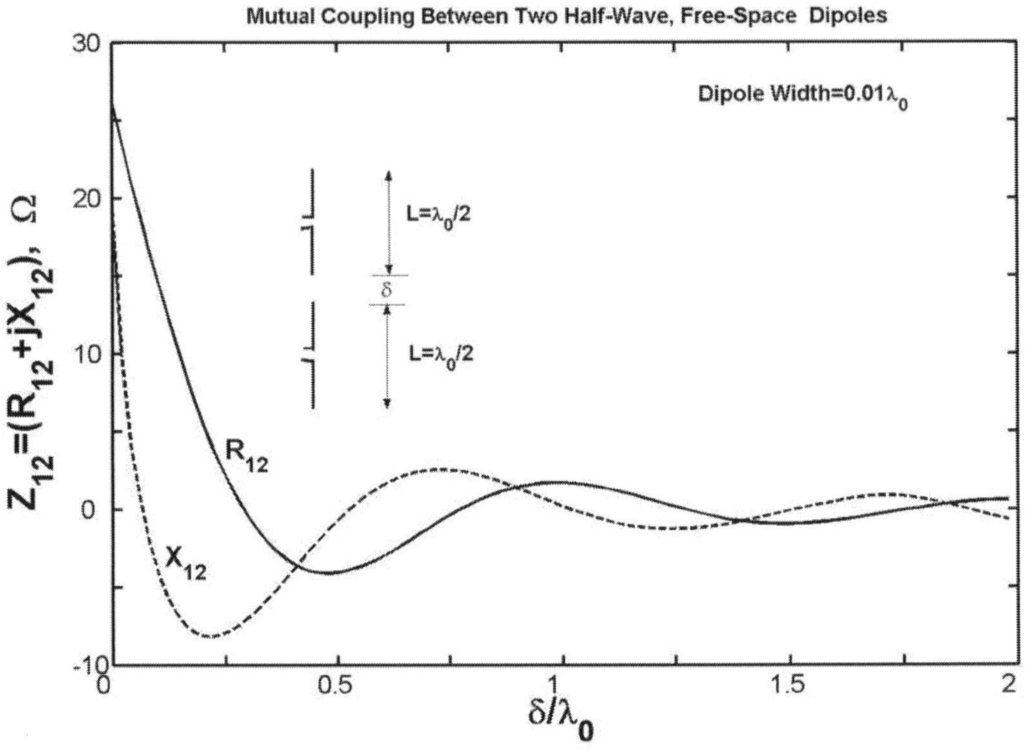

FIGURE 17

The nth segment is described by a given distribution function $f_n(z)$, with an unknown amplitude I_n. Each $f_n(z)$ is expressed by a common function $f(z)$, the same function we used in (9.39) to describe the current over the entire length L, except its length L is substituted with ΔL, and the z coordinate of its center point is moved to z_{0n}. The overlaping of the segments, as shown in Fig.18, ensures that the total current $I(z)$ is continuous over the entire length l. Mathematically, any arbitrary current function $I(z)$ can always be expressed by the segmentation of (9.67), with sufficiently large number of segments N. In practice, only a moderate number of segments may be required to provide reasonble accuracy.

The above segmentation model is equivalent to having N number of overlapping smaller antennas, each of length ΔL. The current distribution $f_n(z)$ on each smaller antenna is assumed given, but its current amplitude I_n remains unknown. The current distribution $I(z)$ on the total antenna is unknown, but can be expressed as a superposition of the currents $I_n f_n(z)$ on each smaller antenna, once the individual current amplitudes I_n can be solved.

The center point of segment n, at $z = z_{0n}$, maybe viewed as the input port of the nth antenna element. Accordingly, the N small antennas can be modeled as a N port circuit, with port currents I_n and voltages V_n. Only one of the N small antennas, identified as

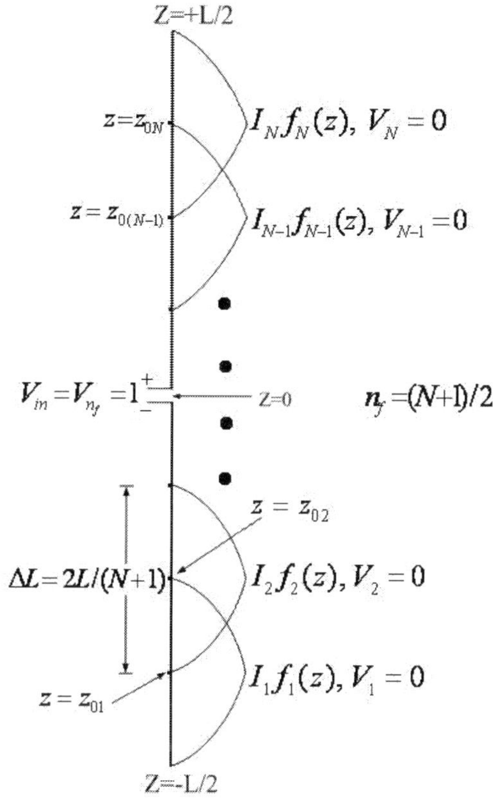

FIGURE 18

the feed element $n = n_f$, is connected to the antenna input with $V_{n_f} = V_{in} = 1V$, while all others are short-circuited with $V_n = 0$.

Under general conditions of V_n, the port currents I_n's can be related to V_n's through a mutual-impedance matrix. This is similar to mutual-impedance modeling in the last section, but extended in general to N elements.

$$\sum_{j=1}^{N} I_j Z_{nj} = V_n, \; n = 1, \dots, N,$$
$$[Z_{ij}]_{N\times N} [I_n]_{N\times 1} = [V_n]_{N\times 1},$$
$$Z_{ij} = Z_m(L \equiv \Delta L, x_0 = y_0 = 0, z_0 \equiv \Delta z_{oij}),$$
$$\Delta z_{oij} = |z_{oi} - z_{oj}|. \tag{9.68}$$

Z_{ij} is the mutual impedance between ith and jth elements, which is equal to the mutual impedance Z_m between two identical antenna elements of length ΔL and distance Δz_{oij}, as derived in the last section 3.3.

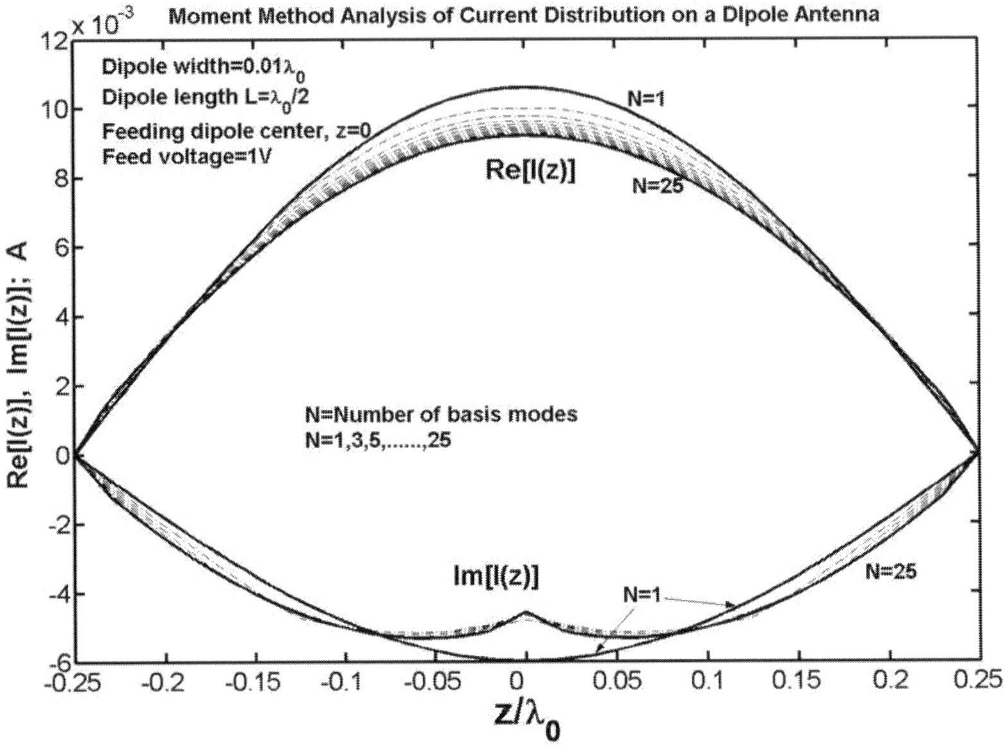

FIGURE 19

Consider now the specific condition of Fig.8 with one feed point, located at the center of the antenna. n_f is the index of the element connected to the antenna input (see Fig.18). In the present case of a center-fed antenna, with odd number of segments N, we have $n_f = (N+1)/2$. Accordingly, the input terminals of all antenna elements except $n = n_f$ are short circuited, with $V_n = 0$, $n \neq n_f$; $V_n = V_{in} = 1$, $n = n_f$. With this voltage excitation, the matrix relation of (9.68) can be solved for the input currents I_n. The current amplitudes I_n can then be used in (9.67) to obtain the total current distribution $I(z)$. Now, the input current I_{in} of the antenna is equal to I_{n_f}. Accordingly, the input impedance Z_{in} of the antenna can be obtained as the ratio of the $V_{in} = 1$ and the $I_{in} = I_{n_f}$ we just solved for.

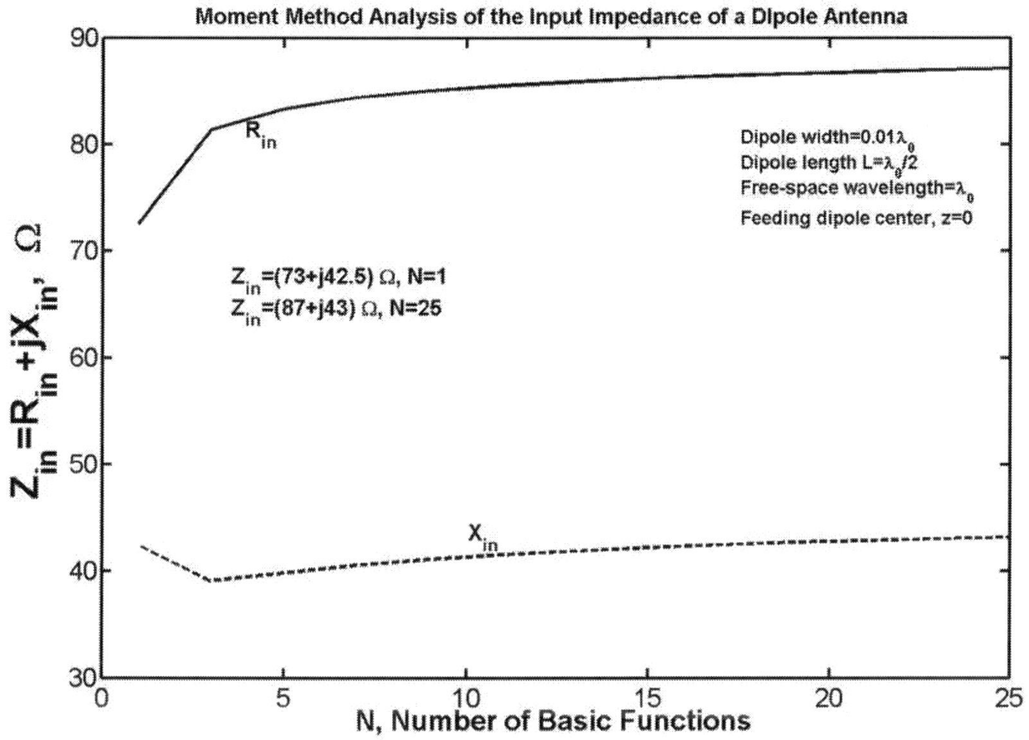

FIGURE 20

$$[I] = \begin{bmatrix} I_1 \\ I_2 \\ . \\ . \\ . \\ . \\ I_N \end{bmatrix}, \; [V] = \begin{bmatrix} 0 \\ 0 \\ . \\ 1 \\ . \\ 0 \\ 0 \end{bmatrix},$$

$$\sum_{j=1}^{N} I_j Z_{nj} = 0, \; n \neq n_f; \; = V_{in} = 1, \; n = n_f,$$

$$[I] = [Z]^{-1} [V],$$

$$Z_{in} = \frac{V_{n_f}}{I_{n_f}} = \frac{V_{in} = 1}{I_{n_f}}. \tag{9.69}$$

Simple variations in the above method can be implemented to move the input terminal from the element $n = n_f = (N+1)/2$ to any other element, or even to have multiple feed terminals with arbitrary relative excitations. This is possible by suitable choice of the voltage excitation matrix $[V]$. That clearly would not have been possible in the simple models of the last sections 3.1-3.3. More significant variations are possible to model a wire antenna of an arbitrary curved shape, by breaking it into a large number of small, linear elements, that are oriented in different directions. The formulation described above would still bc applicable, except we need mutual impedance $Z_{ij} = Z_m$ between different elements of arbitrary relative orientation. The mutual coupling derivation in section 3.3 was particularly formulated for colinear elements, but it could be similary extended for elements of arbitrary orientation, as needed for the general modeling. In any event, it should be clear that the basic principle of the formulation we have developed here covers a much broader scope of modeling arbitrary configurations of thin wire antennas.

Boundary Condition and Field Moments: It would be useful to view the above modeling in terms of the tangential electic field on the surface of the antenna. The antenna is physically made of a strip of conducting material. Assuming that the material used is a perfect conductor, the tangential electric field E_z needs to be zero at every point on the antenna surface. We did not explicitly enforce this boundary condition in our analysis. However, it has been implicitly enforced in (9.69) through the terminal voltges of the antenna segments. That is, the terminal voltage V_n is enforced to be zero for all terminals, except that of the central feed element.

Consider two general elements n and j. The mutual impedance $Z_m = Z_{nj}$, as seen at the element n, contributed due to coupling from the element j, is defined in the last section 3.3 by (9.66). $Z_m = Z_{nj}$ is related to the complex source power P_{nj} seen by the element n, contributed due to the electric field E_{zj} produced from element j.

$$P_{nj} = \frac{1}{2} I_j I_n^* Z_{nj} = -\frac{1}{2} \int\limits_z \int\limits_x E_{zj} J_{zn}^* dxdz = -\frac{1}{2} \int\limits_z \int\limits_x E_{zj}(x,z) I_n^* f_n(z) h(x) dxdz,$$

$$I_j Z_{nj} = \frac{2P_{nj}}{I_n^*} = -\int\limits_z \int\limits_x E_{zj}(x,z) f_n(z) h(x) dxdz. \tag{9.70}$$

The above relation for the mutual impedance Z_{nj} can be used in (9.69) to express the terminal voltage V_n in terms of the total electric field E_z.

$$\begin{aligned} V_n = \sum_{j=1}^{N} I_j Z_{nj} &= -\int\limits_z \int\limits_x \sum_{j=1}^{N} E_{zj}(x,z) f_n(z) h(x) dxdz \\ &= -\int\limits_z \int\limits_x E_z(x,z) f_n(z) h(x) dxdz. \end{aligned} \tag{9.71}$$

The above expression for the terminal voltage V_n may now be equated to zero for all $n \neq n_f$. The equivalent enforcement in terms of the electric field E_z maybe expressed as follows.

$$V_n = 0, \int_z \int_x E_z(x,z) f_n(z) h(x) dx dz = 0, \; n \neq n_f,$$
$$V_n = V_{in} \neq 0, \int_z \int_x E_z(x,z) f_n(z) h(x) dx dz \neq 0, \; n = n_f. \quad (9.72)$$

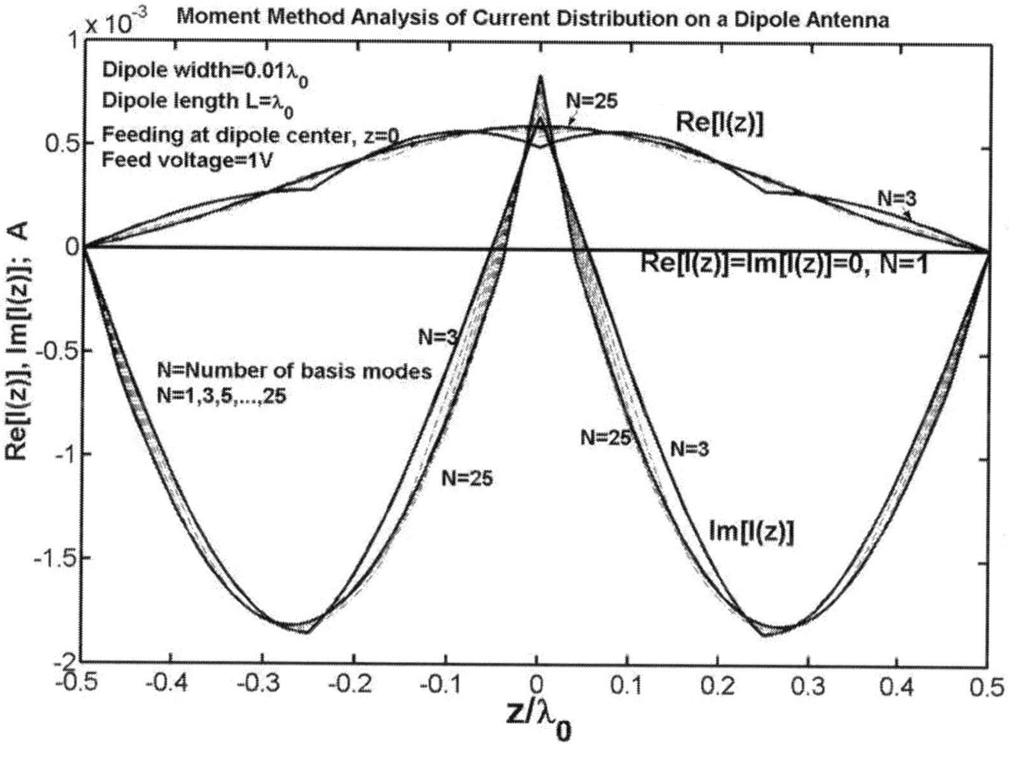

FIGURE 21

We have the integration of E_z, multiplied by the distribution function $f_n(z)h(x)$ - referred to as the *moment* of E_z on the *weighting function* $f_n(z)h(x)$ - equal to zero, for all $n \neq n_f$. Accordingly, the method is referred to as the "Method of Moments." Essentially, instead of enforcing the tangential field E_z to be zero at all points of the antenna, as rigorously required, we are enforcing it in an integrated sense over N segments, other than the feed segment. In the limit with large or infinite N, the length of each segment $\Delta L = L/(N+1)$ would be small, approaching zero. In this limit, the above enforcement through field-moments over very small or zero-length segments would effectively enforce zero tangential fields everywhere on the antenna, except near the feed terminal. as required for a rigorous solution. In other words, a moment method model with ideally infinite number of segments would lead to a rigorous solution. In practice, however, the result may be quite close to the rigorous condition for moderately large N.

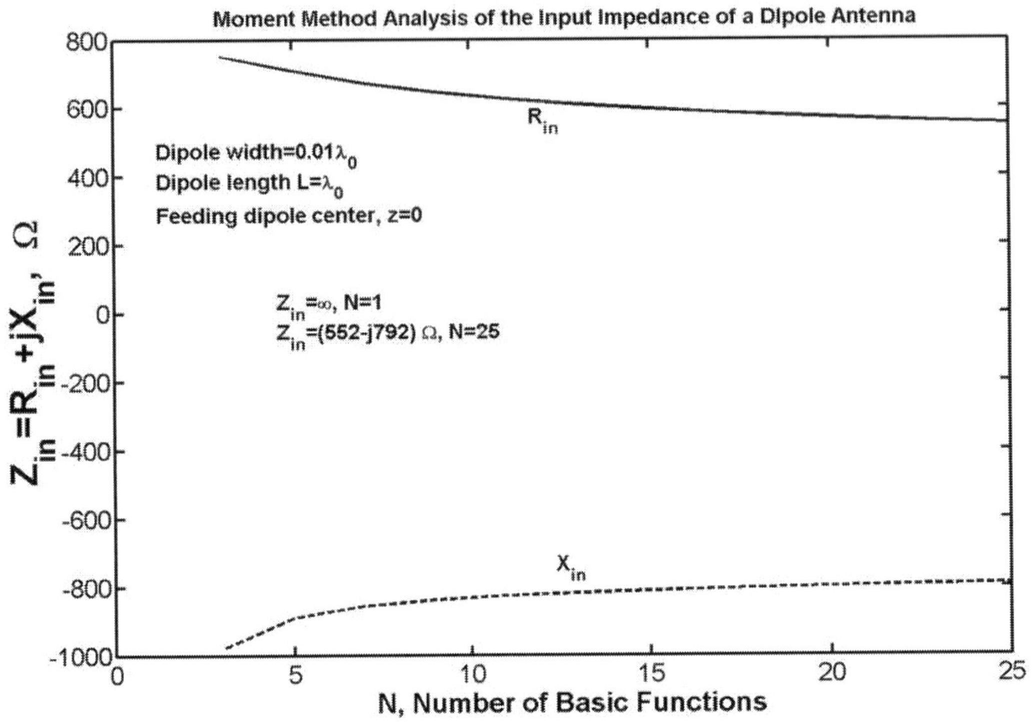

FIGURE 22

Fig.19 plots the real and imaginary parts of the antenna current distribution $I(z)$ for $L = \lambda_0/2$, computed by the method of moments presented above, with different number of segments N. The reference input voltage for the computation is $V_{in} = 1$. The corresponding input impedance $Z_{in} = R_{in} + jX_{in}$ is plotted in Fig.20 as a function of the segmentation number N. The current distribution $I(z)$ with $N = 1$ refers to the simple current distribution we used in sections 3.1-3.3. As seen in Fig.19, this simple distribution with $N = 1$ looks quite similar to more accurate solutions with increased N, except for the imaginary part of the current near the feed point. The input impedance in Fig.20 changes moderately from $Z_{in} = (73 + j42.5)\Omega$ for $N = 1$ to $Z_{in} = (87 + j43)\Omega$ for $N = 25$. Results similar to Figs.19,20 are shown in Figs.21,22 for $L = \lambda_0$ and in Figs.23,24 for $L = 0.25\lambda_0$.

Input impedance Z_{in} calculated with $N = 25$, for different values of length $0 < L < 2.0\lambda_0$, are plotted in Fig.25, as compared with the simple model with $N = 1$. Results for antenna lengths around the resonant values at about $L = 0.5\lambda_0, 1.5\lambda_0$ do not change much for a more accurate model with $N = 25$, as compared to the simple result for $N = 1$. However, the results for $N = 25$ deviate significantly from the simple results for $N = 1$ for operating lengths away from the above resonant points. Particulary near length $L = 1.0\lambda_0, 2.0\lambda_0$, where the simple model with $N = 1$ predicts an idel infinite impedance, the results with $N = 25$ deviate significantly from those ideal values, as one would expect.

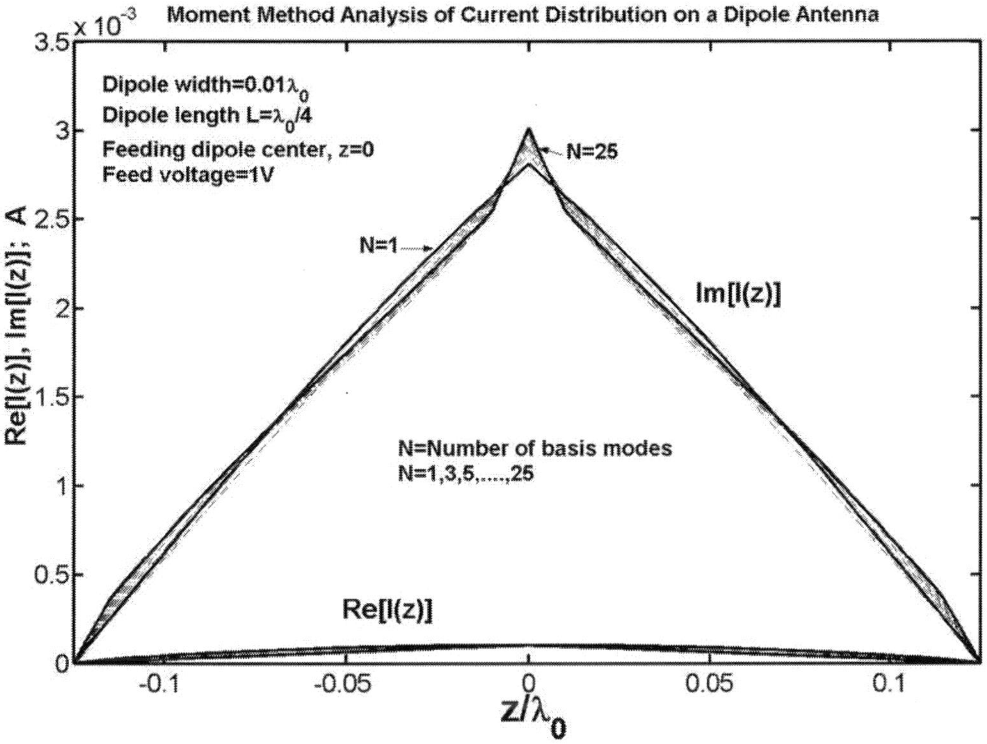

FIGURE 23

4. Folded Dipole Antenna

Figure 26 shows the geometry of a folded-dipole antenna. It consists of a source (V_{in}), connected to the ends of a wire of total length $2L$, folded once as a rectangular loop of height L and width $\delta << L$. The length L can have any value, but is often designed to be about $\lambda_0/2$ for a resonant operation. This antenna configuration can be viewed as a superposition of an "odd mode," and an "even mode," having dual excitations at the center of the two arms of the antenna with even and odd symmetry. The odd mode represents a transmission line mode of operation, without any radiation. Here the top and bottom sections are two identical transmission-line stubs of length $L/4$, connected in series, as shown in Fig.26. The input current I_t in this mode can then be related to V_{in} as follows.

$$I_t = \frac{V_{in}}{2Z_t} = \frac{V_{in}}{2jZ_0 \tan(k_0 L/2)}, \tag{9.73}$$

where Z_0 is the characteristic impedance of a two-wire transmission line having the same wire diameter as that of the folded dipole, and line-to-line separation δ. In contrast, the even mode represents an antenna mode of operation, with $I = 0$ at the top and bottom ends to satisfy the symmetry condition. The antenna mode operates very much as two

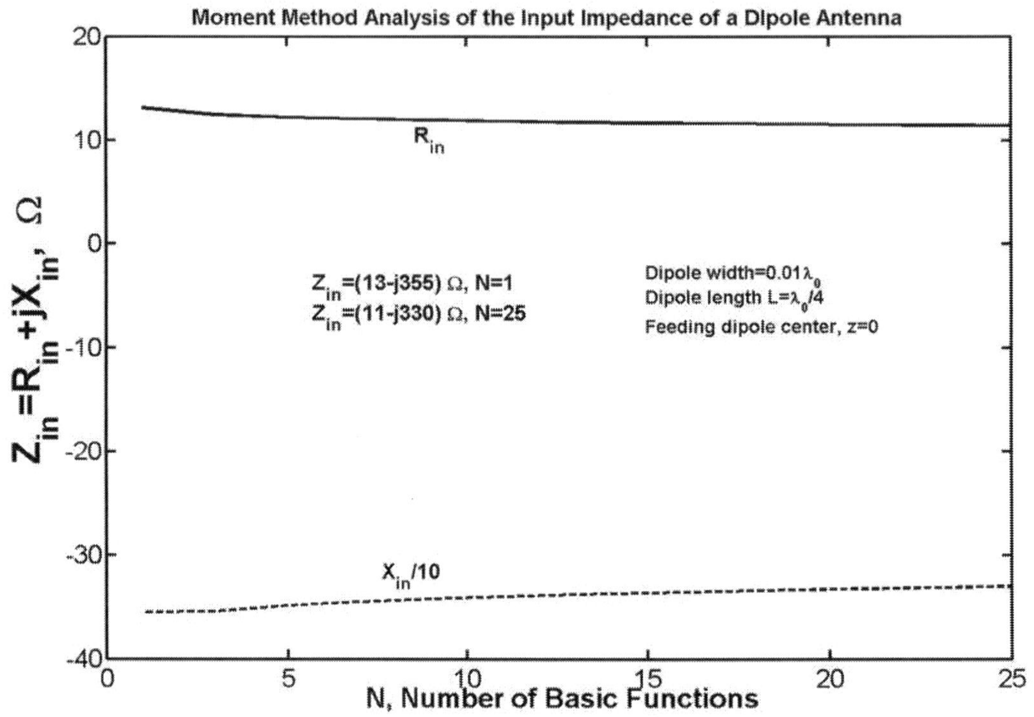

FIGURE 24

dipole antennas of length L each, placed close to each other. The current distribution along the wire is the same as that for an equivalent dipole antenna, and therefore has the same radiation pattern and gain. The input current I_a for the antenna mode can be related to the input voltage V_{in} using the input impedance Z_{dipole} of an equivalent dipole antenna of length L.

$$Z_{dipole} = \frac{V_{in}/2}{2I_a}, \quad I_a = \frac{V_{in}}{4Z_{dipole}}. \tag{9.74}$$

Now, using superposition, the input impedance of the folded dipole can be expressed as a parallel combination of two parts.

$$I_{in} = I_a + I_t = \frac{V_{in}}{4Z_{dipole}} + \frac{V_{in}}{2jZ_0 \tan k_0 L/2}, \tag{9.75}$$

$$Z_{in} = \frac{V_{in}}{I_{in}} = \frac{1}{\frac{1}{4Z_{dipole}} + \frac{1}{2jZ_0 \tan k_0 L/2}} = (4Z_{dipole}) \parallel (2jZ_0 \tan k_0 L/2). \tag{9.76}$$

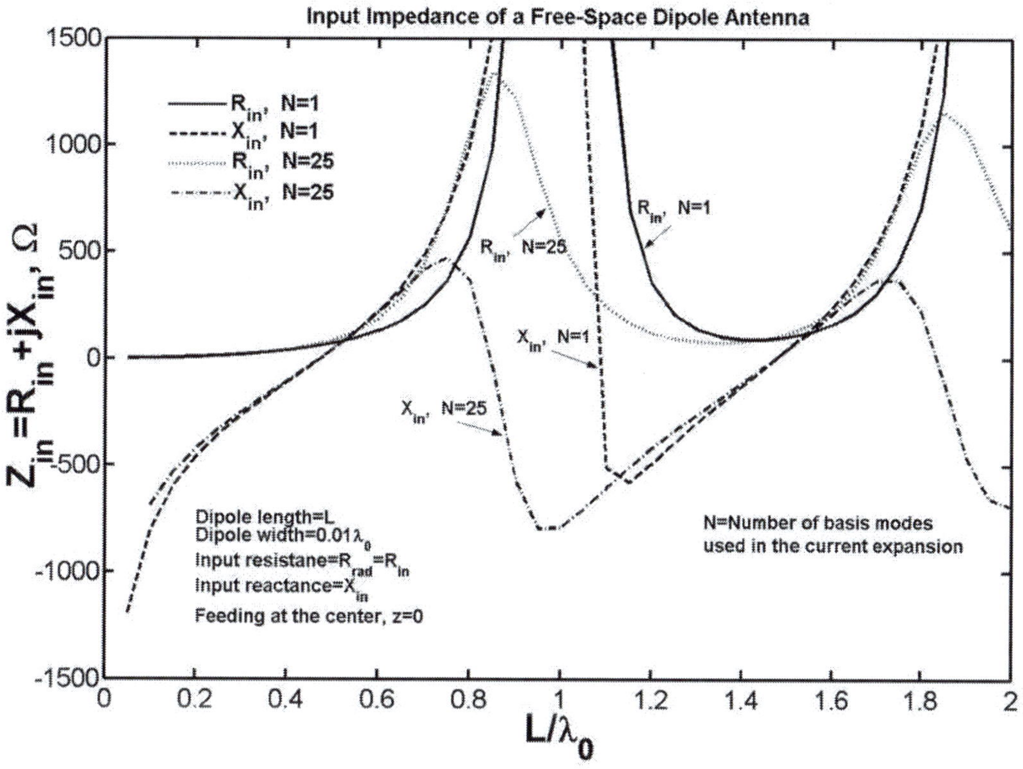

FIGURE 25

If the length L is a half wavelength, or $k_0L/2 = \pi/2$, then the input impedance of the folded dipole antenna is four times that of a half-wave dipole antenna.

$$Z_{in}(L = \lambda_0/2) = 4Z_{dipole}(L = \lambda_0/2) = 4 \times 73 = 292\Omega. \tag{9.77}$$

So, a folded-dipole antenna is a derivative of a regular dipole antenna, with the same radiation pattern, polarization and directivity, except with a different input impedance. For the half-wave case, the folded-dipole antenna simply produces an impedance transformation by a factor of four. The resulting 292Ω impedance is convenient to match to 300Ω systems, used in some broadcast applications.

5. Monopole Antennas

Monopole antennas, as shown in Fig.27, is a class of derivatives of dipole antennas. Here only half of the dipole antenna is needed for operation. A metal ground plane (ideally of infinite size) is used, with respect to which the excitation voltage is applied to the half structure. The half-structure for a regular dipole antenna is called a monopole antenna, in reference to the presence of only one physical side. A similar half structure for a folded-dipole antenna is called a folded-monopole antenna. The presence of the ground plane

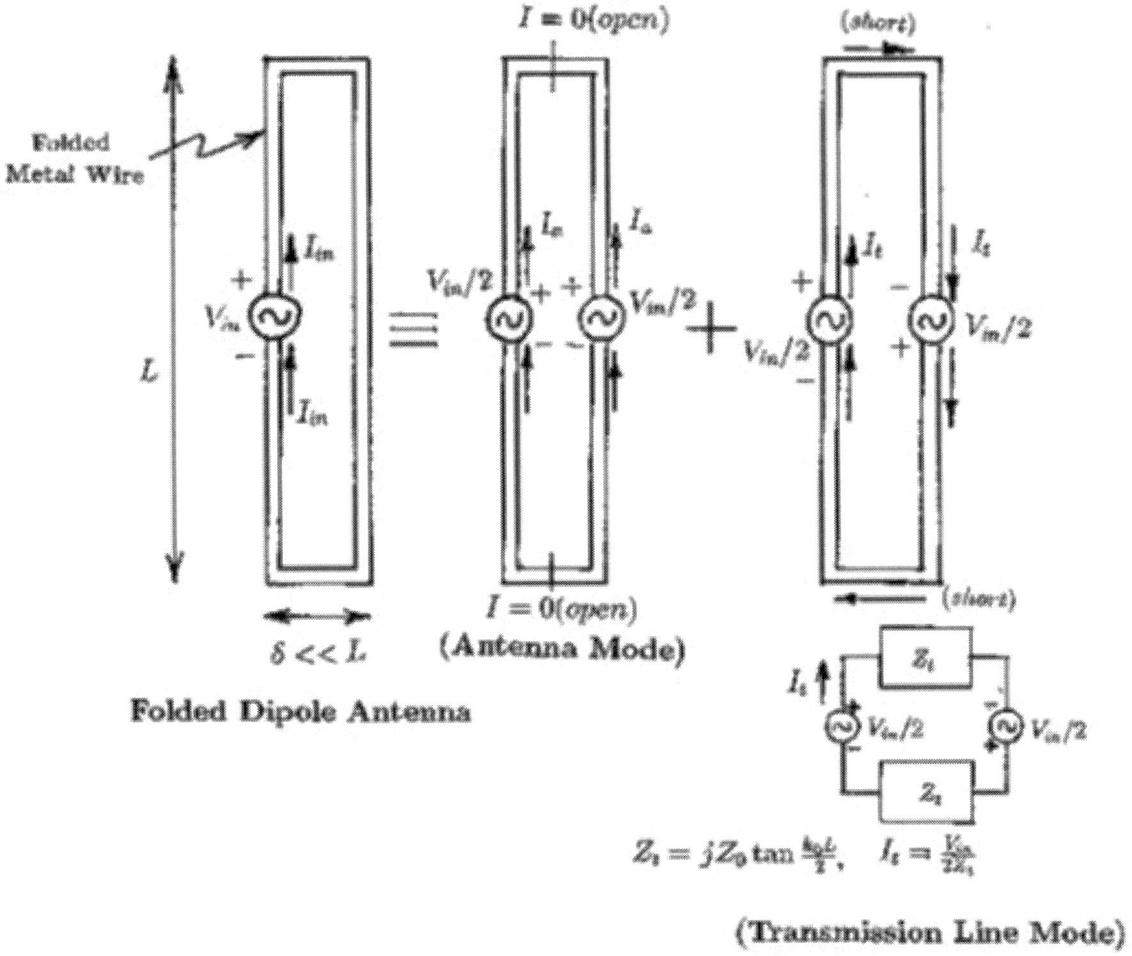

FIGURE 26

allows the monopole antenna to operate as electrically equivalent to a dipole antenna. The ground plane equivalently replaces the lower half by an imaging principle, in analogy to creating an optical image through a mirror. Notice in Fig.27 that in order that the currents in the monopole and dipole structures are the same, one needs the source voltage of the equivalent dipole antenna to be twice as that of the monopole antenna. As a result, the input impedance of the monopole structure is half of that of the corresponding dipole structure.

$$Z_{monopole} = \frac{1}{2} Z_{dipole}; \; Z_{monopole} = \frac{V_{in}}{I_{in}}, \; Z_{dipole} = \frac{2V_{in}}{I_{in}}. \tag{9.78}$$

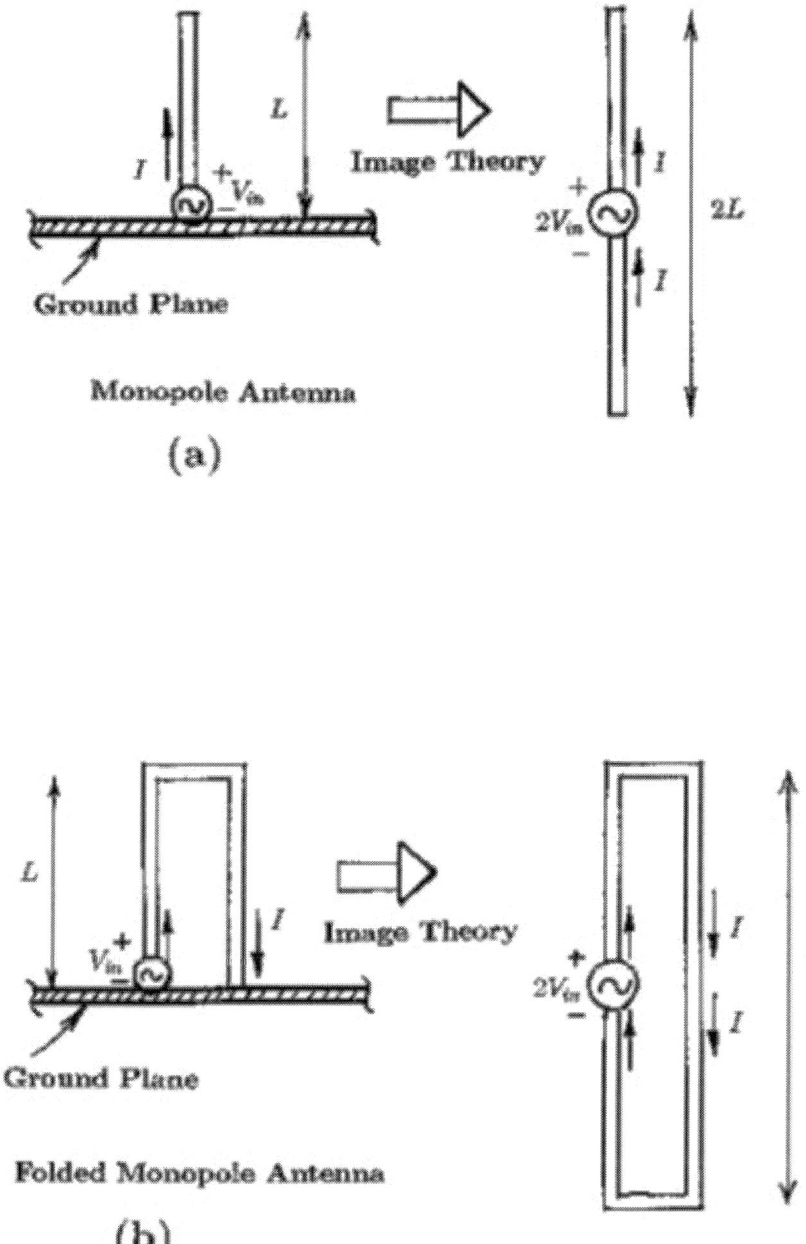

FIGURE 27

For example, if the length of the monopole antenna $L = \lambda_0/4$, such that the corresponding length of the equivalent dipole antenna is $\lambda_0/2$, we have the following values of the radiation impedances.

$$Z_{monopole} = \frac{1}{2} Z_{dipole} = 36.5\Omega; \; Z_{foldedmonopole} = \frac{1}{2} Z_{foldeddipole} = 146\Omega. \tag{9.79}$$

Due to above imaging principle, the polarization of radiation and radiation patterns of a monopole antenna is the same as that of its equivalent dipole antenna. Except, the monopole antenna has field only in the top half space, having zero radiation below the ground plane. In contrast, the equivalent dipole structure has fields in both sides, with the radiation to the bottom side symmetric to that above. Under this situation, the expressions of the directive gain functions (9.9,9.10,9.33,9.34) maybe reevaluated to show that the directivity D of a monopole antenna is twice that of its equivalent dipole structure.

$$D_{monopole} = 2D_{dipole}; \; D_{short\ monopole} = 3.0, \; D_{half-wave\ monopole} = 3.28. \tag{9.80}$$

6. Wire-Loop Antenna

A loop antenna consists of a wire loop of arbitrary shape, fed by an input current I_{in}. A circular loop, as shown in Fig.28, is the shape most commonly used. The radius of the loop, R, could be electrically small or large. If the current distribution along the wire loop is known, or can at least be closely approximated based on some physical insight, then the radiated fields of the loop antenna can be derived analytically, or computed numerically. This is possible by treating the loop as a superposition of small current elements with known amplitudes, physical location and orientation. The far-field quantities for a small dipole antenna, as presented in the section 1, with suitable transformation of its coordinates to account for different physical location and orientation, can be used here to model the elemental radiation.

6.1. Uniform Current, Small Loop. If the current loop is small ($R << \lambda_0$), the far-field quantities can be derived into a simple final form. In this case, the current along the loop $I(\phi')\hat{\phi}'$ can be assumed to be independent of $\hat{\phi}'$, with magnitude equal to the input current I_{in}. With reference to Fig.28, consider the small current element of length $R\Delta\phi'$, carrying a current along the direction $\hat{\phi}'$. The radiation fields due to this current element can be expressed using results from section 1, equations (9.2,9.3). However, one must properly adjust (9.2,9.3) in order to account for the orientation of the current element along $\hat{\phi}'$, and its location at (R, ϕ') defined by a position vector $\overline{R} = R\hat{R}$.

The equation (9.3) maybe used to express the radiating magnetic field $\overline{H}$ due to the current element. .

$$\Delta\overline{H}(r, \theta, \phi; \phi') = \frac{jI_{in}R\Delta\phi'}{2\lambda_0} \frac{e^{-jk_0 r'}}{r'} (\hat{\phi}' \times \hat{r}), \tag{9.81}$$

$$\hat{\phi}' = \cos(\phi - \phi')\hat{\phi} + \cos\theta \sin(\phi - \phi')\hat{\theta} + \sin\theta \sin(\phi - \phi')\hat{r}, \tag{9.82}$$

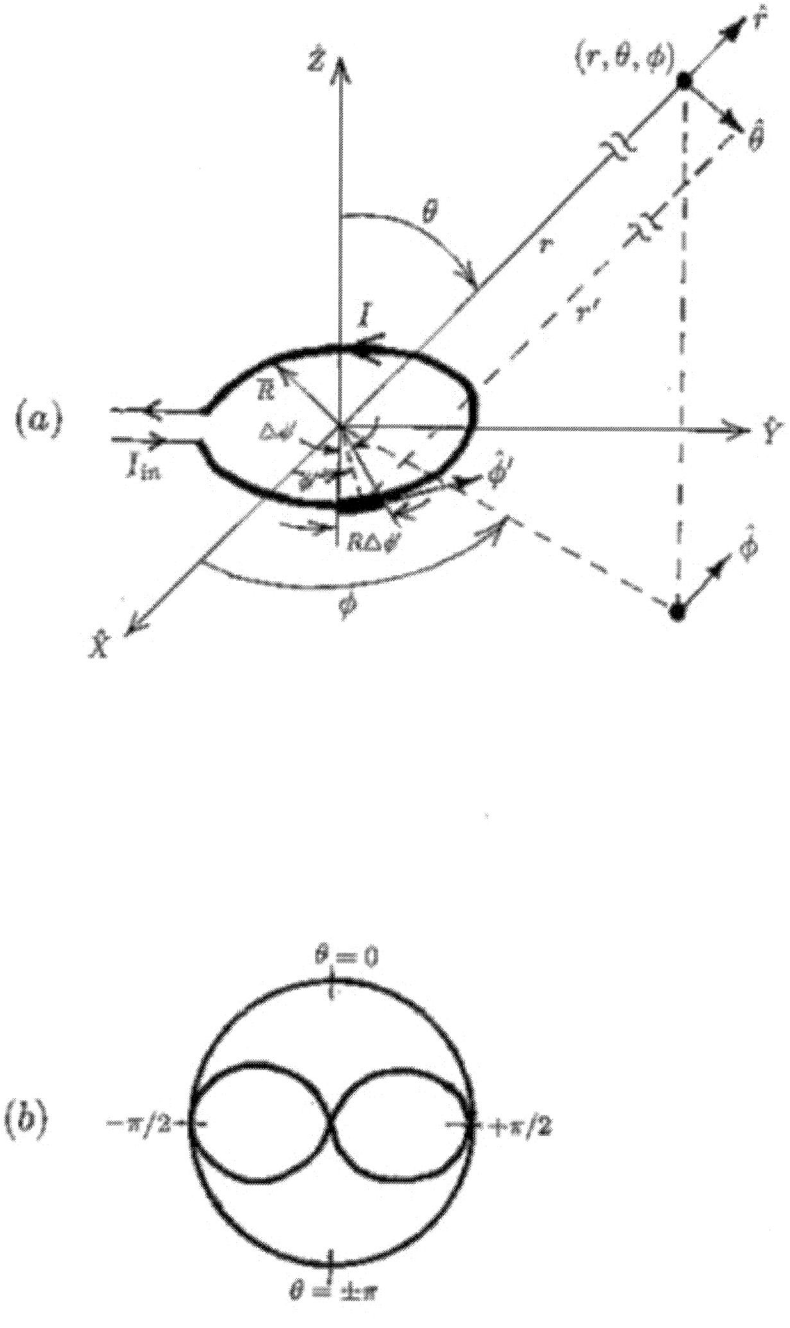

FIGURE 28

$$\overline{R} = R\hat{R} = R[\sin\theta\cos(\phi - \phi')\hat{r} + \cos\theta\cos(\phi - \phi')\hat{\theta} - \sin(\phi - \phi')\hat{\phi}], \tag{9.83}$$

$$r' = r - \overline{R} \cdot \hat{r} = r - R \sin\theta \cos(\phi - \phi'), \tag{9.84}$$

$$(\hat{\phi}' \times \hat{r}) = \cos(\phi - \phi')\hat{\theta} - \cos\theta \sin(\phi - \phi')\hat{\phi}. \tag{9.85}$$

Using the co-ordinate relationships (9.82,9.83,9.84,9.85) we can rewrite (9.81) in the following form.

$$\begin{aligned} \Delta\overline{H}(r,\theta,\phi;\phi') \;\simeq\; & \frac{jI_{in}R\Delta\phi'}{2\lambda_0} \frac{e^{-jk_0 r} e^{jk_0 R \sin\theta \cos(\phi-\phi')}}{r} \\ & [\cos(\phi - \phi')\hat{\theta} - \cos\theta \sin(\phi - \phi')\hat{\phi}]. \end{aligned} \tag{9.86}$$

In the far-field region we have $r >> R$. This condition has been used in (9.86) to approximate r' in the denominator by r. In addition, for a small loop we have $R << \lambda_0 = \frac{2\pi}{k_0}$, which is equivalent to $Rk_0 << 1$. This condition can be used in order to further simplify (9.86) by approximating one of the exponential functions in the numerator.

$$\begin{aligned} \Delta\overline{H}(r,\theta,\phi;\phi') \;\simeq\; & \frac{jI_{in}R\Delta\phi'}{2\lambda_0} \frac{e^{-jk_0 r}[1 + jk_0 R \sin\theta \cos(\phi - \phi')]}{r} \\ & [\cos(\phi - \phi')\hat{\theta} - \cos\theta \sin(\phi - \phi')\hat{\phi}]. \end{aligned} \tag{9.87}$$

The total magnetic field can be obtained by integrating (9.87) over $0 < \hat{\phi}' < 2\pi$, to cover the entire current loop. The $\hat{\phi}$ component of (9.87) can be shown to integrate to zero, whereas the $\hat{\theta}$ component will be non-zero.

$$\begin{aligned} \overline{H}(r,\theta,\phi) &= \int_{\phi'=0}^{2\pi} \Delta\overline{H}(r,\theta,\phi;\hat{\phi}') = \frac{jI_{in}R}{2\lambda_0} \frac{e^{-jk_0 r}}{r} (jk_0 R \sin\theta)(\pi\hat{\theta}) \\ &= -\frac{I_{in}(\pi R^2)}{\lambda_0^2} \frac{e^{-jk_0 r}}{r} (\pi \sin\theta)\hat{\theta} = -\frac{e^{-jk_0 r}}{4\pi r} I_{in} S k_0^2 \sin\theta\ \hat{\theta};\ S = \pi R^2. \end{aligned} \tag{9.88}$$

Once the total radiating magnetic field is known, the radiating electric field can be obtained from it using their far-field relationship.

$$\overline{E}(r,\theta,\phi) = \hat{\phi} \frac{\eta_0 k_0^2 (I_{in} S) \sin\theta e^{-jk_0 r}}{4\pi r} = \eta_0 \overline{H}(r,\theta,\phi) \times \hat{r}. \tag{9.89}$$

If there are N number of loops in the form of a coil, instead of a single loop shown in Fig.28, then the corresponding field quantities would have to be multiplied by N. One may view the far field of a small-loop antenna as "*dual*" to that of a small dipole antenna. This is because the loop antenna has H_θ and E_ϕ far-field components, as shown in (9.88,9.89), in contrast to which a small dipole antenna has E_θ and H_ϕ components. Both the small loop and the small dipole antenna have the same $\sin\theta$ field pattern, thus have the same directive gain $D_g(\theta,\phi) = 1.5 \sin^2\theta$ and directivity $D = 1.5$. Once the radiation fields are known, using a procedure similar to that used for a small dipole antenna in section 1, one can derive the total radiated power, P_{rad}, and the radiation resistance, R_{rad}.

$$P_{rad} = \frac{\eta_0 k_0^4 (S|I_{in}|)^2}{12\pi}, \quad R_{rad} = \frac{2P_{rad}}{|I_{in}|^2} = \frac{\eta_0 k_0^4 S^2}{6\pi} = 20 k_0^4 S^2. \tag{9.90}$$

It may be noted, the above equation (9.90) needs to be multiplied by N^2 for a N-loop coil antenna.

6.2. Uniform Current, General Loop Size. In the above derivation, when the loop radius R is not small enough compared to the wavelength λ_0, the magnetic field in (9.86) can not be simplified into (9.87). However, if the current is still assumed to be uniform over the entire loop, (9.86) can be integrated without the approximation of (9.87), using Bessel-funciton relations (see Appendix).

$$\overline{H}(r,\theta,\phi) = \int_{\phi'=0}^{2\pi} \Delta\overline{H}(r,\theta,\phi;\hat{\phi}') = \frac{jI_{in}R}{2\lambda_0}\frac{e^{-jk_0 r}}{r}\int_{\phi'=0}^{2\pi} e^{jk_0 R\sin\theta\cos(\phi-\phi')}$$

$$\left[\cos(\phi-\phi')\hat{\theta} \quad \cos\theta\sin(\phi-\phi')\hat{\phi}\right]d\phi' = -\frac{I_{in}R\pi}{\lambda_0}\frac{e^{-jk_0 r}}{r} J_1(k_0 R\sin\theta)\hat{\theta},$$

$$\int_{\phi'=0}^{2\pi} \cos(\phi-\phi')\ e^{j\alpha\cos(\phi-\phi')}d\phi' = \int_{\triangle t=2\pi} \cos t\ e^{j\alpha\cos t}dt = 2\pi j J_1(\alpha),$$

$$\int_{\phi'=0}^{2\pi} \sin(\phi-\phi')\ e^{j\alpha\cos(\phi-\phi')}d\phi' = \int_{\triangle t=2\pi} \sin t\ e^{j\alpha\cos t}dt = 0, \tag{9.91}$$

where $J_1(.)$ is the Bessel function of the first kind, of order zero. The above expression maybe verified to approach (9.88) in the limit of small $R \to 0$, by approximating $J_1(\alpha \to 0) = \alpha/2$.

The radiating electric field $\overline{E}$ can be obtained from the above magnetic field using their far-field relationships. The radiated power P_{rad} can be derived by integrating the power density function $P_d = |\overline{E}|^2/(2\eta_0) = |\overline{H}|^2\eta_0/2$ over a sphere.

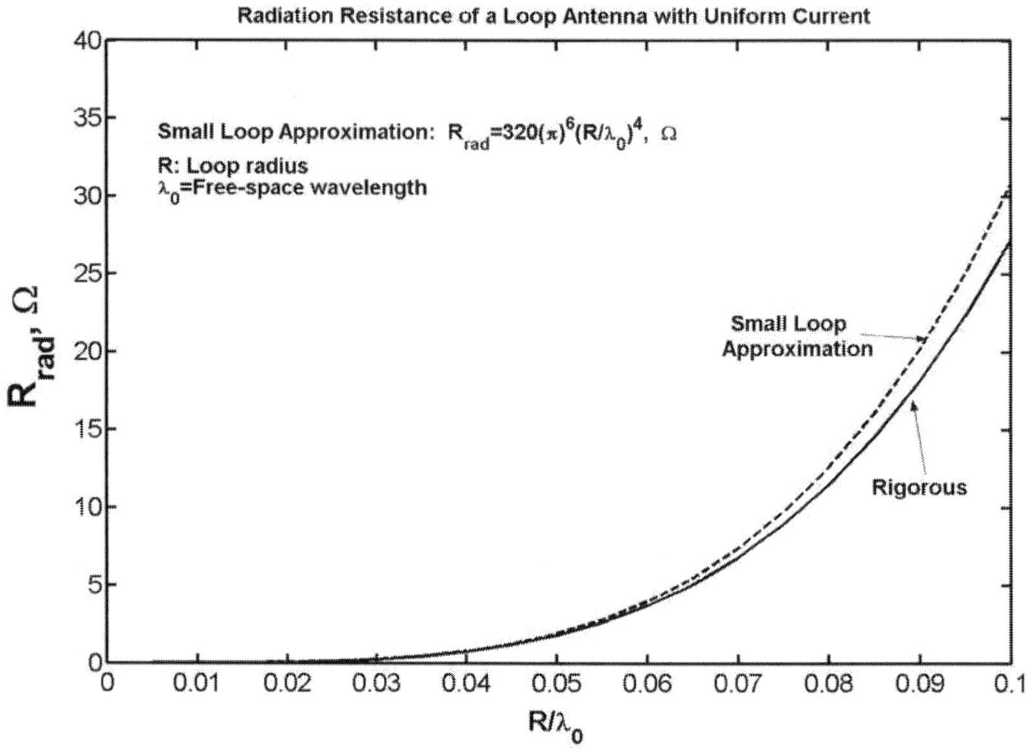

FIGURE 29

$$\overline{E}(r,\theta,\phi) = \frac{I_{in}R\pi\eta_0}{\lambda_0}\frac{e^{-jk_0r}}{r}J_1(k_0R\sin\theta)\hat{\phi} = \eta_0\overline{H}(r,\theta,\phi)\times\hat{r},$$

$$P_d(r,\theta,\phi) = \frac{|\overline{E}(r,\theta,\phi)|^2}{2\eta_0} = \frac{|I_{in}|^2R^2\pi^2\eta_0}{2\lambda_0^2r^2}J_1^2(k_0R\sin\theta),$$

$$P_{rad} = \int\limits_{\theta=0}^{\pi}\int\limits_{\phi=0}^{2\pi}P_d(r,\theta,\phi)r^2\sin\theta d\phi d\theta$$

$$= \frac{|I_{in}|^2R^2\pi^3\eta_0}{\lambda_0^2}\int\limits_{\theta=0}^{\pi}J_1^2(k_0R\sin\theta)\sin\theta d\theta,$$

$$R_{rad} = \frac{2P_{rad}}{|I_{in}|^2} = \frac{2R^2\pi^3\eta_0}{\lambda_0^2}\int\limits_{\theta=0}^{\pi}J_1^2(k_0R\sin\theta)\sin\theta d\theta. \tag{9.92}$$

The radiation resistance R_{rad} calculated using (9.92) is compared in Fig.29 with the simple results of (9.90) for a small-loop approximation. The two results in Fig.29 agree for small R/λ_0, but gradually deviate from each other as R/λ_0 increases, as expected.

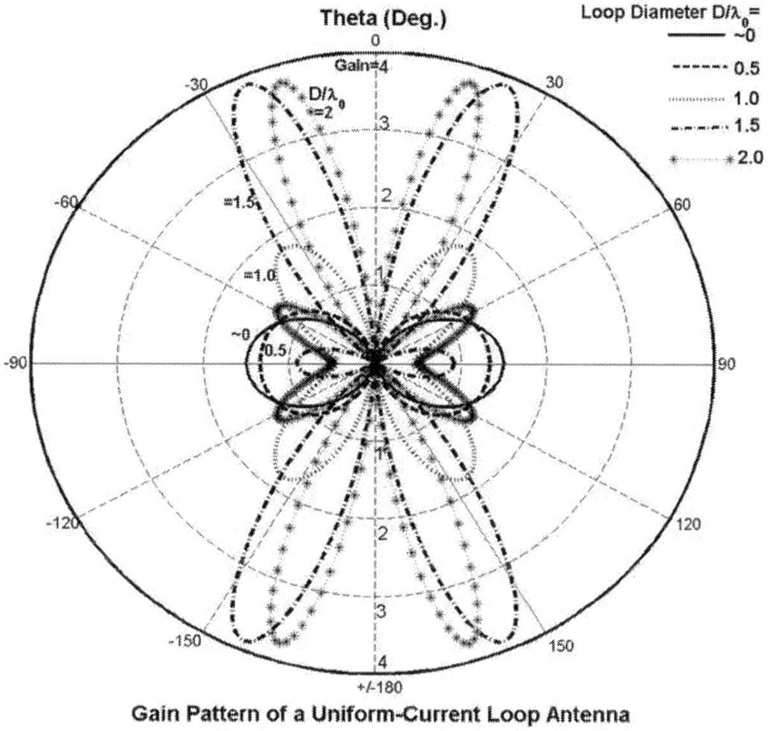

FIGURE 30

The directive gain function $D_g(\theta, \phi)$ can be obtained using the power density function P_d and the radiated power P_{rad}, and expressed in terms of the radiation resistance R_{rad}.

$$D_g(\theta, \phi) = \frac{4\pi r^2 P_d(r, \theta, \phi)}{P_{rad}} = \frac{8\pi r^2 P_d(r, \theta, \phi)}{|I_{in}|^2 R_{rad}} = \frac{4R^2\pi^3\eta_0}{\lambda_0^2 R_{rad}} \, J_1^2(k_0 R \sin\theta). \tag{9.93}$$

The directive gain function D_g, obtained for a uniform loop antenna using (9.93), is plotted in Fig.30 as a function of the elevation angle θ (it is independent of ϕ) in polar form, for selected loop sizes. The directivity D, which is the maximum value of the above directive gain function $D_g(\theta, \phi)$, and the corresponding elevation angle θ_{max} at which the above maximum occurs, are plotted in Fig.31 as a function of the loop radius.

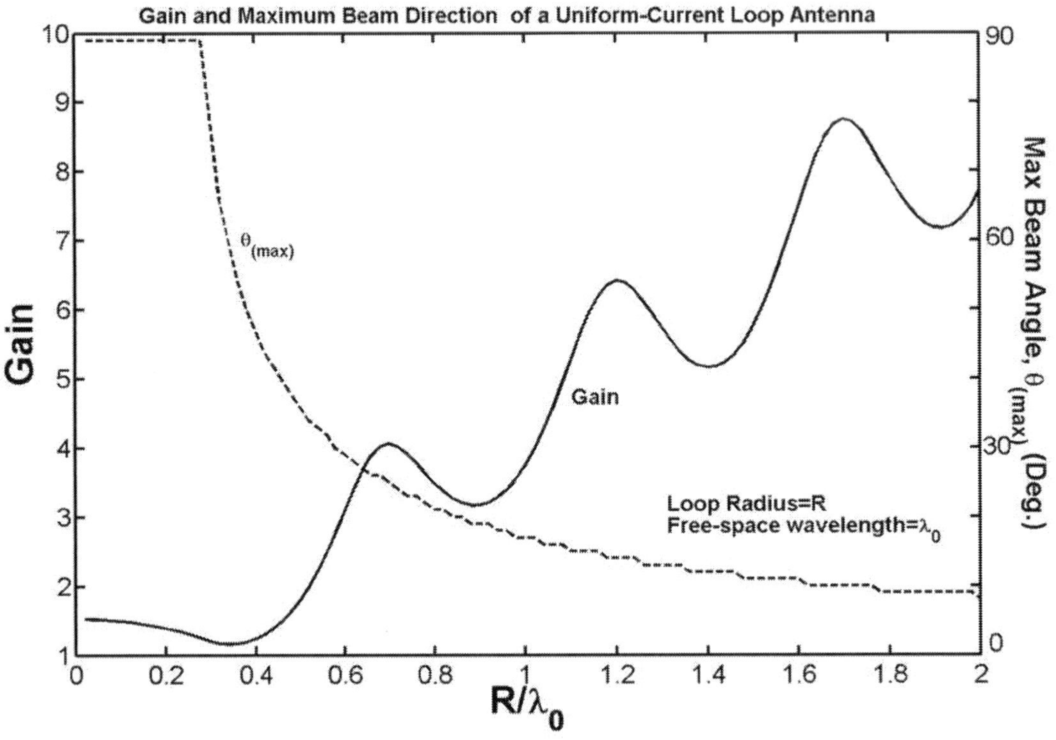

FIGURE 31

6.3. General $\cos(n\phi)$ Current. We will derive the radiation fields ($r \to \infty$) for current distributions of the form $I(\phi') = I_n \cos(n\phi')$. The loop current for general situations can be expressed as a superposition of such functions, and accordingly the resulting radiation fields can be expressed as superpositions of the formulas we will derive here.

$$I(\phi') = I_n \cos(n\phi'). \tag{9.94}$$

The elemental magnetic magnetic field for the above current distribution can be expressed by replacing the constant current I_{in} in (9.86) by the new current distribution. The total magnetic field can then be derived by integrating the elemental field over ϕ'.

$$\overline{H}(r,\theta,\phi) = \int_{\phi'=0}^{2\pi} \frac{jI_n\cos(n\phi')R}{2\lambda_0}\,\frac{e^{-jk_0 r}e^{jk_0 R\sin\theta\cos(\phi-\phi')}}{r}$$
$$\left[\cos(\phi-\phi')\hat{\theta} - \cos\theta\sin(\phi-\phi')\hat{\phi}\right]d\phi'. \tag{9.95}$$

In order to simplify the above integral expression, we establish the following Bessel-function relations.

$$\int_{\phi'=0}^{2\pi} e^{\pm jn\phi'} e^{j\alpha\cos(\phi'-\phi)} d\phi' = e^{\pm jn\phi} \int_{\triangle t=2\pi} e^{\pm jnt}\, e^{j\alpha\cos t} dt$$
$$= 2\pi j^n e^{\pm jn\phi} J_n(\alpha);\ (t = \phi' - \phi),$$

$$\int_{\phi'=0}^{2\pi} [\cos(n\phi'), \sin(n\phi')] e^{j\alpha\cos(\phi'-\phi)} d\phi' = 2\pi j^n [\cos(n\phi), \sin(n\phi)] J_n(\alpha), \quad (9.96)$$

$$\int_{\triangle t=2\pi} \cos(nt)\, e^{j\alpha\cos t} dt = 2\pi j^n J_n(\alpha), \quad \int_{\triangle t=2\pi} \sin(nt)\, e^{j\alpha\cos t} dt = 0,$$

$$\int_{\phi'=0}^{2\pi} \cos(n\phi') e^{j\alpha\cos(\phi'-\phi)} d\phi' = 2\pi j^n \cos(n\phi) J_n(\alpha) = p(\alpha, \phi),$$

$$\int_{\phi'=0}^{2\pi} \cos(n\phi')\cos(\phi'-\phi)\, e^{j\alpha\cos(\phi'-\phi)} d\phi' = \frac{1}{j}\frac{\partial p(\alpha,\phi)}{\partial\alpha} = 2\pi j^{n-1}\cos(n\phi) J_n'(\alpha), \quad (9.97)$$

$$\int_{\phi'=0}^{2\pi} \cos(n\phi')\sin(\phi'-\phi)\, e^{j\alpha\cos(\phi'-\phi)} d\phi' = \frac{1}{j\alpha}\frac{\partial p(\alpha,\phi)}{\partial\phi} = 2\pi n j^{n+1}\sin(n\phi)\frac{J_n(\alpha)}{\alpha}. \quad (9.98)$$

(9.98) can now be used in (9.95) to simplify and express the magnetic field in terms of Bessel functions of first kind, $J_n(.)$. Other fields and radiation characteristics, such as the electric field $\overline{E}$, power density P_d, radiated power P_{rad}, radiation resistance R_{rad}, directive gain function $D_g(\theta, \phi)$, can then be derived from the magnetic field, using the similar steps as in the last section.

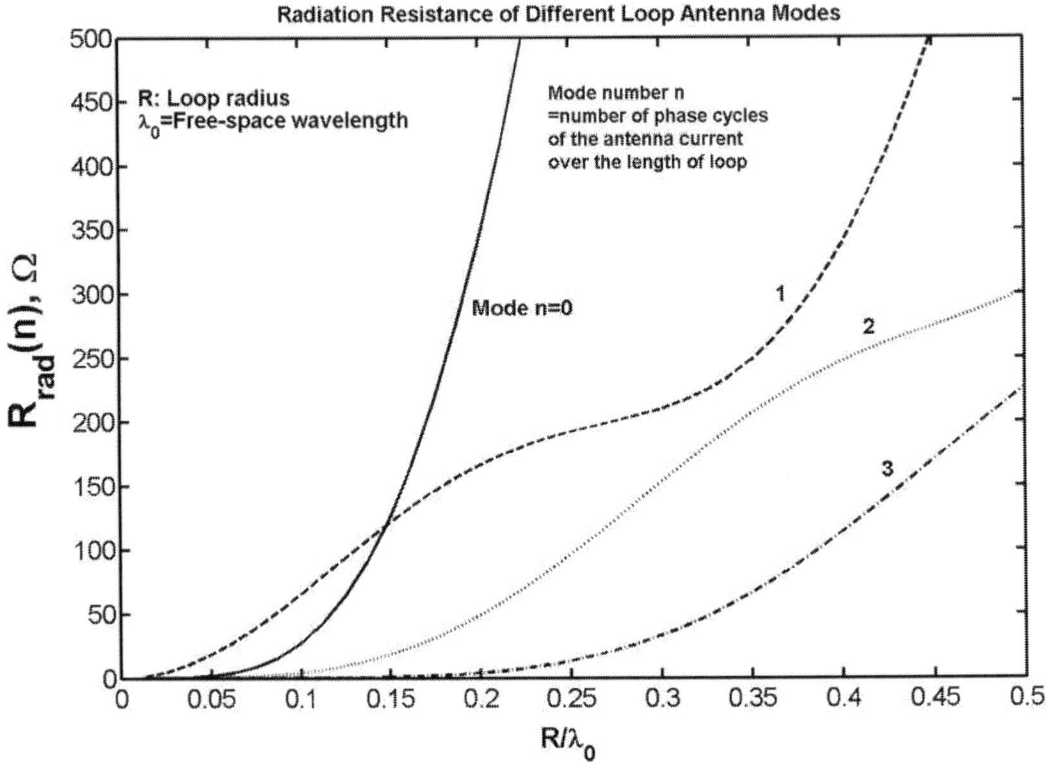

FIGURE 32

$$\overline{H}(r,\theta,\phi) = \frac{I_n R k_0 j^n}{2}\ \frac{e^{-jk_0 r}}{r}[\cos(n\phi){J'}_n(k_0 R\sin\theta)\hat{\theta}$$

$$-n\cos\theta\sin(n\phi)\frac{J_n(k_0 R\sin\theta)}{(k_0 R\sin\theta)}\hat{\phi}],\ .$$

$$\overline{E}(r,\theta,\phi) = \frac{I_n R k_0 \eta_0 j^n}{2}\ \frac{e^{-jk_0 r}}{r}[-\cos(n\phi){J'}_n(k_0 R\sin\theta)\hat{\phi}$$

$$-n\cos\theta\sin(n\phi)\frac{J_n(k_0 R\sin\theta)}{(k_0 R\sin\theta)}\hat{\theta}],$$

$$R_{rad} = \frac{\eta_0 R^2 k_0^2 \pi\delta_n}{2}\int\limits_{\theta=0}^{\pi}[J_n'^2(k_0 R\sin\theta) + n^2\cos^2\theta\frac{J_n^2(k_0 R\sin\theta)}{(k_0 R\sin\theta)^2}]\sin\theta d\theta,$$

$$D_g(\theta,\phi) = D_{g\phi}(\theta,\phi) + D_{g\theta}(\theta,\phi),$$

$$D_{g\phi}(\theta,\phi) = \frac{\eta_0 R^2 k_0^2\pi\delta_n}{R_{rad}}[J_n'^2(k_0 R\sin\theta)],$$

$$D_{g\theta}(\theta,\phi) = \frac{\eta_0 R^2 k_0^2\pi\delta_n}{R_{rad}}[n^2\cos^2\theta\frac{J_n^2(k_0 R\sin\theta)}{(k_0 R\sin\theta)^2}],$$

$$\delta_n = 1,\ n=0;\ = 1/2,\ n>0. \qquad (9.99)$$

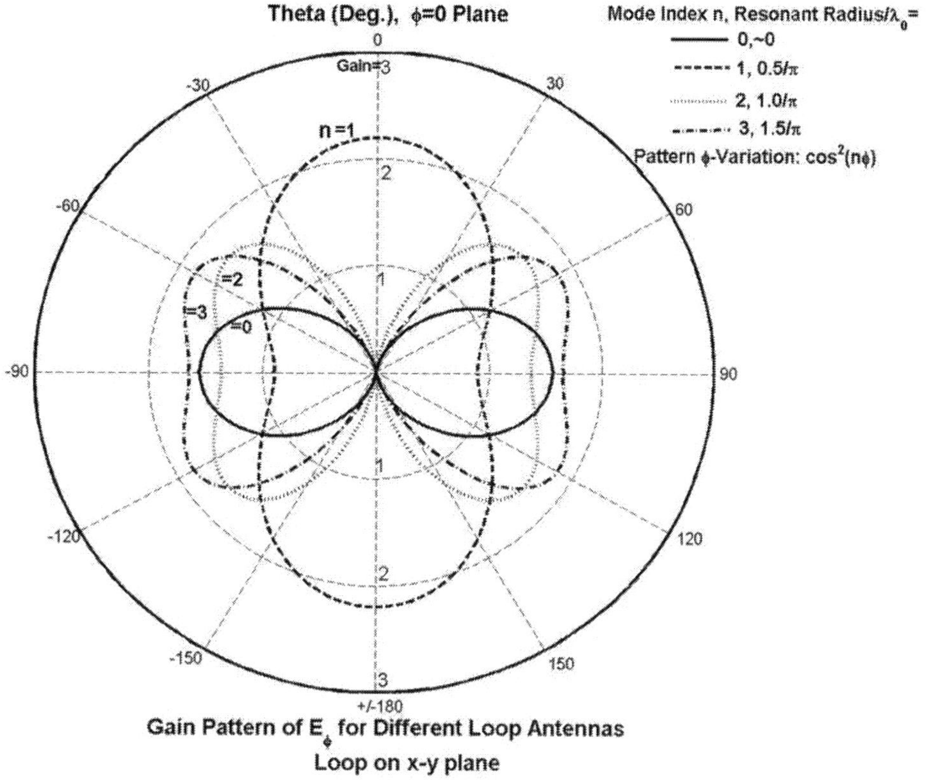

FIGURE 33

The radiation resistance R_{rad} computed by the above formula is plotted in Fig.32 for different mode indices n. The results in Fig.32 for $n = 0$ maybe verified with those in Fig.29, as derived in (9.92) specifically for $n = 0$. The separate parts, $D_{g\phi}$ and $D_{g\theta}$, of the total gain pattern D_g, contributed by the ϕ and θ components of the electric field, are plotted in Fig.33 and Fig.34, respectively, for different mode indices n. The loop sizes in these plots, for the different indices n, are selected close to the respective resonance conditions. The resonance conditions would be studied in the following section.

6.4. Rigorous Field Analysis, Uniform Current. The far-field analyses of last sections 6.1-6.3 do not provide the input reactance of the antenna. A rigorous field analysis, which includes fields close to the source would be developed here. A wire-loop antenna is modeled as a planar current strip, as shown in Fig.35, referred as a strip ring or a strip loop.

We will first model a simple case, with uniform current along the loop, and then extend the analysis to a general current distribution of the form $I_n \cos(n\phi)$ in the following section. For the simple case, the surface current on the strip loop of Fig.35 is expressed as,

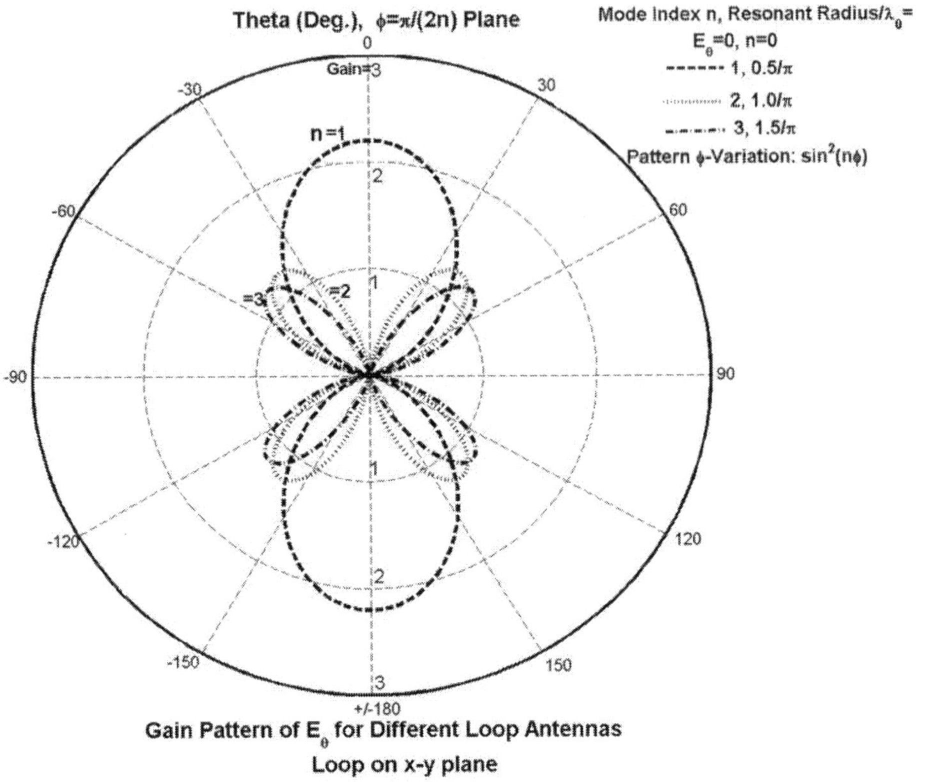

FIGURE 34

$$\begin{aligned}\bar{J}_s(\rho,\phi) &\cong \hat{\phi} I_{in}\frac{R}{w\rho} = (-\hat{x}sin\phi + \hat{y}cos\phi) I_{in}\frac{R}{w\rho},\ |\rho - R| < w/2;\\ &= 0,\ |\rho - R| > w/2.\end{aligned} \tag{9.100}$$

The current is uniform along the loop, with no ϕ dependence. A $1/\rho$ radial dependence is assumed, so that the product of the current at a given ρ and its path length $2\pi\rho$ along the curent over the complete circular loop, is the same for all locations ρ. In a practical situation with a non-zero resistance of the conducting strip, this would result in equal voltage V_{in} along any circular current path on the strip, When the strip width w is small, the current is essentially uniform over the width, with only a small variation contributed due to the $1/\rho$ factor.

$$\tilde{\tilde{\bar{J}}}_s = \int_y \int_x \bar{J}_s e^{-jk_x x} e^{-jk_y y} dx dy = \int_{\rho=0}^{\infty} \int_{\phi=0}^{2\pi} \bar{J}_s e^{-jk_x x} e^{-jk_y y} \rho d\phi d\rho$$

$$= \int_{\rho=0}^{\infty} \int_{\phi=0}^{2\pi} \bar{J}_s e^{-j(k_\rho cos\phi' \rho cos\phi + k_\rho sin\phi' \rho sin\phi)} \rho d\phi d\rho$$

$$= \int_{\Delta\rho} \int_{\phi=0}^{2\pi} I_{in} \frac{R}{w\rho} (-\hat{x} sin\phi + \hat{y} cos\phi) e^{-jk_\rho \rho cos(\phi - \phi')} \rho d\phi d\rho$$

$$= I_{in} \frac{R}{w} \int_{\Delta\rho} (-\hat{x} sin\phi' + \hat{y} cos\phi')(-j2\pi J_1(k_\rho \rho)) d\rho \qquad (9.101)$$

$$= I_{in} \frac{j2\pi R}{w} (\frac{k_y}{k_\rho} \hat{x} - \frac{k_x}{k_\rho} \hat{y}) \int_{\Delta\rho} J_1(k_\rho \rho) d\rho = I_{in} \frac{j2\pi R}{w k_\rho^2} \triangle J_0(k_\rho R)(k_y \hat{x} - k_x \hat{y}),$$

$$k_x = k_\rho \cos\phi', \; k_y = k_\rho \sin\phi',$$

$$\triangle\rho : R_- < \rho < R_+, \; R_+ = R + \frac{w}{2}, \; R_- = R - \frac{w}{2},$$

$$\frac{dJ_0(\alpha)}{d\alpha} = -J_1(\alpha), \; \triangle J_0(k_\rho R) \; = \; J_0(k_\rho R_-) - J_0(k_\rho R_+). \qquad (9.102)$$

$J_n(.)$ refers to the Bessel function of the first kind, of order n. The relationship (9.96) is used in (9.101), with $n = 1$ and the ϕ and ϕ' variables switched (ϕ' refer to different coordinates in (9.96) and (9.101)).

The potential function ψ_{ze} can be expressed using the derivation of chapter 6, section 5, with proper substitution for the source-current transform.

$$\bar{J}_s = \frac{1}{4\pi^2} \int \int \tilde{\tilde{\bar{J}}}_s \, e^{jk_x x} e^{jk_y y} dk_x dk_y$$

$$= \frac{1}{4\pi^2} \int \int I_{in} \frac{j2\pi R}{k_\rho^2 w} \triangle J_0(k_\rho R)(k_y \hat{x} - k_x \hat{y}) e^{+jk_x x} e^{+jk_y y} dk_x dk_y,$$

$$\psi_{ze}(x, y, z) = \frac{1}{4\pi^2} \int \int I_{in} \frac{j2\pi R}{k_\rho^2 w} \triangle J_0(k_\rho R)(\frac{-j\omega\mu}{2k_z}) e^{jk_x x} e^{jk_y y} e^{\mp jk_z z} dk_x dk_y. \qquad (9.103)$$

The electric field $\overline{E}$ can be derived from the above potential using the field-potential relations of (5.35).

$$\overline{E} = -\hat{x} \frac{\partial \psi_{ze}}{\partial y} + \hat{y} \frac{\partial \psi_{ze}}{\partial x}$$

$$= I_{in} \frac{R\omega\mu}{4\pi w} \int \int \frac{\triangle J_0(k_\rho R)}{k_\rho^2 k_z} (-jk_y \hat{x} + jk_x \hat{y}) e^{jk_x x} e^{jk_y y} e^{\mp jk_z z} dk_x dk_y. \qquad (9.104)$$

The source power can be expressed using the electric field of (9.104) and the current transform of (9.102), from which the input impedance is obtained.

$$
\begin{aligned}
P = -\frac{1}{2}\int\int \bar{E}(z=0)\cdot \bar{J}^* ds = \frac{1}{2}|I_{in}|^2 Z_{in},\ Z_{in} = \frac{2P}{|I_{in}|^2},, \\
2P = -I_{in}\frac{R\omega\mu}{4\pi w}\int\int \frac{\triangle J_0(k_\rho a)}{k_\rho^2 k_z}(-jk_y\hat{x} + jk_x\hat{y})\cdot(\tilde{\tilde{J}}\ (k_x,k_y))^* dk_x dk_y \\
= |I_{in}|^2 \frac{R^2\omega\mu}{2w^2}\int\int \frac{|\triangle J_0(k_\rho R)|^2}{k_\rho^2 k_z} dk_x dk_y, \\
Z_{in} = \frac{2P}{|I_{in}|^2} = \frac{R^2 k\eta}{2w^2}\int_{k_\rho=0}^{\infty}\int_{\phi'=0}^{2\pi} \frac{|\triangle J_0(k_\rho R)|^2}{k_\rho^2 k_z} k_\rho d\phi' dk_\rho \\
= \frac{R^2 k\eta}{2w^2}\int_{k_\rho=0}^{\infty}\int_{\phi'=0}^{2\pi} \frac{\left|J_0(k_\rho(R-\frac{w}{2}) - J_0(k_\rho(R+\frac{w}{2})\right|^2}{k_\rho k_z} d\phi' dk_\rho \\
= \frac{R^2 k\eta\pi}{w^2}\int_{k_\rho=0}^{\infty} \frac{\left|J_0(k_\rho(R-\frac{w}{2})) - J_0(k_\rho(R+\frac{w}{2}))\right|^2}{k_\rho k_z} dk_\rho = R_{in} + jX_{in}.
\end{aligned}
\tag{9.105}
$$

The real part of the impedance, R_{in}, is contributed by integration over the range $0 < k_\rho < k$, where k_z is real. Assuming no material loss, R_{in} is equal to the radiation resistance R_{rad}. The resistance values computed from the above expression are verified with those in Fig.29 or Fig.32 (mode number $n = 0$), obtained from the far-field analyses in sections 6.3, 6.2. The imaginary part of the impedance, X_{in}, is plotted in Fig.36 (mode index $n = 0$), together with that for other current variations to be derived in the following section. As can be seen from (9.105) and Fig.36, X_{in} for a uniform loop is always inductive ($X_{in} > 0$).

For a small ($R << \lambda_0$) and thin ($w << \lambda_0$) loop, the following simplification is possible. This would lead to a simple expression for the input resistance $R_{in} = R_{rad}$, which maybe checked with that derived in section 6.1 from the far-field analysis.

$$
\begin{aligned}
J_0(k_\rho(R-\frac{w}{2})) - J_0(k_\rho(R+\frac{w}{2})) = -\frac{dJ_0(k_\rho\rho)}{d\rho}\Big|_{\rho=R} \times w \\
= -\frac{dJ_0(k_\rho\rho)}{d(k_\rho\rho)}\Big|_{\rho=R} \times k_\rho w = J_1(k_\rho R)k_\rho w \simeq \frac{k_\rho R}{2}k_\rho w
\end{aligned}
\tag{9.106}
$$

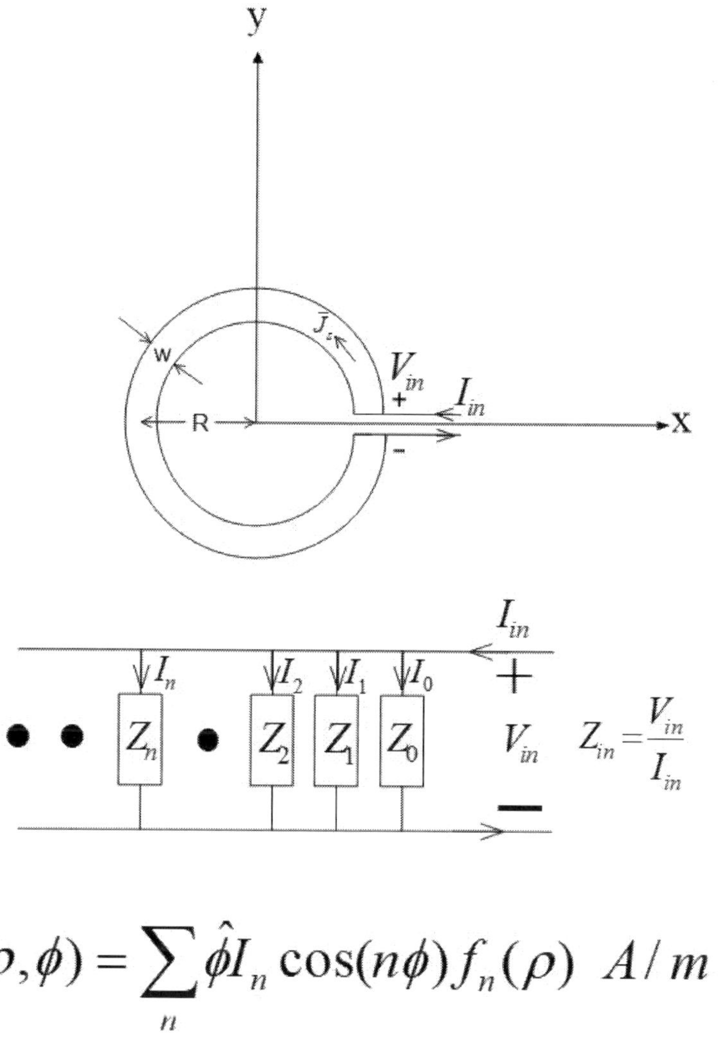

FIGURE 35

$$
\begin{aligned}
R_{in}(R, w << \lambda) &= \frac{R^2 k\eta\pi}{w^2} \int\limits_{k_\rho=0}^{k} \frac{R^2 w^2}{4} k_\rho^4 \frac{1}{k_z k_\rho} dk_\rho, \ (k_z = \sqrt{k^2 - k_\rho^2}) \\
&= \frac{R^4 k\eta\pi}{4} \times k^3 \int\limits_{t=0}^{\frac{\pi}{2}} \frac{\sin^3 t \times k\cos t}{k\cos t} dt \ (k_\rho = k\sin t) \\
&= \frac{R^4 k^4 \eta\pi}{4} \times \frac{4}{6} = S^2 k^4 \times \frac{\eta}{6\pi} = 20 S^2 k^4 = R_{rad} \, . \qquad (9.107)
\end{aligned}
$$

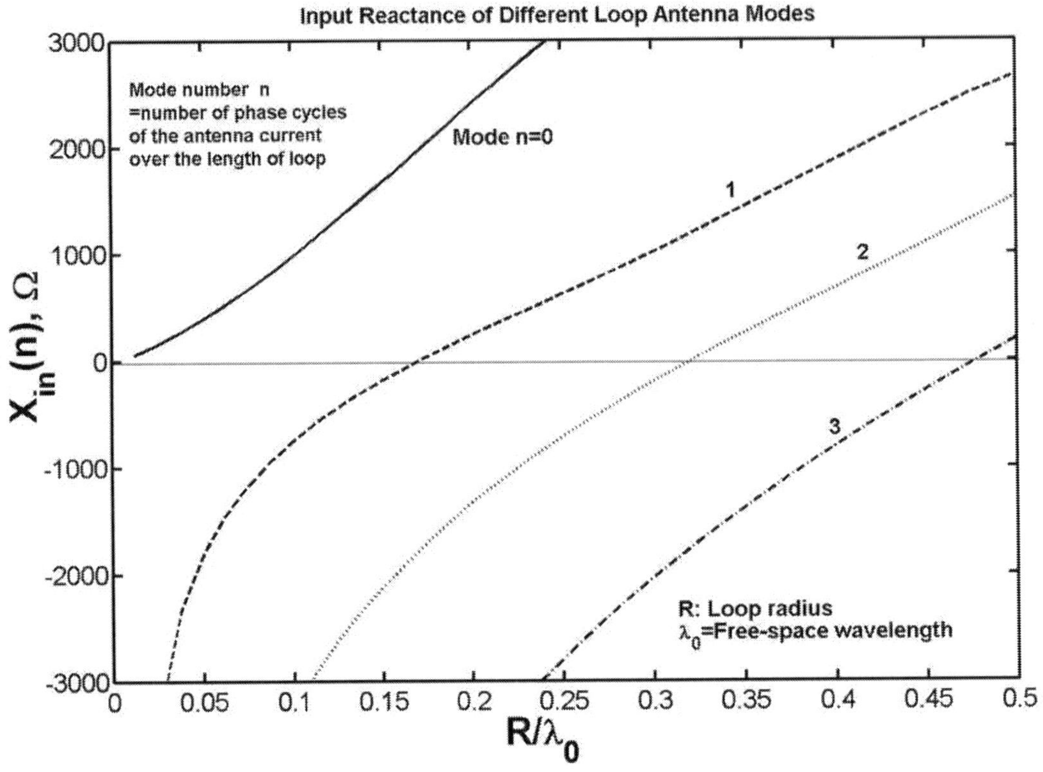

FIGURE 36

6.5. Rigorous Field Analysis, General $\cos(n\phi)$ **Current.** In general, the surface current $\overline{J}_s$ on the strip ring of Fig.35 is expressed as a superposition of $\cos(n\phi)$ distributions.

$$\bar{J}_s(\rho,\phi) = \sum_{n=0}^{N} \hat{\phi} I_n \cos(n\phi) f_n(\rho) = \sum_{n=0}^{N} \bar{J}_{sn}(\rho,\phi),$$

$$\begin{aligned} \bar{J}_{sn}(\rho,\phi) &= \hat{\phi} I_n \cos(n\phi) f_n(\rho) = (-\sin\phi\hat{x} + \cos\phi\hat{y}) I_n \cos(n\phi) f_n(\rho) \\ &= \hat{x} J_{snx}(\rho,\phi) + \hat{y} J_{sny}(\rho,\phi) \equiv \hat{x} J_{snx}(x,y) + \hat{y} J_{sny}(x,y). \end{aligned} \quad (9.108)$$

The radial distribution $f_n(\rho)$ of each mode n is normalized such that the mode current over the strip width w, measured at $\phi = 0$ (at the input terminal) is I_n.

We will analyze the fields produced by each modal distribution $\overline{J}_{sn}$. The analysis is the same as that developed specifically for the simple case of a uniform distribution (mode index $n = 0$) in the last section. However, in the general case of $n \neq 0$, the fields would include both TM_z and TE_z fields, in contrast to only TE_z fields for the simple case of $n = 0$. We would first derive the Fourier transforms of the current components J_{snx}

and J_{sny}, The current transform of J_{snx} is expressed in the following, in terms of Bessel functions J_n of the first kind.

$$
\begin{aligned}
\tilde{J}_{snx} &= \int\int J_{snx}(x, y)e^{-jk_x x}e^{-jk_x x}dxdy \\
&= \int_{\phi=0}^{2\pi}\int_{\rho=R-W/2}^{R+W/2} J_{snx}(\rho, \phi)e^{-j(k_\rho \rho \cos\phi' \cos\phi + k_\rho \rho \sin\phi' \sin\phi)}\rho d\rho d\phi \\
&= \int_{\phi=0}^{2\pi}\int_{\rho=R-W/2}^{R+W/2} -I_n \sin\phi \cos(n\phi) f_n(\rho)e^{-jk_\rho \rho \cos(\phi-\phi')}\rho d\rho d\phi \\
&= \int_{\rho=R-W/2}^{R+W/2}\int_{\phi=0}^{2\pi} -\frac{I_n}{2}(\sin((n+1)\phi) - \sin((n-1)\phi)) \\
&\qquad e^{-jk_\rho \cos(\phi-\phi')}d\phi f_n(\rho)\rho d\rho \\
&= \int_{\rho=R-W/2}^{R+W/2} -\pi j^{-(n+1)} I_n[\sin((n+1)\phi')J_{n+1}(k_\rho \rho) + \sin((n-1)\phi') \\
&\qquad J_{n\;1}(k_\rho \rho)] f_n(\rho)\rho d\rho. \qquad (9.109)
\end{aligned}
$$

(k_x, k_y) and (k_ρ, ϕ') are the transform variables in the rectangular and polar coordinates, respectively. The Bessel-function relations established in (9.96), with suitable adjustment of variables, are used in the above simplification. Additional Bessel-function relations (see Appendix) can be used in order to reduce the ρ integration. A radial distribution function of the form $f_n(\rho) = R/(W\rho)$ or $f_n(\rho) = R^{n-1}/(W\rho^{n-1})$ is selected, as suitable for mathematical simplification. For $W << R$, either of the forms would essentially represent a uniform distribution over the strip width,

$$
\begin{aligned}
\tilde{J}_{snx} = & \int_{\rho=R-W/2}^{R+W/2} -\pi j^{-(n+1)} I_n[\sin(n\phi')\cos\phi'(J_{n+1}(k_\rho\rho) + J_{n-1}(k_\rho\rho)) \\
& + \cos(n\phi;)\sin\phi'(J_{n+1}(k_\rho\rho) - J_{n-1}(k_\rho\rho))] f_n(\rho)\rho d\rho \\
& = \int_{\rho=R-W/2}^{R+W/2} -\pi j^{-(n+1)} I_n[\sin(n\phi')\cos\phi' \frac{2n}{k_\rho\rho} J_n(k_\rho\rho) \\
& - \cos(n\phi')\sin\phi' \; 2J_n'(k_\rho\rho)] f_n(\rho)\rho d\rho (see Appendix) \\
& = -2\pi j^{-(n+1)} I_n[\sin(n\phi')\cos\phi' \; nt_0^{n-2}(\frac{R}{Wk_\rho}) \int_{t=t_1}^{t_2} \frac{J_n(t)}{t^{n-1}} dt \\
& - \cos(n\phi')\sin\phi'(\frac{R}{Wk_\rho}) \int_{t=t_1}^{t_2} J_n'(t) dt] \\
& = -2\pi j^{-(n+1)} I_n \frac{R}{Wk_\rho}[\sin(n\phi')\cos\phi' \; nt_0^{n-2}(\frac{J_{n-1}(t_1)}{(t_1)^{n-1}} \\
& - \frac{J_{n-1}(t_2)}{(t_2)^{n-1}}) - \cos(n\phi')\sin\phi'(J_n(t_2) - J_n(t_1))] (see Appendix), \\
& f_n(\rho) = \frac{R}{W\rho} \; or \; \frac{R^{n-1}}{W\rho^{n-1}}, \; t = k_\rho\rho, \; d\rho = \frac{dt}{k_\rho}.
\end{aligned} \tag{9.110}
$$

The variables t_0, t_1 and t_2 are defined in (9.113).

The transform of the y component of the current J_{sny} can be similarly derived.

$$
\begin{aligned}
\tilde{J}_{sny} = & \int_{\rho=R-W/2}^{R+W/2} \int_{\phi=0}^{2\pi} \frac{I_n}{2}(\cos((n+1)\phi) + \cos((n-1)\phi)) \\
& e^{-jk_\rho\rho\cos(\phi-\phi')} d\phi f_n(\rho)\rho d\rho \\
= & \int_{\rho=R-W/2}^{R+W/2} \pi j^{-(n+1)} I_n[\cos((n+1)\phi')J_{n+1}(k_\rho\rho) - \cos((n-1)\phi') \\
& J_{n-1}(k_\rho\rho)] f_n(\rho)\rho d\rho \\
& = -2\pi j^{-(n+1)} I_n \frac{R}{Wk_\rho}[\sin(n\phi')\sin\phi' \; nt_0^{n-2}(\frac{J_{n-1}(t_1)}{(t_1)^{n-1}} \\
& - \frac{J_{n-1}(t_2)}{(t_2)^{n-1}}) + \cos(n\phi')\cos\phi'(J_n(t_2) - J_n(t_1))].
\end{aligned} \tag{9.111}
$$

The transform of the total mode current $\bar{J}_{sn}$ is described by the above transforms of the two componens J_{snx} and J_{sny}. The spatial distribution of the currents is then expressed as a Fourier summation or integration over the transform variables.

$$
\begin{aligned}
\tilde{\tilde{J}}_{sn} &= \hat{x}\tilde{J}_{snx} + \hat{y}\tilde{J}_{sny} \\
&= (\hat{x}\cos\phi' + \hat{y}\sin\phi')2\pi I_n \sin(n\phi')C_{1n} + (-\hat{x}\sin\phi' + \hat{y}\cos\phi')2\pi I_n \cos(n\phi')C_{2n}, \\
\bar{J}_{sn} &= \frac{1}{4\pi^2}\int\int \tilde{\tilde{J}}_{sn}\, e^{jk_x x}e^{jk_y y}dk_x dk_y \\
&= \frac{I_n}{2\pi}\int\int(\hat{x}\cos\phi' + \hat{y}\sin\phi')\sin(n\phi')C_{hn}e^{jk_x x}e^{jk_y y}dk_x dk_y \\
&+\frac{I_n}{2\pi}\int\int(-\hat{x}\sin\phi' + \hat{y}\cos\phi')\cos(n\phi')C_{en}e^{jk_x x}e^{jk_y y}dk_x dk_y, \quad (9.112)
\end{aligned}
$$

where the paramters C_{hn} and C_{en} are defined in the following..

$$
\begin{aligned}
&t_0 = k_\rho R,\ t_1 = k_\rho(R - W/2),\ t_2 = k_\rho(R + W/2), \\
&C_{hn} = j^{-(n+1)}\frac{R}{Wk_\rho}nt_o^{n-2}\Delta(\frac{J_{n-1}(t_o)}{t_o^{n-1}}),\ C_{en} = -j^{-(n+1)}\frac{R}{Wk_\rho}\Delta J_n(t_o), \\
&\Delta(\frac{J_{n-1}(t_0)}{t_0^{n-1}}) = \frac{J_{n-1}(t_2)}{t_2^{n-1}} - \frac{J_{n-1}(t_1)}{t_1^{n-1}},\ \Delta J_n(t_0) = J_n(t_2) - J_n(t_1). \quad (9.113)
\end{aligned}
$$

The first and second parts of the above current represent sources for TM_z and TE_z radiation, respectively. The respective potential functions can be expressed following the derivation in chapter 6, with proper substitution of the current transforms. The respective fields can then be derived from the potential functions using the field-potential relations of (5.34,5.35). The TM_z and TE_z fields and the associated paramters are represented with subscripts h and e respectively.

$$
\begin{aligned}
\psi_{nzm} &= \pm\frac{I_n}{2\pi}\int\int\sin(n\phi')C_{hn}(\frac{-j}{2k_\rho})e^{jk_x x}e^{jk_y y}e^{\mp jk_z z}dk_x dk_y, \\
\psi_{nze} &= \frac{I_n}{2\pi}\int\int\cos(n\phi')C_{en}(\frac{j\omega\mu_o}{2k_z k_\rho})e^{jk_x x}e^{jk_y y}e^{\mp jk_z z}dk_x dk_y, \\
k_z &= \sqrt{k^2 - k_x^2 - k_y^2},\ Re(k_z) > 0,\ Im(k_z) < 0. \quad (9.114)
\end{aligned}
$$

$$
\begin{aligned}
\overline{E}_{hn} &= \frac{1}{j\omega\epsilon}(\frac{\partial^2}{\partial x\partial z}\hat{x} + \frac{\partial^2}{\partial y\partial z}\hat{y} + (\frac{\partial^2}{\partial z^2} + k^2)\hat{z})\psi_{nzm} \\
&= \frac{I_n}{2\pi}\int\int\sin(n\phi')C_{hn}(-\frac{k_z}{2\omega\epsilon})(\frac{k_x}{k_\rho}\hat{x} + \frac{k_y}{k_\rho}\hat{y} + \frac{k_\rho}{k_z}\hat{z})e^{jk_x x}e^{jk_y y}e^{\mp jk_z z}dk_x dk_y, \\
\overline{E}_{en} &= (-\hat{x}\frac{\partial}{\partial y} + \hat{y}\frac{\partial}{\partial y})\psi_{nze} \\
&= \frac{I_n}{2\pi}\int\int\cos(n\phi')C_{en}(-\frac{\omega\mu_o}{2k_z})(-\frac{k_y}{k_\rho}\hat{x} + \frac{k_x}{k_\rho}\hat{y})e^{jk_x x}e^{jk_y y}e^{\mp jk_z z}dk_x dk_y. \quad (9.115)
\end{aligned}
$$

The sign of the square-root expression for k_z in (9.114) is properly selected, in order that it represents a valid physical situation with waves propagating or decaying away from the source.

The power delivered from each mode is obtained from the above fields and the current transforms of (9.112). The power is then related to an equivalent input impedance $Z_{in}(n) = Z_n$ for the particular mode n.

$$P_{hn} = -\frac{1}{2}\int\int_{z=0} \overline{E}_{hn}\cdot \bar{J}^*_{sn} ds = \frac{1}{2}|I_n|^2 Z_{hn},\ Z_{hn} = \frac{2P_{hn}}{|I_n|^2},$$

$$2P_{hn} = -\frac{I_n}{2\pi}\int\int \sin(n\phi')C_{hn}(-\frac{k_z}{2\omega\epsilon})(\frac{k_x}{k_\rho}\hat{x} + \frac{k_x}{k_\rho}\hat{y} + \frac{k_\rho}{k_z}\hat{z})\cdot(\tilde{\tilde{J}}_{sn}\ (k_x, k_y))^* dk_x dk_y$$

$$= -|I_n|^2\int\int \sin(n\phi')C_{hn}(-\frac{k_z}{2\omega\epsilon})\sin(n\phi')C^*_{hn} dk_x dk_y$$

$$= |I_n|^2\int_{k_\rho=0}^{\infty}\int_{\phi'=0}^{2\pi} \sin^2(n\phi')|C_{hn}|^2\frac{k_z}{2\omega\epsilon} d\phi' k_\rho dk_\rho,$$

$$Z_{hn} = \frac{2P_{hn}}{|I_n|^2} = \pi\eta\delta_n\int_0^{\infty}|C_{hn}|^2\frac{k_z k_\rho}{k}dk_\rho;\ \delta_n = 0,\ n=0;\ = \frac{1}{2},\ n>0 \quad (9.116)$$

$$P_{en} = -\frac{1}{2}\int\int_{z=0} \overline{E}_{en}\cdot \bar{J}^*_{sn} ds = \frac{1}{2}|I_n|^2 Z_{en},\ Z_{en} = \frac{2P_{en}}{|I_n|^2},$$

$$2P_{en} = -\frac{I_n}{2\pi}\int\int \cos(n\phi')C_{en}(-\frac{\omega\mu_o}{2k_z})(-\frac{k_y}{k_\rho}\hat{x} + \frac{k_x}{k_\rho}\hat{y})\cdot(\tilde{\tilde{J}}_{sn}\ (k_x, k_y))^* dk_x dk_y$$

$$= -|I_n|^2\int\int \cos(n\phi')C_{en}(-\frac{\omega\mu_o}{2k_z})\cos(n\phi')C^*_{en} dk_x dk_y$$

$$= |I_n|^2\int_{k_\rho=0}^{\infty}\int_{\phi'=0}^{2\pi} \cos^2(n\phi')|C_{2n}|^2\frac{\omega\mu_o}{2k_z} d\phi' k_\rho dk_\rho,$$

$$Z_{en} = \frac{2P_{en}}{|I_n|^2} = \pi\eta\delta'_n\int_0^{\infty}|C_{en}|^2\frac{k_\rho k}{k_z}dk_\rho,\ \delta'_n = 1,\ n=0;\ = \frac{1}{2},\ n>0. \quad (9.117)$$

The total impedance Z_n for the mode n is the sum of the two parts, Z_{hn} and Z_{en}, contributed from the individual TM_z and TE_z fields, respectively.

$$Z_n = \frac{2P_n}{|I_n|^2} = \frac{2(P_{hn}+P_{en})}{|I_n|^2} = \frac{2P_{hn}}{|I_n|^2} + \frac{2P_{en}}{|I_n|^2} = Z_{hn} + Z_{en}. \quad (9.118)$$

It may be noted, that if the field $\overline{E}$ and current $\overline{J}_s$ in the derivation of (9.116,9.117), correspond to different modal indices, the resulting complex power would be zero. In other words, the power delivered from a particular current mode $n = i$, contributed by the fields produced from a different mode $n \neq i$, would be zero. This is due the orthogonal nature of the ϕ variations of the currents and interacting fields from the different modes, resulting in a zero power integral. Accordingly, the total power delivered from the strip-ring of Fig.35 is the sum of the powers from each individual modes, with no cross-coupling

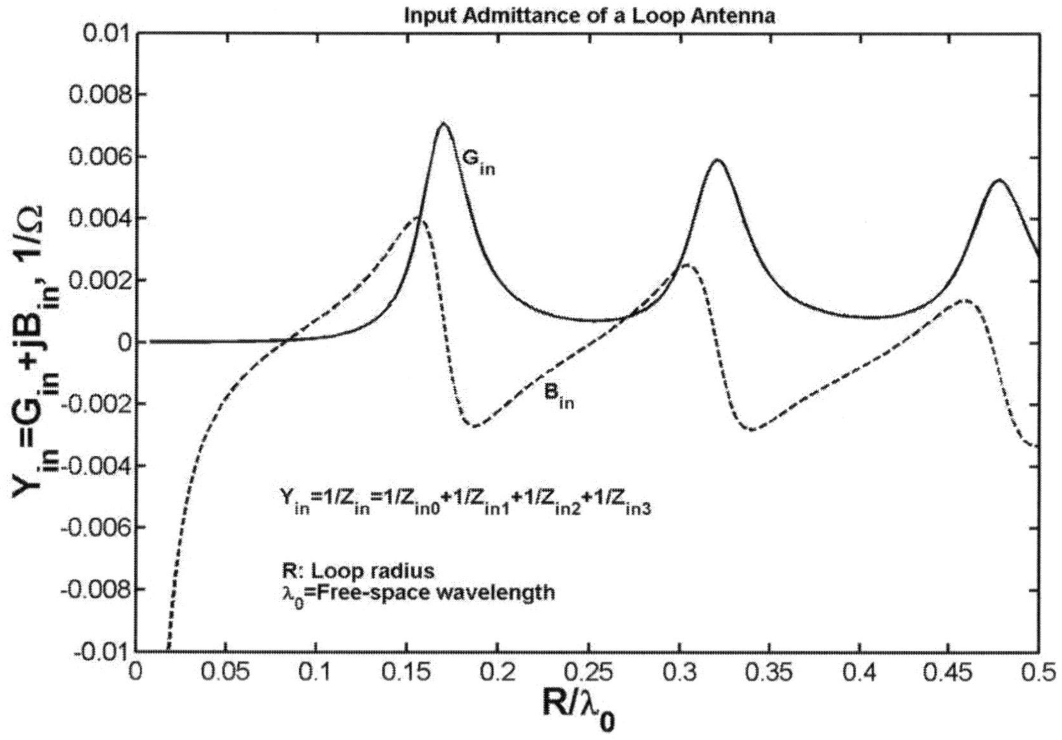

FIGURE 37

contribution. Further, each current mode n sees the same input voltage V_{in}, as shown in Fig.35. Therefore, the orthogonal current modes can be represented as equivalent impedances across the input terminal, all connected in parallel to each other.

$$P_{in} = \frac{|V_{in}|^2}{Z_{in}^*} = \sum_n P_n = \sum_n \frac{|V_{in}|^2}{Z_n^*}, \quad \frac{1}{Z_{in}} = \sum_n \frac{1}{Z_n} = \sum_n \frac{1}{(Z_{hn} + Z_{hn})}. \tag{9.119}$$

The real and imaginary parts of the modal impedance $Z_n = Z_{in}(n)$ are plotted in Figs.32,36, for different indices n, as a function of the loop size. The results for $n = 0$ (uniform current) are verified with those from the last section 6.4, specifically derived for the simple case. The reactance for a uniform ring is seen in Fig.36 to be always inductive, which may also be analytically verified from (9.116,9.117). For all other modes $n \neq 0$, the reactance $X_n = X_{in}(n)$ changes sign from negative to positive, as the size of the loop is increased. The resonant radius R_{0n} of the loop, where the modal reactance X_n in Fig.36 is zero, is about $n\lambda_0/(2\pi)$. This corresponds to having the circular loop-length equal to the n times the operating wavelength λ_0. For operation close to the above resonance condition for a mode n, the modal impedance Z_n would be small compared to that for all other non-resonating modes. Accordingly, the total input impedance Z_{in} in (9.119)

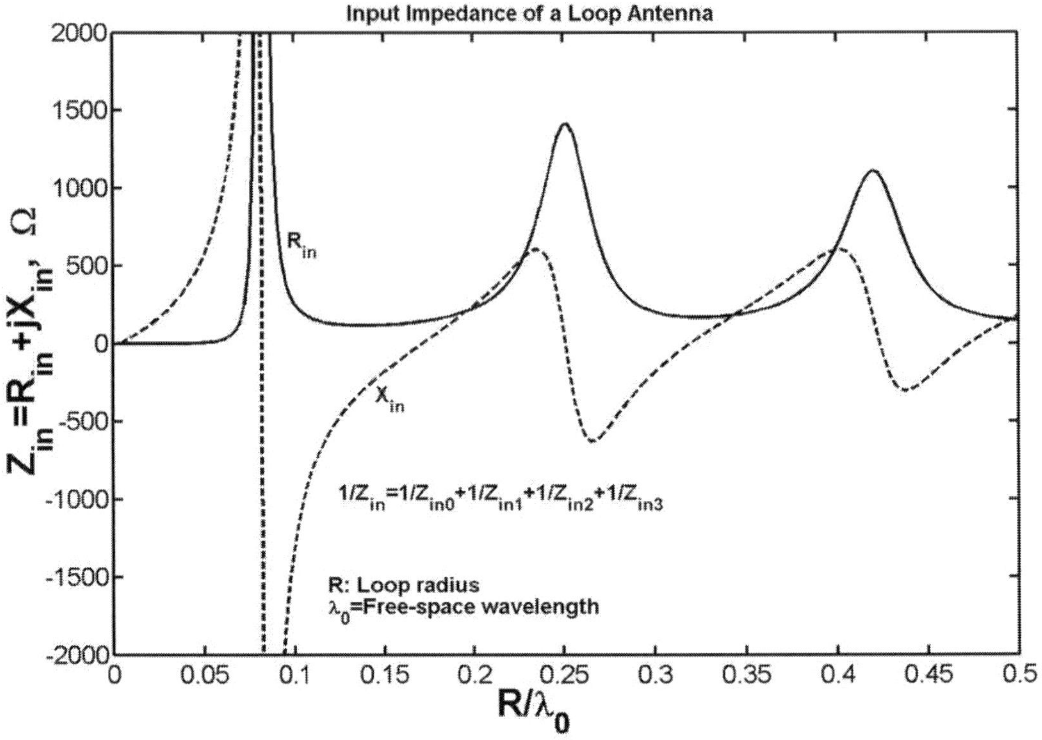

FIGURE 38

can be approximately modeled only by the impedance Z_n of the resonant mode. with negligible contributions from all other modes.

The total impedance Z_{in} and admittance Y_{in}, computed with the selected modes using (9.119), are shown in Figs.38,37, respectively.

Mutual Impedance Between Two Loops

When two elements are operating in the presence of each other, the mutual interaction between them is modeled in terms of mutual impedance Z_m. The general concept and formulation is similar to the mutual impedance modeling for strip-dipole antennas in section 3.3. Consider two current modes with modal indices $n = i$ and $n = l$, which are in general different, represented by their strip currents $\overline{J}_{si}$ and $\overline{J}_{sl}$. The respective radii R_i and R_l, and strip widths w_i and w_l, may also be different, in general. Their center reference points are displaced from each other, as shown in Fig.39. Fore convenience, one of the elements ($n = i$, "source element") is placed at the reference origin, while the other ($n = l$, "receiving element") is assumed to be parallel to the source element placed on the $z = z_0$ plane, centered at (x_0, y_0, z_0). We also assume for convenience, that the two modes are rotationally displaced with each other, with their reference ports aligned along the x axis. Interchanging the two elements would produce the same mutual impedance, with $Z_m = Z_{il} = Z_{li}$, as per the reciprocity principle (see chapter 2, section 4.3.) .

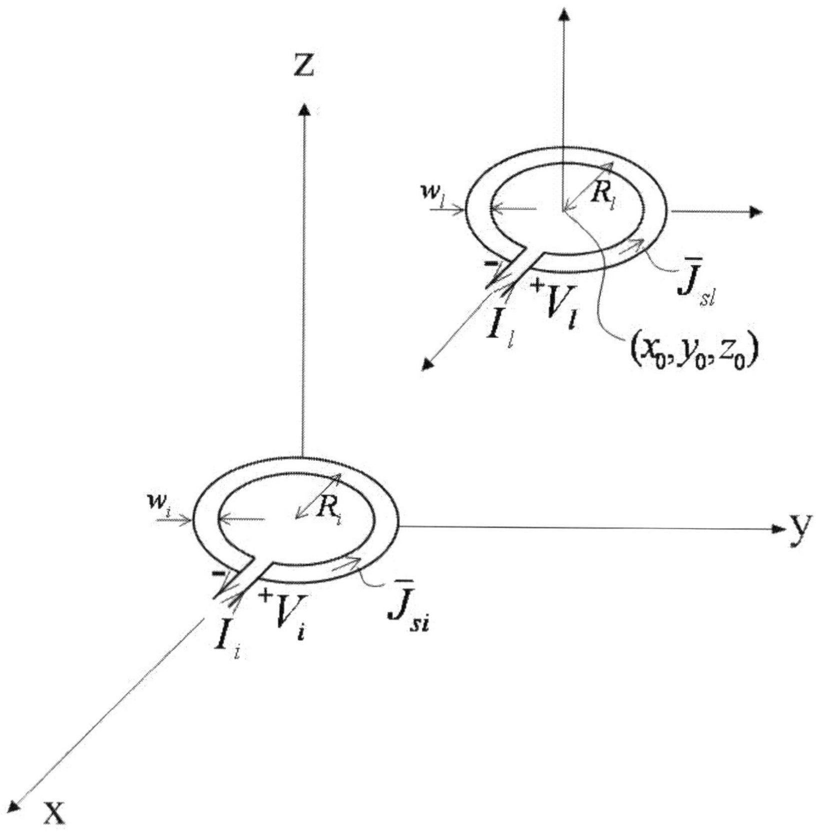

FIGURE 39

The derivation in (9.116-9.118) of the self-impedance Z_n can be extended to the mutual impedance Z_{il}, with suitable adjustment and substitution of variables. In the derivation of (9.116-9.118), we used the same current as a source for the electric field, as well as the receiving or interacting current to evaluate the power integration. For mutual impedance calculation, we need to substitute the index of the source current as $n = i$, that of the receiving or interacting current as $n = l$, and use different physical paramters for the modes accordingly. In addition, the plane of evaluation of the electric field should be $z = z_0$, and the center of the $n = l$ mode be moved to $x = x_0$ and $y = y_0$. This is in contrast to having both modes co-located at the origin, as was the case in the derivation of (9.116-9.118). With the above changes, the mutual impedance Z_{il} can be expressed as follows.

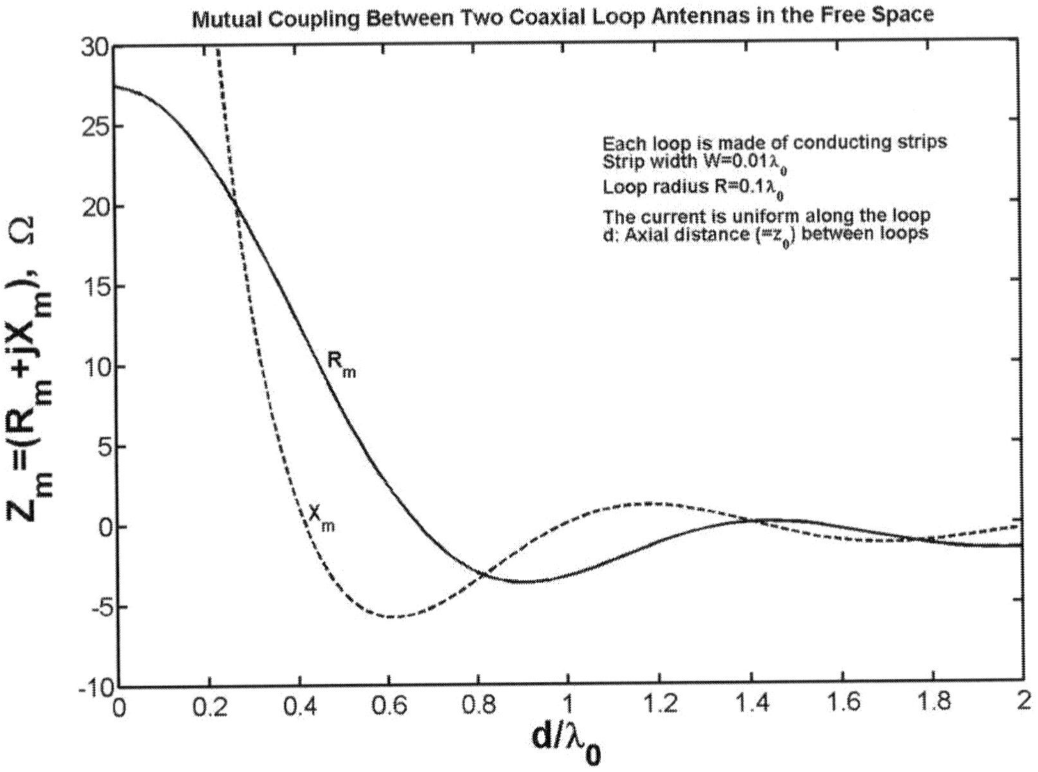

FIGURE 40

$$P_{hil} = -\frac{1}{2}\int\int_{z=z_0} \overline{E}_{hi}(x_0, y_0, z_0)\cdot \bar{J}^*_{sl} ds = \frac{1}{2} I_i I^*_l Z_{hil},$$

$$Z_{hil} = \frac{2P_{hil}}{I_i I^*_l} = \int\int \sin(i\phi')C_{hi}(\frac{k_z}{2\omega\epsilon})\sin(l\phi')C^*_{hl}e^{jk_x x_0}e^{jk_y y_0}e^{\mp jk_z z_0}dk_x dk_y,$$

$$P_{eil} = -\frac{1}{2}\int\int_{z=0} \overline{E}_{ei}(x_0, y_0, z_0)\cdot \bar{J}^*_{sl} ds = \frac{1}{2} I_i I^*_l Z_{eil},$$

$$Z_{eil} = \frac{2P_{eil}}{I_i I^*_l} = \int\int \cos(i\phi')C_{ei}(\frac{\omega\mu_o}{2k_z})\cos(l\phi')C^*_{el}e^{jk_x x_0}e^{jk_y y_0}e^{\mp jk_z z_0}dk_x dk_y,$$

$$Z_{il} = \frac{2P_{il}}{I_i I^*_l} = \frac{2(P_{hil}+P_{eil})}{I_i I^*_l} = \frac{2P_{hil}}{I_i I^*_l} + \frac{2P_{en}}{I_i I^*_l} = Z_{hil} + Z_{eil}. \quad (9.120)$$

where the paramters C_{hi}, C_{hl}, C_{ei} and C_{el} are expressed in (9.113) with index n substituted by i or l, and the radius R by R_i or R_l, and the strip width w by w_i or w_l, respectively.

For general relative positioning, there will be a non-zero mutual impedance between two modes of different indices $i \neq l$. That is in contrast with a zero mutual impedance

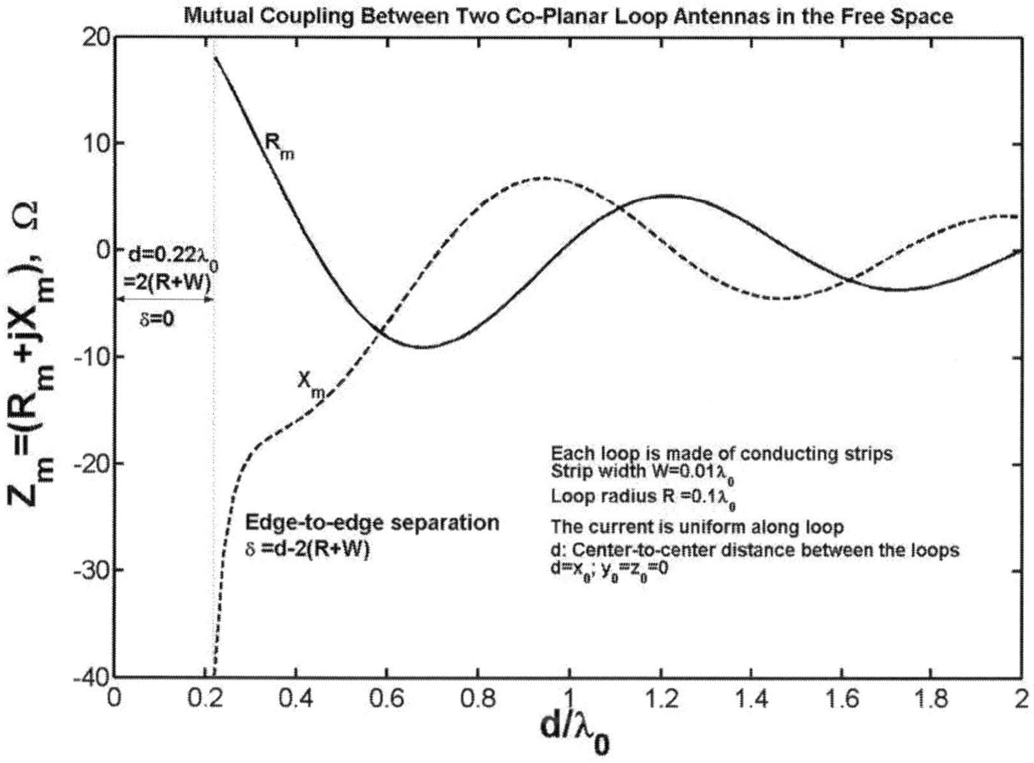

FIGURE 41

between the two different modes, when they are co-located or coaxial. Fig.40 plots the mutual impedance between two idential loops, with uniform currents ($n = 0$), coaxially displaced from each other along the z axis. Whereas, Fig.41 plots the mutual impedance between the same loops when thay are displaced on the plane of the loop, along the x axis.

7. Slot Antenna

A slot antenna, as shown in Fig.42, is considered another basic radiating element. It is realized by removing a small area of metal from an infinite ground plane. For a practical design, however, the ground plane may be finite but large in size. The open area, or the "slot," can be of any shape and size, but is shown in Fig.42 to be rectangular as often used for simplicity. As shown in Fig.1, a slot antenna may be excited by applying a voltage source across the slot.

Babinet's Principle

One may view a slot antenna to be physically "complementary" to a strip-dipole antenna, consisting of a metal strip of the same shape and size as that of the slot. The complementary strip-dipole is to be excited by a "voltage gap," breaking across the strip at the same position where the slot antenna is excited. If the slot-antenna is superimposed

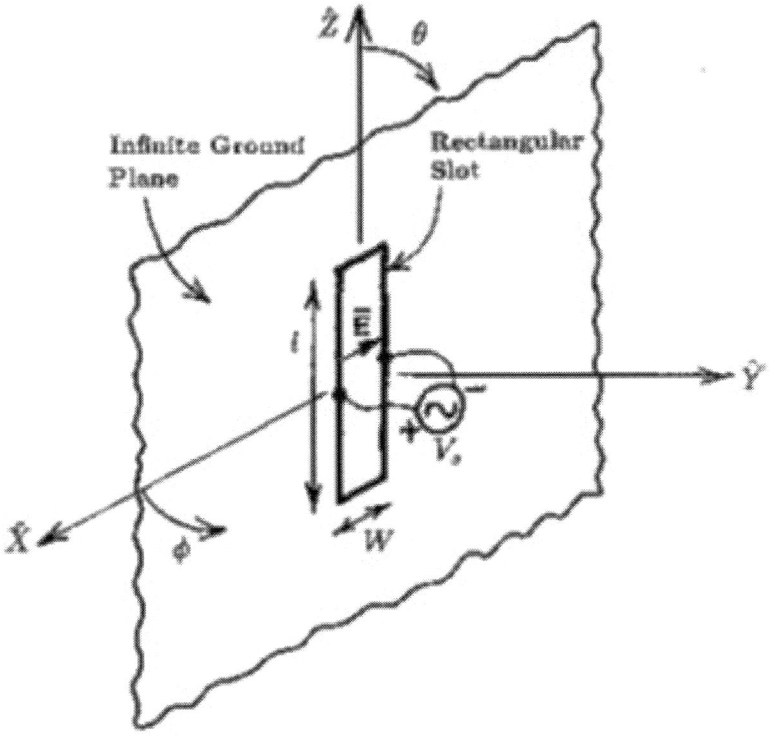

FIGURE 42

on its complementary dipole-antenna structure, one would obtain an infinite conducting plane. Interestingly, employing certain symmetry in Maxwell's equations, and the complementary nature of the above two structures, it can be shown that the fields of a slot antenna and that of a complementary dipole antenna are related to each other in a simple manner. This theory is called *Babinet's principle* [**27**], which consequently relates the various transmit/receive characteristics, such as the pattern, gain and input impedance, of the two complementary antennas.

According to the *Babïnet's principle*, the fields produced by a slot antenna can be obtained directly from the corresponding complementary strip-dipole antenna using the following substitutions:

• (1) for the front side of the slot antenna of ($0 < \phi < \pi$, see Fig.42,) replace $\overline{E}$ of the dipole with $-\overline{H}$ for the slot antenna, $\overline{H}$ with $\overline{E}$, ϵ_0 with μ_0, μ_0 with ϵ_0, and η_0 with $1/\eta_0$, but replace the input current I_{in} of the strip dipole with twice the source voltage V_s for the slot antenna.

• (2) For the back side ($\pi < \phi < 2\pi$) one needs the same substitutions as for the front side, except the signs for the field expressions are reversed.

The above substitutions would also lead to a simple relationship between the complex input impedances of the complementary antennas. If Z_{sl} is the complex input impedance as seen across the input of an arbitrary slot antenna, and Z_{cd} is that of its complementary dipole antenna, then using the substitutions prescribed in the Babinet's principle one can show,

$$4\frac{Z_{cd}}{\eta_0} = \frac{\eta_0}{Z_{sl}}, \; Z_{cd}Z_{sl} = \frac{\eta_0^2}{4}. \tag{9.121}$$

7.1. Small Slot Antenna. As an example, let us apply the Babinet's principle to a small rectangular slot antenna, whose complementary structure is a small dipole antenna studied in section 1. Employing the substitutions prescribed above, the far-field expressions for the small slot antenna can be obtained from those in (9.2,9.3) for a short dipole antenna.

$$\overline{H}(r,\theta,\phi) = \begin{cases} -\hat{\theta}V_s\frac{e^{-jk_0 r}}{r}\frac{jl}{\eta_0\lambda_0}\sin\theta; \; 0 < \phi < \pi, \\ \hat{\theta}V_s\frac{e^{-jk_0 r}}{r}\frac{jl}{\eta_0\lambda_0}\sin\theta; \; \pi < \phi < 2\pi, \end{cases} \tag{9.122}$$

$$\overline{E}(r,\theta,\phi) = \begin{cases} \hat{\phi}V_s\frac{e^{-jk_0 r}}{r}\frac{jl}{\lambda_0}\sin\theta; \; 0 < \phi < \pi, \\ -\hat{\phi}V_s\frac{e^{-jk_0 r}}{r}\frac{jl}{\lambda_0}\sin\theta; \; \pi < \phi < 2\pi. \end{cases} \tag{9.123}$$

Using the above the field expressions, one can obtain the antenna parameters for the small slot antenna following the similar steps in section 1 for its complementary dipole antenna.

$$P_d(r,\theta,\phi) = \frac{|V_s|^2}{2\eta_0 r^2}(\frac{l}{\lambda_0})^2\sin^2\theta, \; p_d(\theta,\phi) = \frac{|V_s|^2}{2\eta_0}(\frac{l}{\lambda_0})^2\sin^2\theta, \tag{9.124}$$

$$\underline{p}_d(\theta,\phi) = \frac{p_d(\theta,\phi)}{p_{dmax}} = \sin^2\theta, \tag{9.125}$$

$$P_{rad} = \int_{\phi=0}^{2\pi}\int_{\theta=0}^{\pi} p_d(\theta,\phi)\sin\theta\, d\theta\, d\phi = \frac{|V_s|^2 4\pi}{3\eta_0}(\frac{l}{\lambda_0})^2 = \frac{|V_s|^2}{2R_{rad}}, \tag{9.126}$$

$$G_{rad} = \frac{1}{R_{rad}} = \frac{8\pi}{3n_0}(\frac{l}{\lambda_0})^2 = \frac{1}{45}(\frac{l}{\lambda_0})^2, \tag{9.127}$$

$$D_g(\theta,\phi) = \frac{3\sin^2\theta}{2}; \; D = 1.5. \tag{9.128}$$

Comparing the radiation resistances of a small slot in (9.127) to that of a small dipole antenna in (9.8), we see that $R_{rad}(dipole) \times R_{rad}(slot) = \frac{\eta_0^2}{4}$, which is consistent with the general impedance relationship of (9.121). As shown in (9.125), a narrow slot antenna (see Fig.42) has the same $\sin^2\theta$ power pattern as that of a small dipole antenna. The small slot antenna has E_ϕ and H_θ radiation fields, in contrast with the E_θ and H_ϕ radiation fields of its complementary small dipole antenna.

Similar procedure can be extended to a half-wave slot antenna, or in general to a rectangular slot antenna of any arbitrary length. This will allow derivation of the far-fields as well as radiation/impedance parameters of a long slot antenna, by relating to those of the complementary dipole structures in section 2.

8. Microstrip Antenna Printed on a Dielectric Substrate

Figs.43, 52 show geometries of a strip antenna - also referred to as a microstrip antenna - printed on a dielectric substrate. The basic structure of a microstrip antenna consists of a dielectric substrate of thickness d, having a complete metalization on one of its surfaces, and a thin metal strip printed on the opposite surface. The substrate is usually thin compared to the operating wave length ($d << \lambda_0$). The metal strip on the top surface can have various shapes, though a rectangular shape as shown in the Figs.43, 52 is commonly used. The strip width w may be narrow, comparable to or less than the substrate thickness d, in which case the antenna is commonly referred to as a printed-dipole or a microstrip-dipole element. On the other hand, a microstrip antenna with a wide width $w >> d$ is casually referred to as a "patch" antenna, in reference to its physical appearance as a square or rectangular patch of conductor covering the dielectric substrate.

The microstip antenna can be excited using various methods [**37, 38**]. We will model two basic configurations of input excitations. The antenna of Fig.43 is shown to be excited from an ideal delta-gap voltage source at the center of the dipole. An alternate feeding arrangement using a coaxial probe, which is normally positioned off the center of the antenna, is shown in Fig.52. In the following sections 8.1-8.6, we will model the basic configuration of Fig.43 with a delta-gap feed, using a formulation which is an extension of that used in the study of a small printed-dipole element in chapter 6, section 6. The characteristics of a coaxial feeding configuration of Fig.52 are then related in section 8.7 to those of the delta-gap feeding configuration of Fig.43, using a impedance transformation formula.

The microstrip antenna is a relatively modern invention. It was introduced in order to allow convenient integration of antenna elements and associated driving circuitry of a communication system, together on a common printed-circuit board or a semiconductor chip [**39, 37**]. In addition to the above convenience and resulting economic advantages, the integrated-circuit technology for the antenna fabrication also allowed high dimensional accuracy, which was otherwise difficult to achieve in a traditional antenna fabrication.

A practical microstrip antenna of Fig.43 or 52, printed on a thin dielectric substrate, is expected to produce maximum radiation in the broadside (perpendicular to the substrate, $\theta = 0^0$) direction, and ideally no radiation in the end-fire (along the surface of the substrate, $\theta = 90^0$) direction. The size of the antenna is usually designed such that the antenna resonates at the operating frequency, producing a real input impedance. For a rectangular microstrip antenna, this requires the length of the antenna, L, to be close to a

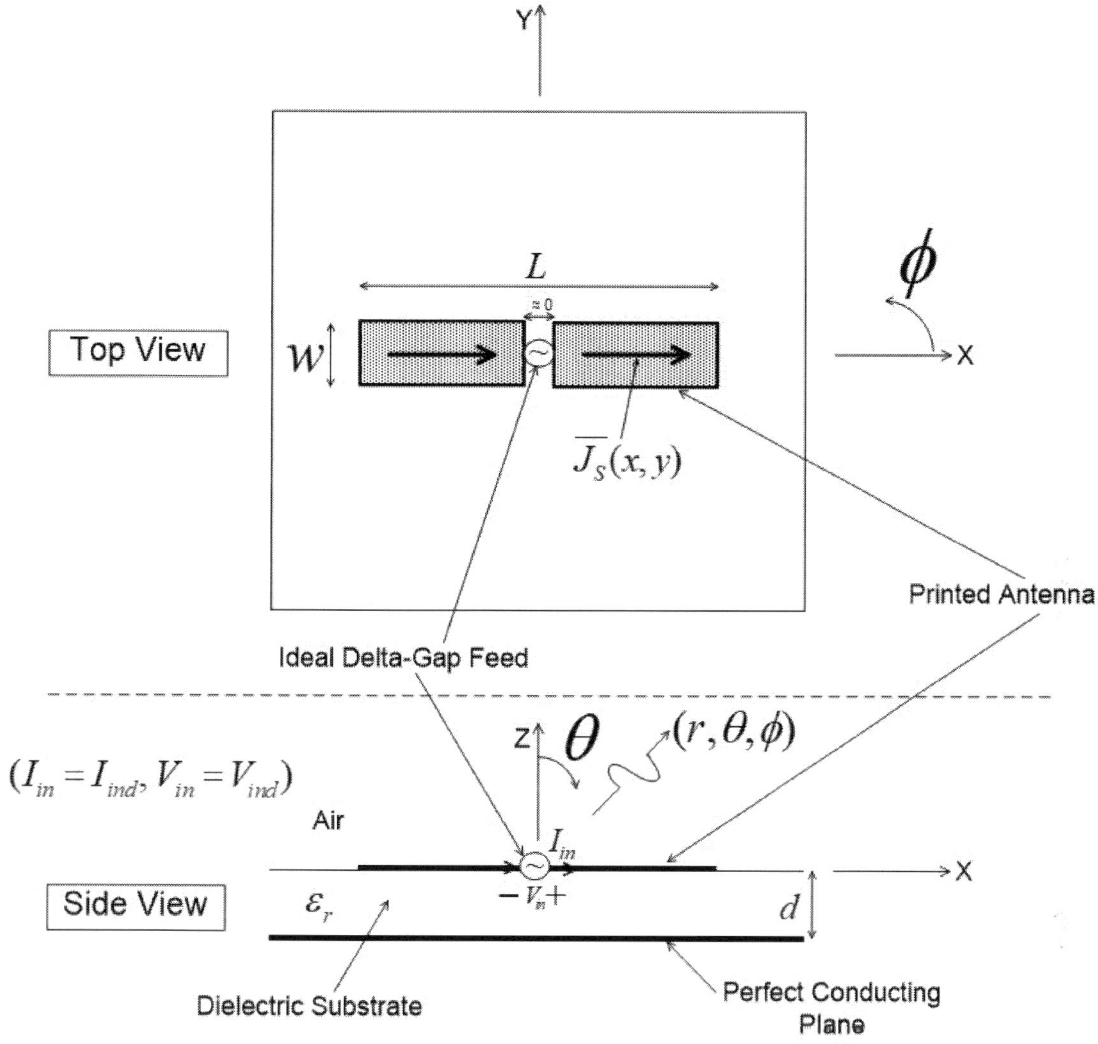

FIGURE 43

half wave-length in the dielectric medium of the substrate. The width of the antenna, w, and the position of the input probe, on the other hand, determine the level of the input impedance. We will model the essential radiation and impedance characteristics in the following sections. For a basic physical understanding, the microstrip antenna of Figs.43, 52 may be conceptualized as a rectangular cavity ($L \times w \times d$), with closed conducting walls on top and bottom but physically open in the other four sides. Most of the antenna fields would be confined in the dielectric medium enclosed by the cavity. Only a part of the cavity fields would fringe through the open side walls into the free-space medium above the substrate, and these fringing fields would be responsible for radiation into the free-space. However, for a practical thin substrate the structure operates principally as a resonant cavity, with only limited fringing radiation. Therefore, the bandwidth of the antenna radiation for such practical designs is expected to be relatively poor, as compared to other common antennas studied in this chapter. The small bandwidth would, however, be adequate in a large class of communication applications.

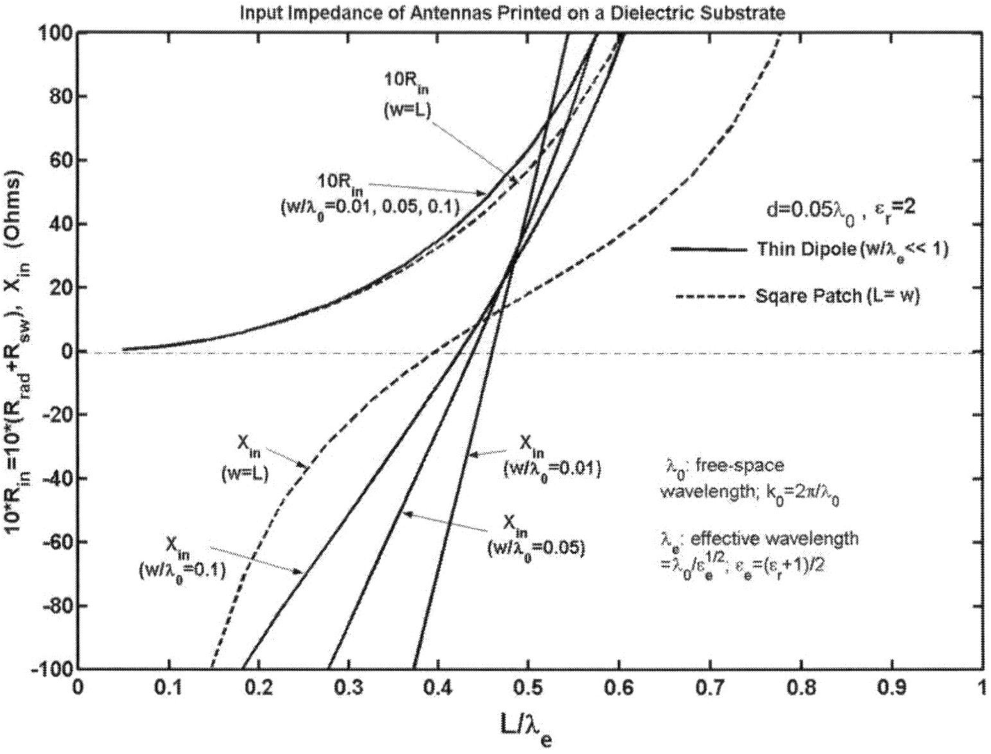

FIGURE 44

8.1. Strip Current Distribution. The current distribution $\bar{J}_s$ for a printed antenna of Fig.43 is expressed in a similar functional form as that for a free-space strip-dipole antenna of section 3.1. Except, the wave number k_0 for the current distribution on the free-space dipole antenna of section 3.1 needs to be substituted by an effective wave number k_e for a printed antenna of Fig.43, in order to properly model the standing-wave behavior of the antenna current on the dielectric substrate. It is reasonable to approximate $k_e \simeq k_0\sqrt{(\epsilon_r + 1)/2}$ representing an effective medium with a relative dielectric constant $\epsilon_e = (\epsilon_r+1)/2$, which is the average of the dielectric constants of the antenna substrate and the air medium. This assumes that the fields are spread about equally in the dielectric and the free-space regions. The ϵ_e could be closer to ϵ_r for antennas with a wide strip $(w >> d)$ and a large dielectric constant $(\epsilon_r >> 1)$. In this case, the fields would be mostly confined in the dielectric medium between the parallel conductors of the wide patch antenna and the ground plane, with only minimal fields fringing into the air medium at the patch edges. We choose $\epsilon_e = (\epsilon_r + 1)/2$ for a general analysis. Any fine adjustment for ϵ_e is not expected to significantly alter the radiation power and fields far from the antenna. It may, however, have relatively more effect on the antenna reactance associated with the stored energy and fields near the antenna. With the above consideration in mind, the current distribution on a general printed antenna element, in the space (x, y) and Fourier (k_x, k_y) domains, is expressed as follows.

$$\bar{J}_s = \hat{x}J_x(x,y) = \hat{x}I(x)h(y), \; I(x) = \int\limits_{y=-w/2}^{+w/2} J_x(x,y)dy, \; \int\limits_{y=-w/2}^{+w/2} h(y)dy = 1,$$

$$h(y) = \frac{1}{w}, \; |y| < \frac{w}{2}; \; = 0, \; |y| > \frac{w}{2},$$

$$I(x) = I_{in}\frac{\sin[k_e(\frac{L}{2} - |x|)]}{\sin(\frac{k_eL}{2})}, \; |x| < L/2; \; = 0, \; |x| > L/2,$$

$$k_e = k_0\sqrt{\epsilon_e}; \; \epsilon_e = \frac{\epsilon_r + 1}{2}, \quad I(x) = I_{in}f(x), \; I(x=0) = I_{in}, \; f(x=0) = 1. \quad (9.129)$$

$$\tilde{J}_x(k_x, k_y) = \int\limits_{x=-L/2}^{L/2}\int\limits_{y=-w/2}^{y/2} J_x(x,y)e^{-jk_xx}e^{-jk_yy}dxdy = I_{in}F(k_x)H(k_y), \quad (9.130)$$

$$H(k_y) = \int\limits_{y=-w/2}^{w/2} h(y)e^{-jk_yy}dy = \frac{1}{w}(\frac{e^{-jk_y\frac{w}{2}} - e^{+jk_y\frac{w}{2}}}{-jk_y}) = \frac{\sin(k_y\frac{w}{2})}{(k_y\frac{w}{2})}, \quad (9.131)$$

$$F(k_x) = \int\limits_{x=-L/2}^{L/2} f(x)e^{-jk_xx}dx = 2\int_0^{\frac{L}{2}} f(x)\cos(k_xx)dx$$

$$= \frac{2}{\sin(\frac{k_eL}{2})}\int_0^{\frac{L}{2}} \sin[k_e(\frac{L}{2} - x)]\cos(k_xx)dx$$

$$= \frac{1}{\sin(\frac{k_eL}{2})}\int_0^{\frac{L}{2}} [\sin(\frac{k_eL}{2} - k_ex + k_xx) + \sin(\frac{k_eL}{2} - k_ex - k_xx)]dx$$

$$= \frac{1}{\sin(\frac{k_eL}{2})}\Big[\frac{\cos(\frac{k_eL}{2} - (k_e - k_x)x)}{(k_e - k_x)} + \frac{\cos(\frac{k_eL}{2} - (k_e + k_x)x)}{(k_e + k_x)}\Big]_0^{\frac{L}{2}}$$

$$= \frac{1}{\sin(\frac{k_eL}{2})}[\frac{\cos(\frac{k_xL}{2}) - \cos(\frac{k_eL}{2})}{(k_e - k_x)} + \frac{\cos(\frac{k_xL}{2}) - \cos(\frac{k_eL}{2})}{(k_e + k_x)}]$$

$$= \frac{2k_e}{\sin(\frac{k_eL}{2})}[\frac{\cos(\frac{k_xL}{2}) - \cos(\frac{k_eL}{2})}{(k_e^2 - k_x^2)}], \quad (9.132)$$

$$\bar{J}_s = \hat{x}\frac{1}{4\pi^2}\int_{-\infty}^{\infty}\int_{-\infty}^{\infty}\tilde{J}_x(k_x,k_y)e^{jk_x x}e^{jk_y y}dk_x dk_y$$
$$= \hat{x}\frac{I_{in}}{4\pi^2}\int_{-\infty}^{\infty}\int_{-\infty}^{\infty}F(k_x)H(k_y)e^{jk_x x}e^{jk_y y}dk_x dk_y. \qquad (9.133)$$

8.2. Fields and Input Impedance. Modeling the fields of a general printed antenna of Fig.43 will essentially follow the basic analysis for a small printed-dipole element, presented in chapter 6, section 6, with appropriate changes for the antenna currents. The amplitude factor Il in (6.95, 6.123), which is the Fourier transform of the current distribution for a small printed-dipole element, needs to be substituted by the current transform $\tilde{J}_x(k_x, k_y)$ of (9.130-9.133) for a general printed antenna. The amplitude factor Il appears outside of the field integral (6.123), because it is constant independent of the transform variables k_x and k_y. In contrast, the current transform $\tilde{J}_x(k_x, k_y)$ is a general function of the transform variables k_x and k_y. Therefore, when the field integral (6.123) is revised for a general printed antenna, the transform $\tilde{J}_x(k_x, k_y)$, substituting for the factor Il, needs to be included inside the integral.

$$E_x(x,y,z) = \frac{1}{4\pi^2}\int_{-\infty}^{\infty}\int_{-\infty}^{+\infty}(-Z^t_{TM}\cos^2\phi' - Z^t_{TE}\sin^2\phi')\tilde{J}_x(k_x,k_y)$$
$$\times e^{+jk_x x}e^{+jk_y y}e^{-jk_{z0}z}dk_x dk_y, \ z>0$$
$$= \frac{1}{4\pi^2}\int_{-\infty}^{\infty}\int_{-\infty}^{+\infty}(-Z^t_{TM}\cos^2\phi' - Z^t_{TE}\sin^2\phi')I_{in}F(k_x)H(k_y)$$
$$\times e^{+jk_x x}e^{+jk_y y}e^{-jk_{z0}z}dk_x dk_y, \ z>0. \qquad (9.134)$$

A field integral of the above form can not be integrated and reduced to a simple expression, that would be rigorously valid at all points of observation. However, simple field expressions that are approximately valid only at large distances can be derived. Such approximations, with separate contributions from the free-space and surface-wave radiations, are presented in the following sections 8.5, 8.6.

The complex power P drawn from the antenna current can be expressed by integrating the elemental power $dP = (1/2)\triangle V\triangle I^* = -(1/2)E_x(z=0)J_x^* dxdy$ over the strip surface. Using the E_x expression of (9.134) and the current expressions of (9.129-9.132), the above power integral in the space variables x and y can be simplified as an integral over the transform variables k_x and k_y.

$$dP = \tfrac{1}{2}\triangle V\triangle I^* = \tfrac{1}{2}(-E_x(z=0)dx)(J_x^* dy)$$
$$= -\tfrac{1}{2}E_x(z=0)J_x^* dxdy, \qquad (9.135)$$

$$P = \int_y \int_x dP = -\frac{1}{2} \int_y \int_x E_x(x, y, z = 0) J_x^*(x, y) dxdy$$

$$= \frac{1}{8\pi^2} \int_{k_y=-\infty}^{+\infty} \int_{k_x=-\infty}^{+\infty} I_{in} F(k_x) H(k_y) (Z_{TM}^t \cos^2 \phi' + Z_{TE}^t \sin^2 \phi')$$

$$\times \Big(\int_y \int_x I_{in}^* f(x) h(y) e^{+jk_x x} e^{+jk_z y} dxdy \Big) dk_x dk_y$$

$$= \frac{1}{8\pi^2} \int_{k_y=-\infty}^{+\infty} \int_{k_x=-\infty}^{+\infty} I_{in} F(k_x) H(k_y) (Z_{TM}^t \cos^2 \phi' + Z_{TE}^t \sin^2 \phi')$$

$$\times I_{in}^* F(-k_x) H(-k_y) dk_x dk_y$$

$$= \frac{|I_{in}|^2}{8\pi^2} \int_{k_y=-\infty}^{+\infty} \int_{k_x=-\infty}^{+\infty} F^2(k_x) H^2(k_y) (Z_{TM}^t \cos^2 \phi' + Z_{TE}^t \sin^2 \phi') dk_x dk_y. \tag{9.136}$$

The source power is equated with the input power at the antenna feed terminals, expressed in terms of input current I_{in} and input impedance Z_{in}. Accordingly, the input impedance of the antenna can be expressed using the power integral of (9.44) and the current transform of (9.39-9.42).

$$P = \frac{1}{2} V_{in} I_{in}^* = \frac{1}{2} Z_{in} |I_{in}|^2, \; Z_{in} = 2P/|I_{in}|^2,$$

$$Z_{in} = \frac{1}{4\pi^2} \int_{k_y=-\infty}^{+\infty} \int_{k_x=-\infty}^{+\infty} F^2(k_x) H^2(k_y) (Z_{TM}^t \cos^2 \phi' + Z_{TE}^t \sin^2 \phi') dk_x dk_y. \tag{9.137}$$

It can be convenient to implement the above integration in polar coordinates k_ρ and ϕ', where $k_x = k_\rho \cos \phi'$ and $k_y = k_\rho \sin \phi'$.

$$Z_{in} = R_{in} + jX_{in} = \frac{1}{4\pi^2} \int_{k_\rho=0}^{\infty} \int_{\phi'=0}^{2\pi} F^2(k_x = k_\rho \cos \phi') H^2(k_y = k_\rho \sin \phi')$$

$$\times (Z_{TM}^t(k_\rho) \cos^2 \phi' + Z_{TE}^t(k_\rho) \sin^2 \phi') k_\rho d\phi' dk_\rho. \tag{9.138}$$

$$
\begin{aligned}
Z_{in}^{TM} &= \frac{1}{4\pi^2}\int\limits_{k_\rho=0}^{\infty}\int\limits_{\phi'=0}^{2\pi} F^2(k_x = k_\rho\cos\phi') \\
\times H^2(k_y &= k_\rho \sin\phi')\cos^2\phi' d\phi' Z_{TM}^t(k_\rho)k_\rho dk_\rho, \\
Z_{in}^{TE} &= \frac{1}{4\pi^2}\int\limits_{k_\rho=0}^{\infty}\int\limits_{\phi'=0}^{2\pi} F^2(k_x = k_\rho\cos\phi') \\
\times H^2(k_y &= k_\rho \sin\phi')\sin^2\phi' d\phi' Z_{TE}^t(k_\rho)k_\rho dk_\rho, \\
Z_{in} &= Z_{in}^{TM} + Z_{in}^{TE}.
\end{aligned} \tag{9.139}
$$

The results for the input resistance R_{in} and reactance X_{in}, as computed from the above integral (9.138), are plotted in Fig.44 for a specific $\epsilon_r = 2$. Results having similar trends with respect to antenna width w and length L are expected for other values of ϵ_r. The resistance R_{in} in Fig.44 is not as strongly dependent on the antenna width w as the reactance X_{in}. The normalized resonant length L/λ_e in Fig.44, corresponding to a zero reactance $X_{in} = 0$, is seen to be less than 0.5, with some variation due to the antenna width w. A narrower width w results in a stronger variation of the reactance with L/λ_e. This is equivalent to having a stronger variation of the reactance with the operating frequency or wavelength, for a given antenna length L, as the antenna width is reduced. This leads to a narrower antenna bandwidth for a narrower antenna width w. The resistance R_{in} consists of the radiation resistance R_{rad}, contributed from power radiated into the free-space medium, and the surface-wave resistance R_{sw}, contributed from power radiated as surface waves in the dielectric substrate. Expressions for the individual parts are presented in the following sections.

It may be noticed, the impedance integral in (9.138) for a general printed antenna is in a similar form as that in (6.124) for a small printed-dipole element. The difference is, the additional transform factors $|F(k_x)|^2$ and $|H(k_y)|^2$ appear inside the integral (9.138), substituting for the factor l^2 appearing outside the integral (6.124). Accordingly, all subsequent analyses for impedance and radiation power for a small printed-dipole element in chapter 6, sections 6.5.1, 6.5.2, which were established upon the integral (9.138), can be repeated for a general printed antenna, by simply implementing the above substitution.

8.3. Free-Space Radiation Power. Like for a small printed-dipole element in chapter 6, section 6.5.1, the real part of the Z_{in} integral (9.138) for a general printed antenna, when evaluated over the range $k_\rho < k_0$, would result in the radiation resistance R_{rad}. However, unlike a small printed-dipole, the ϕ' integration for a general printed-dipole antenna can not be performed analytically, and needs to be evaluated computationally.

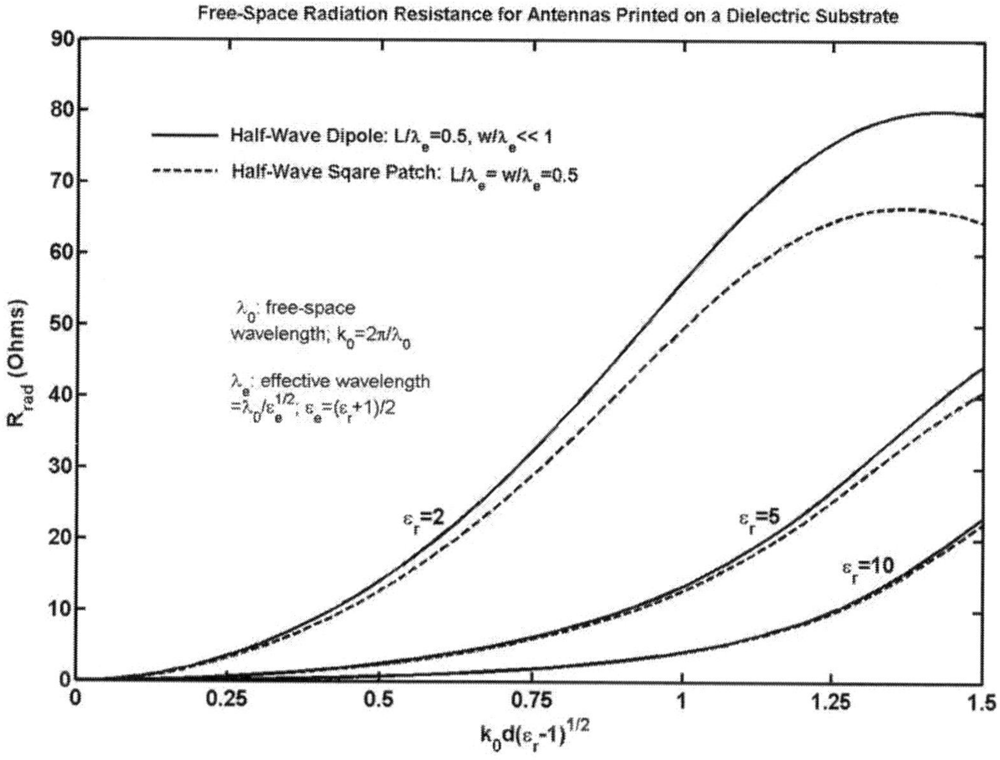

FIGURE 45

$$R_{rad} = Real[\frac{1}{4\pi^2} \int_{k_\rho=0}^{k_0} \int_{\phi'=0}^{2\pi} (Z_{TM}^t \cos^2\phi' + Z_{TE}^t \sin^2\phi') F^2(k_x = k_\rho \cos\phi')$$

$$\times H^2(k_y = k_\rho \sin\phi') k_\rho d\phi' dk_\rho],$$

$$P_{rad} = \frac{1}{2}|I|^2 R_{rad}. \qquad (9.140)$$

Because Z_{TM}^t and Z_{TE}^t are functions of only k_ρ, independent of the angle ϕ', the ϕ' integration may not include these functions. The R_{rad} can be decomposed into R_{rad}^{TM} and R_{rad}^{TE}, as contributed from the TM_z and TE_z parts of the above integration, respectively.

$$\begin{aligned}
R_{rad}^{TM} &= \frac{1}{4\pi^2}\int_{k_\rho=0}^{k_0}\int_{\phi'=0}^{2\pi} \cos^2\phi' F^2(k_x = k_\rho\cos\phi') \\
&\times H^2(k_y = k_\rho\sin\phi')d\phi' Real(Z_{TM}^t)k_\rho dk_\rho, \\
R_{rad}^{TE} &= \frac{1}{4\pi^2}\int_{k_\rho=0}^{k_0}\int_{\phi'=0}^{2\pi} \sin^2\phi' F^2(k_x = k_\rho\cos\phi') \\
&\times H^2(k_y = k_\rho\sin\phi')d\phi' Real(Z_{TE}^t)k_\rho dk_\rho, \\
P_{rad}^{TM} &= \frac{1}{2}|I_{in}|^2 R_{rad}^{TM},\ P_{rad}^{TE} = \frac{1}{2}|I_{in}|^2 R_{rad}^{TE}, \\
P_{rad} &= P_{rad}^{TM} + P_{rad}^{TE},\ R_{rad} = R_{rad}^{TM} + R_{rad}^{TE}.
\end{aligned} \tag{9.141}$$

The radiation resistance R_{rad}, as computed using (9.140), is plotted in Fig.45 as a function of the antenna substrate thickness d, for different dielectric constant ϵ_r and antenna width w as parameters. The R_{rad} increases with the substrate thickness d, and is not strongly dependent on the antenna width w.

8.4. Surface-Wave Radiation Power. The same procedure of chapter 6, section 6.5.2 for a small printed-dipole element can be repeated for a general printed antenna. The starting point of the analysis would be the integral expression for the input impedance Z_{in} in (9.138), instead of that in (6.124, 6.127) for a small printed-dipole element. The main difference in the analysis would be the ϕ' integration, which can no longer be evaluated analytically for a general printed antenna. Accordingly, the final result would be expressed in terms of a ϕ' integration, which may be evaluated computationally.

$$\begin{aligned}
Z_{sw}^{TM} &= \frac{-ja'}{8\pi}\int_{\phi'=0}^{2\pi} F^2(k_x = k_{\rho p}\cos\phi')H^2(k_y = k_{\rho p}\sin\phi')\cos^2\phi' d\phi' \\
&= R_{sw}^{TM},\ P_{sw}^{TM} = \frac{1}{2}|I_{in}|^2 R_{sw}^{TM}, \\
Z_{sw}^{TE} &= \frac{-jb'}{8\pi}\int_{\phi'=0}^{2\pi} F^2(k_x = k_{\rho p}\cos\phi')H^2(k_y = k_{\rho p}\sin\phi')\sin^2\phi' d\phi' \\
&= R_{sw}^{TE},\ P_{sw}^{TE} = \frac{1}{2}|I_{in}|^2 R_{sw}^{TE}.
\end{aligned} \tag{9.142}$$

The total surface-wave resistance R_{sw}, and the associated total surface-wave power $P_{sw} = (1/2)|I_{in}|^2 R_{sw}$, are computed by adding individual contributions from all TE_z and TM_z modes supported by the substrate structure. The propagation constant $k_{\rho p}$, and the residue term a' or b', for individual surface-wave modes, which are required in the evaluation of (9.142), are available in chapter 6, section 6.4.

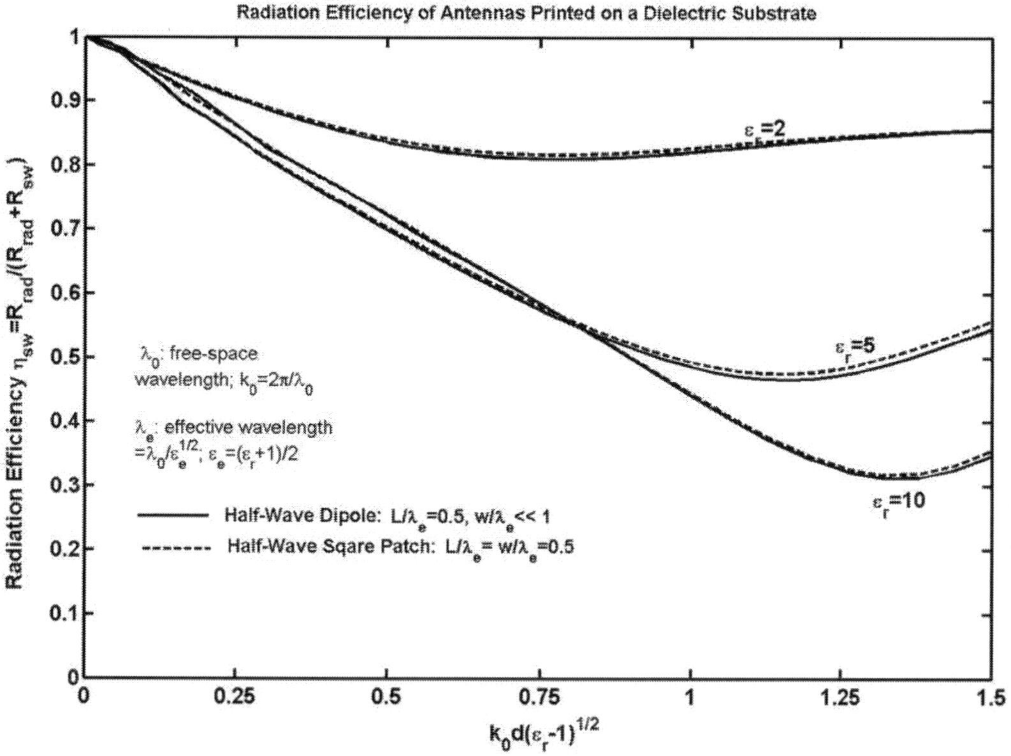

FIGURE 46

The power P_{sw} radiated in the form of surface waves would not be utilized for signal communication into the free-space medium. Accordingly, the power P_{sw} is normally considered to be undesirable in an antenna application. The surface-wave efficiency η_{sw} of a printed antenna is defined as the ratio of the useful power P_{rad} derived in (9.141), which is radiated into the free-space medium, to the total antenna power $P_{in} = P_{rad} + P_{sw}$ which is sum of the P_{rad} and the surface-wave power P_{sw} derived in (9.142).

$$\eta_{sw} = \frac{P_{rad}}{P_{in}} = \frac{P_{rad}}{P_{rad} + P_{sw}} = \frac{R_{rad}}{R_{rad} + R_{sw}} = \frac{R_{rad}}{R_t}. \tag{9.143}$$

The surface-wave efficiency η_{sw} as defined above is plotted in Fig.46 as a function of the antenna substrate thickness d, for different dielectric constant ϵ_r and antenna width w as parameters. Higher η_{sw} is achieved for a thinner antenna substrate, with η_{sw} approaching unity as the substrate thickness d approaches zero.

It may be useful to plot the surface-wave resistance R_{sw} of a printed antenna, contributed due to any given surface-wave mode, properly normalized with respect to the corresponding surface-wave resistance for a small printed-dipole element, as presented in Fig.47. The normalized results of the Fig.47 can be used together with data for the propagation constant β_{sw} of the particular surface-wave mode and the corresponding R_{sw} of a

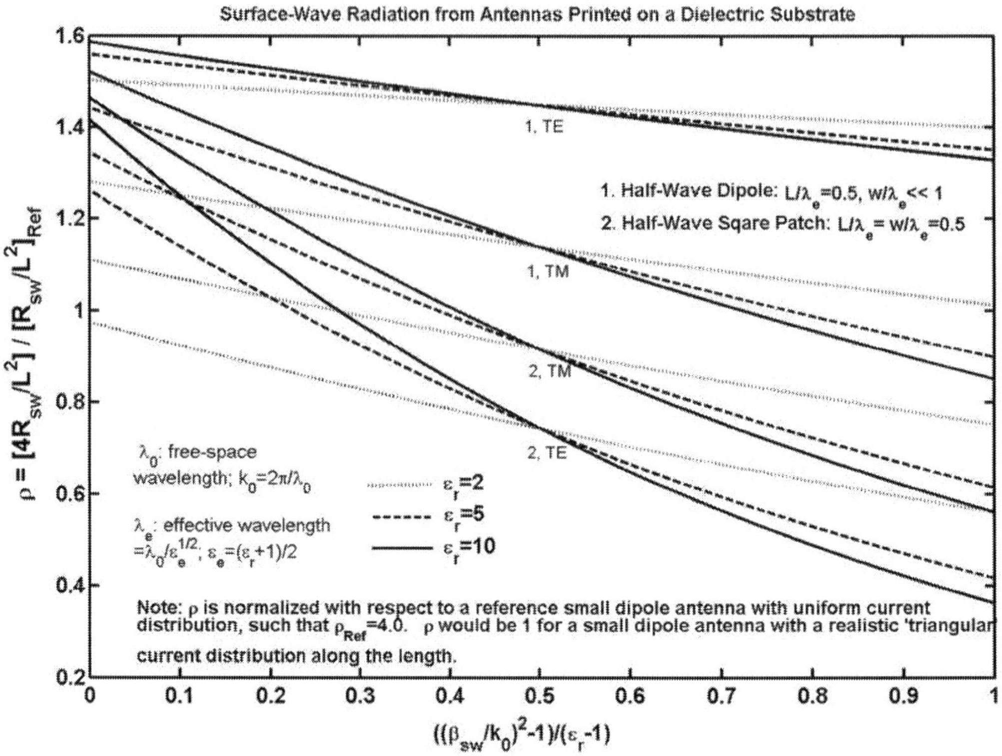

FIGURE 47

small printed-dipole element, respectively presented in Figs.9, 10, chapter 6, in order to obtain the actual value of the R_{sw} for the printed antenna due to the given surface-wave mode. Individual results for each different surface-wave mode supported in a given design (Fig.9, chapter 6) then need to be added in order to obtain the total R_{sw}. The total input resistance $R_t = R_{rad} + R_{sw}$ of a printed antenna, normalized with respect to that for a small printed-dipole element, is plotted in Fig.48.

8.5. Free-Space Radiation Fields. The procedure in chapter 6, section 6.6.1 for a small printed-dipole element can be repeated here for a general printed antenna. The starting point would be a field integral of the form in (9.134), instead of that in (6.123, 6.135) for a small printed-dipole element. The final results can be obtained by substituting the factor Il in those of chapter 6, section 6.6.1 with the current transform $\tilde{J}_x(k_x, k_y) = I_{in}F(k_x)H(k_y)$, evaluated at $k_x = -k_0 \sin\theta \cos\phi$ and $k_y = -k_0 \sin\theta \sin\phi$.

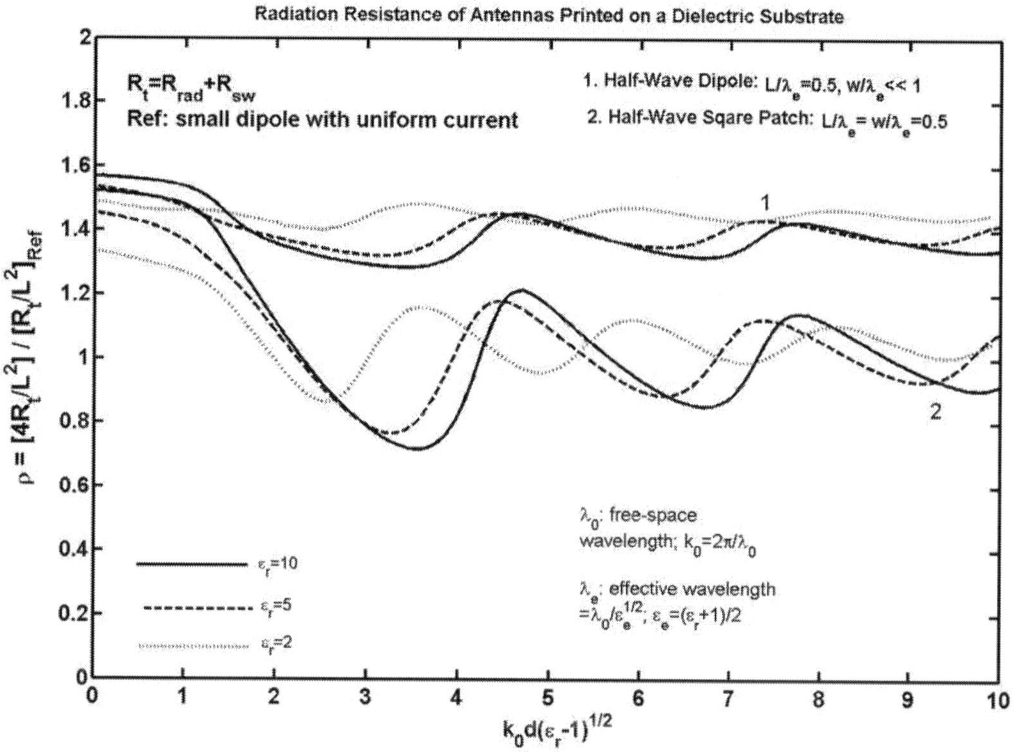

FIGURE 48

$$
\begin{aligned}
E_x(r,\theta,\phi)_{r\to\infty} \simeq -\frac{1}{4\pi^2}&[Z^t_{TM}(k_\rho = k_0\sin\theta)\cos^2\phi + Z^t_{TE}(k_\rho = k_0\sin\theta)\sin^2\phi] \\
&\times \tilde{J}_x(k_x = -k_0\sin\theta\cos\phi, k_y = -k_0\sin\theta\sin\phi) \\
&\times k_0^2\sin\theta\cos\theta \times e^{-jk_0r} \times \sqrt{\frac{2\pi}{-jk_0r}} \times \sqrt{\frac{2\pi}{-jk_0r\sin^2\theta}} \\
= -jI_{in}&\frac{e^{-jk_0r}}{\lambda_0 r}F(k_x = -k_0\sin\theta\cos\phi)H(k_y = -k_0\sin\theta\sin\phi) \\
\times\cos\theta&[Z^t_{TM}(k_\rho = k_0\sin\theta)\cos^2\phi + Z^t_{TE}(k_\rho = k_0\sin\theta)\sin^2\phi], \quad (9.144)
\end{aligned}
$$

$$E_x^{TM}(r,\theta,\phi)_{r\to\infty} \simeq -jI_{in}\frac{e^{-jk_0r}}{\lambda_0 r}F(k_x=-k_0\sin\theta\cos\phi)H(k_y=-k_0\sin\theta\sin\phi)$$
$$\times Z_{TM}^t(k_\rho=k_0\sin\theta)\cos\theta\cos^2\phi,$$
$$E_x^{TE}(r,\theta,\phi)_{r\to\infty} \simeq -jI_{in}\frac{e^{-jk_0r}}{\lambda_0 r}F(k_x=-k_0\sin\theta\cos\phi)H(k_y=-k_0\sin\theta\sin\phi)$$
$$\times Z_{TE}^t(k_\rho=k_0\sin\theta)\cos\theta\sin^2\phi, \quad (9.145)$$

$$E_\theta(r,\theta,\phi)_{r\to\infty} = \frac{E_x^{TM}(r,\theta,\phi)_{r\to\infty}}{\cos\theta\cos\phi} \simeq -jI_{in}\frac{e^{-jk_0r}}{\lambda_0 r}F(k_x=-k_0\sin\theta\cos\phi)$$
$$\times H(k_y=-k_0\sin\theta\sin\phi)Z_{TM}^t(k_\rho=k_0\sin\theta)\cos\phi,$$
$$E_\phi(r,\theta,\phi)_{r\to\infty} = -\frac{E_x^{TE}(r,\theta,\phi)_{r\to\infty}}{\sin\phi} \simeq jI_{in}\frac{e^{-jk_0r}}{\lambda_0 r}F(k_x=-k_0\sin\theta\cos\phi)$$
$$\times H(k_y=-k_0\sin\theta\sin\phi)Z_{TE}^t(k_\rho=k_0\sin\theta)\cos\theta\sin\phi. \quad (9.146)$$

The power radiation pattern, which is the square of the field pattern as expressed above, normalized to unity maximum value, are plotted in Figs.49-51 for printed antennas operating near resonance with $L/\lambda_e = 0.5$, for three different values of ϵ_r equal to 2, 5 and 10. The corresponding patterns for a small printed-dipole element are also plotted for reference, showing essentially similar shapes. As compared to a small printed-dipole element, the angular spread of the pattern is somewhat reduced in the $\phi = 0^0$ plane due to a non-zero antenna length $L/\lambda_e = 0.5$, and in the $\phi = 90^0$ plane due to a non-zero antenna width w.

8.6. Surface-Wave Radiation Fields. The field analysis for the surface-wave radiation in chapter 6, section 6.6.2 for a small printed-dipole element can be repeated here for a general printed antenna. The starting point would be a field integral of the form in (9.134), instead of that in (6.123, 6.141) for a small printed-dipole element. The final results can be obtained by substituting the factor Il in those of chapter 6, section 6.6.2 with the current transform $\tilde{J}_x(k_x,k_y) = I_{in}F(k_x)H(k_y)$, evaluated at $k_x = -k_{\rho p}\cos\phi$ and $k_y = -k_{\rho p}\sin\phi$.

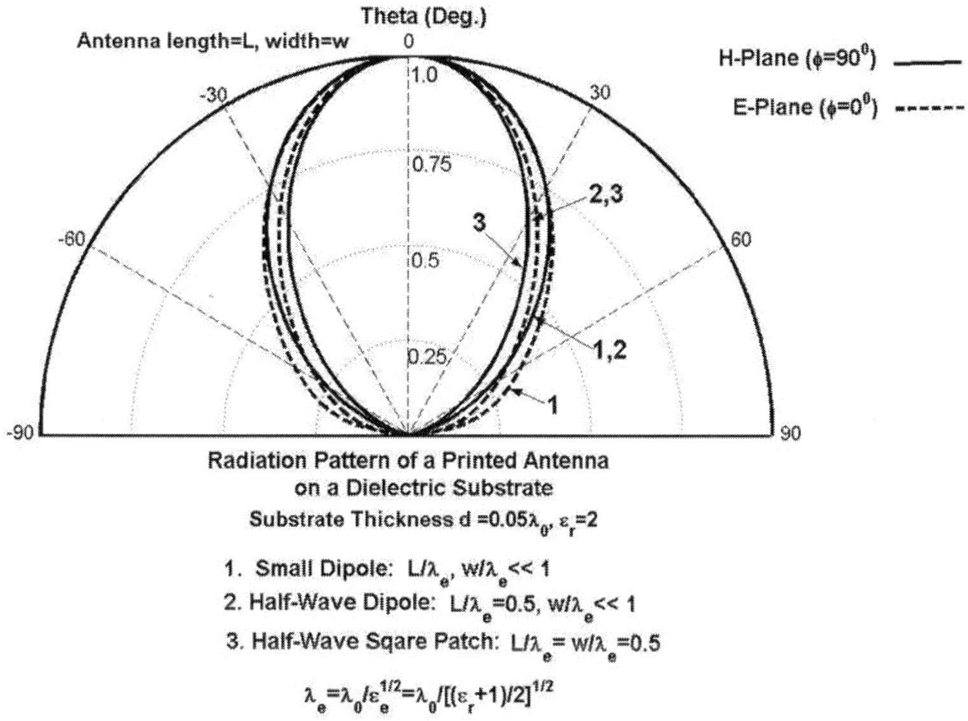

FIGURE 49

$$
\begin{aligned}
E_x^{TM}(\rho,\phi;z=0)_{\rho\to\infty} &\simeq \tilde{J}_x(k_x=-k_{\rho p}\cos\phi, k_y=-k_{\rho p}\sin\phi) \\
&\qquad \times\frac{a'}{2}\sqrt{\frac{1}{2\pi\rho k_{\rho p}}}\cos^2\phi e^{+j3\pi/4}e^{-jk_{\rho p}\rho} \\
= I_{in}F(k_x=-k_{\rho p}\cos\phi)H(k_y=-k_{\rho p}\sin\phi)&\frac{a'}{2}\sqrt{\frac{1}{2\pi\rho k_{\rho p}}}\cos^2\phi e^{+j3\pi/4}e^{-jk_{\rho p}\rho}, \\
E_x^{TE}(\rho,\phi;z=0)_{\rho\to\infty} &\simeq \tilde{J}_x(k_x=-k_{\rho p}\cos\phi, k_y=-k_{\rho p}\sin\phi) \\
&\qquad \times\frac{b'}{2}\sqrt{\frac{1}{2\pi\rho k_{\rho p}}}\sin^2\phi e^{+j3\pi/4}e^{-jk_{\rho p}\rho} \\
= I_{in}F(k_x=-k_{\rho p}\cos\phi)H(k_y=-k_{\rho p}\sin\phi)&\frac{b'}{2}\sqrt{\frac{1}{2\pi\rho k_{\rho p}}}\sin^2\phi e^{+j3\pi/4}e^{-jk_{\rho p}\rho}. \quad (9.147)
\end{aligned}
$$

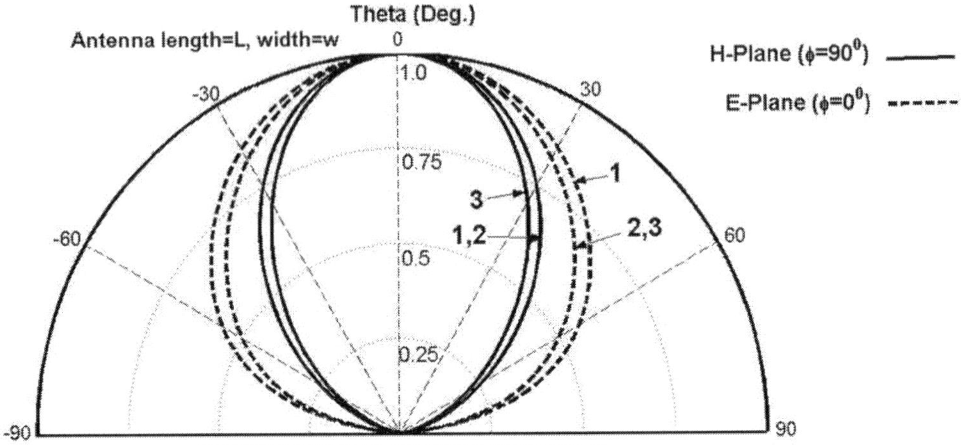

Radiation Pattern of a Printed Antenna on a Dielectric Substrate

Substrate Thickness $d = 0.05\lambda_0$, $\varepsilon_r = 5$

1. Small Dipole: L/λ_e, $w/\lambda_e << 1$
2. Half-Wave Dipole: $L/\lambda_e = 0.5$, $w/\lambda_e << 1$
3. Half-Wave Sqare Patch: $L/\lambda_e = w/\lambda_e = 0.5$

$\lambda_e = \lambda_0/\varepsilon_e^{1/2} = \lambda_0/[(\varepsilon_r+1)/2]^{1/2}$

FIGURE 50

8.7. Printed Antenna with a Coaxial Input. The impedance characteristics for a printed antenna with a coaxial feeding can be related to that with a delta-gap feeding, with the antenna dimensions and substrate parameters same in both cases. The current I_{ind}, voltage V_{ind} and impedance Z_{ind} defined at the delta-gap input terminals (port 1) can be related to the respective parameters I_{inp}, V_{inp} and Z_{inp} at the coaxial-probe input terminals (port 2), by using two-port circuit parameters. With the feeding only across the probe input, the delta-gap terminals are to be closed with a short circuit, with voltage V_{ind} equal to zero. The circuit parameter Z_{21} is the mutual impedance from the delta-gap port to the probe port, which is equal to the reverse mutual impedance Z_{12} from the probe port to the delta-gap port, assuming a reciprocal device (see chapter 2, section 4.3, or appendix G). The Z_{12} or Z_{21} is derived in (9.150) using a transmission line model for the standing-wave current distribution on the antenna strip. The parameter Z_{22} is equal to the impedance Z_{ind} as seen at the input of the delta-gap feed, analyzed in the last sections. The parameter Z_{11} is the self impedance seen by the coaxial probe, expressed in (9.149).

$$V_{inp} = I_{inp}Z_{11} + I_{ind}Z_{12},$$
$$V_{ind} = I_{inp}Z_{21} + I_{ind}Z_{22} = 0,$$
$$\frac{I_{ind}}{I_{inp}} = -\frac{Z_{21}}{Z_{22}} = -\frac{Z_{21}}{Z_{ind}}, \; Z_{22} = Z_{ind},$$
$$Z_{inp} = \frac{V_{inp}}{I_{inp}} = Z_{11} + \frac{I_{ind}}{I_{inp}}Z_{12} = Z_{11} - \frac{Z_{12}Z_{21}}{Z_{ind}} = Z_{11} - \frac{Z_{12}^2}{Z_{ind}}. \quad (9.148)$$

The Z_{11} can be obtained from a modeling of the fields produced due to the current on the coaxial probe. The current on the probe is assumed to be uniform over the length of the probe, which is an accurate description for a small length d. The fields of a coaxial probe of length d, with its uniform current placed between the two parallel plates of a printed patch and the ground plane, are equivalent to the fields of a uniform line-current source of infinite length. This equivalence can be established by using imaging of the probe current over the two parallel plates. Modeling of this equivalent problem is available in chapter 6, section 4, and the results can be readily used for the present study. Accordingly, the probe impedance Z_{11} is equated to the equivalent impedance over length d, seen by a uniform line-current source of infinite length. The radius r_p of the probe, which is equal to that of the equivalent line current source, is assumed to be practically small. This leads to a simple approximation for the impedance, resulting in a mostly imaginary impedance $Z_{11} \simeq jX_{11}$. The small real part of Z_{11} can be ignored, which is equivalent to ignoring any direct radiation from the probe. Any strong radiation is generated through indirect coupling with the dominant radiating current expressed in (9.129, 9.150). It is this indirect coupling we are now trying to model using the two-port parameters.

$$Z_{11} \simeq -j\frac{\omega\mu d}{2\pi} ln(0.89k_0\sqrt{\epsilon_r}r_p) = -j\frac{\eta_0 d}{\lambda_0} ln(0.89k_0\sqrt{\epsilon_r}r_p); \; r_p << \frac{\lambda_0}{\sqrt{\epsilon_r}}. \quad (9.149)$$

The current distribution $I(x)$ on the printed antenna, and the corresponding voltage distribution $V(x)$ across the antenna patch and the conducting ground plane, can be expressed as a standing-wave with an effective wave number k_e. The probe feeding is normally used for a wide printed antenna ($w >> d$), where the fields are mostly confined in the dielectric medium of the substrate, with only minimal fields fringing into the air medium at the antenna edges. Therefore, for such practical designs, the wave number k_e in the following standing-wave modeling can be approximated as $k_0\sqrt{\epsilon_r}$. We assume that the antenna is operating near resonance, where the length L is close to a half wave length in the effective dielectric medium. The current is expressed by a cosine distribution, which is zero at the two ends $x = \pm L/2$, and is of a maximum value I_{ind} at the center $x = 0$. On the other hand, the voltage is expressed as a sine distribution with maximum amplitudes of $\pm V_0$ at the two ends $x \pm L/2$, but zero at the center $x = 0$, and is equal to V_{inp} as seen by a open-circuited ($I_{inp} = 0$) probe at location $x = -\delta$. The characteristic impedance of the strip transmission line is $Z_c = \eta d/w$, as modeled in chapter 4, section 10.1. Further, the impedance $Z(x) = V(x)/I(x)$ at a location x is equal to $-jZ_c \cot(k_e(L/2 - x))$, which is the impedance of a open-circuited stub of length $(L/2 - x)$, as modeled in chapter 4, section 9.4.

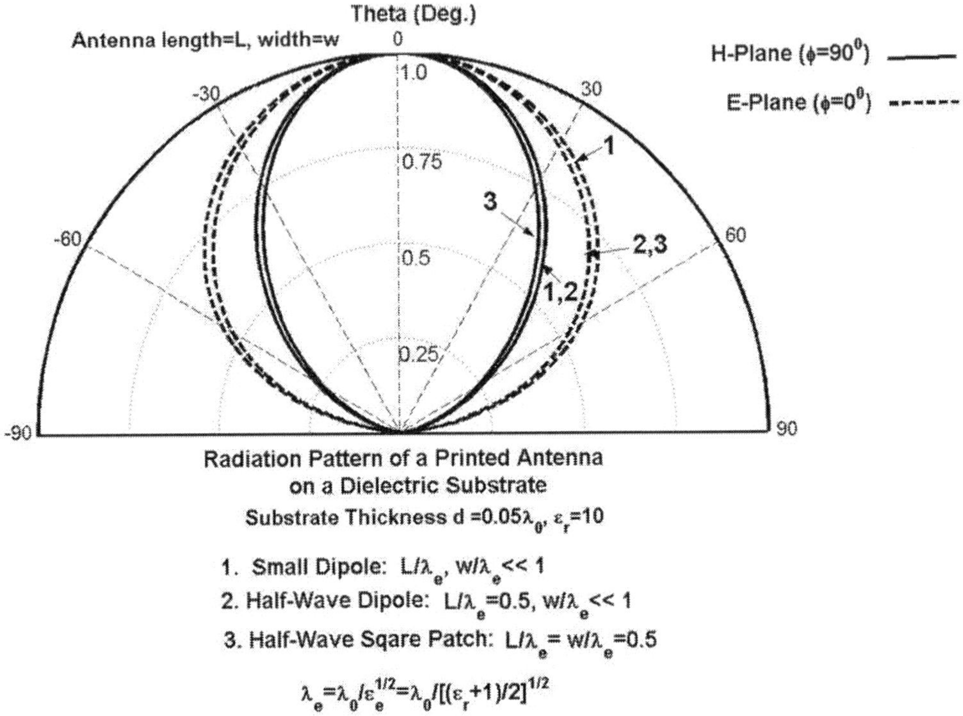

FIGURE 51

$$
\begin{aligned}
I(x) &= I_{ind}\cos(k_e x) = I_{ind}\cos(k_0\sqrt{\epsilon_r}x), \\
V(x) &= V_0\sin(k_e x) = -V_{inp}\frac{\sin(k_0\sqrt{\epsilon_r}x)}{\sin(k_0\sqrt{\epsilon_r}\delta)}\ (I_{inp}=0), \\
Z(x) = \frac{V(x)}{I(x)} &= -\frac{V_{inp}}{I_{ind}}\frac{\tan(k_0\sqrt{\epsilon_r}x)}{\sin(k_0\sqrt{\epsilon_r}\delta)} = -jZ_c\cot(k_0\sqrt{\epsilon_r}(\frac{L}{2}-x)) \\
&= -jZ_c\tan(k_0\sqrt{\epsilon_r}x),\ L = \frac{\lambda_0}{2\sqrt{\epsilon_r}},\ k_0\sqrt{\epsilon_r}\frac{L}{2} = \frac{\pi}{2}, \\
\frac{V_{inp}(I_{inp}=0)}{I_{ind}} &= jZ_c\sin(k_0\sqrt{\epsilon_r}\delta) = Z_{21} = Z_{12},\ Z_c = \frac{\eta d}{w} = \frac{\eta_0 d}{\sqrt{\epsilon_r}w}.
\end{aligned}
\tag{9.150}
$$

Using the mutual impedance Z_{12} of (9.150) and the probe impedance Z_{11} of (9.149) in (9.148), we obtain an expression for the input impedance Z_{inp} seen at the probe input, and also relate the probe current I_{inp} to the current I_{ind} on the printed strip.

$$Z_{inp} = Z_{11} - \frac{Z_{12}^2}{Z_{ind}} = Z_{11} + \frac{Z_c^2}{Z_{ind}} \sin^2(k_0\sqrt{\epsilon_r}\delta) =$$
$$-j\frac{\eta_0 d}{\lambda_0} ln(0.89 k_0 \sqrt{\epsilon_r} r_p) + \frac{\eta_0^2 d^2 \sin^2(k_0\sqrt{\epsilon_r}\delta)}{\epsilon_r w^2} \frac{1}{Z_{ind}},$$
$$\frac{I_{ind}}{I_{inp}} = -j\frac{Z_c}{Z_{ind}} \sin(k_0\sqrt{\epsilon_r}\delta) = -j\frac{\eta_0 d \sin(k_0\sqrt{\epsilon_r}\delta)}{\sqrt{\epsilon_r} w} \frac{1}{Z_{ind}}. \quad (9.151)$$

The radiating strip-current $I(x)$ in (9.150) is assumed to be uniformly distributed over the width (y dimension) of the strip. Accordingly, the associated electric field $E_z(x, y) = E_z(x)$ (see Fig.52), and the voltage $V(x) = -E_z(x)d$ used in the transmission-line modeling of (9.150), are also independent of y. This is an adequate description for a practical patch antenna with a wide strip ($w >> d$). The above model results in the voltage $V_{inp} = V(x = -\delta)$ at the probe input, and the Z_{21} as derived in (9.150), to be independent of any probe offset in the y dimension. This consequently leads to (9.151), having no dependence of the input impedance Z_{inp} or current I_{inp} on the y-dimensional probe offset.

Once we have the current I_{inp} related to the current I_{ind} by (9.151), all results for the printed antenna derived in the last sections, which were expressed in terms of the input current $I_{in} = I_{ind}$, can now be expressed in terms of the probe current I_{inp}. In other words, all results for a printed antenna with a delta-gap voltage feed would be applicable for a probe-fed element as well, by expressing the current $I_{in} = I_{ind}$ at the delta-gap input in terms of the current I_{inp} at the probe input.

Effect of the Probe Offset δ.

Based on the modeling approach prescribed above, we can deduce some basic characteristics with respect to the probe location. As derived in (9.151), the I_{ind} would be zero and Z_{inp} equal to Z_{11}, when the offset distance δ of the probe is zero. Substituting the resulting $I_{ind} = 0$ in the analysis of a printed antenna with a delta-gap feed, presented in the last sections 8.3-8.6, we get the radiation power, the associated radiation resistance seen at the antenna input, and the fields far from the antenna, to be zero. For this case with $\delta = 0$, the source at the input of Fig.52 would see only the probe structure, centered between the parallel plates of the patch and the ground plane, with a mostly reactive or imaginary impedance $Z_{inp} = Z_{11} \simeq jX_{11}$ for a practical thin probe ($r_p << \lambda_0/\sqrt{\epsilon_r}$). The mostly imaginary value for Z_{11} means there is very little radiation originated directly from the probe, as also mentioned before.

The above design with a centered probe-feed ($\delta = 0$), which would produce essentially no radiation power as we explained, would clearly not serve any practical purpose as a useful antenna. Accordingly, in a practical antenna design, the probe needs to be placed off the center of the patch, as suggested in the introduction of section 8. Placing the probe farther off the patch center would result in increased radiation power and in a larger value for the associated input radiation resistance. As expressed in (9.151), when the offset distance δ of the probe in the x dimension is increased from zero at the antenna center to $L/2$ at the edge of the antenna, the radiating current I_{ind} would increase with a $\sin(k_0\sqrt{\epsilon_r}\delta)$ dependence. This is for a given value of the probe current I_{inp}, independent of any probe offset in the y dimension. The result would lead to an increase of the radiation power, and

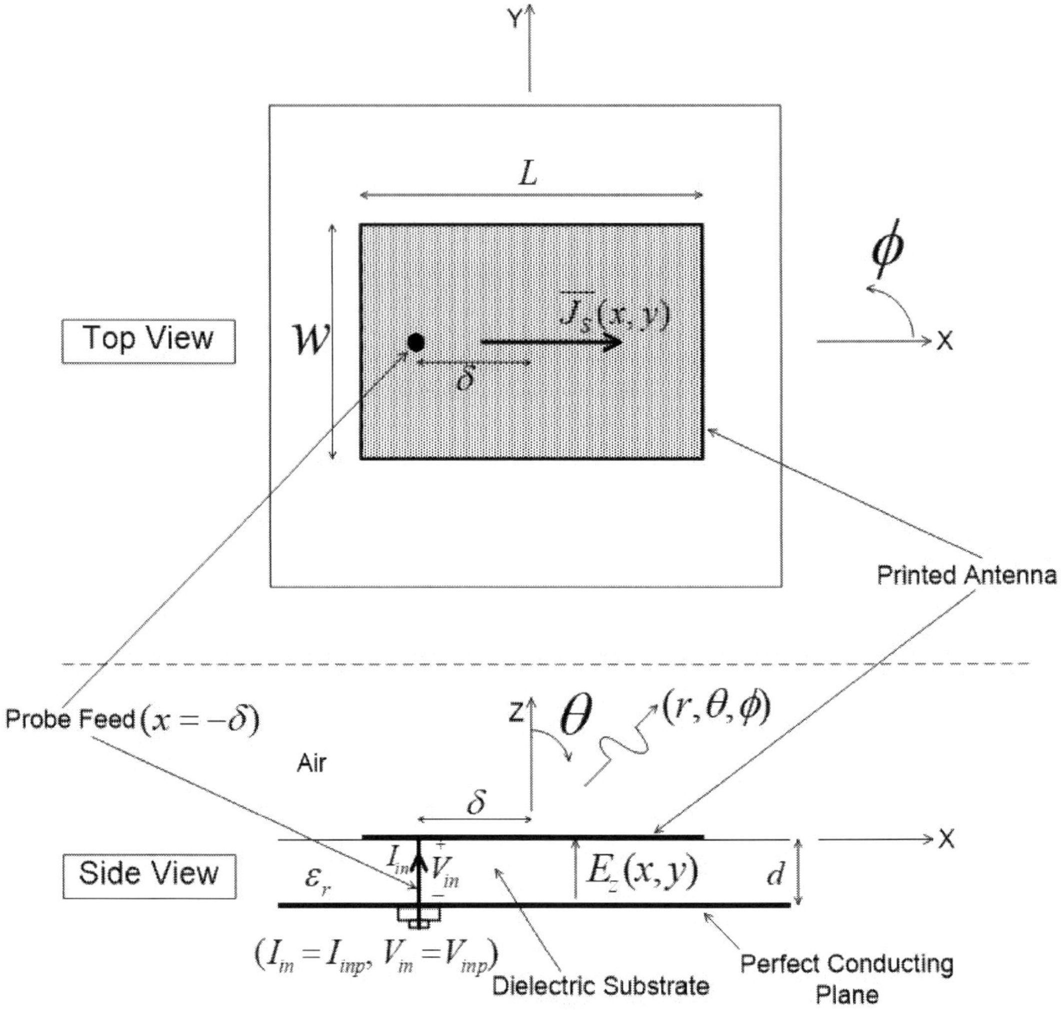

FIGURE 52

of the associated input radiation resistance, with a $\sin^2(k_0\sqrt{\epsilon_r}\delta)$ dependence. The total input impedance Z_{inp}, as per (9.151), consists of the probe's self impedance Z_{11} which is essentially reactive, plus an additional part having the $\sin^2(k_0\sqrt{\epsilon_r}\delta)$ dependence with δ. This basic trend can be utilized for impedance matching, where the input impedance can be conveniently adjusted to a desired level by adjusting the placement of the feeding probe, all other parameters remaining unchanged.

It may be noted, that our analysis and the above discussion considers coupling only to the radiating patch current that flows in the x direction (see Fig.43). Coupling to any y directed patch current can be similarly studied by interchanging between parameters for the x and the y dimensions. This would be particularly important when the patch is nearly square in shape ($L \simeq w$), in which case both the x and y currents can be strongly excited because they would be resonant at about the same operating frequency. However, one of the radiating currents could be suppressed, if desired, by proper placement of the probe. The $x-$ or the $y-$directed patch current would be zero if the probe is placed centered about the x or the y dimension, respectively. In other words, any non-zero offset of the probe from the patch center in only the x or the y dimension would excite the radiating patch current flowing only in the x or the y direction, respectively, while the other current component remains zero. Both the current components can be simultaneously excited, with individual amplitudes adjusted by proper selection of the probe offsets in each dimension. This can be useful for adjusting the polarization of the antenna radiation, particularly for achieving a circularly polarized radiation.

9. Other Common Antenna Geometries

There are many possible antenna geometries in practical use today, each having some desirable properties suited for specific applications. We have only discussed in this chapter selected basic radiating elements. The reader may be referred to antenna text books [**27**]-[**32**] for a more elaborate set of practical antenna geometries. A few of other antenna geometries in common use are shown in Fig.53: (a) a horn antenna, (b) a Yagi-Uda array, and (c) a spiral plate antenna. A horn antenna is distinguished by its feeding from a input waveguide, that can be modeled as a slot or aperture antenna with suitable aperture-field distribution. The Yagi-Uda array is popularly used as a TV antenna, having a high-gain radiation along the end-fire direction, over a broad bandwidth. Much broader bandwidth of radiation can be achieved by a class of "frequency-independent" antennas, such as a spiral-plate antenna shown in Fig.53(c). In principle, one needs an infinite-sized spiral structure in Fig.1(c) for an ideal, frequency-independent operation. However, a practical spiral antenna will have to be truncated at a finite spiral length. This would result in a finite but very large bandwidth. Other frequency-independent antennas include (i) conical spiral and (ii) log-periodic structures.

Another common class of antennas make use of additional material structures, such as a focusing reflector, or a dielectric lens structure, surrounding a basic source antenna, in order to achieve focusing of the antenna radiation. This results in increasing the antenna gain. Though these structures are not really a part of the fundamental radiation mechanism, their designs are often critical for the over-all performance of the total antenna.

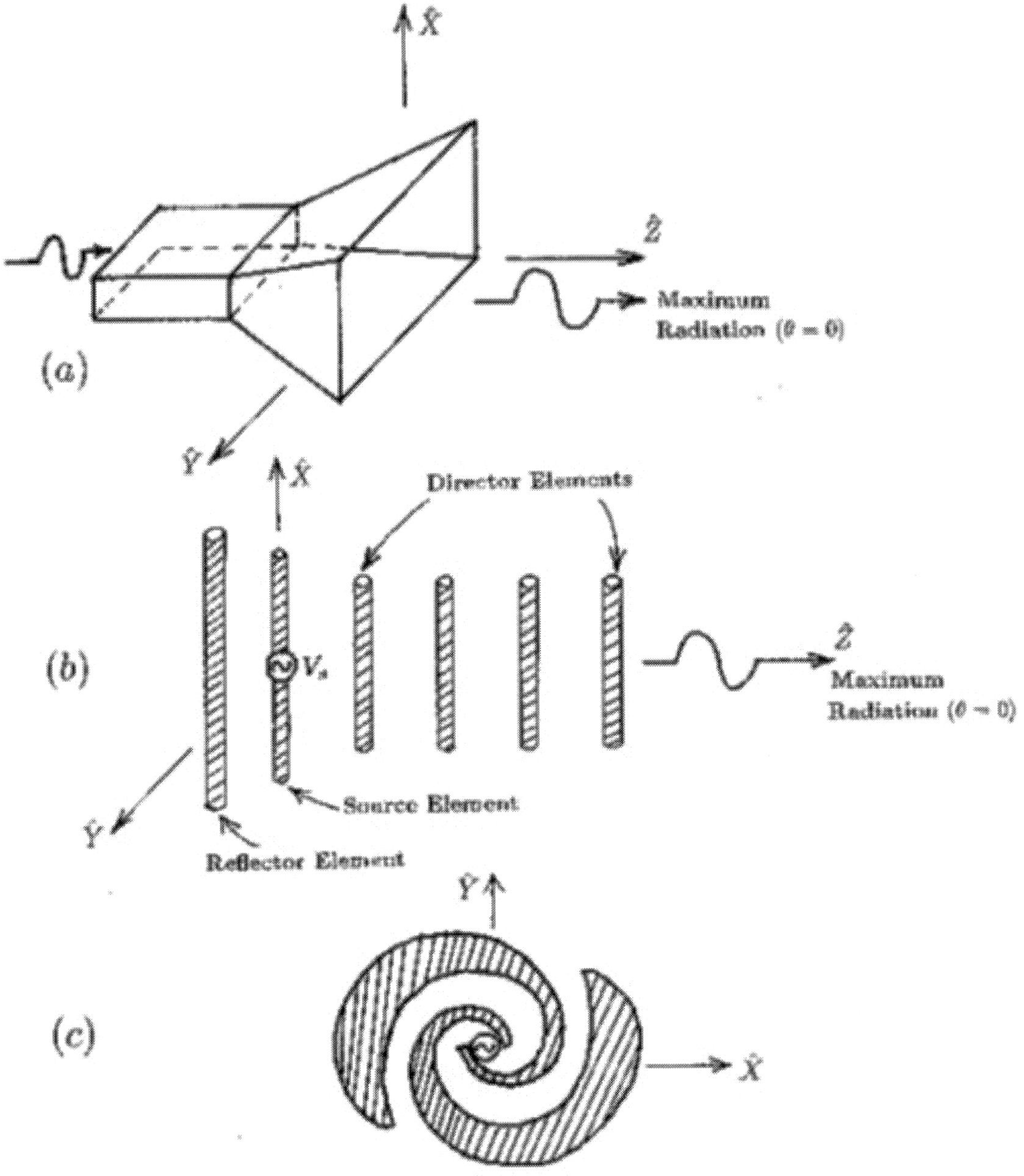

$\hat{X}$
$\hat{Z}$
Maximum Radiation ($\theta = 0$)
(a)
$\hat{Y}$
$\hat{X}$
Director Elements
(b)
V_s
$\hat{Z}$
Maximum Radiation ($\theta = 0$)
Source Element
$\hat{Y}$
Reflector Element
$\hat{Y}$
(c)
$\hat{X}$

Figure 53

ANTENNA TYPE		POLARIZATION	POWER PATTERN $\underline{P}_d(\theta,\phi)$	DIRECTIVITY D	RADIATION RESISTANCE, $R_{rad}(\Omega)$
Dipole Antennas (current along $\hat{z}$)	Short Dipole (Length=l)	$E_\theta(TM_z)$	$\sin^2\theta$	1.5	$80\pi^2(l/\lambda_0)^2$
	Half-Wave Dipole (Length= $\lambda_0/2$)	$E_\theta(TM_z)$	$\frac{\cos^2(\pi/2\cos\theta)}{\sin^2\theta}$	1.64	73
	Half-Wave Folded Dipole (Length= $\lambda_0/2$)	$E_\theta(TM_z)$	$\frac{\cos^2(\pi/2\cos\theta)}{\sin^2\theta}$	1.64	292
Monopole Antennas (current along $\hat{z}$)	Short Monopole (Length=l)	$E_\theta(TM_z)$	$\sin^2\theta$ (only half space, $0\le\theta\le\pi/2$)	3.0	$160\pi^2(l/\lambda_0)^2$
	Half-Wave Monopole (Length= $\lambda_0/4$)	$E_\theta(TM_z)$	$\frac{\cos^2(\pi/2\cos\theta)}{\sin^2\theta}$ (only half space, $0\le\theta\le\pi/2$)	3.28	36.5
	Half-Wave Folded Monopole (Length= $\lambda_0/4$)	$E_\theta(TM_z)$	$\frac{\cos^2(\pi/2\cos\theta)}{\sin^2\theta}$ (only half space, $0\le\theta\le\pi/2$)	3.28	146
Loop Antenna (Loop on x-y plane)	Small Loop Antenna (Loop Area=S)	$E_\phi(TE_z)$	$\sin^2\theta$	1.5	$320\pi^4(S/\lambda_0^2)^2$
Slot Antennas (slot length along $\hat{z}$)	Short Slot Antenna (Length=l)	$E_\phi(TE_z)$	$\sin^2\theta$	1.5	$45(\lambda_0/l)^2$
	Half-Wave Slot Antenna (Length= $\lambda_0/2$)	$E_\phi(TE_z)$	$\frac{\cos^2(\pi/2\cos\theta)}{\sin^2\theta}$	1.64	485
Microstrip Antenna (Antenna current along $\hat{x}$, patch on the x-y plane)	Rectangular, Edge Fed (Length= $\lambda_0/(2\sqrt{\varepsilon_r})$	Both E_θ and E_ϕ (TM_z)	No simple analytical formulas. See microstrip antenna section.		

Table I: Summary of basic antenna parameters of common antenna types. Refer to appropriate sections in the text for the geometry of the specific antennas.

CHAPTER 10

ANTENNA ARRAYS

An *antenna array* is made of a number of individual antenna elements, each located at a different position in space, with independent excitation to each input. Such an array may provide many practical advantages over a single antenna element. By properly designing for the individual locations and input excitations, one can achieve radiation characteristics that might not be normally feasible using a single antenna. Also, if one can control the input excitations to the individual elements in the real time, the radiation characteristics of the array, such as the pointing direction or radiation pattern, can be changed in a dynamic manner. This would allow significant flexibility in order to adjust to a dynamically changing situation. We will analyze the basic theory of a general antenna array, and derive results for specific situations of (1) a one-dimensional array and (2) a two-dimensional array with uniform as well as non-uniform amplitudes of input excitations.

An arbitrary configuration of an array of N identical antennas is shown in Fig.1. The center of the $i-$th antenna is defined by a position vector $\overline{R}_i = \hat{x}x_i + \hat{y}y_i + \hat{z}z_i$, and the input current to the $i-$th antenna is I_i. The electric field of the array in the far-field region can be expressed as the superposition of the individual contributions.

$$\overline{E}_i = I_i \frac{e^{-jk_0 r_i}}{r_i} \overline{e}_i(\theta, \phi), \ \overline{E} = \sum_{i=1}^{N} \overline{E}_i = \sum_{i=1}^{N} I_i \frac{e^{-jk_0 r_i}}{r_i} \overline{e}_i(\theta, \phi), \tag{10.1}$$

$$\overline{r}_i = \overline{r} - \overline{R}_i = \hat{r} r_i, \ r_i = r - \overline{R}_i \cdot \hat{r} = r - \triangle_i. \tag{10.2}$$

In the far field region, we have $r \gg \triangle_i$. Under this situation, the r_i's in the denominator of (10.1) may be approximated as $r_i = r - \triangle_i \simeq r$. However, such approximation is not appropriate for the propagation factor in the numerator. This is due to the oscillating nature of the propagation factor, which is sensitive to the small difference $\triangle_i$ between r_i and r.

$$\overline{E} = \sum_{i=1}^{N} I_i \frac{e^{-jk_0(r-\triangle_i)}}{r - \triangle_i} \overline{e}_i(\theta, \phi) \simeq \frac{e^{-jk_0 r}}{r} \sum_{i=1}^{N} I_i e^{jk_0 \triangle_i} \overline{e}_i(\theta, \phi). \tag{10.3}$$

Often it may be assumed, with good approximation, that the transmit properties of each antenna is unaffected by the presence of the other elements. This assumption is valid when the *mutual-coupling* effects between the antennas have negligible effects on the radiation characteristics of the individual elements. Under this practical approximation, the total radiation from the array can be further simplified.

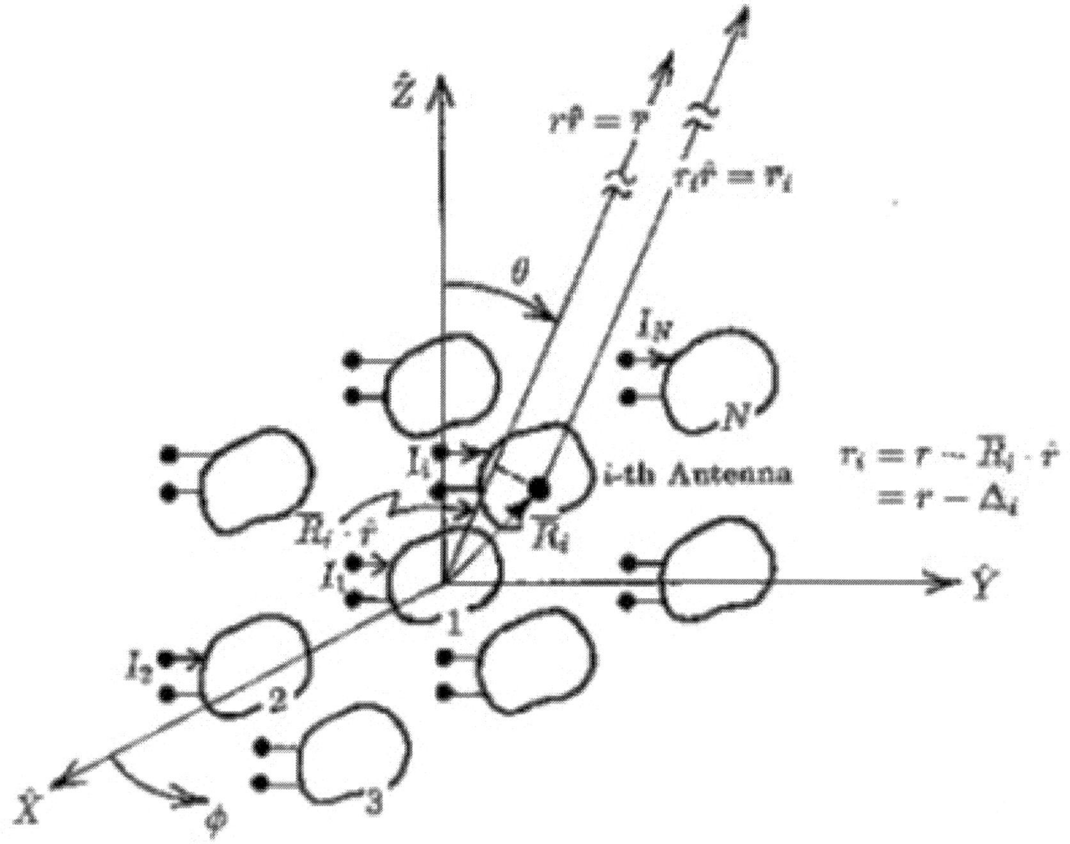

FIGURE 1

$$\overline{E} \simeq \overline{e}(\theta,\phi)\frac{e^{-jk_0 r}}{r}\sum_{i=1}^{N} I_i e^{jk_0\triangle_i};\ \overline{e}_i(\theta,\phi) = \overline{e}(\theta,\phi), \tag{10.4}$$

$$\overline{E} = (EF)\times(AF);\ EF = \overline{e}(\theta,\phi)\frac{e^{-jk_0 r}}{r},\ \ AF = \sum_{i=1}^{N} I_i e^{jk_0\triangle_i}. \tag{10.5}$$

The total radiation is thus separated into two independent factors. The element factor EF is the contribution due to the individual antenna elements, where as the array factor AF may be seen as the effect due to the array configuration. It allows to study and design the array, independent of the specific antenna elements to be used. When the antenna input currents are allowed to have different phases α_i, as well as different amplitudes $|I_i|$, the array factor (AF) may be expressed in the following compact form.

$$AF = \sum_{i=1}^{N} |I_i| e^{j\alpha_i} e^{jk_0\triangle_i} = \sum_{i=1}^{N} |I_i| e^{j\psi_i}; \ \psi_i = \alpha_i + k_0\triangle_i = \alpha_i + k_0(\hat{r} \cdot \overline{R}_i). \tag{10.6}$$

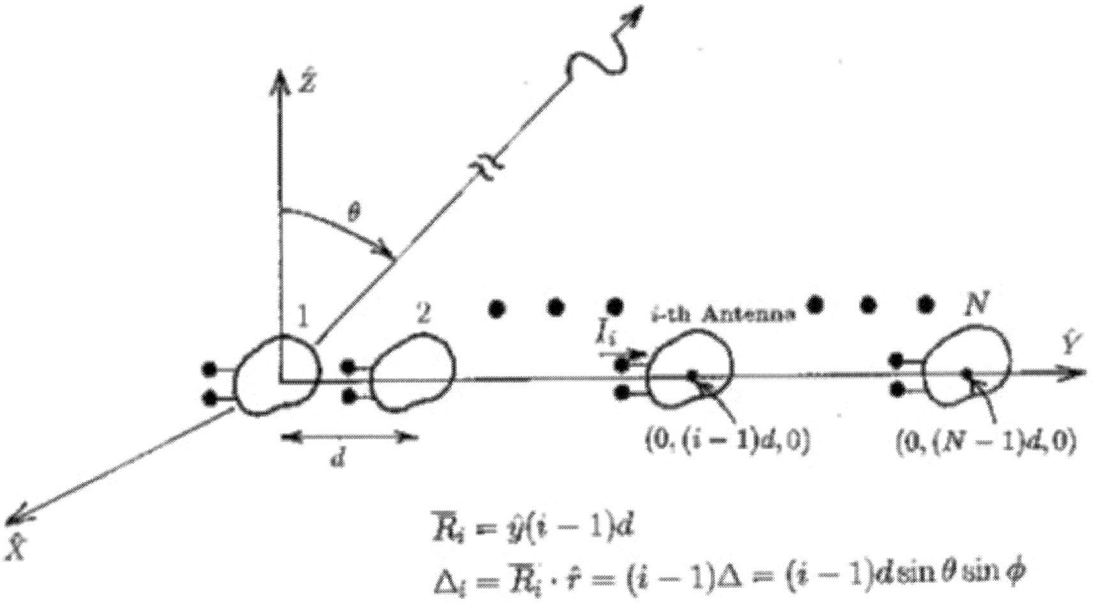

FIGURE 2

1. One-Dimensional, Uniform Array

As an example, consider an array of N antenna elements, equally spaced along the $\hat{y}$ axis, as shown in Fig.2. Also assume that each antenna is excited with the same amplitude, but phased in a linearly progressive manner. The array factor for this case can be expressed as follows.

$$I_i = |I_i| e^{j(i-1)\alpha} = I e^{j(i-1)\alpha}, \tag{10.7}$$

$$\overline{R}_i = \hat{y}(i-1)d, \ \triangle_i = \hat{r} \cdot \overline{R}_i = (i-1)d \sin\theta \sin\phi = (i-1)\Delta, \tag{10.8}$$

$$AF = I \sum_{i=1}^{N} e^{j\psi_i}; \ \psi_i = (i-1)\psi = (i-1)(\alpha + k_0\triangle) = (i-1)(\alpha + k_0 d \sin\theta \sin\phi), \tag{10.9}$$

$$|AF| = I|\sum_{i=1}^{N} e^{j(i-1)\psi}| = I|\frac{1-e^{jN\psi}}{1-e^{j\psi}}| = I|\frac{\sin\frac{N\psi}{2}}{\sin\frac{\psi}{2}}|. \tag{10.10}$$

It may be useful to normalize the array factor, with respect its maximum value NI.

$$AF_n(\theta,\phi) = \frac{|AF(\theta,\phi)|}{|AF(\theta,\phi)|_{max}} = |\frac{\sin\frac{N\psi}{2}}{N\sin\frac{\psi}{2}}| = |f(\psi)|;\ AF_n(\theta,\phi) = |f(\psi)| \leq 1. \tag{10.11}$$

The above array factor is plotted in Fig.3, as a function of the array variable ψ, for different number of elements N. Notice that the beam width gets progressively narrower with increasing number of elements N. The beam width approaches zero for $N \to \infty$, and the array in this limiting case would produce non-zero radiation in only one specific direction in space, with no radiation in all other directions.

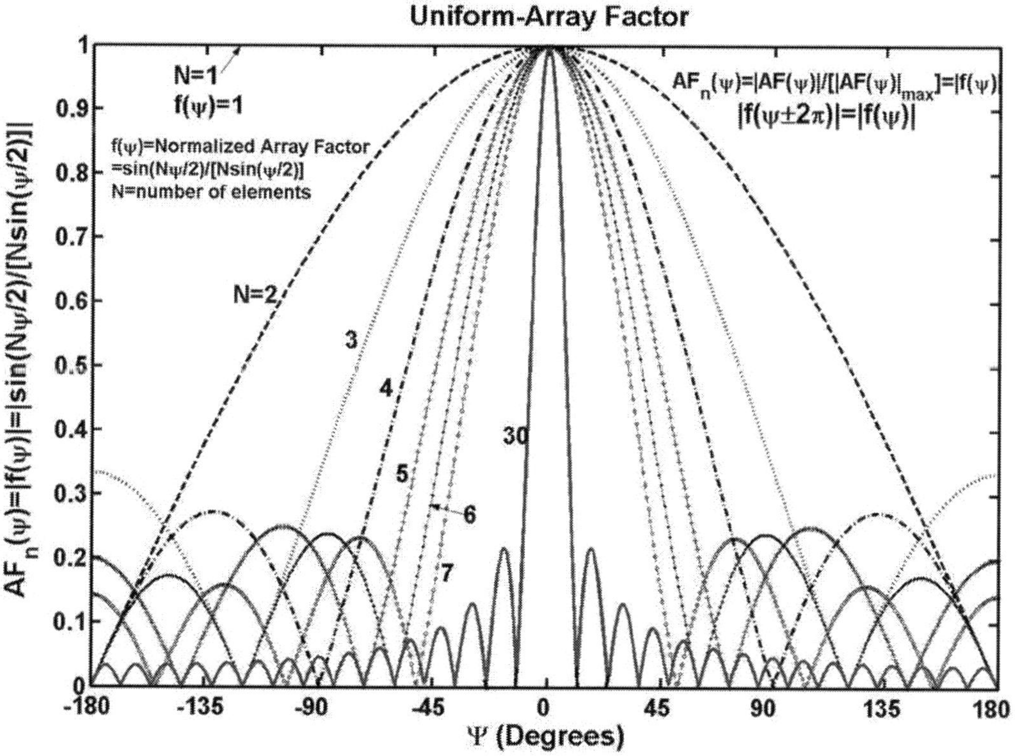

FIGURE 3

Array Side Lobe Level

The peak value of the normalized array factor $AF_n(\theta, \phi) = |f(\psi)|$ is unity, which occurs at $\psi = 0$ corresponding to the main direction of radiation. The next peak of AF_n in Fig.3 occurs approximately at $N\psi/2 = 1.5\pi$, where $AF_n = \frac{1}{N \sin \frac{1.5}{N}}$. This corresponds to the first side lobe level (SLL) of the array factor. For an array with large number of elements N, this level can be shown to be independent of N.

$$SLL = \frac{1}{N \sin \frac{1.5\pi}{N}} \simeq \frac{1}{1.5\pi} \cong 13dB. \tag{10.12}$$

Assuming that the element factor EF does not change significantly from the main beam to the side lobe of the array, the above side lobe level (10.12) is approximately equal to the SLL of the total radiation pattern. Remember that this SLL is derived for any large array ($N \to \infty$), while assuming that the amplitudes of excitation are uniform over the array. However, one can achieve a lower value of SLL by using a non-uniform amplitude distribution, to be discussed in the following sections 2, 3.

Graphical Method of Sketching Array Factor

Notice that the array factor in (10.10) is represented as a function of $\psi = \alpha + k_0 d \sin\theta \sin\phi$. This is relatively easy to sketch, on the ψ plane, as shown in Fig.3. However, ultimately one needs to sketch the array factor as a function of the space angles (θ, ϕ). This mapping may be done in a graphical manner, as demonstrated in Fig.4 for a four-element array. In Fig.4, the mapping from the ψ (upper sketch) to the θ plane (lower polar sketch), is demonstrated for the specific case of $\phi = \pi/2$, and $\alpha = \pi/2$. The reasoning for the method, as illustrated in Fig.4, should be self explanatory, and accordingly extended for other values of ϕ and α, as well as for any other general functional forms of the array factor $AF(\psi)$.

Notice that the array factor will be maximum ($= NI$) when ψ is zero or even multiples of π. This will determine the angle of the maximum radiation θ_{max} for a given value of ϕ, or vice versa.

$$|AF(\theta, \phi)|_{max} = I \left| \frac{\sin \frac{N\psi_{max}}{2}}{\sin \frac{\psi_{max}}{2}} \right| = NI, \tag{10.13}$$

$$\psi_{max} = \alpha + k_0 d \sin\theta_{max} \sin\phi = 2n\pi, \;\; n = 0, \pm 1, \pm 2, \cdots . \tag{10.14}$$

For example, if we observe on the $\phi = \pi/2$ plane,

$$\psi_{max} = \alpha + k_0 d \sin\theta_{max} = 2n\pi, \;\; n = 0, \pm 1, \pm 2, \cdots ; \; \phi = \frac{\pi}{2}. \tag{10.15}$$

Grating Lobes

In general, depending on the spacing d, there can be more than one value of n for which one can find a solution for θ_{max}. The radiation lobe corresponding to $n = 0$ is called the main lobe, whereas all other lobes corresponding to all n other than $n = 0$ are called the *grating lobes* (borrowing the name from similar situation in the study of periodic gratings.) For the main-lobe radiation, we have,

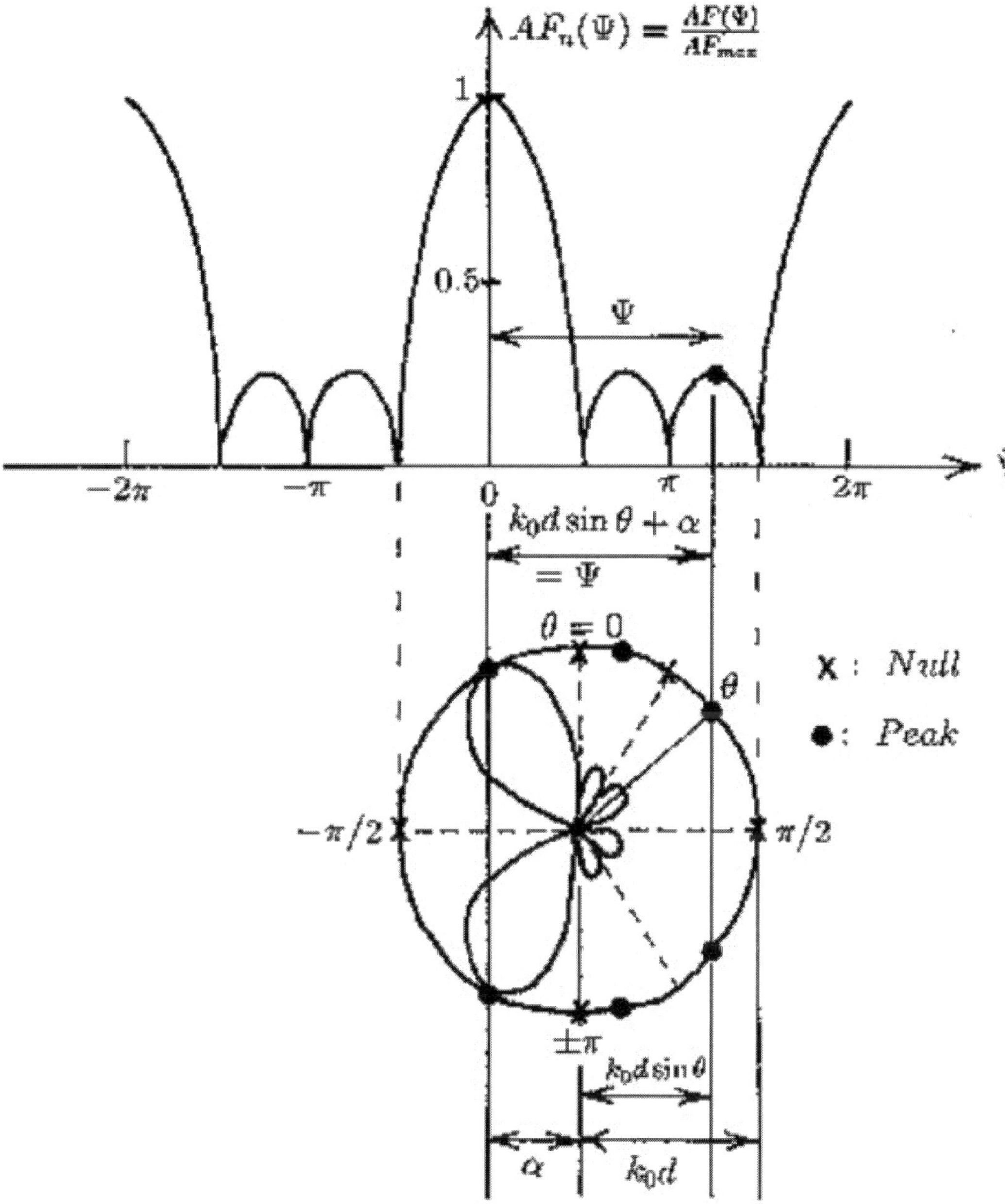

FIGURE 4

$$\theta_{max} = \arcsin \frac{-\alpha}{k_0 d}; \; n = 0, \; \phi = \pi/2. \tag{10.16}$$

Accordingly, the value of θ_{max} can be steered by controlling the inter-element phasing α. However, it may be desirable to design d such that no grating lobe is allowed to exist, for all possible values of α. As can be shown from (10.16), this is achieved if $k_0 d \leq \pi/2$, requiring the inter-element spacing to be less than half a wavelength.

2. One-Dimensional, Non-Uniform Array: Binomial Distribution

The above analysis of section 1 is extended here for a one-dimensional array of Fig.2, with a general distribution of input amplitudes $|I_i|$, which may not be a constant I for each element. As a specific case, consider a distribution with $|I_i|$ proportional to the $(i-1)$th coefficient of a binomial power series of order $(N-1)$. That is, $|I_i| = (N-1)!/[(N-i)!(i-1)!]$.

$$I_i = |I_i| e^{j(i-1)\alpha} = {}^{(N-1)}C_{(i-1)} e^{j(i-1)\alpha} = \frac{(N-1)}{(N-i)!(i-1)!} e^{j(i-1)\alpha},$$

$$AF = \sum_{i=1}^{N} |I_i| e^{j(i-1)\psi} = \sum_{i=1}^{N} {}^{(N-1)}C_{(i-1)} e^{j(i-1)\psi} = (1+e^{j\psi})^{(N-1)},$$

$$\psi = (\alpha + k_0 \triangle) = (\alpha + k_0 d \sin\theta \sin\phi),$$

$$|AF| = \left|(1+e^{j\psi})^{(N-1)}\right| = 2^{(N-1)} \cos^{(N-1)}(\psi/2). \tag{10.17}$$

The array factor is normalized with respect its maximum value of $2^{(N-1)}$.

$$AF_n(\theta,\phi) = \frac{|AF(\theta,\phi)|}{|AF(\theta,\phi)|_{max}} = |\cos^{(N-1)}(\psi/2)| = |f(\psi)|,$$

$$f(\psi) = \cos^{(N-1)}(\psi/2), \;\; AF_n(\theta,\phi) = |f(\psi)| \leq 1. \tag{10.18}$$

The above normalized array factor is plotted in Fig.5, for different number of elements N. Notice that unlike the array factor for a uniform array in Fig.3, there is no side-lobe in in the case of a binomial amplitude distribution. However, the beam width is relatively larger for the binomial distribution, as compared to the uniform distribution, when the number N of array elements is the same in both cases. This is illustrated in Fig.7.

The main-beam or any grating-lobe direction, which corresponds to the maximum value for the array factor AF, is determined when the array variable ψ is zero or any integral multiples of 2π. However, the variables ψ, and consequently the above beam condition $\psi = 0$, is independent the specific functional form of the array factor $AF(\psi)$, which is determined by the distribution of the input amplitudes $|I_i|$. Accordingly, formulas (10.13-10.16) for the main-beam or any grating-lobe direction would remain the same for a binomial distribution, as for a uniform, a Tchebychev, or any other specific distribution.

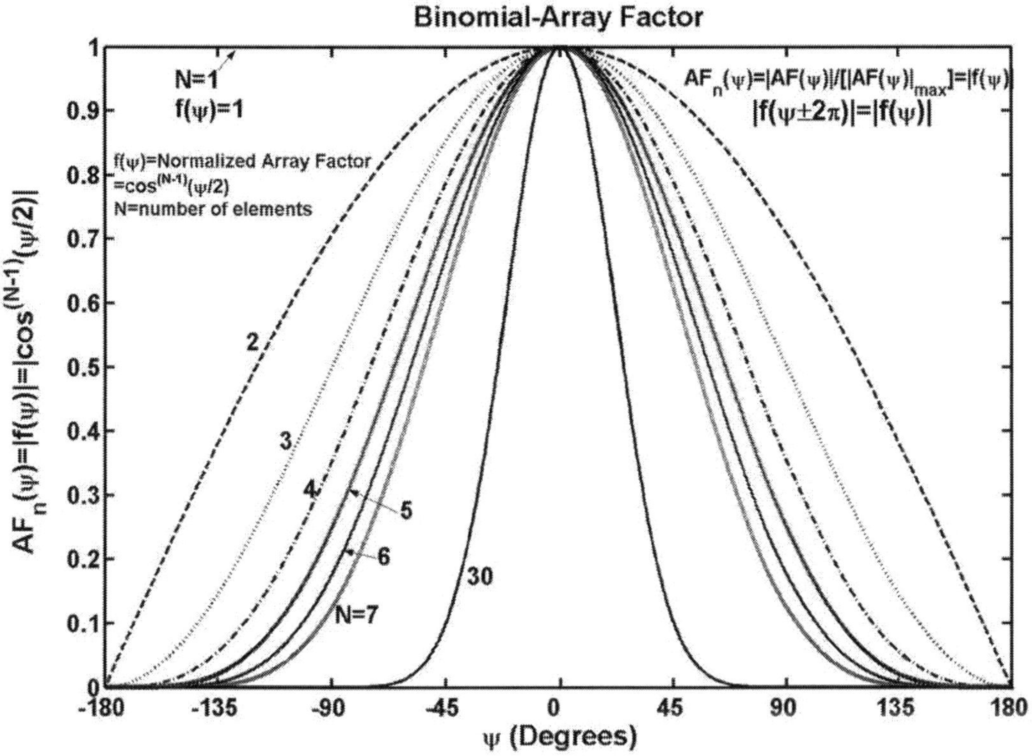

FIGURE 5

3. One-Dimensional, Non-Uniform Array: Tchebychev Distribution

We will study another useful amplitude distribution, which is based on special types of polynomial functions, referred to as the Tchebychev polynomials. Let us first study the functional forms of such special polynomials in the following, before we can use them for the array application.

3.1. Tchebychev Polynomials. The Tchebychev polynomial $T_N(z)$ is based on expansion of the function $\cos(Nt)$ in terms of the powers of $\cos(t)$.

$$T_N(z = \cos t) = \cos(Nt), 1 \leq z = \cos t \leq 1. \tag{10.19}$$

The function $T_N(z)$ would oscillate in the range $-1 \leq T_N(z) \leq 1$, $-1 \leq z \leq 1$. The functional form of the polynomial of z, when it is established by substituting $z = \cos t$ in the expansion of $\cos(Nt)$ in terms of $z = \cos t$, explicitly applies in the range $1 \leq z = \cos t \leq 1$. The same functional form would remain valid also outside the range $-1 \leq z \leq 1$, and therefore can be applicable as well for any real value of $-\infty \leq z \leq \infty$. However, the characteristic oscillatory behavior would not be maintained outside the range $-1 \leq z \leq 1$.

An iterative algorithm is established to generate the polynomial for any order N, starting from simple known forms for $N = 0$ and $N = 1$.

$$
\begin{aligned}
T_0(z) = \cos(0t) = 1,\ T_1(z) = \cos(t) = z,\\
\cos((N-2)t) = \cos((N-1)t - t) = \cos((N-1)t)\cos t + \sin((N-1)t)\sin t,\\
\cos(Nt) = \cos((N-1)t + t) = \cos((N-1)t)\cos t - \sin((N-1)t)\sin t,\\
\cos((N-2)t) + \cos(Nt) = 2\cos((N-1)t)\cos t,\\
T_{(N-2)}(z) + T_N(z) = 2T_{(N-1)}(z)z,\ \ T_N(z) = 2T_{(N-1)}(z)z - T_{(N-2)}(z)\ . \quad (10.20)
\end{aligned}
$$

The same iterative relation would be valid as well if the cosine functions are substituted by cosine-hyperbolic functions. Therefore, the above Tchebychev functions may also be expressed in terms of cosine-hyperbolic functions, valid for $z = \cosh t > 1$.

$$
T_N(z = \pm\cosh t) = (\pm 1)^N \cosh(Nt),\ z = \pm\cosh t,\ |z| > 1. \quad (10.21)
$$

The coefficient of the z^i term of the general polynomial $T_N(z)$ is equal to a_{Ni}, $0 \le i \le N$. Following the above iteration algorithm, the coefficient a_{Ni} is related to $a_{(N-1)(i-1)}$ and $a_{(N-2)i}$. a_{Ni} is zero for all odd or all even values of i, respectively when the order number N is odd or even. Equivalently, $T_N(z)$ is an odd or even function of z, for an odd or even value of N, respectively.

$$
\begin{aligned}
T_N(z) = \sum_{i=0}^{N} a_{Ni} z^i;\ a_{Ni} = 0,\ i > N,\\
T_N(z) = \sum_{i=0}^{N} a_{Ni} z^i = 2zT_{N-1}(z) - T_{N-2}(z)\\
= 2z \sum_{i=0}^{N-1} a_{(N-1)i} z^i - \sum_{i=0}^{N-2} a_{(N-2)i} z^i,\\
a_{Ni} = 2a_{(N-1)(i-1)} - a_{(N-2)i}\ . \quad (10.22)
\end{aligned}
$$

Using the above iteration formula, Polynomial expressions for $T_N(z)$ are derived for all $N \le 10$, as listed in the following.

$$
\begin{aligned}
T_0(z) &= 1,\\
T_1(z) &= z,\\
T_2(z) &= 2z^2 - 1,\\
T_3(z) &= 4z^3 - 3z,\\
T_4(z) &= 8z^4 - 8z^2 + 1,\\
T_5(z) &= 16z^5 - 20z^3 + 5z,\\
T_6(z) &= 32z^6 - 48z^4 + 18z^2 - 1,\\
T_7(z) &= 64z^7 - 112z^5 + 56z^3 - 7z,\\
T_8(z) &= 128z^8 - 256z^6 + 160z^4 - 32z^2 + 1,\\
T_9(z) &= 256z^9 - 576z^7 + 432z^5 - 120z^3 + 9z,\\
T_{10}(z) &= 512z^{10} - 1280z^8 + 1120z^6 - 400z^4 + 50z^2 - 1. \qquad (10.23)
\end{aligned}
$$

$T_{N-1}(z)$ is plotted in Fig.6 for several values of N, which we try to fit to an array factor with N antenna elements.

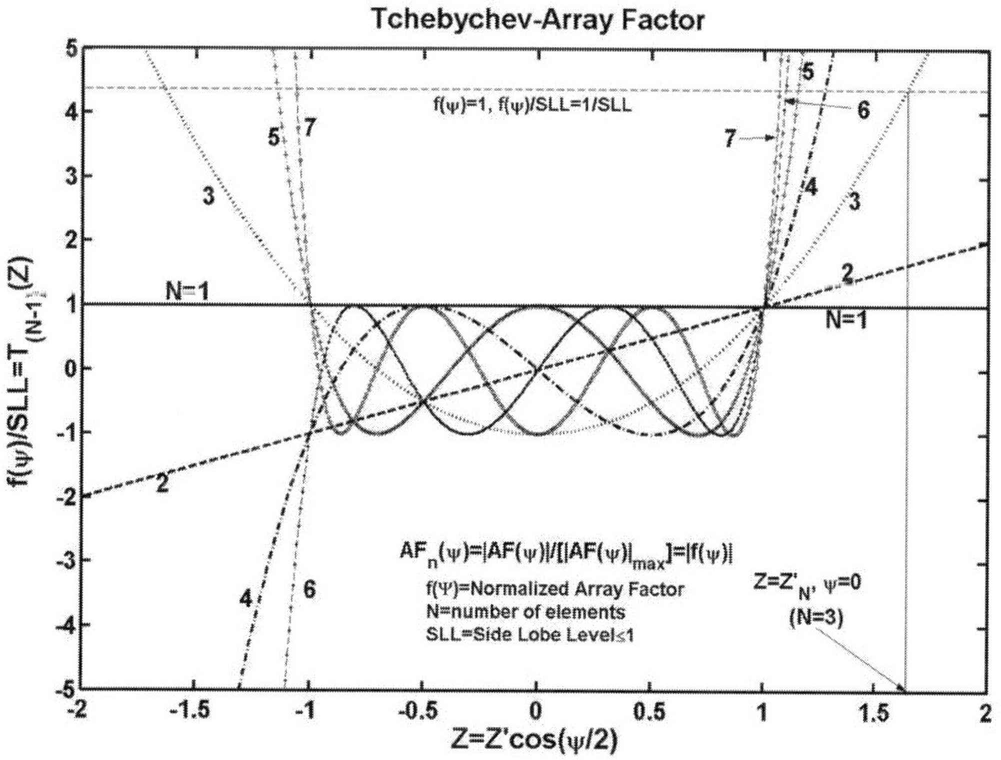

FIGURE 6

For notational convenience, let us change the indexing of antenna elements in Fig.2 from $1 \leq i \leq N$ to $i' = 2i - (N+1)$, $(N-1) \leq i' \leq (N-1)$, having only odd or only even indices i', respectively for even or odd values of N. Unlike only positive integer indices i, we have now both positive and negative integer indices i'. Let us also assume a symmetric distribution of the antenna input amplitudes about the center of the array. That is, we have $|I'_{i'}| = |I'_{-i'}|$. We normalize the input amplitudes such that the sum of all amplitudes is unity.

$$AF = \sum_{i=1}^{N} |I_i| e^{j(i-1)\psi} = e^{j(N-1)\psi/2} \sum_{i'=-(N-1)}^{(N-1)} |I'_{i'}| e^{ji'(\psi/2)},$$

$$I'_{i'} = |I_i|, \; i' = 2i - (N+1),$$

$$|AF(\theta,\phi)| = \left| \sum_{i'=-(N-1)}^{(N-1)} |I'_{i'}| e^{ji'(\psi/2)} \right| = |f(\psi)| = \frac{|AF(\theta,\phi)|}{|AF(\theta,\phi)|_{max}} = AF_n(\theta,\phi),$$

$$|AF|_{max} = \left| \sum_{i'=-(N-1)}^{(N-1)} |I'_{i'}| \right| = 1,$$

$$f(\psi) = \sum_{i'=-(N-1)}^{(N-1)} |I'_{i'}| e^{ji'(\psi/2)} = \sum_{i'=1}^{(N-1)} 2|I'_{i'}| \cos(i'(\psi/2)), \; (\text{even } N, \text{odd } i');$$

$$= |I'_{i'=0}| + \sum_{i'=2}^{(N-1)} 2|I'_{i'}| \cos(i'(\psi/2)), \; (\text{odd } N, \text{even } i'). \quad (10.24)$$

We can now express the normalized array factor $f(\psi)$ using a Tchebychev polynomial: $f(\psi) = (SLL) \times T_{N-1}(z = z' \cos(\psi/2))$, where $z' \geq 1$ is a design parameter to be determined. $f(\psi)$ would exhibit an oscillatory (equal ripple) behavior in the range $|z| \leq 1; |\cos(\psi/2)| \leq (1/z')$, which is equivalent to having all side lobes of the array at an equal level. The parameter $z' = z'_N$ is dependent on the number of elements N and the desired side-lobe level ($SLL \leq 1$). z'_N may be determined using the hyperbolic-cosine form of the Tchebychev function in (10.21).

$$\begin{aligned} f(\psi) &= (SLL) T_{N-1}(z = z' \cos(\psi/2)), \\ f(\psi = 0) = f(\psi)_{max} &= 1 = (SLL) T_{N-1}(z'), \\ T_{N-1}(z' = \cosh t > 1) &= \frac{1}{SLL} = \cosh((N-1)t), \\ z' = \cosh t &= \cosh[\frac{1}{(N-1)} \cosh^{-1}(\frac{1}{SLL})]. \end{aligned} \quad (10.25)$$

With the above known value of z'_N, the expression for $f(\psi) = (SLL) \times T_{N-1}(z = z' \cos(\psi/2))$ may be expanded in terms of power functions $\cos^p(\psi/2)$, $0 \leq p \leq (N-1)$, with known coefficients, using the Tchebychev polynomials of (10.23).

Now, the normalized array factor $f(\psi)$ of (10.24) consists of a series of functions $\cos(i'(\psi/2), 0 \leq i' \leq (N-1)$, each of which can be expanded as a series of power functions $\cos^p(\psi/2)$, $0 \leq p \leq i'$ using the Tchebychev polynomial $T_{i'}(\cos(\psi/2))$ of (10.23). Accordingly, the total function $f(\psi)$ is also expressed as a series of power functions $\cos^p(\psi/2)$, $0 \leq p \leq (N-1)$, with its unknown coefficients linearly related to the unknown antenna input amplitudes $I'_{i'}$. This expansion may be equated with the known expansion of the $f(\psi)$, as explained earlier, with known coefficients for $\cos^p(\psi/2)$. The input amplitudes $I'_{i'}$ can now be determined by solution of a complete set of linear equations.

3.2. Design Procedure for a Tchebychev Array. We will demonstrate the above procedure for a specific design, with $N = 10$ number of antenna elements, and a desired side-lobe level $SLL = 0.23$, $SLL(dB) = 20\log(SLL) \sim 13dB$. This is intended for comparison with a $N-$element uniform array, which is shown in (10.12) to have about the same side-lobe level $(SLL \sim 2/(3\pi))$.

(1) First, solve for z' from (10.25):

$$z' = \cosh[\frac{1}{9}\cosh^{-1}(1/0.23)] = 1.0287. \tag{10.26}$$

(2) Express $f(\psi)$ using (10.25,10.23):

$$\begin{aligned} f(\psi) &= (SLL)T_{N-1}(z = z'\cos(\psi/2)) \\ &= (SLL)[256z'^9\cos^9(\psi/2) - 576z'^7\cos^7(\psi/2) + 432z'^5\cos^5(\psi/2) \\ &\quad -120z'^3\cos^3(\psi/2) + 9z'\cos(\psi/2)]. \end{aligned} \tag{10.27}$$

(3) Express $f(\psi)$ using (10.24,10.23):

$$\begin{aligned} f(\psi) &= 2[I'_1\cos(\psi/2) + I'_3\cos(3(\psi/2)) + I'_5\cos(5(\psi/2)) \\ &\quad + I'_7\cos(7(\psi/2)) + I'_9\cos(9(\psi/2))] \\ &= 2(256I'_9)\cos^9(\psi/2) + 2(-576I'_9 + 64I'_7)\cos^7(\psi/2) \\ &\quad + 2(432I'_9 - 112I'_7 + 16I'_5)\cos^5(\psi/2) + 2(-120I'_9 + 56I'_7 - 20I'_5 + 4I'_3)\cos^3(\psi/2) \\ &\quad + 2(9I'_9 - 7I'_7 + 5I'_5 - 3I'_3 + I'_1)\cos(\psi/2). \end{aligned} \tag{10.28}$$

(4) Equate the two expressions (10.27,10.28) for $f(\psi)$, by equating the individual coefficients of each power function $\cos^p(\psi/2)$, $1 < p < (N-1)$ in the two expressions. Solve the resulting linear equations for all $I'_{i'}$. It may be solved manually. Alternately, a matrix inversion program may be used, which would be particularly helpful for large N.

$$\begin{aligned}
2(256I'_9) &= (SLL)256z'^9, \\
2(-576I'_9 + 64I'_7) &= (SLL)(-576z'^7), \\
2(432I'_9 - 112I'_7 + 16I'_5) &= (SLL)432z'^5, \\
2(-120I'_9 + 56I'_7 - 20I'_5 + 4I'_3) &= (SLL)(-120z'^3), \\
2(9I'_9 - 7I'_7 + 5I'_5 - 3I'_3 + I'_1) &= (SLL)(9z'),
\end{aligned} \tag{10.29}$$

$$[A]\,[I'] = (SLL)\,[B]\,, \tag{10.30}$$

$$[A] = \begin{bmatrix} 1 & -3 & 5 & -7 & 9 \\ 0 & 4 & -20 & 56 & -120 \\ 0 & 0 & 16 & -112 & 432 \\ 0 & 0 & 0 & 64 & -576 \\ 0 & 0 & 0 & 0 & 256 \end{bmatrix}, \quad [I'] = \begin{bmatrix} I'_1 \\ I'_3 \\ I'_5 \\ I'_7 \\ I'_9 \end{bmatrix}, \quad [B] = \begin{bmatrix} 9z' \\ -120z'^3 \\ 432z'^5 \\ -576z'^7 \\ 256z'^9 \end{bmatrix}, \tag{10.31}$$

$$[I'] = (SLL)\,[I']^{-1}\,[B]\,. \tag{10.32}$$

We will solve the amplitudes I' from the linear equations (10.29) by manual calculation.

$$\begin{aligned}
I'_9 &= (SLL)z'^9/2 = 0.1484, \\
I'_7 &= [(SLL)(-576z'^7)/2 + 576I'_9]/64 = 0.0739, \\
I'_5 &= [(SLL)432z'^5/2 - 432I'_9 + 112I'_7]/16 = 0.0874, \\
I'_3 &= [(SLL)(-120z'^3)/2 + 120I'_9 - 56I'_7 + 20I'_5]/4 = 0.0987, \\
I'_1 &= (SLL)(9z')/2 - 9I'_9 + 7I'_7 - 5I'_5 + 3I'_3 = 0.1055.
\end{aligned} \tag{10.33}$$

The resulting array factor (10.24,10.25) is compared in Fig.7 with that for a uniform (10.10) and a binomial (10.17) array with the same number of elements ($N = 10$). For this design with a $SLL = 0.23 \sim 13dB$, the shape of the main beam for the Tchebychev and the uniform array are about the same, whereas the binomial array exhibits a larger beamwidth. Note that the side lobe level (SLL) could be changed in a new design of a Tchebychev array to any desired level, by proper design of the input amplitudes, which would trade a higher (SLL) with a narrower beam width, or vice versa. In contrast, a uniform array always produces about a fixed side lobe level ($SLL = 0.23 \sim 13dB$), and a binomial array exhibits a gradual reduction in the array factor away from the main beam at $\psi = 0$ without any side lobe.

As discussed in section 2 for a binomial array, the main-beam or any grating-lobe direction is independent of any specific distribution of the input amplitudes $|I_i|$. The formulas (10.13-10.16) for the main-beam or any grating-lobe direction would remain the same for a Tchebychev distribution, as for a binomial or a uniform distribution.

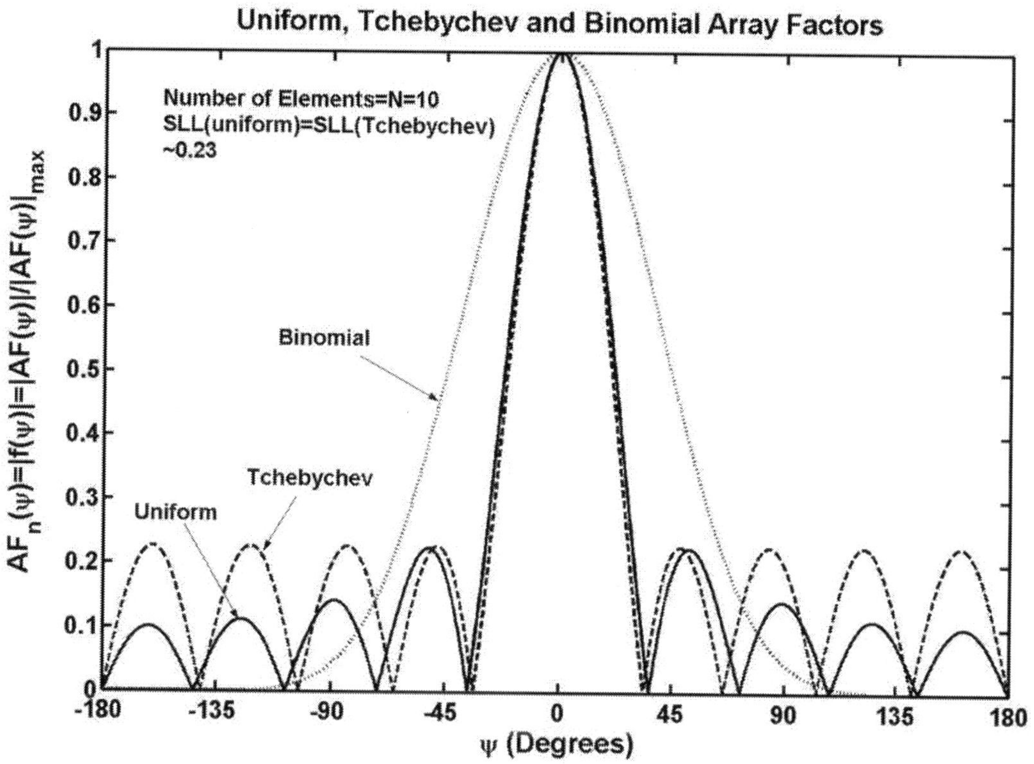

FIGURE 7

4. Two-Dimensional, Uniform Array

A one-dimensional array has the freedom of steering the main beam only in one plane, θ or ϕ. The concept may be extended to an array in two-dimensions, as shown in Fig.8, in order to allow beam steering both in θ and ϕ. As shown in Fig.8, let us consider an array of $(m \times n)$ elements, placed on the $x-y$ plane, with uniform amplitudes of excitation. The inter-element spacing is d_x along the x-axis, and d_y along the y-axis, which are in general different from each other. The inter-element phasing is α_x and α_y, respectively along the x and y axes, which can be independently controlled in order to allow steering in both θ and ϕ.

The derivation of the array factor for the two-dimensional array, as we present below, follows the same procedure used for a one-dimensional array. Same terminology is also used for equivalent parameters. We now need double-indexing of the elements along x and y, and accordingly need double summations for the total fields, in stead of a single summation for a one-dimensional array.

$$I_{pq} = |I_{pq}| e^{j[(p-1)\alpha_x + (q-1)\alpha_y]} = I e^{j[(p-1)\alpha_x + (q-1)\alpha_y]}, \tag{10.34}$$

$$\overline{R}_{pq} = \hat{x}(p-1)d_x + \hat{y}(q-1)d_y, \tag{10.35}$$

$$\triangle_{pq} = \hat{r}\cdot\overline{R}_{pq} = (p-1)d_x \sin\theta\cos\phi + (q-1)d_y \sin\theta\sin\phi = (p-1)\triangle_x + (q-1)\triangle_y, \tag{10.36}$$

$$AF = I\sum_{p=1}^{m}\sum_{q=1}^{n} e^{j\psi_{pq}} = I\sum_{p=1}^{m} e^{j(p-1)\psi_x}\sum_{q=1}^{n} e^{j(q-1)\psi_y} = I(AF_x)(AF_y), \tag{10.37}$$

$$\psi_{pq} = (p-1)(\alpha_x + \triangle_x) + (q-1)(\alpha_x + \triangle_y) = (p-1)\psi_x + (q-1)\psi_y, \tag{10.38}$$

$$|AF_x| = |\sum_{p=1}^{m} e^{j(p-1)\psi_x}| = |\frac{1-e^{jm\psi_x}}{1-e^{j\psi_x}}| = |\frac{\sin\frac{m\psi_x}{2}}{\sin\frac{\psi_x}{2}}|, \tag{10.39}$$

$$|AF_y| = |\sum_{q=1}^{n} e^{j(q-1)\psi_y}| = |\frac{1-e^{jn\psi_y}}{1-e^{j\psi_y}}| = |\frac{\sin\frac{n\psi_y}{2}}{\sin\frac{\psi_y}{2}}|, \tag{10.40}$$

$$|AF| = I|AF_x||AF_x| = I|\frac{\sin\frac{m\psi_x}{2}}{\sin\frac{\psi_x}{2}}||\frac{\sin\frac{n\psi_y}{2}}{\sin\frac{\psi_y}{2}}|, \tag{10.41}$$

$$|AF(\theta,\phi)|_{max} = I|\frac{\sin\frac{m\psi_{x(max)}}{2}}{\sin\frac{\psi_{x(max)}}{2}}||\frac{\sin\frac{n\psi_{y(max)}}{2}}{\sin\frac{\psi_{y(max)}}{2}}| = mnI, \tag{10.42}$$

$$AF_n(\theta,\phi) = \frac{|AF(\theta,\phi)|}{|AF(\theta,\phi)|_{max}} = (\frac{\sin\frac{m\psi_x}{2}}{m\sin\frac{\psi_x}{2}})(\frac{\sin\frac{n\psi_y}{2}}{n\sin\frac{\psi_y}{2}}). \tag{10.43}$$

In general, it may be possible to have more than one value of $\psi_{x(max)}$ and $\psi_{y(max)}$, for which the array factor is maximum. All such values, other than the dominant ones ($\psi_{x(max)} = 0,\ \psi_{y(max)} = 0$), correspond to the grating lobes of the array structure.

$$\psi_{x(max)} = 0, \pm 2\pi, \pm 4\pi \ldots;\ \psi_{y(max)} = 0, \pm 2\pi, \pm 4\pi \ldots. \tag{10.44}$$

As for a one-dimensional array, no grating lobe will occur, when one designs with $d_x,\ d_y \leq \lambda_0/2$. For the dominant lobe of radiation, which is of particular practical interest, we have,

$$\alpha_x + k_0 d_x \sin\theta_{max}\cos\phi_{max} = 0,\ \alpha_y + k_0 d_y \sin\theta_{max}\sin\phi_{max} = 0, \tag{10.45}$$

$$\theta_{max} = \arcsin\left(\sqrt{(\frac{\alpha_x}{k_0 d_x})^2 + (\frac{\alpha_y}{k_0 d_y})^2}\right),\ \phi_{max} = \arccos\left(\frac{-\alpha_x}{k_0 d_x \sin\theta_{max}}\right). \tag{10.46}$$

Accordingly, for given spacings d_x and d_y, the maximum directions θ_{max} and ϕ_{max} may be varied by controlling the inter-element phasing α_x and α_y.

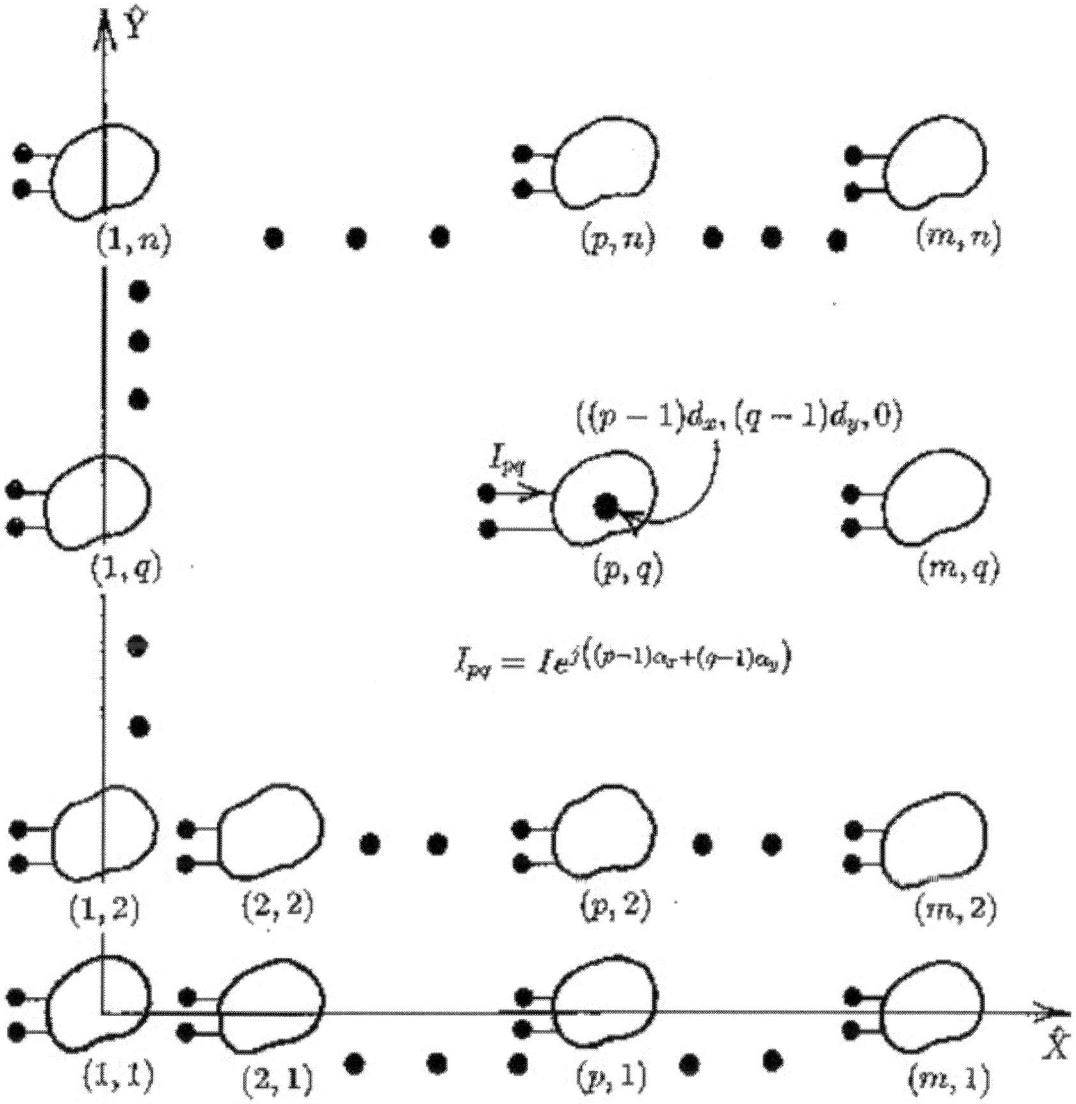

FIGURE 8

5. Two-Dimensional, Non-Uniform Array

We extend the above analysis of section 4 for a two-dimensional array of Fig.8, but with a general distribution of input amplitudes $|I_{pq}|$, which is not a constant I for each element. For convenience, let us assume that the sum of current amplitudes of all elements is normalized to unity.

$$\sum_{q=1}^{n}\sum_{p=1}^{m}|I_{pq}| = 1. \tag{10.47}$$

Further, we will consider a useful class of distributions, where the normalized current amplitude $|I_{pq}|$ can be expressed as a product of two normalized amplitudes I_p and $I_q|$, each determined by only one index, either p or q, and each is normalized to unity when summed over all values of its own index.

$$|I_{pq}| = |I_p||I_q|, \ \sum_{q=1}^{n}|I_q| = 1, \ \sum_{p=1}^{m}|I_p| = 1,$$
$$\sum_{q=1}^{n}\sum_{p=1}^{m}|I_{pq}| = \sum_{q=1}^{n}\sum_{p=1}^{m}|I_p||I_q| = \sum_{q=1}^{n}|I_q|\sum_{p=1}^{m}|I_p| = 1. \tag{10.48}$$

The derivations in section 4 may now be modified for a non-uniform array with the above conditions (10.47,10.48), resulting in the following expression for the array factor.

$$I_{pq} = |I_{pq}|e^{j[(p-1)\alpha_x+(q-1)\alpha_y]} = |I_p||I_q|e^{j[(p-1)\alpha_x+(q-1)\alpha_y]}, \tag{10.49}$$

$$AF = \sum_{p=1}^{m}\sum_{q=1}^{n}|I_{pq}|e^{j\psi_{pq}} = \sum_{p=1}^{m}|I_p|e^{j(p-1)\psi_x}\sum_{q=1}^{n}|I_q|e^{j(q-1)\psi_y} = (AF_x)(AF_y),$$
$$AF_x(\psi_x) = \sum_{p=1}^{m}|I_p|e^{j(p-1)\psi_x}, \ AF_y(\psi_y)\sum_{q=1}^{n}|I_q|e^{j(q-1)\psi_y},$$
$$|AF|_{max} = |AF_x(\psi_x)|_{max}|AF_y(\psi_y)|_{max}$$
$$= |AF_x(\psi_x = 0)||AF_y(\psi_y = 0) = \sum_{p=1}^{m}|I_p|\sum_{q=1}^{n}|I_q| = 1, \tag{10.50}$$

where the variables ψ_{pq}, ψ_x and ψ_y remain unchanged from those for a uniform array in section 4.

$$\psi_{pq} = (p-1)(\alpha_x + \triangle_x) + (q-1)(\alpha_x + \triangle_y) = (p-1)\psi_x + (q-1)\psi_y. \tag{10.51}$$

As for a uniform array in section 4, the above array factor AF for a non-uniform array is also expressed as a product of two factors $AF_x(\psi_x)$ and $AF_y(\psi_y)$. However, the specific expressions for the two factors would now be different. $AF_x(\psi_x)$ and $AF_y(\psi_y)$ essentially have the same expressions as the array factor AF for a one-dimensional array (see (10.50,10.17,10.24)) having m or n number of elements, but with the array parameter ψ for the one-dimensional array replaced by ψ_x or ψ_y, respectively. Accordingly, the normalized input amplitudes $|I_p|$ or $|I_q|$ may be designed to be equal to those for a specific type of one-dimensional array, respectively with m or n elements. Then, the factor $AF_x(\psi_x)$ or $AF_y(\psi_y)$ would have the same functional form as the array factor $AF(\psi)$ for the respective one-dimensional array. If $|I_p|$ and $|I_q|$ are designed both with a binomial, or both with a Tchebychev, type distribution (see sections 2, 3), then the particular two-dimensional array may be referred to as a planar binomial or Tchebychev array. However, the individual factors AF_x and AF_y may as well be designed independently with a different type of distribution - a binomial, a Tchebychev, a uniform, or any other suitable distribution, in order to achieve different pattern behavior in the orthogonal ($\phi = 0$ and $\phi = \pi/2$) planes.

The main-beam or any grating-lobe direction, which corresponds to the maximum value for the array factor AF, or equivalently of the factors AF_x and AF_y, is determined when the array variables ψ_x and ψ_y are zero or any integral multiples of 2π. However, the variables ψ_x and ψ_y, and consequently the above beam conditions $\psi_x = 0$ and $\psi_y = 0$, are independent of any functional forms of the array factors $AF_x(\psi_x)$ or $AF_y(\psi_y)$, which are determined by the distribution of the input amplitudes $|I_p|$ and $|I_q|$. Accordingly, formulas (10.44-10.46) for the main-beam or any grating-lobe direction would remain the same for a binomial or a Tchebychev array, as for a uniform array.

6. Mutual Coupling

As mentioned before, the analyses discussed earlier in this section are based on a critical assumption - that there is negligible mutual interactions between the antenna elements. This allowed us to make the key simplification of (10.4). Under situations when this assumption is no longer valid, one needs to use the rigorous equation (10.3). In this rigorous formulation, $\overline{e}_i(\theta, \phi)$ is different from that of an isolated antenna. Instead, $\overline{e}_i(\theta, \phi)$ is the transmit vector when only the i-th element is excited, (other elements open circuited, with zero input currents), while all other elements are still physically present in their respective locations. In other words, $\overline{e}_i(\theta, \phi)$ is the superposition of the primary radiation from the i-the antenna (when isolated), with secondary radiations from all the antenna elements passively excited through mutual coupling.

In addition to the radiated fields $\overline{e}_i(\theta, \phi)$, the voltages at the antenna inputs will also change due to the mutual-coupling effects. An N-element array needs to be rigorously modeled as a N-port network, using a $[Z]_{N\times N}$ impedance matrix.

$$[V] = [Z][I]; \; V_i = \sum_{k=1}^{N} I_k Z_{ik}, \; i = 1, N. \tag{10.52}$$

The input voltage of an antenna element, operating in an array environment, is dependent upon the input currents of all other elements. The linearity constant relating V_i and I_k is called the mutual impedance Z_{ik}. When one excites only the i-th element, with all other currents zero (but all other elements physically present,) from (10.52) we have $V_i = I_i Z_{ii}$. Z_{ii} is called the self impedance of the ith element, operating in the array environment. Note that the corresponding value, when the i-th antenna is allowed to operate in an isolated manner, is $V_i = I_i Z_{in}$. Z_{in} is, in general, different from Z_{ii}, and they would be equal to each other only in the limit of zero mutual coupling.

7. Infinite Antenna Array

A large periodic array of antenna elements may be modeled as an ideal array with infinite number of elements, allowing significant simplicity in analysis and understanding. This assumes useful operating conditions, where the elements are excited by signals of equal magnitude and the signal phase is progressing linearly with the position of the element. An ideal infinite array is simply a periodic continuation of the actual array to infinite distances. When the number of elements in a finite-sized array is sufficiently large, the environment "seen" by individual elements would closely resemble the ideal infinite array. That may be true for most elements, except the ones that are close to the outer edges of the finite array. Accordingly, a modeling for the ideal infinite array is intended to be a reasonable approximation for the characteristics of only the central elements of a finite-sized array.

Fig.9 shows the arrangement of an infinite array in two dimensions, where the individual elements are positioned on the $y = 0$ plane, periodically arranged in x and z directions. Due to the physical as well as excitation symmetry in the Fig.9, the mutual coupling seen by any selected element, produced due to radiation from all other elements, would be equal independent of the position of specific element in the array. The self-coupling is also the same for each element, because each element in the array is assumed to be physically identical to each other. Accordingly, the total input impedance seen by each element in the infinite array of Fig.9 would be the same for each element. On the other hand, if the array was finite in size, each element would strictly see a different environment, resulting in different input impedances for individual elements. The differing impedances for each element in the finite array could, in principle, be calculated by adding individual mutual-impedance contributions from each other element, but that would require a computationally complex procedure. In contrast, the input impedance for the infinite array of Fig.9 can be modeled by exploiting the physical symmetry and periodic nature of the electromagnetic fields in the structure, resulting in significant analytical simplicity. We will demonstrate such a modeling in the following, specifically for strip dipole elements, employing a Fourier analysis.

The antenna input impedance in an array environment would be in general different from the input impedance seen by a single, isolated element. The array impedance is expected to be a function of the inter-element spacing and the direction of radiation from the array. A simplified analysis of an infinite-array structure is expected to model such essential characteristics.

7.1. Infinite Array of Strip-Dipole Elements. Consider that each element in the infinite array of Fig.9 is a strip dipole element, conforming to the $y = 0$ plane, which is also the plane of the array. The longitudinal direction of each element is aligned with the z axis, with only z-directed antenna currents. They are periodically arranged with element spacings a and b in the x and z directions, respectively. The input of each element is excited with currents of equal magnitude, but having linear phase progression. α_x and α_z are the inter-element phase difference in the x and z directions, respectively. Accordingly, the distribution of the z directed current $J_{zmn}(x, z)$ on the (m, n) element may be related to that $J_{z00}(x, z)$ on the $(0, 0)$ element at the origin. The current distribution $J_z(x, z)$ we have used in (9.39) for an isolated antenna element may also be used here to describe the current distribution $J_{z00}(x, z)$ for the central element.

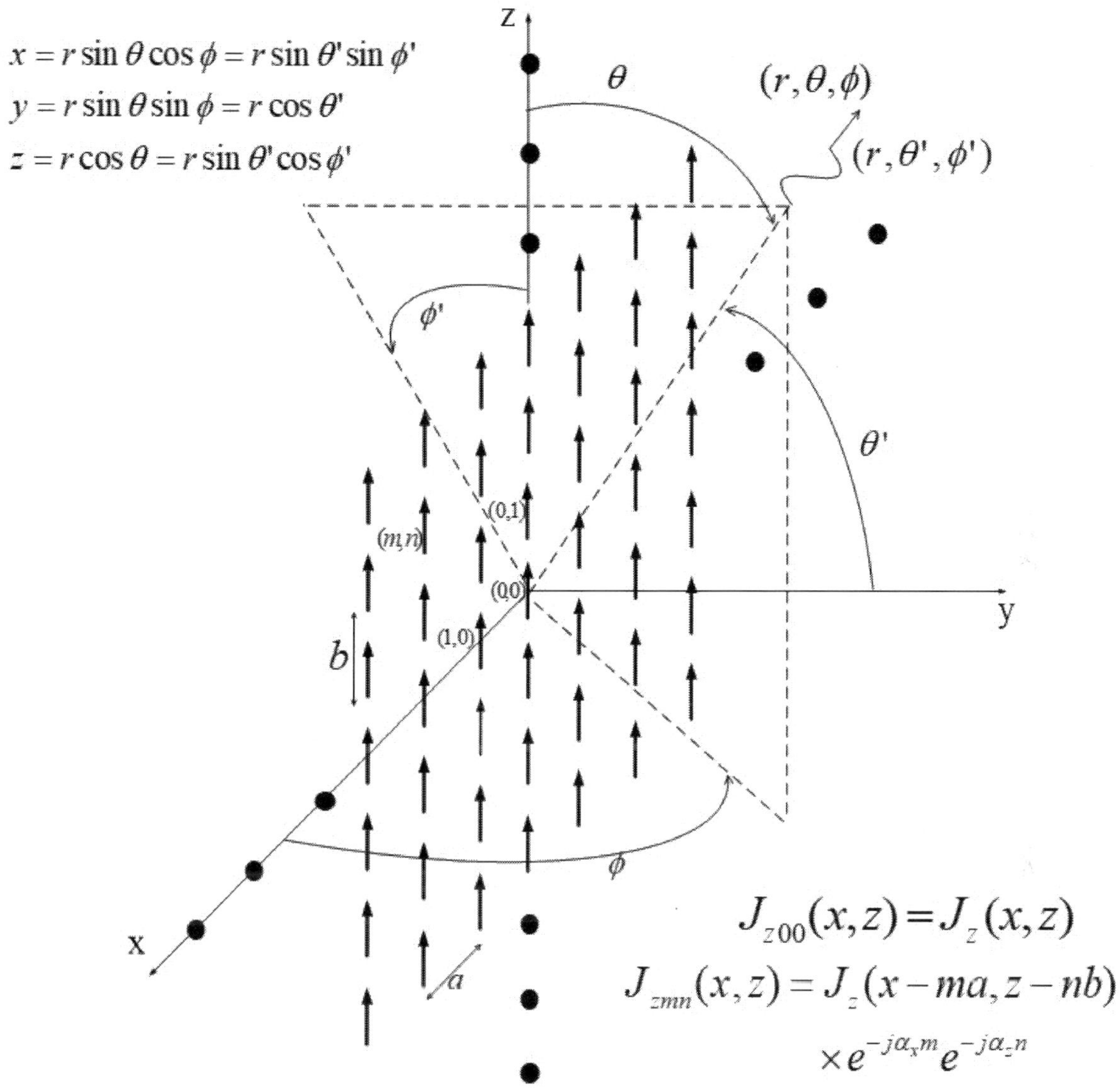

FIGURE 9

$$
\begin{aligned}
J_{zmn}(x,z) =&\ e^{-jm\alpha_x} e^{-jn\alpha_z} J_{z00}(x-ma, z-nb) \\
= &\ J_{z00}(x,z) * [e^{-j\alpha_x m} e^{-j\alpha_z n} \delta(x-ma)\delta(z-nb)], \\
&\ J_{z00}(x,z) = J_z(x,z),
\end{aligned} \tag{10.53}
$$

where $*$ represents a convolution operation. Now, the current distribution $J_{za}(x,z)$ for the entire infinite-array structure can be expressed as a superposition of the elemental currents $J_{zmn}(x,z)$.

$$J_{za}(x,z) = \sum_n \sum_m J_{zmn}(x,z) = \sum_n \sum_m e^{-jm\alpha_x} e^{-jn\alpha_z} J_{z00}(x-ma, z-nb)$$
$$= \sum_n \sum_m J_z(x,z) * [e^{-jm\alpha_x} e^{-jn\alpha_z} \delta(x-ma)\delta(z-nb)]$$
$$= J_z(x,z) * \sum_n \sum_m e^{-jk_{x0}ma} e^{-jk_{z0}nb} \delta(x-ma)\delta(z-nb)$$
$$= J_z(x,z) * [e^{-jk_{x0}x} e^{-jk_{z0}z} \sum_n \sum_m \delta(x-ma)\delta(z-nb)], \; k_{x0} = \frac{\alpha_x}{a}, \; k_{z0} = \frac{\alpha_z}{a}. \quad (10.54)$$

The periodic delta functions in (10.54) may be expressed as a Fourier series. Equivalently, the Fourier transform in the x and z coordinates, as defined in (6.28), of the periodic delta functions may also be expressed as a periodic delta functions in the (k_x, k_z) wave-number space.

$$\sum_n \sum_m \delta(x-ma)\delta(z-nb) = \sum_n \sum_m a_{mn} e^{j\frac{2\pi}{a}mx} e^{j\frac{2\pi}{b}ny}, \; a_{mn} = \frac{1}{ab},$$
$$\Im_{xz}[\sum_n \sum_m \delta(x-ma)\delta(z-nb)] = \frac{4\pi^2}{ab} \sum_n \sum_m \delta(k_x - \frac{2\pi}{a}m)\delta(k_z - \frac{2\pi}{b}n),$$
$$\Im_{xz}[e^{j\frac{2\pi}{a}mx} e^{jz\frac{2\pi}{b}nz}] = \Im_x[e^{j\frac{2\pi}{a}mx}]\Im_z[e^{j\frac{2\pi}{b}nz}],$$
$$\Im_x[e^{j\frac{2\pi}{a}mx}] = 2\pi\delta(k_x - \frac{2\pi}{a}m), \; \Im_z[e^{j\frac{2\pi}{b}nz}] = 2\pi\delta(k_z - \frac{2\pi}{b}n). \quad (10.55)$$

$\Im_{xz}$ represents a two-dimensional Fourier transform in the x and z directions, as we have defined in (6.28), and $\Im_x$ and $\Im_z$ represent the corresponding one-dimensional transforms in the x and z directions, respectively.

Now, using the Fourier theory for convolution functions, we can express the Fourier transform $\widetilde{J}_{za}(k_x, k_z)$ for the array current $J_{za}(x,z)$ of (10.54) in terms of the transform $\widetilde{J}_z(k_x, k_z)$ for the current $J_z(x,z)$ on the central element. The result can then be simplified using (10.55) and the shifting property of the Fourier transform.

$$\widetilde{J}_{za}(k_x, k_z) = \frac{4\pi^2}{ab} \widetilde{J}_z(k_x, k_z) \Im_{xz}[e^{-jk_{x0}x} e^{-jk_{z0}z} \sum_n \sum_m \delta(x-ma)\delta(z-nb)]$$
$$= \frac{4\pi^2}{ab} \widetilde{J}_z(k_x, k_z) \sum_n \sum_m \delta(k_x + k_{x0} - \frac{2\pi}{a}m)\delta(k_z + k_{z0} - \frac{2\pi}{b}n)$$
$$= \frac{4\pi^2}{ab} \sum_n \sum_m \widetilde{J}_z(k_{xm}, k_{zn})\delta(k_x - k_{xm})\delta(k_z - k_{xm}),$$
$$k_{xm} = -k_{x0} + \frac{2\pi}{a}m = -\frac{\alpha_x}{a} + \frac{2\pi}{a}m, \; k_{zn} = -k_{z0} + \frac{2\pi}{b}n = -\frac{\alpha_z}{b} + \frac{2\pi}{b}n. \quad (10.56)$$

The above transform consists of contributions from discrete values of wavenumbers, that are periodically spaced in the (k_x, k_z) wave-number space. Equivalently, the array current $J_{za}(x, z)$ may be expressed as a Fourier inverse of its transform $\widetilde{J}_{za}(k_x, k_z)$. The result would effectively form a Fourier-series decomposition of the array current into infinite number of Fourier current sheets.

$$J_{za}(x,z) = \frac{1}{4\pi^2} \int_{k_z=-\infty}^{\infty} \int_{k_x=-\infty}^{\infty} \widetilde{J}_{za}(k_x, k_z) e^{jk_x x} e^{jk_z z} dk_x dk_z$$
$$= \frac{1}{ab} \sum_n \sum_m \widetilde{J}_z(k_{xm}, k_{zn}) e^{jk_{xm}x} e^{jk_{zn}z}. \quad (10.57)$$

7.1.1. *Floquet Mode Representation.* Each current sheet in the above decomposition of the array current may be seen as a source of a plane wave. The resulting plane wave may be propagating in a given direction in the surrounding free space, or be evanescent in the normal direction (y). This would depend, respectively, if the sum of k_{xm}^2 and k_{zn}^2 is less or more than k_0^2, where k_0 is the wave number in the free space. The corresponding field components may be expressed using the source-field relations for a plane-wave in Ch.6.2.4. Such decomposition of the fields and currents in a periodic structure into plane-wave components is referred to as a Floquet-mode decomposition. Each Fourier component with wave numbers (k_{xm}, k_{zn}), that are periodically spaced in the (k_x, k_z) wave-number space, are called the Floquet modes. Fig.10 shows the relative positions of different Floquet modes on the (k_x, k_z) plane. Such a diagram, referred to as a Floquet-mode diagram, would be useful for array design and understanding.

It would be practically desirable to design an array that radiates only one plane wave, propagating in a given direction $(\theta = \theta_0, \phi = \phi_0)$, or $(\theta' = \theta'_0, \phi' = \phi'_0)$ (see Fig.9). This is equivalent to having the $(m = 0, n = 0)$ Floquet mode to propagate or radiate, with $(k_{xo}^2 + k_{z0}^2) < k_0^2$, while all other Floquet modes are evanescent, with $(k_{xm}^2 + k_{zm}^2) > k_0^2$, $(m, n) \neq (0, 0)$.

7.1.2. *Design of Inter-Element Spacing.* The above design condition is to be established using the expressions of k_{xm} and k_{zn} in (10.56). With this objective, it may be sufficient to ensure that the two lowest-order Floquet modes $(m = 1, n = 0)$ and $(m = 0, n = 1)$ are evanescent (see Fig.10), for all possible values of $(k_{xo}^2 + k_{z0}^2) < k_0^2$ (all radiation angles), with $-k_0 < k_{x0} < k_0$ and $-k_0 < k_{z0} < k_0$. Once these two modes are ensured to be evanescent, then the Floquet modes $(m = -1, n = 0)$ and $(m = 0, n = -1)$ that are symmetric to the above two modes would also be evanescent. All other Floquet modes with higher indices can be shown to be evanescent as well. Consider first the evanescent condition for the $(m = 1, n = 0)$ mode.

$$k_{xm}^2 + k_{zn}^2 = k_{x1}^2 + k_{z0}^2 = (-k_{x0} + \frac{2\pi}{a})^2 + k_{z0}^2 > k_0^2,$$
$$(-k_0 + \frac{2\pi}{a})^2 > k_0^2, \ -k_0 + \frac{2\pi}{a} > k_0, \ \frac{2\pi}{a} > 2k_0, \ a < \frac{\pi}{k_0} = \frac{\lambda_0}{2}. \quad (10.58)$$

This imposes an upper limit for the design of the element spacings ($a < \lambda_0/2$). Similarly, the evanescent condition for the ($m = 0, n = 1$) mode can also be shown to impose an upper limit for the element spacing ($b < \lambda_0/2$).

7.1.3. *Design of Array Phasing.* The wave numbers k_{x0} and k_{z0} of the $(0,0)$ Floquet mode in (10.57), assumed to be the only propagating mode, may be related to the associated propagation direction ($\theta = \theta_0, \phi = \phi_0$), or equivalently ($\theta' = \theta'_0, \phi' = \phi'_0$) (Fig.9). The basic plane-wave relations (5.26) are used. This provides design formulas for the inter-element phases α_x and α_z, necessary for a desired angle of radiation.

$$\begin{aligned} k_{x0} &= k_0 \sin\theta_0 \cos\phi_0 = \frac{\alpha_x}{a} = k_0 \sin\theta'_0 \sin\phi'_0, \\ \alpha_x &= k_0 a \sin\theta_0 \cos\phi_0 = k_0 a \sin\theta'_0 \sin\phi'_0, \\ k_{z0} &= k_0 \cos\theta_0 = \frac{\alpha_z}{b} = k_0 \sin\theta'_0 \cos\phi'_0, \\ \alpha_z &= k_0 b \cos\theta_0 = k_0 b \sin\theta'_0 \cos\phi'_0. \end{aligned} \qquad (10.59)$$

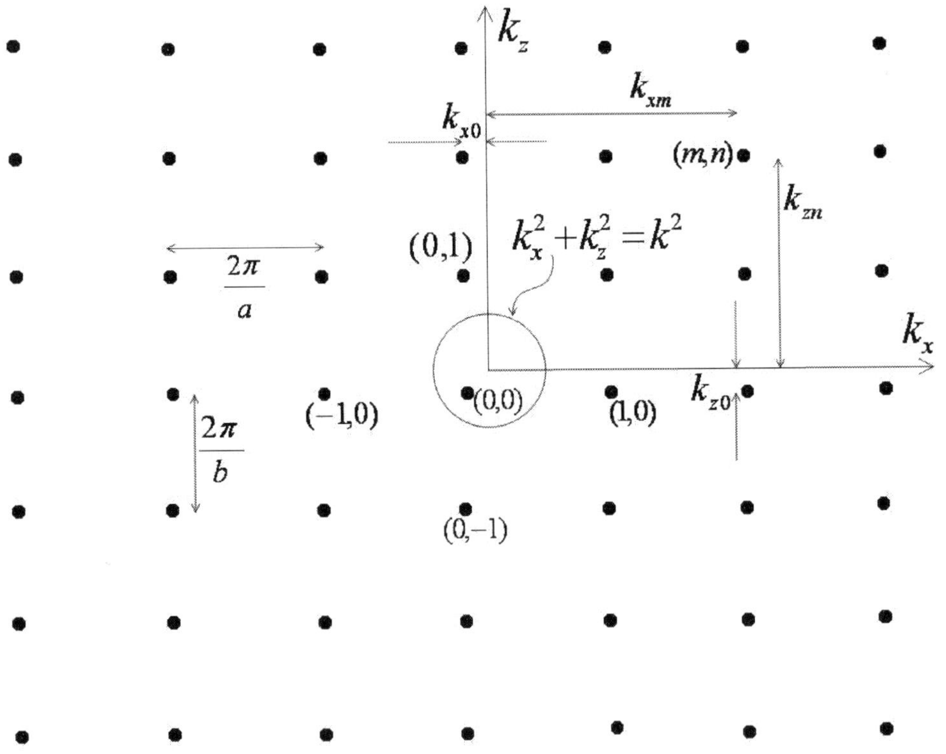

FIGURE 10

7.1.4. *Array Field and Impedance Modeling.* Let us now express the electric field E_z produced due to the array current in (10.56,10.57). We will use the same relation (9.38) we used for an isolated strip dipole, except the current transform $\widetilde{J}_z(k_x, k_z)$ for the isolated element is now replaced by $\widetilde{J}_{az}(k_x, k_z)$ for the entire array. The complex input impedance seen by the central element, which would be identical to that seen by any other array element as discussed before, is equated to the complex source power delivered from the current $\widetilde{J}_z(k_x, k_z)$ on the central element. This assumes that the current I_{in} at the input terminal of the central element is unity. The complex impedance or power is expressed as a surface integral of the product of the electric field $E_z(x, z)$ and element current $J_z(x, z)$ over the surface of the central element. This can then be equivalently expressed as a Fourier integral in the transform domain. This procedure is also similar to that used for an isolated strip dipole element in (9.44), except the electric field is now produced by the total array instead of only one isolated element.

$$E_z(x,y,z) = \frac{1}{j\omega\epsilon_0}\frac{1}{4\pi^2}\int_{k_z}\int_{k_x}\frac{\widetilde{J}_{za}(k_x,k_z)}{2jk_y}(k_0^2 - k_z^2)e^{jk_x x}e^{jk_z z}e^{\mp jk_y y}dk_x dk_z,$$

$$Z_{in} = -\int_z\int_x E_z(y=0)J_z^* ds \ (I_{in}=1)$$

$$= -\frac{1}{j\omega\epsilon_0}\frac{1}{4\pi^2}\int_{k_z}\int_{k_x}\frac{\widetilde{J}_{za}(k_x,k_z)\widetilde{J}_z^*(k_x,k_z)}{2jk_y}(k_0^2 - k_z^2)dk_x dk_z. \quad (10.60)$$

ϵ_0 is the permittivity of the free-space medium. Using (10.56), the integral expression of (10.60) can be expresses as a summation over k_{xm} and k_{zn}.

$$Z_{in} = \frac{1}{2\omega\epsilon_0} \times \frac{1}{ab}\sum_n\sum_m \frac{\left|\widetilde{J}_z(k_{xm},k_{zn})\right|^2}{k_{ymn}}(k_0^2 - k_{zn}^2),$$

$$k_{ymn} = \sqrt{k_0^2 - k_{xm}^2 - k_{zn}^2},\ k_{y00} = k_0\cos\theta_0' = k_0\sin\theta_0\sin\phi_0. \quad (10.61)$$

The real part R_{in} of the above input impedance would be contributed only by the $(m = 0, n = 0)$ Floquet mode, and the imaginary part jX_{in} would be contributed by all other Floquet modes. This is because k_{ymn} would be positive only for the $(0, 0)$ mode. This assumes that the array is designed with only one radiating direction, as discussed before.

$$R_{in} = \frac{1}{ab} \times \frac{1}{2\omega\epsilon_0} \times \frac{\left|\widetilde{J}_z(k_{x0}, k_{z0})\right|^2}{k_{y00}}(k_0^2 - k_{z0}^2)$$

$$= \frac{1}{ab} \times \frac{1}{2\omega\epsilon_0} \times \frac{\left|\widetilde{J}_z(k_{x0}, k_{z0})\right|^2}{k_0 \sin\theta_0 \sin\phi_0} k_0^2 \sin^2\theta_0$$

$$= \frac{1}{ab} \times \frac{1}{2\omega\epsilon_0} \times \frac{\left|\widetilde{J}_z(k_{x0}, k_{z0})\right|^2}{k_0 \cos\theta'_0}(k_0^2 - k_0^2 \sin^2\theta'_0 \cos^2\phi'_0)$$

$$= \frac{1}{ab} \times \frac{\eta_0}{2} \times \frac{\sin\theta_0}{\sin\phi_0} \left|\widetilde{J}_z(k_{x0}, k_{z0})\right|^2$$

$$= \frac{1}{ab} \times \frac{\eta_0}{2} \times \frac{(1 - \sin^2\theta'_0 \cos^2\phi'_0)}{\cos\theta'_0} \left|\widetilde{J}_z(k_{x0}, k_{z0})\right|^2. \quad (10.62)$$

$\eta_0 = k_0/(\omega\epsilon_0)$ is the wave impedance of the free-space medium, which is equal to $120\pi \simeq 377\Omega$.

7.1.5. *Half-Wave Dipole Elements.* The above analysis is valid for general situations for any antenna length, width and array element spacing. Let us specifically calculate the array input impedance for dipole elements that are half-wavelength long. This antenna length is close to practical designs often used, allowing the antenna to operate near a resonance. Further, we will select inter-element separations $a = b = \lambda_0/2$, which are at the upper bound of the design condition in (10.58). Results for R_{in} and X_{in} are shown in Fig.11 and Fig.12, respectively, for different angles of radiation. For radiation normal to the array surface ($\theta'_0 = 0$), the array input resistance is calculated to be about 76.4Ω. This value may be compared with about 73Ω resistance for an isolated half-wave dipole element, derived in (9.35).

$$R_{in}(l = \lambda_0/2) = \frac{1}{ab} \times \frac{\eta_0}{2} \times \frac{\sin\theta_0}{\sin\phi_0} \left|\widetilde{J}_z(k_{x0}, k_{z0})\right|^2 = \frac{240}{\pi} \simeq 74.6\Omega,$$

$$\widetilde{J}_z(k_{x0}, k_{z0}) = \frac{2}{k_0} = \frac{\lambda_0}{\pi}, \; k_{x0} = k_{z0} = 0,$$

$$\theta_0 = \phi_0 = \frac{\pi}{2}, \; \theta'_0 = 0, \; a = b = \frac{\lambda_0}{2}. \quad (10.63)$$

The strip current transform of (9.40,9.42) is used for the above result, with strip width w assumed to be much smaller than the wave length. Note that the array impedance depends on unit cell dimensions as well as radiation angles. It may be observed from Figs.11,Fig.12 that for radiation along the $\phi'_0 = 0$ plane (the E-plane) the array impedance tends to reduce with increasing θ'_0, leading to a zero impedance for $\theta'_0 = \pi/2$. On the other hand, for $\phi'_0 = \pi/2$ (the H-plane) the impedance increases with θ'_0, leading to an infinite impedance for $\theta'_0 = \pi/2$. There is comparatively moderate degree of variation of the impedance with θ'_0 (at least for lower values of θ'_0,) for radiation in the $\phi'_0 = \pi/2$ plane (the D-plane).

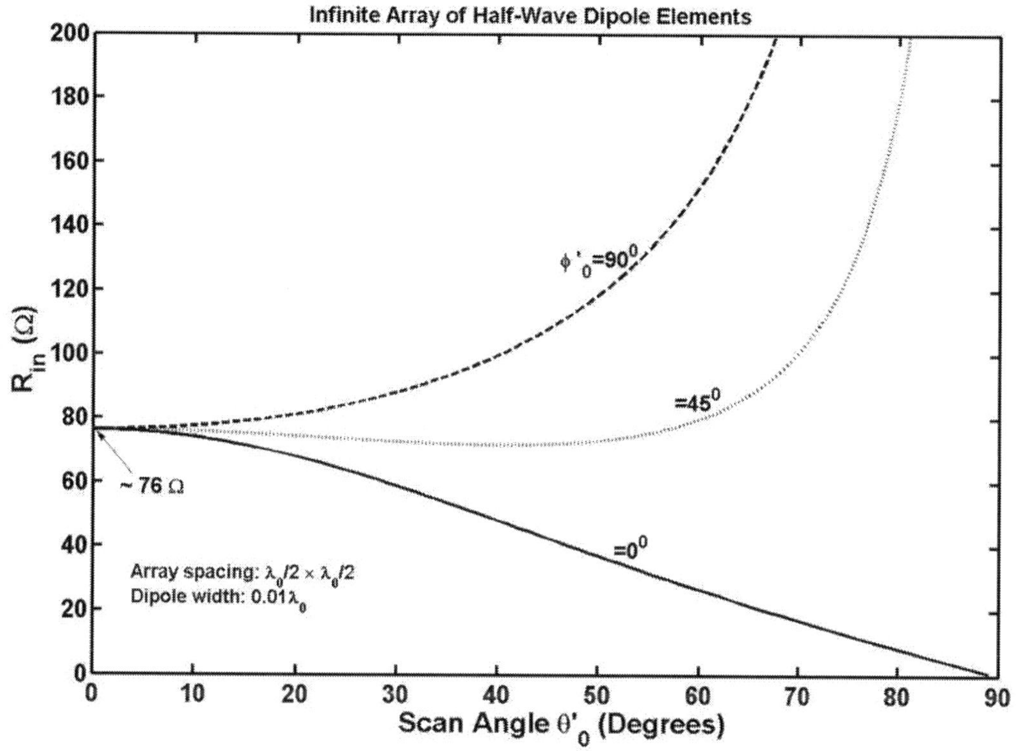

FIGURE 11

7.1.6. *Small Dipole Elements.* If the length l of the dipole elements is sufficiently small ($l << \lambda_0$), the input resistance expression in (10.62) may be simplified.

$$J_z = l\delta(x)\delta(z), \ \widetilde{J}_z = l,$$

$$R_{in} = \frac{l^2}{ab} \times \frac{\eta_0}{2} \times \frac{\sin\theta_0}{\sin\phi_0} = \frac{l^2}{ab} \times \frac{\eta_0}{2} \times \frac{(1-\sin^2\theta_0'\cos^2\phi_0')}{\cos\theta'_0}. \tag{10.64}$$

For radiation normal to the array plane ($\theta_0 = \phi_0 = \frac{\pi}{2}$, $\theta_0' = 0$,) and for half-wavelength array spacing ($a = b = \frac{\lambda_0}{2}$),

$$R_{in} = \frac{l^2}{\lambda_0^2}2\eta_0 = 240\pi(\frac{l}{\lambda_0})^2; \ \theta_0 = \phi_0 = \frac{\pi}{2}, \ \theta_0' = 0, \ a = b = \frac{\lambda_0}{2}. \tag{10.65}$$

Compare this with $R_{in} = 80\pi^2\frac{l^2}{\lambda_0^2}$ we derived in (9.8) for an isolated small dipole element. The array input resistance for radiation along the principal axes may be of particular

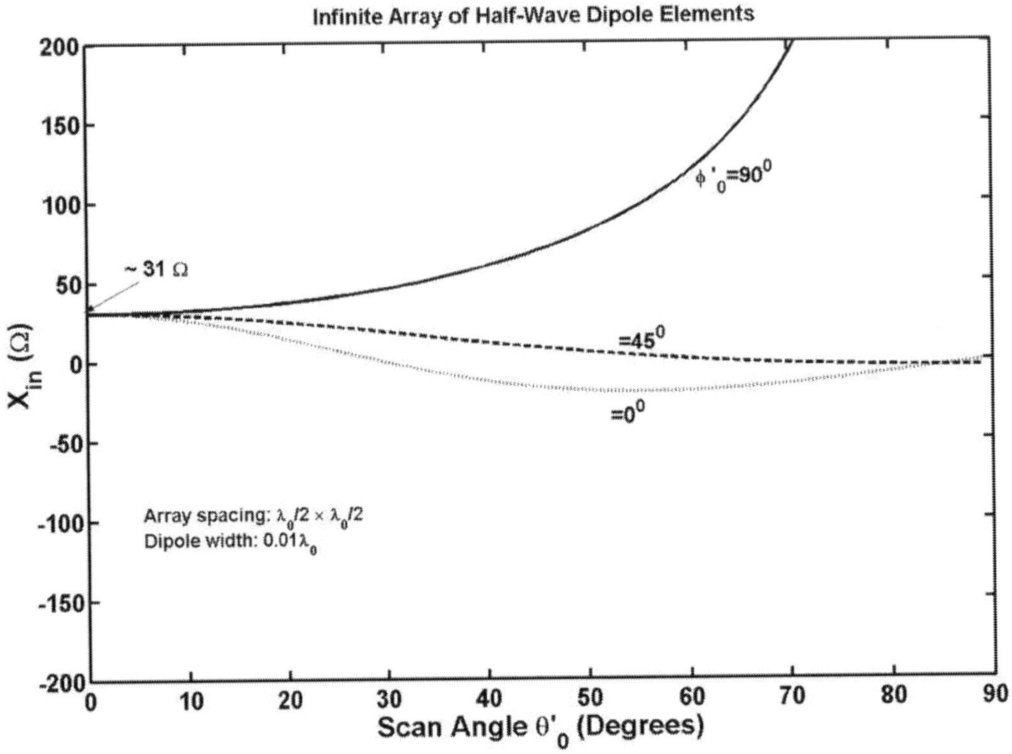

FIGURE 12

interest: (a) along the z-axis where $\theta_0 = 0$ or $(\theta_0' = \pi/2, \phi_0' = 0)$, and (b) the x-axis where $(\theta_0 = \pi/2, \phi_0 = 0)$ or $(\theta_0' = \pi/2, \phi_0' = \pi/2)$.

$$R_{in}(\theta_0 = 0;\ \theta_0' = \pi/2, \phi_0' = 0) = 0,$$
$$R_{in}(\theta_0 = \pi/2,\ \phi_0 = 0;\ \theta'_0 = \pi/2,\ \phi'_0 = \pi/2) = \infty. \tag{10.66}$$

It is also of interest to know the dependence of array resistance for radiation along specific planes. The plane with $\phi_0' = 0$, which is parallel to the dipole elements, or equivalently parallel to the electric field of radiation, is referred to as the E-plane. On the other hand, the plane with $\phi_0' = \pi/2$, which is orthogonal to the dipole elements, or equivalently parallel to the magnetic field of radiation, is referred to as the H-plane. The plane with $\phi_0' = \pi/4$, which is mid way between the E and H- planes, is referred to as the diagonal- or the D-plane.

$$E - plane,\ \phi_0' = 0:\ R_{in} = \frac{l^2}{ab} \times \frac{\eta_0}{2} \times (\cos\theta_0'),$$
$$H - plane,\ \phi_0' = \pi/2:\ R_{in} = \frac{l^2}{ab} \times \frac{\eta_0}{2} \times \frac{1}{(\cos\theta_0')},$$
$$H - plane,\ \phi_0' = \pi/4:\ R_{in} = \frac{l^2}{ab} \times \frac{\eta_0}{4} \times (\cos\theta_0' + \frac{1}{\cos\theta_0'}). \quad (10.67)$$

The above trends of array input resistance for small dipole elements are compared in Fig.13 for half-wave dipoles. The resistances in Fig.13 are normalized with respect to the value for radiation normal to the array ($\theta_0' = 0$). As may be observed from Fig.13, the trends for such normalized array resistances are generally alike for small as well as half-wave dipole arrays. The trends for the small and half-wave dipoles are identical to each other for the H-plane.

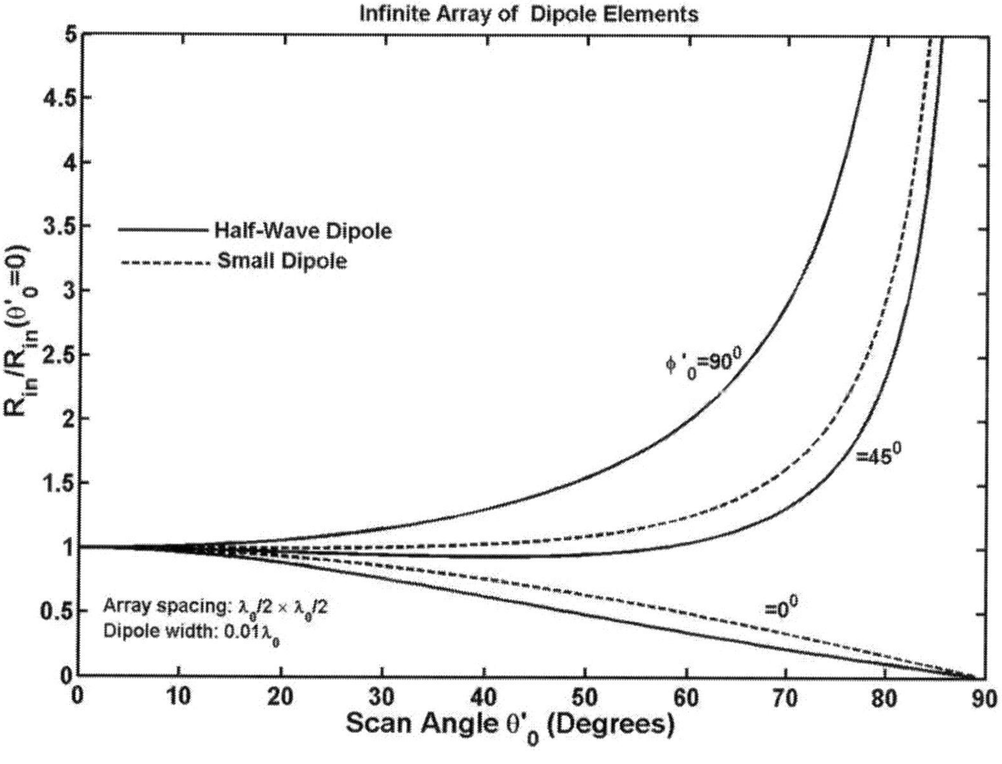

FIGURE 13

APPENDIX A

Differential Operators in 3-Dimensional Space

1. Ingtegration Over Curved Paths or Surfaces

1.1. Line or Contour Integral. An integral of a given vector function $\overline{A}$, over a given contour or curved line C_{12}, is defined in the following limiting form.

$$\int_{C_{12}} \overline{A} \cdot \overline{dl} = \lim_{\substack{\Delta l_i \to 0, \\ N \to \infty}} \sum_{i=1}^{N} \overline{A}_i \cdot \overline{\Delta l_i} \ . \tag{A.1}$$

In a general situation (shown in Fig.1), when the contour has two distinct starting and end points, the line integral is referred to as a open-line integral. In a more specific situation, where the end point (2) joins back at the starting point (1), the open contour C_{12} would turn into a closed contour or loop C, and the associated line integral would be referred to as a closed-line integral. In this case, the final integral is independent of the actual starting point of integration. Accordingly, it is not necessary to define a starting point in a closed-line integral. A closed-line integral maybe distinguished from an open-line integral, by using a special integral symbol shown with a circular loop. The circular loop represents the closed-loop nature of the associated integration path C.

$$\int_{\substack{Loop \\ C}} \overline{A} \cdot \overline{dl} \equiv \oint_{C} \overline{A} \cdot \overline{dl} \ . \tag{A.2}$$

A particular line integral is described by three specific parameters, as listed in the following. (a) A continuous path or contour, that is C_{12} for a open contour or C for a closed contour. (b) A specific direction along the contour, representing the orientation of the path elements $\overline{\Delta l_i}$. For integral along a open contour C_{12}, the contour direction is implicitly defined through specifying the starting and end points (1) and (2), respectively. However, for an integral along a closed contour C, where the start and end points are the same, it is necessary to specify the orientation along the contour. The orientation of the path elements $\overline{\Delta l_i}$ would otherwise be ambiguous, affecting the sign of the final integral. (c) A given vector field $\overline{A}$, defined at every point along the contour.

1.2. Surface Integral. An integral of a given vector function $\overline{A}$, over a given surface S, is defined in the following limiting form.

$$\int\!\!\int_{S} \overline{A} \cdot \overline{ds} = \lim_{\substack{\Delta S_i \to 0, \\ N \to \infty}} \sum_{i=1}^{N} \overline{A}_i \cdot \overline{\Delta S}_i. \tag{A.3}$$

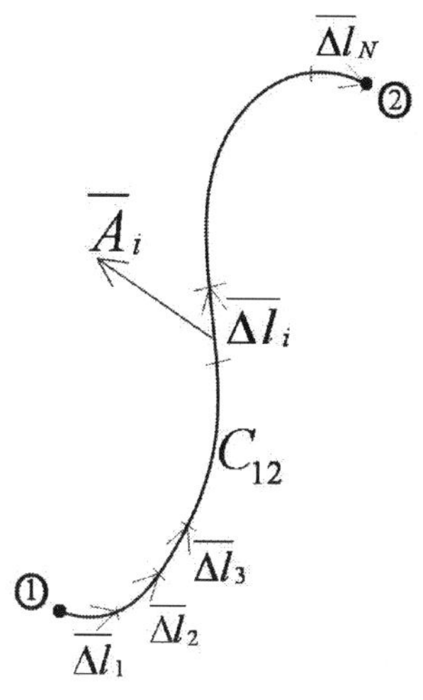

FIGURE 1

When the surface S completely encloses a given three-dimensional space of volume V, the associated surface integral is referred to as a closed-surface integral. When that is not the case (the general situation shown in Fig.2), the surface integral may be referred to as a open-surface integral. A closed-surface integral maybe distinguished from an open-surface integral, by using a special double-integral symbol shown with a circular loop. The circular loop is supposed to represent a closed spherical surface, which in turn represents the closed nature of the associated integration surface S. The double-integral notation, which is used for both open- and closed-surface integrals, indicates integration in two spatial dimensions that needs to be performed over the associated surface S.

$$\int\int_{\substack{Closed\\S}} \overline{A}\cdot\overline{ds} \equiv \oiint_S \overline{A}\cdot\overline{ds}. \tag{A.4}$$

A particular surface integral is described by three specific parameters, as listed in the following. (a) A given surface S is specified, over which the integration is performed. The surface may be a planar configuration, described on a two dimensional space. The surface, in this case, would not enclose any non-zero volume, and therefore can only be a open type of surface. In general, the surface is defined in a three dimensional space, which may either be a open or a closed type of surface. (b) Orientation of the normal directions $\hat{n}_i$ need to be specified for each surface element ΔS_i. The normal orientation has two possibilities, one of which will have to be definitively selected in reference to the entire surface S. The two possibilities may be informally referred to as either pointing "out of" or "into" the surface S. One of these possibilities is selected, and the specific choice is maintained uniformly over the entire surface S. That is, if the outward normal direction is selected, the orientation for $\hat{n}_i$ is directed "out of" the surface at every point on the surface S. Or else, if the inward normal direction is selected, the orientation for $\hat{n}_i$ is directed "into" the surface at every point on the surface S. (c) A given vector field $\overline{A}$ is defined at every point over the surface S. The component of the vector $\overline{A}$, along a normal (as defined above) to the surface S, contributes to the total integral.

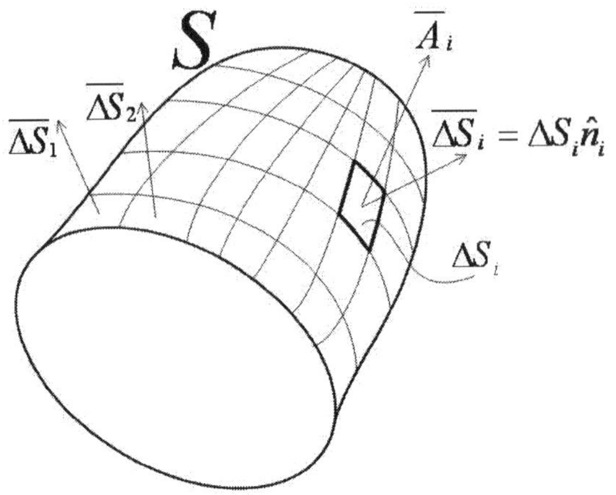

FIGURE 2

1.3. Volume Integral. An integral of a given scalar function ψ, over a given volume τ, is defined in the following limiting form.

$$\int\int\int_{\tau} \psi d\tau = \lim_{\substack{\Delta\tau_i \to 0, \\ N\to\infty}} \sum_{i=1}^{N} \psi_i \Delta\tau_i \ . \tag{A.5}$$

A particular volume integral is described by two parameters: (a) a given region in the three-dimensional space with volume τ, and (b) a given scalar function ψ defined at every point inside the volume region. Unlike the line and surface integrals defined above, that are performed for a vector function, the volume integral defined here is performed for a scalar function. The volume integral maybe expressed using a triple-integral operation, which indicates integration in three spatial dimensions that needs to be performed over the associated volume τ.

2. Gradient, Curl and Divergence in General Curved Coordinates

Gradient, Curl and Divergence are special types of derivatives in a three-dimensional space, operating on a scalar or vector function. These derivatives may be referred to as derivative functions or operators. In the following, we will define each of the derivative operators, and express them in a general curved coordinate system. Expressions in the rectangular, spherical and cylindrical coordinate systems, which are special cases of the general curved coordinate system, can then be obtained by simply substituting appropriate paramaters for the specific case.

In a general curved coordinate system, as shown in Fig.3, a point in a three-dimensional space is defined with three variables (u, v, w). At every point, three curved lines can be described by changing one of the variables, while keeping the other two variables constant. These curves may be referred to as the three "curved axes u, v and w," at the particular point. The axis name refers to the only variable that changes along the particular curved axis. The curved coordinates are assumed to be orthogonal to each other, in the sense that the tangents to the three curved axes are orthogonal to each other at every point in space. The three unit vectors, $\hat{u}$, $\hat{v}$ and $\hat{w}$, defined along tangents to the respective curved axes, constitute the three orthogonal vector directions. At any given point, as one given coordinate variables is incremented by one unit, while keeping the other two unchanged, the corresponding length displacement Δl is referred as the "length factor" for the particular coordinate, at the particular point. The length factors for the three coordinates u, v, and w are represented by the variables α_u, α_v and α_w, respectively.

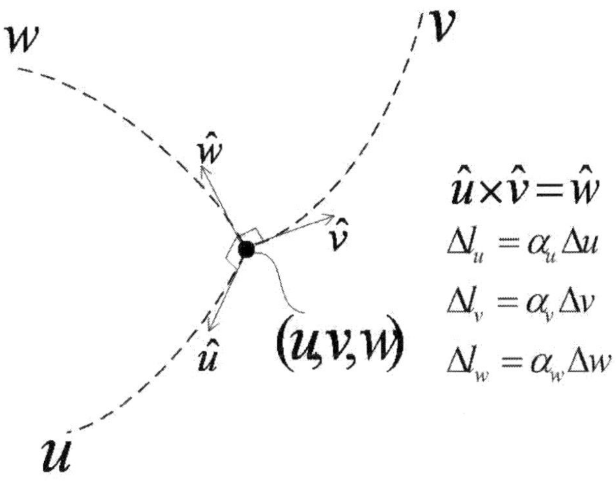

FIGURE 3

2.1. The Gradient Function $\overline{\nabla}\psi$. The gradient is a derivative function, which operates on a scalar function ψ as its variable, but results in a vector function as its outcome. Accordingly, the gradient is a vector function, represented by an operator notation $\overline{\nabla}\psi$, and the resulting vector is referred to as a gradient vector of the function ψ.

The gradient vector in a three-dimensional space maybe defined by defining its component along any particular direction. The component of the gradient vector along a given direction $\hat{l}$ is defined as the derivative of the associated scalar function ψ, per unit length $\triangle l$ along the given direction.

$$\overline{\nabla}\psi \cdot \hat{l} = \lim_{\triangle l \to 0} \frac{\triangle \psi}{\triangle l},$$
$$\lim_{\triangle l \to 0} \overline{\nabla}\psi \cdot \overline{\triangle l} = \triangle \psi; \; \overline{\triangle l} = \hat{l}(\triangle l) \; . \tag{A.6}$$

Like any vector, the gradient vector in a three-dimentional space is completely described by its individual components along three orthogonal directions $\hat{u}$, $\hat{v}$ and $\hat{w}$. For example, the component of $\overline{\nabla}\psi$ along the direction $\hat{u}$ is expressed using the basic definition of (A.6), as follows.

$$\overline{\nabla}\psi \cdot \hat{u} = \lim_{\Delta u \to 0} \frac{\psi_2 - \psi_1}{\Delta l} = \lim_{\Delta u \to 0} \frac{\psi(u + \frac{\Delta u}{2}, v, w) - \psi(u - \frac{\Delta u}{2}, v, w)}{\alpha_u \Delta u} = \frac{1}{\alpha_u}\frac{\partial \psi}{\partial u}. \quad (A.7)$$

Components along other directions can be similarly expressed.

$$\overline{\nabla}\psi \cdot \hat{v} = \frac{1}{\alpha_v}\frac{\partial \psi}{\partial v}, \quad \overline{\nabla}\psi \cdot \hat{w} = \frac{1}{\alpha_w}\frac{\partial \psi}{\partial w}. \quad (A.8)$$

The three orthogonal components, as expressed above, may now be combined to get the complete expression for the gradient $\overline{\nabla}\psi$.

$$\begin{aligned} \overline{\nabla}\psi &= \hat{u}(\overline{\nabla} \cdot \hat{u}) + \hat{v}(\overline{\nabla}\psi \cdot \hat{v}) + \hat{w}(\overline{\nabla}\psi \cdot \hat{w}) \\ &= \hat{u}\frac{1}{\alpha_u}\frac{\partial \psi}{\partial u} + \hat{v}\frac{1}{\alpha_v}\frac{\partial \psi}{\partial v} + \hat{w}\frac{1}{\alpha_w}\frac{\partial \psi}{\partial w}. \end{aligned} \quad (A.9)$$

2.2. The Curl Function $\overline{\nabla} \times \overline{A}$. The curl is a derivative function, which operates on a vector function $\overline{A}$ as its variable, and results in a vector function as its outcome. The resulting vector is called the curl vector of $\overline{A}$, and in reference to the nature of the outcome the curl is called a vector function. It maybe represented by the operator notation $\overline{\nabla} \times \overline{A}$, in recognition of the final expression for the curl, which appears similar (although not always equal) to a cross-product of the gradient operator $\overline{\nabla}$ with the vector function $\overline{A}$.

Because the curl is a vector function, it may be defined by defining its component along any particular direction. The component of the curl vector along a given direction $\hat{n}$ is defined as the path integration of the associated vector function $\overline{A}$, evaluated over an elemental closed contour ΔC located on a plane perpendicular to the given direction, then divided by the elemental area ΔS enclosed by the contour ΔC.

$$\begin{aligned} (\overline{\nabla} \times \overline{A}) \cdot \hat{n} &= \lim_{\Delta S \to 0} \frac{\int_{\substack{Loop \\ \Delta C}} \overline{A} \cdot \overline{dl}}{\Delta S}, \\ \lim_{\Delta S \to 0} (\overline{\nabla} \times \overline{A}) \cdot \overline{\Delta S} &= \int_{\substack{Loop \\ \Delta C}} \overline{A} \cdot \overline{dl}; \; \overline{\Delta S} = \hat{n}\Delta S. \end{aligned} \quad (A.10)$$

As required, the orientation of the path elements along the contour ΔC are to be properly defined. This is done in reference to the given direction $\hat{n}$, which is normal to plane of the contour ΔC, using a standard right-hand rule. As per the rule, let the thumb of the right hand be aligned along the normal direction. Then, the orientation of the other fingers of the right hand, curved in a "grip" form, would refer to the orientation of the path elements along ΔC.

Like any vector, the curl vector may be completely described by its individual components along three orthogonal directions $\hat{u}$, $\hat{v}$ and $\hat{w}$. For example, the component of $\overline{\nabla} \times \overline{A}$ along the direction $\hat{w}$ may be derived using the basic definition of (A.10), as follows (see Fig.4).

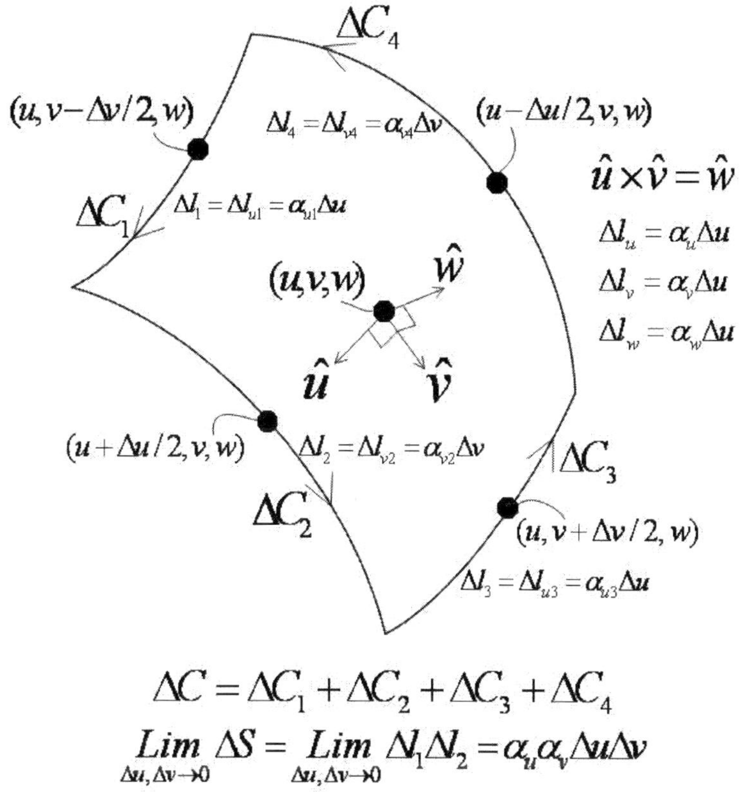

FIGURE 4

$$(\overline{\nabla}\times\overline{A})\cdot\hat{w} = \lim_{\substack{\triangle S\to 0\\ \Delta u,\Delta v\to 0}} \frac{\int_{\substack{Loop\\ \Delta C}} \overline{A}\cdot\overline{dl}}{\Delta S}; \ \triangle C \equiv \triangle C_1 + \triangle C_2 + \triangle C_3 + \triangle C_4,$$

$$\int_{\Delta C_1} \overline{A}\cdot\overline{dl} = +A_{u1}\Delta l_1 = +A_{u1}\alpha_{u1}\Delta u = +A_{lu1}\Delta u,$$

$$\int_{\Delta C_3} \overline{A}\cdot\overline{dl} = -A_{u3}\Delta l_3 = -A_{u3}\alpha_{u3}\Delta u = -A_{lu3}\Delta u,$$

$$\frac{\int_{\Delta C_1+\Delta C_3} \overline{A}\cdot\overline{dl}}{\Delta S} = -\frac{(A_{lu3}-A_{lu1})\Delta u}{\alpha_u\alpha_v\Delta u\Delta v} = -\frac{[A_{lu}(u, v+\frac{\Delta v}{2}, w) - A_{lu}(u, v-\frac{\Delta v}{2}, w)]}{\alpha_u\alpha_v\Delta v},$$

$$\lim_{\substack{\triangle S\to 0\\ \Delta u,\Delta v\to 0}} \frac{\int_{\Delta C_1+\Delta C_3} \overline{A}\cdot\overline{dl}}{\Delta S} = -\frac{1}{\alpha_u\alpha_v}\frac{\partial A_{lu}}{\partial v} = -\frac{\alpha_w}{\alpha_u\alpha_v\alpha_w}\frac{\partial A_u\alpha_u}{\partial v},$$

$$\int_{\Delta C_2} \overline{A} \cdot \overline{dl} = A_{v2}\Delta l_2 = A_{v2}\alpha_{v2}\Delta v = A_{lv2}\Delta v,$$

$$\int_{\Delta C_4} \overline{A} \cdot \overline{dl} = -A_{v4}\Delta l_4 = -A_{v4}\alpha_{v4}\Delta v = -A_{lv4}\Delta v,$$

$$\frac{\int_{\Delta C_2+\Delta C_4} \overline{A} \cdot \overline{dl}}{\Delta S} = \frac{(A_{lv2} - A_{lv4})\Delta v}{\alpha_u \alpha_v \Delta u \Delta v} = \frac{[A_{lv}(u+\frac{\Delta u}{2}, v, w) - A_{lv}(u - \frac{\Delta u}{2}, v, w)]}{\alpha_u \alpha_v \Delta u},$$

$$\lim_{\substack{\Delta S \to 0 \\ \Delta u, \Delta v \to 0}} \frac{\int_{\Delta C_2+\Delta C_4} \overline{A} \cdot \overline{dl}}{\Delta S} = -\frac{1}{\alpha_u \alpha_v}\frac{\partial A_{lv}}{\partial u} = -\frac{\alpha_w}{\alpha_u \alpha_v \alpha_w}\frac{\partial A_v \alpha_v}{\partial u},$$

$$(\overline{\nabla} \times \overline{A}) \cdot \hat{w} = \lim_{\substack{\Delta S \to 0 \\ \Delta u, \Delta v \to 0}} \frac{\int_{\Delta C_1+\Delta C_2+\Delta C_3 \Delta C_4} \overline{A} \cdot \overline{dl}}{\Delta S} = \frac{\alpha_w}{\alpha_u \alpha_v \alpha_w}(\frac{\partial A_v \alpha_v}{\partial u} - \frac{\partial A_u \alpha_u}{\partial v}). \quad \text{(A.11)}$$

Components of the curl along other directions can be similarly derived. However, the final expressions for the other components may be directly obtained by proper interchange of the indices u, v and w in the above expression. Interchanging $u \to v$, $v \to w$, and $w \to u$ in (A.11), we get the expression for the $\hat{u}$ component of $\overline{\nabla} \times \overline{A}$.

$$(\overline{\nabla} \times \overline{A}) \cdot \hat{u} = \frac{\alpha_u}{\alpha_u \alpha_v \alpha_w}[\frac{\partial A_w \alpha_w}{\partial v} - \frac{\partial A_v \alpha_v}{\partial w}] \, . \quad \text{(A.12)}$$

Similarly, interchanging again $u \to v$, $v \to w$, and $w \to u$ in (A.12), we get the expression for the $\hat{v}$ component of $\overline{\nabla} \times \overline{A}$.

$$(\overline{\nabla} \times \overline{A}) \cdot \hat{v} = \frac{\alpha_v}{\alpha_u \alpha_v \alpha_w}[\frac{\partial A_u \alpha_u}{\partial w} - \frac{\partial A_w \alpha_w}{\partial u}] \, . \quad \text{(A.13)}$$

The three orthogonal components of the curl, as expressed above, may now be combined to get the complete expression for $\overline{\nabla} \times \overline{A}$.

$$\overline{\nabla} \times \overline{A} = \hat{u}(\overline{\nabla} \times \overline{A} \cdot \hat{u}) + \hat{v}(\overline{\nabla} \times \overline{A} \cdot \hat{v}) + \hat{w}(\overline{\nabla} \times \overline{A} \cdot \hat{w})$$
$$= \frac{1}{\alpha_u \alpha_v \alpha_w}[\alpha_u \hat{u}(\frac{\partial}{\partial v}A_w \alpha_w - \frac{\partial}{\partial w}A_v \alpha_v) + \alpha_v \hat{v}(\frac{\partial}{\partial w}A_u \alpha_u - \frac{\partial}{\partial u}A_w \alpha_w)$$
$$+\alpha_w \hat{w}(\frac{\partial}{\partial u}A_v \alpha_v - \frac{\partial}{\partial v}A_u \alpha_u)]. \quad \text{(A.14)}$$

The above expression may be written in a compact manner using a matrix notation.

$$\overline{\nabla} \times \overline{A} = \frac{1}{\alpha_u \alpha_v \alpha_w} \begin{bmatrix} \alpha_u \hat{u} & \alpha_v \hat{v} & \alpha_w \hat{w} \\ \frac{\partial}{\partial u} & \frac{\partial}{\partial v} & \frac{\partial}{\partial w} \\ \alpha_u A_u & \alpha_v A_v & \alpha_w A_w \end{bmatrix} . \quad \text{(A.15)}$$

The above matrix expression can be shown to be equivalent to its required form in (A.14), by expanding the determinant form as if it is a determinant of a conventional matrix consisting of numbers or variables. Except, in this case the unit vectors and derivative operators would substitute for respective numbers or variables in a conventional matrix. In this process, multiplication of a derivative operator with variables or functions is to be interpreted as evaluating the particular derivative of the given variables or functions.

2.3. The Divergence Function $\overline{\nabla} \cdot \overline{A}$. The divergence is a derivative function, which operates on a vector function $\overline{A}$ as its variable, but results in a scalar function as its outcome. In reference to the outcome, the divergence is called a scalar function, represented by the operator notation $\overline{\nabla} \cdot \overline{A}$. The symbolic notation is in recognition of the final expression for the divergence operator, which appears similar (not always equal) to a dot-product of the gradient operator $\overline{\nabla}$ with the vector function $\overline{A}$.

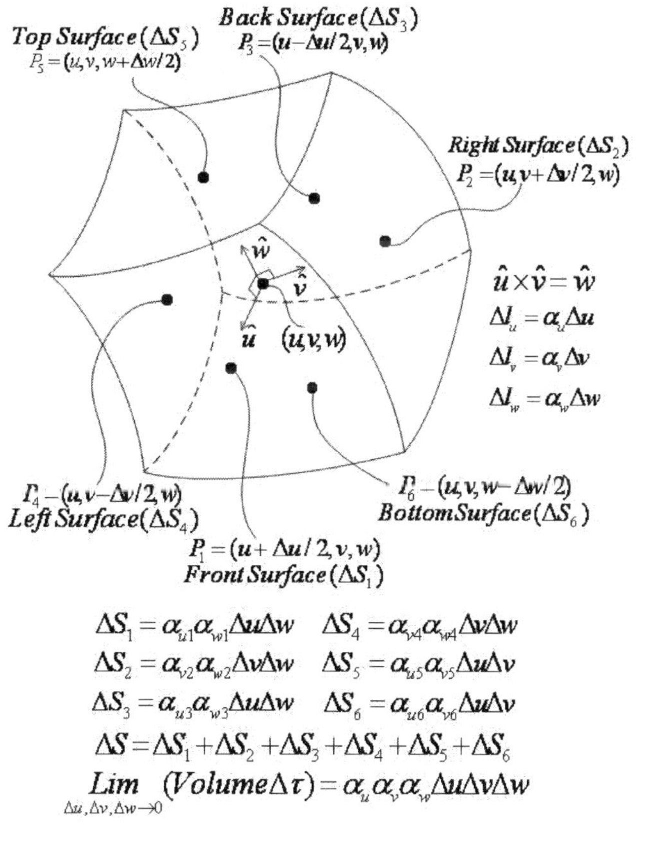

FIGURE 5

The divergence of a given vector function, at a given point in space, is defined as the surface integration of the particular vector function $\overline{A}$, evaluated over an elemental closed surface ΔS around the reference point, then divided by the elemental volume $\Delta\tau$ enclosed by the closed surface ΔS.

$$\overline{\nabla} \cdot \overline{A} = \lim_{\Delta\tau \to 0} \frac{\frac{\int\int_{Closed}\overline{A}\cdot\overline{ds}}{\Delta S}}{\Delta\tau},$$
$$\lim_{\Delta\tau \to 0} (\overline{\nabla} \cdot \overline{A})\Delta\tau = \int\int_{\substack{Closed\\ \Delta S}} \overline{A} \cdot \overline{ds} \ . \tag{A.16}$$

Using the above basic definition, in the following we will derive the expression of the divergence function in a general curved, orthogonal coordinate system (see Fig.5).

$$\overline{\nabla} \cdot \overline{A} = \lim_{\substack{\Delta\tau \to 0\\ \Delta u, \Delta v, \Delta w \to 0}} \frac{\frac{\int\int_{Closed}\overline{A}\cdot\overline{ds}}{\Delta S}}{\Delta\tau},$$
$$\Delta\tau = \Delta l_u \Delta l_v \Delta l_w = \alpha_u \Delta u \alpha_v \Delta v \alpha_w \Delta w,$$
$$\Delta S \equiv \Delta S_1 + \Delta S_2 + \Delta S_3 + \Delta S_4 + \Delta S_5 + \Delta S_6. \tag{A.17}$$

The surface ΔS encloses a closed box of volume $\Delta\tau$, and can be decomposed into six open surfaces on the six sides of the closed box. The total surface integral over the entire closed surface ΔS may be evaluated in three steps. In each step, the surface integral is evaluated over a pair of surfaces that are located on the opposite sides of the closed box. First, the integration over the opposing surfaces ΔS_1 and ΔS_2 are evaluated as follows.

$$\int\int_{\Delta S_1} \overline{A} \cdot \overline{ds} = A_{u1}\Delta S_1 = A_{u1}\alpha_{v1}\alpha_{w1}\Delta v \Delta w = A_{su1}\Delta v \Delta w,$$
$$\int\int_{\Delta S_3} \overline{A} \cdot \overline{ds} = -A_{u3}\Delta S_3 = -A_{u3}\alpha_{v3}\alpha_{w3}\Delta v \Delta w = -A_{s3}\Delta v \Delta w,$$
$$\frac{\int\int_{\Delta S_1 + \Delta S_3} \overline{A} \cdot \overline{ds}}{\Delta\tau} = \frac{(A_{su1} - A_{su3})\Delta v \Delta w}{\alpha_u \alpha_v \alpha_w \Delta u \Delta v \Delta w} = \frac{1}{\alpha_u \alpha_v \alpha_w} \frac{(A_{su1} - A_{su3})}{\Delta u},$$
$$\lim_{\substack{\Delta\tau \to 0\\ \Delta u, \Delta v, \Delta w \to 0}} \frac{\int\int_{\Delta S_1 + \Delta S_3} \overline{A} \cdot \overline{ds}}{\Delta\tau} = \frac{1}{\alpha_u \alpha_v \alpha_w} \lim_{\Delta u \to 0} \frac{A_{su1} - A_{su3}}{\Delta u}$$
$$= \frac{1}{\alpha_u \alpha_v \alpha_w} \lim_{\Delta u \to 0} \frac{A_{su}(u + \frac{\Delta u}{2}, v, w) - A_{su}(u - \frac{\Delta u}{2}, v, w)}{\Delta u}$$
$$= \frac{1}{\alpha_u \alpha_v \alpha_w} \frac{\partial A_{su}}{\partial u} = \frac{1}{\alpha_u \alpha_v \alpha_w} \frac{\partial A_u \alpha_v \alpha_w}{\partial u}. \tag{A.18}$$

Similarly, integration over the opposing faces ΔS_2 and ΔS_4 may be derived. However, the resulting expression may also be directly obtained by interchanging $u \to v$, $v \to w$, and $w \to u$ in the final expresison of (A.18).

$$\lim_{\substack{\Delta\tau \to 0\\ \Delta u, \Delta v, \Delta w \to 0}} \frac{\int\int_{\Delta S_2 + \Delta S_4} \overline{A} \cdot \overline{ds}}{\Delta\tau} = \frac{1}{\alpha_u \alpha_v \alpha_w} \frac{\partial A_{sv}}{\partial v} = \frac{1}{\alpha_u \alpha_v \alpha_w} \frac{\partial A_v \alpha_u \alpha_w}{\partial v}. \tag{A.19}$$

Integration over the remaining faces ΔS_5 and ΔS_6 may also be similarly derived, or directly obtained by interchanging $u \to v$, $v \to w$, and $w \to u$ in the expression of (A.19).

$$\lim_{\substack{\Delta\tau\to 0\\ \Delta u,\Delta v,\Delta w\to 0}} \frac{\int\int_{\Delta S_5+\Delta S_6} \overline{A}\cdot\overline{ds}}{\Delta\tau} = \frac{1}{\alpha_u\alpha_v\alpha_w}\frac{\partial A_{sw}}{\partial w} = \frac{1}{\alpha_u\alpha_v\alpha_w}\frac{\partial A_w\alpha_u\alpha_v}{\partial w}. \tag{A.20}$$

Now, the individual parts in (A.18,A.19,A.20) can be combined in (A.17) to obtain the expression for the divergence in the curved coordinate frame (u, v, w).

$$\begin{aligned}\overline{\nabla}\cdot\overline{A} &= \lim_{\substack{\Delta\tau\to 0\\ \Delta u,\Delta v,\Delta w\to 0}} \frac{\int\int_{\Delta S_1+\Delta S_2+\Delta S_3+\Delta S_4+\Delta S_5+\Delta S_6} \overline{A}\cdot\overline{ds}}{\Delta\tau}\\ &= \frac{1}{\alpha_u\alpha_v\alpha_w}\left[\frac{\partial A_u\alpha_v\alpha_w}{\partial u} + \frac{\partial A_v\alpha_u\alpha_w}{\partial v} + \frac{\partial A_w\alpha_u\alpha_v}{\partial w}\right].\end{aligned} \tag{A.21}$$

3. Differential Operators in Specific Coordinate Systems

The general results, derived in the last section for a general, curved, orthogonal coordinate system, may be expressed specifically for rectangular, cylindrical and spherical coordinates. This is accomplished by substituting the coordinate variables (u, v, w) and the length coefficients $(\alpha_u, \alpha_v, \alpha_w)$ with their respective paramters for the specific coordinate system.

3.1. Rectangular Coordinates.

A rectangular coordinate system, represented by coordinate variables (x, y, z), and the respective unit vectors $(\hat{x}, \hat{y}, \hat{z})$, is shown in Fig.6.

$$\begin{aligned}(\hat{u}, \hat{v}, \hat{w}) = (\hat{x}, \hat{y}, \hat{z}),\ (u, v, w) &= (x, y, z),\\ \Delta l_x &= \alpha_x\Delta x = \Delta x,\ \alpha_x = 1 = \alpha_u,\\ \Delta l_y &= \alpha_y\Delta y = \Delta y,\ \alpha_y = 1 = \alpha_v,\\ \Delta l_z &= \alpha_z\Delta z = \Delta z,\ \alpha_z = 1 = \alpha_w,\\ \overline{\nabla}\psi &= \hat{x}\frac{\partial\psi}{\partial x} + \hat{y}\frac{\partial\psi}{\partial y} + \hat{z}\frac{\partial\psi}{\partial z},\\ \overline{\nabla}\times\overline{A} &= \begin{bmatrix}\hat{x} & \hat{y} & \hat{z}\\ \frac{\partial}{\partial x} & \frac{\partial}{\partial y} & \frac{\partial}{\partial z}\\ A_x & A_y & A_z\end{bmatrix},\\ \overline{\nabla}\cdot\overline{A} &= [\frac{\partial A_x}{\partial x} + \frac{\partial A_y}{\partial y} + \frac{\partial A_Z}{\partial z}].\end{aligned} \tag{A.22}$$

3.2. Cylindrical Coordinates.

A cylindrical coordinate system, represented by coordinate variables (ρ, ϕ, z), and the respective unit vectors $(\hat{\rho}, \hat{\phi}, \hat{z})$, is shown in Fig.7.

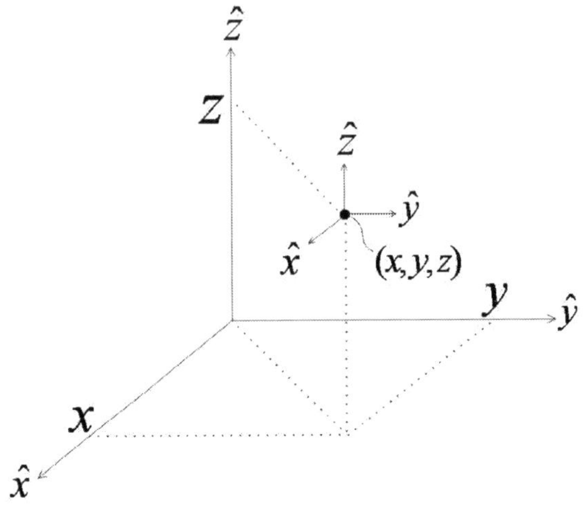

Figure 6

$$
\begin{aligned}
(\hat{u}, \hat{v}, \hat{w}) = (\hat{\rho}, \hat{\phi}, \hat{z}),\ (u, v, w) = (\rho, \phi, z),\\
\Delta l_\rho = \alpha_\rho \Delta\rho = \Delta\rho,\ \alpha_\rho = 1 = \alpha_u,\\
\Delta l_\phi = \alpha_\phi \Delta\phi = \rho\Delta\phi,\ \alpha_\phi = \rho = \alpha_v,\\
\Delta l_z = \alpha_z \Delta z = \Delta z,\ \alpha_z = 1 = \alpha_w,\\
\overline{\nabla}\psi = \hat{\rho}\frac{\partial\psi}{\partial\rho} + \hat{\phi}\frac{1}{\rho}\frac{\partial\psi}{\partial\phi} + \hat{z}\frac{\partial\psi}{\partial z},\\
\overline{\nabla}\times\overline{A} = \frac{1}{\rho}\begin{bmatrix} \hat{\rho} & \rho\hat{\phi} & \hat{z} \\ \frac{\partial}{\partial\rho} & \frac{\partial}{\partial\phi} & \frac{\partial}{\partial z} \\ A_\rho & \rho A_\phi & A_z \end{bmatrix},\\
\overline{\nabla}\cdot\overline{A} = \frac{1}{\rho}[\frac{\partial A_\rho \rho}{\partial\rho} + \frac{\partial A_\phi}{\partial\phi} + \frac{\partial A_z \rho}{\partial z}]\\
= \frac{1}{\rho}\frac{\partial A_\rho \rho}{\partial\rho} + \frac{1}{\rho}\frac{\partial A_\phi}{\partial\phi} + \frac{\partial A_z}{\partial z}.
\end{aligned}
\tag{A.23}
$$

3.3. Spherical Coordinates.

A spherical coordinate system, represented by coordinate variables (r, θ, ϕ), and the respective unit vectors $(\hat{r}, \hat{\theta}, \hat{\phi})$, is shown in Fig.8.

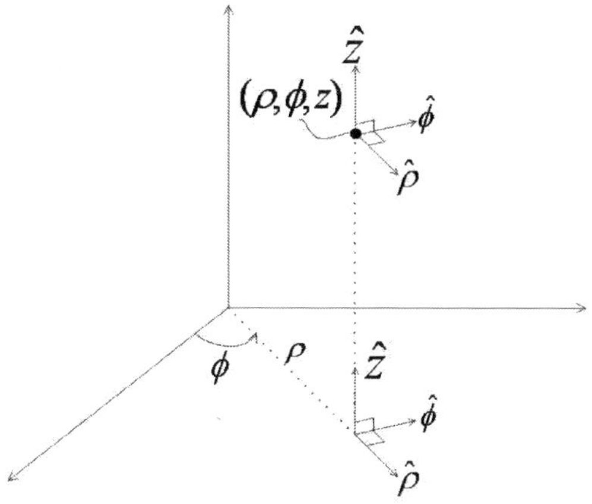

FIGURE 7

$$
\begin{aligned}
(\hat{u},\hat{v},\hat{w}) = (\hat{r},\hat{\theta},\hat{\phi}),\ (u,v,w) = (r,\theta,\phi),\\
\Delta l_r = \alpha_r \Delta r = \Delta r,\ \alpha_r = 1 = \alpha_u,\\
\Delta l_\theta = \alpha_\theta \Delta\theta = r d\theta,\ \alpha_\theta = r = \alpha_v,\\
\Delta l_\phi = \alpha_\phi \Delta\phi = r\sin\theta d\phi,\ \alpha_\phi = r\sin\theta = \alpha_w,\\
\overline{\nabla}\psi = \hat{r}\frac{\partial\psi}{\partial r} + \hat{\theta}\frac{1}{r}\frac{\partial\psi}{\partial\theta} + \hat{\phi}\frac{1}{r\sin\theta}\frac{\partial\rho}{\partial\phi},\\
\overline{\nabla}\times\overline{A} = \frac{1}{r^2\sin\theta}\begin{bmatrix} \hat{r} & r\hat{\theta} & r\sin\theta\hat{\phi} \\ \frac{\partial}{\partial r} & \frac{\partial}{\partial\theta} & \frac{\partial}{\partial\phi} \\ A_r & rA_\theta & r\sin\theta A_\phi \end{bmatrix},\\
\overline{\nabla}\cdot\overline{A} = \frac{1}{r^2\sin\theta}[\frac{\partial A_r r^2\sin\theta}{\partial r} + \frac{\partial A_\theta r\sin\theta}{\partial\theta} + \frac{\partial A_\phi r}{\partial\phi}]\\
= \frac{1}{r^2}\frac{\partial A_r r^2}{\partial r} + \frac{1}{r\sin\theta}\frac{\partial A_\theta\sin\theta}{\partial\theta} + \frac{1}{r\sin\theta}\frac{\partial A_\phi}{\partial\phi}.
\end{aligned}
\tag{A.24}
$$

4. Useful Relations and Theorems

4.1. $\overline{\nabla}\cdot\overline{\nabla}\times\overline{A} = 0$.

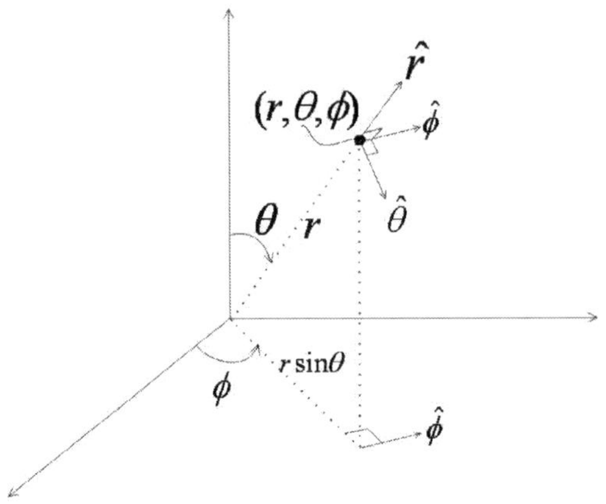

FIGURE 8

$$\begin{aligned}
\overline{\nabla}\times\overline{A} &= \frac{1}{\alpha_u\alpha_v\alpha_w}[\alpha_u\hat{u}(\frac{\partial A_w\alpha_w}{\partial v}-\frac{\partial A_v\alpha_v}{\partial w})+\alpha_v\hat{v}(\frac{\partial A_u\alpha_u}{\partial w}-\frac{\partial A_w\alpha_w}{\partial u}) \\
&\quad +\alpha_w\hat{w}(\frac{\partial A_v\alpha_v}{\partial u}-\frac{\partial A_u\alpha_u}{\partial v})]=\overline{A}', \\
\overline{\nabla}\cdot\overline{\nabla}\times\overline{A} &= \overline{\nabla}\cdot\overline{A}'=\frac{1}{\alpha_u\alpha_v\alpha_w}[\frac{\partial A'_u\alpha_v\alpha_w}{\partial u}+\frac{\partial A'_v\alpha_u\alpha_w}{\partial v}+\frac{\partial A'_w\alpha_u\alpha_v}{\partial w}] \\
&= \frac{1}{\alpha_u\alpha_v\alpha_w}[\frac{\partial}{\partial u}(\frac{\partial A_w\alpha_w}{\partial v}-\frac{\partial A_v\alpha_v}{\partial w})+\frac{\partial}{\partial v}(\frac{\partial A_u\alpha_u}{\partial w}-\frac{\partial A_w\alpha_w}{\partial u})+\frac{\partial}{\partial w}(\frac{\partial A_v\alpha_v}{\partial u}-\frac{\partial A_u\alpha_u}{\partial v})] \\
&= \frac{1}{\alpha_u\alpha_v\alpha_w}[\frac{\partial^2 A_w\alpha_w}{\partial u\partial v}-\frac{\partial^2 A_v\alpha_v}{\partial u\partial w}+\frac{\partial^2 A_u\alpha_u}{\partial v\partial w}-\frac{\partial^2 A_w\alpha_w}{\partial u\partial v}+\frac{\partial^2 A_v\alpha_v}{\partial u\partial w}-\frac{\partial^2 A_u\alpha_u}{\partial v\partial w}] \\
&= 0. \qquad \text{(A.25)}
\end{aligned}$$

4.2. $\overline{\nabla}\times(\overline{\nabla}\psi)=0.$

$$\begin{aligned}
\overline{A} &= \overline{\nabla}\psi=\hat{u}\frac{1}{\alpha_u}\frac{\partial\psi}{\partial u}+\hat{v}\frac{1}{\alpha_v}\frac{\partial\psi}{\partial v}+\hat{w}\frac{1}{\alpha_w}\frac{\partial\psi}{\partial w}, \\
\overline{\nabla}\times\overline{A} &= \frac{1}{\alpha_u\alpha_v\alpha_w}[\alpha_u\hat{u}(\frac{\partial A_w\alpha_w}{\partial v}-\frac{\partial A_v\alpha_v}{\partial w})+\alpha_v\hat{v}(\frac{\partial A_u\alpha_u}{\partial w}-\frac{\partial A_w\alpha_w}{\partial u}) \\
&\quad +\alpha_w\hat{w}(\frac{\partial A_v\alpha_v}{\partial u}-\frac{\partial A_u\alpha_u}{\partial v})] \\
= \overline{\nabla}\times(\overline{\nabla}\psi) &= \frac{1}{\alpha_u\alpha_v\alpha_w}[\alpha_u\hat{u}(\frac{\partial}{\partial v}(\frac{\partial\psi}{\partial w})-\frac{\partial}{\partial w}(\frac{\partial\psi}{\partial v}))+\alpha_v\hat{v}(\frac{\partial}{\partial w}(\frac{\partial\psi}{\partial u})-\frac{\partial}{\partial u}(\frac{\partial\psi}{\partial w})) \\
&\quad +\alpha_w\hat{w}(\frac{\partial}{\partial u}(\frac{\partial\psi}{\partial v})-\frac{\partial}{\partial v}(\frac{\partial\psi}{\partial u}))]=0. \qquad \text{(A.26)}
\end{aligned}$$

4.3. $\overline{\nabla}\cdot(\overline{A}\times\overline{B})=\overline{B}\cdot\overline{\nabla}\times\overline{A}-\overline{A}\cdot\overline{\nabla}\times\overline{B}.$

$$\overline{A} = \hat{u}A_u + \hat{v}A_v + \hat{w}A_w,\ \overline{B} = \hat{u}B_u + \hat{v}B_v + \hat{w}B_w,$$
$$\overline{A}\times\overline{B} = \hat{u}(A_vB_w - B_vA_w) + \hat{v}(B_uA_w - A_uB_w) + \hat{w}(A_uB_v - B_uA_v),$$
$$\nabla\cdot(\overline{A}\times\overline{B}) = \tfrac{1}{\alpha_u\alpha_v\alpha_w}[\tfrac{\partial}{\partial u}(A_v\alpha_vB_w\alpha_w - B_v\alpha_vA_w\alpha_w)$$
$$+\tfrac{\partial}{\partial v}(B_u\alpha_uA_w\alpha_w - A_u\alpha_uB_w\alpha_w) + \tfrac{\partial}{\partial w}(A_u\alpha_uB_v\alpha_v - B_u\alpha_uA_v\alpha_v)]$$
$$= \tfrac{1}{\alpha_u\alpha_v\alpha_w}[\tfrac{\partial}{\partial u}(A_{lv}B_{lw} - B_{lv}A_{lw}) + \tfrac{\partial}{\partial v}(B_{lu}A_{lw} - A_{lu}B_{lw}) + \tfrac{\partial}{\partial w}(A_{lu}B_{lv} - B_{lu}A_{lv})],$$
$$A_{lu} = \alpha_uA_u,\ A_{lv} = \alpha_vA_v,\ A_{lw} = \alpha_wA_w,$$
$$B_{lu} = \alpha_uB_u,\ B_{lv} = \alpha_vB_v,\ B_{lw} = \alpha_wB_w. \tag{A.27}$$

$$\overline{B}\cdot\nabla\times\overline{A} = \tfrac{1}{\alpha_u\alpha_v\alpha_w}[B_u\alpha_u(\tfrac{\partial A_w\alpha_w}{\partial v} - \tfrac{\partial A_v\alpha_v}{\partial w}) + B_v\alpha_v(\tfrac{\partial A_u\alpha_u}{\partial w} - \tfrac{\partial A_w\alpha_w}{\partial u})$$
$$+B_w\alpha_w(\tfrac{\partial A_v\alpha_v}{\partial u} - \tfrac{\partial A_u\alpha_u}{\partial v})]$$
$$= \tfrac{1}{\alpha_u\alpha_v\alpha_w}[B_{lu}(\tfrac{\partial A_{lw}}{\partial v} - \tfrac{\partial A_{lv}}{\partial w}) + B_{lv}(\tfrac{\partial A_{lu}}{\partial w} - \tfrac{\partial A_{lw}}{\partial u}) + B_{lw}(\tfrac{\partial}{\partial u}A_{lv} - \tfrac{\partial}{\partial v}A_{lu})]$$
$$= \tfrac{1}{\alpha_u\alpha_v\alpha_w}[(B_{lw}\tfrac{\partial}{\partial u}A_{lv} - B_{lv}\tfrac{\partial}{\partial u}A_{lw}) + (B_{lu}\tfrac{\partial}{\partial v}A_{lw} - B_{lw}\tfrac{\partial}{\partial v}A_{lu})$$
$$+B_{lv}\tfrac{\partial}{\partial w}A_{lu} - B_{lu}\tfrac{\partial}{\partial w}A_{lv}]. \tag{A.28}$$

Similarly, by interchanging A and B in the above relation,

$$\overline{A}\cdot\nabla\times\overline{B} = \tfrac{1}{\alpha_u\alpha_v\alpha_w}[(A_{lw}\tfrac{\partial}{\partial u}B_{lv} - A_{lv}\tfrac{\partial}{\partial u}B_{lw}) + (A_{lu}\tfrac{\partial}{\partial v}B_{lw} - A_{lw}\tfrac{\partial}{\partial v}B_{lu})$$
$$+(A_{lv}\tfrac{\partial}{\partial w}B_{lu} - A_{lu}\tfrac{\partial}{\partial w}B_{lv})]. \tag{A.29}$$

Now, subtract (A.29) from (A.28), and the result can be verified to be equal to (A.27).

$$\overline{B}\cdot\nabla\times\overline{A} - \overline{A}\cdot\nabla\times\overline{B} = \tfrac{1}{\alpha_u\alpha_v\alpha_w}[\tfrac{\partial}{\partial u}(A_{lv}B_{lw} - A_{lw}B_{lv}) + \tfrac{\partial}{\partial v}(B_{lu}A_{lw} - B_{lw}A_{lu})$$
$$+\tfrac{\partial}{\partial w}(B_{lv}A_{lu} - B_{lu}A_{lv})] = \nabla\cdot(\overline{A}\times\overline{B}). \tag{A.30}$$

4.4. Path Independence of Gradient Integral.

$$d\psi = \tfrac{\partial\psi}{\partial u}du + \tfrac{\partial\psi}{\partial v}dv + \tfrac{\partial\psi}{\partial w}dw$$
$$= \tfrac{1}{\alpha_u}\tfrac{\partial\psi}{\partial u}\alpha_udu + \tfrac{1}{\alpha_v}\tfrac{\partial\psi}{\partial v}\alpha_vdv + \tfrac{1}{\alpha_w}\tfrac{\partial\psi}{\partial w}\alpha_wdw$$
$$= \left(\tfrac{1}{\alpha_u}\tfrac{\partial\psi}{\partial u}\hat{u} + \tfrac{1}{\alpha_v}\tfrac{\partial\psi}{\partial v}\hat{v} + \tfrac{1}{\alpha_w}\tfrac{\partial\psi}{\partial w}\hat{w}\right)\cdot(\alpha_udu\hat{u} + \alpha_vdv\hat{v} + \alpha_wdw\hat{w})$$
$$= \overline{\nabla}\psi\cdot(dl_u\hat{u} + dl_v\hat{v} + dl_w\hat{w}) = \overline{\nabla}\psi\cdot\overline{dl} = \overline{\nabla}\psi\cdot(dl\hat{l});\ \tfrac{d\psi}{dl} = \overline{\nabla}\psi\cdot\hat{l},$$
$$\textstyle\int_{C_{12}}\overline{\nabla}\psi\cdot\overline{dl} = \int_{C_{12}}d\psi = \int_1^2 d\psi = \psi_2 - \psi_1 = \Delta\psi. \tag{A.31}$$

The above result is independent of any specific path C_{12}, connecting two given end points 1 and 2 (see Fig.1). The line integral of the gradient vector $\overline{\nabla}\psi$ over a path C_{12} is simply the difference between the values of the function ψ at the end points 1 and 2.

Consider a closed contour C, which is a special case for an open contour C_{12} with its end points 1 and 2 joining together as one. As per the above principle under the special condition, the line integral of a gradient vector $\overline{\nabla}\psi$ over the closed contour C would be zero. This property of a gradient function may also be alternately derived by applying Stokes' theorem (A.36) to the curl relation in (A.26).

4.5. Divergence Theorem. Consider a general volume region V, enclosed by its boundary surface S. The total volume V may be decomposed into small elements ΔV_j, each bordered by elemental closed surface ΔS_j (see Fig.9). The integral of a vector function $\overline{A}$ over each closed-surface element ΔS_j can be equated to the volume integral of the divergence function $\overline{\nabla}\cdot\overline{A}$ over the volume element ΔV_j. This is based on the definition of a divergence function in A.16.

$$(\overline{\nabla}\cdot\overline{A})_j = \lim_{\Delta V_j\to 0}\frac{\int\int_{\substack{Closed\\ \Delta S_j}}\overline{A}\cdot\overline{ds}}{\Delta V_j},$$
$$\int\int_{\substack{Closed\\ \Delta S_j}}\overline{A}\cdot\overline{ds} = (\overline{\nabla}\cdot\overline{A})_j\Delta V_j = \sum_{k=1}^{6}\int\int_{\Delta S_{kj}}\overline{A}\cdot\overline{ds}. \tag{A.32}$$

The above results can be added over all elements j. The volume integrals over all the elements j would be added to have the volume integral over the total volume V. On the other hand, in the process of adding the surface integrals over all ΔS_j, the contributions from common surface segments ΔS_{jk} from neighboring surface elements ΔS_j would cancel. This is because the normals to the common surface segments are oriented in opposite directions. This would leave contributions only from the surface segments ΔS_{jk} that are on the outer surface S. Accordingly, the surface integrals over all closed-surface elements ΔS_j would be added to result in the surface integrals over the entire outer surface S.

$$\int\int\int_V(\overline{\nabla}\cdot\overline{A})dV = \sum_j\int\int\int_{\Delta V_j}(\overline{\nabla}\cdot\overline{A})dV = \sum_j(\overline{\nabla}\cdot\overline{A})_j\Delta V_j$$
$$= \sum_j\sum_{k=1}^{6}\int\int_{\Delta S_{kj}}\overline{A}\cdot\overline{ds} = \sum_{\substack{all\ k\\ on\ S}}\overline{A}_k\cdot\overline{\Delta S}_k = \int\int_{\substack{Closed\\ S}}\overline{A}\cdot\overline{ds}, \tag{A.33}$$

$$\int\int_{\substack{Closed\\ S}}\overline{A}\cdot\overline{ds} = \oint\!\!\!\oint_S\overline{A}\cdot\overline{ds} = \int\int\int_V(\overline{\nabla}\cdot\overline{A})dV. \tag{A.34}$$

The above useful result is referred to as the divergence theorem, which may be stated as follows. The integral of the divergence of a given vector function $\overline{A}$ evaluated over a given volume V, is equal to the integral of the vector function $\overline{A}$ evaluated over the surface S that encloses the given volume region V. The reference normal directions at different points on the surface S are assumed to be directed "outward" from the volume region V.

4.6. Stokes' Theorem. Consider a continuous, open surface S, which may be non-planar in general, bordered by a closed loop C. which may also be non-planar in general. The total surface S may be decomposed into small surface elements ΔS_k, each bordered by elemental closed loop ΔC_k (see Fig.10). The integral of a vector function $\overline{A}$ over each loop element ΔC_k can be equated to the surface integral of the curl function $\overline{\nabla}\times\overline{A}$ over the surface element ΔS_k. This is based on the definition of a curl function in A.10.

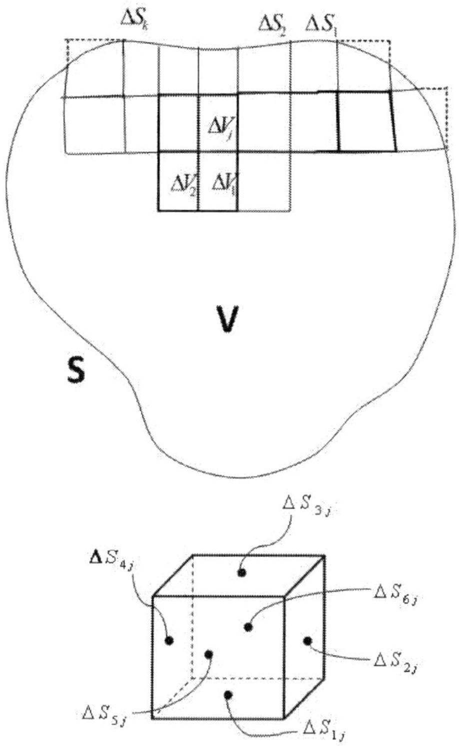

Decomposition of a closed surface element ΔS_j into six sides

FIGURE 9

$$(\overline{\nabla} \times \overline{A})_k \cdot \hat{n}_k = \lim_{\Delta S_j \to 0} \frac{\int_{\substack{Closed \\ \Delta C_k}} \overline{A} \cdot \overline{dl}}{\Delta S_k},$$

$$\int_{\substack{Closed \\ \Delta C_k}} A \cdot d\bar{l} = \sum_{\substack{Loop \\ \Delta C_k}} \overline{A}_{ik} \cdot \overline{\Delta l}_{ik} = (\overline{\nabla} \times \overline{A})_k \cdot \hat{n}_k \Delta S_k = (\overline{\nabla} \times \overline{A})_k \cdot \overline{\Delta S}_k. \quad \text{(A.35)}$$

The above results can be added over all elements k. The surface integrals over all the elements k would be added to have the surface integral over the total surface S. On the other hand, in the process of adding the loop integrals over all ΔC_k, the contributions from common line segments from neighboring loop elements would cancel. This is because the common segments are oriented in the opposite directions. This would leave contributions only from the line segments along the border loop C. Accordingly, the line integrals over all loop elements ΔC_k would be added to result in the line integrals over the border loop C.

$$\int \int_S (\overline{\nabla} \times \overline{A}) \cdot \overline{ds} = \sum_k (\nabla \times \overline{A})_k \cdot \overline{\Delta S}_k = \sum_k \sum_{\substack{Loop \\ \Delta C_k}} \overline{A}_{ik} \cdot \overline{\Delta l}_{ik}$$
$$= \sum_{\substack{All\ i\ on \\ Loop\ C}} (\overline{A}_i \cdot \overline{\Delta l}_i) = \int_{\substack{Loop \\ C}} \overline{A} \cdot \overline{dl}, \quad \text{(A.36)}$$

$$\int_{\substack{Loop \\ C}} \overline{A} \cdot \overline{dl} = \oint_C \overline{A} \cdot \overline{dl} = \int \int_S (\overline{\nabla} \times \overline{A}) \cdot \overline{ds}. \quad \text{(A.37)}$$

The above result is referred to as Stokes' theorem, which may be stated in words as follows. The integral of the curl of a vector function $\overline{A}$, evaluated over a given open surface S, is equal to the integral of the vector function $\overline{A}$ evaluated over the closed contour C that borders the surface S. The reference normal directions to the surface S near the border contour C, and the orientation of the path elements along C, are to be properly referenced with respect to each other using a standard right-hand rule. As per the rule, let the thumb of the right hand be aligned along the normal to a surface element near the border contour C. Then, the orientation of the other fingers of the right hand, curved in a "grip" form and positioned along the border contour C, would refer to the proper orientation of the path elements along ΔC (see Fig.10).

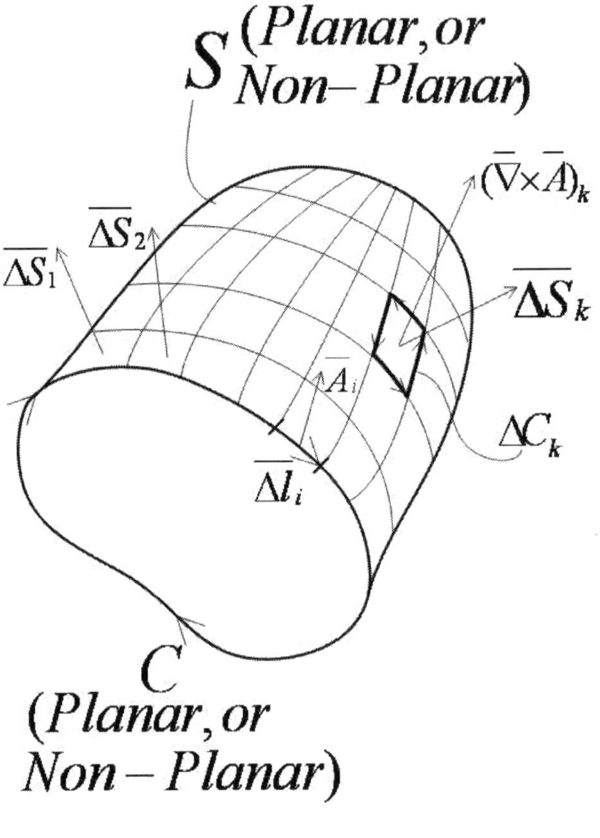

FIGURE 10

APPENDIX B

Basic Phasor Concept for Harmonic Signals

The phasor analysis is commonly used in electrical circuit analysis for harmonic signals, It is also used in the analysis of harmonic waves. We present here some basic phasor concepts, and establish notational principles which we would use in different analyses. .

1. Definition

Let us define a complex phasor in functional and notational terms. A time-harmonic signal, for example a voltage signal $V(t)$, is represented by its complex phasor V_p. The peak amplitude of the time-harmonic signal $V(t)$ is equal to the magnitude $|V_p|$ of the phasor, and the phase of $V(t)$ is equal to the phase $\arg(V_p)$ of the phasor.

$$V(t) = |V_p| \cos(\omega t + \arg(V_p)) = Re(|V_p| e^{j(wt+\arg(V_p))}) = Re(V_p e^{j\omega t}). \tag{B.1}$$

j is the positive imaginary number, which is the square-root of -1. Notationally, the variable used for a complex phasor is the same variable used for the time function the phasor represents, but with a subscript p. This distinction would be useful, particularly when the time function and its phasor variable are used together in a common context. When there is no obvious confusion, for the sake of notational economy, we may use the same variable to interchangeably refer to the time function as well as its phasor. Accordingly, the time function $V(t)$ may be represented by a phasor V or V_p. The phasor V is a complex number, which is not a function of time t. Whereas, the time function $V(t)$ is usually shown with the time dependence. The distinctions between the time function and its complex phasor may often be self explanatory, based on proper context of an analysis or expression.

2. Sum and Difference of Phasors

The sum or difference between two harmonic signals, each with same frequency ω, is also a harmonic signal with the same frequency. And. the phasor of the sum or difference function may be shown to be the sum or difference of the individual phasors, respectively.

$$V_1(t) = Re(V_{1p} e^{jwt}),\ V_2(t) = Re(V_{2p} e^{jwt}),\ V(t) = Re(V_p e^{jwt}),$$
$$V(t) = V_1(t) \pm V_2(t) = Re(V_{1p} e^{jwt}) \pm Re(V_{2p} e^{jwt}) = Re((V_{1p} \pm V_{2p}) e^{j\omega t}),$$
$$V_p = V_{1p} \pm V_{2p}. \tag{B.2}$$

Further, two or more complex phasors, each represented with a magnitude and a phase, or with real and imaginary parts, can be added and subtracted using complex algebra. This procedure is also equivalent to adding or subtracting two dimensional vectors, as shown in Fig.1. Accordingly, complex algebra or vector analysis may be conveniently used in adding or subtracting harmonic functions. In other words, trigonometric signals may be treated as complex-algebraic or vector variables.

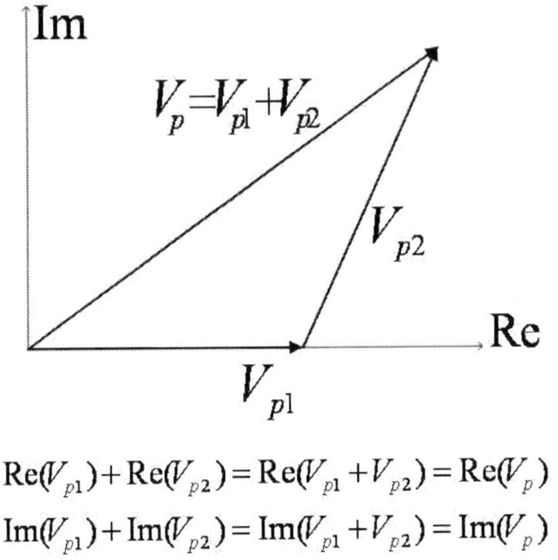

FIGURE 1

3. Phasor and Time-Derivative

The time derivative of a harmonic signal is also a harmonic signal with the same frequency. It may be shown that the phasor of the time derivative of a harmonic function is obtained by multiplying the phasor of the harmonic function by $j\omega$. This would allow time-differential operators to be transformed into algebraic multiplications, leading to simpler analysis.

$$\frac{dV(t)}{dt} = -\omega|V_p|\sin(\omega t + \arg(V_p)) = \omega|V_p|\cos(\omega t + \pi/2 + \arg(V_p)),$$

$$\frac{dV(t)}{dt}\Big|_p = \omega|V_p|e^{j(\pi/2+\arg(V_p))} = \omega|V_p|e^{j\pi/2}e^{j\arg(V_p)} = j\omega V_p,$$

$$\frac{\partial}{\partial t} \equiv j\omega, \ \frac{\partial^2}{\partial t^2} \equiv (j\omega)^2 = -\omega^2. \tag{B.3}$$

4. Dot Product of Two Phasors

A complex phasor may be treated as a two-dimensional vector. The real and imaginary parts of the phasor are treated as two vector components. The magnitude and phase of the phasor are treated as the magnitude and the angle with respect to the real axis of the equivalent vector, respectively. Accordingly, a dot product of two phasors may be defined in a similar manner as the dot product of two vectors.

$$V_{p1} \cdot V_{p2} = |V_{p1}||V_{p2}|\cos(\phi_1 - \phi_2),\ V_{p1} = |V_{p1}|e^{j\phi_1},\ V_{p2} = |V_{p2}|e^{j\phi_2}. \tag{B.4}$$

The dot product of the phasors may also be equivalently expressed using complex algebra and complex-conjugate operation (*).

$$\begin{aligned} V_{p1} \cdot V_{p2} &= |V_{p1}||V_{p2}|\cos(\phi_1 - \phi_2) = Re(|V_{p1}||V_{p2}|e^{j(\phi_1-\phi_2)}) \\ &= Re(|V_{p1}|e^{j\phi_1}|V_{p2}|e^{-j\phi_2}) = Re(V_{p1}V_{p2}^*). \end{aligned} \tag{B.5}$$

We will see what this dot product would mean, in relations to the harmonic functions $V_1(t)$ and $V_2(t)$ which the individual phasors represent. Consider the product of the individual harmonic functions, and take the time average of the product.

$$\begin{aligned} &< V_1(t)V_2(t) >= \frac{1}{T}\int_{t=0}^{T} |V_{p1}|\cos(\omega t + \phi_1)|V_{p2}|\cos(\omega t + \phi_1)dt \\ &= \frac{|V_{p1}||V_{p2}|}{2T}\int_{t=0}^{T}[\cos(2\omega t + \phi_1 + \phi_2) + \cos(\phi_1 - \phi_2)]dt = \frac{|V_{p1}||V_{p2}|}{2}\cos(\phi_1 - \phi_2), \\ &V_{p1} \cdot V_{p2} = |V_{p1}||V_{p2}|\cos(\phi_1 - \phi_2) = Re(V_{p1}V_{p2}^*) = 2 < V_1(t)V_2(t) > . \end{aligned} \tag{B.6}$$

where, the operator $< . >$ represents time average. We see that the dot product of two phasors is equal to twice the time average of the product of the respective time-harmonic functions. The two time harmonic functions are assumed to have the same frequency ω and time period $T = 2\pi/\omega$. Otherwise, the above dot-product relation would not be valid or meaningful.

Consider a special case, when the two harmonic functions are identical.

$$\begin{aligned} &V_{p1} = V_{p2} = V_p,\ V_1(t) = V_2(t) = V(t),\ \phi_1 = \phi_2 = \phi, \\ &< V^2(t) >= V_{rms}^2 = \frac{1}{T}\int_0^T |V_p|^2\cos^2(\omega t + \phi)dt = \frac{|V_p|^2}{2}, \\ &|V_p|^2 = 2 < V^2(t) >= 2V_{rms}^2,\ |V_p| = V_{peak} = \sqrt{2}V_{rms}. \end{aligned} \tag{B.7}$$

The square-root of the time average (or the mean value) of the square of the harmonic function $V(t)$ is referred to as the root-mean-square or the RMS value, V_{rms}, of the harmonic function. In other words, $< V^2(t) >$ is equal to V_{rms}^2. Using this definition, we see that the phasor magnitude, which is defined to be the peak value of the harmonic function, is equal to $\sqrt{2}$ times V_{rms}.

5. Peak or RMS Phasor

In all the above representations we have defined the peak amplitude of a harmonic signal to be equal to the magnitude of its complex phasor. However, it may sometimes be useful to define the root-mean-square (RMS) value of the harmonic signal to be equal to the magnitude of the complex phasor. Accordingly, a phasor may be referred to as a RMS-phasor or a peak-phasor, in order to clarify which definition is used in a particular situation. The peak value of a harmonic signal, as per (B.7), is larger than the RMS value by a factor $\sqrt{2}$. Accordingly, when a RMS-phasor is used, the peak value or a harmonic signal would be $\sqrt{2}$ times the magnitude of its phasor. In this case, .

$$V(t) = \sqrt{2}|V_p|\cos(\omega t + \arg(V_p)) = Re(\sqrt{2}|V_p|e^{j(\omega t + \arg(V_p))}) = Re(\sqrt{2}V_p e^{j\omega t}),$$
$$|V_p| = V_{rms},\ V_{peak} = \sqrt{2}V_{rms} = \sqrt{2}|V_p|. \quad \text{(B.8)}$$

The time dependence, expressed in (B.1) using a peak phasor, and that in (B.8) using a RMS-phasor simply differ by a factor $\sqrt{2}$.

The dot-product relation of (B.6), which was expressed assuming peak phasors, would also have to be properly modified when RMS phasors are used. The new relation is obtained by replacing in (B.6) the functional expressions of the harmonic functions in terms of the respective RMS phasors, as defined above in (B.8).

$$V_{p1} \cdot V_{p2} = |V_{p1}||V_{p2}|\cos(\phi_1 - \phi_2) = Re(V_{p1}V_{p2}^*),$$
$$< V_1(t)V_2(t) >= \frac{1}{T}\int_{t=0}^{T} \sqrt{2}|V_{p1}|\cos(\omega t + \phi_1)\sqrt{2}|V_{p2}|\cos(\omega t + \phi_1)dt$$
$$= \frac{|V_{p1}||V_{p2}|}{T}\int_{t=0}^{T} [\cos(2\omega t + \phi_1 + \phi_2) + \cos(\phi_1 - \phi_2)]dt = |V_{p1}||V_{p2}|\cos(\phi_1 - \phi_2),$$
$$V_{p1} \cdot V_{p2} = |V_{p1}||V_{p2}|\cos(\phi_1 - \phi_2) = Re(V_{p1}V_{p2}^*) =< V_1(t)V_2(t) > . \quad \text{(B.9)}$$

Accordingly, the dot product of two RMS phasors is equal to the time average of the product of the respective harmonic functions. Note that dot-product relation in (B.6) using peak phasor, and that in (B.9) using RMS phasors, simply differ by a factor of 2.

The sum and difference relation in (B.2), and the equivalence of a time-derivative operation to multiplication by $j\omega$ shown in (B.3), would remain valid, without any modification, for RMS or peak phasors. This assumes that all phasors uniformly use the same definition.

APPENDIX C

Delta Function and Useful Limiting Forms

Delta functions in one and two dimensions are defined. Some useful limiting forms of the delta functions are analyzed.

1. Definition

1.1. Delta Function in One Dimension.

$$\begin{aligned} \delta(x) &= 0;\ x \neq 0, \\ &\neq 0;\ x = 0, \end{aligned}$$

$$\int_{-\infty}^{\infty} \delta(x)dx = 1 = 2\int_{0}^{\infty} \delta(x)dx, \tag{C.1}$$

$$f(x)\delta(x) = f(0)\delta(x),$$

$$\int_{-\infty}^{\infty} f(x)\delta(x)dx = \int_{-\infty}^{\infty} f(0)\delta(x)dx = f(0), \tag{C.2}$$

where $f(x)$ is a general "well-behaved" function of x. Equation (C.2) may be referred to as the "sampling theorem" of delta function.

1.2. Delta Function in Two Dimensions.

$$\begin{aligned} \delta(x)\delta(y) &= 0;\ x, y \neq 0, \\ &\neq 0;\ x, y = 0, \end{aligned}$$

$$\int\int_{-\infty}^{\infty} \delta(x)\delta(y)dxdy = \iint \delta(x)\delta(y)ds = 1,\ ds = dxdy. \tag{C.3}$$

A similar principle as in (C.2) may be extended for two dimensions

$$f(x, y)\delta(x)\delta(y) = f(0, 0)\delta(x)\delta(y),$$

$$\int\int_{-\infty}^{\infty} f(x, y)\delta(x)\delta(ydxdy = \int\int_{-\infty}^{\infty} f(0, 0)\delta(x)\delta(y)dxdy = f(0, 0), \tag{C.4}$$

where $f(x, y)$ is a general "well-behaved" function of x, which is finite and continuous for all values of (x, y).

1.2.1. *Alternate Polar Form.*

$$\delta_p(\bar{\rho}) \equiv \delta_p(\rho, \phi) \triangleq \frac{\delta(\rho)}{\pi\rho}, \text{ for all } \phi, \tag{C.5}$$

where $\bar{\rho} \equiv (\rho, \phi)$ are standard polar coordinates for (x, y), related as $x = \rho\cos\phi$ and $y = \rho\sin\phi$. $\delta(\rho)$ is a one-dimensional delta function of ρ.

$$\delta_p(\bar{\rho}) = 0;\ \rho \neq 0,$$
$$\neq 0;\ \rho = 0,$$

$$\iint \delta_p(\bar{\rho})ds = \int_0^\infty \int_{\phi=0}^{2\pi} \frac{\delta(\rho)}{\pi\rho}\rho d\phi d\rho = \int_0^\infty 2\frac{\delta(\rho)}{\rho}\rho d\rho = 1,\ ds = \rho d\phi d\rho. \tag{C.6}$$

Comparing (C.6) with (C.3),

$$\delta_p(\bar{\rho}) = \frac{\delta(\rho)}{\pi\rho} = \delta(x)\delta(y). \tag{C.7}$$

$\delta_p(\bar{\rho}) \equiv \delta_p(\rho, \phi)$ is an alternate form of the two dimensional delta function $\delta(x)\delta(y)$ in the polar form.

2. Useful Limiting Relations

Often it is useful to define a delta function in the form of a particular regular function, as it approaches a limit with respect to a parameter. In the following we will study some useful limiting relations.

(I)

$$f_1(x, \Delta) = \frac{\Delta}{(x^2 + \Delta^2)}. \tag{C.8}$$

Examine the above function as Δ approaches zero.

$$f_1(x, \triangle \to 0) = 0;\ x \neq 0,$$
$$\neq 0;\ x = 0. \tag{C.9}$$

Now evaluate the integration of the function $f_1(x, \Delta)$ with respect to x .

$$\int_{-\infty}^{\infty} f_1(x; \Delta)dx = \int_{-\infty}^{\infty} \frac{\Delta}{x^2 + \Delta^2}dx$$

$$= \int_{\mp\frac{\pi}{2}}^{\pm\frac{\pi}{2}} \frac{\Delta}{\Delta^2 \sec^2 t}\Delta \sec^2 t\ dt = \pm\pi,\ \triangle \geq 0; \tag{C.10}$$

$$x = \Delta \tan t,\ dx = \Delta \sec^2 t dt.$$

Compare (C.9,C.11) with the definition of $\delta(x)$ in (C.1).

$$f_1(x, \Delta \to 0_\pm) \equiv \pm\pi\delta(x). \tag{C.11}$$

(I.1)

Consider another function $g(x, \Delta)$, which is a general "well-behaved" function of x and Δ (finite and continuous for all values of x and Δ) .

$$\begin{aligned} \lim_{\triangle \to 0_\pm} (g(x, \triangle) f_1(x, \triangle)) &= g(x, \triangle = 0)(\lim_{\triangle \to 0_\pm} f_1(x, \triangle)) \\ &= \pm\pi\delta(x) g(x, \triangle = 0) = \pm\pi\delta(x) g(x = 0, \triangle = 0). \end{aligned} \tag{C.12, C.13}$$

This is similar to the sampling theorem in (C.2), but now applicable in a limiting form.

$$\lim_{\triangle \to 0_\pm} \int_{-\infty}^{\infty} \frac{\triangle}{\triangle^2 + x^2} g(x, \triangle) dx = \pm\pi g(0, 0). \tag{C.14}$$

(II)

$$f_2(\rho, \phi, \Delta) = \frac{\Delta}{(\Delta^2 + \rho^2)^{\frac{3}{2}}}. \tag{C.15}$$

Examine the above function as $\Delta \to 0$.

$$\begin{aligned} f_2(\rho, \phi, \triangle \to 0) &= 0;\ \rho \neq 0, \\ &\neq 0;\ \rho = 0. \end{aligned} \tag{C.16}$$

Evaluate the integration of $f_2(\rho, \phi, \Delta)$ in two dimensions.

$$\begin{aligned} \iint f_2(\rho; \Delta) ds &= \int_{\rho=0}^{\infty} \int_{\phi=0}^{2\pi} \frac{\Delta}{(\Delta^2 + \rho^2)^{\frac{3}{2}}} \rho d\phi d\rho \\ &= 2\pi \int_0^{\pm\frac{\pi}{2}} \frac{\Delta}{\Delta^3 \sec^3 t} \Delta \tan t \Delta \sec^2 t dt \\ &= 2\pi \int_0^{\pm\frac{\pi}{2}} \sin t dt = \pm 2\pi,\ \Delta \geq 0; \\ \rho &= \Delta \tan t,\ d\rho = \Delta \sec^2 t dt. \end{aligned} \tag{C.17}$$

Comapre (C.16,C.17) with (C.6,C.7).

$$f_2(\rho, \phi, \Delta \to 0_\pm) = \pm 2\pi\delta_p(\rho) = \pm 2\frac{\delta(\rho)}{\rho} = \pm 2\pi\delta(x)\delta(y). \tag{C.18}$$

(II.1)

Consider another function $g(\rho, \phi, \Delta)$, which is a general "well-behaved" function of ρ, ϕ and Δ (finite and continuous everywhere) .

$$\begin{aligned}\lim_{\triangle\to 0\pm}(g(\rho,\phi,\triangle)f_2(\rho,\phi,\triangle)) &= g(\rho,\phi,\triangle=0)(\lim_{\triangle\to 0\pm} f_2(\rho,\phi,\triangle)) \\ &= \pm 2\pi\delta_p(\bar{\rho})g(\rho,\phi,\triangle=0) = \pm 2\pi\delta_p(\bar{\rho})g(0,0,0). \end{aligned} \tag{C.19}$$

This is similar to the sampling theorem in (C.4), applied in a limiting form.

$$\lim_{\triangle\to 0\pm}\iint \frac{\triangle}{(\triangle^2+\rho^2)^{3/2}}g(\rho,\phi,\triangle)ds = \pm 2\pi g(0,0,0). \tag{C.20}$$

(III)

$$f_3(\rho,\phi,\Delta) = \frac{\Delta a}{\Delta^2+\rho^2}e^{-ja\sqrt{\Delta^2+\rho^2}},\ a>0. \tag{C.21}$$

Examine the above function as $\Delta \to 0$.

$$\begin{aligned} f_3(\rho,\phi,\Delta\to 0) &= 0;\ \rho\neq 0, \\ &\neq 0,\ \rho = 0, \end{aligned} \tag{C.22}$$

$$\begin{aligned}\iint f_3(\rho;\Delta)ds &= \int_{\rho=0}^{\infty}\int_{\phi=0}^{2\pi}\frac{\Delta a}{\Delta^2+\rho^2}e^{-ja\sqrt{\Delta^2+\rho^2}}\rho d\phi d\rho \\ &= 2\pi\int_{\rho=0}^{\infty}\frac{\Delta a}{\Delta^2+\rho^2}e^{-ja\sqrt{\Delta^2+\rho^2}}\rho d\rho \\ &= 2\pi\int_{|\Delta|a}^{\infty}\frac{\Delta a}{t^2}e^{-jt}tdt; \\ t^2 &= a^2(\Delta^2+\rho^2). \end{aligned} \tag{C.23}$$

Consider $\int_x^\infty \frac{e^{-jt}}{t}dt = h(x)$ exists for $x > 0$.

$$\begin{aligned} h(x) &= \int_x^\infty \frac{e^{-jt}}{t}dt, \\ h'(x) = \frac{dh}{dx} &= -\frac{e^{-jx}}{x},\ h'(x\to 0) = -\frac{1}{x}, \\ h(x\to 0) &= -\ln x + c, \end{aligned} \tag{C.24}$$

where c is a finite constant. From (C.23) and (C.24),

$$\iint f_3(\rho;\Delta\to 0)ds = 2\pi\lim_{\Delta\to 0}[\Delta a\{-\ln(|\Delta|a)+c\}] = 0, \tag{C.25}$$

where we use the limit relation $\lim_{x\to 0}(x\ln|x|) = 0$. Note that both f_3 and f_2 may have similar general behavior around $\rho = 0$ as Δ approaches zero. However, unlike the integration of f_2, the integration of f_3 is zero as Δ approaches zero. This is true for any value of a, including a approaching zero. Therefore, the function f_3 in its limit as Δ approaches zero may be considered a delta function with zero amplitude, for all non-zero values of a. Note the trivial situation when the function f_3 is identically equal to zero if $a = 0$.

The role of the oscillating factor $e^{-ja\sqrt{\Delta^2+\rho^2}}$ is important for the above limiting expression (C.23-C.25) of the function f_3 to exist and be equal to zero. Without the factor, the function f_3 would approach $1/\rho^2$ for large values of ρ, for all $\Delta \neq 0$. Therefore, the polar integration of the f_3 in (C.23) with respect to $\rho d\rho$ would be infinite for all $\Delta \neq 0$, and the limit of the integration (C.23-C.25) as $\Delta \to 0$ can be shown not to exist.

3. Fourier Transform

$$\Im_x \delta(x) \triangleq \int_{-\infty}^{\infty} \delta(x) e^{-jk_x x} dx = 1. \tag{C.26}$$

This uses the sampling theorem of (C.2).

$$\begin{aligned}\Im_{xy} \delta(x)\delta(y) &\triangleq \iint_{-\infty}^{\infty} \delta(x)\delta(y) e^{-jk_x x} e^{-jk_y y} dx dy = 1 \\ = \Im_{xy} \delta_p(\bar{\rho}) &= \iint_{-\infty}^{\infty} \frac{\delta(\rho)}{\pi\rho} e^{-jk_x x} e^{-jk_y y} dx dy.\end{aligned} \tag{C.27}$$

This uses the sampling theorem of (C.4).

APPENDIX D

Wave Function $\frac{1}{r}\exp(-jkr)$

The basic characteristics of the elementary wave function $\frac{1}{r}\exp(-jkr)$ in the spherical coordinates will be analyzed. Functional and Fourier-transform relations of the wave function will be derived.

1. Elementary Spherical Wave Function

$$f(\bar{r}) \equiv f(r,\theta,\phi) = \frac{1}{r}e^{-jkr}, \text{ for all } \theta \text{ and } \phi, \qquad \text{(D.1)}$$
$$r = \sqrt{x^2+y^2+z^2} = \sqrt{\rho^2+z^2}.$$

$f(\bar{r})$ is the elementary spherical wave function, which is independent of θ and ϕ. (r,θ,ϕ), (ρ,ϕ,z), and (x,y,z) are the standard spherical, cylindrical and rectangular coordinates, respectively.

1.1. Two-Dimensional Fourier Transform. A two-dimensional Fourier Transform of the wave function $f(\bar{r})$ with respect to x and y, and its inversion formula may be defined in a conventional way.

$$\Im_{xy} f(\bar{r}) \equiv \tilde{f}(k_x,k_y,z) \triangleq \int_{y=-\infty}^{\infty}\int_{x=-\infty}^{\infty} \frac{e^{-jkr}}{r} e^{-jk_x x} e^{-jk_y y}\, dx\, dy, \qquad \text{(D.2)}$$

where the operator $\Im_{xy}$ or $\sim$ overlay alternately refers to Fourier transformation.

$$f(\bar{r}) = \frac{1}{4\pi^2}\int_{k_y=-\infty}^{\infty}\int_{k_x=-\infty}^{\infty} \tilde{f}(k_x,k_y,z)e^{+jk_x x}e^{+k_y y}dk_x dk_y. \qquad \text{(D.3)}$$

Transform and inversion in any other two rectangular coordinates, with the third dimension as a variable, can be similarly defined.

2. First Derivatives of $f(\bar{r})$ as a Limiting Delta Function

The first derivatives with respect to x, y and z are symmetric with respect to each other.

$$\frac{\partial f}{\partial x} = \frac{-x}{r^3}e^{-jkr} - j\frac{kx}{r^2}e^{-jkr}$$
$$\frac{\partial f}{\partial y} = \frac{-y}{r^3}e^{-jkr} - j\frac{ky}{r^2}e^{-jkr}$$
$$\frac{\partial f}{\partial z} = \frac{-z}{r^3}e^{-jkr} - j\frac{kz}{r^2}e^{-jkr}$$
$$\frac{\partial f}{\partial u} = \frac{-u}{r^3}e^{-jkr} - j\frac{ku}{r^2}e^{-jkr}; \; u = x,\ y,\ z. \tag{D.4}$$

Examine the first derivative $(\partial f/\partial z)$. We will see that it behaves as a limiting form of a two-dimensional delta function, as $z \to 0$. This limiting behavior is illustrated in Fig.1, where we plot $(\partial f/\partial z)\rho'/k^2$ with respect to the normalized polar variable $\rho' = k\rho$, for different values of the normalized paramter $z' = zk$.

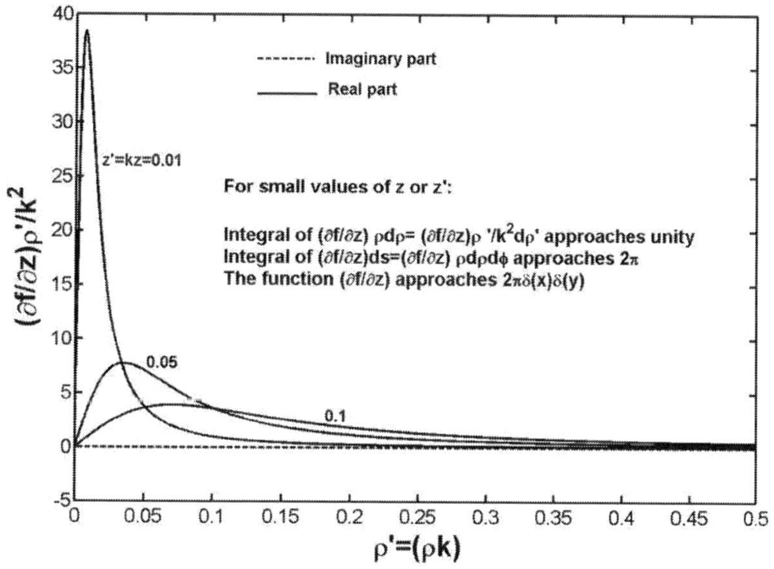

FIGURE 1

First, as z approaches zero, the derivative funciton $(\partial f/\partial z)$ in (D.4) can be shown to be localized around the origin $\rho = 0$. This is an essential behavior for a delta function.

$$\left.\frac{\partial f(\bar{r})}{\partial z}\right|_{z\to 0} \begin{aligned} &= 0, \;\; \rho \neq 0, \\ &\neq 0, \;\; \rho = 0. \end{aligned} \tag{D.5}$$

Then, let us evaluate the two-dimensional integration of $(\partial f/\partial z)$ in the (x, y) or (ρ, ϕ) plane, as $z \to 0$.

$$\lim_{z\to 0\pm}\iint \frac{\partial f}{\partial z}ds$$
$$= \lim_{z\to 0\pm}\iint (-\frac{z}{r^3}e^{-jkr} - j\frac{kz}{r^2}e^{-jkr})ds$$
$$= \mp 2\pi - j0 = \mp 2\pi. \tag{D.6}$$

We use the limiting relations established in Appendix C for the above evaluation. For the first part of the integration in (D.6), we use the limiting relation for the functon $f_2(\rho,\phi,\Delta)$ in Appendix C, with $\Delta = z$ and $g(\rho,\Delta) = e^{-jkr}$ substituted in (C.20). For the second part of the integration in (D.6), we use the limiting relation for the functon $f_3(\rho,\phi,\Delta;a)$ in Appendix C, with $\Delta = z$ and $a = k$ substituted in (C.25).

Using (D.5,D.6) we can establish the following delta-function equivalence for $(\partial f/\partial z)$.

$$\lim_{z\to 0\pm}\frac{\partial f(\bar{r})}{\partial z} \equiv \mp 2\pi\delta_p(\rho) = \mp 2\frac{\delta(\rho)}{\rho} = \mp 2\pi\delta(x)\delta(y). \tag{D.7}$$

Note that the above limiting relation is valid independent of k. As a useful special case as $k \to 0$, the above relation applies to function $\frac{1}{r}$ as well. In other words, the first derivative of $\frac{1}{r}$ with respect to z also behaves as a two-dimensional delta function, as z approaches zero.

Once the derivative is proved to be a delta function as $z \to 0$, its Fourier transform in the limiting case can be simplified.

$$\Im_{xy}[\lim_{z\to 0\pm}\frac{\partial f}{\partial z}] = \lim_{z\to 0\pm}\Im_{xy}\frac{\partial f}{\partial z} = \Im_{xy}[\mp 2\pi\delta(x)\delta(y)] = \mp 2\pi. \tag{D.8}$$

3. Second Derivatives and Wave Equation

Second derivatives can be obtained by differentiating (D.4) again with respect to x, y or z. One can then check that the second derivatives of the elementary wave function would satisfy a standard wave equation, as expected.

$$\frac{\partial^2 f(\bar{r})}{\partial u^2} = -\frac{e^{-jkr}}{r^3} - \frac{jk}{r^2}e^{-jkr} + 3\frac{jku^2}{r^4}e^{-jkr} - \frac{k^2u^2}{r^3}e^{-jkr} + \frac{3u^2}{r^5}e^{-jkr};\ u = x, y, z, \tag{D.9}$$
$$\frac{\partial^2 f(\bar{r})}{\partial x^2} + \frac{\partial^2 f(\bar{r})}{\partial y^2} + \frac{\partial^2 f(\bar{r})}{\partial z^2} = -k^2\frac{e^{-jkr}}{r} = -k^2 f(\bar{r}). \tag{D.10}$$

Using the inverse transform expression (D.3) for $f(\bar{r})$ in both sides of (D.10) we get an equivalent wave equation for the transform $\tilde{f}$.

$$\frac{\partial^2 \tilde{f}}{\partial z^2} + (k^2 - k_x^2 - k_y^2)\tilde{f} = 0. \tag{D.11}$$

Solving the above wave equation (D.11),

$$\tilde{f}(k_x, k_y, z) = \tilde{f}_0(k_x, k_y) e^{\mp jk_z z};$$
$$k_z = \sqrt{k^2 - k_x^2 - k_y^2}. \tag{D.12}$$

The uper(negative) and lower(positive) signs of the dual-sign notation used above are selected for $z > 0$ and $z < 0$, respectively. This notation would be consistently used in other places. In order to have finite value of the function at $z \to \pm\infty$, we also need to select proper sign for the square-root expression in the above equation, such that $Re(k_z) > 0,\ Im(k_z) < 0$.

From the z-dependent transform expression (D.12) one can also derive an equivalent expression for the transform of the derivative, which is equal to the derivative of the transform.

$$\Im_{xy}(\frac{\partial f(\bar{r})}{\partial z}) = \frac{\partial}{\partial z}\Im_{xy} f(\bar{r}) = \frac{\partial}{\partial z}\tilde{f}(k_x, k_y, z) = \mp jk_z \tilde{f} = \mp jk_z \tilde{f}_0 e^{\mp jk_z z}. \tag{D.13}$$

4. Fourier Transform Expression for $f(\bar{r})$

The basic properties of the wave funciton we derived above in terms of its derivatives and transforms can be employed to deduce a simple form for its Fourier transform. Using (D.13) in (D.8) we can solve for $\tilde{f}_0$. Then using the expression of $\tilde{f}_0$ in (D.12) one can find the final expression for the Fourier Transform of the wave function $f(\bar{r})$.

$$\Im_{xy} \lim_{z \to 0\pm} \frac{\partial f(\bar{r})}{\partial z} = \lim_{z \to 0\pm} \Im_{xy} \frac{\partial f(\bar{r})}{\partial z} = \mp jk_z \tilde{f}_0(k_x, k_y) = \mp 2\pi, \tag{D.14}$$

$$\tilde{f}_0 = \frac{2\pi}{jk_z}, \tag{D.15}$$

$$\Im_{xy} f(\bar{r}) = \Im_{xy}(\frac{1}{r} e^{-jkr}) = \tilde{f}(k_x, k_y, z) = \tilde{f}_0(k_x, k_y) e^{\mp jk_z z} = \frac{2\pi}{jk_z} e^{\mp jk_z z}. \tag{D.16}$$

4.1. Special Cases.

One may obtain a useful transform expression as a special case of (D.16) by taking the limit as $z \to 0$.

$$\Im_{xy}(\frac{1}{\rho} e^{-jk\rho}) = \frac{2\pi}{jk_z}. \tag{D.17}$$

Other useful transforms would be obtained by taking a limit as $k \to 0$ in both (D.16) and (D.17).

$$\Im_{xy}\left(\frac{1}{r}\right) = \frac{2\pi}{k_\rho} e^{\mp k_\rho z}, \tag{D.18}$$

$$\Im_{xy}\left(\frac{1}{\rho}\right) = \frac{2\pi}{k_\rho}, \tag{D.19}$$

$$k_\rho = jk_z|_{k=0} = \sqrt{k_x^2 + k_y^2} > 0.$$

As in the general expression of the transform, the negative and positive signs in the dual-sign notation are used for $z > 0$ and $z < 0$ respectively.

4.2. Alternate Forms of Transform. We have used transforms in the xy plane. Similar transforms can be expressed over other two planes with suitable change of variables and definitions.

$$\Im_{xz} f(\bar{r}) = \tilde{f}(k_x, k_z, y) \triangleq \int_{z=-\infty}^{\infty} \int_{x=-\infty}^{\infty} f(\bar{r}) e^{-jk_x x} e^{-jk_z z} dx dz$$

$$= \frac{2\pi}{jk_y} e^{\mp jk_y y}; \tag{D.20}$$

$$k_y = \sqrt{k^2 - k_x^2 - k_z^2}; \; Re(k_y) > 0, \; Im(k_y) < 0.$$

APPENDIX E

Bessel Functions of Integer Order

A general expression in the Fourier-integral form, and associated differential equation, for Bessel Functions are presented. Useful special integral forms, differential and iterative relations, and approximate expressions for small and large arguments are given. The results are derived, starting from the general Fourier-integral expression of the Bessel Functions. One may like the derivations for a complete understanding, or one may treat the different final results and relations as useful properties or identities of the Bessel Functions, for ready reference.

1. General Fourier-Integral Expression

$$f(\rho)e^{jn\phi} = \frac{j^{-n}}{\pi} \int_{-\infty}^{+\infty} \frac{(k_x + jk_y)^{\mp n}}{k_y} e^{jk_x x} e^{\mp jk_y y} dk_x;$$

$$(-)\ for\ y > 0,\ 0 < \phi < \pi;\ (+)\ for\ y < 0,\ \pi < \phi < 2\pi;$$

$$x = \rho\cos\phi,\ y = \rho\sin\phi;\ k_y = \sqrt{1 - k_x^2}. \qquad \text{(E.1)}$$

Note that the parameter $k_y = \sqrt{1 - k_x^2}$ has two possible values, one of which should be properly selected. Over the integration domain $1 < k_x < \infty$ and $-\infty < k_x < -1$, the two possible values of k_y are either positive or negative imaginary, of which only the negative imaginary value needs to be selected. This would ensure that the integrand parameter $e^{\mp jk_y y}$ is not infinite as y approaches $\pm\infty$, such that the resulting integral (E.1) and the associated function $f(\rho)$ do exist and are finite. On the other hand, in the integration domain $-1 < k_x < +1$ the parameter k_y is real, but it may be selected with either a positive or negative sign. These choices lead to two valid cases for the function $f(\rho)$:

$$f(\rho) = H_n^{(2)}(\rho) = J_n(\rho) - jY_n(\rho);\ Re(k_y) > 0, Im(k_y) < 0, \qquad \text{(E.2)}$$

$$= -H_n^{(1)}(\rho) = -J_n(\rho) - jY_n(\rho);\ Re(k_y) < 0, Im(k_y) < 0. \qquad \text{(E.3)}$$

$H_n^{(1)}$ and $H_n^{(2)}$ are referred to as the Hankel Functions of order n, of the first and second kind, respectively. Subtracting (E.3) from (E.2), we obtain an expression for the function $J_n(\rho)$, referred to as the Bessel Function of order n, of the first kind.

$$J_n(\rho) = \frac{1}{2}[H_n^{(2)}(\rho) + H_n^{(1)}(\rho)] = Re(f(\rho))$$

$$= \frac{j^{-n}}{\pi} e^{-jn\phi} \int_{-1}^{+1} \frac{\cos(n \sin^{-1} k_y + k_y y)}{k_y} e^{jk_x x} dk_x; \; Re(k_y) > 0. \tag{E.4}$$

Similarly, adding (E.2) and (E.3), we would obtain an expression for the function $Y_n(\rho)$, referred to as the Bessel Function of order n, of the second kind.

$$Y_n(\rho) = \frac{1}{-2j}[H_n^{(2)}(\rho) - H_n^{(1)}(\rho)] = -Im(f(\rho))$$

$$= -\frac{j^{-n-1}}{\pi} e^{-jn\phi} [\int_{-1}^{+1} -j \sin(n \sin^{-1} k_y + k_y y)$$

$$+\{\int_{-\infty}^{-1} + \int_{1}^{+\infty}\} e^{-j(n \sin^{-1} k_y + k_y y)}] \frac{e^{jk_x x}}{k_y} dk_x;$$

$$Re(k_y) > 0, Im(k_y) < 0, \; y > 0, 0 < \phi < \pi. \tag{E.5}$$

The J_n and Y_n are real functions, but H_n's are complex functions.

2. Special Integral Forms

The above expressions (E.4) and (E.5) are valid for different choices of ϕ. Consider useful special situations:

(I) $\phi = 0, y = 0, x = \rho$.

By substituting the above variables in (E.4), we have,

$$J_n(\rho) = \frac{j^{-n}}{\pi} \int_{-1}^{+1} \frac{\cos(n \sin^{-1} k_y)}{k_y} e^{jk_x\rho} dk_x \;(Re(k_y) > 0)$$

$$= \frac{j^{-n}}{\pi} \int_{0}^{\pi} \cos(n\phi') e^{j\rho\cos\phi'} d\phi' \;(k_x = \cos\phi', k_y = \sin\phi')$$

$$= \frac{j^{-n}}{2\pi} \int_{0}^{2\pi} \cos(n\phi') e^{j\rho\cos\phi'} d\phi'$$

$$= \frac{j^{-n}}{2\pi} \int_{0}^{2\pi} [\cos(n\phi') + j\sin(n\phi')] e^{j\rho\cos\phi'} d\phi'$$

$$= \frac{j^{-n}}{2\pi} \int_{0}^{2\pi} e^{j(n\phi' + \rho\cos\phi')} d\phi'. \tag{E.6}$$

The following integral relation is assumed in the above derivation, which is simply a result of the odd symmetry of the associated integrand about $\phi' = \pi$.

$$\int_{0}^{2\pi} \sin(n\phi') e^{j\rho\cos\phi'} d\phi' = 0. \tag{E.7}$$

Equivalent expressions for $J_n(\rho)$ can also be obtained by taking complex-conjugate of the expressions in (E.6).

$$J_n^*(\rho) = J_n(\rho) = \frac{(-j)^{-n}}{\pi} \int_{0}^{\pi} \cos(n\phi') e^{-j\rho\cos\phi'} d\phi'$$

$$= \frac{j^n}{\pi} \int_{0}^{\pi} \cos(n\phi') e^{-j\rho\cos\phi'} d\phi' = \frac{j^n}{2\pi} \int_{0}^{2\pi} \cos(n\phi') e^{-j\rho\cos\phi'} d\phi'$$

$$= \frac{j^n}{2\pi} \int_{0}^{2\pi} e^{j(n\phi' - \rho\cos\phi')} d\phi'. \tag{E.8}$$

Relations between $J_n(-\rho)$, $J_{-n}(\rho)$ and $J_n(\rho)$ maybe established using (E.6), (E.8).

$$J_n(-\rho) = (-1)^n J_n(\rho) = J_{-n}(\rho). \tag{E.9}$$

Now, substituting $\phi = 0$, $y = 0$, $\rho = x$ in the $Y_n(\rho)$ expression of (E.5), we have,

$$Y_n(\rho) = \frac{j^{-n-1}}{\pi}[\int_0^{\pi} j\sin(n\phi') - \{\int_{\pi}^{\pi+j\infty} + \int_{-j\infty}^{0}\}e^{-jn\phi'}]e^{j\rho\cos\phi'}d\phi'$$

$$= \frac{j^{-n}}{\pi}\int_0^{\pi}\sin(n\phi')e^{j\rho\cos\phi'}d\phi' - \frac{j^{-n}}{\pi}\{\int_{-\infty}^{0} + \int_{-\pi j}^{-\pi j+\infty}\}e^{nt}e^{j\rho\cosh t}dt \ (t = \phi'/j)$$

$$= \frac{j^{-n}}{\pi}[\int_0^{\pi}\sin(n\phi')e^{j\rho\cos\phi'}d\phi' - \int_{-\infty}^{0} e^{nt}e^{j\rho\cosh t}dt - \int_0^{+\infty} e^{nt}(-1)^n e^{-j\rho\cosh t}dt]$$

$$= \frac{j^{-n}}{\pi}[\int_0^{\pi}\sin(n\phi')e^{j\rho\cos\phi'}d\phi' - \int_0^{\infty}\{e^{-nt}e^{j\rho\cosh t} + e^{+nt}(-1)^n e^{-j\rho\cosh t}\}dt]. \qquad \text{(E.10)}$$

$$jIm(Y_n(\rho)) = 0 = \frac{j^{-n}}{\pi}[\int_0^{\pi}\sin(n\phi')e^{j\rho\cos\phi'}d\phi']$$

$$-\frac{j^{-n}}{\pi}[\int_0^{\infty}\{e^{-nt} + e^{+nt}(-1)^n\}\cos(\rho\cosh t)dt]; \ n = odd,$$

$$= \frac{j^{-n}}{\pi}[\int_0^{\pi}\sin(n\phi')e^{j\rho\cos\phi'}d\phi'] - \frac{j^{-n}}{\pi}[$$

$$j\int_0^{\infty}\{e^{-nt} + e^{+nt}(-1)^{n+1}\}\sin(\rho\cosh t)dt]; \ n = even. \qquad \text{(E.11)}$$

$$Re(Y_n(\rho)) = Y_n(\rho) = -\frac{j^{-n}}{\pi}[j\int_0^{\infty}\{e^{-nt} + e^{+nt}(-1)^{n+1}\}\sin(\rho\cosh t)dt]$$

$$= \frac{2j^{-n-1}}{\pi}\int_0^{\infty}\cosh(nt)\sin(\rho\cosh t)dt; \ n = odd,$$

$$= -\frac{j^{-n}}{\pi}[\int_0^{\infty}\{e^{-nt} + e^{+nt}(-1)^n\}\cos(\rho\cosh t)dt]$$

$$= -\frac{2j^{-n}}{\pi}\int_0^{\infty}\cosh(nt)\cos(\rho\cosh t)dt; \ n = even. \qquad \text{(E.12)}$$

Relations between $Y_n(-\rho)$, $Y_{-n}(\rho)$ and $Y_n(\rho)$ maybe established from (E.12).

$$Y_n(-\rho) = (-1)^n Y_n(\rho) = Y_{-n}(\rho). \tag{E.13}$$

Combining the above relationship for $Y_n(\rho)$, and (E.9) for $J_n(\rho)$, with (E.2,E.3), similar relations for the Hankel Functions may also be established.

$$\begin{aligned} H_n^{(1)}(-\rho) &= (-1)^n H_n^{(1)}(\rho) = H_{-n}^{(1)}(\rho), \\ H_n^{(2)}(-\rho) &= (-1)^n H_n^{(2)}(\rho) = H_{-n}^{(2)}(\rho). \end{aligned} \tag{E.14}$$

It maybe useful to express the fundamental Bessel Functions $J_0(\rho)$ and $Y_0(\rho)$ by substituting $n = 0$ in the general expressions of (E.6) and (E.12), respectively.

$$\begin{aligned} J_0(\rho) &= \frac{1}{\pi}\int_0^{\pi} e^{j\rho\cos\phi'}d\phi' = \frac{1}{2\pi}\int_0^{2\pi} e^{j\rho\cos\phi'}d\phi' \\ &= \frac{1}{\pi}\int_0^{\pi} \cos(\rho\cos\phi')d\phi' = \frac{1}{2\pi}\int_0^{2\pi} \cos(\rho\cos\phi')d\phi', \end{aligned} \tag{E.15}$$

$$Y_0(\rho) = -\frac{2}{\pi}\int_0^{\infty} \cos(\rho\cosh t)dt. \tag{E.16}$$

(II) $\phi = \frac{\pi}{2}, x = 0, y = \rho$.

By substituting the above variables in the $J_n(\rho)$ expressions of (E.4), we have,

$$\begin{aligned} J_n(\rho) &= \frac{(-1)^n}{\pi}\int_{-1}^{+1} \frac{\cos(n\sin^{-1}k_y + k_y\rho)}{k_y}dk_x \ (Re(k_y) > 0) \\ &= \frac{(-1)^n}{\pi}\int_0^{\pi} \cos(n\phi' + \rho\sin\phi')d\phi'. \end{aligned} \tag{E.17}$$

An alternative expression may also be obtained using (E.17) and (E.9).

$$J_n(\rho) = \frac{1}{\pi}\int_0^{\pi} \cos(\rho\sin\phi' - n\phi')d\phi'. \tag{E.18}$$

Similarly, with the above substitution of variables in the $Y_n(\rho)$ expression of (E.5), we have,

$$Y_n(\rho) = \frac{(-1)^n}{\pi}\int_0^{\pi}\sin(n\phi' + \rho\sin\phi')d\phi' - \frac{(-1)^n}{\pi j}\{\int_{\pi}^{\pi+j\infty} + \int_{-j\infty}^{0}\}e^{-jn\phi'}e^{-j\rho\sin\phi'}d\phi'$$

$$= \frac{(-1)^n}{\pi}\int_0^{\pi}\sin(n\phi' + \rho\sin\phi')d\phi' - \frac{(-1)^n}{\pi}\{\int_{-\infty}^{0} + \int_{-j\pi}^{-j\pi+\infty}\}e^{nt}e^{\rho\sinh t}dt \ (\phi' = jt)$$

$$= \frac{(-1)^n}{\pi}\int_0^{\pi}\sin(n\phi' + \rho\sin\phi')d\phi' - \frac{(-1)^n}{\pi}[\int_{-\infty}^{0} e^{nt}e^{\rho\sinh t}dt + \int_0^{\infty} e^{nt}(-1)^n e^{-\rho\sinh t}dt]$$

$$= \frac{(-1)^n}{\pi}[\int_0^{\pi}\sin(n\phi' + \rho\sin\phi')d\phi' - \int_0^{\infty} e^{-nt}e^{-\rho\sinh t}dt - \int_0^{\infty} e^{nt}(-1)^n e^{-\rho\sinh t}dt]$$

$$= \frac{(-1)^n}{\pi}[\int_0^{\pi}\sin(n\phi' + \rho\sin\phi')d\phi' - \int_0^{\infty}(e^{-nt} + e^{nt}(-1)^n)e^{-\rho\sinh t}dt]. \quad (E.19)$$

An alternative expression may also be obtained using (E.19) and (E.13).

$$Y_n(\rho) = \frac{1}{\pi}[\int_0^{\pi}\sin(\rho\sin\phi' - n\phi')d\phi' - \int_0^{\infty}(e^{nt} + e^{-nt}(-1)^n)e^{-\rho\sinh t}dt]. \quad (E.20)$$

The fundamental Bessel Functions $J_0(\rho)$ and $Y_0(\rho)$ maybe expressed by substituting $n = 0$ in the general expressions of (E.18) and (E.20), respectively.

$$J_0(\rho) = \frac{1}{\pi}\int_0^{\pi}\cos(\rho\sin\phi')d\phi', \quad (E.21)$$

$$Y_0(\rho) = \frac{1}{\pi}\int_0^{\pi}\sin(\rho\sin\phi')d\phi' - \frac{2}{\pi}\int_0^{\infty}e^{-\rho\sinh t}dt. \quad (E.22)$$

3. Bessel Differential Equation

$$(\rho^2\frac{\partial^2}{\partial\rho^2} + \rho\frac{\partial}{\partial\rho} + \rho^2 - \frac{\partial^2}{\partial\phi^2})B(\rho,\phi) = 0. \quad (E.23)$$

The validity of the above differential equation may first be checked for the space dependent part $e^{jk_x x}e^{\mp jk_y y} = e^{j\rho\cos(\phi'\pm\phi)}$ of (E.1), where $k_x = \cos\phi'$, $k_y = \sin\phi'$, $x = \rho\cos\phi$, $y = \rho\sin\phi$. Then, (E.23) should be satisfied for any function that can be expressed in a general integral from as (E.1), which contains the integrand factor $e^{j\rho\cos(\phi'\pm\phi)} = e^{jk_x x}e^{\mp jk_y y}$. For the specific case in (E.1), where $B(\rho,\phi) = f(\rho)e^{jn\phi}$, (E.23) may be simplified with $\frac{\partial^2}{\partial\phi^2} = -n^2$.

$$(\rho^2\frac{\partial^2}{\partial\rho^2} + \rho\frac{\partial}{\partial\rho} + \rho^2 - n^2)f(\rho) = 0, \tag{E.24}$$

where $f(\rho) = B_n(\rho)$ is any of the specific Bessel Functions $J_n(\rho)$, $Y_n(\rho)$, $H_n^{(2)}(\rho)$ or $H_n^{(1)}(\rho)$. (E.24) may be solved for B_n by assuming a series expansion of B_n with unknown coefficients, and then matching coefficients by applying (E.24). Solutions for the lowest orders $n = 0$ and $n = 1$ maybe used to obtain solutions for higher orders $n > 1$ by using suitable iterative relations. For reference, plots of the Bessel Functions $J_n(\rho)$ and $Y_n(\rho)$, for a limited range of variable $0 < \rho < 12$ and order numbers $n \leq 3$, are shown in Figs.1,2.

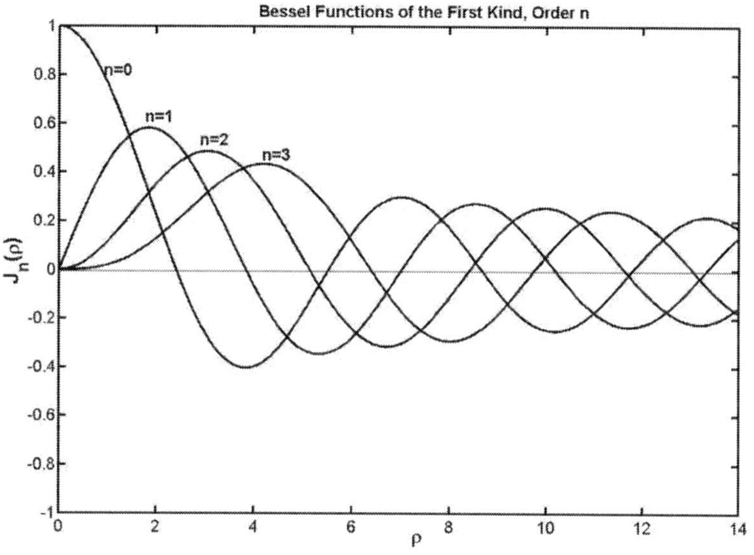

FIGURE 1

4. Useful Differential and Iterative Relations

$$\begin{aligned} B_{n-1}(\rho) + B_{n+1}(\rho) &= \frac{2n}{\rho}B_n(\rho), \\ B_{n-1}(\rho) - B_{n+1}(\rho) &= 2B_n'(\rho); \\ B_n &= J_n, Y_n, H_n^{(2)}, H_n^{(1)}. \end{aligned} \tag{E.25}$$

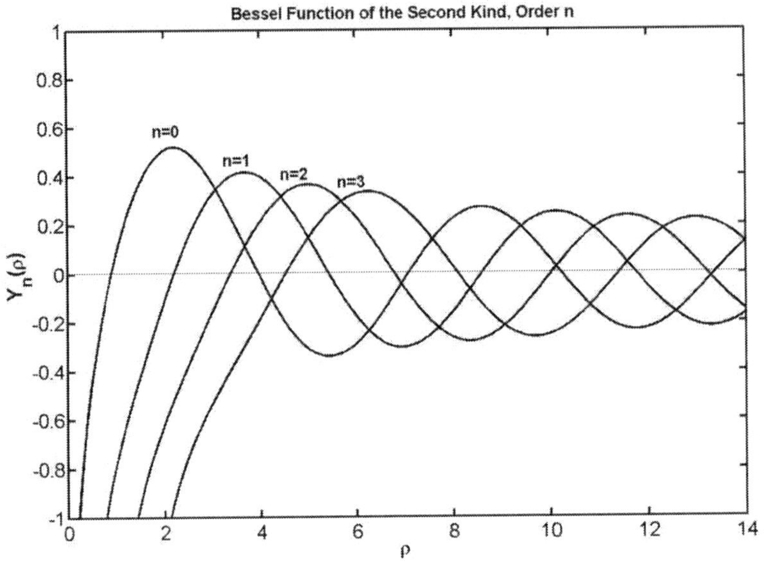

FIGURE 2

(E.25) maybe proved for the Bessel Functions $J_n(\rho)$ and $Y_n(\rho)$ by substitution of the respective expressions (E.18) and (E.20) and suitable simplification. Proof of (E.25) for $J_n(\rho)$ and $Y_n(\rho)$ may then be extended for $H_n^{(2)}(\rho)$ and $H_n^{(1)}(\rho)$ by superposition.

(E.25) is useful to iteratively calculate B_{n+1} from B_n and B_{n-1}, starting from B_1 and B_0.

Two useful expressions for the derivative $B_n'(\rho)$ may be obtained from (E.25).

$$B_n'(\rho) = B_{n-1}(\rho) - \frac{n}{\rho}B_n(\rho) \tag{E.26}$$

$$= \frac{n}{\rho}B_n(\rho) - B_{n+1}(\rho). \tag{E.27}$$

(E.27) may also be expressed in another useful form.

$$-B_{n+1}(\rho) = -\frac{n}{\rho}B_n(\rho) + B_n'(\rho),$$

$$-\frac{B_{n+1}(\rho)}{\rho^n} = -\frac{n}{\rho^{n+1}}B_n(\rho) + \frac{B_n'(\rho)}{\rho^n} = \frac{d}{d\rho}(\frac{B_n(\rho)}{\rho^n}),$$

$$-\frac{B_n(\rho)}{\rho^{n-1}} = \frac{d}{d\rho}(\frac{B_{n-1}(\rho)}{\rho^{n-1}}). \tag{E.28}$$

5. Approximation for Small Arguments

By finding the lth derivative of the $J_n(\rho)$ expression in (E.6) with respect to ρ, and then taking the limit as $\rho \to 0$, it can be shown,

$$\left.\frac{d^l J_n(\rho)}{d\rho^l}\right|_{\rho=0} = \frac{j^{l-n}}{\pi} \int_0^{\pi} \cos(n\phi') \cos^l(\phi') d\phi'$$
$$= 0,\ l < n; \ = \frac{1}{2^n},\ l = n. \tag{E.29}$$

Expanding $J_n(\rho)$ by a Taylor-series in ρ, and then using (E.29), the following limiting expression can be established.

$$J_n(\rho \to 0) \simeq \frac{1}{n!}(\frac{\rho}{2})^n. \tag{E.30}$$

Taking the limit of the $Y_n(\rho)$ expression in (E.20) for $n > 0$, as $\rho \to 0$, it maybe shown,

$$Y_n(\rho \to 0, n > 0) \simeq -\frac{1}{\pi} \int_0^{\infty} e^{nt} e^{-\frac{\rho e^t}{2}} dt = -\frac{1}{\pi} \int_{\frac{1}{2}}^{\infty} t'^{(n-1)} 2^n e^{-\rho t'} dt';$$
$$t' = \frac{e^t}{2},\ dt' = e^t \frac{dt}{2} = t' dt. \tag{E.31}$$

The above expression can be integrated in parts,

$$Y_n(\rho \to 0, n > 0) \simeq -\frac{1}{\pi}[2^n t'^{(n-1)} \frac{e^{-\rho t'}}{-\rho} \Big|_{\frac{1}{2}}^{\infty} + \int_0^{\infty} \frac{e^{-\rho t'}}{\rho} 2^n (n-1) t'^{(n-2)} dt']. \tag{E.32}$$

The first part in the above expression is proportional to $1/\rho$. Whereas, the second part can be seen to be significantly larger in magnitude than the first part, with higher powers of ρ in the denominator. Accordingly, for approximations as ρ approaches zero, we need to keep only the second part. This second part can be similarly integrated in parts again, and the process successively repeated, with the following result.

$$Y_n(\rho \to 0, n > 0) \simeq -\frac{1}{\pi} \int_0^{\infty} \frac{e^{-\rho t'}}{\rho^{n-1}} 2^n (n-1)! dt'$$
$$= -\frac{1}{\pi} \frac{(n-1)!}{\rho^{n-1}} 2^n \frac{e^{-\rho t'}}{-\rho} \Big|_0^{\infty} = -\frac{2^n}{\pi} \frac{(n-1)!}{\rho^n}. \tag{E.33}$$

The above limiting expresison for $Y_n(\rho \to 0)$ is valid for $n > 0$. The limiting expression for $n = 0$ can be derived from that for $n = 1$, by using the differential relationship (E.27).

$$\frac{dY_0(\rho \to 0)}{d\rho} = -Y_1(\rho \to 0) \simeq \frac{2}{\pi\rho},$$
$$Y_0(\rho \to 0) \simeq \frac{2}{\pi} \ln \rho. \tag{E.34}$$

6. Approximation for Large Arguments

As ρ approaches ∞ for any given ϕ, or equivalently as x, y approach $\pm\infty$, the integrand for $H_n^{(2)}(\rho)$ in (E.1,E.2) may need to evaluated only over the domain $-1 < k_x < +1$. Outside this domain, that is over $-\infty < k_x < -1$ and $+1 < k_x < +\infty$, the paramter k_y is selected to be negatively imaginary. This results in the integrand paramter $e^{\mp jk_y y}$, and therefore the associated integration, to approach zero as y approaches $\pm\infty$. Specifically consider the case in (E.1,E.2) with $0 < \phi < \pi,\ \rho \to \infty$, or equivalently $y \to +\infty$.

$$H_n^{(2)}(\rho \to \infty) e^{jn\phi} \simeq \frac{j^{-n}}{\pi} \int_{-1}^{+1} \frac{e^{-jn\phi'}}{k_y} e^{j\rho\cos(\phi'+\phi)} dk_x \ (0 < \phi < \pi,\ k_x = \cos\phi')$$

$$= \frac{j^{-n}}{\pi} \int_0^{\pi} e^{-jn\phi'} e^{j\rho\cos(\phi'+\phi)} d\phi' = \frac{j^{-n}}{\pi} e^{jn\phi} \int_{t=\phi}^{\pi+\phi} e^{-jnt} e^{j\rho\cos t} dt \ (t = \phi' + \phi). \tag{E.35}$$

Now, the above integration can be evaluated using the method of stationary-phase approximation.

$$H_n^{(2)}(\rho \to \infty) e^{jn\phi} \simeq \frac{j^{-n}}{\pi} e^{jn\phi} e^{-j\pi n} e^{-j\rho} \sqrt{\frac{2\pi}{-j\rho}},$$
$$H_n^{(2)}(\rho \to \infty) \simeq \sqrt{\frac{2}{\pi\rho}} j^{n+\frac{1}{2}} e^{-j\rho} = \sqrt{\frac{2}{\pi\rho}} e^{-j(\rho - \frac{n\pi}{2} - \frac{\pi}{4})},$$
$$H_n^{(1)}(\rho \to \infty) = H_n^{*(2)}(\rho \to \infty) \simeq \sqrt{\frac{2}{\pi\rho}} e^{+j(\rho - \frac{n\pi}{2} - \frac{\pi}{4})},$$
$$J_n(\rho \to \infty) \simeq \sqrt{\frac{2}{\pi\rho}} \cos(\rho - \frac{n\pi}{2} - \frac{\pi}{4}),$$
$$Y_n(\rho \to \infty) \simeq \sqrt{\frac{2}{\pi\rho}} \sin(\rho - \frac{n\pi}{2} - \frac{\pi}{4}). \tag{E.36}$$

7. $J_n(\rho)$ and $Y_n(\rho)$ Relationship

From (E.1)

$$H_0^{(2)}(\rho) = \frac{1}{\pi}\int_{-\infty}^{+\infty} \frac{1}{k_y} e^{+jk_x x} e^{\mp jk_y y} dk_x;$$

$$(-)\ y > 0,\ (+)\ y < 0;\ k_y = \sqrt{1-k_x^2},\ Re(k_y) > 0,\ Im(k_y) < 0. \qquad \text{(E.37)}$$

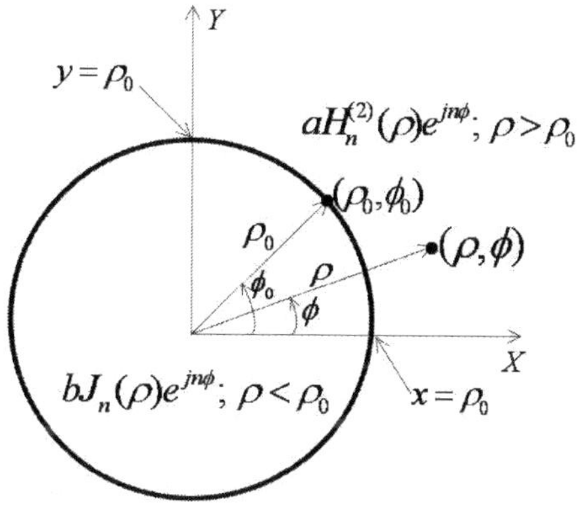

FIGURE 3

One may synthesize $H_n^{(2)}(\rho)$ by superposition of $H_0^{(2)}(x - x_0, y - y_0)$ (see Fig.3). As may be seen from symmetry and by enforcing continuity of $H_0^{(2)}(\rho)$:

$$\begin{gathered} f(\rho,\phi) = \tfrac{1}{2\pi}\int_0^{2\pi} H_0^{(2)}(x - x_0,\ y - y_0) e^{jn\phi_0} d\phi_0 \\ = aH_n^{(2)}(\rho)e^{jn\phi}, \rho > \rho_0;\ = bJ_n(\rho)e^{jn\phi}, \rho < \rho_0,\ ; \\ aH_n^{(2)}(\rho_0) = bJ_n(\rho_0), \\ x_0 = \rho_0\cos\phi_0, y_0 = \rho_0\sin\phi_0, \\ H_0^{(2)}(x,y) \triangleq H_0^{(2)}(\rho = \sqrt{x^2+y^2}). \end{gathered} \qquad \text{(E.38)}$$

From (E.37),

$$\frac{1}{2\pi}\int_{0}^{2\pi} H_0^{(2)}(x-x_0,\ y-y_0)e^{jn\phi_0}d\phi_0$$

$$= \frac{1}{2\pi}\int_{0}^{2\pi}\frac{1}{\pi}\int_{-\infty}^{+\infty}\frac{1}{k_y}e^{+jk_x(x-x_0)}e^{-jk_y(y-y_0)}e^{jn\phi_0}d\phi_0\ (y>\rho_0)$$

$$= \frac{1}{\pi}\int_{-\infty}^{+\infty}\frac{1}{k_y}e^{+jk_xx}e^{-jk_yy}(\frac{1}{2\pi}\int_{0}^{2\pi}e^{-jk_xx_0}e^{+jk_yy_0}e^{jn\phi_0}d\phi_0)dk_x$$

$$= \frac{1}{\pi}\int_{-\infty}^{+\infty}\frac{1}{k_y}e^{+jk_xx}e^{-jk_yy}(\frac{1}{2\pi}\int_{0}^{2\pi}e^{-j\rho_0\cos(\phi'+\phi_0)}e^{jn\phi_0}d\phi_0)dk_x\ (k_x=\cos\phi',\ k_y=\sin\phi')$$

$$= \frac{1}{\pi}\int_{-\infty}^{+\infty}\frac{1}{k_y}e^{+jk_xx}e^{-jk_yy}e^{-jn\phi'}(\frac{1}{2\pi}\int_{\substack{\Delta t=2\pi\\ t=\phi'+\phi}}e^{jnt}e^{-j\rho_0\cos t}dt)dk_x\ (t=\phi'+\phi)$$

$$= \frac{1}{\pi}\int_{-\infty}^{+\infty}\frac{e^{-jn\phi'}}{k_y}e^{+jk_xx}e^{-jk_yy}j^{-n}J_n(\rho_0)dk_x = J_n(\rho_0)H_n^{(2)}(\rho)e^{jn\phi},\ y>\rho_0. \quad \text{(E.39)}$$

Comparing (E.38), (E.39) for $y=\rho_{0+}$,

$$a = J_n(\rho_0),\ b = H_n^{(2)}(\rho_0). \quad \text{(E.40)}$$

Examine the following operation on the right-hand side of (E.38).

$$\frac{\partial f}{\partial\rho}\Big|_{\rho=\rho_{0+}} - \frac{\partial f}{\partial\rho}\Big|_{\rho=\rho_{0-}} = [J_n(\rho_0)H_n'^{(2)}(\rho_0) - H_n^{(2)}(\rho_0)J_n'(\rho_0)]e^{jn\phi}$$
$$= [J_n(\rho_0)J_n'(\rho_0) - jJ_n(\rho_0)Y_n'(\rho_0) - J_n(\rho_0)J_n'(\rho_0) + jY_n(\rho_0)J_n'(\rho_0)]e^{jn\phi}$$
$$= -j[-Y_n(\rho_0)J_n'(\rho_0) + J_n(\rho_0)Y_n'(\rho_0)]e^{jn\phi}. \quad \text{(E.41)}$$

Now attempt to perform the same operation on the left-hand side of (E.38). From (E.30,E.34) and (E.2), with $\rho\to 0$,

$$
\begin{aligned}
H_0^{(2)}(\rho) &\simeq -jY_0(\rho) \sim -j\frac{2}{\pi}ln(\rho), \\
\frac{\partial H_0^{(2)}(\rho)}{\partial x} &\sim -\frac{2j}{\pi\rho}\frac{\partial\rho}{\partial x} = -\frac{2j}{\pi\rho}\frac{x}{\rho} = -\frac{2j}{\pi}\frac{x}{x^2+y^2} \\
&= \mp 2j\delta(y),\ x = 0_\pm\ ; \\
\frac{1}{\pi}\frac{x}{x^2+y^2} &= 0,\ y \neq 0,\ x \to 0;\ \ \neq 0,\ y = 0, x \to 0,
\end{aligned}
$$

$$
\begin{aligned}
\frac{1}{\pi}\int\limits_{-\infty}^{+\infty}\frac{x^2}{x^2+y^2}dy = \frac{2}{\pi}\int\limits_{0}^{\infty}\frac{x}{x^2+y^2}dy = \frac{2}{\pi}\int\limits_{t=0}^{\pm\frac{\pi}{2}}\frac{x\times x\sec^2 t}{x^2\sec^2 t}dt \\
= +1,\ x > 0;\ \ = -1.\ x < 0.
\end{aligned}
$$

In general,

$$
\frac{\partial H_0^{(2)}(\rho_0)}{\partial x'} = \mp 2j\delta(y'),\ x' = 0_\pm\ , \tag{E.42}
$$

where (x', y') are any orthogonal axes, rotated about (x,y).

$$
\begin{aligned}
\frac{\partial H_0^{(2)}(x-x_0,\ y-y_0)}{d\rho} &= \mp 2j\delta[\rho_0(\phi-\phi_0)] = \mp 2j\frac{1}{\rho_0}\delta(\phi-\phi_0), \\
&\rho = \rho_{0+},\ x' = \rho-\rho_0;\ y' = \rho_0(\phi-\phi_0), \\
\frac{\partial H_0^{(2)}(x-x_0,\ y-y_0)}{d\rho}\Big|_{\rho=\rho_{0+}} &- \frac{\partial H_0^{(2)}(x-x_0,\ y-y_0)}{d\rho}\Big|_{\rho=\rho_{0-}} \\
&= -4j\frac{1}{\rho_0}\delta(\phi-\phi_0).
\end{aligned} \tag{E.43}
$$

Use (E.41,E.43) in (E.38).

$$
\begin{aligned}
\frac{1}{2\pi}\int\limits_{0}^{2\pi} -4j\frac{1}{\rho_0}\delta(\phi-\phi_0)e^{jn\phi_0}d\phi_0 = \frac{-2j}{\pi\rho_0}e^{jn\phi} \\
= -j[J_n(\rho_0)Y_n'(\rho_0) - J_n'(\rho_0)Y_n(\rho_0)]e^{jn\phi} \\
J_n(\rho_0)Y_n'(\rho_0) - J_n'(\rho_0)Y_n(\rho_0) = \frac{2}{\pi\rho_0}.
\end{aligned}
$$

Because ρ_0 is arbitary, the above equation should be valid for any $\rho = \rho_0$

$$J_n(\rho)Y_n'(\rho) - J_n'(\rho)Y_n(\rho) = \frac{2}{\pi\rho}. \tag{E.44}$$

APPENDIX F

Method of Stationary-Phase Integration

Fourier-type integrals involving wave functions, when evaluated at sufficiently large distances, can be approximated using a special technique. Such an integral, evaluated over a definite range of variable(s), is mostly contributed from integration around specific points where the derivative of the phase of thc integrand is zero. These points are referred to as the "stationary-phase" points. Accordingly, the total integral is approximated by evaluating the integration around the specific points using analytical formulas.

We will first establish analytical formulas for certain standard integrals. These formulas would then be used in the analytical evaluation of the integrals around the stationary-phase points. The procedure is demonstrated for selected functions in one and two dimensions,

1. Useful Integrals

$$\int_{-\infty}^{+\infty} e^{-x^2} dx = \int_{-\infty}^{+\infty} e^{-y^2} dy = A,$$

$$A^2 = \int_{-\infty}^{+\infty} e^{-x^2} dx \int_{-\infty}^{+\infty} e^{-y^2} dy = \int_{-\infty}^{+\infty}\int_{-\infty}^{+\infty} e^{-(x^2+y^2)} dxdy$$

$$= \int_{\phi=0}^{2\pi}\int_{\rho=0}^{\infty} e^{-\rho^2} \rho d\rho d\phi = 2\pi \int_{0}^{\infty} e^{-\rho^2} \rho d\rho$$

$$= \pi \int_{0}^{\infty} e^{-t} dt = \pi, \ (\rho^2 = x^2 + y^2 = t; 2\rho d\rho = dt),$$

$$A = \sqrt{\pi}. \tag{F.1}$$

The above integrand is a real exponential function. Similar procedure may be extended to exponential funcitons with complex arguments, by suitable substitution and deformation of the integration contour in the complex plane (see Fig.1).

$$\int_{-\infty}^{+\infty} e^{-j\theta^2} d\theta = \int_{-\sqrt{j}\infty, C_2}^{+\sqrt{j}\infty} e^{-t^2} \frac{dt}{\sqrt{j}} = \int_{-\infty, C_1}^{+\infty} e^{-t^2} \frac{dt}{\sqrt{j}} = \sqrt{\frac{\pi}{j}}$$
$$(j\theta^2 = t^2,\ t = \theta\sqrt{j},\ dt = d\theta\sqrt{j}), \tag{F.2}$$

$$\int_{-\infty}^{+\infty} e^{-j\alpha\theta^2} d\theta = \int_{-\infty}^{+\infty} e^{-t^2} \frac{dt}{\sqrt{j\alpha}} = \sqrt{\frac{\pi}{j\alpha}}. \tag{F.3}$$

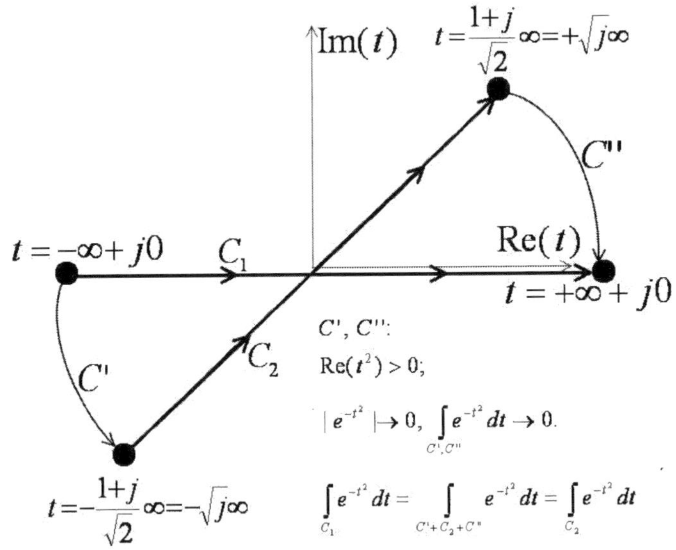

FIGURE 1

2. Integrals with Stationary-Phase Points

We will first introduce the concept of approximating a highly oscillating integral, based on its stationary-phase behavior. The basic concept is established for a one-dimensional integral, with a simple phase function having one stationary-phase point. This basic form can then be extended to more general phase functions for integration in one and two dimensions, with any number of stationary-phase points.

2.1. Basic Integral Form in One Dimension. It would be useful first to examine the characteristic of the function $e^{-j\alpha\theta^2}$, when α is a large real positive number. Fig.2 plots the function for $\alpha = 20\pi$, showing rapid oscillatory behavior with variation of θ, except in the vicinity of $\theta = 0$. As *alpha* would increase, the oscillation away from $\theta = 0$ would increase indefinitely, while maintaining minimal oscillation near $\theta = 0$. This above function may be compared with $e^-j\alpha\theta$, which is also an oscillating funciton, with oscillation frequency increasing indefinitely as α increases. However, unlike $e^{-j\alpha\theta^2}$, the funciton $e^-j\alpha\theta$ would oscillate equally rapidly for all values of θ. The phase function of $e^{-j\alpha\theta^2}$ is referred to as "stationary" at $\theta = 0$, which means the derivative of the phase function is zero at $\theta = 0$. In contrast, the phase of $e^-j\alpha\theta$ is non-stationary at all locations, because the derivative of the phase in this case is non-zero.

Now, consider a new function which is a product of $e^{-j\alpha\theta^2}$ and another "smooth" function $f(\theta)$ with no singular behavior. Like $e^{-j\alpha\theta^2}$, the new function would also be highly osillatory when α is large, except near $\theta = 0$. Accordingly, for a large value of α, the integration of the new product function, over a region that includes the stationary point, would be dominantly contributed from the immediate vicinity of the stationary-phse point at $theta = 0$. Integrations from regions away from the stationary-phase point would be negligible due the strong osillations, because the equally positive and negative regions of oscillation would cancel each other. Therefore, the total integration would essentially remain unchanged if the function $f(\theta)$ is everywhere replaced by its value at the stationary-phase point $\theta = 0$. As a result, the integral of the product function would be approximately equal to the product of $f(\theta)$, evaluated at the stationary-phase point $\theta = 0$, and the integral of $e^{-j\alpha\theta^2}$ as developed in (F.3). Note that the limits of integration would not be critical, as long as the integration range cover the stationary point. Accordingly, the limits may be selected to be from $-\infty$ to $+\infty$ as in (F.3), for analytical simplicity.

$$\lim_{\alpha \to \infty} \int_{-\infty}^{+\infty} f(\theta) e^{-j\alpha\theta^2} d\theta \simeq \int_{-\infty}^{+\infty} f(0) e^{-j\alpha\theta^2} d\theta$$
$$= f(0) \int_{-\infty}^{+\infty} e^{-j\alpha\theta^2} d\theta = f(0)\sqrt{\frac{\pi}{j\alpha}}. \tag{F.4}$$

In contrast to the above case with a stationary-phase point, the integral of another product function $f(\theta)e^{-j\alpha\theta}$, which has no stationary-phase point, would be negligible for a large value of α. Without any stationary-phase point, the integral would oscillate rapidly at all locations, leading to the negligible total integration due to cancellation of the negative and positive parts of oscillation.

2.2. General Integral Form in One Dimension. The above concept may now be extended for a general phase function $g(\theta)$. The stationary point for the phase function can be located by equating the derivative of $g(\theta)$ to be zero. The phase function can be approximated around the stationary point using a Taylor series. The integration method developed above can then be applied. In the following derivation, we would assume that there is only one stationary-phase point. In case there are multiple stationary-phase points, the process needs to be repeated for integration around each point, with individual results similarly expressed by substituting for the respective parameters. The total integral would simply be the sum of the individual results.

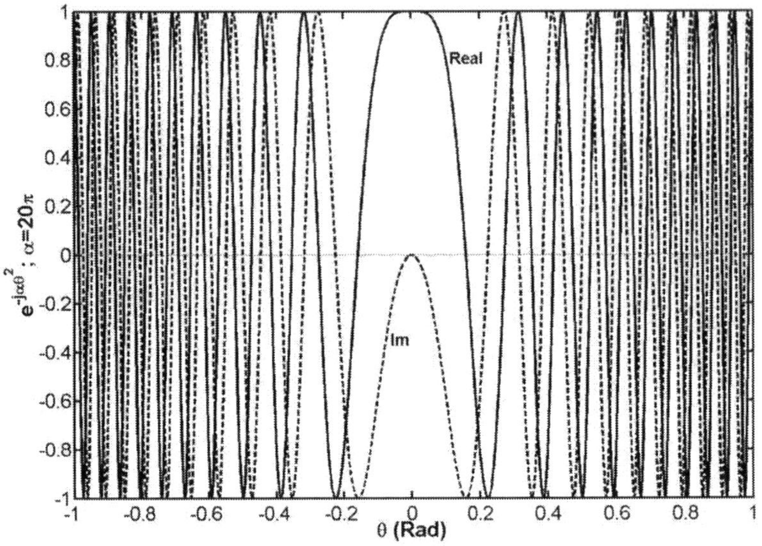

FIGURE 2

$$\lim_{\alpha\to\infty}\int_{-\infty}^{+\infty} f(\theta)e^{-jg(\theta)\alpha}d\theta \simeq \lim_{\alpha\to\infty}\int_{-\infty}^{+\infty} f(\theta)e^{-j\alpha[g(\theta_0)+(\theta-\theta_0)g'(\theta_0)+\frac{(\theta-\theta_0)^2}{2}g''(\theta_0)]}d\theta$$

$$= \lim_{\alpha\to\infty} e^{-j\alpha g(\theta_0)}\int_{-\infty}^{+\infty} f(\theta)e^{-j\alpha g''(\theta_0)\frac{(\theta-\theta_0)^2}{2}}d\theta \;(g'(\theta_0)=0)$$

$$= e^{-j\alpha g(\theta_0)}\int_{-\infty}^{+\infty} f(\theta_0)e^{-j\alpha g''(\theta_0)\frac{(\theta-\theta_0)^2}{2}}d\theta = e^{-j\alpha g(\theta_0)}f(\theta_0)\int_{-\infty}^{+\infty} e^{-j\alpha g''(\theta_0)\frac{(\theta-\theta_0)^2}{2}}d\theta$$

$$= e^{-j\alpha g(\theta_0)}f(\theta_0)\int_{-\sqrt{j}\infty}^{+\sqrt{j}\infty} e^{-t^2}\frac{dt}{\sqrt{\frac{j\alpha g''(\theta_0)}{2}}} = e^{-j\alpha g(\theta_0)}f(\theta_0)\sqrt{\frac{2}{j\alpha g''(\theta_0)}}\int_{-\infty}^{+\infty} e^{-t^2}dt$$

$$= e^{-j\alpha g(\theta_0)}f(\theta_0)\sqrt{\frac{2\pi}{j\alpha g''(\theta_0)}}$$

$$(j\alpha g''(\theta_0)\frac{(\theta-\theta_0)^2}{2} = t^2,\; t=\sqrt{\frac{j\alpha g''(\theta_0)}{2}}(\theta-\theta_0),\; dt=\sqrt{\frac{j\alpha g''(\theta_0)}{2}}d\theta). \quad \text{(F.5)}$$

2.3. General Integral Form in Two Dimensions. The above procedure for one-dimensional integration may be extended to an integration in two dimensions.

$$\begin{aligned}
&\lim_{\alpha \to 0} \int_{-\infty}^{+\infty}\int_{-\infty}^{+\infty} f(\theta,\phi)e^{-j\alpha g(\theta,\phi)} d\theta d\phi \\
&\simeq \lim_{\alpha\to\infty} \int_{-\infty}^{+\infty}\int_{-\infty}^{+\infty} f(\theta,\phi)e^{-j\alpha[g(\theta_0,\phi_0)+(\theta-\theta_0)\frac{\partial g(\theta_0,\phi_0)}{\partial\theta}+(\phi-\phi_0)\frac{\partial g(\theta_0,\phi_0)}{\partial\phi}]} \\
&\times e^{-j\alpha[(\theta-\theta_0)(\phi-\phi_0)\frac{\partial^2 g(\theta_0,\phi_0)}{\partial\theta\partial\phi}+\frac{(\theta-\theta_0)^2}{2}\frac{\partial^2 g(\theta_0,\phi_0)}{\partial\theta^2}+\frac{(\phi-\phi_0)^2}{2}\frac{\partial^2 g(\theta_0,\phi_0)}{\partial\phi^2}.....]} d\theta d\phi \\
&\simeq \int_{-\infty}^{+\infty}\int_{-\infty}^{+\infty} f(\theta_0,\phi_0)e^{-j\alpha g(\theta_0,\phi_0)} e^{-j\alpha\frac{(\theta-\theta_0)^2}{2}\frac{\partial^2 g(\theta_0,\phi_0)}{\partial\theta^2}} e^{-j\alpha\frac{(\phi-\phi_0)^2}{2}\frac{\partial^2 g(\theta_0,\phi_0)}{\partial\phi^2}} d\theta d\phi \\
&(\frac{\partial^2 g}{\partial\theta\partial\phi} = \frac{\partial g}{\partial\theta} = \frac{\partial g}{\partial\phi} = 0;\ \theta=\theta_0, \phi=\phi_0) \\
&= f(\theta_0,\phi_0)e^{-j\alpha g(\theta_0,\phi_0)} \sqrt{\frac{2\pi}{j\alpha\frac{\partial^2 g(\theta_0,\phi_0)}{\partial\theta^2}}}\sqrt{\frac{2\pi}{j\alpha\frac{\partial^2 g(\theta_0,\phi_0)}{\partial\phi^2}}} \\
&= f(\theta_0,\phi_0)e^{-hj\alpha g(\theta_0,\phi_0)} \frac{2\pi}{j\alpha}\sqrt{\frac{1}{\frac{\partial^2 g(\theta_0,\phi_0)}{\partial\theta^2}\frac{\partial^2 g(\theta_0,\phi_0)}{\partial\phi^2}}}. \qquad \text{(F.6)}
\end{aligned}$$

3. Examples

3.1. TM_z Potential for a z-directed, Uniform Line Current ($\rho \to \infty$). The TM_z potential of a uniform line current along the z axis is expressed in (6.40). We will evaluate the potential function for a sufficiently large distance $\rho \to \infty$, using the stationary-phase method of integration we developed above.

$$\psi_{zm} = \frac{1}{4\pi j}\int\limits_{-\infty}^{\infty} \frac{1}{k_y} e^{jk_x x} e^{-jk_y y} dk_x$$

$$(k_y = \sqrt{k^2 - k_x^2},\ Im(k_y) < 0,\ Re(k_y) > 0,\ y > 0)$$

$$\simeq \frac{1}{4\pi j}\int\limits_{-k}^{k} \frac{1}{k_y} e^{j(k_x x - k_y y)} dk_x = \frac{1}{4\pi j}\int\limits_{-k}^{k} \frac{1}{k_y} e^{j[k\cos\phi'\rho\cos\phi - k\sin\phi'\rho\sin\phi]} dk_x$$

$$(k_x = k\cos\phi',\ k_y = k\sin\phi',\ x = \rho\cos\phi,\ y = \rho\sin\phi)$$

$$= \frac{1}{4\pi j}\int\limits_{\phi'=0}^{\pi} \frac{1}{k\sin\phi'} e^{jk\rho\cos(\phi+\phi')} k\sin\phi' d\phi'\ (0 < \phi < \pi)$$

$$= \frac{1}{4\pi j}\int\limits_{t=\phi}^{\phi+\pi} e^{jk\rho\cos t} dt\ (t = \phi + \phi'). \tag{F.7}$$

Now, we use (F.5) with $f(\theta) = f(t) = \frac{1}{4\pi j}$, $g(\theta) = g(t) = -\cos t, \alpha = \rho k$, $g'(t) = \sin t = 0,\ t_0 = \pi,\ \phi < t < \phi + \pi)$, and $g''(t_0) = \cos t_0 = -1,\ g(t_0) = 1$.

$$\psi_{zm}(\rho \to \infty) \simeq \frac{e^{-jk\rho}}{4\pi j}\sqrt{\frac{2\pi}{-jk\rho}} = \frac{1}{4j}\sqrt{\frac{2j}{\pi k\rho}} e^{-jk\rho}. \tag{F.8}$$

The above result may be compared with the rigorous expression (6.48), with approximation used for $H_0^{(2)}(k\rho)$ when $(\rho \to \infty)$ from Appendix E.

$$\psi_{zm} = \frac{1}{4j} H_0^{(2)}(k\rho) \simeq \frac{1}{4j}\sqrt{\frac{2j}{\pi k\rho}} e^{-jk\rho}. \tag{F.9}$$

Approximate expressions for different field components can be derived from the potential expression of (F.9) using field-potential relations (6.49). For example, the z-directed electric field E_z may be expressed as follows.

$$E_z(\rho \to \infty) = \frac{\omega\mu}{j}\psi_{zm}(\rho \to \infty) = -\frac{\omega\mu}{4}\sqrt{\frac{2j}{\pi k\rho}} e^{-jk\rho}. \tag{F.10}$$

3.2. E_z of a Small Dipole $(Il\hat{z})$, $(r \to \infty)$. From (6.30), and applying the field-potential relation (5.34), we have

$$E_z(x,y,z) = -\frac{1}{4\pi^2}\left(\frac{Il}{j\omega\epsilon}\right)\int_{-\infty}^{\infty}\int_{-\infty}^{\infty}\frac{(k^2-k_z^2)}{2jk_y}e^{jk_x x}e^{jk_z z}e^{-jk_y y}dk_x dk_z$$

$$(y>0;\ 0<\theta<\pi,\ 0<\phi<\pi)$$

$$\simeq \frac{-Il}{8\pi^2\omega\epsilon}\int\int_{(k_x^2+k_z^2<k^2)}\frac{(k^2\sin^2\theta')}{k\sin\theta' sin\phi'}e^{jkr(\sin\theta'\cos\phi'\sin\theta\cos\phi+\cos\theta'\cos\theta)}$$

$$\times e^{jkr(-\sin\theta'\sin\phi'\sin\theta\sin\phi)}k^2\sin^2\theta'\sin\phi' d\theta' d\phi'$$

$$= \frac{-Il}{8\pi^2\omega\epsilon}k^3\int_{\phi'=0}^{\pi}\int_{\theta'=0}^{\pi}\sin^3\theta' e^{jkr(\sin\theta'\sin\theta\cos(\phi+\phi')+\cos\theta'\cos\theta)}d\theta' d\phi'$$

$$(k_x = k\sin\theta'\cos\phi',\ k_z = k\cos\theta',\ k_y = \sqrt{k^2-k_x^2-k_z^2}$$
$$= k\sin\theta'\sin\phi';\ Re(k_y)>0,\ Im(k_y)<0). \tag{F.11}$$

We may use the results of section 2.3, with the following substitution. Note that the integration variables in (F.11) are θ' and ϕ', with θ and ϕ used as specific parameters, whereas the integration variables in section 2.3 are θ and ϕ.

$$f(\theta',\phi') = -\frac{Il}{8\pi^2\omega\epsilon}\sin^3\theta',\ \alpha = kr,$$
$$g(\theta'+\phi') = -[\sin\theta'\sin\theta\cos(\phi+\phi')+\cos\theta'\cos\theta],$$
$$\frac{\partial g}{\partial\theta'} = -[\cos\theta'\sin\theta\cos(\phi+\phi')-\sin\theta'\cos\theta] = 0,$$
$$\frac{\partial q}{\partial\phi'} = [\sin\theta'\sin\theta\sin(\phi+\phi')] = 0,$$
$$\theta_0' = \pi-\theta,\ \phi_0' = \pi-\phi\ (0<\theta,\phi,\theta',\phi'<\pi),$$
$$g(\theta_0',\phi_0') = +1,$$
$$\frac{\partial^2 g(\theta_0',\phi_0')}{\partial\theta'^2} = -[-\sin\theta_0'\sin\theta\cos(\phi+\phi_0')-\cos\theta_0'\cos\theta] = -1,$$
$$\frac{\partial^2 g(\theta_0',\phi_0')}{\partial\phi'^2} = \sin\theta_0'\sin\theta\cos(\phi+\phi_0') = -\sin\theta_0'\sin\theta = -\sin^2\theta. \tag{F.12}$$

$$E_z(r,\theta,\phi)_{r\to\infty} \simeq \frac{-Ilk^3}{8\pi^2\omega\epsilon}\sin^3\theta\times e^{-jkr}\times\sqrt{\frac{2\pi}{-jkr}}\times\sqrt{\frac{2\pi}{-jkr\sin^2\theta}}$$
$$= -\frac{Ilj}{4\pi}\frac{k^2}{\omega\epsilon}\sin^2\theta\frac{e^{-jkr}}{r},$$
$$E_\theta(r,\theta,\phi) \simeq \frac{-E_z}{\sin\theta} = Ilj\frac{k\eta}{4\pi}\sin\theta\frac{e^{-jkr}}{r} = jI\frac{\eta}{2}\frac{l}{\lambda}\sin\theta\frac{e^{-jkr}}{r}. \tag{F.13}$$

The above result may be compared with (6.33), which was obtained by algebraic approximation of the rigorous expression of (6.32).

For a general z-directed strip-dipole element, with width W and length L, its current transform $\widetilde{J}_z(k_x, k_z)$ would replace Il for a small-dipole element. The derivation of the electric field $E_z(r \to \infty)$ in the general case would follow similar steps as for a small-dipole element. Except, the integration function $f(\theta', \phi')$, and accordingly $f(\theta'_0, \phi'_0)$, would be different. This would amount to substituting Il by $\widetilde{J}_z(k \sin\theta'_0 \cos\phi'_0, k\cos\theta'_0)$ in the final expression.

$$f(\theta', \phi') = -\frac{\widetilde{J}_z(k\sin\theta'\cos\phi', k\cos\theta')}{8\pi^2\omega\epsilon}\sin^3\theta',$$

$$\begin{aligned} E_z(r,\theta,\phi)_{r\to\infty} &\simeq \widetilde{J}_z(k\sin\theta'_0\cos\phi'_0, k\cos\theta'_0) \times j\frac{k\eta}{4\pi}\sin\theta\frac{e^{-jkr}}{r} \\ &= \widetilde{J}_z(-k\sin\theta\cos\phi, -k\cos\theta)j\frac{k\eta}{4\pi}\sin\theta\frac{e^{-jkr}}{r}. \end{aligned} \tag{F.14}$$

APPENDIX G

Reciprocity Relationship For a Two-Port Circuit

Consider a general circuit consisting of M current branches and N voltage nodes, as shown in Fig.1. There are two pairs of external terminals, defined as ports 1 and 2. Fig.1 specifically shows the circuit excited by a voltage source V_{1e} across the port 1, carrying an input current I_{1e}, while the port 2 remains open with current $I_{2e} = 0$ and voltage V_{2e}. In this form, the port 1 may be considered as an additional current branch with $I_{M+1} = I_{1e}$, Accordingly, the excited configuration of Fig.1 is effectively a $(M + 1)$-branch circuit.

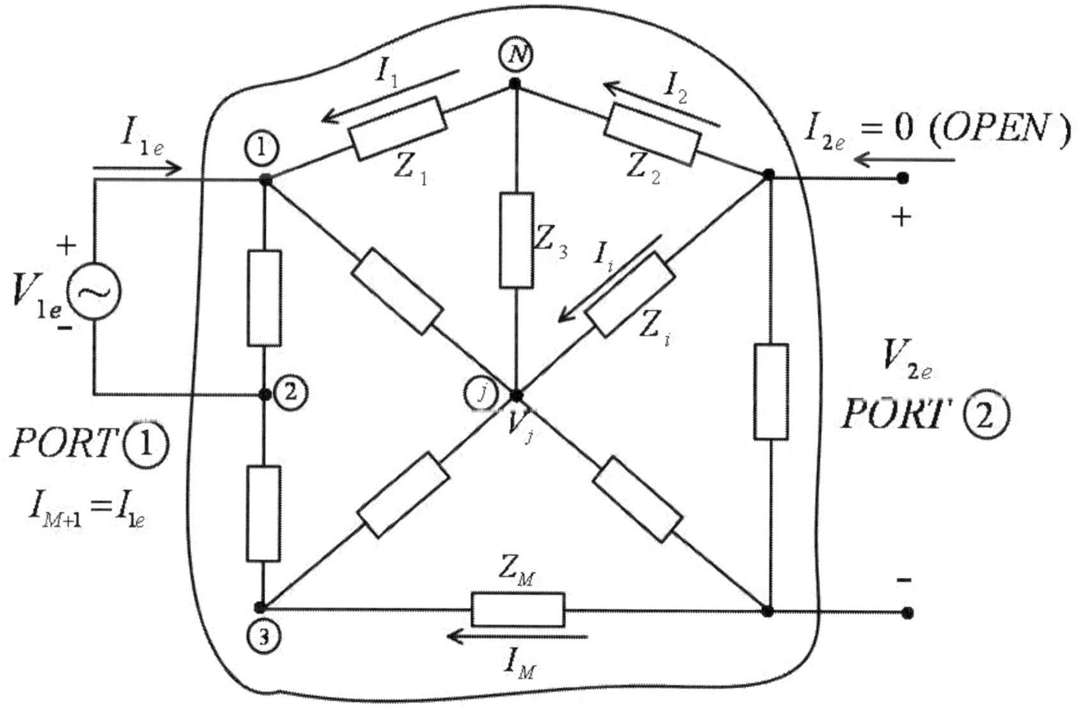

FIGURE 1. A general two-port network with N nodes and M branches (excluding the input branch at port 1), networked in an arbitrary manner, excited at the port 1. The ith arm is represented by an impedance Z_i, carrying a current I_i, and the voltage at the jth node is V_j.

For an arbitrary branch i, define i^+ as the node toward which the current I_i is directed, and i^- as the node current I_i flows from. For example, in Fig.1 we have $2^+ = N$, $i^+ = j$, and $1^- = N$. Consider an arbitrary node k. Kirchoff's Current Law (KCL) applied at the node k would require that the sum of all incoming currents to node k be equal to that of all outgoing currents. Using the above notation we have,

$$\sum_{\substack{all\ q \\ q^+=k}} I_q = \sum_{\substack{all\ q \\ q^-=k}} I_q \ ; \ for\ all\ k = 1, N. \tag{G.1}$$

Let us assign an arbitrary set of numbers (voltages) $a_j, j = 1, N$, to all the nodes. Note that these voltages may not be the same as the actual voltages of the particular circuit of Fig.1. The voltage a_k may be multiplied with KCL relationship of (G.1), and then summed over all nodes $k = 1, N$, to obtain the following relations.

$$a_k \Big[\sum_{\substack{all\ q \\ q^+=k}} I_q - \sum_{\substack{all\ q \\ q^-=k}} I_q \Big] = 0; \ for\ all\ k = 1, N, \tag{G.2}$$

$$\sum_{k=1}^{N} a_k \Big[\sum_{\substack{all\ q \\ q^+=k}} I_q - \sum_{\substack{all\ q \\ q^-=k}} I_q \Big] = 0, \tag{G.3}$$

$$\sum_{k=1}^{N} a_k \sum_{\substack{all\ q \\ q^+=k}} I_q - \sum_{k=1}^{N} a_k \sum_{\substack{all\ q \\ q^-=k}} I_q = 0. \tag{G.4}$$

Let us examine the first double summation in (G.4). Here, the inner summation is first performed over all incoming current branches to a particular node k, together multiplied by the nodal a_k, and the result is then summed over all nodes $k = 1, N$. The order of the summation may be interchanged as follows.

$$\sum_{k=1}^{N} a_k \sum_{\substack{all\ q \\ q^+=k}} I_q = \sum_{q=1}^{M+1} I_q \sum_{\substack{all\ k \\ k=q^+}} a_k. \tag{G.5}$$

However, for a given branch q, there is only one unique node $k = q^+$ to which the current I_q is directed. Therefore, the inner summation in (G.5) consists of only one element a_k=a_{q^+}. As a result, the double summation is simplified into a single summation over all branches.

$$\sum_{k=1}^{N} a_k \sum_{\substack{all\ q \\ q^+=k}} I_q = \sum_{q=1}^{M+1} I_q \sum_{\substack{all\ k \\ k=q^+}} a_k = \sum_{q=1}^{M+1} I_q a_{q^+}. \tag{G.6}$$

Similarly, the second double summation in (G.4) may also be simplified into a single summation over all branches.

$$\sum_{k=1}^{N} a_k \sum_{\substack{all\ q \\ q^-=k}} I_q = \sum_{q=1}^{M+1} I_q \sum_{\substack{all\ k \\ k=q^-}} a_k = \sum_{q=1}^{M+1} I_q a_{q^-}. \tag{G.7}$$

Using (G.6) and (G.7) in (G.4) we get,

$$\sum_{q=1}^{M+1} I_q a_{q^+} - \sum_{q=1}^{M+1} I_q a_{q^-} = 0, \tag{G.8}$$

$$\sum_{q=1}^{M+1} I_q \left(a_{q^+} - a_{q^-}\right) = 0, \tag{G.9}$$

$$\sum_{q=1}^{M=1} I_q \left(\Delta a_q\right) = 0; \quad \Delta a_q = a_{q^-} - a_{q^+}. \tag{G.10}$$

As mentioned, the nodal parameters $a_j; j = 1, N$, used in (G.1-G.10) can be any arbitrary set of numbers associated with the N nodes. Now, consider the same basic circuit of Fig.1, but excited from the port 2 with voltage V_{2e} and current I_{2e}, while the port 1 remains open with $I_{1e} = 0$ and V_{1e}, as shown in Fig.2. We may choose to equate the a_j's in (G.1-G.10) with the actual voltages $V_j'; j = 1, N$ of Fig.2.

$$\sum_{q=1}^{M+1} I_q \left(V'_{q^-} - V'_{q^+}\right) = \sum_{q=1}^{M+1} I_q \Delta V'_q = 0, \tag{G.11}$$

$$\Delta V'_q = V'_{q^-} - V'_{q^+} = I'_q Z_q, \tag{G.12}$$

$$I_{M+1} = I_{1e}, \tag{G.13}$$

$$\Delta V'_{M+1} = -V'_{1e} = V'_{(M+1)^-} - V'_{(M+1)^+}, \tag{G.14}$$

$$\sum_{q=1}^{M} I_q \Delta V'_q - I_{1e} V'_{1e} = 0, \tag{G.15}$$

$$\sum_{q=1}^{M} I_q I'_q Z_q = I_{1e} V'_{1e}. \tag{G.16}$$

Now, the entire derivation of (G.1-G.16) maybe repeated, starting with the analysis of the $(M+1)$-branch circuit of Fig.2. In this analysis, the nodal parameters a'_j may be equated to the actual voltages V_j of Fig.1.

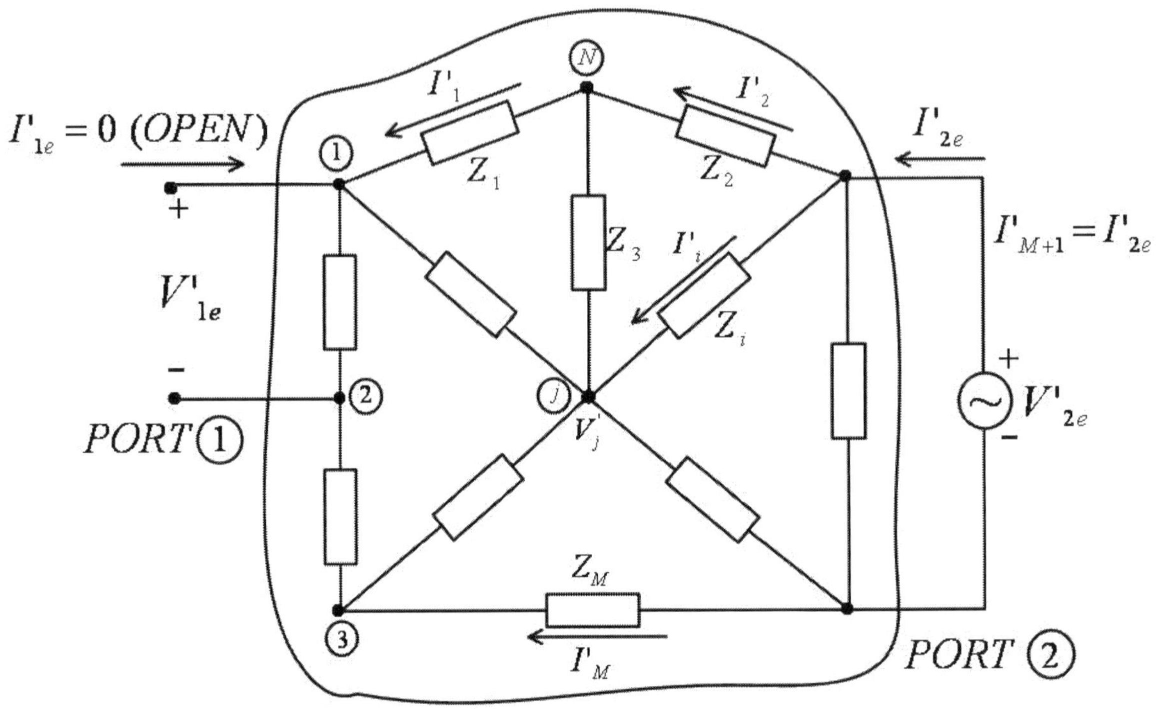

FIGURE 2. The same two-port circuit as in Fig.1, but excited at the port 2. The voltage at the jth node is V_j' and the current in the ith branch is I_i'.

$$\sum_{q=1}^{M+1} I_q'(V_{q^-} - V_{q^+}) = \sum_{q=1}^{M+1} I_q'(\Delta V_q), \tag{G.17}$$

$$\Delta V_q = V_{q^-} - V_{q^+} = I_q Z_q, \tag{G.18}$$

$$I'_{M+1} = I'_{2e}, \tag{G.19}$$

$$\Delta V_{M+1} = V_{(M+1)^-} - V_{(M+1)^+} = -V_{2e}, \tag{G.20}$$

$$\sum_{q=1}^{M} I_q' \Delta V_q - I_{2e}' V_{2e} = 0, \tag{G.21}$$

$$\sum_{q=1}^{M} I'_q I_q Z_q - I'_{2e} V_{2e} = 0. \tag{G.22}$$

$$\sum_{q=1}^{M} I'_q I_q Z_q = I'_{2e} V_{2e}. \tag{G.23}$$

Comparing (G.23) and (G.16), we get the following relationship.

$$I_{1e} V'_{1e} = I'_{2e} V_{2e}, \tag{G.24}$$

$$\frac{V'_{1e}}{I'_{2e}} = \frac{V_{2e}}{I_{1e}}. \tag{G.25}$$

In reference to Figs.1,2, V'_{1e} and V_{2e} are measured across open-circuit terminals. Therefore we may express (G.25) with a more explicit description for the measuring conditions.

$$\left.\frac{V'_{1e}}{I'_{2e}}\right|_{I'_{1e}=0} = \left.\frac{V_{2e}}{I_{1e}}\right|_{I_{2e}=0}. \tag{G.26}$$

The impedance parameter in the left-hand side of the above relation is defined as the mutual impedance Z_{12}, as measured at the port 1 with excitation from the port 2. Similarly, the impedance parameter in the right-hand side is defined as the mutual impedance Z_{21}, as measured at the port 2 with excitation from the port 1.

$$Z_{12} = Z_{12}. \tag{G.27}$$

This relationship (G.26), (G.27) is referred to as the reciprocity relationship for a two-port circuit.

Extension to a General n-Port Circuit

The derivation and final result of the above reciprocity relationship would also apply to a circuit with more than two ports, in general n ports, that are defined across arbitrary pairs of nodes of the particular circuit. We may choose any two, i and j, of the n ports at one instance, while all other ports are open-circuited, and treat the configuration as a two-port circuit. The reciprocity relationship (G.27) for a two-port circuit would apply to the above specific configuration, with i and j as its two specific ports. Accordingly, a reciprocity relation of (G.27) may be extended to a generel n-port circuit, as expressed in the following form.

$$Z_{ij} = \left.\frac{V_{ie}}{I_{je}}\right|_{I_{ke}=0,\ all\ k\neq j} = Z_{ji} = \left.\frac{V_{je}}{I_{ie}}\right|_{I_{ke}=0,\ all\ k\neq i}. \tag{G.28}$$

Extension to a Circuit with Inter-Coupled Branches

Following similar steps, the above relation may also be extended to a much more general situation. In this general case, any arm/branch of the circuit may consist of not simply a single, isolated impedence, but the branch maybe coupled to one or more branches (in general all branches), which can be modeled using sub-circuits. The inter-coupling sub-circuits are assumed to be reciprocal, that can be described by suitable impedance parameters which satisfy reciprocity conditions similar to (G.28). Under this condition, the voltage V'_q in (G.12) maybe expressed in a more general form, leading to revision of (G.16).

$$\Delta V'_q = V'_{q^-} - V'_{q^+} = \sum_{p=1}^{M} I'_p Z_{qp};$$

$$Z_{qp} = \left.\frac{\Delta V'_q}{I'_p}\right|_{I'_k=0,\ all\ k\neq p} = Z_{pq} = \left.\frac{\Delta V'_p}{I'_q}\right|_{I'_k=0,\ all\ k\neq q}. \tag{G.29}$$

$$\sum_{q=1}^{M} I_q \sum_{p=1}^{M} I'_p Z_{qp} = \sum_{q=1}^{M}\sum_{p=1}^{M} I_q I'_p Z_{qp} = I_{1e} V'_{1e}. \tag{G.30}$$

Similarly, equations (G.18,G.23) need also to be revised.

$$\Delta V_q = V_{q^-} - V_{q^+} = \sum_{p=1}^{M} I_{qp} Z_{qp}, \tag{G.31}$$

$$\sum_{q=1}^{M} I'_q \sum_{p=1}^{M} I_p Z_{qp} = \sum_{q=1}^{M}\sum_{p=1}^{M} I'_q I_p Z_{qp} = I'_{2e} V_{2e}. \tag{G.32}$$

Inter-changing indices and sequence of summation in (G.32), and then comparing with (G.30), the same relationship (G.24) is obtained. This would lead to the same reciprocity condition (G.27) for a two-port circuit, or equivalently the same condition (G.28) for a general n port circuit.

$$I'_{2e} V_{2e} = \sum_{q=1}^{M}\sum_{p=1}^{M} I'_q I_p Z_{qp} = \sum_{q=1}^{M}\sum_{p=1}^{M} I'_p I_q Z_{pq}, \tag{G.33}$$

$$I_{1e} V'_{1e} = \sum_{q=1}^{M}\sum_{p=1}^{M} I_q I'_p Z_{qp} = I'_{2e} V_{2e} = \sum_{q=1}^{M}\sum_{p=1}^{M} I_q I'_p Z_{pq};\ Z_{pq} = Z_{qp}. \tag{G.34}$$

With the above form of inter-coupling between different branches, we have a much more complex circuit. Any general electromagnetic field can be viewed and modeled as a general, inter-connected and inter-coupled circuit, if the material medium would equivalently confirm to the reciprocal behavior. Such materials are referred to as reciprocal materials, and the associated electromagnetic fields are referred to as "reciprocal fields." Accordingly, the reciprocity relation of (G.28) would as well apply to such reciprocal electromagnetic fields, when suitable port terminals and associated currents and voltages are specified. Such a reciprocity relation is also derived in chapter 2, section 4.3, using field relations in an electromagnetic field, directly from Maxwell's equations.

Bibliography

[1] D. K. Cheng. *Fundamentals of Engineering Electromagnetics.* Addison-Wesley Publishing, 1993.

[2] Jr. D. H. Hayt. *Engineering Electromagnetics.* McGraw-Hill, New York, 1995.

[3] N. N. Rao. *Elements of Engineering Electromagnetics.* Prentice Hall, New Jersey, 2000.

[4] D. Halliday, R. Resnick, and J. Walker. *Fundamentals of Physics, Volume 2, 10th Edition.* John Wiley and Sons, 2015.

[5] E. M. Purcell. *Electricity and Magnetism (Berkeley Physics Course, Vol.2, 2 Ed.).* Prentice Hall, New Jersey, 1984.

[6] Richarrd P. Feynman, Robert B. Leighton, and Mathew Sands. *Lectures on Physics, Vol.II.* Addision Wesley, 1964.

[7] S. Ramo, J. R. Whinnery, and T. Van Duzer. *Fields and Waves in Communication Electronics, 3rd Edition.* John Wiley and Sons, 1993.

[8] J. D. Jackson. *Classical Electrodynamics, 2 Ed.* John Wiley and Sons, New York, 1975.

[9] R. F. Harrington. *Time Harmonic Electromagnetic Fields.* John Wiley and Sons; IEEE Press, 2001.

[10] James Clark Maxwell. *A Treatise on Electricity and Magnetism, Vol. I and II (Reprint from 1873).* Dover Publications, 2007.

[11] V. A. Valkenburg. *Network Analysis, 3rd Edition.* Prentice-Hall, Inc., 1980.

[12] W. H. Hayt, J. E. Kemmerly, and S. M. Durbin. *Engineering Circuit Analysis, 7th Edition.* McGraw-Hill, New York, 2007.

[13] C. K. Alexander and M. N. O. Sadiku. *Fundamentals of Electric Circuits, 5th Edition.* McGraw-Hill, New York, 2012.

[14] Richarrd P. Feynman, Robert B. Leighton, and Mathew Sands. *Lectures on Physics, Vol.I, Ch.15.* Addision Wesley, 1964.

[15] R. S. Elliott. *Electromagnetics: History, Theory and Applications.* Wiley-IEEE Press, 1999.

[16] Ray Skinner. *Relativity for Scientists and Engineers.* Dover Publications, Inc., 1982.

[17] Albert Einstein. Zur Elektrodynamik bewegter Körper (On the Electrodynamics of Moving Bodies). *Annalen der Physik*, 322(10):891–921, 1905.

[18] Albert Einstein and Anna Beck (English Translator). *The Collected Papers of Albert Einstein, Volume 2: The Swiss Years: Writings 1900-1909 (see Documents 23 and 24).* Princeton University Press, 1989.

[19] D. M. Pozar. *Microwave Engineering, 4th Edition.* John Wiley and Sons, 2012.

[20] R. E. Collin. *Foundations for Microwave Engineering.* McGraw-Hill, Inc., 1992.

[21] A. V. Oppenheim, A. S. Willsky, and S. H. Nawab. *Signals and Systems, 2nd Edition.* Pearson Publishing, 1996.

[22] C. L. Phillips, J. M. Parr, and E. A. Riskin. *Signals, Systems and Transforms, 5th Edition.* Pearson Publishing, 2012.

[23] B. P. Lathi. *Linear Systems and Signals, 2nd Edition.* Oxford University Press, 2004.

[24] L. B. Felsen and N. Marcuvitz. *Radiation and Scattering of Waves (Ch.4).* Prentice Hall, 1973.

[25] D. M. Pozar. *Microwave Engineering, Ch.10 (4th Edition).* John Wiley and Sons, 2012.

[26] Max Planck. On the Law of Distribution of Energy in the Normal Spectrum. *Annalen der Physik*, 309(3):553–563, 1901.

[27] C. A. Balanis. *Antenna Theory: Analysis and Design, 2nd Edition.* John Wiley and Sons, Inc., 1997.

[28] J. D. Kraus and D. A. Fleisch. *Electromagnetics with Applications.* McGraw-Hill, New York, 1999.

[29] J. D. Kraus. *Antennas.* McGraw-Hill, New York, 1950.

[30] J. Jasik (Editor). *Antenna Engineering Handbook.* McGraw-Hill, New York, 1961.

[31] R. E. Collin and F. J. Zucker (Editors). *Antenna Theory, Parts I, II.* McGraw-Hill, New York, 1969.

[32] S. A. Schelkunoff and H. T. Friis. *Antennas,Theory and Practice.* John Wiley and Sons, Inc., 1952.

[33] A. Popoulis and S. U. Pillai. *Probability, Random Variables and Stochastic Processes, 4th Edition.* McGraw-Hill, New York, 2002.

[34] K. Pahlavan and A. H. Levesque. *Wireless Information Networks, 2nd Edition.* John Wiley and Sons, 2005.

[35] T. S. Rappaport. *Wireless Communications: Principles and Practice, 2nd Edition.* Prentice Hall, 2002.

[36] K. Fujimoto. *Mobile Antenna Systems Handbook, 3rd Edition.* Artech House Publishers, 2008.

[37] D. M. Pozar. Microstrip Antennas. *IEEE Proceedings*, 80:79–91, 1992.

[38] D. M. Pozar and D. H. Schaubert. *Microstrip Antennas.* IEEE Press, New York, 1995.

[39] K. R. Carver and J. W. Mink. Microstrip Antenna Technology. *IEEE Transactions on Antennas and Propagation*, AP-29(1):2–24, January 1981.

Index

Made in the USA
Monee, IL
05 November 2026

67985670R00324